MECHANICS OF COMPOSITE MATERIALS

OFFICE OF NAVAL RESEARCH STRUCTURAL MECHANICS SERIES

Already published in this series:

STRUCTURAL MECHANICS, Proceedings of the First Symposium on Naval Structural Mechanics.

Edited by J. NORMAN GOODIER AND NICHOLAS J. HOFF, 1960

PLASTICITY, Proceedings of the Second Symposium on Naval Structural Mechanics.

Edited by E. H. LEE AND P. S. SYMONDS, 1960

HIGH TEMPERATURE STRUCTURES AND MATERIALS, Proceedings of the Third Symposium on Naval Structural Mechanics.

Edited by A. M. FREUDENTHAL, B. A. BOLEY AND H. LIEBOWITZ, 1964

MECHANICS AND CHEMISTRY OF SOLID PROPELLANTS, Proceedings of the Fourth Symposium on Naval Structural Mechanics.

Edited by A. C. ERINGEN, H. LIEBOWITZ, S. L. KOH AND J. M. CROWLEY, 1967

OFFICE OF NAVAL RESEARCH
STRUCTURAL MECHANICS SERIES

MECHANICS OF COMPOSITE MATERIALS

PROCEEDINGS OF THE FIFTH SYMPOSIUM
ON NAVAL STRUCTURAL MECHANICS

Held at the Marriott Motor Hotel,
Philadelphia, Pennsylvania
May 8–10, 1967

SPONSORED BY THE OFFICE OF NAVAL RESEARCH
AND THE GENERAL ELECTRIC COMPANY, SPACE SCIENCES LABORATORY

Edited by

F. W. WENDT
H. LIEBOWITZ
N. PERRONE

PERGAMON PRESS

OXFORD · NEW YORK · TORONTO · SYDNEY · BRAUNSCHWEIG

Pergamon Press Ltd., Headington Hill Hall, Oxford
Pergamon Press Inc., Maxwell House, Fairview Park, Elmsford, New York, 10523
Pergamon of Canada Ltd., 207 Queen's Quay West, Toronto 1
Pergamon Press (Aust.) Pty. Ltd., 19a Boundary Street,
Rushcutters Bay, N.S.W. 2011, Australia
Vieweg & Sohn GmbH, Burgplatz 1, Braunschweig

First edition 1970
Library of Congress Catalog Card No. 78–96061

Made in Hungary
08 006421 3

PREFACE

A LOW-WEIGHT, high-strength, high-stiffness material which can be tailored to meet the structural requirement—these are some of the virtues which scientists and engineers in the field of materials and structures have perennially been searching for in a structural material. The use of the multiphase—or composite—material promises to fulfill these needs and open up broad, new horizons whose ultimate consequences regarding any phase of structural engineering cannot yet be fully imagined.

The fabrication of fibrous composites, together with the achievement of high strength-to-weight and stiffness-to-weight ratios, is a task of utmost significance—both with respect to small-scale specimens developed to determine material properties experimentally, and to full-size structural elements.

Many valid questions arise, especially at a time when more and more emphasis is being placed on specific applications of composites. For example:

What factors influence the achievement of these high ratios of strength- and stiffness-to-weight?

How does concentration and orientation of the reinforcement affect strength and stiffness under various loading conditions?

What needs to be done to tailor the material so that optimum performance can be achieved for a given requirement in a given structure?

How can these materials best be utilized to meet the requirements of a structural system?

Several hundred scientists and engineers from the United States and seven foreign countries, who are particularly concerned with these aspects of composite development, gathered in Philadelphia, Pennsylvania, from May 8 to May 10, 1967, to attend the International Conference on the Mechanics of Composite Materials, co-sponsored by the Office of Naval Research and the General Electric Company's Space Sciences Laboratory. The international gathering offered a rare and unique opportunity for people to exchange their ideas concerning past, present, and future work in this field.

A highlight of the conference was the International Forum, in which experts in composites from six foreign nations presented summaries of research and development work being done in their respective countries. These talks were of extreme interest in that they illuminated the particular areas of research and interest outside of this country.

The papers presented at the conference served to intimately couple the structural requirements of a system, whether it has naval, aerospace, or reentry applications, to the studies that have been carried out and are currently under way into the fundamental mechanical properties of composite materials and structural elements made of composite materials, as well as into their behavior under various environmental conditions.

Although the speakers represented a diversity of scientific and engineering disciplines and their areas of interest ranged from outer space to inner space, it was generally agreed that there are major shortcomings that remain to be eliminated if composites are to achieve their full potential. The needs are basically the following:

1. Reliable non-destructive testing techniques.
2. Improved methods for quality control in fabrication.
3. Better design data.
4. Improved test techniques to determine basic properties.
5. Joints and penetration procedures.
6. More research into the basic mechanisms of composite materials (micromechanics).
7. Lower cost.

It was generally felt that the science of composites is still in its infancy and that the problems cited will be solved.

In summary, the outstanding properties of composites, high strength and stiffness, low density, ease of formability and the ability to tailor design, to name a few, will have a dramatic, if not revolutionary, effect on structures in the years to come.

Following is the complete program of the conference.

Monday, May 8

Opening Session

Welcoming Address:

Dr. Leo Steg, Manager, Space Sciences Laboratory, General Electric Company.

Session I—Requirements

Chairman: Dr. Peter King, Chief Scientist, Office of Naval Research, Department of the Navy

Dr. John P. Craven, Chief Scientist, Special Projects Office, Department of the Navy, Missile and Aircraft Systems Constraints and Operational Requirements.

Capt. Laurence L. Jackson, Jr., Commanding Officer, Atlantic Undersea Test and Evaluation Center, Department of the Navy, Ship Systems Constraints and Operational Requirements.

Mr. Norris F. Dow, Consulting Engineer, Space Sciences Laboratory, General Electric Company, Materials, and Engineering Problems.

Session II—Applications

Chairman: Capt. Arthur M. Blamphin, Staff Member, Materials Advisory Board, National Academy of Sciences

Prof. Albert G. H. Dietz, Department of Civil Engineering, Massachusetts Institute of Technology, Structural Uses of Composites.

Capt. Samuel R. Heller, Jr., Director, Hull Systems and Weapons Support Division, Naval Ship Engineering Center, Department of the Navy, The Use of Composite Materials in Naval Ships.

Mr. Herbert S. Schwartz, Assistant to Chief, Nonmetallic Materials Division, Air Force Materials Laboratory, Department of the Air Force, Applications of Reinforced Plastics in Aircraft.

Dr. John D. Stewart,† Manager, Research and Engineering, and Mr. A. M. Garber, Manager, Structural Mechanics Laboratory Operation, Re-entry and Environmental Systems Division, General Electric Company, Applications of Composite Materials in Re-entry Vehicles.

Mr. Richard R. Heldenfels, Chief, Structures Research Division, Langley Research Center, National Aeronautics and Space Administration, Applications of Composite Materials in Space Vehicle Structures.

International Forum

Mr. James E. Gordon, Explosives Research and Development Establishment, Ministry of Technology, United Kingdom, Composites Research in England.

M. Jacques Duflos, Directeur Commercial, Pierre Genin & Cie, France, French Developments in the Field of Thermal-resistant and Mechanical Composites.

Prof. Ulrich Hütter, Institut für Flugzeugbau, Technische Hochschule Stuttgart, Germany, Composites Research in Germany.

Tuesday, May 9

Session III—Micromechanics

Chairman: Prof. Daniel C. Drucker, Engineering Division, Brown University

Prof. Zvi Hashin, Division of Engineering Mechanics, University of Pennsylvania, Theory of Composite Materials.

Prof. Mark J. Beran, Towne School of Civil and Mechanical Engineering, University of Pennsylvania, Statistical Theory of Heterogeneous Media.

† Dr. Stewart was not able to present his paper at the conference; however, the full text of his paper is included in these Proceedings.

Prof. Maurice A. Brull, Division of Engineering Mechanics, University of Pennsylvania, Stress–Strain Relations for Composite Materials in Shells of Arbitrary Geometry.

Prof. August J. Durelli,† Prof. V. J. Parks, Mr. H. C. Feng, Dr. F. P. Chiang, Civil Engineering and Mechanics Department, The Catholic University of America, and now Assistant Professor at the State University of New York (Stonybrook), Strains and Stresses in Matrices with Inserts. Prof. George Herrmann† and Prof. Jan D. Achenbach, Department of Civil Engineering, Northwestern University, Wave Propagation in Laminated and Fiber-reinforced Composites.

Prof. Rene B. Testa† and Prof. Bruno A. Boley, Department of Civil Engineering and Engineering Mechanics, Columbia University, Basic Thermoelastic Problems in Fiber-reinforced Materials. Prof. Severino L. Koh, School of Aeronautics, Astronautics and Engineering Sciences, Purdue University, Continuum Theories for Composite Materials.

Session IV—Structural Mechanics

Chairman: Dr. Nicholas Perrone, Acting Head, Structural Mechanics Branch, Office of Naval Research, Department of the Navy

Dr. John Hedgepeth,† Vice-President, Engineering, Astro Research Corporation (formerly Assistant Director for Engineering, The Martin Company), and Dr. Donald F. Haskell, Staff Consultant, Fairchild Hiller, Space and Electronic Systems Division (formerly Staff Engineer, Martin Marietta Corporation), Structural Mechanics of Fiber-reinforced Composites.

Mr. Max E. Waddoups† and Mr. James R. Blacklock, Senior Structures Engineers, General Dynamics Corporation, The Application of Finite Element Stiffness Matrix Analysis for Composite Structures.

Mr. Kenneth Hom, Head, Structures Evaluation Branch, Submarine Structures Division, Naval Ship Research and Development Center, Fiber-reinforced Plastics for Hydrospace Applications.

Dr. Kenneth R. Berg, Chief, Structural and Mechanical Engineering, Whittaker Research & Development Division, Problems in the Design of Joints and Attachments.

Mr. Louis A. Riedinger,† Manager, Structures Division, Mr. Murat H. Kural and Mr. George W. Reed, Jr., Structures Division, Lockheed Missiles and Space Company, Evaluation of the Potential Structural Performance of Composites.

Prof. Ulrich Hütter, Institut für Flugzeugbau, Technische Hochschule Stuttgart, Germany, Optimization of Shell Structures Under Bending and Torsion Loads.

† Indicates speaker.

Prof. Lucien A. Schmit, Jr., Professor-in-Charge, Solid Mechanics, Structures and Mechanical Design, Case Institute of Technology, The Structural Synthesis Concept and Its Potential Role in Design with Composites.

Dr. Hans Schuerch, President, Astro Research Corporation, Advanced Concepts for Composites.

Wednesday, May 10

Session V—Failure Mechanics

Chairman: Dr. George R. Irwin, Supervisor, Mechanics Division, U.S. Naval Research Laboratory, Department of the Navy

Dr. B. Walter Rosen, Consulting Engineer, Space Sciences Laboratory, General Electric Company, Strength of Uniaxial Fibrous Composites.

Mr. Phillip W. Mast, Mechanics Division, U.S. Naval Research Laboratory, Department of the Navy, Fracture Volume Changes in Laminated Orthotropic Materials.†

Dr. George A. Cooper‡ and Prof. Anthony Kelly, Department of Metallurgy, University of Cambridge, England, The Contribution to the Work of Fracture of a Composite Material of "Pull-out" of Fibers.

Prof. Sol R. Bodner‡ and Prof. J. M. Lifshitz, Department of Mechanical Engineering, Technion – Israel Institute of Technology, Experimental Investigations on the Dynamic Strength of Composites.

Dr. William J. Douglas,‡ Senior Analyst, Keystone Computer Associates, and Prof. Walter Jaunzemis, Department of Engineering Mechanics, Pennsylvania State University, Stability of Prestrained Laminated Media.

Dr. Robert Heller, Professor of Engineering Mechanics, Virginia Polytechnic Institute (formerly Deputy Director, Institute for the Study of Fatigue and Reliability, Columbia University), Mechanical Properties of Wire-reinforced Grains.

Dr. Larry S. Shu, Research Scientist, Space Sciences Laboratory, General Electric Company, The Evaluation of the Strength of Fiber-reinforced Composites.

Dr. Frederick W. Schmiedeshoff, Director of Research, Benet Research & Engineering Laboratories, Watervliet Arsenal, U.S. Army Weapons Command, Elastic Interface.

Session VI—Environmental Interaction

Chairman: Mr. James E. Gordon, Explosives Research and Development Establishment, Ministry of Technology, United Kingdom

Dr. Stephen W. Tsai, Professor and Director of Materials Research

† This paper is not available for publication in these Proceedings.

‡ Indicates speaker.

Laboratory, Washington University, Environmental Factors in the Design of Composite Materials.

Dr. Frederick L. Schuyler,† Principal Scientist, Theoretical Fluid Physics, Space Sciences Laboratory, and Dr. Edward Friedman, Solid Mechanics Laboratory Operation, Re-entry Systems, General Electric Company, High-temperature Ablation Interaction.

Dr. Sam Brelant,‡ RESEP Project Engineer, Aerospace Corporation, and Mr. Ira Petker, Senior Engineer, Composites Technology Section, Aerojet-General Corporation, Fabrication and Environmental Interaction Effects of Filament-wound Composites.

Mr. Nathan Fried, Head, High Strength Plastics Program, U.S. Naval Applied Science Laboratory, Department of the Navy, Degradation of Composite Materials: The Effect of Water on Glass-reinforced Plastics.

International Forum

Mr. Tadashi Tao, Chief Chemical Scientist, Nagoya Aircraft Factory, Mitsubishi Heavy Industries, Ltd., Japan, Composites Research in Japan.

Mr.Karl-Axel Olsson, Research Engineer, Division of Aircraft Structure, Royal Institute of Technology, Sweden, Composites Applications and Research in Sweden.

The editors wish to express their sincere appreciation for the splendid cooperation received from the authors and session chairmen, not only in preparation of the presentations, but also in subsequent revisions and editorial polishing of the proceedings. A special note of thanks to the participants from abroad who helped to make the conference a memorable event with fresh viewpoints and highlights of composite material application in their respective countries.

Few words can express the dedication to minute detail and tireless follow-up that was so evident in the smooth handling of conference correspondence and other secretarial work by Mrs. Barbara Walker and Mrs. Dolores Novak —all the way from the planning stage to documentation of the proceedings. A special word of thanks also belongs to Mr. Ted Ryan and his capable staff of the Laboratory Publications and Information office within the Space Sciences Laboratory, among them Mr. Charles De Julius for behind-the-scene organizational efforts and certainly the editorial support from Mr. Jack Wood, Mr. Paul Klatt, Miss Betty Keeley and Mr. Abner Harrison.

FREDERICK W. WENDT

HAROLD LIEBOWITZ

NICHOLAS PERRONE

† Dr. Schuyler was not able to present his paper at the conference; however, the full text of his paper is included in these Proceedings.

‡ Indicates speaker.

CONTENTS

WELCOMING ADDRESS

DR. LEO STEG

Space Sciences Laboratory, General Electric Company,
Philadelphia, Pennsylvania

IT is my pleasure to greet you in the name of the General Electric Company. I feel particularly close to this meeting because the topic of structures is within my own professional interests. I see a number of colleagues in the audience. The Structural Mechanics Group of the Office of Naval Research seems to have remarkable insight in electing very promising topics for its international conferences. I doubt that the structures community could discuss a more consequential topic than the present one. The idea that an almost arbitrarily anisotropic material can be put together and used for structure design permits removal of the age-old constraint of materials properties; one can now truly achieve optimization. If any of you have ever tried optimization without this constraint, you will find it stimulating. The impact on structures, whether they be aircraft or naval structures, is very impressive. There is an F-111B flying about with a significant amount of boron filaments in load-bearing structures. It is doing very well—with a substantial weight saving—and I just wonder how much structural design has gone into this aircraft, particularly with respect to the boron filaments. When I have talked with the people who build these airplanes they say that they look for major contributions from the structural community furnishing the design criteria. In fact, I think that before the year is out, composite materials will be in rather wide supply at much lower cost. It is very appropriate that the limelight and the pressure now shift from the materials community onto the structures community and that the question as to how to design with these materials be thoroughly explored. So I don't think I could open this meeting with a better thought than that I think we are due for a renaissance in the structures field. I'm pleased to be here and have G.E. join the Navy in welcoming you to this meeting.

Thank you very much.

SESSION I

REQUIREMENTS

MISSILE AND AIRCRAFT SYSTEMS CONSTRAINTS AND OPERATIONAL REQUIREMENTS

DR. JOHN P. CRAVEN

Chief Scientist, Special Projects Office,
Department of the Navy

IN ITS broadest sense a review of naval requirements for composite materials would encompass nearly all of the hardware employed by the Navy in the conduct of its missions. For in the most general sense, nearly all modern materials are composites or mixtures of a number of substances where the admixture is chosen to improve some functional behavior of the material in its particular application. Uniqueness of functional requirement thus creates the need for unique material characteristics which in turn requires the development of a new material or in the alternative the construction of a composite having the net characteristics.

The most that can be accomplished in this introductory talk, therefore, is to cite a number of naval applications which place unusual requirements on materials and to indicate those solutions which are now employed or which are projected as possible. Examples will be chosen primarily in the area of the author's experience with the Deep Submergence Systems Project and the Polaris/Poseidon–Fleet Ballistic Missile Program. It is hoped that these few examples will stimulate further ingenuity in the development of composites and in the inevitable feedback which generates new opportunities and requirements.

It is convenient therefore to begin at the bottom with naval requirements in deep ocean technology and to proceed to the free surface and then to the requirements of missiles and aircraft. The central materials problem in deep ocean technology is the development of high strength-to-weight materials which can be fabricated and employed in configurations which result in efficient structures for the deep ocean. Other major problems are in the development of buoyancy materials, material combinations whose moduli are so matched as to permit numerous cycling at depth, materials which resist biological fouling, materials which are optically transparent, structures which are resistant to effects of nearly implosion or which implode with a low peak pressure, materials which are corrosion resistant, and structures with carefully controlled conductive paths.

In the initial developments of deep running conventional submarines, the major structural form which has been employed is the ring-stiffened cylinder. Spacing and depth of stiffener and thickness of plate have been carefully tailored to provide a balanced structure in terms of modes of failure at collapse depth. Increase in depth has been obtained primarily by increase in material strength, by increasingly sophisticated fabrication techniques, and by elimination of welds. This approach, which has been best illustrated by Pellini of N.R.L., results in structures which closely approach the characteristics of the parent material. This technique has been realized to a large extent in the commercially built submersible *Aluminaut* (made by Reynolds Metals Company). Here the rings are integral with the plate and were machined from forged aluminum rings. Welds are avoided by bolting the rings together. The full depth capability of *Aluminaut* has not yet been attained, however, presumably because of uncertainties with respect to the end caps whose forming has been a difficult manufacturing process. The experience with *Aluminaut* and other military deep conventional submersibles suggests the application of material combinations to compensate for the cost and difficulty associated with fabrication of the higher stress levels. Extension of this form of technology suggests sandwich construction for greater section modules at lower weight; the use of composites where ring structures are designed with a material and shape which optimizes its role as stiffening member, and plates are designed with material and shape which optimizes its role as an enclosing surface; the use of multi-layers of differing material so that a portion of the structure is under prestressing tension at the surface and is balanced over its full range of capability over the full range of depth; and the use of composites such as glass-reinforced plastic in order to attain requisite strength and toughness. As yet, developments in this regard have been restricted to the laboratory and these will be covered in some detail in this symposium by papers from the David Taylor Model Basin.

The alternate direction to the attainment of a deep submersible capability is in the use of spherical hulls or, as in the case of the Navy Deep Submergence Research Vehicle, the use of nested spherical hulls. At deep depths the added structural efficiency of the sphere permits a great depth capability for more conventional materials. The limited volume associated with this structure is such that it is highly desirable to make as much of the machinery as is possible free-flooded and with a net positive buoyancy. This can only be accomplished with the use of flotation materials. It is in this regard that composites look most attractive for this class of vehicle. Only a limited number of flotation materials are available ranging from the lighter fluids such as the gasoline employed in the *Trieste* bathyscaphe to the composites of glassy materials whose compressive strength permits them to survive in the form of foams of hollow spheres or other buoyant structures. The leading contender is now the syntactic foams which consist of microballoons of glass in an epoxy resin. Although the material has little structural strength, it provides a buoyancy of approximately 27 lb/ft^3 and with suitable development could achieve buoyancies of 30–40 lb/ft^3. It appears ideal as a filler material in sandwich structures and is so employed in the superstructure of the Lockheed *Deep Quest* and was considered for similar application for the Navy DSRV.

In the extension of submersibles to the deepest part of the ocean, the syntactic foams themselves will not be adequate if extremely large and cumbersome vehicles are to be avoided. The preliminary trade-off study of the DSRV has shown that a 20,000 ft depth capability can be achieved by use of pressure hulls of either high-strength steel or titanium, provided that they are supplemented with buoyancy spheres of a high-strength material. The possibilities in this regard include a number of composites. Indeed if implosion resistance requirements are to be met, some shock mitigating material is essential as part of each buoyancy sphere. The major contenders at present are ceramics such as alumina, the glasses, pyrex, pyroceram prestressed glass, the fiberglasses in conventional "motor case" configuration, and fiberglass composites of glued spherical segments with fibers oriented normal to the sphere surface. This latter configuration has been successfully constituted in 10 in. spheres and with a weight-to-displacement ratio of 0.4 and has been tested to a depth of 44,000 ft.

The extension of this technology to the manned pressure hull of a deep submersible should add two desirable functional attitudes to the deep submersible capability-transparency, or high visibility and low cost. The front running contender in this regard is a composite of massive glass and plastic. As a parent material the massive glass has the requisite compressive weight ratio. Although its impact resistance of the structure increases as a result of the preloading associated with depth, it is still inadequate against sharp high intensity loading. This characteristic has been investigated by Perry at N.O.L. Spherical segments of surface compressed glass have been coated with various materials to determine the effect on impact resistance. Polyurethane and rubber have proven effective in reducing the impace bonding strain for thicknesses in excess of 1/2 in. More dramatic results have been obtained in explosion-resistance tests. Here the "no damage" standoff distance for a 1 lb pentalite charge was determined as a function of depth for a number of glass configurations with and without elastomeric coatings. Critical dynamic pressures of more than 14,000 psi were achieved at depths of 14,000 to 22,000 ft for 10 in. glass spheres coated with 1/8 in. plastic. With this promising beginning the clear limitation to the use of glass in composite form is in the fabrication of hatch openings and penetrations and techniques for non-destructive testing.

The Naval Ordnance Test Station is building as an advanced development a small 54 in. diameter sphere constructed of an appropriately coated glass. No penetrations will be employed and the chief structural problem centers about the mating surfaces and the glass-to-metal configuration. The design of this joint is of critical importance requiring as it does a metal of approximately the same modulus as the glass and restraint not only in the direction of principle stress but centering restraint against lateral tension inducing forces. Doubtless other considerations will dictate more complex composite structures when larger or non-spherical hulls are required. This is particularly true for large vehicles because of practical limitation in the size of glass sections which can be produced. This suggests the use of segmented geodesics in metallic matrices to insure no glass-to-glass contact.

Such structures will highlight a problem already generic to deep ocean technology. This is the problem of differential compression with depth. The need to match structural stiffness so that the working of the structure with continuous cycling is not destructive requires careful design. An old example is illustrative of the point. The Model Basin has proposed a structure with steel jacket and aluminum rings. Prior to its initial pressurization both jacket and rings are in effectively zero stress. Initially (because of its higher modulus) the steel absorbs the load until the deformation induces stress in the aluminum. As depth is increased during initial pressurization the stress on both increases as a composite until the steel passes its yield at which point it deforms around the aluminum ring.

After the permanent deformation of the first dive the return to the surface results in a pretensioning of the steel and an effective prestressed composite structure. This beneficial type of environmental working can prove deleterious for other material combinations and configurations unless carefully matched in the design process.

Other generic material problems which we aggravated or are probably aggravated in the deep ocean environment are stress corrosion resulting from the compression and concentration of gas under high pressure, biological fouling and electrolytic protection. This latter is of particular concern when employing composites since the conductivity of sea water permits gradients to be established either as a result of the natural potential of the material or as the result of electrical equipments and machinery.

A greater variety of composite materials requirements are posed by the shallower water programs of saturated diving and salvage. The design problems may not be as stringent but the applications will be perhaps more unique. This naval capability derives from the Man-in-the-Sea program which is more popularly identified with the SEALAB. The technique of saturation diving is one in which the swimmer has established equilibrium between the gases which are dissolved in his tissues and the depth at which he is working. The tools which the diver employs, the structures in which he will work as transported, the protective equipments he will use present unique design problems. Mobile SEALAB structures may be designed as internally pressurized structures, or as structures which are at times internally or externally pressurized. In the former instances the use of internal pressure composites such as fiberglass or even prestressed reinforced concrete seems attractive. The vehicles which are employed to transfer the diver from the habitat to the surface and to the deck decompression chamber are also internally pressurized structures for which light weight is a distinct asset. Unique transport vehicles may also be employed such as the diver transport vehicle which has an externally pressurized sphere for pilot and copilot and an internally pressurized sphere for the swimmer or aquanauts.

At the surface of the sea, unique materials requirement result from the slamming and wave slap characteristics of the free surface and the corrosion problems associated with continuous wetting and drying. A number of unique craft are now required by the Navy in the form of hydrofoil and surface effect craft. For the hydrofoil boat, a major structural problem is associated with the foils which have an extremely high lift per unit area, which may be subjected to cavitation and at the same time must be shock resistant should obstacles be encountered. The use of high modulus fibre composites with elastomeric coatings is a possible application which might meet this lightweight high-strength application. Superstructures of hydrofoil, gencraft, and more conventional ships are clearly candidates for the application of fiber composites.

Another unique functional requirement for surface and near surface craft are found in sonar domes as sonic devices which must be structurally sound and be transparent to underwater sound or must have especially tailored characteristics, or in the case of mounting material must be highly acoustically absorbent. The design of composites to meet this wide range of condition is particularly difficult because of the compression of the elastomal which is usually employed in such application.

Recent attention has also been focused on composites for armor for small craft, river craft, landing craft, etc. The requirements here are for light weight, nonfragmentation and resistance to high velocity impact.

Before proceeding to the missile and aircraft requirements of the Navy, it is important to introduce a category which is generic to all environments, that of power and propulsion. Although nuclear power is beyond the purview or classification of this paper it should be noted that nuclear fuel elements, curves and vector structure provide an important environment for composites which can supply requisite structural strength, resist high temperature, and survive in the radiation environment. This requirement is aggravated if one looks to advanced developments such as direct conversion which will require the incorporation of thermionic emitting materials such as cesium, or thermoelectric materials, or fluids useful for magnetohydrodynamic propulsion.

Beyond the prime energy source, the characteristics of present and future process fluids ranging from steam, ammonia, sodium, or gas, etc., place a requirement for non-corrosive and chemically inert materials incorporated in the turbine sections. Equally important are the turbine blades and other rotating machinery. Examples include a proposal to employ beryllium wire in a metallic composite to benefit from its high strength-to-weight capabilities without the penalty of its low modulus.

The majority of missile and aircraft material problems derive from the high accelerations of modern systems, the high temperatures associated with high re-entry velocity or high Mach number, the participation of the fuel as a part of the structure, the out-of-atmosphere radiation environment, the internal pressures associated even with atmosphere or less than atmosphere living spaces at high altitude, and the weapons environment.

The Navy's requirement in the ballistic missile field have been centered in the Poseidon development program. The history of transition metal motor cases to fiber glass motor cases as early as the A-2 missile development is well known. These developments and their subsequent modification in the A-3 and Poseidon programs are such that the major problems in motor case development are well in hand. It should be noted that the motor case and solid propellant are a true composite in the sense that the bond between motor case and liner and between liner and propellant has required careful development and a detailed understanding of the mechanics of shrinkage in the fabrication and coring process and the mechanics of the structure during flight. In the long range programs associated with Polaris, attention has been paid to reinforced grains and to grains which contribute to the structural integrity of the case. In such composites a careful design of the burning surface to insure that the internal pressure matches the remaining strength of the case and that the booster, like the smile on the Cheshire Cat, fades away at the completion of burning. Developments in reinforced propellant grains are still of interest in the smaller air-to-air, air-to-surface advanced development missile programs of the Navy and they may again be of interest when successors for the Poseidon missile are required.

A recent application of composites in the Polaris program is of interest in demonstrating the unusual ways in which the needs for composites are manifest. The increase of diameter of the Poseidon missile over its predecessors greatly limits the space available for shock integration and protection in the missile tube. As a result it was desired that this protection be provided by a material meeting specified stress compression limits. The unusual characteristics suggested the use of a structural form which would yield initially until the statistically generated honeycomb had been compressed and the continued yielding would then be a function of the parent material. This failed to give the desired stress-deformation response at high stress levels. The final solution was the use of a sandwich structure with notched V-shaped structures. In this composite the stress characteristics of the initial structure are maintained until initial buckling takes place. After buckling, the desired stress structure curve is restored when the notches on the V are closed and a uniformly thick buckling resistant matrix is established throughout the structure.

Both missile re-entry vehicles and aircraft and associated weapons are subject to high temperature environments. The re-entry vehicles of the missile may be subjected to additional environmental effects. Classification precludes discussion of this most interesting field but it suffices to note that the postulated high radiation defense environments make re-entry vehicles an important field for composites. However, limits can be set on the materials capabilities and their environments with regard to specific application. Predictions made by the Air Systems Command show the types of composites and their hoped-for characteristics as a function of temperature. As anticipated, refractory composites are the only suitable materials above 1200° F, whereas below that level a larger number of possibilities are presented.

In achieving composites as indicated many unique techniques will be employed. The high potential of whiskers and of thin films is such that many techniques have been and will be attempted to efficiently embed them in a matrix which guarantees structural integrity. The mixing of alloys and metals in intimate but structurally segregated relationship will also call for unusual techniques as in the controlled solidification of an aluminum–aluminum metal eutectic to produce a truly laminated section or through the random lamination of ceramics and metal to minimize slip associated with highly specified layers of ceramic and metal. The Air Systems Command has prepared a predicition of results and possible attainments with representative composites. There is, however, a long way yet to go to achieve these projections.

In summary it has been shown in a Whitman's Sampler of Navy requirements that there is a need for a wide variety of composite materials to meet a wider spectrum of Navy needs and requirements. Many more could be cited but they would only reinforce the need for the profession to be prepared to construct a material to meet as yet unforeseen and peculiar naval applications on short notice and with high precision.

SHIP SYSTEMS CONSTRAINTS AND OPERATIONAL REQUIREMENTS

CAPT. L. L. JACKSON, JR., U.S.N.

Commanding Officer, Atlantic Undersea Test and Evaluation Center

TO AID in presenting the fairly complex subjects of constraints and requirements as simply and directly as possible, the subject matter is divided as follows:

- The definition of a Navy ship.
- The mission of the Navy; that is, what is expected and required of the Navy in its defense of this country.
- A brief discussion of environment, since environmental influence certainly must be taken into account when some specific material is being considered for use in a Navy ship.
- The requirements which are placed on materials. Obviously, these are of critical importance if a ship is to do efficiently the job for which it was designed and built.
- Finally, a summary of the foregoing points by, perhaps giving you something to think about in terms of future needs—at least as I see these needs at this time.

I. NAVY SHIP

I have never encountered anywhere a clear, handy definition of what is a Navy ship. However, with apologies to Noah Webster, here is my considered definition of what constitutes a Navy ship. It is a sizeable structure, self-propelled, self-sufficient and manned, which in a given time can efficiently convey a payload from one point in the water environment to another; it can direct forceful action against an enemy and can successfully resist the effects of attack. Size differentiates the ship from a boat or craft, and by calling it a structure we emphasize the fact that a ship is constructed—not merely assembled or formed or produced or manufactured, as are other artifacts, such as vessels or vehicles or shapes of various kinds.

To elaborate, the self-sufficiency of a ship is important in that it has a long-term connotation as compared with aircraft, land vehicles, boats or other means of moving payloads. A ship is expected to remain at sea for months, and during that time to rely entirely upon its own resources. The fact that it is self-propelled distinguishes it from a barge, float or any other sea-mounted structure, which might be encountered at sea.

We must not forget that ships are manned—that the trained and dedicated men on board give the ship purpose and flexibility of action, without which it would not be worthy of the name "ship." Another good reason on the emphasis that ships are manned is the fact that designers, constructors and producers of equipment to be used in them, sometimes neglect the human element and the essential relation between man and structure.

In moving a payload, speed, endurance and maneuverability—including the ability to change speed and direction with minimum delay—all are essential characteristics of a Navy ship. Certainly, the efficiency of this mode of transport is apparent when we consider that ships carry 98% of the goods moved across the world ocean. Further, the types of payloads that are moved by this means are almost infinitely variable. They include just about everything from the projectiles, shells and bullets of the battleship, to minutely fabricated parts or tremendously large quantities of petroleum carried in mammoth tankers. Literally, a ship can carry anything from rare porcelain to 50-ton tanks, ready for battle.

But, to return to purely Navy ships, on some attack submarines the payload might be restricted to torpedoes, while on fast deployment logistics ships, large numbers of soldiers can be carried, together with sufficient fighting equipment to establish and secure a beachhead far across the ocean.

The capabilities which enable a Navy ship to take forceful action against the enemy while simultaneously resisting the effects of an attack are the attributes that differentiate the Navy ship from one used in commercial transport. Naturally, the provision of such capabilities always will result in considerably higher cost compared to the ship used only in commercial transport, but without these capabilities, it is not a Navy ship.

Lastly, but by no means least, a ship must be able to float. Seemingly this is an obvious statement. However I do not refer to flotation under ideal conditions. I refer to the ability of a Navy ship to remain under control and afloat in the water or under water, where environment may be far from passive. The ship must be able to maintain flotation and stability while under abnormal and perhaps violent conditions resulting from an attack (Figs. 1 and 2).

This definition of what constitutes a Navy ship should make us better able to understand what we're talking about when addressing the subject of ship system constraints and operational requirements.

FIG. 1. Surface ship

FIG. 2. Underwater ship

II. NAVY MISSION

Navy missions, or the manner in which a ship is expected to perform when ordered to carry out a given task, consist of three main types.

Most important of these is the long-term mission where ships may undertake a strike against enemy ships or shore installations within range of guns and projectiles. Strike warfare, however, includes the ability to carry aircraft, which in turn are equipped with missiles. It also includes the ability to launch missiles directly from the ship, on the surface or under the sea against enemy targets in the air, on the sea, on the shore or even far inland (Fig. 3).

FIG. 3. Strike warfare mission

The second type of mission involves the launching of torpedoes from surface ships or submarines against other submarines. This is the anti-submarine warfare mission, but is by no means confined to the firing of torpedoes to kill the enemy submarine. The vital prerequisites are to detect, locate and identify the enemy submarine preparatory to sinking it. Not only submarines are involved in such missions but also mines, aircraft and various types of equipment often play a role (Fig. 4).

A third Navy mission is the conduct of amphibious warfare. The need for this capability has been highlighted in Southeast Asia and other sensitive areas, but it was emphasized most particularly during our World War II operations in the Pacific. Amphibious warfare denotes the Navy's capability to transport a fully equipped armed force across an ocean to the enemy's

FIG. 4. Anti-submarine warfare mission (artist's concept)

location, to land that force on a designated beach and to give it continuous support in the conduct of operations immediately after the forces have landed.

There are still other types of missions, not all of them necessarily aggressive. For example, there is the all-important matter of logistics, where the Navy must provide supplies and support to overseas bases and installations, or to other ships in the Fleet. Then there is what might be called sea-based strategic deterrence warfare—essentially the function of the Polaris and Poseidon missiles in the FBM weapon system. Such a mission may not involve the actual firing of missiles; it is enough if the enemy knows that the Navy is poised, ready to deter a potential aggressor from nuclear attack, and practically invulnerable to enemy counterattack.

III. ENVIRONMENT

Certain environmental effects must be taken into account in considering ship system constraints.

The environmental conditions imposed by the enemy obviously can affect the ability of a ship to function effectively. If the enemy presents an environment of nuclear weapons, a submarine will be subjected to tremendous shock loading if the explosion is underwater. A surface ship, depending upon the distance, may encounter a pulverizing air blast and mountainous waves. The heat wave, the lethal radiation and fallout combine to transform the normal environment into an extremely dangerous one. In addition, the shock effect of such a blast, whether undersea or on the surface, must be absorbed by the ship's hull and her installed equipment. Even without nuclear weapons, ships and submarines must have reasonable resistance against torpedoes, bombs, projectiles and missiles.

However, the enemy is not alone in producing environmental factors capable of imposing severe constraints on our ships. The motion of the waves, for example, as a ship rises upon the crest of one gigantic peak and plunges into the trough of the next, produces severe multiple stresses. Because she may be stressed at different points, the ship obviously must be designed and built with an ample margin of strength to withstand such punishment (Fig. 5). In the case of submarines, and to a much lesser extent, the bottoms of surface ships, hydrostatic pressure must be taken into account. As we go deeper into the ocean environment, the pressure exerted against the ship's structure increases at the rate of 44 psi for each additional 100 ft of depth. This means that at 20,000 ft—admittedly much deeper than we can now go—the hydrostatic loading on a submarine's hull would be 8800 psi. Under such loadings, composite materials can soon become "decomposited".

Again, sea water, by reason of its salt content, is a highly corrosive chemical which relentlessly corrodes almost any material it encounters, particularly metals. In fact, the corrosive properties of salt water make some materials practically useless in an ocean environment. Among metals, magnesium, usually so valuable because of its exceptional strength-to-weight ratio, is perhaps the most vulnerable to salt water corrosion.

Temperature variations at sea cannot be forgotten in the consideration of environmental effects. These may range from well below zero in the polar

FIG. 5. The natural environment

regions to over 100° F in tropical regions. Undersea, the variation is not so great, since temperatures rarely drop below 27° F in the coldest parts of the ocean and rise to not more than 90° F in the warmest parts. Nevertheless, these temperature extremes are highly significant in relation to certain types of materials and cannot be ignored (Fig. 6).

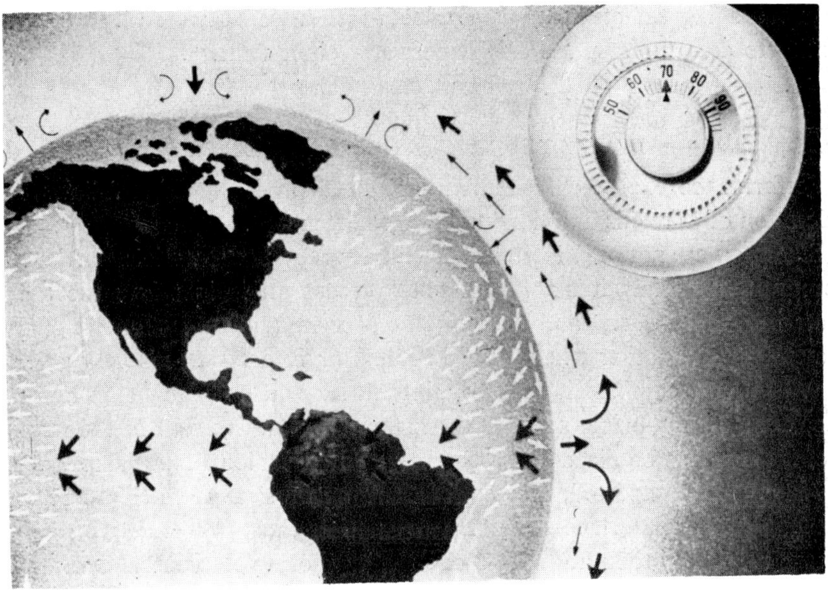

FIG. 6. The thermal environment

Still another environmental factor in the ocean is the biological fouling of the ship's bottom. This can present as much of a problem today as it did in the old days of wooden hull sailing ships. Barnacles and other encrustations that attach to the exposed parts of a ship or submarine can reduce the speed and the efficiency of the propulsion by a significant margin. We have yet to find an effective antidote to this condition that has plagued us for centuries. Perhaps composite materials can provide a good solution.

Lastly, we must face an environmental problem resulting from encounters within the interfaces of the oceans. These interfaces are the shore and the seabed, together with effective winds and even other objects (including shipwrecks).

The environmental problems facing our ships are generally understood, though perhaps not nearly as well as the Navy—and those of us directly concerned with ships and submarines—would like to have them understood. It is for this that, we have major programs underway dedicated to the study of the oceans, to learning how best to meet the many natural hazards to which ships are exposed almost from the time they are launched, and to the vital subject of engineering for ocean use.

My position as Commanding Officer of the Atlantic Undersea Test and Evaluation Center gives me some insight into these environmental hazards, because I have installed in the "Tongue of the Ocean" a variety of systems deep underwater, exposed to environmental effects. I am interested, therefore, in encouraging the study of these effects with a view toward improving our ability to care for and use this underwater equipment.

IV. MATERIALS REQUIREMENTS

Performance expected from materials used on Navy ships follows consideration of missions and the operating environment.

Since you are not only interested in but are expert in materials, particularly composite materials, it is appropriate to present an insight into the use of materials by the Navy, and into why the Navy may show preference for one material over another when viewed against a given task. It is very possible that I may be emphasizing some points which will be brought up by other speakers, but my prime purpose is to summarize my version of the requirements that must be placed on any material applicable to Navy ships.

As might be expected, the requirements of the Navy are the result of a rather complex interface between two parties, those who analyze possible future threats to the nation (and can thereby formulate general operational needs and long range objectives) and those who must be prepared to meet these needs effectively, yet economically.

Unfortunately, this process or interface is not simply a timephase sequence of procedural steps with a definite starting point and an equally clear terminal point. A number of highly qualified individuals take a careful collective look at the future. They do this, of course, with one objective in mind; to ensure that our Navy is fully capable of defending the United States and our Allies any-where—above or below the seas—well into the future.

The problem is that each of these individuals is educated in a different profession; each therefore is bound by a different discipline. Each may also be imbued with different concepts as to what constitutes the best material for a given purpose. Research and development teams throughout the Navy study, analyze, test and experiment with materials, then report the results for the benefit of the rest of the Navy. Experienced and trained naval officers study threats and mission scenarios to evolve long-range requirements for new ships and systems. Engineers and naval architects convert these requirements into preliminary designs and proposed technical approaches. Finally, the Navy will produce a specific operational requirement. This requirement is in the form of a statement of the characteristics which the ship is expected to have.

Given such an operational requirement, the system developer or ship designer can produce detailed specifications, and it is these which determine what material will be selected. Obviously, any designer or system engineer worthy of the name will recommend those materials whose use and application will result in the greatest number of benefits. But here again, no matter how exotic the material, or how dramatic a break-through it represents, it must still meet certain fundamental needs from which there can be no departure.

It must have the properties required for the purpose. These include strength, elasticity, toughness, durability, weight-to-displacement, also fatigue, corrosion and creep resistance.

The designer must be able to design with this material, that is, to fashion it to his needs with the assurance that everything is known beforehand about how the material actually performs under a wide variety of environmental conditions and not merely how it *may* react, or what its performance *could* be. Without this knowledge about it, the material will not be accepted for use in the ship's system.

The builder must be able to work and form the material in reasonable time and cost without causing the material to change its properties in an unexpected fashion.

Assuming that the material has met these preliminary criteria, it must be amenable to simple repair and maintenance. By this, I mean much more than a new bolt hole or a fresh coat of paint. I mean material which can actually be *reworked:* from which portions can be cut out and replaced, or to which additional material may be added.

Surprisingly often, for example, it becomes necessary to cut out entire sections of submarines, remove obsolete or damaged equipment and replace it with new equipment, then reassemble or reconstruct the separated portions of the submarine. The hull structure even of submarines must be reworkable.

The material must be compatible with other materials which are necessary to the structure or support of the ship. By compatible, I mean that such differing materials must not be susceptible to any harmful interaction when in close proximity. When you consider that a choice of materials might range all the way from normally corrosive or destructive gases to fluids whose chemical qualities could prove ruinous to certain metals, you begin to grasp the magnitude of the problem. It is perhaps a comforting thought that designers of rocket propellant tanks, fuel lines and rocket engines have to wage a ceaseless battle against the corrosive and destructive interplay of incompatible materials!

Again, our Navy materials must be durable in use, and this durability should be capable of reliable verification by some straightforward quality assurance test, preferably of the non-destructive kind. Tests to destruction not only are costly but also time-consuming. I lay stress on non-destructive, foolproof quality checks because it is necessary that Navy constructors, operators and overhaulers reassure themselves frequently that the material is, in fact, durable, that it has lost none of the essential properties for which it was chosen in the first place.

A vital requisite in regard to the choice of Navy materials has not been forgotten, but rather, saved till last. The material must be cost competitive. Other things being equal, there is no factor more likely to influence the selection of a particular material than basic cost comparison. These costs take into account any development test and evaluation required, the cost of fabricating with it, and the cost of maintaining it over a long period of time.

V. SUMMARY

I have attempted to define for you what constitutes a Navy ship, to outline some of the warfare missions it may be directed to undertake, to discuss the restraints that environment may impose upon a ship, and last, to underscore the vital importance of the correct choice of materials in constructing a ship or its equipment.

It only remains for us to take a joint look at the future and consider briefly those operational requirements which will have a direct influence on the choice of materials for use in Navy ships. As a submarine officer, now engaged in undersea test and evaluation of advanced systems, I feel strongly that *the most significant goal* toward which we must work in the future is the development of submarines able to operate throughout the hydrosphere. By comparison with the ocean depth, today's deepest diving submarines operate only just below the surface. It is much the same thing as with aircraft, the highest flying of which, by comparison with the stratosphere, is no more than skimming the earth's surface!

When we consider that the oceans reach to an effective depth of 20,000 ft, and that the hydrosphere covers 70.8% of the entire globe, one can wonder that the Navy did not years ago recognize a destiny to enter that new world and claim it as the *natural* medium for Navy warfare.

It is, perhaps, difficult to visualize the ocean bed as a series of valleys, plains, hills and mountain ranges, complete with passes and defiles, but that is preci ely what we would see if we could go that deep. Only the forms of life and the medium in which they live would differ. So far, mankind has barely begun to exploit the untold riches which the sea has to offer, not only food, but also valuable chemicals and minerals. Ahead lie untapped, virtually limitless resources to which oceanography and its allied sciences hold the key. The Navy must go ahead of or with those who wish to exploit these resources.

If I have digressed a little, perhaps you will put it down to my enthusiasm for this subject.

As for general Navy requirements, *my second thought* is that we must acquire greatly increased speeds for surface ships. During the past one hundred and fifty years, the fastest surface ships have perhaps doubled in speed, but even that is not saying much. It is, in fact, saying practically nothing by comparison with aircraft, which have increased their speed by more than twenty times in the past fifty years, and their payload by about a thousand times. Requirements for the rapid surface transportation of many types of payloads will make it almost mandatory, for ships to be able to double their present speeds before long. As you know, experiments with hydrofoils, air cushion vehicles and shipboard use of gas turbines are making headway in this direction. In turn, we will need new approaches to materials, new methods of propulsion, new control systems, and with them, no doubt, a new concept affecting fundamental requirements such as efficiency, durability and endurance.

FIG. 7. Deep diving submersible

To sum up, then, this look at the future suggests that we would benefit immeasurably from deep-diving submarines (Fig. 7) in the true Jules Verne sense of the term, and from surface ships of vastly greater speed than we presently have.

It is, perhaps, not too much to ask that you gentlemen bear in mind these significant operational needs as you pursue your studies of composite materials.

MATERIALS, AND ENGINEERING PROBLEMS

NORRIS F. DOW

Space Sciences Laboratory,
General Electric Company

ABSTRACT

Methods of characterizing material performance both for use in designs and in evaluations are reviewed, and inadequacies of methods developed for homogeneous materials for use with composites are considered. It is concluded that: (1) better methods are needed for the determination of even the static mechanical properties such as tensile strength, compressive and shear strengths and (2) better design approaches are needed to utilize the high unidirectional properties of advanced filaments. Progress toward improved techniques for material characterizations is described, and the "lumped reinforcement" concept for utilization of unidirectional properties is discussed.

I. INTRODUCTION

Materials both solve and create engineering problems. In the early days of application, new materials as a rule seem to create more problems than they solve. Composite materials are no exception to this rule. Indeed with their many unique characteristics composite materials are truly engineering problems in themselves, but especially in their application to design do they create problems for the engineer.

In the past engineers have built up techniques for the evaluation of homogeneous materials which lead to success in application. Ultimate strengths, yield strengths, elongations, etc., have been defined which measure the material properties and, while in some instances the measures have not been well defined (for example, elongation values require shrewd interpretation), the engineer has generally learned to use them so as to apply the materials to advantage.

For non-homogeneous materials like composites, however, complications arise in the employment of traditional measures both of material properties and of end efficiencies of application. Complications arise both because the properties in many cases are difficult to measure and so the values obtained have been adulterated by improper testing (for example, the determination of the tensile strength of a composite requires sophisticated techniques, as will be shown), and because single values no longer even approximately describe properties in all material directions.

This paper therefore addresses itself to a discussion of some of the problems associated with the engineering application of composites. As will be evident, solutions to many of these problems depend in large measure on advances in understanding of the mechanics involved; in some areas contributions of this nature are described—in others needed further contributions are identified.

II. TENSILE STRENGTH

A. *Introduction*

The ultimate tensile stress for a material has long been one of the prime measures of its applicability for engineering design. In the past, with homogeneous materials, the need for test data to establish this stress has been taken for granted. Analytical correlations between tensile strengths and constituent properties for alloys are beyond the present state-of-the-art for mechanics, and the empirical approach is relied upon. For filamentary composites, on the other hand, almost the reverse situation applies. Progress has been made with an analytical model(1) while reliable experimental measurements are difficult enough so that truly satisfactory standard tensile tests have yet to be established. Perhaps the first problem created by composite materials is that one must learn how to test them, even for the simple measure of tensile strength.

This problem has arisen in part because of an unfortunate reliance that has been placed by material developers upon the flexure test. The flexure test is indeed simple and, when intelligently used, is valid enough for comparisons among fabrication procedures, resins, and the like.(2) For engineering design data, however, "flex strengths" are essentially valueless, of course, as pointed out in detail by Hackman and Foye.(3)

Unfortunately, available tensile test data for filamentary composites are in many cases hardly more characteristic of the real material properties than flexure test data. Little is to be gained by citing the variety of inadequate tensile tests that have been proposed and tried. Suffice it to say here that the problem is the usual one of obtaining a uniform state of simple, uniaxial tensile stress over a representative test length, particularly unaccompanied by unrepresentative discontinuities in filaments or cross-section. Rather than exploring faults of bad tests, let us examine in some detail one of the better approaches—the N.O.L. ring "split-dee" test(4) with a view to progress toward a better standard which can yield reliable design data.

B. *The N.O.L. Ring, "Split-dee" Tensile Test*

The N.O.L. ring "split-dee" tensile test, shown schematically in Fig. 1, has the advantages that the specimen may readily be fabricated by filament winding and tested in a conventional universal testing machine without special fixtures. It has the disadvantages that no test section is available at which

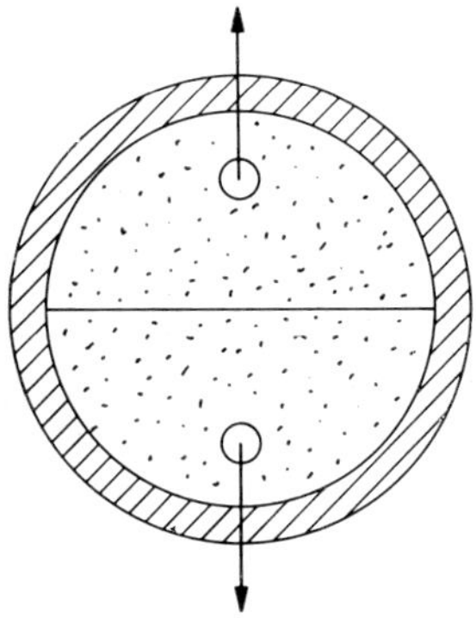

FIG. 1. Schematic of N.O.L. ring, "split-dee" tensile test (from ref. 4)

strain gages may be mounted to measure the stress–strain properties, and, more seriously, the test introduces substantial bending moments in the ring where the split occurs between the two dees. These bending moments extend above and below the split, as shown in Fig. 2. Actually, because the membrane stress in the ring causes it to increase in diameter as the load is increased, the ring tends to pull away from the corners of the dee, and the maximum bending moment is that associated with the load times the deflection to the center of the ring cross-section near the corners of the dees. The magnitude of this maximum bending moment is substantial (Fig. 3), reaching a value of 35% of the membrane stress when that stress reaches 100,000 psi for a typical glass-reinforced epoxy, 0.060 in. thick, for example.

Curiously, the magnitude of the bending moment is relatively insensitive to the composite properties. For example, the increment in stress in a hypothetical boron/epoxy ring with a Young's modulus of 60,000,000 psi is essentially the same as that in a glass/epoxy ring with a modulus of 7,500,000 psi at the same load (Fig. 4). The increased bending stiffness of the boron/epoxy is compensated for by the decreased diametral expansion of the ring under load, so that the bending stresses remain essentially constant as E varies. The boron/epoxy will be more sensitive to initial differences in diameter between the ring and the dees, however, and greater scatter in the test data may be expected for the stiffer rings for this reason alone.

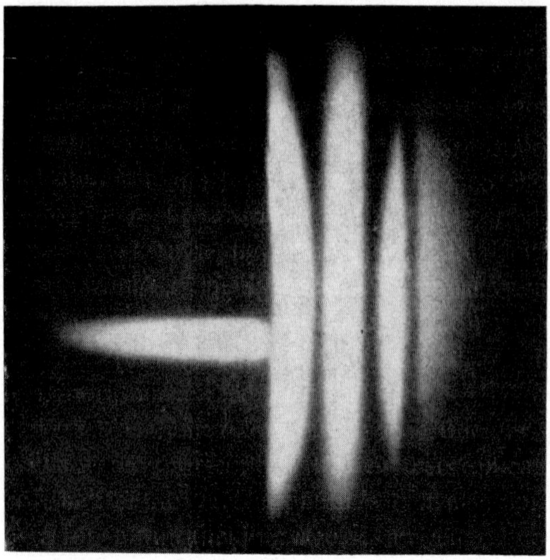

FIG. 2. Photoelastic study of stresses in the vicinity of the split between the dees in an N.O.L. ring "split-dee" tensile test

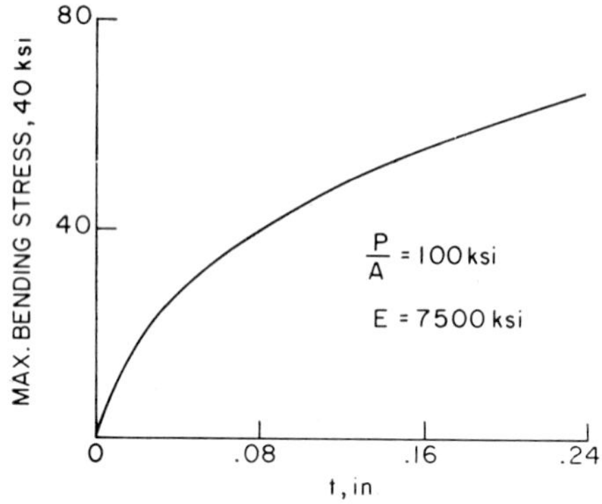

FIG. 3. Results of analysis of maximum bending moment in "split-dee" test of a glass-filament-reinforced epoxy N.O.L. ring

GLASS/EPOXY	BORON/EPOXY
E = 7,500,000 psi	E = 60,000,000 psi
σ_{Ave} = 100,000 psi	σ_{Ave} = 100,000 psi
$\sigma_{Bending}$ = 35,400 psi	$\sigma_{Bending}$ = 36,400 psi
σ_{Max} = 135,400 psi	σ_{Max} = 136,400 psi
σ_{Min} = 64,600 psi	σ_{Min} = 63,600 psi

FIG. 4. Comparison of maximum stresses in boron-reinforced and glass-reinforced epoxy N.O.L. rings during "split-dee" test

C. *The "Race-track" Specimen*

In order to obtain test data more representative of simple axial tension, plainly a better test is needed than the "split-dee" ring test. A specimen design which yields smaller maximum bending moments than a circular ring is the "race-track" specimen shown schematically in Fig. 5. Results of analyses of the maximum bending moments for a "race-track" specimen with 1 in. long straightaways and for a circular ring are compared in Fig. 6. While the

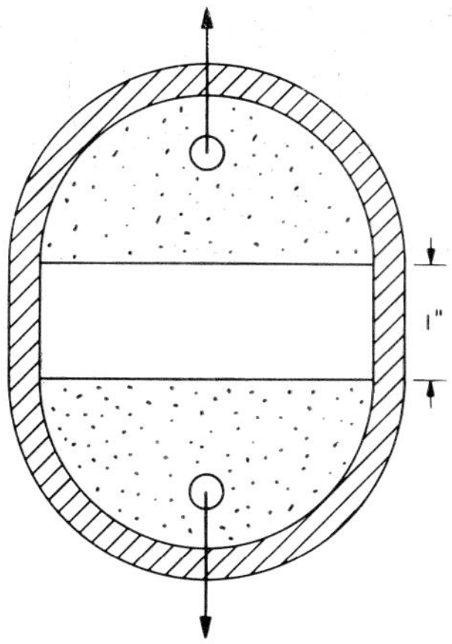

FIG. 5. "Race-track", filament-wound tensile specimen

CIRCULAR	RACE-TRACK (1" straight)
σ_{Ave} =100,000 psi	σ_{Ave} =100,000psi
$\sigma_{Bending}$ =35,400 psi	$\sigma_{Bending}$ =13,200psi

FIG. 6. Comparison of maximum bending moments for circular and "race-track" specimens in "split-dee" test

bending is not completely eliminated by the straightaway, it is substantially reduced in magnitude. Furthermore, a test section is now available for strain gages for the measurement of the stress–strain characteristics during test.

The mechanics of the reduction of the maximum bending moment are suggested by the schematic sketch of Fig. 7. Under load the mid-point of the straightaway deflects inward toward the center of the track, so that even though the curved portions pull away from the dees the eccentricity of the tensile load at mid-span is small. The maximum moment accordingly is still that near the corner of the dee—not at mid-span. Qualitative confirmation of the deflection analysis is given by the photoelastic specimen shown in Fig. 8.

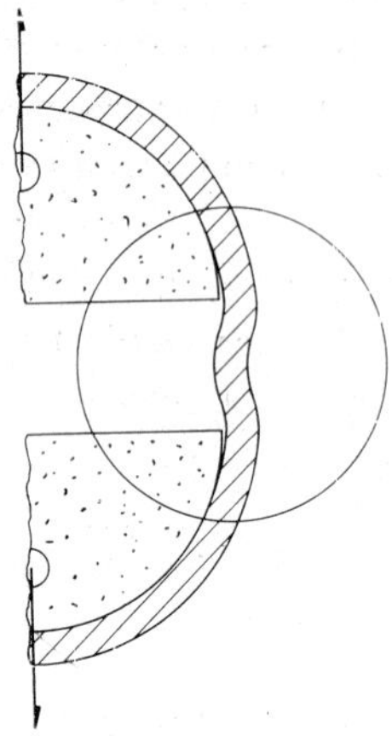

FIG. 7. Schematic representative of mechanics of deflection of "race-track" specimen

FIG. 8. Photoelastic study of stresses in the straightaway of a "race-track" specimen

D. *Application to Design*

Even if a good tensile strength measurement is available for filamentary composites, its applicability for design is open to question. Because of the nonductile nature of the composites, they are likely to exhibit a similar notch sensitivity to non-ductile metals. Not surprisingly in view of the scarcity of good data for simple tension for composites, data which will permit correlations for notch sensitivities are virtually non-existent. Thus it is evident that engineering design of filamentary composites for tension loading applications must still be accompanied with prototype construction and test. Attempted and continued correlations of the results of prototype tests with good tensile data are needed to clarify the unknowns in this area. Particularly needed first, however, are good tensile data.

III. COMPRESSIVE STRENGTH

A. *Introduction*

The identification of an engineering number representative of the compressive "strength" of a material is less straightforward, even for homogeneous materials, than the corresponding tension value. Compressive failures for most structural applications are associated with instabilities which are usually more appropriately related to some sort of yield stress than an ultimate stress. Difficulties arise in the definition of yield stress, however. A value is needed which measures both the overall and detailed shape of the material stress–strain curve, as by some function of stress, tangent modulus, and secant modulus, out to strains somewhat greater than the unit shortenings corresponding to the structural failure of interest. Various yield stresses have been defined—the 0.2% offset value, the value at which the tangent modulus is one half of the secant modulus,[5] etc. These values were relatively successful for correlating instabilities with material properties for the usual structural metals. When the material stress–strain curve was extraordinary, however, the correlation was less satisfactory.

Compared to the usual structural metals, filamentary composites have extraordinary stress–strain curves. In order to evaluate at least qualitatively the problems for engineering design introduced by the new character of these stress–strain properties, let us reconsider some of the problems of correlation of the compressive strengths of structures and the compressive strengths of materials. We can then discuss more appropriately the problems of measurement of the required properties for composite materials, and after that relate these once more to the engineering design problem.

B. *Problems of Definition of Compressive Strength*

In 1954 the NACA conducted a series of compression tests on identically proportioned stiffened panels of a variety of materials.[6] Shortly after the conclusion of the tests the investigators concerned all left the NACA, and extensive analyses of the data have not been carried out. The test results are of interest from the standpoint that they provide some clues about the definition of compressive strength of structural elements which may help in the definition of a "compressive strength" appropriate for the type of stress–strain characteristics encountered in filamentary composite materials. For present purposes only a brief study is made of these old data, with a view to determining whether some guidelines are indicated suitable for extension to composites.

Three groups of materials were tested.

(a) A family of aluminum alloys having the same Young's modulus but varying 0.2% offset yield stress (Fig. 9).
(b) A family of materials having essentially the same 0.2% offset yield stress but varying Young's modulus (Fig. 10).
(c) A family of materials having essentially the same 0.2% offset yield strain but varying moduli and yield stresses (Fig. 11).

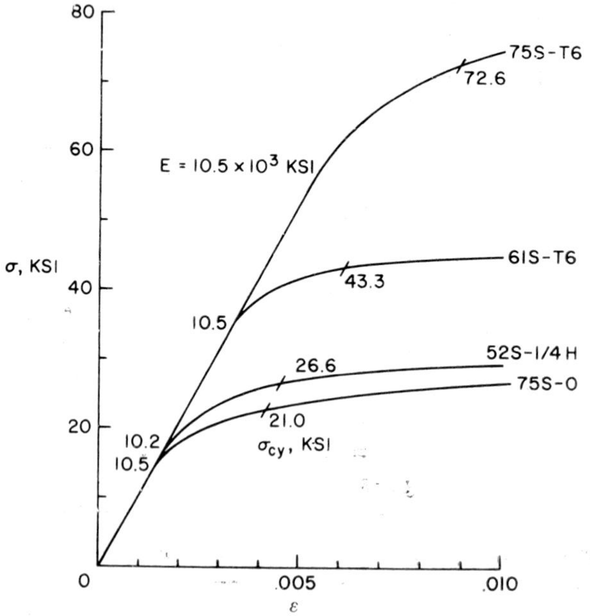

FIG. 9. Stress–strain curves for panels having materials of the same Young's modulus (from ref. 6)

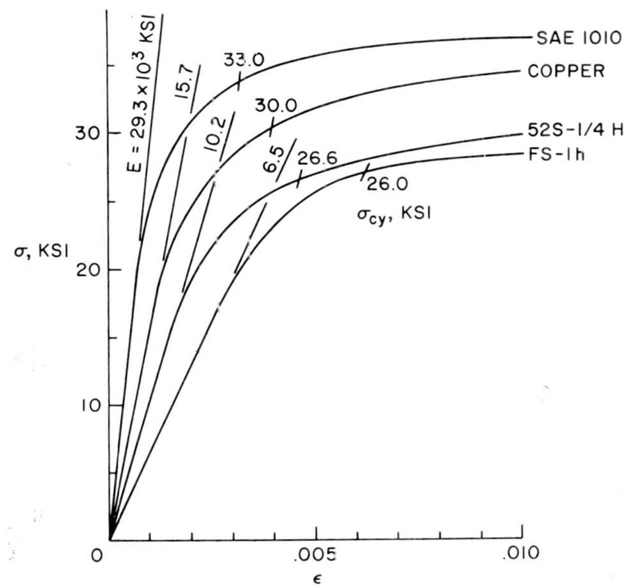

FIG. 10. Stress–strain curves for panels having materials of essentially the same 0.2% offset yield stress (from ref. 6; the unformed materials in this case had closer values of σ_{cy},—forming the Zees raised the properties of the steel and copper substantially)

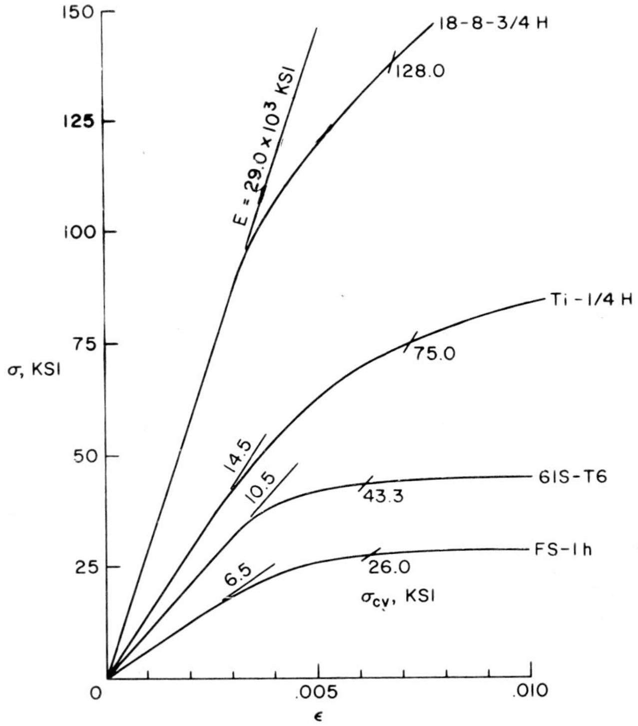

FIG. 11. Stress–strain curves for panels having materials of essentially the same 0.2% offset yield strain (from ref. 6)

If the 0.2% offset yield stress were a good measure of the compressive strength, a reasonable assumption would be that for a sturdy stiffened panel —that is, a short specimen with small width/thickness ratios (b/ts) for all plate elements—the failure stress would be close to that yield stress. Actually measured ratios of these values are given in Table 1, and indeed the correla-

TABLE 1. FAILURE STRESSES AND 0.2% OFFSET YIELD STRESSES FOR STURDY STIFFENED PANELS OF VARIOUS MATERIALS (FROM REF. 6)

Material	Yield stress, σ_{cy}, ksi	Failure stress, $\bar{\sigma}_f$, ksi	$\bar{\sigma}_f/\sigma_{cy}$
18.8	128	116.2	0.91
Ti	75	73.3	0.98
7075-T6	72.6	60.4	0.83
61S-T6	43.3	40.0	0.92
1010	33	35.9	1.09
Cu	30	30.1	1.00
52S-$\frac{1}{4}$ H	26.6	25.8	0.97
FS-1h	26.0	23.1	0.89
75S-0	21.0	20.9	1.00

tion is reasonable. There seems to be a weak additional correlation with yield strain. If the 1/6th power of yield strain were incorporated in the measure of σ_{cy}, the correlation would be within 7%.

If a non-sturdy panel is used as a basis for correlation (by non-sturdy is meant one having large values of b/t for the plate elements), the results are as shown in Tables 2 and 3. For the non-sturdy panels the correlation with yield strain is stronger than for the sturdy panels, and the incorporation of a 1/2 power of the yield strain as a factor would effect a correlation within 9%.

TABLE 2. FAILURE STRESSES AND 0.2% OFFSET YIELD STRESSES FOR NON-STURDY STIFFENED PANELS OF VARIOUS MATERIALS (FROM REF. 6)

Material	Yield stress, σ_{cy}, ksi	Failure stress, $\bar{\sigma}_f$, ksi	$\bar{\sigma}_f/\sigma_{cy}$
18-8	128	58.2	0.455
Ti	75	37.5	0.500
7075-T6	72.6	29.8	0.410
61S-T6	43.3	23.2	0.536
1010	33	22.8	0.691
Cu	30	19.2	0.640
52S-$\frac{1}{4}$ H	26.6	16.5	0.620
FS-1h	26.0	14.0	0.538
75S-0	21.0	11.6	0.552

TABLE 3. COMPARISON OF NORMALIZED RATIOS $\bar{\sigma}_f/\sigma_{cy}$ FOR STURDY AND NON-STURDY STIFFENED PANELS OF VARIOUS MATERIALS (FROM REF. 6) ($\bar{\sigma}_f/\sigma_{cy}$ FOR COPPER TAKEN AS UNITY)

Material	$\bar{\sigma}_f/\sigma_{cy}$ Sturdy	$(\bar{\sigma}_f/\sigma_{cy})$ Norm Non-sturdy
18-8	0.91	0.711
Ti	0.98	0.781
7075-T6	0.83	0.641
61S-T6	0.92	0.838
1010	1.09	1.080
Cu	1.00	1.00
52S-$\frac{1}{4}$ H	0.97	0.969
FS-1h	0.89	0.841
75S-0	1.00	0.863

Thus it appears that a function of the conventional 0.2% offset yield stress and strain could be derived to correlate short panel compressive strengths for both sturdy and non-sturdy panels within about 10%. That is, the old measure of a "compressive yield stress" to which engineers and designers are accustomed, while crude, is truly a useful design number for customary materials.

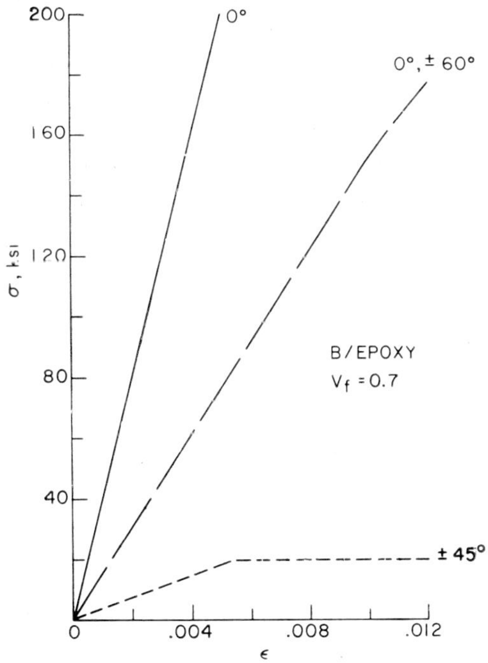

FIG. 12. Representative stress–strain curves for boron/epoxy composite laminates

For many filamentary composites, unfortunately, no 0.2% offset yield stress exists at all. Stress–strain curves for composites vary from essentially straight lines for unidirectional reinforcement, to bilinear functions for 0°, ±60° quasi-isotropic laminates, to flat-topped functions for ±45° configurations (Fig. 12). Only the latter can be considered to have a 0.2% offset yield stress at any strain of engineering interest. A new definition of material compressive strength is called for.

C. *Plate Tests*

It would appear that perhaps a better definition of compressive strength can be derived from a basic structural element like a plate than from an assembly of plates like a stiffened panel. In order to investigate this possibility, the Space Sciences Laboratory conducted a series of tests on two types of boron/epoxy plates and 7075-T6 aluminum alloy plates in V-groove fixtures. The results suggest not only that further study is required to devise an engineering measure of compressive strength for composites but also that some of the past experiments on metal plates may have been misleading. These results have sufficiently important implications for the design of composites that they need to be considered in some detail.

Stress–strain properties of the three types of plate tested are shown in Fig. 13. The 7075-T6 aluminum-alloy (the solid curve on the figure) was used as a calibration of the test techniques. The "isotropic" boron/epoxy—the long dash curve—was used as representative of a material which does not exhibit a "yield stress" below 2% strain. The ±45° boron/epoxy—the dotted curve—represented a non-strain-hardening material.

First results were those for the 7075-T6 material. The ultimate strengths of these plates are plotted as the ×-points in Fig. 14 and compared with the results in the literature.[5] As can be seen, the agreement is good in the intermediate range of stresses—that is, it appears that the V-groove fixture corresponds to that from which previous correlations of plate strengths and material properties have been made, at least for the intermediate range of width/thickness ratios.

Results for the boron/epoxy plates are also plotted in Fig. 14 and compared with the aluminum-alloy results. The ±45° plates, represented by the ⊡-points, not surprisingly achieved the 20 ksi stresses representative of the flat top of the material stress–strain curve. The "isotropic" boron/epoxy (the ⊙-points), however, despite its high stress–strain curve, gave results below those for the 7075-T6 aluminum alloy. A clue to these low stresses may be given by the fact that the aluminum-alloy tests at the low b/t ratios are also low compared to the old data. Thus the V-groove fixture may be inadequate for highly stressed plates, and clearly the definition of an engineering yield

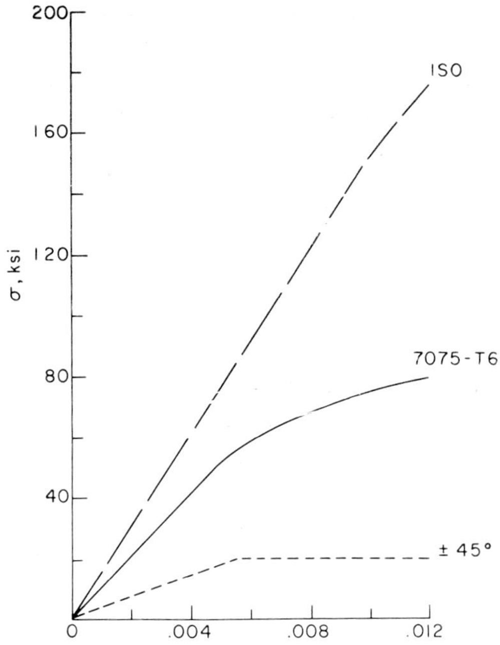

FIG. 13. Stress–strain properties of plates tested in V-groove fixture

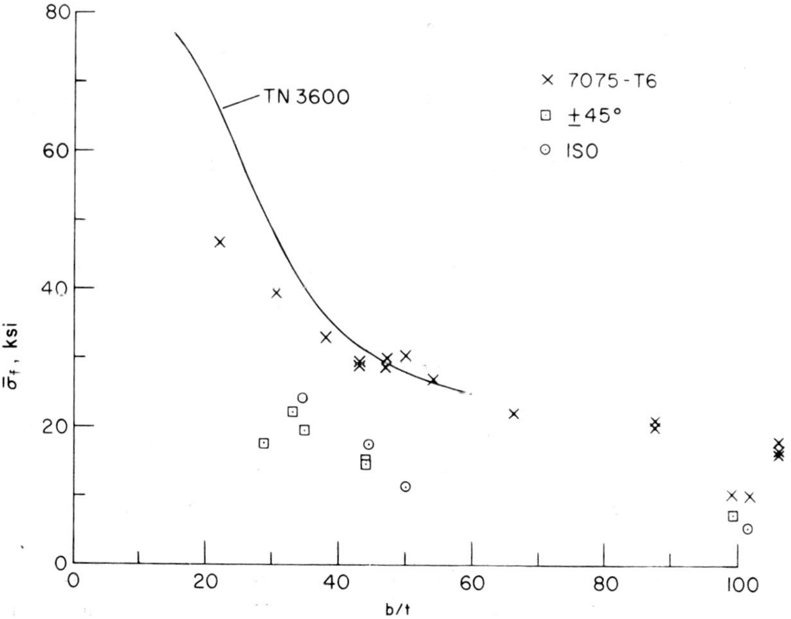

FIG. 14. Average stress at failure for plates tested in V-groove fixtures

stress for the correlation of these stresses is still in question. Certainly all indications are that properly supported isotropic boron/epoxy plates should give much higher stresses.

D. *Ultimate Compressive Strength*

In view of the low plate strengths measured for the isotropic reinforcement pattern, the question naturally arises as to whether the high ultimate strengths indicated by the stress–strain curves of Fig. 13 are really accessible. In other words, the question arises as to whether the low plate strengths are just indications of low ultimate strengths. For the isotropic configuration, data are available in the literature[7] which do indeed confirm the strength values shown. Perhaps more important for design implications, however, is the question whether high compressive strengths are truly achievable in a *unidirectionally* reinforced configuration, or whether binder weaknesses limit the attainment of the strengths inherent in advanced filaments.

Because of the importance to design, studies have been carried out in the Space Sciences Laboratory—in collaboration with the Re-entry Systems Department which fabricated the specimens—of the ultimate strength characteristics of unidirectionally reinforced composites.

Two aspects of this study bear on the problem of engineering design: (a) if the test specimen is well supported at the ends so that premature "brooming" type failures are not encountered, simple unidirectionally reinforced composites can indeed achieve high stresses (a photograph of a test fixture found to give satisfactory results by casting the specimen ends in a low melting point alloy is shown in Fig. 15); (b) the binder properties are important to the attainment of high compressive strengths, as discussed below.

Employing the theory developed by Rosen,[8] calculations were made of the compressive strengths of a three-constituent composite comprising axially oriented boron filaments and randomly oriented chopped glass filaments. The results of the calculations are presented in Fig. 16. The figure shows that, with the total volume fraction of boron reinforcement constant, the compressive strength should increase rapidly as the glass content is increased from zero to about 18%, because of the enhanced strength and stiffness of the binder. Test results for three-component composites are indicated by the points on the figure, and while there are low points (attributable to resin-free areas in the specimens, caused by the learning process for manufacturing the three-component system) the better specimens do show increases in strength.

The importance of such binder reinforcement for design applications will be discussed further under the heading "Evaluations" (see the "lumped reinforcement concept" pp. 44, 45).

FIG. 15. Top and end views of compression specimen cast in cerrobend end fixture

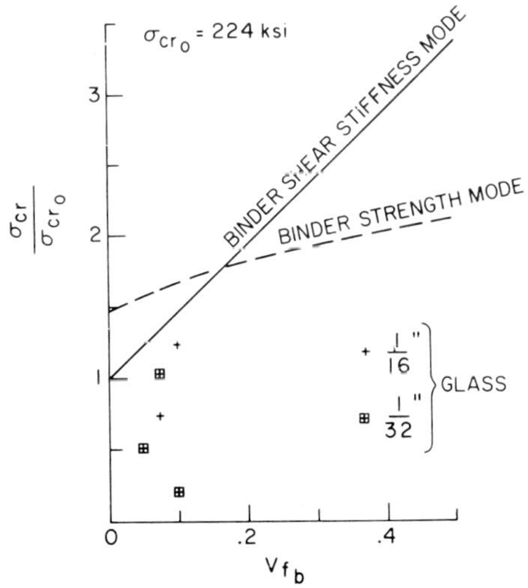

FIG. 16. Calculated and measured compressive strengths of boron-filament-reinforced composites with chopped glass added to the binder

E. *Application to Design*

The applicability to engineering design of measures of compressive strength of composite materials is obviously, from the foregoing discussion, in even a less well-defined state than for the tension case. Further testing of plate specimens and assemblies is needed to clarify the problems of correlations. Testing of plates of systematically related mechanical properties, as in the NACA panel tests, would probably provide the most valuable data.

The V-groove test of plates also needs to be critically re-examined. As noted previously, the data for the 7075-T6 aluminum-alloy plates tested in the Space Sciences Laboratory agreed with reported NACA results in the intermediate range of b/ts (Fig. 14). At the large width/thickness ratio ($b/t = 100$), no previous results were reported, but the new tests appear low relative to an extrapolation of the old curve. The one new test at a small b/t is also low, and the new data for initial buckling stresses in general correspond to lower values than would be calculated for simple support.

Particularly in view of the anisotropic nature of filamentary composites, much work must yet be done to establish engineering criteria for compressive design.

IV. SHEAR

A. *Introduction*

If the relationships between mechanical properties and design appear confusing for tension and compression in filamentary composites, then the situation for shear can only be described by the word “chaotic”. Because of the extreme difference in properties between filaments and binders, even in the early days of fiberglass reinforcement a tendency to shear failure in the binder was recognized. Unfortunately, these failures have acquired the adjective “interlaminar”, and either in consequence of this misnomer, or again because of the need for simple, comparative tests for material development purposes, progress toward engineering design values of composite shear strengths has been minimal or negative. Accordingly, in this section the discussion will be limited to some remarks about the problems of measurement of shear properties, and some comments about the implications both for engineering design and future development of interlaminar shear weakness.

B. *Measurement of Shear Strengths*

The design of a specimen having a test section subjected to a state of uniform shearing stress is essentially impossible. The closest practical approach is probably a thin-walled circular cylinder loaded in torsion. Unfortunately this is a difficult specimen to build and test compared to the simple, cheap "interlaminar shear" test beams of common practice.

About all that can be said about the common "interlaminar shear" test is that it does not measure interlaminar shear. The state of stress varies throughout the volume of the specimen, the direction of shear reverses under the point of load application, and, particularly, the stress state on one side of the cross-section is combined shear and tension and on the other combined shear and compression, either of which may be critical in some combination. Determination of the real failure mode and its relation to shear strength in such a test is a challenge to the data analyst.

B. W. Rosen[9] has proposed, and the author concurs, that a better test of shear strength is a block compression test transverse to the filament direction. In such a test the state of stress is uniform over a reasonable specimen volume, and the compression is in a direction unlikely to affect pronouncedly the shear failure. The shear character of failure encountered in such a test is

FIG. 17. Typical transverse compression failures of unidirectionally reinforced fibrous composites (from ref. 10)

illustrated in Fig. 17 (taken from ref. 10), and the type of data obtainable is indicated in Fig. 18.

Data like those of Fig. 18 are not available for torsion tests of circumferentially wound circular cylinders. *A priori* it appears that such torsion tests would provide true engineering measures of composite shear strengths in the plane of the filaments.

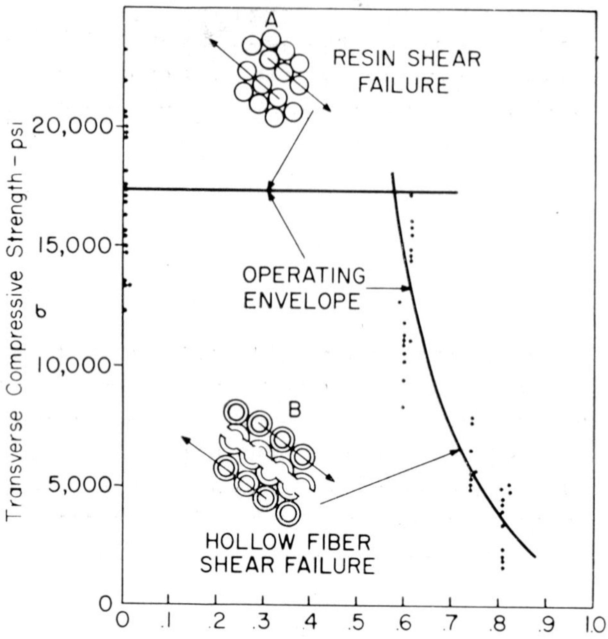

FIG. 18. Typical data obtained from transverse compression of unidirectionally reinforced composites (from ref. 10)

C. *Three-dimensional Reinforcement*

Reinforcement transverse to the filaments should increase the in-plane shear strength. Actual attempts to achieve such reinforcement by binder additions, however, have been rather disappointing,[11] in that substantial improvements have not been achieved. Three-dimensional (orthogonal filaments) reinforcement also has not yielded marked improvement.[12] How much these failures to achieve advances are functions of fabrication or test techniques are not clear. Proper three-dimensional reinforcement should lead to enhancement of shear as well as compressive strengths, as noted before, and continued efforts in this direction are appropriate.

D. *Application to Design*

In the present state-of-the-art practical application to design is impractical. Design values having engineering significance for shear strengths of filamentary composites do not exist. At best the designer must design away from shear stresses, as by multidirectional laminates, such that some layers of the laminate have filaments directed to pick up the loads as tension or compression in the filaments. The design of laminates to avoid shear failures under all conditions is not always possible, however, and shear in the plane of filaments is certainly one of the unresolved engineering problems of filamentary composites.

V. EVALUATIONS, AND THEIR IMPLICATIONS

A. *Introduction*

As was initially stated, new materials create as well as solve problems. Since the foregoing three sections have been concerned more with created than solved problems for composites, it is only just that some attention be devoted also to the advances which composites make available.

That advanced composites do make substantial improvements accessible has been amply demonstrated. Great potential improvements have been indicated for highly loaded structures as in aircraft.(13) Improvements have also been predicted for more lightly loaded components such as boost-vehicle shells,(14) but here sophisticated design is essential, and competition with better metals, like a good beryllium, is intense; that is, if the engineering design is not well done, the apparent advantage of the composites may be lost.

The present section therefore attempts to recapitulate somewhat to emphasize guidelines found effective for composite design, and to indicate engineering problems involved (over and beyond the basic ones already mentioned). As a mechanism for this recapitulation, a study will first be described of the application of composite materials to enhance the efficiency of deep-submergence shell structures. This is chosen as a suitable example because previous efficiency studies of such shells have been somewhat deficient in just the areas vital to best composite design, and so have not presented the composites to best advantage, and because the nature of the application is such as to bring out clearly the types of problems generally encountered in efficient composite design.

B. *Efficiency of Deep Submergence Shells*

The efficiency of deep-submergence shells is generally determined relative to their buoyancy. For shallow depths and with dense materials such that the difference between inner and outer diameters of the shell are relatively small, the evaluation is a simple measure of shell weight required per unit displacement to carry the external pressure. If the shell volume is large compared to the payload volume, however, as it tends to become with composite materials for the greater depths, the evaluation is complicated by the loss in payload volume available associated with the thick shell wall. Accordingly, for the latter cases a fair comparison of efficiencies must take account of the shell volume as well as the payload volume.

The consequences of the importance of shell volume for deep submergence are illustrated in Figs. 19 and 20. In Fig. 19 the efficiencies of steel and boron/epoxy shells are compared on the simplest basis, namely that of pure mono-

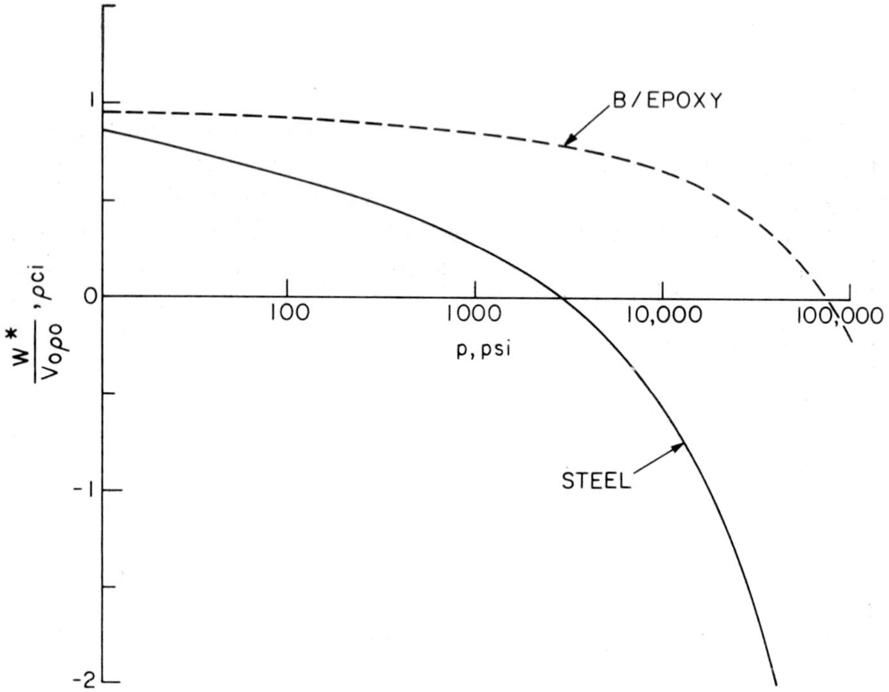

FIG. 19. Comparative efficiencies of deep-submergence cylindrical shells of steel and boron/epoxy when the outer diameter is not critical

coque construction with solid shells of the two materials. The buoyancy ratio in the form

$$\frac{W^*}{V_0\varrho_0} = \frac{\text{Total weight displaced minus shell weight}}{\text{Payload volume} \times \text{density of water}}$$

is plotted against the external pressure to be carried. The steel shell seems to have a negative buoyancy above about 3000 psi pressure; the boron/epoxy still exhibits a substantial buoyancy margin at that pressure.

If the proportions required are considered, however, and if sandwich construction is permitted so that the steel can be reasonably effectively utilized, the picture is different (Fig. 20). Comparing shells of the same overall diameters, one finds that the composite shows a substantial improvement only for the very deep submergence range. The great gains at low pressure apparent in Fig. 14 were the results of the use of thicker shell walls for the composites.

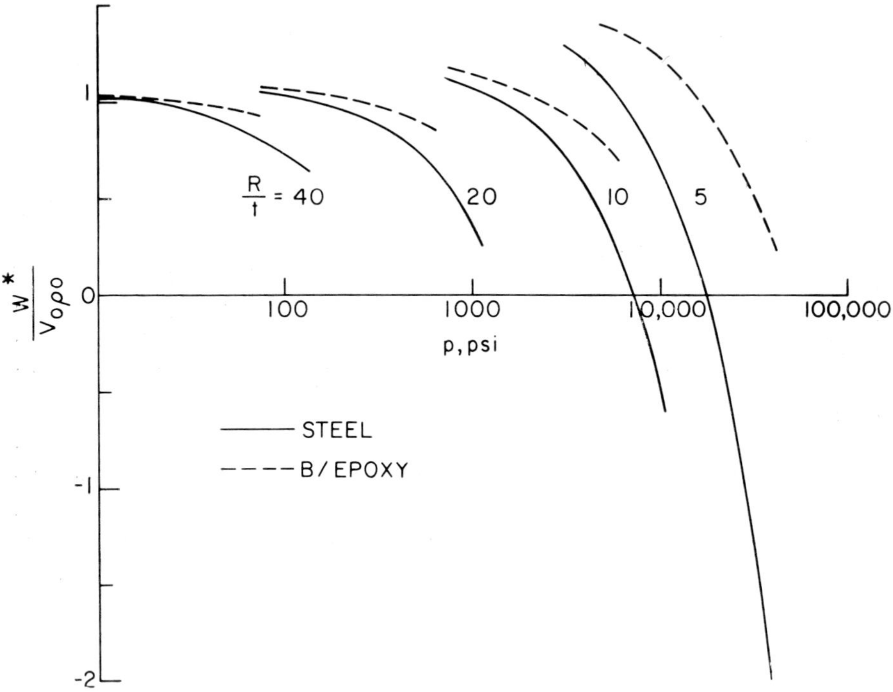

FIG. 20. Comparative efficiencies of deep-submergence cylindrical shells of steel and boron/epoxy for specified outer diameters

Considered from the standpoint of potential improvement, however, it must be recognized that while the gains from the use of the composite do not appear as significant as the simple evaluation would suggest, still they are gains, and gains that are difficult if not impossible to attain by other than the material improvements that advanced composites make accessible. For the deep-submergence application, as for most applications, sensational improvements in structural efficiency are not usually accessible. Optimization of material and configuration is required and proper optimizations generally tend to diminish rather than magnify differences among the various designs possible.

C. *The Importance of Density and Unidirectionality*

The gains in efficiency for the composite shells indicated in Figs. 19 and 20 confirm again the two outstanding conclusions to be drawn from efficiency studies for aircraft[13] and boost vehicles,[14] namely,

(1) The most effective improvement of the many that have been achieved in composite properties is the reduction in material density.
(2) The great gains in unidirectional strength and stiffness possible with advanced filamentary composites are diminished or even lost if the structure is required to have multidirectional properties.

The first of these conclusions is of primary importance to further materials development. Emphasis directed toward the achievement of filaments of lower density will pay greater dividends (other factors being constant) than further increases in strength and stiffness.

The second conclusion is of importance to the designer. To capitalize upon the high unidirectional properties, the designer should make the structure to the greatest degree possible unidirectional. Among the first indicators of the merits of capitalizing upon the unidirectionality were the stiffened panels of ref. 13 incorporating heavy unidirectionally reinforced composite stiffeners on thin, $\pm 45°$ reinforced composite skin. This "lumped reinforcement" concept was found far more efficient than other configurations.

The lumped reinforcement concept can be extended to a variety of structural forms, including those like plates that require biaxial properties. For example, a composite plate may be made with beryllium faces and unidirectionally reinforced boron/epoxy core material to combine the high elastic

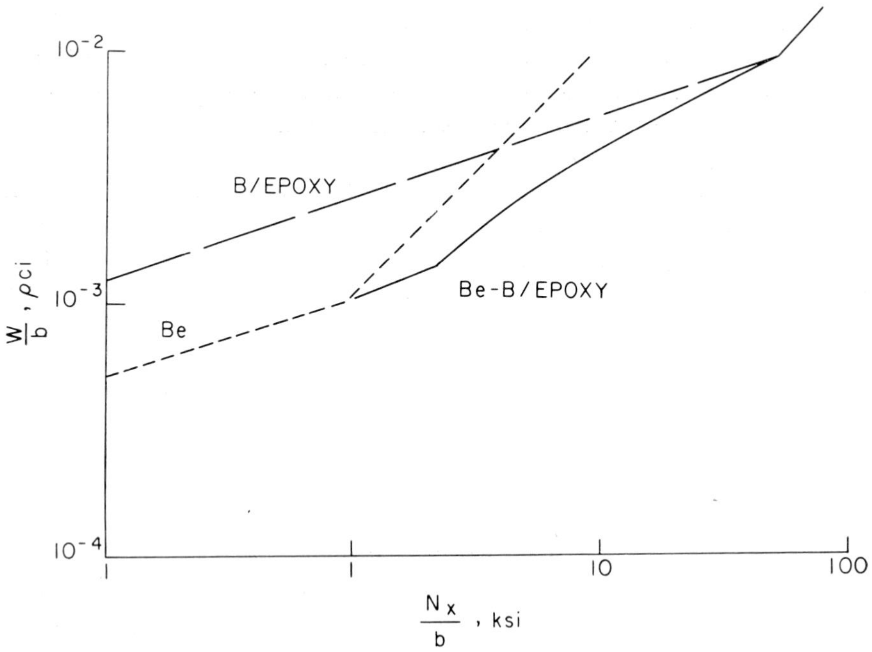

FIG. 21. Efficiencies of composite and beryllium flat plates in compression

plate-buckling resistance of beryllium with the high unidirectional strength of the boron/epoxy. The efficiency of this combination is shown in Fig. 21.

In Fig. 21 the efficiency of beryllium plates is shown strength limited at low load intensities—compressive load divided by plate width. The combination beryllium, 0° boron/epoxy core sandwich plate extends the strength while maintaining the buckling efficiency provided by the beryllium faces.

Similar applications of the lumped reinforcement concept to combine strength and stiffness properties in new structural configurations will naturally suggest themselves. For example, the use of unidirectionally reinforced composite rings at the intersections of intersecting sphere pressure vessels is an obvious possibility for the deep-submergence problem.

VI. CONCLUDING DISCUSSION

After reviewing some of the problems confronting the engineers attempting the translation of the potential gains associated with advanced composites into actually improved designs, one must draw mixed conclusions. All the evidence indicates that the gains are there, but even many of the most elementary tools necessary for the designer are still inadequate. Many of these inadequacies are exactly those to which the First International Conference. on the Mechanics of Composite Materials is addressed; others will require further attention. Specifically the designer presently needs:

(1) Better methods for the characterization of the properties of composites Tensile, compressive, and shear strength properties particularly need to be defined in adequate terms for translation to design. The part of the definition deriving from mechanical property tests is the one that needs most improvement.

(2) Better methods for capitalizing upon the high, unidirectional strength and stiffness properties made accessible by filamentary composites. The "lumped reinforcement" concept is one approach toward this goal. Binder reinforcement may contribute to its successful achievement.

Curiously this review has not gone beyond the problems of static strength. Dynamic, fatigue, and thermal loadings; problems of joints, and all the other problems besetting the designer have not been considered here, not because they have not received research attention, but only because the static problems are of evident major concern.

ACKNOWLEDGEMENTS

Researches described in this paper were supported by the Office of Naval Research (Contract N00014-66-C0167), the Air Force (Contract AF 33(615)-2447), and the NASA (Contract NASw-1377).

REFERENCES

1. ROSEN, B. W. Tensile Failure of Fibrous Composites, *AIAA Journal*, November 1964.
2. COLE, L. F. and MULVANEY, W. P. Mechanical Property Correlations between High Performance Composites and Cast Epoxy Resin Data, SPI 22nd Annual Conference, Washington, D.C., February 1967.
3. HACKMAN, L. and FOYE, R. Data Requirements for Design with Filamentary Composite Materials, AFML-TR-66-274, September 1966.
4. FRIED, N. Survey of Methods of Test for Parallel Filament Reinforced Plastics, in ASTM STP No. 327, June 1962.
5. ANDERSON, R. A. and ANDERSON, M. S. Correlation of Crippling Strength of Plate Structures with Material Properties, NACA TN 3600, January 1956.
6. DOW, N. F., HICKMAN, W. A. and ROSEN, B. W. Data on the Compressive Strength of Skin-Stringer Panels of Various Materials, NACA TN 3064, January 1954.
7. COHEN, L. Measurement of Boron Filamentary Composite Behaviour Relative to Mechanical Design of Re-entry Vehicle Structures, SPI 22nd Annual Technical Conference, Washington, D.C., February 1967.
8. DOW, N. F., ROSEN, B. W. and HASHIN, Z. Studies of Mechanics of Filamentary Composites, NASA CR-492, June 1966.
9. ROSEN, B. W. Personal communication, February 1967.
10. ROSEN, B. W. and KETLER, A. E., JR. Hollow Glass Fiber Reinforced Plastics, Final Report on Contract NOw-61-0613-d, November 1962.
11. PROSEN, S. P., MUELLER, C. E. and BARNET, F. R. Means for Increasing Interlaminar Shear Resistance in Filament Wound Structures, AFML-TR-66-274, September 1966.
12. JUDGE, J. F. NASA, Air Force Funding Avco Multidirectional Reinforced Plastics, *Technology Week*, October 31, 1966.
13. DOW, N. F., ROSEN, B. W. and KINGSBURY, H. B. Evaluation of the Potential of Advanced Composite Materials for Aircraft Structures, AFML-TR-66-144, May 1966.
14. ROSEN, B. W. and DOW, N. F. Influence of Constituent Properties upon the Structural Efficiency of Fibrous Composite Shells, presented at AIAA Sixth Structures and Materials Conference, Palm Springs, Calif., April 1965.

SESSION II

APPLICATIONS

STRUCTURAL USES OF COMPOSITES

ALBERT G. H. DIETZ†

Massachusetts Institute of Technology, Cambridge, Massachusetts

I. INTRODUCTION

Earthbound static structures such as buildings in some ways represent simpler engineering design problems than do space, deep submergence, or transport vehicles generally, but in other ways the restrictions may be more severe. Among these are cost competition with other, often well-entrenched materials that are capable of performing satisfactorily, and the long life expected of buildings, calling for durability over periods far in excess of those demanded of many other applications.

Composites are not new in buildings and their appurtenances. They include plain concrete; reinforced concrete, a composite of the next higher order; plywood, particleboards and fiberboards generally; cement-asbestos board; plastics-based laminates; and sandwiches of several of these combined. The ancient Egyptians made plywood; the Romans had concrete.

This paper will concentrate principally on the newer fiber-based composites and their use in structures, drawing mainly upon building applications for examples. As a practical matter this almost always means glass fiber reinforced unsaturated polyesters because of the required combination of reasonable cost and relatively simple fabrication processes.

Glass fiber reinforced plastics have some advantages and some limitations when compared with other structural materials. Among the most important advantages are:

1. FORMABILITY. Reinforced plastics have no inherent shape. They must be formed, and this can be done on relatively simple molds. Shapes can therefore be selected that are most advantageous for the given application. The designer is largely freed from the limitations on shape imposed by some other materials.

2. STRENGTH AND LIGHTNESS. Reinforced plastics can provide relatively great strength, and in proportion to their weight are among the strongest structural materials. Because of their resistance to cracking they can be thin, as compared with that other plastic material, concrete, for example, which must be appreciably thick and heavy under even the most favorable condi-

† Professor of Building Engineering, Department of Civil Engineering, Department of Architecture.

tions. The strength properties of reinforced plastics can be tailored to fit the imposed stresses, a feature not true of all structural materials.

3. LIGHT TRANSMISSION. Of all the structural materials available, reinforced plastics alone can transmit light. They can therefore provide the unique combination of structure, enclosure, and light transmission.

Among the most serious limitations are:

1. LOW STIFFNESS. Compared with steel and aluminum, reinforced plastics possess low stiffness. Even the most densely-packed unidirectional materials have a low elastic modulus; in most instances they are no better than concrete or wood, and may be lower. Design is consequently often limited by deflection rather than strength. Advantage must therefore be taken of the formability of these materials to select shapes such as three-dimensional curved plates and shells, ribbed and corrugated structures, and sandwiches, to overcome the deficient stiffness.

2. COST. Compared with other major structural materials, reinforced plastics are not cheap. Every pound must be stretched to the utmost. This is another reason for selecting inherently efficient structural shapes. It continues to be a major limitation in construction.

3. LIMITED SIZE. No good way of fabricating reinforced plastics directly in place has yet been found, although some experimentation has been carried on. This means that parts must be fabricated in the shop to sizes that can be transported through the streets. This, however, is in line with the strong trend toward offsite fabrication generally. Nevertheless, the problem of joining the prefabricated sections arises and it can be serious with reinforced plastics. Much needs to be done to find fast, reliable, inexpensive ways of using adhesives for field connections.

4. UNCERTAIN DURABILITY. There is no long history of use of reinforced plastics in building and similar structures. Compared with wood, masonry, concrete, steel, and even aluminum, these materials are infants. The oldest applications are perhaps 15 years old, a mere beginning as buildings go. Some boats are reported to be older and still in good shape, but even here the limit must be about 25 years. Lacking reliable short-term tests to predict long-time behavior, the durability of these materials must be rated as uncertain. It *is* certain that poorly-designed and fabricated materials may deteriorate rapidly and become worthless in a few years. Well-designed and fabricated reinforced plastics building components are, however, giving a good account of themselves and show good possibilities of long survival. Moreover, better formulations appear to be coming along, and methods of renewing deteriorated surfaces in the field are available and improving. Nevertheless, the prudent designer will so design his components that they can, if necessary, be removed and replaced easily.

5. FIRE. Any plastic, including reinforced plastics, can be destroyed by fire. Some formulations support their own combustion, others are self-extinguishing. Increased fire-resistance may mean decreased weatherability. Design must therefore be such as to minimize the fire hazard by making possible ready egress of inhabitants, and to make certain that fire damage does not lead to catastrophic failure. In all fairness it must be said that incombustible structural materials can lead to disastrous failures in fires. The fairly recent cases of large incombustible structures that collapsed in a matter of minutes when combustible contents caught fire emphasize the point.

II. EXAMPLES

A. *Radomes*

The radars of the warning network often need to be protected from inclement weather in relatively inaccessible places.[1,2] Any radome cover must be transparent to electromagnetic radiation, withstand high winds with wind-driven sand and ice particles (150 m.p.h. usually specified), and the parts often must be flown into remote areas and assembled with a minimum of equipment by relatively unskilled personnel. Many such radomes (Fig. 1a) are constructed of glass fiber reinforced plastics. The widely-used 55-ft diameter spherical radome, for example, consists of membranes of 1/16 in. thick random chopped glass fiber in a polyester resin matrix, the membranes formed integrally with edge ribs approximately 1/4 to 3/8 in. thick and 3 to 4 in. deep. When assembled by field bolting, the sections form a network of ribs supporting the membrane. Design must take into account the positive wind load on the windward side and the negative load caused by reduced air pressure on the leeward side. Under compressive stress ribs must not become unstable and allow the shell to collapse inward.

Larger radomes up to 150 ft in diameter either have a network of metal ribs supporting reinforced plastics sheets (Fig. 1b) or are hexagonal 6 in. thick sandwiches with 1/16 in. thick reinforced plastics facings bonded to a resin-impregnated honeycomb core (Fig. 1c). Radomes 500 or more feet in diameter are contemplated.

B. *Moscow Pavilions (Fig. 2)*

At the time of the United States Exhibition in Moscow in 1959 it was decided to build three pavilions to protect some of the exhibits of photography, art, and architecture.[3] The design chosen was a series of "umbrellas", each consisting of a central shaft 20 ft high surmounted by a hexagonal canopy

FIG. 1a. Radome of random chopped, glass fiber reinforced, polyester membranes molded integrally with ribs. Membranes are approximately 1/16 in. thick; ribs approximately 1/4 to 3/8 in. thick by 3 to 4 in. deep

FIG. 1b. Radome 150 ft in diameter with metal ribs supporting reinforced plastic skins

FIG. 1c. Radome 150 ft in diameter consisting of 6 in. thick sandwich panels with 1/6 in. thick reinforced plastics facings bonded to resin-impregnated honeycomb core. (Lincoln Laboratories)

16 ft across. Ninety such umbrellas were clustered in various configurations to form the three pavilions. The requirements were for resistance to 60 m.p.h. winds in irregular gusts, combined with maximum translucence in the canopies for natural daytime lighting and artificial overhead lighting at night. To meet these rather stringent requirements, advantage was taken of the formability of reinforced plastics to provide a 1/4 in. thick hollow shaft flaring from 6 in. in diameter at the base to 3 ft at the top, surmounted by six 1/16 in. thick doubly-curved "petals" with 1/4 in. thick edge ribs to form the hexagonal canopy. The very thin canopies provided ample light transmission but their double curvature provided the strength and rigidity to withstand the specified wind loading.

Although the pavilions were designed to be strictly temporary and no thought was given to weather resistance, the Soviets re-erected the units at Yalta where they are still standing.

C. *Rearrangeable Schoolhouse (Fig. 3)*

Changing requirements often make it desirable to rearrange spaces in a school, or to change its location. To provide such flexibility a prototype design was carried out utilizing downward-sloping lightweight hyperbolic paraboloids, 8 ft square in plan, four of them attached to the top of a hollow

FIG. 2. Pavilions at United States Exhibition in Moscow. Canopies 16 ft across consist of doubly-curved 1/16 in. thick reinforced plastics skins formed integrally with supporting ribs 1/4 in. thick. Supporting columns 20 ft high are 1/4 in. thick

FIG. 3. Prototype rearrangeable schoolhouse. Roof panels are hyperbolic paraboloid sandwiches 2 in. thick with 1/16 in. thick glass fiber reinforced plastics on foamed-in-place polyurethane foam cores. (M.I.T. Department of Architecture and Department of Civil Engineering)

column. To the outer edges of these paraboloids identical but upward-sloping units were attached to provide rooms 32 ft square. Hypars were sandwiches of glass fiber reinforced polyester facings with either honeycomb or foamed-in-place polyurethane cores. They were light enough to be put in place by men without mechanical hoisting equipment. After standing three years to observe their behavior under snow and wind loads and normal exposure, the prototype was dismantled and re-erected in a neighborhood where it is now occupied by a Head-Start program.

D. *Apollo Building*

To provide light, one wall of the Apollo moonshot assembly building(4) is largely faced with sandwich panels consisting of 1/16 in. thick random chopped, glass fiber reinforced, polyester bonded to a grid core of small I-shaped

FIG. 4. Wall of Apollo Assembly building faced with sandwich panels of reinforced plastics bonded to aluminum grid core. Outer facing is covered with thin transparent film of polyvinyl fluoride for increased weather resistance. (Filon)

aluminum alloy ribs (Fig. 4). For increased weather resistance the outer facing is covered with a film of polyvinyl fluoride laminated to the reinforced plastic. The sandwich panels must resist the wind loads at the same time that they transmit light.

E. *House of the Future (Fig. 5a)*

In this house built in 1957, four cantilever wings extend 16 ft from a central core to support the living quarters.[5] To provide sufficient stiffness the 8 ft wide by 16 ft long floor and roof sections of the U-shaped bents are monocoque box girders utilizing 0.30 in. thick shells of heavy woven glass fiber rovings and polyester resin (Fig. 5b). The upper surface of the 4.5 in. thick floor section is a sandwich with 0.30 in. thick upper facing, 0.10 in. thick lower facing, and a phenolic resin impregnated kraft paper honeycomb core. To provide the monocoque action the floor sandwich is epoxy-bonded along its edges to the lower shell. A similar construction is employed in the roof.

Design loads are:

Dead load, approximately 10 psf
Live loads, 50 psf floor, 30 psf roof
90 m.p.h. winds
Earthquake (Los Angeles code)

Stiffness rather than strength, except at connections, is the controlling

FIG. 5a. *(Legend, see page 57.)*

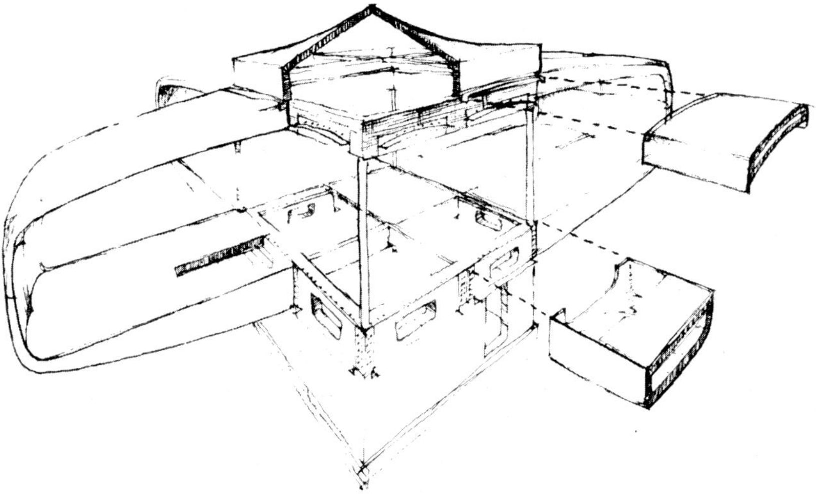

FIG. 5b.

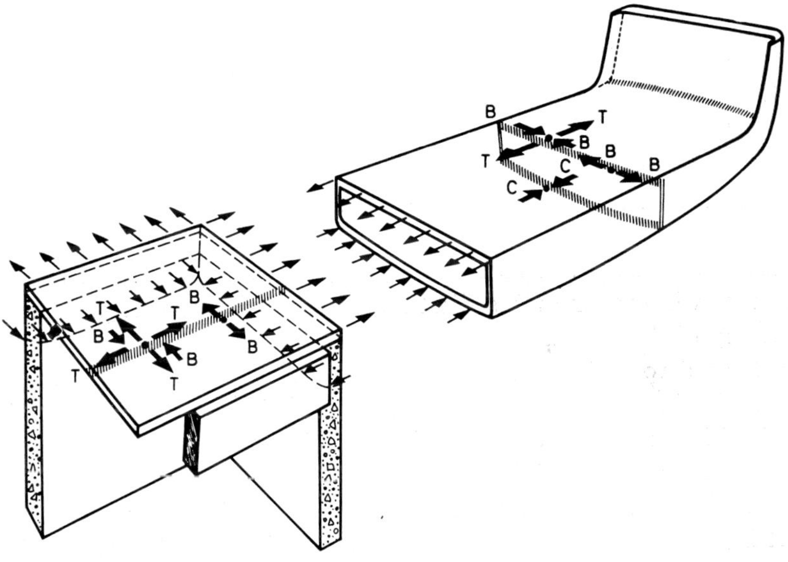

FIG. 5c.

FIG. 5a. House of the future. General view showing molded doubly-curved shells forming cantilever wings. FIG. 5b. Exploded diagrammatic view showing monocoque cantilevers with outer shells bonded along edges to floor and ceiling sandwiches of woven glass rovings reinforced facings bonded to resin-impregnated paper honeycomb cores. FIG. 5c. Diagrammatic representation of stresses in floor bents; longitudinal compression in shell; longitudinal tension in floor sandwich facings; transverse bending in floor sandwich causing compression in upper facing and tension in lower facing. (Monsanto, Hamilton and Goody) factor in this house design. Under full live load, to avoid a feeling of shakiness, deflection at the outboard ends of the bents is limited to 3/4 in.

Because the bents act as cantilevers, longitudinal compressive stresses are set up in the shell of the lower monocoque section which is doubly curved to provide resistance to buckling (Fig. 5c). Longitudinal tensile stresses are set up in the facings of the floor sandwich. The floor sandwich also acts in bending in the transverse direction, setting up transverse compressive stresses in the upper facing and transverse tensile stresses in the lower facing. Sandwich facings are therefore subjected to combined stresses. Furthermore, strain-gage measurements made on a test unit show that because of shear-lag across the 8 ft width of the bent, longitudinal stresses in the bent vary by about 50%. All stresses, however, are well within allowable limits.

The house was originally expected to stand about three years, but in the ten years it has been on display in Disneyland, approximately 2,000,000 people per year have gone through it, it has experienced the expected high winds, and several minor to moderate seismic shocks have occurred. Careful periodic measurements reveal no discernible creep or distortion, and examinations of structural details show no signs of distress. It appears to be as sound as when erected.

III. STRESS ANALYSIS

The combination of stresses often occurring in structures, as exemplified by the House of the Future, raises questions respecting the internal stress response of laminated composite materials such as are often employed. No exhaustive treatment will be attempted here because other papers in this conference will examine the subject more deeply, but the following rather elementary discussion may serve to introduce the problem.

On the macro scale of structural design rather than the micro scale of micromechanics, a composite consisting of a thin layer of random fibers embedded in a matrix is essentially isotropic in its own plane, and has one value each for elastic modulus, shear modulus, and Poisson's ratio, with the usual relationships among them. A composite consisting of parallel fibers in a matrix, or of a 90° woven fabric is orthotropic[6] and has two main or "natural" axes at right angles to each other (Fig. 6). In a thin, essentially two-dimensional layer the twelve elastic constants of three-dimensional orthotropy reduce to:

E_L longitudinal elastic modulus,
E_T transverse elastic modulus,
G_{LT} shear modulus in LT plane,
ν_{LT} Poisson's ratio, relating transverse strain to longitudinal strain,
ν_{TL} Poisson's ratio, relating longitudinal strain to transverse strain.

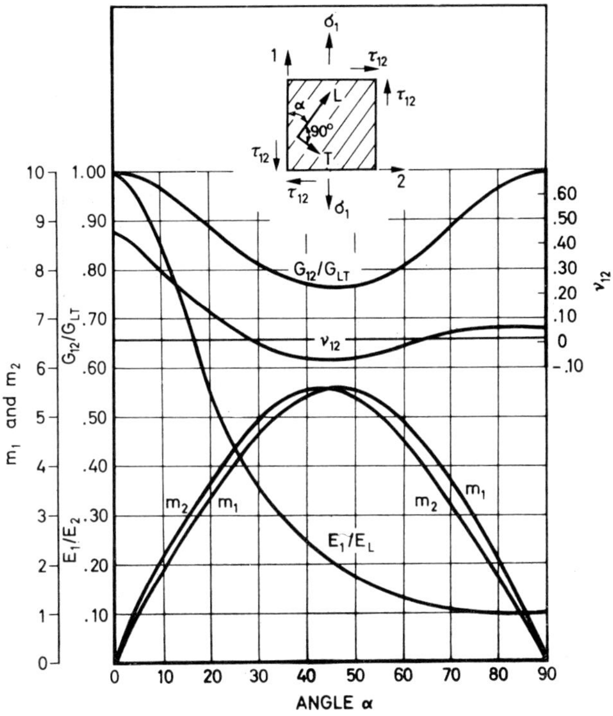

FIG. 6. Variation of elastic modulus E, shear modulus G, Poisson's ratio ν, and distortion factors m_1 and m_2 for orthotropic material

Moreover,

$$\frac{\nu_{LT}}{\nu_{TL}} = \frac{E_L}{E_T}.$$

In directions other than parallel to the $L-T$ or natural axes, these elastic constants vary between the L and T values. The relationships may be expressed in various ways derived from the behavior of orthotropic systems. The equations given below are taken from Hoff.[7] The subscripts 1 and 2 refer to axes 1 and 2 making an angle α with L and T respectively.

$$\frac{E_L}{E_1} = \cos^4\alpha + \frac{E_L}{E_T}\sin^4\alpha + \frac{1}{4}\left(\frac{E_L}{G_{LT}} - 2\nu_{LT}\right)\sin^2 2\alpha \tag{1}$$

$$\nu_{12} = \frac{E_1}{E_L}\left\{\nu_{LT} - \frac{1}{4}\left(1 + 2\nu_{LT} + \frac{E_L}{L_T} - \frac{E_L}{G_{LT}}\right)\sin^2 2\alpha\right\} \tag{2}$$

$$\frac{G_{LT}}{G_{12}} = \frac{G_{LT}}{E_L}\left\{\left(1 + 2\nu_{LT} + \frac{E_L}{E_T}\right) - \left(1 + 2\nu_{LT} + \frac{E_L}{E_T} - \frac{E_L}{G_{LT}}\right)\cos^2 2\alpha\right\}. \tag{3}$$

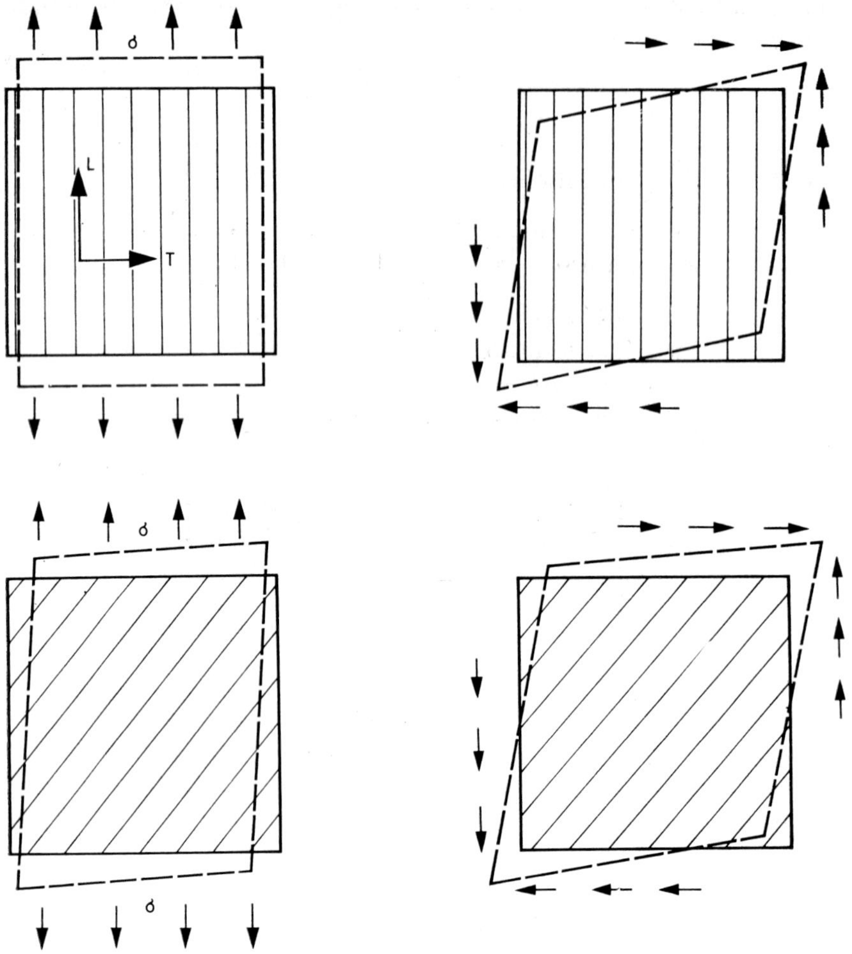

FIG. 7. *Upper*. Direct stresses in direction of natural axes cause direct longitudinal and transverse strains, and shear stresses cause shear distortion. *Lower*. Direct stresses in directions not parallel to natural axes cause shear distortion, and shear stresses cause longitudinal and transverse strains

A direct stress σ_L applied parallel to the L direction causes only a direct strain $\varepsilon_L = \sigma_L/E_L$ in that direction, a corresponding Poisson's strain $\varepsilon_T = -\nu_{LT}\varepsilon_L$ at right angles, and conversely for the T direction. Similarly, a shear stress applied in the $L-T$ directions causes only an angular distortion, $\gamma_{LT} = \tau_{LT}/G_{LT}$ (Fig. 7).

In any direction other than L or T, a direct stress σ_1, causes not only a direct strain $\varepsilon_1 = \sigma_1/\varepsilon_1$ and a Poisson's strain $\varepsilon_2 = -\nu_{12}\varepsilon_1$ but a shear distortion as well (Fig. 7), associated with a distortion factor m_1 so that the relationship may be written $\gamma_{12} = -m_1\sigma_1/E_L$. In a similar fashion, a shear stress τ_{12} causes not only shear distortion $\gamma_{12} = \tau_{12}/G_{12}$ but direct strains in the 1 and 2 directions $\varepsilon_1 = -m_1\tau_{12}/E_L$ and $\varepsilon_2 = m_2\tau_{12}/E_L$. Using Hoff's derivations

$$m_1 = \sin 2\alpha \left\{\nu_{LT}+\frac{E_L}{E_T}-\frac{1}{2}\frac{E_L}{G_{LT}}-\cos^2\alpha\left(1+2\nu_{LT}+\frac{E_L}{E_T}-\frac{E_L}{G_{LT}}\right)\right\} \quad (4)$$

$$m_2 = \sin 2\alpha \left\{\nu_{LT}+\frac{E_L}{E_T}-\frac{1}{2}\frac{E_L}{G_{LT}}-\sin^2\alpha\left(1+2\nu_{LT}+\frac{E_L}{E_T}-\frac{E_L}{G_{LT}}\right)\right\}. \quad (5)$$

Figure 8 shows the variations of E_1, G_{12}, ν_{12}, m_1, and m_2 for a material having the following elastic constants in the L–T directions:

$$E_L = 3\times 10^6 \text{ psi}$$
$$E_T = 0.15\times 10^6 \text{ psi}$$
$$G_{LT} = 0.25\times 10^6 \text{ psi}$$
$$\nu_{LT} = 0.40, \text{ from which } \nu_{TL} = 0.40\times\frac{0.15}{3.00} = 0.02.$$

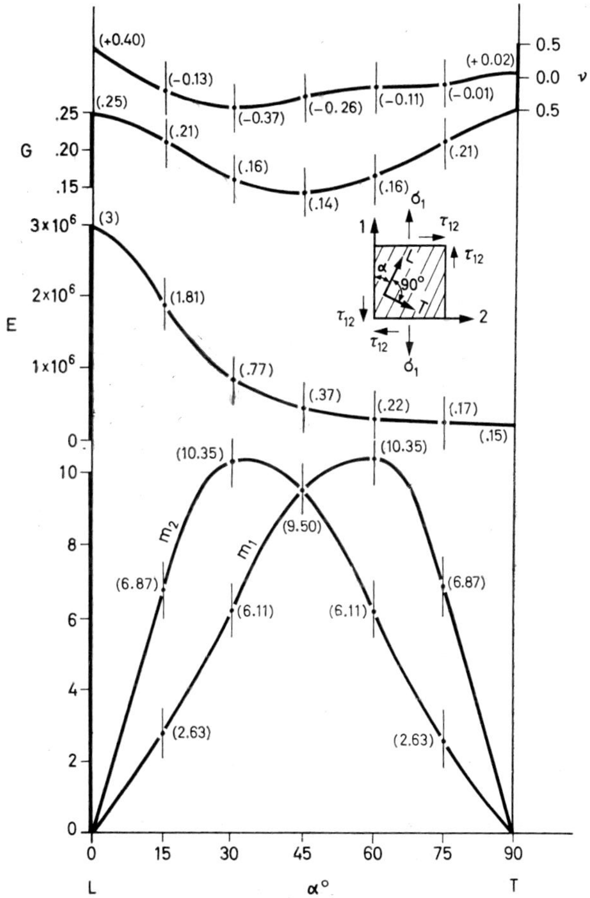

FIG. 8. Variation of E_1, G_{12}, ν_{12}, m_1, and m_2 for planar orthotropic material having E_L = 3,000,000 psi, E_T = 150,000 psi, G_{LT} = 250,000 psi, and ν_{LT} = 0.40

These are simply arbitrary values not necessarily typical of any actual material but selected to show wide variation between the L and T directions, such as might be characterized by relatively high-modulus fibers laid parallel in a low-modulus matrix.

When a laminate is made up of bonded layers of orthotropic materials with the natural axes of the layers not parallel to each other, and a direct stress is applied in the plane of the laminate, each layer will attempt to distort according to its own elastic characteristics (Fig. 9), but being bonded together,

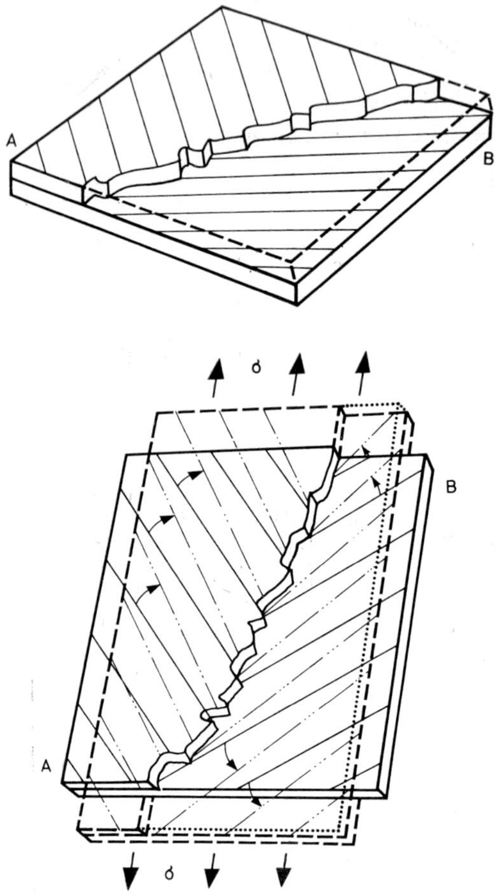

FIG. 9. In a laminate with natural axes not parallel, imposed stresses set up different stresses in each layer and interlaminar shear between layers

they are constrained to act together. This sets up not only different stresses in the different layers but interlaminar shear stresses. The relationships may be developed, for a two-component laminate consisting of layers a and b of orthotropic material. Stresses σ_1, σ_2 and τ_{12} are applied as shown in Fig. 10.

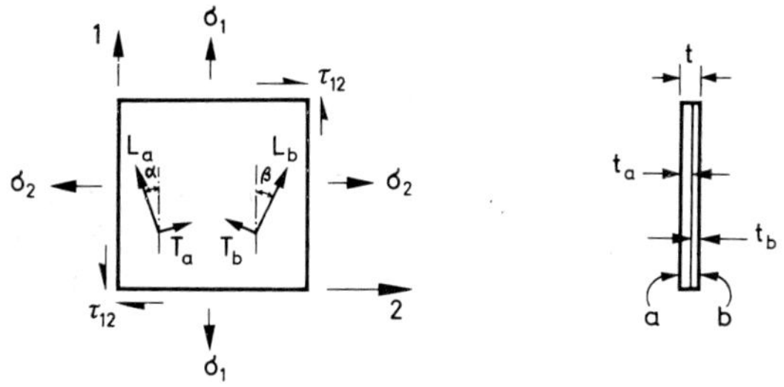

FIG. 10. Particle of laminate consisting of layers of orthotropic materials a and b with stresses applied in 1 and 2 directions. Natural axes of layers are at angles α and β to 1-direction

Two basic assumptions are made:

1. Strains in all layers are equal:

$$\varepsilon_{1a} = \varepsilon_{1b} = \varepsilon_1 \tag{6a}$$

$$\varepsilon_{2a} = \varepsilon_{2b} = \varepsilon_2 \tag{6b}$$

$$\gamma_{12a} = \gamma_{12b} = \gamma_{12}. \tag{6c}$$

2. The sums of the internal stresses in the layers of the laminate equal the imposed stresses:

$$\sigma_{1a}t_a + \sigma_{1b}t_b = \sigma_1 t;\ \sigma_{1b} = \frac{\sigma_1 t - \sigma_{1a}t_a}{t_b} \tag{7a}$$

$$\sigma_{2a}t_a + \sigma_{2b}t_b = \sigma_2 t;\ \sigma_{2b} = \frac{\sigma_2 t - \sigma_{2a}t_a}{t_b} \tag{7b}$$

$$\tau_{12a}t_a + \tau_{12b}t_b = \tau_{12} t;\ \tau_{12b} = \frac{\tau_{12} t - \tau_{12a}t_a}{t_b}. \tag{7c}$$

Equations (6) can be written for the individual layers:

$$\varepsilon_1 = \begin{cases} \varepsilon_{1a} = \dfrac{\sigma_{1a}}{E_{1a}} - \nu_{21a}\dfrac{\sigma_{2a}}{E_{2a}} - m_{1a}\dfrac{\tau_{12a}}{E_{La}} \\ \varepsilon_{1b} = \dfrac{\sigma_{1b}}{E_{1b}} - \nu_{21b}\dfrac{\sigma_{2b}}{E_{2b}} - m_{1b}\dfrac{\tau_{12b}}{E_{Lb}} \end{cases} \tag{8a}$$

$$\varepsilon_2 = \begin{cases} \varepsilon_{2a} = -\nu_{12a}\dfrac{\sigma_{1a}}{E_{1a}} + \dfrac{\sigma_{2a}}{E_{2a}} - m_{2a}\dfrac{\tau_{12a}}{E_{La}} \\ \varepsilon_{2b} = -\nu_{12b}\dfrac{\sigma_{1b}}{E_{1b}} + \dfrac{\sigma_{2b}}{E_{2b}} - m_{2b}\dfrac{\tau_{12b}}{E_{Lb}} \end{cases} \tag{8b}$$

$$\gamma_{12} = \begin{cases} \gamma_{12a} = -m_{1a}\dfrac{\sigma_{1a}}{E_{La}} - m_{2b}\dfrac{\sigma_{2a}}{E_{La}} + \dfrac{\tau_{12a}}{G_{12b}} \\ \gamma_{12b} = -m_{1b}\dfrac{\sigma_{1b}}{E_{Lb}} - m_{2b}\dfrac{\sigma_{2b}}{E_{Lb}} + \dfrac{\tau_{12b}}{G_{12b}}. \end{cases} \tag{8c}$$

Solution of these equations leads to the following:

$$A_{11}\sigma_{1a}+A_{12}\,\sigma_{2a}+A_{13}\tau_{12a} = \frac{t}{t_a t_b}\left(\frac{\sigma_1}{E_{1b}} - \nu_{21b}\frac{\sigma_2}{E_{2b}} - m_{1b}\frac{\tau_{12}}{E_{Lb}}\right) \tag{9a}$$

$$A_{21}\sigma_{1a}+A_{22}\,\sigma_{2a}+A_{23}\tau_{12a} = \frac{t}{t_a t_b}\left(-\nu_{12b}\frac{\sigma_1}{E_{1b}} + \frac{\sigma_2}{E_{2b}} - m_{2b}\frac{\tau_{12}}{E_{Lb}}\right) \tag{9b}$$

$$A_{31}\sigma_{1a}+A_{32}\,\sigma_{2a}+A_{33}\tau_{12a} = \frac{t}{t_a t_b}\left(-m_{1b}\frac{\sigma_1}{E_{Lb}} - m_{2b}\frac{\sigma_2}{E_{Lb}} + \frac{\tau_{12}}{G_{12b}}\right) \tag{9c}$$

in which

$$A_{11} = \frac{1}{E_{1a}t_a} + \frac{1}{E_{1b}t_b}, \quad A_{12}^{*} = -\frac{\nu_{21a}}{E_{2a}t_a} - \frac{\nu_{21b}}{E_{2b}t_b}, \quad A_{13} = -\frac{m_{1a}}{E_{La}t_a} - \frac{m_{1b}}{E_{Lb}t_b}$$

$$A_{21}^{*} = -\frac{\nu_{12a}}{E_{1a}t_a} - \frac{\nu_{12b}}{E_{1b}t_b}, \quad A_{22} = \frac{1}{E_{2a}t_a} + \frac{1}{E_{2b}t_b}, \quad A_{23} = -\frac{m_{2a}}{E_{La}t_a} - \frac{m_{2b}}{E_{Lb}t_b}$$

$$A_{31} = A_{13}, \qquad A_{32} = A_{23}, \qquad A_{33} = \frac{1}{G_{12a}t_a} + \frac{1}{G_{12b}t_b}$$

$^{*}A_{21} = A_{12}$, numerically.

These equations may be applied, for example, to finding the internal stresses in the layers of the laminated particle in Fig. 11, made of the material shown in Fig. 8. The L-directions of layers a and b make alternate left and right hand angles of 15° with the direction of stress σ_1. A shear stress is also applied in the 1–2 directions.

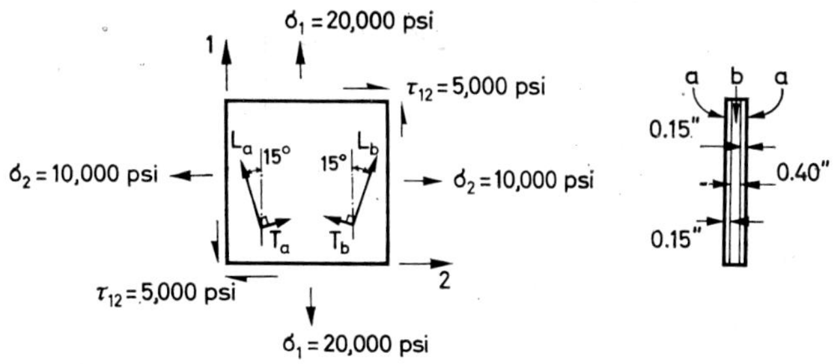

FIG. 11. Particle of laminate with layers a and b oriented as shown with respect to imposed stresses. Layers a and b composed of same material as Fig. 8

It is instructive first to determine the variation of the average stresses σ_1, σ_2 and τ_{12} at various directions, and particularly to find the values and directions of the principal stresses (where shear stress disappears) and the value of maximum shear. This is conveniently done by Mohr's circles. The result is shown at A in Fig. 12, in which the average values of σ_1', σ_2', and τ_{12}' are plotted as functions of the angle θ with respect to the 1-direction.

Equations (9) can now be employed to determine the stresses σ_{1a}, σ_{2a} and τ_{12a} in the a layers associated with the 1–2 directions. Equations (7) then solve for the stresses σ_{1b}, σ_{2b}, and τ_{12b} in the b layers associated with the 1–2 directions. Again it is instructive to determine the variations of the stresses in the individual layers a and b at various angles θ with respect to the 1-direction. Using Mohr's circles these are found to be as shown at B and C in Fig. 12. These are carried only to 180°; beyond this the curves repeat.

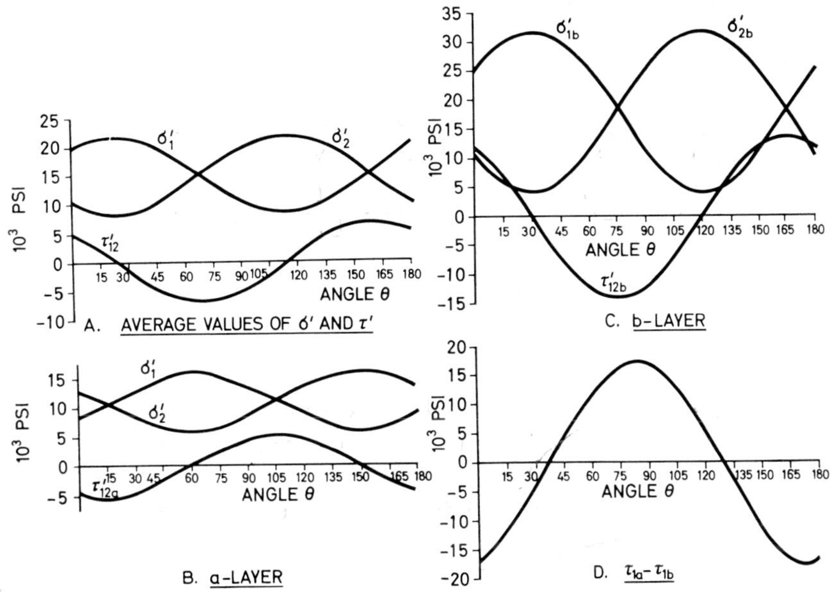

FIG. 12. A. Variation of average direct and shear stresses in particle of Fig. 11 at various angles θ with respect to 1 and 2 directions. B and C. Variations of direct and shear stresses in layers a and b at various angles θ with respect to 1 and 2 directions. D. Difference between values of shear stress in layers a and b

Examination of Fig. 12A shows that the average principal stresses attain values of 22,000 psi and 7800 psi at 22.5° and 112.5° and maximum shear stress of 7000 psi at 67.7° and 157.7° as compared with the stresses in the 1–2 directions of 20,000 psi, 10,000 psi, and 5000 psi respectively.

In the a-layers, as shown in Fig. 12B, the magnitudes of the stresses σ'_{1a}, σ'_{2a}, and τ'_{12a} are less than the average and attain maximum and minimum values of 16,000, 5700 and 5100 psi respectively. The maximum and minimum principal stresses, however, are attained at 60° and 150° and maximum shear at 15° and 105°.

In the b-layers, as shown in Fig. 12C, the range is much greater, and the maximum and minimum principal stresses go to 32,000 psi and 4400 psi at 30° and 120°, while maximum shear reaches 13,900 psi at 75° and 165°.

As this example illustrates, stresses in the layers of a laminate composed of orthotropic materials can differ widely from the average, and design based upon average values can be seriously in error. Moreover, the peak stresses in the layers do not necessarily occur in the same directions as the averages.

The positive and negative values of shear indicate that shear distortions are either to the right or to the left. As curves *B* and *C* of Fig. 12 show, shear may be directed in opposite directions at the same values of θ in layers *a* and *b*. The difference between them is an indication of interlaminar shear and is plotted at *D* in Fig. 12.

The arrangement of layers in Fig. 11 is not likely to be the best. Different arrangements will result in different stresses in and among the layers, with some of them more, and some of them less, favorable.

The same procedure can be applied to multi-layer laminates but the solution becomes laborious. Here, evidently, is an excellent opportunity for computers to carry out the tedious calculations necessary to find the best orientation of the various layers.

IV. CONCLUSIONS

As structural materials, fiber-reinforced plastics have the advantages of formability, strength and lightness, ability to tailor strength to stress, and light transmission. They have the limitations of low stiffness, cost, uncertain durability over long time-periods, and susceptibility to fire. They may be essentially isotropic or orthotropic. In laminates of orthotropic materials the stresses in the individual layers may differ widely from the average imposed stresses, and significant interlaminar shear may be set up between layers of the laminate.

ACKNOWLEDGEMENTS

Information respecting radomes was furnished by Edward Murphy, Lincoln Laboratories. George Nelson was the architect for the Moscow pavilions. Hamilton and Goody were the architects for the House of the Future. Joseph Schiffer and Marvin Goody were the architects for the Rearrangeable Schoolhouse. Engineering for the Moscow pavilions, the House of the Future, and the schoolhouse was carried out by F. J. Heger, H. R. Simpson, F. J. McGarry, R. P. Whittier, and the author.

REFERENCES

1. ORABONA, J. F. Description of Large Radomes Now in Use, CAMROC Technical Memorandum No. 12, Prepared for Cambridge Observatory Committee, M.I.T. Lincoln Laboratory, Lexington, Massachusetts, 1966.
2. Private communication, Edward Murphy, M.I.T. Lincoln Laboratory, Lexington, Massachusetts.
3. U.S. Pavilion in Moscow, *Modern Plastics*, December 1959.
4. Erecting the Moon-Rocket Building, *Engineering News-Record*, January 1965.
5. DIETZ, A. G. H., GOODY, M. E., HEGER, F. J., MCGARRY, F. J. and WHITTIER, R. P. Engineering the Plastics House of the Future, *Modern Plastics*, June and July 1957.
6. LOVE, A. E. H., *A Treatise on the Mathematical Theory of Elasticity*, Dover Publications, New York, 1944.
7. HOFF, N. J., The Strength of Laminates and Sandwich Structural Elements, in *Composite Engineering Laminates*, ed. A. G. H. Dietz, M.I.T. Press, Cambridge, Mass., 1969.

THE USE OF COMPOSITE MATERIALS IN NAVAL SHIPS†

S. R. HELLER, JR.‡

Naval Ship Engineering Center, Washington, D.C.

I. INTRODUCTION

"Lighter than aluminum! Stronger than steel! Corrosion-resistant! Rot-proof! Dielectric! Non-magnetic! Thermally insulating!" The torrent of superiorities claimed gushes on as endlessly as the glib chant of the carnival barker until one can no longer resist the Lorelei call and furtively slides his coin forward in exchange for ogling the "Miracle of the Midway", this modern version of Little Egypt. There, inside the tent basking in the spotlight, is the nonpareil, as unprepossessing as another miracle substance, penicillin.

What is this paragon of materials? A plethora of proprietary tradenames has been used to identify it. Equally incomprehensible is the polysyllabic chemical term, characteristic of complex organic compounds. For the purpose of this paper, it can be dubbed *glass-reinforced plastic*.

A. *Definition*

Glass-reinforced plastic, or GRP, is a specific kind of *composite material*. Indeed, it is that particular version of composite material which is most widely used in the construction and outfitting of naval ships. To put these terms in the proper perspective, they must be defined.

1. A *composite material* is a combination of metal alloys or plastics with built-in strengthening agents, usually in the form of filaments, foils, flakes, or whiskers of a stronger material. The resulting material has properties which are combinations of those of its constituents.

2. *Glass-reinforced plastic* is a particular kind of composite material in which a synthetic resin is strengthened by glass particles.

† The opinions expressed herein are those of the author alone and should not be construed as reflecting the official views of the Department of the Navy or the Naval service at large.

‡ Captain, U.S.N. Director, Hull Systems and Weapons Support Division.

B. *Attractiveness of Glass-reinforced Plastic*

The partial listing of superiorities claimed for glass-reinforced plastic is sufficient to whet the appetite of the most jaded structural designer. Despite his innate skepticism toward the flamboyant claims of hucksters, his professionalism demands that he sift fact from fiction. He has heard the claims, but he requires incontrovertible facts.

1. MECHANICAL PROPERTIES. Of the several superiorities claimed, that alluding to a high strength-to-weight ratio is the one which first attracts attention and interest. Table 1 is a brief summary of the strength and weight properties of typical structural materials used in naval ships and craft. The superiority of GRP in this comparison is clear. What makes these data compelling is that the properties are the *minimums* specified.

Were the strength-to-weight ratio the sole criterion of mechanical properties, GRP would be *the* logical choice as the structural material for naval ships. Indeed, if the strength-to-weight ratio were the sole criterion, we would be hard put to explain the shift from wood to iron a century ago!

TABLE 1. WEIGHT AND STRENGTH OF TYPICAL STRUCTURAL MATERIALS USED IN NAVAL SHIPS

Material	Specification	Density (lb/in^3)	Yield strength (psi)	Strength/Density (in.)	Order of merit
Medium steel	MIL-S-16113 Grade M	0.283	33,000	116,600	1.00
High tensile steel	MIL-S-16113 Grade HT	0.283	47,000	166,100	1.42
High yield steel	MIL-S-16216 Grade HY 80	0.283	80,000	282,700	2.42
Aluminum (5456-H321)	QQ-A-250	0.0975	33,000	338,500	2.90
High yield steel	MIL-S-16216 Grade HY 100	0.283	100,000	353,400	3.03
Oak	MIL-W-15154	0.0226†	8,400‡	372,200	3.19
Alaska yellow cedar	MIL-L-2594	0.0162†	7,100‡	438,300	3.76
Southern yellow pine	MIL-W-24126	0.0208†	9,300‡	446,500	3.83
Douglas fir	MIL-W-2038	0.0174†	7,800‡	449,300	3.85
GRP	MIL-P-17549 Grade 1	0.0606	45,000††	742,600	6.37

† Average value from *Wood Manual* NAVSHIPS 250–336.

‡ Average value for bending fiber stress at proportional limit from *Wood Manual* NAVSHIPS 250–336 for 12% moisture content.

†† Flexural yield strength in wet condition as determined by testing in accordance with Fed. Std. LP-406b.

Indeed, as Verweij[1] points out, to consider the strength-to-weight ratio is "an oversimplification to such an extent that it becomes partially incorrect." Verweij[1] goes on to develop the relationships for comparison on an equal weight basis:

(a) For tensile strength, the ratio of tensile strength to specific gravity.

(b) For flexural strength, the ratio of flexural strength to the *square of the specific gravity*.

(c) For rigidity, the ratio of elastic modulus to the *cube of the specific gravity*.

Accordingly, Table 2 has been prepared to show the relative merits of the typical structural materials insofar as flexural strength and rigidity are concerned. Again, the superiority of GRP over the *metals* is clear, but the superiority of the various types of wood over GRP now emerges. Table 3, an abbreviated consolidation of both Tables 1 and 2, summarizes the comparison of these mechanical properties on an equal weight basis. The superiority of wood over GRP is again apparent. Such a comparison, even though vastly better than that of Table 1 is still incomplete, for it neglects the inhomogeneity of wood as compared with metals or GRP. The strengths and rigidity of wood have been taken for the grain *parallel* to the length of the beam. When the grain is *perpendicular* to the length of the beam (either in the width or depth direction), the properties are substantially *less* than those given. Also the known weakness of wood fastenings is not taken into consideration in Tables 1–3 inclusive.

In any event, properties in addition to strength and rigidity must be considered in evaluating any potential structural material.

2. DURABILITY. Several of the superiorities claimed may be loosely linked into the category, durability. The technical literature is replete with subjective evaluations of durability.

(a) In 1952, Alfers and Bushey[2] stated, "The resins and processes evolved made possible the fabrication of large laminates having exceptional strength combined with light weight and outstanding resistance to heat, chemicals, abrasion, and fatigue." In comparing GRP boats with their wood counterparts, Alfers and Bushey cite superiority of GRP in dimensional stability, distortion, deterioration including dry rot, and resistance to marine borers.

(b) Alfers and Graner[3] reported freedom from marine-borer damage, dry rot, and corrosion based on a bit more than 10 years of experience. Maintenance costs are reported to have been "low in cost compared with repair cost for wooden hulls."

(c) Buermann and Della Rocca[4] repeated the more commonly quoted advantages, "resistance to corrosion and dry rot, and reduced maintenance ...". Resistance of GRP to most acids, oils, and solvents at normal temperatures is also reported by them.

TABLE 2. FLEXURAL STRENGTH AND RIGIDITY OF TYPICAL STRUCTURAL MATERIALS ON AN EQUAL WEIGHT BASIS

Material	Specific gravity (γ)	Flexural strength (σ_f in ksi)	$\sigma_f \gamma^{-2}$ (ksi)	Order of merit for flexural strength	Flexural modulus (E in ksi)	$E\gamma^{-3}$ (ksi)	Order merit for rigidity
Medium steel	7.84	33.0	0.537	1.00	30,000	62.255	1.00
HTS	7.84	47.0	0.765	1.42	30,000	62.255	1.00
HY 80	7.84	80.0	1.302	2.42	30,000	62.255	1.00
Aluminum	2.70	33.0	4.527	8.43	10,000	508.053	8.16
HY 100	7.84	100.0	1.627	3.03	30,000	62.255	1.00
Oak	0.63†	8.4†	21.164	39.41	1,810†	7238.639	116.27
Alaska cedar	0.44†	7.1†	36.674	68.29	1,420†	16669.797	267.77
Southern pine	0.58†	9.3†	27.646	51.48	1,990†	10199.27	163.83
Douglas fir	0.48†	7.8†	33.854	63.04	1,950†	17632.378	283.23
GRP	1.68	45.0‡	15.944	29.69	2,300‡	485.065	7.79

† Average value from *Wood Manual* NAVSHIPS 250–336.

‡ Determined in wet condition in accordance with Fed. Std. LP-406b.

TABLE 3. COMPARISON OF MECHANICAL PROPERTIES OF TYPICAL STRUCTURAL MATERIALS ON AN EQUAL WEIGHT BASIS

Material	Tensile strength	Flexural strength	Rigidity
GRP	1.0	1.00	1.0
Douglas fir	0.6	2.12	36.3
Southern yellow pine	0.6	1.73	21.0
Alaska yellow cedar	0.6	2.30	34.4
Oak	0.5	1.33	14.9
HY 100 steel	0.48	0.10	0.1
Aluminum	0.46	0.28	1.04
HY 80 steel	0.38	0.08	0.1
High tensile steel	0.22	0.05	0.1
Medium steel	0.16	0.03	0.1

As experience has accumulated, more objective evaluations of durability have become available.

(d) Alfers[5] reported several specific applications of GRP *where especially serious corrosion problems had previously been encountered.* These applications included piping for washdown of radioactive material, submarine battery exhaust ventilation, submarine fairwaters and superstructures, destroyer deckhouses, submarine sonar domes, and propeller shaft coverings. After service ranging from 5 to 7 years, the GRP was still essentially free from maintenance.

(e) Cobb[6] reported on the condition of three 40-ft Coast Guard patrol boats after 10 years of service. The three boats had been in continuous use in a variety of waters, both salt and fresh, with operating hours totaling 18,700 hours. All three boats were, at the time of the survey, operating in the Houston Ship Canal. They had been selected for this duty because of the highly corrosive attack (sulphuric acid waste dumped from oil and gas refineries) on the steel boats. The GRP hulls showed no effect from this exposure although one boat had been in this service for eight years. Testing of samples cut from the hulls revealed no penetration of the laminate by oil, water, or dirt. Moreover, the mechanical properties were found from testing of these samples to be more than adequate for the service. Service records indicated maintenance costs for the GRP hulls were only 20% of the maintenance costs for corresponding steel hulls.

(f) Spaulding[7] reported the cost of maintaining a GRP amphibious landing craft to be 25% of the cost of its wooden counterpart. It was also reported[7] that the "down-time" for repairs of the GRP hull was only 10% of that for the wooden hull.

(g) The most completely documented evaluation of the performance of a large glass-reinforced plastic structure is that presented by Fried and Graner.[8]

TABLE 4. COMPARATIVE DATA FROM GRP FAIRWATER U.S.S. *Halfbeak* (SS352)

Property	Condition	Original data†	Current data (11 years service)			Specification requirement‡
			Panel 1	Panel 2	Average	
Flexural strength (psi)	Dry	52,400	51,900	51,900	51,900	50,000
	Wet††	54,300	46,400	47,300	46,900	45,000
Flexural modulus (psi $\times 10^6$)	Dry	2.54	2.62	2.41	2.52	2.50
	Wet	2.49	2.45	2.28	2.37	2.30
Compressive strength (psi)	Dry	Not determined	40,200	38,000	39,100	33,000
	Wet		35,900	35,200	35,600	28,000
Barcol hardness	Dry	55	53	50	52	—
Specific gravity	Dry	1.68	1.69	1.66	1.68	—
Resin content (%)	Dry	47.6	47.4	48.2	47.8	35—43

† Average of three panels.
‡ MIL-P-17549, Grade 1.
†† Two-hour boil.

The structure involved was a large GRP fairwater installed in U.S.S. *Halfbeak* (SS352). During the fabrication and installation, samples of the laminates were continuously monitored by the U.S. Naval Applied Science Laboratory to assure compliance with the specification, MIL-P-17549, Grade 1. Five years after installation, a small sample was tested and no serious degradation was noted. Indeed, Verweij[1] quotes a flexural strength of 88% and flexural modulus of 91% of original values. Eleven years after installation when a modification to the structure was approved for military reasons, samples were again taken for testing. The results are given in Table 4 (table 2 of ref. 8).

FIG. 1. Marine fouling on GRP test panels after 14-month immersion in sea water

Table 4 furnishes evidence that the mechanical properties of GRP which has been properly fabricated (i.e. reinforcement well "wetted out" by the resin, low void content, and meets specification requirements in all respects) are not degraded with time.

In addition to their extensive treatment of the mechanical properties of the GRP fairwater for *Halfbeak*, Fried and Graner[8] also reported results of a separate, but related, investigation of extended sea water immersion effects. Pertinent to this discussion are three of their findings:

(1) After a 5-year immersion period, a laminate, similar to that used in the *Halfbeak* fairwater, retained over 95% of original strength and stiffness characteristics.

(2) The properties of properly fabricated GRP materials are not affected seriously by long-term exposure to weather.

(3) Samples were covered with marine growth as shown in Fig. 1 (fig. 7 of ref. 8), but were easily cleaned by mild scraping as shown in Fig. 2 (fig. 8 of ref. 8). Further, it appears to be feasible to provide an anti-fouling agent directly with the resinous matrix to discourage marine fouling.

3. DIELECTRICITY. Alfers and Graner[3] cite extensive use of plastic in general as electrical insulating material. Buermann and Della Rocca[4] provide a rationale for the insulating properties by exploring the composition of the glass used for reinforcing. They conclude that the glass, a lime-alumina

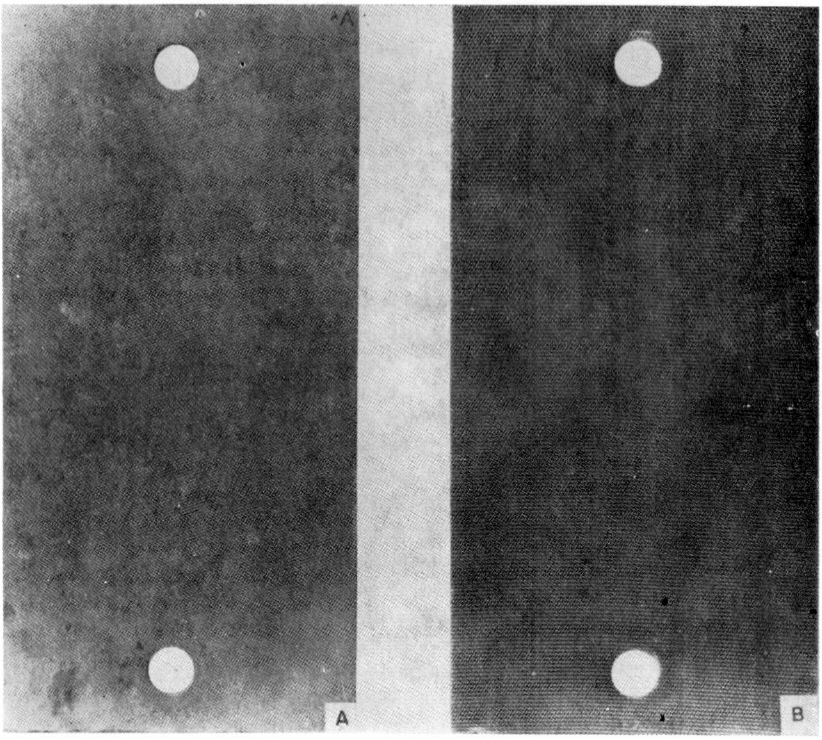

FIG. 2. Test panels in Fig. 1 after cleaning

borosilicate of low alkali content and high moisture resistance, was originally developed for electrical insulating application—hence the term "E-glass". The really telling blow, however, is provided by Alfers[9] who cites the specific choice of GRP as the material for constructing large masts for the command ship, U.S.S. *Wright* (CC2), for enhanced communication performance because of its electrical insulation properties.

4. NONMAGNETISM. It is virtually self-evident that non-ferrous materials are non-magnetic. Hence, GRP fully qualifies.

5. THERMAL INSULATION. Alfers[5] cites the replacement of a steel sea cabin in U.S.S. *Southerland* (DDR743) with a GRP structure of sandwich construction, the core of which was a light-weight phenolic-impregnated paper honeycomb. Alfers[5] further states, "No additional thermal insulation was used, since the honeycomb panel was found to be the equivalent of 1-inch thick fiberglass blanket... It has been reported to be cooler in hot climates...." Della Rocca[10] confirms the adequacy of the thermal insulation properties of GRP sandwich construction. Cobb[6] reports that, during a serious tanker fire at the port of Houston, "the crew of the fiberglass boat

TABLE 5. HEAT CONDUCTIVITY OF TYPICAL STRUCTURAL MATERIALS USED IN NAVAL SHIPS

Material	Heat conductivity (Btu/ft/ft²/hr/°F)
GRP (single-skin)	0.12
Core (polyvinyl chloride foam)	0.03
Oak	0.14
Fir	0.09
Pine	0.14
Steel	40
Aluminum	140

were able to operate closer to the intense flames and for a longer period of time than with steel hulls because of the low heat conductive nature of the fiber-glass boat. It was, furthermore, pointed out that while the topside paint was scorched, the laminate did not ignite nor was it otherwise damaged." Table 5, an abridgement of data reported by Verweij,(1) compares the heat conductivity of typical structural materials. The superiority of GRP over metals in this respect is clear. Moreover, in the single-skin construction mode, GRP is equivalent to wood in heat conductivity; in the sandwich mode, GRP is superior to wood.

II. USE IN SURFACE SHIPS

According to Alfers and Bushey,(2) the U.S. Navy first evinced interest in structural plastics by a contract in 1942 for 12 laminated plastic surf boats with Western Plastic of Los Angeles, California. The boats, to be laminated of cotton cloth bonded with a thermoplastic material in a volatile solvent, were for evaluation. Because the hull shape was questionable, the design was modified in 1945. Although a production contract for 300 boats was awarded to the same company in July 1945, the end of World War II shortly thereafter caused cancellation. During the course of these initial boatbuilding contracts, polyester thermosetting resins were developed and found immediate application in the manufacture of parts for military aircraft. Because of the availability of thermosetting resins which require no heat in their cure, the thermoplastic materials, which do require heat for manufacture, were virtually discarded as boatbuilding materials.

Aircraft experience during World War II with reinforced plastics including glass laminates demonstrated that these materials were specially adaptable to mass production with people of *little skill*. Moreover, the raw ingredients were *noncritical* in a wartime environment. Furthermore, man-hours for production with these synthetic materials were only about one-fourth of those required for conventional material.

A. *Early Use in Boats*

In spite of the sad experience of the two aborted boat contracts, the great potential of the thermosetting resins was sufficiently attractive that the U.S. Navy's Bureau of Ships, in June 1946, placed two development contracts for 28-ft personnel boats of laminated plastic construction. One of the contracts, that with Winner Manufacturing Co., was executed by the "bag molding" method; i.e., a single female mold was used to receive the laminate and a bladder was placed on top to exert pressure during the cure. The other contract, with Marco Chemical, Inc., was executed by the "injection method", i.e., a pair of mating molds was filled with the reinforcing material, the resin was forced between them by drawing a vacuum,[7] and the laminate was then cured with the molds in place. It is of historical interest to note that, when the first hull was removed from its mold on 30 December 1947 under the Winner contract, it was, at that time, the largest monolithic plastic structure ever built.

Although the bag molding method was used with some success commercially, the three boats built by Winner with this method were the only U.S. Navy craft so constructed because of its expense and complexity. On the other hand, the injection method was widely used for Navy craft for nearly a decade. The continuing problem with the injection method was resin content. Throughout the laminate there were dry spots with no resin and other spots with overconcentration of resin. At best, the resin content was about 75% with correspondingly low mechanical properties.

Spaulding[7] has said that 1950 was a turning point. In that year, the U.S. Navy commenced "production" (as opposed to "developmental") contracts for plastic boats. Also in that year, the U.S. Navy began serious consideration of "sandwich" construction (as opposed to "single skin"). In addition, 1950 marked the start of laying up laminates by hand using glass mat and cloth (Fig. 3) in various combinations.

Figure 3 also shows glass woven roving. When this low cost reinforcement became available in 1955, it sounded the death knell of the vacuum injection method. With it came laminates of low resin content but with high strength, produced rapidly by the "hand lay-up" process. Heretofore, hand lay-up had not produced laminates which were improvements over those made by the "injection" method.

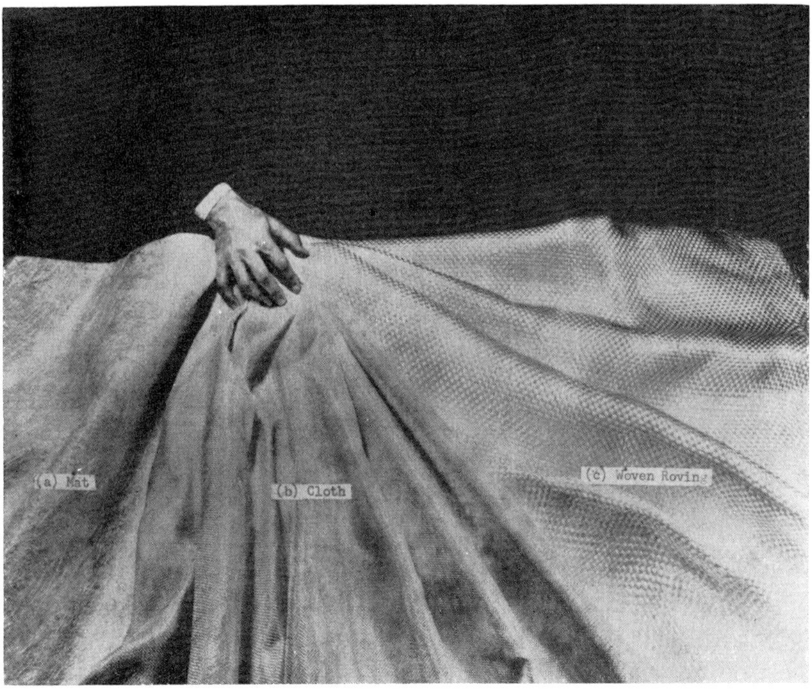

FIG. 3. Typical examples of glass mat, cloth and woven roving

Although the early Navy uses of sandwich construction were unsuccessful and the method is not now used for hulls, the concept deserves more attention than mere passing mention. In 1951, a contract was awarded for the development and construction of a GRP minesweeping boat (MSB), 57 ft long. After five years of intensive research and development, the boat was delivered. The midship section of the MSB, shown in Fig. 4 (fig. 16 of ref. 3), featured a 3/16 in. outer skin, an 1/8 in. inner skin, and a honeycomb core of resin-impregnated cotton cloth weighing 4.7 lb/ft^3 and $1\frac{3}{4}$ in. thick. The boat has been in use at the Mine Defense Laboratory at Panama City, Florida as a general purpose workboat for 10 years. Although the honeycomb core has become filled because of water absorption, the hull is sound structurally and has been evaluated as "satisfactory".

Between 1955 and 1962, 32 Navy craft from 33 to 50 ft in length were constructed by the "core mold" process. This process uses a number of wooden forms in place of a mold (Fig. 5) over which low density foam core boards are planked. When the core planking is complete, one skin (usually the outer using a female mold) is laid up. This is followed by the installation of bulkheads and girders. Then, the hull is inverted, the wooden forms removed, and the inner skin laid up. Although the process was conceived as a means

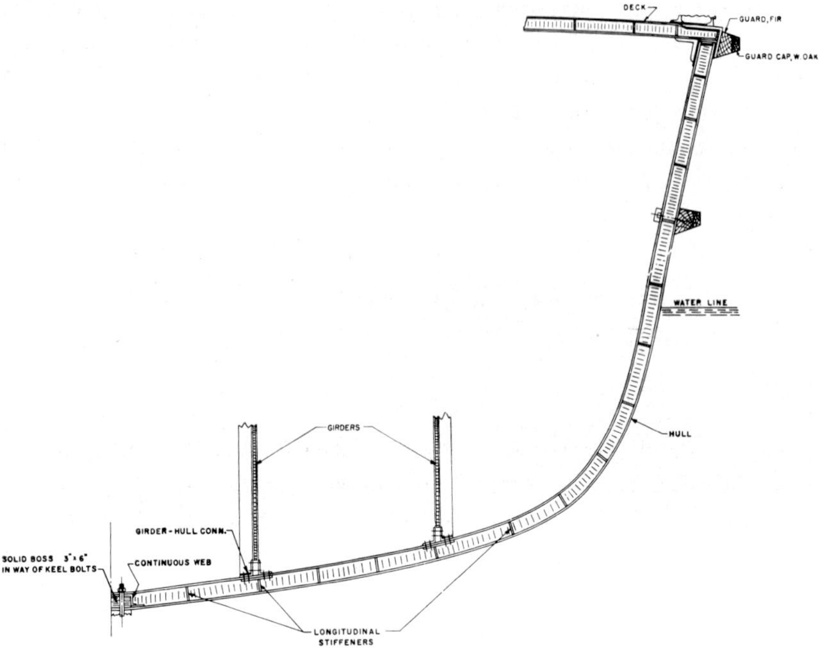

FIG. 4. Midship section of MSB

FIG. 5. Core mold construction of boats

of minimizing mold costs, the system proved to be expensive. Moreover, the finished hull was inferior—often unsatisfactory structurally—because of fastening, joint, and core bonding problems as well as poor impact resistance associated with thin skins and brittle cores.

Despite the lack of success in hull construction, the sandwich method has been successfully used for bulkheads, flats, decks, and cabin tops. Polyurethane and cellular cellulose acetate (CCA) weighing 6 to 8 lb/ft^3 and endgrain balsa have given satisfactory service. Generally speaking, honeycomb cores continue to be plagued by water absorption, poor face bonds, and lack of resilience.

B. *Early Use in Ships*

In the section *Attractiveness of Glass-reinforced Plastic*, a number of early applications were cited in support of superiorities claimed for GRP. It is pertinent now to consider these and other applications in greater depth.

1. PIPING. Alfers and Graner[3] have described the extensive laboratory evaluation of GRP pipe and the pilot installation in U.S.S. *Keller* (DE419) of glass-reinforced-polyester pipe and fittings. Unfortunately, this pilot installation performed unsatisfactorily in service. Hindsight—always 20/20—pinpoints the difficulty as "too much, too soon". The installation preceded proper material development.

Offsetting this poor performance of early GRP pipe are two later (1960) developments:

(a) The report by Zimmie, in discussing,[4] that the American Bureau of Shipping had, at that time, approved the use of certain grades of polyvinyl chloride piping for limited application. Zimmie also reported the installation of over 2100 ft of such piping in a Great Lakes ore carrier.

(b) The reports by Alfers[5] of highly successful use of polyvinyl chloride piping in a wide variety of naval ships in the washdown system (Fig. 6). Alfers also furnished comparative cost and weight data which is repeated here in Table 6.

TABLE 6. COMPARISON OF WEIGHT AND COST OF TYPICAL PIPING MATERIALS

Material	Relative weight	Relative cost
Polyvinyl chloride	1.00	1.00
Aluminum	1.12	0.98
Steel	2.54	0.47

FIG. 6. Washdown system using plastic pipe

FIG. 7. GRP sea cabin for U.S.S. *Southerland* (DDR743)

A related application is that of ventilation ducts. Alfers[5] reports that about 15% of the ducts in U.S.S. *Bittern* (MHC43), a wooden minesweeper, were made of polyvinyl chloride.

2. SUPERSTRUCTURE. The replacement of a steel sea cabin in U.S.S. *Southerland* in 1955 has previously been mentioned[5] with regard to thermal insulation qualities. The structural aspects are also noteworthy. The dimensions are 9 ft by 7 ft by 6 ft 5 in. high; an overall view is shown in Fig. 7. Sandwich construction was used throughout. The outer skin was 3/16 in. thick; the inner skin, 1/8 in. thick; the overall thicknesses were 2 in. for the decks and $1\frac{3}{4}$ in. for the sides. The core was phenolic impregnated paper honeycomb weighing 4.5 lb/ft^3. The design load was a wind of 130 m.p.h. The weight of the GRP structure was 2085 lb as compared to 6000 lb for the steel structure it replaced—a weight saving of 65%.

A similar but larger GRP structure was installed[5] in U.S.S. *Fechteler* (DDR870) in 1956. This deckhouse extension, 35 ft long by 4 ft wide by 7.5 ft high, was also of sandwich construction. The skin thicknesses were identical to those used in the sea cabin, but the cores were thicker. The sides used cores $3\frac{1}{2}$ in. thick; the deck, 2 in. thick. In addition, there were three internal bulkheads of similar construction. The design problem was somewhat complicated in that there had been numerous cases of storm damage to the steel structures and reinforcement was impossible because weight and moment compensation was not available in those old ships. Consequently, an adequate reinforcement of the steel structure was designed, and then the GRP structure was designed for comparable strength. The reinforced steel structure would have weighed 7500 lb, the GRP replacement weighed only 2400 lb—a weight saving of 68%.

In addition to the weight savings, no additional thermal insulation was required, more usable space and neater appearance were gained by the use of flat sides without internal stiffeners, and the isolation of dissimilar metals to prevent electrolysis (a problem when aluminum is used to replace steel) was avoided.

3. TOPSIDE APPURTENANCES. The specific choice of GRP[9] as the material for constructing large masts for the command ship, U.S.S. *Wright* (CC2), for enhanced communication performance because of its electrical insulation properties has previously been mentioned. These masts, shown in Fig. 8, vary in length from 33 ft to 83 ft. The 83-ft mast (Fig. 9) had an extremely interesting structural aspect: its length above the weather deck, 74 ft, was required to be unsupported (i.e., without guys or stays). The long mast varies in diameter from 5 ft at the base to 2 ft at the top and tapers in thickness from 1 in. at the base to 1/2 in. at the top. Also of interest is the mode of construction. The mast was molded in two longitudinal parts and then joined

by bonding. Despite the large deflections which accompany the large unsupported length, these long masts have performed satisfactorily for nearly three years at sea. A similar set of masts has been installed in U.S.S. *Saipan* (AGMR2). In this case, the mast has been molded in one piece with the longitudinal seams eliminated.

Radomes, protective covers for radar antennas, are, perhaps, the oldest shipboard application of glass-reinforced plastics.[2] For this use, a character-

FIG. 8. GRP mast installation in U.S.S. *Wright* (CC2)

istic which makes GRP unique among structural materials for ship use—minimum reflectivity of electromagnetic radiation—is exploited. Thus, a radome protects the radar antenna from the marine atmosphere, wind, and blast without adversely affecting the equipment's detection capability.

4. TANKS. As has already been stated by Cobb[6], properly fabricated GRP is not penetrated by oil or water. It was natural, therefore, that GRP be tried as the structural material for the internal tanks for wooden minesweepers. Its strength-to-weight advantage, its corrosion resistance, the noncriticality of its raw materials, and its inherent non-magnetic character all combined to make it a logical competitor to non-magnetic metals. A typical example is shown in Fig. 10.

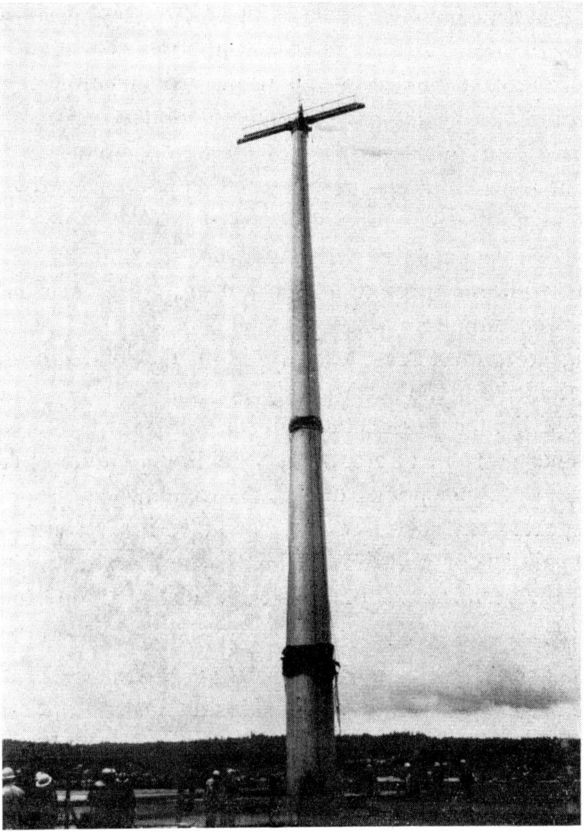

FIG. 9. 83-ft GRP mast for U.S.S. *Wright* (CC2)

FIG. 10. GRP tank for minesweeper

According to Alfers and Graner[3] these tanks were required to pass successfully the following tests:

(a) *Pressure*. No leakage was permitted when filled with water and pressurized for 24 hours to a head equal to twice the tank depth.

(b) *Deflection*. During the pressure test, the maximum deflection must not exceed 1/40 of the minimum span between stiffening members or 1/2 in., whichever is less.

(c) *Sloshing*. No leakage was permitted when the tank was 2/3 full of water and rolled for 40 hours through a total arc of 30 degrees at a frequency of 18 to 20 cycles per minute.

Tanks designed to meet these test criteria in addition to the environmental conditions specified for the specific ship applications were reported[3] to be from 30 to 40% lighter than metal tanks designed for the same service. The weight was reported to be approximately 0.9 lb per gallon of capacity; costs varied from $2 to $3 per pound of finished weight.

Alfers[5] reported extensive use of GRP tanks in capacities ranging from 20 to 20,000 gallons for a variety of boats, small craft, and minesweepers. Furthermore, the performance of these tanks has been reported to be highly successful.

5. MISCELLANEOUS. Alfers[5] reported that polythene sheets of varying thickness are used in nuclear powered ships as neutron radiation shielding.

Also reported by Alfers[5] are trial installations of decorative laminates for interior partitions and bulkheads where the freedom from maintenance offsets the cost of material and installation. It has been shown that, for service lives in excess of only four years, the laminated partitions are cheaper than the conventional painted metal ones.

Over 2000 size 0–type 0 minesweeping floats, 17 ft long, have been in service since 1956 *without failure*. These floats weigh 950 lb each, less than half of their stainless steel counterparts. They are more easily handled because of their lesser weight and have *no* magnetic signature (Fig. 11).

C. *Present Use in Ships*

The unparalleled success in the previously described trial applications has confirmed the superiorities claimed for GRP and has, indeed, established it as a structural material.

1. BOATS. There have been more then 3000 GRP boats built for the U.S. Navy in the past 20 years. Service experience with them including some for as much as 18 years has, in general, been excellent. This experience is summarized by the following:

(a) *Strength*. Spaulding[7] reported the results of full-scale crushing and

FIG. 11. Minesweeping float

bending tests on 26-ft motor whaleboats made of both wood and GRP. The boats were loaded in sagging condition by a hydraulic press. The wooden boat failed at a load of 24,997 lb with a keel deflection of $16\frac{1}{2}$ in.; the plastic boat sustained a load of 41,698 lb with a deflection of $3\frac{1}{8}$ in.

(b) *Resistance to underwater explosions*. Spaulding[7] reported that limited tests against both wooden and GRP 36-ft minesweeping launches (MSL) indicated superior performance of the plastic boats.

(c) *Costs*. Spaulding[7] reported that, when procured in quantities of 10 or more, GRP boats cost about the same as wooden boats. This reflects amortization of the cost of the mold which is not required for wood construction. Material costs average about $0.375 per pound for fire-retardant resin,

$0.49 per pound for woven roving, and $0.44 per pound for the laminate. Labor costs for GRP are drastically reduced from those of wood because of the use of *unskilled* labor for GRP and the need for only about half the corresponding man-hours of labor. As an average, the cost of structure, including labor, material, overhead, and share of the mold proportioned for quantities of 10 or more, for a Navy GRP boat is about $1.50 per pound.

Recent data from the Naval Amphibious Base at Coronado, California on comparative maintenance costs of wooden and GRP amphibious landing craft used to train assault boat coxswains are pertinent. These boats averaged 100 landings per day in light to moderate surf and were always used by inexperienced personnel. There were 18 wooden and 15 plastic LCVP involved; cost data were accumulated for each boat over a 9-month period.

Material costs for the wooden boats totaled $2978.22 for an average of $165.46 per boat. For the plastic boats, the material costs totaled $1818.91 for an average of $121.26 per boat. For the wooden boats, the total man-hours involved were 3349 for an average of 186 per boat; for the plastic boats, a total of 1505 man-hours were involved for an average of 100 per boat.

Although this report stressed the fact that the level of skill required for repairs to wooden boats was distinctly greater than that required for plastic boats and asserted that the training needed for repair personnel was more than six months for wood and only five weeks for GRP, in projecting the labor costs from the man-hours reported the *same pay scale* will be assumed. A daily rate of $50 (equivalent to $6.25 per hour) will be assumed. The comparative repair costs are given in Table 7 from which it can be seen that GRP repairs are about half those of wood. Factors which have not been included in Table 7 are that major repairs to wooden boats require longer time and greater storage area for materials than for GRP boats. Also not included is that wooden repairs require extensive *shop* equipment and tools whereas GRP repairs can be accomplished with *portable* power tools.

Because of this outstanding experience, it is little wonder that Spaulding[(7)] has said, "Fiberglass has replaced wood as a material for new hull construction for Navy small craft to 60 feet in length." Indeed, it can be stated almost

TABLE 7. COMPARATIVE COSTS OF REPAIRS TO WOOD AND GRP BOATS

Material	Material cost†	Man-hours required†	Pay rate‡ ($/MH)	Labor cost	Total cost	Relative cost
Wood	$165.46	186	$6.25	$1162.50	$1327.96	1.00
GRP	$121.26	100	$6.25	$ 625.00	$ 746.26	0.56

† Derived from Commanding Officer, Naval Amphibious Base, Coronado letter serial 3711 of 17 October 1966.
‡ Assumed.

FIG. 12. 31-ft river patrol boat

categorically that all Navy small craft are constructed of GRP. A typical recent example is the 31-ft river patrol boat (PBR), 120 of which were procured in 1966 for use in Vietnam (Fig. 12). Only for exceptionally arduous service would either steel or aluminum be selected in place of GRP.

2. CRAFT HULLS. Although the U.S. Navy has not used GRP for craft longer than 60 ft, there is a large and steadily growing body of foreign experience with commercial craft that are longer. Two 77-ft pilot boats using sandwich construction have been built in Holland[11] and have now had two years of successful experience in the hard pilot service off the Hook of Holland. Della Rocca[10] cites satisfactory service of 67-ft power yachts built in England, an 80-ft passenger boat and a 107-ft, 100-ton tanker built in Russia, and a 74-ft side trawler built in South Africa, but furnishes no details of their design or construction. Spaulding[7] reported the existence of these same craft and indicated a 96-ft trawler was under construction in South Africa. Advertising information indicates that a design of a 130-ft trawler is on hand. The January 1967 issue of *Reinforced Plastics* had a short article describing the construction in GRP of a 34-ft test section of a 160-ft long ship contemplated. The effort was jointly executed by Bristol Aeroplane Plastics, Ltd. and John I. Thornycroft & Co., Ltd. under contract with the British Ministry of Defence (Navy). A sandwich-type construction using a patented box core scheme was employed. The consortium of Bristol-Thornycroft was reported in this article to be ready to take orders for ships up to 200 ft in length. Verweij,[1] however, reflects more caution in setting 120 ft

in length as the upper limit for use of GRP for hull structure. Coincidentally, that is the upper limit set for GRP in the Rules published by Lloyd's Registry of Shipping.

The U.S. Navy has not been dormant in its consideration of the use of GRP for the hull structure of craft longer than 60 ft. In 1966, it requested bids for a 80-ft YP (a yacht-like craft used for seamanship training of midshipmen at the Naval Academy) to be built of GRP using the single-skin construction scheme. Unfortunately, this proved to be a poor choice *since only one unit was desired*. The bids received reflected the high cost of molds and were rejected because of the inordinate expense (about $2.65 per pound of structure) involved.

Although the U.S. Navy was thwarted in its efforts to advance the state of the art by building a 80-ft GRP hull, it is *my* opinion that, if *multiple* units of a *new* craft up to 120 feet in length were now required, these craft would be built of GRP. Certainly, the feasibility studies reported by Spaulding and Della Rocca,[12] the statement by Spaulding[7] "that GRP had essentially replaced wood as a hull construction material except in large minesweepers", and the existence of Lloyd's Rules for craft of that length bolster and lend credibility to *my* opinion.

3. SUPERSTRUCTURE. Prior successful experience with GRP superstructures and deckhouses has been exploited. The use of GRP for this application has been expanded. For example, the shipbuilder for the first several ships of the U.S.S. *Asheville* (PGM84) class of patrol gunboats opted to build the deckhouses, one level 58 ft long by 18 ft wide by 9 ft high on which is mounted another level 27 ft long by $10\frac{1}{2}$ ft wide by 8 ft high, of GRP rather than aluminum (the hulls are aluminum). Lindsay[13] reports that the use of GRP

FIG. 13. Overall view of GRP deck house after test blast

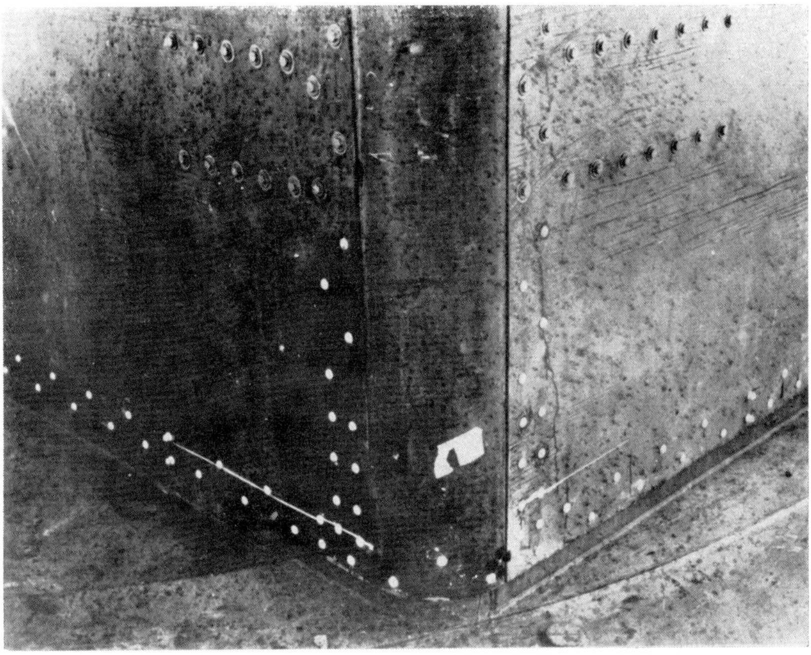

FIG. 14. Close-up of typical minor damage after test blast

significantly reduces radar reflectivity "to the point where even a very experienced and sharp-eyed radar operator has difficulty in detecting the ship." Lindsay[13] also claims a further advantage for GRP in this application since there is no spalling action when attacked by small caliber ammunition. It is pertinent to note, however, that the shipbuilder made several attempts before a successful superstructure was developed and the resulting structure was *slightly heavier* than the original design of aluminum. This experience proves that it is very difficult for GRP to compete with aluminum on weight and cost for *extremely light structure*.

In addition, a small surface ship deckhouse 24 ft long by 17 ft wide by 8 ft high, was designed of GRP to withstand a high level air blast. The structure was built and tested by such a blast along with similar aluminum structures that represented both conventional and blast resistant designs.[9] Although this was the first GRP deckhouse designed to resist blast and the response of such a structure and its fastenings to very high rates of loading was imperfectly understood, the structure withstood the blast acceptably. Figure 13 is an overall view of the deckhouse after the test blast; Fig. 14 is a close-up of representative minor damage.

4. TOPSIDE APPURTENANCES. Radomes continue to be built of GRP. As the antennas have increased in size and complexity, the radomes too have

become larger. Moreover, the recent radomes have been designed, as was the experimental deckhouse previously discussed, to withstand high level air blast.

The design of the new MSO-523 class of ocean minesweepers includes a GRP structure for the combined purposes of a mast and exhaust gas disposal. Selection of GRP for this "mack" was based on weight-saving, corrosion-resistance, low maintenance, and the non-magnetic property of the material.

5. MISCELLANEOUS. The design of the new MSO-523 class of ocean minesweepers also features extensive use of GRP for ventilation ducts, tanks, furniture, and hull fittings such as ladders, lifelines, and minesweeping floats.

In naval ships in general, formed polyvinyl chloride sheets are used for urinal partitions and flat polyester glass laminated sheets are used for shower and water closet partitions.[5] On the contrary, lavatories, waterclosets, and soap dishes which have, in the past, been made of plastic are no longer used because of their poor resistance to abrasion by cleaning and scouring materials. Piping and shielding applications previously described have been continued.

Spherical flasks for storing compressed air for starting gas turbines are now made of structural plastics of very high tensile strength.[5] These flasks are relatively small, up to $3\frac{1}{2}$ ft^3 capacity. They do, however, demonstrate again weight saving and freedom from corrosion. The application requires[3] severe testing: 10,000 cycles from 0 to 3000 psi at room temperature, burst pressure in excess of 6000 psi and to ability withstand the gun-fire test and other requirements of the Military Specification MIL-C-1210 for metal high-pressure gas cylinders.

Lindsay[13] reports the use of a rather sophisticated composite armor for ships operating off the coast of Viet Nam. According to Lindsay, the armor is composed of "a facing of very hard material (aluminum oxide and boron carbide are two materials used) cemented to a back-up laminate of glass fiber woven roving and polyester resin."

The fan blades for the U.S. Navy's first ground effect machine, SKMR-1, were made of GRP skin and internal stiffening bonded to the aluminum hub. Equal strength designs were made for both aluminum and magnesium blades. A weight comparison showed a savings of 37% over the magnesium construction and 54% over aluminum. Over three years of operation have been logged successfully insofar as the fans are concerned.

6. PLASTIC FOAMS. Although the material used for insulation and filling of voids is not glass-reinforced plastic nor, strictly speaking, a composite material, it is a close relative and deserves mention here. Vinyl chloride foam in sheets and foamed-in-place polyurethane are now relatively widely used for thermal insulation. Not only is the thermal conductivity equal or superior

to previously used material, but the time and man-hours of labor required for installation are *an order of magnitude less*.

Similar materials similarly installed are now used to provide buoyancy in small boats. They are also used to fill rudders and bilge keels with attendant weight saving and freedom from maintenance. As with the insulation, the time and man-hours to install are significantly less than the wood and bitumastic previously used.

A special minesweeper has been rendered virtually unsinkable by the use of foamed-in-place plastic buoyancy material. In this application, the buoyancy material is placed in the cargo holds of the merchant hull which was converted.[(5)]

III. USE IN SUBMARINES

Nowhere in the Navy is the environment more severe than in submarines. Yet, at the same time, the pressure for weight-saving in submarines is unremitting. It was natural, therefore, that submarine applications of GRP were among the first found.

A. *Early Use*

In the earlier discussion of *durability* mention was made of the GRP fairwater for *Halfbeak* and several GRP applications where especially serious corrosion problems had previously been encountered. These pilot installations, indeed, heralded more widespread use.

1. FAIRWATERS. The outstanding success of the fairwater installed on *Halfbeak* (Fig. 15) in 1953 led to a series of similar GRP installations as the aluminum fairwaters became uneconomical to repair. Alfers[(5)] cites installations on U.S.S. *Grampus* (SS523) in 1956, on U.S.S. *Cutlass* (SS478) and U.S.S. *Irex* (SS482) in 1957, and on U.S.S. *Tiru* (SS416), U.S.S. *Requin* (SS481), U.S.S. *Pickerel* (SS524), and U.S.S. *Pomfret* (SS341) in 1959 and 1960. Fried and Graner[(8)] state, furthermore, that "they are easily competitive in cost and weight with aluminum construction." Service reports are glowing with respect to reliability and freedom from maintenance.

2. SUPERSTRUCTURES. At about the same time, 1953, that the pilot installation of a GRP fairwater on *Halfbeak* was being made, a corresponding installation of a GRP superstructure (the walking area above the pressure hull) was made on U.S.S. *Croaker* (SSK246). Damage was sustained by contact with tugs and piers during mooring operations. Moreover, according to Alfers[(5)] the plastic flexed and emitted crackling noise. The design was, at best, marginal, so the GRP structure was removed.

FIG. 15. GRP fairwater for U.S.S. *Halfbeak* (SS352) under construction

The successful experience with the GRP fairwater on *Halfbeak* pointed out why the *Croaker* installation was not successful and how to solve the noise problem. In 1956, a new GRP superstructure design was executed and a small test installation was made on U.S.S. *Guavina* (AOSS362). The weight of this GRP installation was 1543 lb; that of the conventional construction it replaced was 1817 lb. The success of the *Guavina* installation led to a more extensive one, about 100 ft long, in *Tiru* in 1960.

3. PIPING AND VENTILATION. Because of the severe corrosion associated with battery exhaust ventilation, a trial GRP installation was made in U.S.S. *Perch* (APSS313) as a replacement of the rubber-covered metal system. Similarly, the rubber-covered copper tubing used in battery cooling water systems has been replaced with polyvinyl chloride piping. Both installations have been eminently successful.

4. SONAR DOMES. Investigation of material properties indicated that good sonar transmission at various frequencies was achievable with glass-reinforced plastics. When this attribute is coupled with weight-saving, freedom from maintenance and corrosion, and ease of fabrication of complex shapes, there is little wonder why there was early interest in using GRP in place of rubber and steel for sonar domes. Alfers(5) also cites the smoothness and fairness of the surface as advantages because of the corresponding reduction in flow noise.

As a result of these numerous apparent advantages, an experimental installation was made in 1954 during the trials of which all were confirmed. Consequently, 54 of these domes were installed in 1955. After four years of service, there had been only one failure reported. Following these installations was that of a larger ribless conformal nose dome in U.S.S. *Albacore* (AGSS 569) the service and performance of which have been excellent.

After several abortive attempts to produce a plastic dome with plastic ribs for U.S.S. *Nautilus* (SSN571), the structural problems were solved by designing and installing a composite dome—steel trusses for ribs with a GRP skin. This composite dome proved to be eminently satisfactory, both structurally and acoustically, during *Nautilus*'s history-making voyage under the North Pole. This success led to several similar installations.

5. Insulation. Foamed-in-place polyurethane was used instead of cork as insulation for the refrigerated storage spaces in U.S.S. *Grayback* (SSG574) in 1957 and U.S.S. *Skipjack* (SSN585) in 1958.

Vinyl chloride sheet has been used in place of cork to insulate escape trunks. This material is fire-resistant and requires no vapor seal. Moreover, it has better impact strength than cork and is easier to install. Alfers(5) states that installed cost, weight, and thermal conductivity are comparable to cork.

Plastic foam, preformed, has also been used in place of cork for pipe insulation for low temperature service. The plastic pipe lagging is, according to Alfers,(5) 40% cheaper than cork, is easier to install because of its flexibility and is only 60% of the weight of cork.

B. *Present Use*

"Nothing succeeds like success" is a shopworn aphorism. Nevertheless, it is true. And because of its truth, it succinctly summarizes the present use of GRP in submarines: extension of that which has been proven.

1. Fairwaters. The continued success of installations patterned after that on *Halfbeak* has led to more and more similar installations. As the existing aluminum fairwaters become uneconomical to repair or when there is to be an extensive change to accommodate new features, the new fairwaters are constructed of GRP.

2. Superstructures. Successful installations in *Guavina* and *Tiru* have led to still other similar installations. Probably the most extensive use of GRP in this sort of external appendage has been in U.S.S. *Dolphin* (AGSS555), a small deep-diving research submarine which, being weight-limited, required extreme weight-saving measures. Not only was the superstructure (dorsal

FIG. 16. Artist's conception of U.S.S. *Dolphin* (AGSS555)

position) of *Dolphin* constructed of GRP, but the pectoral and ventral fairings (Fig. 16) for sonar hydrophones also were made of GRP.

3. PIPING AND VENTILATION. The satisfactory service experience with plastic battery exhaust ventilation and battery cooling water piping has led to continued routine use of plastics for these applications. In addition, polyvinyl chloride piping is now used for electronics cooling water piping.

4. SONAR DOMES. The series of successful GRP sonar domes over the past dozen years has led to the selection of GRP as THE material for this application. Figure 17 shows a typical submarine sonar dome now in use.

5. INSULATION. Foam plastics continue in use for thermal insulation in submarines. Recent tests, reported by Alfers,[(5)] have demonstrated the practicability of spraying urethane foam insulation. Projections of the demonstration indicate the use of only *one-twelfth* the labor involved in installing cork insulation.

6. VOID FILLING. Foamed-in-place plastics have virtually replaced wood packing and bitumastic as the filling of non-functioning voids in submarines just as is the case on surface ships because of weight saving and freedom from maintenance. Approximately 20,000 lb of plastic void filling material of all types is now used in each submarine.

Unicellular urethane foams with densities of 2 to 8 lb/ft^3 and compressive strengths of 20 to 165 psi at 10% deflection are customarily used for voids inside the ship. The fillings, in these cases, serve no structural purpose, but are used only to prevent trash and water from collecting and to protect the metal from corrosion.

The voids in rudders and diving planes, on the other hand, must be filled with a material which will provide structural support to the plating. This

FIG. 17. Typical GRP submarine sonar dome

application, then, requires greater strength than that of the filler used for nonstructural purposes. Syntactic foams, made of phenolic microballoons and either an epoxy or a polyester resin, have been developed for this purpose. Densities vary from about 20 to about 40 lb/ft^3. An extensive program was carried out by the Republic Aviation Corporation for the Portsmouth Naval Shipyard to develop a filler for *Dolphin*. One of the results of this program, Fig. 18, is the variation of ultimate compressive strength with density for filling compounds using epoxy resin. Pourability, water absorption, permanent set, and flammability are considered in addition to strength and density. Generally speaking, epoxy based compounds have better strength-to-weight ratio and pour rates than polyester compounds for the same permanent set, water absorption, and flammability values.

7. RUDDERS. Because of the need for weight saving in *Dolphin*, the rudders were designed and constructed as composites. Each rudder had a steel spar and hub, GRP skin, and epoxy void-filling compound. Figure 19, a cutaway view of one of the rudders, shows the mode of construction.

8. DECKS. The extreme need for weight saving in *Dolphin* led to the use of GRP for internal platform decks. The deck and its supporting structure could not exceed 5 lb/ft^2. As installed, the deck is composed of two thin GRP skins bonded to a GRP honeycomb core.

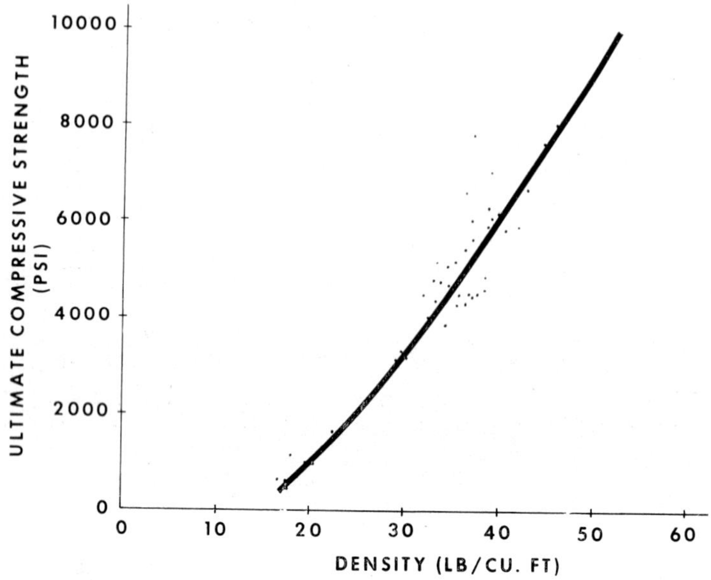

FIG. 18. Variation of ultimate compressive strength of syntactic foam with glass content

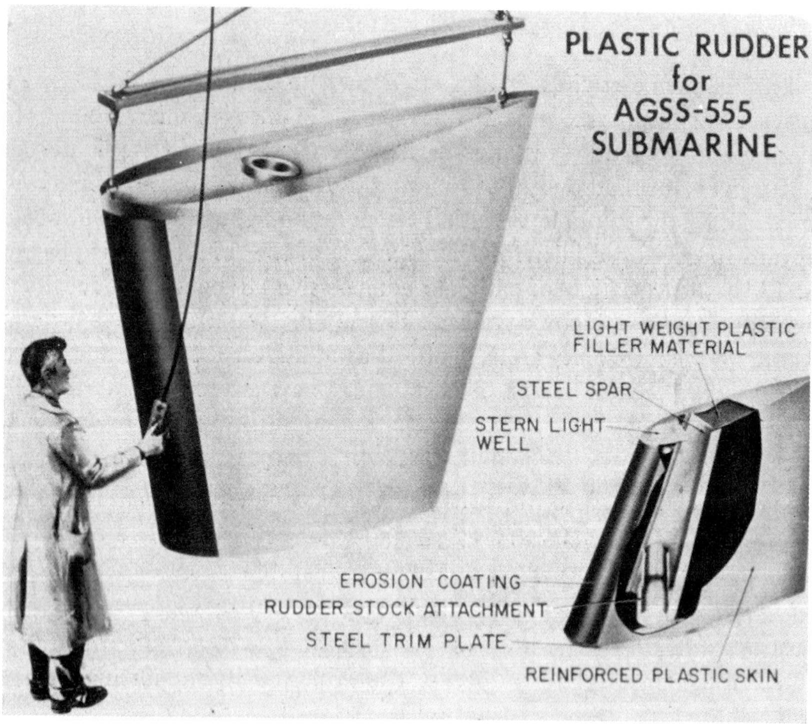

FIG. 19. Cutaway view of composite rudder for U.S.S. *Dolphin* (AGSS555)

FIG. 20. Typical GRP submarine mast fairing

FIG. 21. Schematic construction of syntactic foam

9. MAST FAIRINGS. All submarine masts including periscopes that must be operated while the submarine is partially submerged are usually fitted with streamlined fairings to minimize resistance. Because these fairings are external to the hull, they are continuously wet. In the past they have been constructed of corrosion-resisting metals. Glass-reinforced plastics have recently been adopted because of weight saving, freedom from maintenance, use of non-critical materials, and initial cost saving. Figure 20 shows typical construction. Service experience has been excellent.

10. MISCELLANEOUS. The external skin of *Alvin*,[14] a small deep-diving research submersible, is made of GRP. This skin, which is the hydrodynamic shape of the vehicle and is *not* subject to external pressure, is constructed in a manner similar to the fairwaters and superstructures previously described.

Early deep-diving vehicles, such as *Trieste*, used gasoline for buoyancy. Beginning with *Alvin*, recent research vehicles have used syntactic foam, glass microspheres embedded in a plastic matrix, in place of the gasoline. Figure 21 shows, schematically, the construction of syntactic foam.

IV. POSSIBLE FUTURE USES

Diligent and meticulous research enables the historian to chronicle faithfully the past. No matter how time-consuming is this process, it is far less challenging than the task of foretelling the future. It is to this challenge, however, that I address myself, mindful of the warnings of Pope† and Solomon.‡ At the risk of being wearisome, I must repeat that what follows is purely personal. It will be noted, however, that, in many instances, it is guided by past practice: cautious extension of what has been proven successful.

A. *Surface Ships*

It is axiomatic in any business venture that the least-cost product acceptable to the customers is that which is continued. Shipbuilding is no exception. The principle of "least cost over the life cycle" is belatedly entering the forum where "least first cost" has long held sway. Glass-reinforced plastic construction in a variety of applications, especially small craft, has, by its durability, relative freedom from maintenance, and ease of repair, established its relative overall economy despite some disparity in initial cost. Coupled with this economy is the outstanding service performance of GRP articles and structures. Nothing impresses the shipbuilding industry, notorious for its conservatism, more than economy and performance.

1. SMALL BOATS. Already the primary material for constructing boats up to 60 ft in length for the U.S. Navy and accepted by foreign countries and a large classification society for hulls up to 120 ft in length, GRP can be expected to become the primary hull structural material for U.S. Navy craft up to 120 ft long. Single-skin construction will undoubtedly continue to be the preferred mode. But, improved materials and improved bonding surely will be developed by a vigorous and dynamic technology so that sandwich construction, with its better strength-to-weight and rigidity characteristics,[1]

† "Fools rush in where angels fear to tread."
‡ "A fool's mouth is his destruction."

will gain acceptance despite earlier failures and find application in the longer craft.

2. SHIP HULLS. It is exceedingly difficult to say when a craft stops being a "boat" and becomes a "ship". The U.S. Navy has long used the criterion that a craft lifted by and stowed upon another craft is a "boat" and the "host craft" is a "ship". Moreover, almost by fiat, "boats" have been limited to 100 ft in length. Now is no time to argue the merits of this definition; rather it is convenient to use it.

Although economy and performance are compelling arguments, they are not the *only* arguments for use of GRP. Some of the special characteristics of GRP, such as high strength-to-weight ratio and lack of magnetism, will surely cause its selection for special kinds of ships.

(a) *Minesweepers.* Spaulding and Della Rocca(12) have already studied GRP for use in minesweeper hulls and confirmed its feasibility, at least from a design standpoint. For such use, both the high strength-to-weight ratio and lack of magnetism are important, if not, compelling arguments. I, therefore, expect that future generations of minesweepers will be built with GRP as the hull structural material. Such a prediction is based not only on the light-weight and non-magnetic characteristics of GRP, but also in consideration of the multiple units to be built which will permit early amortization of mold costs and, hence, overall economy. Figure 22 (fig. 12 of ref. 12) demonstrates this aspect. Such a prediction also tacitly assumes that the resistance of GRP to underwater explosions for these large craft will match previous experience(7) with small boats.

(b) *Hovercraft.* The craft that operate either very close to the water but above it, *ground effect machines*, or those that maintain slight contact with the water, *captured air bubble craft*, require extreme weight-saving measures to be able to carry payloads. For these applications, extensive use of GRP including the basic hull structure is almost certain to come.

(c) *Landing Craft.* Recently, steel has begun to give way to aluminum for these amphibious craft because the weight of the vehicles they must carry has increased and the beaching draft must be maintained. The inexorable increase of vehicular weight with passing time is almost certain to continue. Equally certain is that the beaching draft will be required to be kept constant. Some time in the future, then, the hull structure of landing craft will probably shift to GRP. Technically, it is within the state of the art today. Only the resistance to abrasion needs to be improved.

3. SUPERSTRUCTURES. Present and past installations of GRP superstructures have been eminently satisfactory in all respects. Their continued use is assured. In my opinion, however, GRP use for this application will not be universal. Rather, I feel, it will follow certain predictable paths.

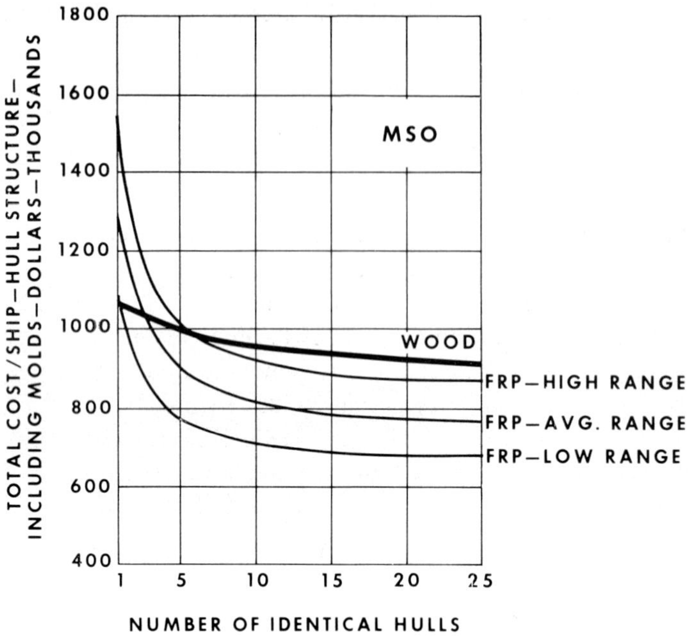

FIG. 22. Estimated cost of hull structure for MSO

(a) For small, high-speed craft, such as the motor patrol gunboats (PGM), hydrofoils of all kinds, and hovercraft, GRP will be used *initially* (when the ships are first built) because of the weight savings possible.

(b) For special ships, such as minesweepers, GRP will be used *initially* because of the non-magnetic property of the material.

(c) For other special ships with extensive topside electronics, such as command ships (CC, AGC) and communication relay ships (AGMR), GRP will be used *initially* because it has been found to minimize intermodulation products generated at nonlinear junctions (the "rusty bolt" effect) and thus enhance electronic performance.

(d) For the remaining ships, steel or aluminum will be used *initially*.

The paths predicted should not be interpreted as reflecting any dissatisfaction with GRP superstructures. On the contrary, these predictions are intended to reflect both prudence and foresight. GRP will be used with prudence initially (and, by inference, throughout the ship's life) for those ships whose mission, environment, and constraints require one or more of GRP's *unique* properties. Equally wisely, but with some foresight, GRP will not be used initially for those ships which do *not* require the unique properties. For these ships, the weight reduction, which occurs when GRP is used as a *replacement* for steel or aluminum, will be held in abeyance by the naval architect for use in the future *when* some new weapons or electronics system must be installed.

It is, perhaps, callous to observe that every *new* or *improved* weapons and electronics system is always bigger and heavier than the one it replaces. It is, however, realistic, and the ship designer should keep some measures to accommodate this growth "up his sleeve".

4. TOPSIDE APPURTENANCES. As with GRP superstructures, present and past installations of GRP topside appurtenances (masts, radomes, and the like) have been eminently satisfactory. Their continued use is also assured. I do, however, foresee an extension of the use of GRP, primarily because of the weight saving and freedom from maintenance. Lifelines and their stanchions appear to be the first hull fittings headed toward GRP. Booms, davits, and ladders are also likely candidates. Bitts, chocks, cleats, padeyes, and sheaves may, if fabrication technology makes the rapid strides now expected, also be made of GRP. As I see it, those items which are widely used (i.e., multiple production is possible) will be made of GRP, because then the benefits will be gained at no increase in cost.

5. VENTILATION. With the exception of minesweepers, where the nonmagnetic property of GRP is so important, little use has been made of GRP for ventilation ducts in surface ships. Just as was forecast for GRP superstructures, I envision a broader use of GRP for ventilation ducts when a ship is modernized or converted. Here, the weight reduction possible appears to be the salient factor.

Based on the successful use of GRP for the fan blades for SKMR-1, I foresee a pronounced effort to make fan blades, fairing pieces, and fan casings themselves from GRP in the future. Here again, the mass production of standard units will keep cost about constant at the same time that weight saving and freedom from maintenance will be gained. Recent conversations with the industry indicate that such a shift in materials is already contemplated.

6. PIPING. In no other area of marine application is the field as fertile for shift to GRP as is that of piping. The limited uses already made of plastic piping have been very successful, notably for the washdown of radioactive material. The growing use ashore of GRP piping in a variety of severe services, such as oil fields and chemical plants, attests to its quality and reliability when material selection is carefully considered. Filament-wound pipe can be made in almost any size and for any operating pressure desired. Acrylic lining material has withstood severe abrasive service both from high flow rates and from slurried material. There is now commercially available a substantial line of pipe fittings made of GRP.

It is true that GRP piping is not readily bent. It is equally true that joints are not as readily made as the present brazed or welded connections. But these apparent defects need not be disqualifying. On the contrary, they can, if the marine engineer does not suffer from the "Not Invented Here" attitude, be transformed into the strongest arguments for the use of GRP piping. By careful design and engineering—done *once* for a class of ships—the piping layout of the systems for which GRP is suitable can *maximize* the number of *straight* runs and *minimize* the number of *field joints*. Such care in design has, in the past few years, been given to submarine piping for the purpose of *eliminating* joints and thereby *improving safety*. Without exception, the *increased* cost of design and engineering has been more than offset by the *decreased* cost of pipe installation, joining, and testing.

To be sure, GRP is *not* suitable for universal use. Its elevated temperature properties are not good. It is doubtful if there would be any weight or cost saving possible in very high pressure, small size applications, such as compressed gases or hydraulic power transmission. But there is no apparent reason, except inertia, why it cannot be used for services at temperatures of not more than about 200°F and pressures of not more than about 150 psi. Even with those limitations, GRP piping could be used for drainage, potable water, salt water cooling, fuel and lubricating oil filling, overflow, and transfer, sounding tubes, air escapes, and plumbing. Weight saving and freedom from maintenance have been demonstrated. Careful engineering should permit a savings in labor costs of installation. This cost saving and that accruing over the life of the ship from reduced maintenance is expected to show an overall dollar saving in life cycle costs. Furthermore, GRP does not involve critical materials such as copper, nickel, and chromium.

I would be foolhardy indeed to predict the widespread use of GRP piping in naval ships in the near future. But the *facts* are available. Some day these will be reanalyzed, and a quixotic young engineer will lead the crusade. Hopefully, that day will not be too long in coming.

7. INSULATION. It is confidently expected that the spraying techniques successfully demonstrated recently will become fully operational in the near future. Then the savings in time and man-hours projected by Alfers[5] will be realized as will the savings in money.

8. MISCELLANEOUS. The use of plastics for the filling of voids, for buoyancy in boats and other special applications, for nonintegral tanks, for sonar domes, for furniture, and for bulkheads and partitions is expected to continue, expand, and be improved as new materials and processes are developed. Not only does there appear to be no retreat, but rather an avid desire to exploit the special properties of plastics in these applications.

9. HYDROFOILS. No projection into the future would be complete without some specific mention of hydrofoils. Casual mention of the probability of use of GRP in the superstructure of hydrofoil craft has already been made.

As with other light-weight, high-speed craft, GRP will become a prime candidate for the hull structure itself as well as for the superstructure. Indeed, for weight-saving reasons, GRP will probably find its widest use in hydrofoil craft. If the craft is to be a combatant, as, for example, the hydrofoil patrol gunboats (PGH) now under construction, it is likely that some form of composite armor, such as that described by Lindsay,[13] would be installed.

An interesting study by Rufolo[15] predicts a weight saving of 44% for the foil by using GRP instead of HY-80 steel now used. Rufolo[15] then analyzes the effect of the weight saved in increasing speed with no other changes, in increasing endurance by using the weight saved for added fuel, and in using the weight saved for increasing propulsion power and thereby speed. These parametric studies clearly provide justification for the development of lightweight structural materials. They also underscore the likelihood that GRP will be extensively used in a wide variety of applications for hydrofoils and other high-performance craft.

B. *Submarines*

Composite materials, especially GRP, have been used with outstanding success in an increasing number of different applications in submarine construction over the past 15 years. For the most part, these successful applications have been selected because of the weight reduction possible with GRP construction or because of its resistance to corrosion. This trend is expected to continue.

Fairwaters and superstructures are expected to be made of GRP not only when replacement of existing metal construction is necessary, but also for initial construction. Mast fairings and sonar domes will, in all likelihood, continue to be made of GRP as will the non-pressure structure of deep submersibles. Syntactic foam will continue to be used for filling voids and, in deep-diving research vehicles, for obtaining buoyancy. As with surface ships, plastic foam insulation is expected to be applied in the future by spraying.

More widespread use of plastics in ventilation systems is expected. Just as for surface ships, both ducts and fans made of GRP are foreseen—and for the same reason. If the GRP decks and rudders now installed in *Dolphin* are as successful as expected, the future use of GRP for these applications seems assured. Similarly, successful development and use of GRP hull fittings (lifelines, stanchions, cleats, ladders, and the like) for surface ships will be followed by use in submarines.

On the other hand, more widespread use of plastic piping is not expected, largely because of the high pressures and small sizes involved. I do, however, expect to see plumbing, fresh water and low-pressure salt water systems join the battery and electronics cooling water systems now made of plastics.

1. HULL STRUCTURE. The attractiveness of GRP as the structural material for submarines, because of its high strength-to-weight ratio is clearly shown in Fig. 23. Indeed, because of this attractiveness, there has been an extensive study of the material. Pellini[16] has summarized the status, projections, and scope of additional research required for a wide variety of candidate materials. Kies[17] has focused attention exclusively on reinforced plastics.

Kies[17] carefully recounts the degradation of glass by moisture while under tension, the susceptibility of glass to static fatigue under tension, the sensitivity of compressive strength of GRP cylinders to the resin system used, the lack

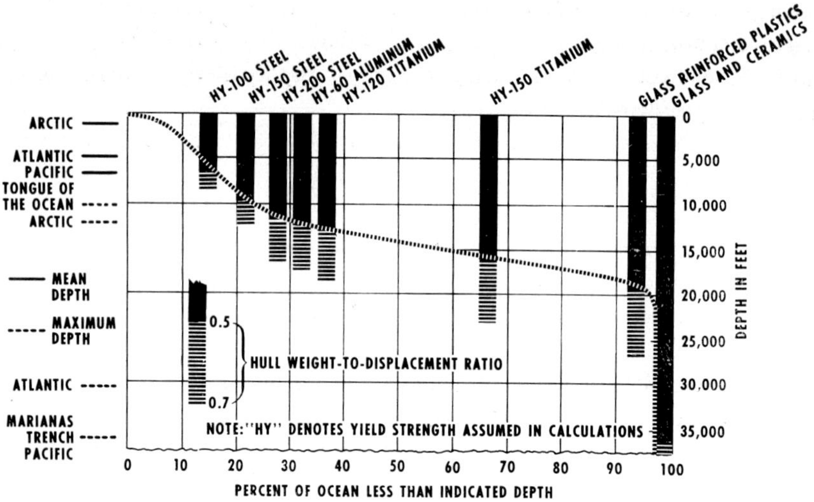

FIG. 23. Depth potential for various materials

of knowledge of fracture toughness of composite materials, the variability of glass–resin interface bonds, the unpredictable effect of lay-up and fabrication on failure by unwinding, buckling, and shear cracking, the loss of compressive strength with cyclic loading; the necessity for minimizing geometric stress concentrations, and the difficulty of making reliable attachments and penetrations. Nevertheless Kies[17] concludes on a confident note: that no type of material has improved as rapidly as reinforced plastic and that, given the necessary funding and research effort, reinforced plastic material suitable for deep research submersibles will be available in five years.

I am not nearly as sanguine as Kies. I wholeheartedly agree that the potential of reinforced plastics for use as the structural material for deep research vehicles exists. But, and it is a big "but", when the performance of a material in a particular application is so susceptible to variability in its fabrication and there is no nondestructive test technique to determine these effects before service use, I submit that it is premature to select such a material for that particular purpose.

2. COMPRESSED AIR FLASKS. The attractiveness of filament-wound air flasks, because of the weight-saving potential (50 to 65%), became evident with the successful use of the technique for rocket motor cases. In spite of the success attained in the aerospace and missile applications, early efforts to build high-pressure air bottles for submarines were frustrating. Obtaining consistency in rupture strength was the first difficulty, but this was, at last, overcome.

The next problem was obtaining a liner which was impervious to air diffusion. If air diffuses into the liner and then the pressure in the bottle is released, bubbles are formed within the liner and it "sponges". After a concerted test and evaluation program involving a number of materials, both polymers and metals, a thin electro-deposited nickel foil was found to be suitable.

The problems still remaining are the adhesion of the liner to the wall of the bottle and the detection of unsealed pinholes in the liner. It is expected that these problems can be solved with modest additional studies.

The suitability of filament-wound GRP for the stowage of high-pressure gases seems to be acknowledged widely. The stowage containers may be GRP with an improved metal foil liner or GRP wound over a thicker metal liner which will be part of the structure itself. It is expected that air and oxygen flasks, industrial gas bottles, and even the ubiquitous carbon dioxide fire extinguishers will soon be made of filament-wound GRP.

V. OBSTACLES TO EXPANDED USE

The list of superiorities claimed for GRP seemed endless. Subsequent evaluation essentially validated the claims. Yet, in the face of these proven characteristics, there was no mass stampede toward adoption. Rather there is a cautious, almost tentative, extension of its use in naval ships.

Although GRP is today widely accepted as the structural material for boats, most people consider building larger surface craft with GRP as the hull structural material to be just short of incredible. This duality of judgment is symptomatic of the apprehension and mistrust with which GRP is viewed. There seems to be a tacit acceptance of the material for appendages, appurtenances, fittings, and secondary structure, but a marked reluctance for its use as primary structure.

Not all of this reluctance can be attributed to the conservatism of the shipbuilding industry. Nor can this subjective opinion be summarily rejected as intuitive, emotional, and unfounded on fact. There is, in the long recital by Kies[17] of difficulties and unknowns, ample cause for hesitancy, doubt, and caution on the part of those who are charged with decisions affecting the safety of others.

A. *Fabrication Problems*

It seems strange to associate the term "fabrication problems" with a material whose ability to be fabricated into complex shapes is unquestioned. Nevertheless, fabrication problems do abound. Some of these "problems" have been solved, but the solutions adversely affect cost. For instance, the temperature and humidity during fabrication must be kept within rather narrow limits. This requires enclosures with proper "air conditioning". The existence of an enclosure requires adequate ventilation to minimize the annoying effects of styrene vapor which is released during the fabrication process.

Moreover, because the material is made at the same time and in the same place that the part is fabricated, molds are required. The requirement of a mold brings with it added cost unless there are multiple units to amortize the cost. GRP is suitable for repetitive production—not for single units.

1. CONTINUITY OF LAY-UP. Far more serious than the annoyance of styrene vapor and the added cost of molds, enclosures, and environmental controls is the present inability to stop and restart the lay-up process. One of the bitter pills of experience learned in boatbuilding with GRP is that, with existing resins, the strength of joints or bonds between two separately fabricated elements of structure is much less than that of either element. This fact has virtually forced the avoidance of such joints, usually called "secondary bonds" and of sandwich construction which is dependent on them. This inability to make strong secondary bonds has virtually forced continuous lay-up which has, in turn, limited the size of structure.

This problem is certainly amenable to solution. Kies[17] has said "Probably no other type of material has been improving so fast as GRP in recent years." Certainly, a vigorous and dynamic industry can be expected to develop resins and techniques to improve secondary bonds—to develop a "plastic weld". But, until it is done, and done reliably, the use of GRP for primary structure must be approached with extreme caution.

2. PENETRATIONS AND ATTACHMENTS. Kies[17] has cited that the problems in obtaining permanence, reliability, and leakproofness of joints between GRP and metal parts are not small. Kies has also observed that such joints will probably always be required for a variety of reasons. In the discussion herein on *Piping*, some of these reasons are given.

Although this problem is akin to and has several aspects in common with "secondary bonds", it is somewhat more complicated by the differences in moduli of elasticity and in coefficients of thermal expansion of the dissimilar materials to be joined. The designer, of course, can help by locating such penetrations and attachments in areas of low stress, where possible. But this is no panacea. What is obviously needed is intensive development of superior attachment details and adhesives. Devices as ingenious and as simple as wire staples and slide fasteners are archetypal.

3. Variability of manual operations. Kies[17] has observed that lay-up and fabrication quality affect the failure of composite materials. Many other investigators have observed that thorough wetting of the reinforcement and low void content of the product are required for good strength. The validity of these observations is well established. The ability to measure the strength of the composite without cutting a coupon is, unfortunately, not now available. Nor is there a device or technique except the human eye to determine the void content in the laminate nondestructively.

A recent example, drawn from the U.S. Navy's boatbuilding program, will highlight the dilemma. In the normal course of production of small boats, a large quantity was recently rejected because of the presence of "white laminate"; i.e., a visible (to the naked eye) indication of concentrated voids. The affected areas were duly removed, tested for strength, and found deficient. All production was stopped while the source of the difficulty was sought. As might be expected, the boat builders blamed the weavers of the roving used, the weavers blamed the suppliers of glass fibers, and the glass suppliers blamed the boat builders' resins. Each participant claimed his process was unchanged and that the fault lay elsewhere. Nevertheless, because of the urgent need for boats, all involved joined in the search for the source of trouble. The trouble was eventually found and corrected—and boat production resumed. Because this incident is still a matter in litigation, it would be premature to discuss it further. Suffice it to say that the error was traced to human variations.

B. *Quality Control*

The possible adverse effect of human variations in the fabrication process just described undermines confidence in reinforced plastics for structural use. Furthermore, this example dramatizes the importance of both quality control and the techniques to assure quality.

1. Raw material monitoring. Kies[17] has observed that recently it has become clear that commercial surface finishes of glass fibers are not necessarily inert and permanent in storage while awaiting impregnation by resin. Thus, it may be desirable to designate storage conditions to be sure that the surface finish desired is realized. It also is important to test samples of each lot of glass reinforcement *before* making a laminate to assure the proper surface finish.

Similarly, it is important to test not only each batch of resin constituents *before* use, but also the resin after mixing the constituents. It is also indicated that the resin should be sampled and tested frequently during the laminating process to assure that all its properties are within the tolerances allowed.

2. NONDESTRUCTIVE TESTING OF FINISHED PRODUCT. It has been noted previously that there is not now available a device or technique for determining strength, wetting of reinforcement, void content of laminate, or bonding of elements nondestructively. This lack is a serious shortcoming. Unlike all other structural materials which are furnished in various forms with definitive properties and then joined to make the finished product, glass-reinforced plastic material is made simultaneously with the finished product. Although knowledge of the properties of the constituents is helpful, it is not sufficient because of the human involvement inherent in the fabrication process.

The seriousness of the lack of a nondestructive test technique has long been appreciated. Indeed, there has been much effort devoted toward developing such a technique. Radiography has been tried without marked success. When glass with a modest amount of lead present was used in a filament winding process, radiography was successful, however, in determining whether or not the winding pattern was as specified. Ultrasonic scanning of laminates for void content has shown better promise than radiography. To date, however, the ultrasonic technique is, at best, a laboratory tool. What is needed is a device or technique suitable for use in the shop and aboard ship.

The reward for the development of a successful nondestructive test technique is well worth the effort involved. Without it, the expansion of the use of GRP will proceed with caution and hesitancy. The time involved to gain acceptance of GRP as a hull structural material will be long and the process arduous. With it—and the ability to demonstrate that the finished product has the required properties—the acceptance of GRP as a full-fledged structural material will be immediate. Surely, this is a sufficient carrot to prod a dynamic industry into vigorous action.

ACKNOWLEDGEMENTS

A survey such as this is based on the work of many people—too many to be listed individually. It is hoped that the authors whose works are included in the list of references will regard this listing as reflecting proper credit for their contributions. Still others will find the recognition of their efforts in similar listings appended to each of the papers enumerated in the list of references.

Specific thanks, however, must go to the several people and activities who generously furnished the illustrations used: to Messrs. J. B. Alfers and W. R. Graner of the Naval Ship Engineering Center for Figs. 4, 6–11, 17,

21, and 23; to Mr. K. B. Spaulding, Jr. also of the Naval Ship Engineering Center for Figs. 5, 12, and 22; to Mr. N. Fried of U.S. Naval Applied Science Laboratory for Figs. 1, 2, and 15; to Mr. B. Cobb, Jr. of Owens–Corning Fiberglass Corp. for Fig. 3; to the Philadelphia Naval Shipyard for Figs. 13 and 14; to the Portsmouth Naval Shipyard for Figs. 18 and 19; and to Mr. J. K. Shiver of Lunn Laminates, Inc. for Fig. 20.

REFERENCES

1. VERWEIJ, D. Comparison Between Plastic and Conventional Boatbuilding Materials, Third FAO Technical Meeting on Fishing Boats, Göteborg, Paper TMB-5/6, Oct. 1965.
2. ALFERS, J. B. and BUSHEY, A. C., JR., Down to Sea in Plastics, *Buships Journal*, **1,** No. 6, Oct. 1952, pp. 5–10.
3. ALFERS, J. B. and GRANER, W. R. Reinforced Plastics — A Structural Material for Marine Applications, *Trans. Soc. Nav. Arch. & Mar. Eng.*, **62,** 1954, pp. 5–29.
4. BUERMANN, T. M. and DELLA ROCCA, R. J. Fiberglass Reinforced Plastics for Marine Structures, *Trans. Soc. Nav. & Mar. Eng.*, **68,** 1960, pp. 138–192.
5. ALFERS, J. B. The Use of Plastics for Ships, Soc. Nav. Arch. & Mar. Eng., Phila. Section, Jan. 15, 1950.
6. COBB, B., JR. A Report on Long Term Durability Fiberglass Boats, *Yachting*, **112,** No. 3, Sept. 1962, pp. 93–94, 122.
7. SPAULDING, K. B., JR. Fiberglass Boats in Naval Service, *Naval Engineers Journal*, **78,** No. 2, April 1966, pp. 333–340.
8. FRIED, N. and GRANER, W. R. Durability of Reinforced-Plastic Structural Materials in Marine Service, *Marine Technology*, **3,** No. 3, July 1966, pp. 321–327.
9. ALFERS, J. B. Plastics in Ships, *Naval Ship Systems Command Technical News*, **15,** No. 9, Sept. 1966, pp. 5–8.
10. DELLA ROCCA, R. J. A 110-Ft Fiberglass Reinforced Plastic Trawler, Third FAO Technical Meeting on Fishing Boats, Göteborg, Paper TMB-5/7, Oct. 1965.
11. VANDEN ERDE, L. Praktische Toepassing van Gewapende Plastics in de Scheepsboww Neemt Toe, *Nautica*, May 1964, pp. 240–243.
12. SPAULDING, K. B., JR. and DELLA ROCCA, R. J. Fiberglass-Reinforced Plastic Minesweepers, *Trans. Soc. Nav. Arch. & Mar. Eng.*, **73,** 1965, pp. 415–449.
13. LINDSAY, E. M. Glass Fiber Reinforced Plastic as a Marine Structural Material, American Institute of Aeronautics and Astronautics, Paper No. 66–699, Aug. 1966.
14. MAVOR, J. W., JR., FROEHLICH, H. E., MARQUET, W. M. and RAINNIE, W. O. JR. *Alvin*, 6000-ft. Submergence Research Vehicle, Soc. Nav. Arch. & Mar. Eng., Paper No. 3, Nov. 1966.
15. RUFOLO, A. Foil Weight Saving and Hydrofoil Performance, *Naval Engineers Journal*, **78,** No. 5, Oct. 1966, pp. 905–913.
16. PELLINI, W. S. Status and Projections of Developments in Hull Structural Materials for Deep Ocean Vehicles and Fixed Bottom Installations, Naval Research Laboratory Report 6107, Nov. 4, 1964.
17. KIES, J. A. Filaments for Reinforcement and the Applicability of Filament-Wound Laminates for Deep-Submergence Vehicles, *Marine Technology*, **3,** No. 1, Jan. 1966, pp. 52–59.

APPLICATIONS OF REINFORCED PLASTICS IN AIRCRAFT

HERBERT S. SCHWARTZ

Air Force Materials Laboratory,
Wright-Patterson Air Force Base

I. INTRODUCTION

Prior to World War II, the principal uses of reinforced plastics in aircraft were for secondary and non-structural uses, such as pulleys, control cable conduits, fairings, and fillets. In most instances these parts were machined from high-pressure molded paper, cotton, or asbestos into phenolic laminate stock shapes, such as sheet, rod or tube. The requirement for expensive high-pressure molding for these phenolic laminate materials, plus the fact that they had only fair mechanical properties, limited their consideration for large area structural parts.

During the early 1940's however, low viscosity polyester resins of the styrene cross-linking type became available. It was found that these resins could be used to laminate glass fabric using pressures ranging from atmospheric down to contact pressure. This permitted use of low cost wood and plaster tooling and also permitted inexpensive fabrication of complex shaped parts such as the ducting shown in Fig. 1.

II. RADOMES

The first major use of low-pressure glass fabric polyester laminates in exterior parts of aircraft was in radomes. Early airborne radomes were made of thermoplastic material and were structurally weak. Early ground based radomes were made of plywood which was adversely affected electrically by moisture absorption. One solution to this problem was to apply a thin overlay of resin-impregnated glass fabric to the outer surfaces of these plywood radomes to serve as a moisture barrier. This was done in 1942, and is believed to be the first use of glass fabric reinforced plastic in radome construction.

The first use of airborne glass fiber reinforced plastic radomes is believed to have occurred in 1943 with the equipping of B-17 bombers with retractable hemispherical radomes. Subsequently, other World War II vintage aircraft were equipped with radar and radomes. These radomes were usually appen-

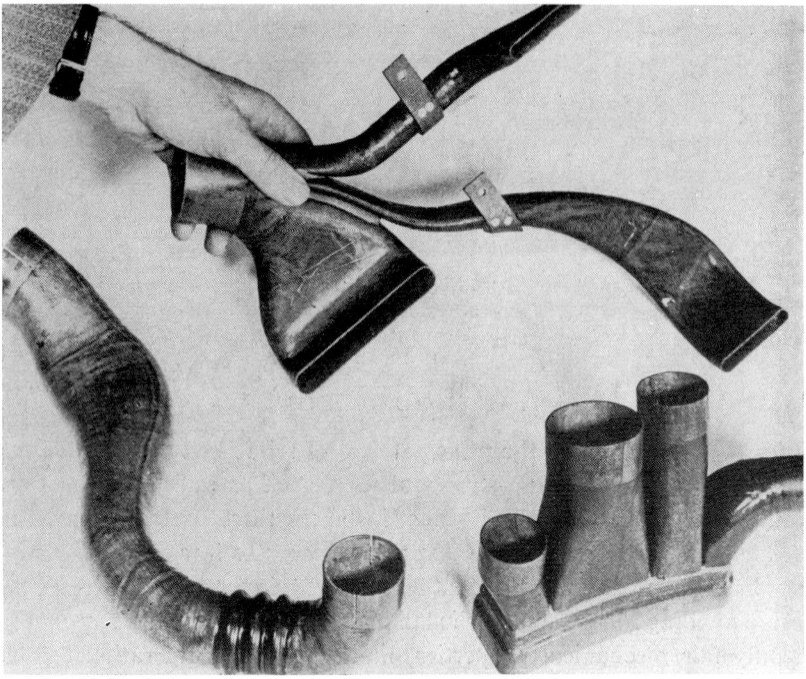

FIG. 1. Reinforced plastic ducting

FIG. 2. Nose radome for F-100

dages which protruded from the aircraft fuselage, since they were not anticipated in the original aircraft design. With the advent of high speed jet aircraft, it was mandatory that radome design be integrated with airframe design to give low drag characteristics. An example of a "faired-in" radome is a research version of the nose radome for the F-100 fighter aircraft, shown in Fig. 2. Here the lower portion of the radome also forms the cowling for the air intake. Another example is the "chin radome" for the B-70 aircraft,

FIG. 3. Chin radome

shown in Fig. 3. Here the radome forms the lower section of the nose of the aircraft, hence the "chin" designation.

Some radomes are heated with bleed air from the engine to prevent formation of ice on the exterior surface. An example of a nose radome of this type is illustrated in Fig. 4. This is a "fluted-core" sandwich construction having inner and outer laminate skins with rectangular cross-section laminate flutes holding the skins together. The flutes act as small ducts to carry the warm air, and the method for making the flutes is somewhat similar to that used to make the conventional sized ducts shown in Fig. 1.

FIG. 4. Fluted core radome

FIG. 5. Grumman E-2A airplane

FIG. 6. Radome for E-2A airplane

There are still some types of operational aircraft which carry "appendage" radomes, but by intent and not through lack of foresight in design. An example is the Grumman E-2A Hawkeye airplane shown in Fig. 5, with a 24 ft diameter radome mounted atop the fuselage. This radome is part of the airborne radar surveillance system of the aircraft. A close-up of the radome during its manufacture is shown in Fig. 6.

III. FUSELAGE AND WINGS

At about the same time that glass fabric polyester laminates were being adopted for use in radomes, these materials were also being used in development of fuselage and wing structures for aircraft. This pioneering work,

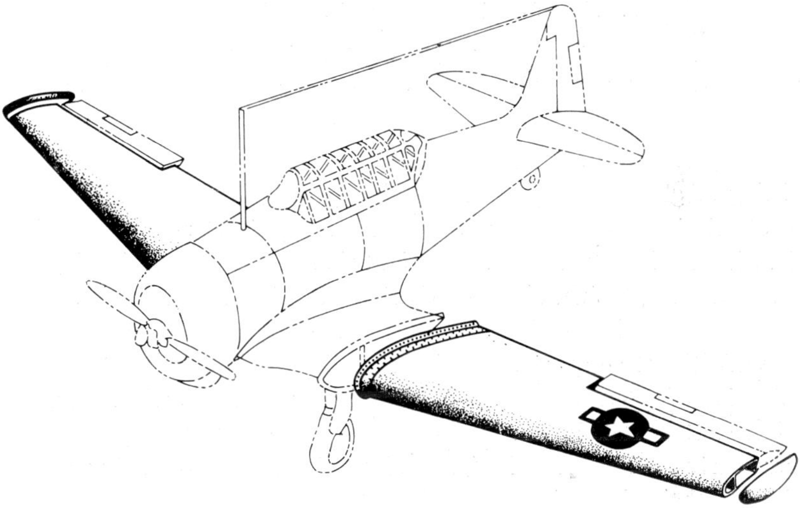

FIG. 7. AT-6C airplane

which was performed at Wright-Patterson Air Force Base, Ohio, included design, fabrication, static testing, assembly, and flight testing of a reinforced plastic fuselage for the BT-15 airplane, and wings for the AT-6C airplane. The BT-15 airplane was a two-place, low wing cantilever monoplane of the basic trainer class. The AT-6C was an advanced trainer airplane, very similar in appearance to the BT-15. Figure 7 is a sketch of the AT-6 airplane showing the general outline of the aircraft in light dotted lines, and the reinforced plastic wing section in dark solid lines.

The fuselage for the BT-15 was of sandwich construction, having reinforced plastic faces and a balsa wood core. Fabrication of the fuselage was completed on November 7, 1943; static tests were conducted November 11–15, 1943; and flight testing was performed on March 24, 1944. On a strength/weight

FIG. 8. Reinforced plastic wing for AT-6C airplane

basis, the reinforced plastic fuselage was 50% stronger than the aluminum fuselage and 80% stronger than the wooden fuselage then in service.

The wings for the AT-6C airplane were also of sandwich construction, consisting of glass fabric polyester laminate skins, and a core consisting of strips of cellular cellulose acetate wrapped with polyester-impregnated glass fabric. Figure 8 is a photograph of a right side reinforced plastic wing panel, prior to attachment of the aileron and other miscellaneous hardware. Fabrication of the first reinforced plastic wing was completed on May 31, 1945. Several additional wings were fabricated and static testing was successfully completed on September 10, 1946. The reinforced plastic wings were flight tested more than 1600 hours.

These early applications of reinforced plastics were reasonably successful, considering their pioneering nature. Why then weren't they quickly followed up by application to subsequent military aircraft? One answer is that the load conditions for high-speed jet aircraft were much greater than for the small trainer aircraft. Also modulus of elasticity of glass-reinforced plastics was too low to satisfy design requirements for primary structures on the newer military aircraft. In addition, the work on the AT-6C plastic wing defined some problems that still plague us today. For example, some quotations from the report covering that work are as follows:

> "The efficiency of the attachment of the metal fittings to the glass laminate of the... wing panel is low. Further work is necessary on the design of glued and bolted joints to improve the efficiency.
>
> "The low elongation of glass fiber laminates has an appreciable effect on the stress distribution in the structure and should be given serious consideration in the design of glass plastic structures.

"The low resistance of glass (fabric) laminates to cleavage should be considered in design of such structures."

How prophetic these statements sound today!

Although the use of reinforced plastics in fuselage and wing structures for military aircraft has been limited, these materials have been used in such structures for small personal civilian aircraft such as the Taylorcraft Model 20, certificated in 1955, and the Piper PA-29.

IV. EMPENNAGE

Some highly stressed reinforced plastic structures which have been used on military aircraft are parts of the empennage, including parts for both the vertical and horizontal tail. One fairly recent example is the vertical tail structure above the horizontal stabilizer on the Grumman E-2A aircraft. This is the same airplane as shown in Fig. 5, which carries the 24 ft diameter radome. The antenna inside the radome rotates continuously, scanning the entire 360° around the aircraft. It was found that the original metallic vertical tail sections interfered with the radar beam, and this led to the development of reinforced plastic vertical tail sections to replace the metal ones. Figure 9

FIG. 9. Reinforced plastic vertical tail of E-2A airplane

FIG. 10. Reinforced plastic leading edge of C-141 horizontal stabilizer

gives a close-up of one of the outboard vertical tail sections made of reinforced plastic (that portion above the horizontal stabilizer). These reinforced plastic tail structures successfully passed static, flight, and radar transmission tests and are now in use on the aircraft.

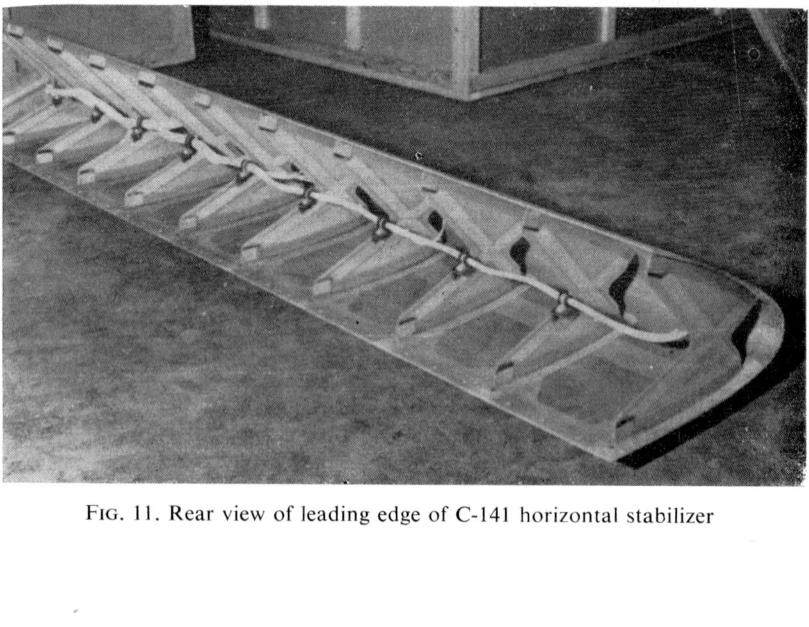

FIG. 11. Rear view of leading edge of C-141 horizontal stabilizer

FIG. 12. F-111 airplane in flight

Another interesting empennage part is the leading edge of the horizontal stabilizer on the C-141 cargo aircraft. The leading edge, which is made of several sections, is electrically heated from the inside surface to provide de-icing capability. An end view of one of the sections is shown in Fig. 10, and a rear view showing the attached electrical cable for heating is shown in Fig. 11.

One of the recent developmental applications of boron fiber reinforced plastic is a midsection of the horizontal stabilizer for the F-111 airplane.

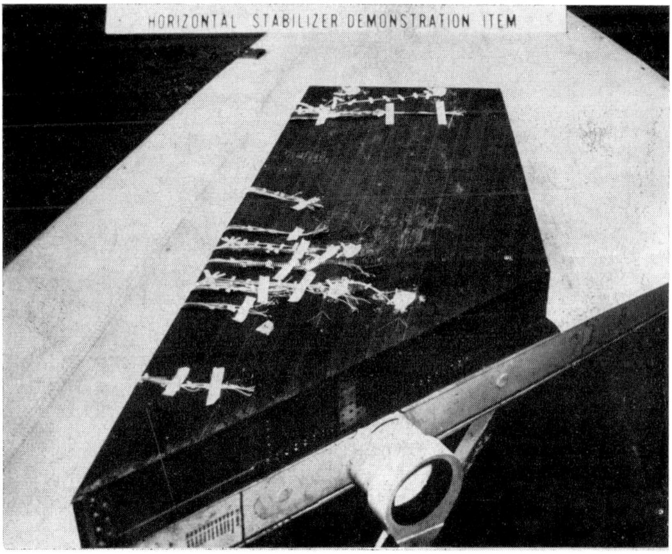

FIG. 13. Experimental boron fiber reinforced plastic section of F-111 horizontal stabilizer

The F-111 is photographed in flight in Fig. 12, and the boron fiber reinforced plastic horizontal stabilizer section (instrumented for structural testing) is shown in Fig. 13. This horizontal stabilizer section is of sandwich construction, and consists of boron fiber epoxy matrix skins, full depth glass fiber plastic honeycomb core, glass fiber plastic closure spars along each edge, and titanium end fittings. Studies showed that a weight saving of 350 lb in an entire F-111 horizontal stabilizer assembly could be realized on a direct substitution basis using the above described configuration.

V. ROTATING PARTS

Rotating parts such as helicopter rotor blades, aircraft propeller blades and jet engine compressor blades are subjected to inertia loads due to their own weight and to oscillatory stresses. The high strength to density ratio of reinforced plastics makes them attractive for resisting inertia loads, and their good damping and fatigue properties make them attractive for resisting the vibrational loads.

Plastics were first used on helicopter rotor blades about 1945 to improve the resistance to abrasion of leading edges of wood blades. Since that time, reinforced plastic rotor blades have been adopted for use on several types of helicopters. One model which uses all reinforced plastic helicopter rotor blades is the Kaman HH-43B shown in Fig. 14. Figure 15 gives a view of the blade from underneath and Fig. 16 a close-up cross-sectional view. Mechanical requirements of this blade are particularly severe, and include design parameters of 60,000 lb centrifugal force at the root ends of the blades, vibratory bending moments of 20,000 in.-lb for normal flight and twice that value for

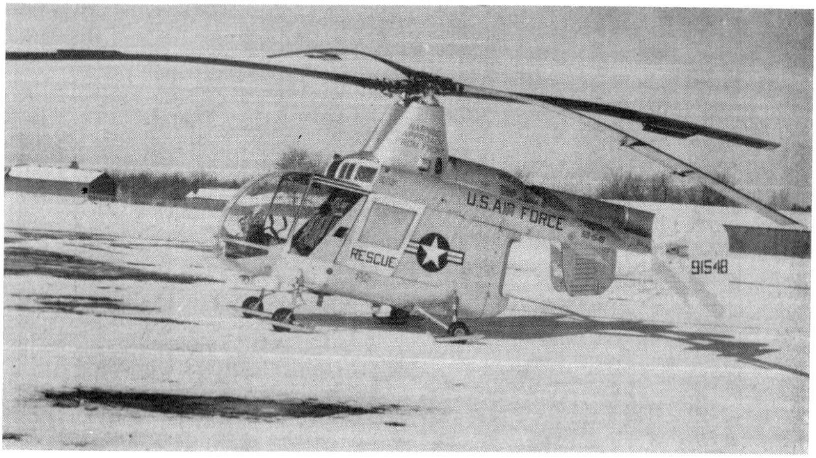

FIG. 14. Kaman HH-43B helicopter

FIG. 15. Reinforced plastic helicopter rotor blade for Kaman HH-43B

sharp maneuvers, torsional loading of 3600 in.-lb, and continuous vibrations between 220 and 260 cycles/min. Full endurance of the plastic rotor blade up to thirty mission cycles or 1000 flight hours is expected.

FIG. 16. Cross-section of reinforced plastic helicopter rotor blade

FIG. 17. X-100 airplane in hovering mode

Reinforced plastic propeller blades for V/STOL aircraft are a highly significant recent development. Two experimental V/STOL aircraft which were equipped with reinforced plastic propeller blades are the X-100 and the X-19, shown in Figs. 17 and 18 respectively. The X-100 was the first airplane of its

FIG. 18. X-19 VTOL aircraft equipped with reinforced plastic propeller blades

type ever flown with reinforced plastic propeller blades. Close-ups of a group of experimental reinforced plastic propeller blades are given in Fig. 19. The X-100 blade, 5 ft long, is at the extreme left, and the X-19 blade, $6\frac{1}{2}$ ft long, is second from the left.

Although reinforced plastic compressor blades for jet engines have been undergoing investigation in the U.S.A. for over fifteen years, there is as yet no known use of these blades in U.S.A. manufactured airborne operational

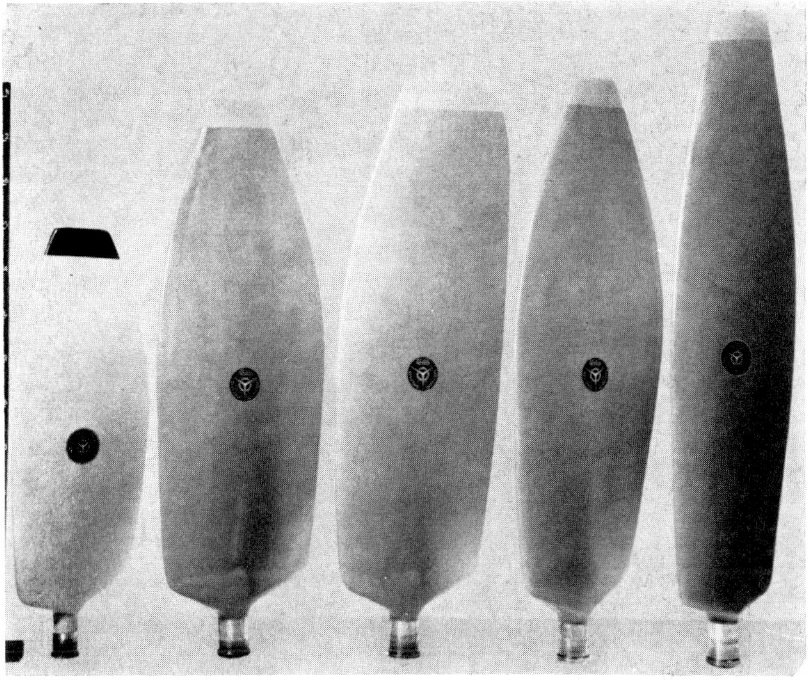

FIG. 19. Experimental reinforced plastic propeller blades

engines. Glass fiber reinforced plastic jet engine compressor blades have been mounted on compressor hubs (see Fig. 20) and subjected to ground testing. Some of the problems encountered were excessive deflection of the blade tips; some anticipated problems are excessive erosion due to sand and rain. However, some of the newer composite materials such as boron fiber reinforced plastics or graphite fiber reinforced plastics may minimize or eliminate these problems.

VI. PRESSURE VESSELS

Although filament wound reinforced plastic pressure vessels have received most of their publicity by being used as rocket motor cases, such pressure vessel constructions are also used in aircraft. Figure 21 is a photograph of filament wound reinforced plastic high-pressure gas storage bottles used in aircraft to actuate hydraulic systems. The plastic bottles, which have an internal rubber bladder, can withstand repeated pressurization to 5000 psi. They are used in military aircraft primarily because they are light in weight and do not shatter if pierced by a projectile.

FIG. 20. Reinforced plastic jet engine compressor blades mounted on hub

VII. SUMMARY

The foregoing discussion has described a number of applications for structural reinforced plastics, some of which have come into operational use, but many of which have not despite superior performance to competing materials. In many instances, reinforced plastic parts are not used in production because of: (a) lack of adequate design data, (b) problems in reproducibility and quality control in raw materials and finished parts, and (c) lack of adequate

FIG. 21. Filament wound high-pressure gas bottles

or reliable non-destructive testing procedures. Fortunately, progress is being made in all these areas, and the future looks bright for significantly expanded use of reinforced plastics in aircraft.

APPLICATIONS OF COMPOSITE MATERIALS IN RE-ENTRY VEHICLES

DR. J. D. STEWART and A. M. GARBER

General Electric Company, Re-entry and Environmental Systems Division, Philadelphia, Pennsylvania

I. INTRODUCTION

Advances in re-entry vehicle technology within the past decade have been remarkable, as man has learned to master the severe environmental conditions associated with hypervelocity flight. In these very high temperature environments, the use of conventional engineering materials is greatly restricted due to thermal degradation. The utilization of reinforced plastic composites as a thermal shield has enabled these materials to serve their intended purpose. The interaction of the high energy environment with the exposed thermal protection material results in a relatively small amount of controlled sacrificial erosion, or ablation. The associated energy-absorption processes which take place control the surface temperature and greatly restrict the flow of heat into the substrate.

Reinforced plastic composites have demonstrated the capability of absorbing, dissipating, and blocking a large amount of thermal energy. They have relatively low thermal conductivities and they can tolerate very high temperatures. In fact, their performance increases with environmental temperature. A wide variety of reinforced plastics are available and their properties and characteristics can be varied by changes in resin and reinforcement material. They are lightweight, costs are acceptable and shapes can be fabricated with existing techniques which include bag molding, high-pressure molding and tape winding.

A principal factor influencing the selection of a plastic-reinforced composite is the heating and loading environment encountered during the re-entry phase of the flight. Consequently, it is convenient to group those vehicles which exhibit similar characteristics into several classes as listed in Table 1. Each class is characterized by specific environmental conditions that play a vital role in selecting the optimum thermal protection system. The technical disciplines involved in the selection of a material system involves the fusion of the sciences of materials, structures, thermodynamics, aerodynamics, flight dynamics, and physics. Consequently, the importance of an integrated systems approach cannot be overemphasized. The selection of a material system

TABLE 1. CLASSES OF RE-ENTRY SYSTEMS

1	BALLISTIC MANEUVERING RE-ENTRY SYSTEMS
2.	BALLISTIC RE-ENTRY SYSTEMS
3.	ORBITING MANEUVERING RE-ENTRY SYSTEMS
4.	SCIENTIFIC SPACE SYSTEMS
5.	INTERPLANETARY SYSTEMS

must be based on an approach that will permit the design of an efficient system. However, it is impossible to speak of a universal best reinforced plastic composite, and it is equally difficult to establish a single best criterion for rating system performance. Each material exhibits its optimum performance in a specific environment and may be unusable in another.

Vehicle configuration or shape will vary among the various classes of re-entry vehicles. A sketch of a typical ballistic re-entry vehicle, with its thermostructural wall consisting of a reinforced plastic composite heat shield bonded to a conventional metal substructure, is shown in Fig. 1.

II. THE ENVIRONMENT

The flight paths of bodies that re-enter the earth's and other planetary atmospheres vary from ballistic missiles to interplanetary landers. The thermal, chemical, and mechanical variables which affect the selection of a rein-

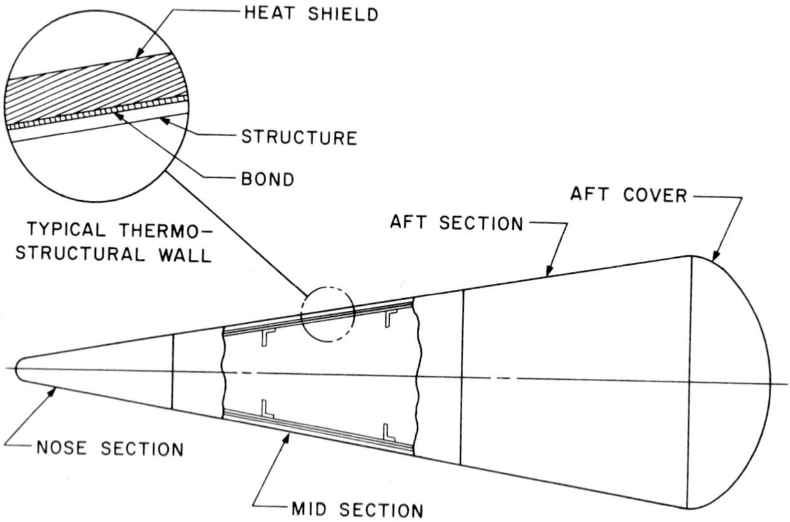

FIG. 1. Typical re-entry vehicle

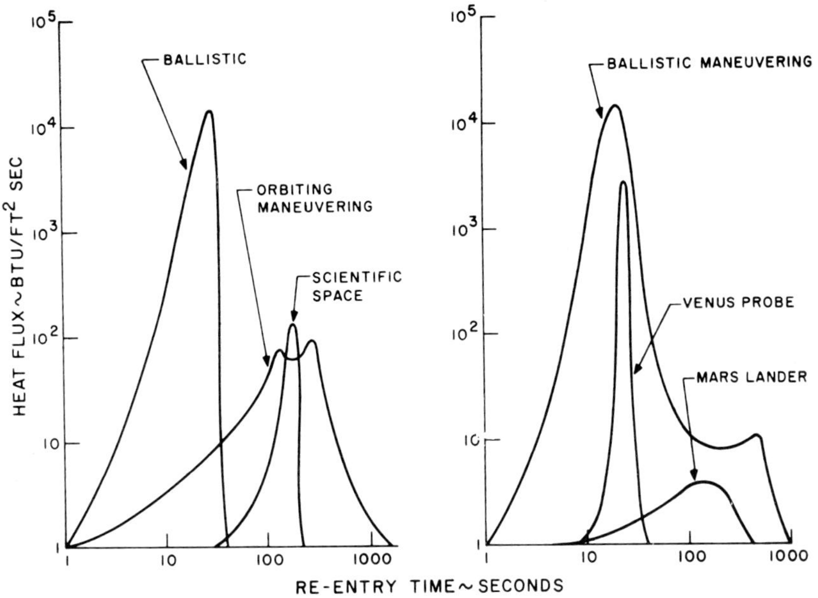

FIG. 2. Heat flux vs. re-entry time

forced composite vary greatly from vehicle to vehicle. This is shown in Figs. 2, 3 and 4 which present heat flux versus re-entry time, altitude versus velocity and stagnation enthalpy, and peak heat flux versus characteristic time for the various classes of re-entry vehicles. Heating times vary from several seconds

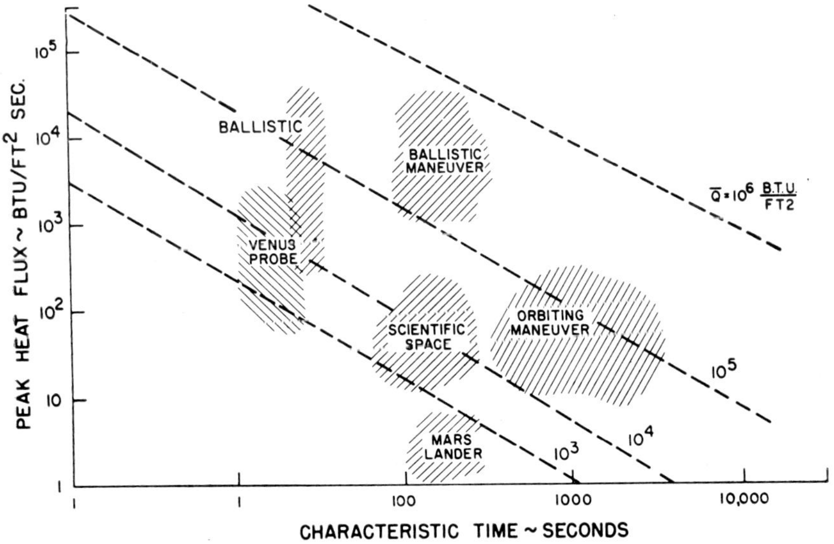

FIG. 3. Typical heat inputs for various re-entry systems

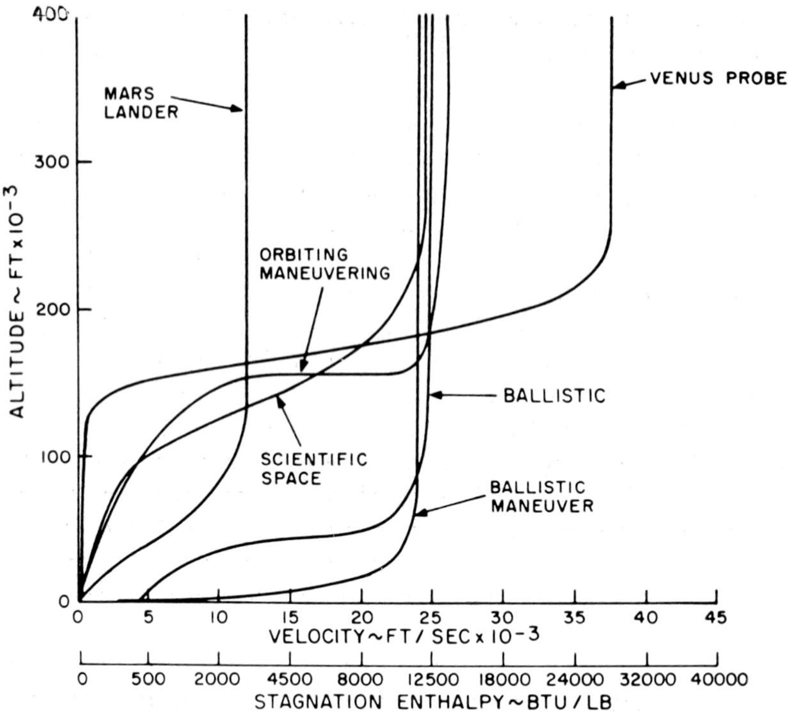

FIG. 4. Altitude vs. velocity

to several thousand seconds. Peak heat fluxes may vary from 10 Btu/ft^2 sec to 10,000 Btu/ft^2 sec.

Other typical re-entry environments are listed in Table 2, which shows that dynamic pressures can vary from less than one up to 200 atm, stagnation

TABLE 2. TYPICAL RE-ENTRY ENVIRONMENTS

VEHICLE SYSTEM		MACH NO.	STAGNATION POINT HEAT FLUX (BTU/ft^2 SEC X 10^{-3})	INTEGRATED HEAT FLUX (BTU/ft^2 X 10^{-4})	ENTRY TIME (SEC X 10^{-2})	DYNAMIC PRESSURE (atm)	STAGNATION PRESSURE (atm)	G's
BALLISTIC MANEUVER		24	37	65	1.2	200	400	300
BALLISTIC		24	15	16.2	.45	55	120	100
ORBITING MANEUVERING		25	.10	15	25	.3	1	30
SCIENTIFIC SPACE		25	.16	2	3.5	.3	1	20
INTERPLANETARY								
	MARS LANDER	12	.005	.07	2.5	.4	1	15
	VENUS PROBE	35	3.3	2.6	.2	3	7	500

pressures from 1 to 400 atm and peak deceleration forces from 15 to 300 g (see Table 2). In addition, shear stresses can be as high as 1000 lb/ft^2 and less than 1 lb/ft^2.

The aerodynamic shape and the stability of the body and its motions must also be considered and will impose additional restrictions. For example, in the case of axially symmetric shapes, convective heating is generally most intense at or very close to the stagnation point. This is where the gas enthalpy is at a maximum and the boundary layer is the thinnest. For very blunt re-entry vehicles like the Apollo, the peak convective heating rate occurs on the rounded corners of the forward heat shield. This is caused by the high local pressure gradient with the occurrence of sonic flow at the corners. Convective heating is generally found less intense at other body stations, because of a decrease in the stream enthalpy and an increase in boundary layer thickness. However, while blunting will lead to reduced convective heating, it also results in high deceleration forces, which may exceed human tolerances or vehicle structural limits. Blunting is also impractical at high velocities due to the intense heating from the radiating gas around the vehicle. By using pointed or conical shapes the shock is brought closer to the vehicle surface, and the radiant heating is diminished.

Reinforced plastic composites traveling at very high speeds can be seriously influenced by mechanical aspects of the re-entry environment, such as pressure, particle impact, acoustical vibration and inertial forces. The general effect of mechanical forces is to cause premature removal of surface material with attendant reduction in material ablation efficiency. At high stagnation pressures, the mechanically weak ablating surface layer may be crushed and blown away. Shear stresses at the ablating surface can be of sufficient magnitude to cause sloughing, run off, particle erosion or premature char failure. The surface char layer formed on certain reinforced plastic composites during ablation serves many useful functions and leads to a significant increase in ablative performance. A number of mechanisms may act to remove this char. For example, the thermally weakened surface material on an ablating reinforced plastic composite may be fractured and removed by acoustic vibrations. Thermal stresses in the char layer and in the virgin material can result in surface cracking of the reinforced plastic. Gaseous products formed in the material substrate can generate large internal stresses in the char.

In addition to re-entry, a reinforced plastic composite heat shield must be capable of withstanding other environments, see Table 3, which are part of the total mission profile. During powered flight, heating due to high launch velocities, intense booster vibrations and high noise levels can be significant. In long duration space environments reinforced plastics must be able to withstand vacuum, cyclic temperature variations, radiation and particle impact without any degradation to the performance of the heat shield during its

TABLE 3. ADDITIONAL ENVIRONMENTAL CONDITIONS

- POWERED FLIGHT
  - VIBRATIONS
  - HEATING
- SPACE
  - VACUUM
  - TEMPERATURE
  - RADIATION
  - PARTICLE IMPACT
- LANDING/RECOVERY
  - AIR
  - LAND
  - WATER
- GROUND HANDLING/STORAGE
  - READINESS
  - TEMPERATURE
  - SHOCK
  - HUMIDITY

final re-entry phase. Many manned or unmanned re-entry vehicles must be capable of performing a landing or recovery operation. This environment will result in additional requirements upon the heat shield system.

Finally, the ground handling/storage environment associated with any re-entry vehicle will impose its own unique demands, which in certain cases can be as significant to the heat shield material as the re-entry environment. Vehicles that have a storage life requirement must be composed of materials that will not degrade in natural environments, such as temperature and humidity extremes.

III. APPLICATIONS

A. *Material Selection*

The performance of a heat protection material is a very involved function of both materials and environmental variables. In addition, design criteria associated with the substructure, in terms of allowable backface temperature, and aerodynamic configuration, in terms of allowable shape change, are very important. Since the total design spectrum of re-entry environments is very large, it is clear that there is no single material that will be optimum in all environments. A material may exhibit good performance in a specific environment and be unacceptable in another condition. Optimum performance can be obtained by tailoring certain material properties and characteristics that are of critical significance for certain classes of re-entry vehicles. Those parameters which are of prime importance compared to others which are desirable but are of secondary importance for each class of vehicle are so indicated in Table 4. The heat of sublimation/vaporation should be as high as possible, to minimize the weight of material required to accommodate the incident flux, and is of primary importance for the ballistic maneuver, ballistic, and Venus probe vehicles where the heat fluxes are very high. A low thermal conductivity is required for insulation purposes and is of prime significance for those vehicles which are subjected to low heat flux levels. Other material properties which require high values are the heat of gasification, decomposition temperature, char strength, and specific heat. The mechanical properties of the material (stiffness, ductility, and strength) should be as high as possible in order to match the mechanical properties of the substrate

TABLE 4. MATERIALS SELECTION AND PERFORMANCE CRITERIA

MATERIAL PARAMETER	DESIRED TREND	BALLISTIC MANEUVER	BALLISTIC	SCIENTIFIC SPACE	ORBITING MANEUVERING	PLANETARY MARS LANDER	PLANETARY VENUS PROBE
HEAT OF SUBLIMATION/VAPORIZATION	HIGH	X	X				X
THERMAL CONDUCTIVITY	LOW			X	X	X	
HEAT OF GASIFICATION	HIGH			X	X	X	
TEMPERATURE OF DECOMPOSITION	HIGH			X	X	X	
CHAR STRENGTH	HIGH	X	X				X
THERMAL DIFFUSIVITY	LOW	X	X	X	X	X	X
SPECIFIC HEAT	HIGH			X	X	X	
DECOMPOSITION PRODUCTS	LOW MOLECULAR WEIGHT			X	X	X	
SURFACE EMISSIVITY	HIGH			X	X	X	
THERMAL EXPANSION	LOW(1)	X	X	X	X	X	X
MECHANICAL PROPERTIES F_{TU}, E, ϵ_{TU}	HIGH(1)	X	X	X	X	X	X

(1) Depends on interaction with other parameters and design criteria

material. In addition, a low thermal expansion would tend to result in better thermal compatibility conditions with the substrate material at elevated temperatures. A low thermal diffusivity is desired in order to minimize the amount of energy conducted into the material and maximize the thermal capacity of the ablating material. A low molecular weight of the decomposition products results in high blocking efficiency.

The composition and construction of a reinforced plastic composite has a profound influence on its performance in a re-entry environment. Important composition and construction parameters include the type of resin, reinforcement and filler, ratio of individual constituents, reinforcement orientation, and fabrication and processing conditions.

Polymeric materials are usually classified as either low or high-temperature ablators. Low-temperature ablators, such as polytetrafluoroethylene, are generally used when the incident flux is low, when a large volume of low molecular weight gas is desired to block convective heating and when maximum insulation is desired. When the environment involves high heating rates, high-temperature resins such as the phenolics, epoxies, and silicones are used.

These cross-linked resins tend to form carbon or other refractory decomposition products during exposure to high temperatures. This surface char or residue formed during ablation serves several useful functions and leads to a significant increase in performance. It acts as a refractory barrier between the virgin reinforced plastic substrate and the environment. Gases formed in the substrate absorb heat as they infiltrate through this porous layer, and it is capable of reaching very high temperature levels and thus dissipating energy by reradiation.

Of the resins available for heat shield composites, phenolics have been widely utilized where good thermal stability and mechanical properties are desired. They combine ease of application with versatility and good processing characteristics. For structural plastics, epoxy resins have been widely used as well. They are very easy to process because of their 100% reactivity, and they can be formulated for reasonable temperature stability.

However, the upper temperature limits to which such resins may be exposed while maintaining useful mechanical properties restrict their applicability for high performance heat shield/structure applications. This consideration is of particular importance in regimes where vehicles may be exposed to a prolonged heat soak prior to or during entry. Newer resins are available which possess greater thermal stability than the phenolics and, in some cases, have improved charring characteristics as well. Figure 5 illustrates the results

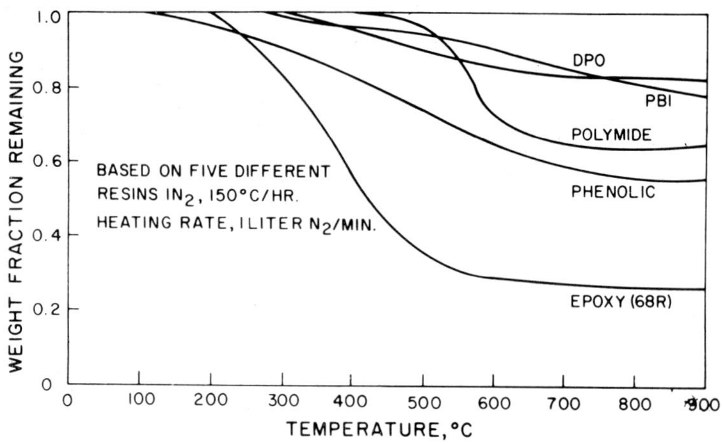

FIG. 5. Thermogravimetric analysis

of thermogravimetric analyses performed on a number of resins in a nitrogen atmosphere, including, for comparison purposes, a phenolic resin used for heat shields and an epoxy resin designed for structural applications. This chart shows that there are several resins displaying a combination of weight retention to very high temperatures and a high yield of char residue. Chemically, these are based on all-carbon or carbon–nitrogen aromatic ring structures. Newer resins, such as the polyphenylenes and paraphenyl phenols, still in the laboratory stage, display even better characteristics.

A wide variety of fibrous materials are used to reinforce the resin to form a strong composite which will perform in an ablation environment. The reinforcements can vary greatly in chemical composition and physical form. Generally, they are classified according to whether they are organic or inorganic. Organic fibers such as nylon are used when the exposure conditions involve low to moderate heating rates with low pressure and shear forces. Other organic fibers that are available include linen, cotton, rayon, orlon, etc. However, nylon has been employed most often. The organic fibers generate a large volume of gaseous products during ablation which serves to thermally insulate the substructure by absorbing and blocking large quantities of heat. However, a basic limitation of the organic fibers is their inability to anchor a surface char layer to the virgin material. The char layer is thus vulnerable to erosion effects associated with moderate pressures and shear forces.

Inorganic fibers such as silica and quartz are widely used. These greatly increase the strength and stiffness of the ablative plastic and perform well in environments which involve high pressure and shear forces. The fibers maintain themselves intact in the hot char layer and, consequently, they can mechanically reinforce the weak degraded surface layer and tie it to the virgin material. Asbestos fiber reinforcing materials have also been used. Refractory fibers such as carbon and graphite, have been developed and used for very high-temperature ablation conditions. These refractory fibers are usually formed from organic fibers, such as rayon or polyacrylonitrile, are nonmelting and have high sublimation temperatures.

Fibers are utilized in many different forms. Yarns in the form of a continuous filament have not been used widely to date for ablation materials. Molded parts using chopped, or tailored, fabric reinforcements have been used to form such parts as nose tips. The most widely used form of fiber are the fabrics and tapes which are prepared by weaving continuous filaments or yarns. These result in highly directional properties which must be tailored to the particular application. Reinforcement dimensions and cloth patterns are very important in designing a reinforced plastic composite heat shield. Particulate fillers have also been used in conjunction with the basic reinforcements to modify ablation and/or mechanical characteristics.

The ablative characteristics may vary significantly with the proportions of the individual material components. A high resin content may be desirable when it is necessary to generate a large volume of gas to handle intense convective heating. When environments include large mechanical pressure and shear forces, it is usually desirable to increase the percentage of reinforcing agent. Systematic orientation of the reinforcement in tape or filament form introduces directional or anisotropic mechanical, thermal, and ablative properties. While orientations parallel to the surface result in maximum in-plane strengths and a minimum of heat conduction, they do subject the composite to increased erosion resulting from pressure shear forces and internal gases trapped within the fibers. On the other hand, orientations perpendicular to the surface offer the greatest resistance to mechanical erosion. However, in-plane strengths are at their lowest levels and substrate temperatures will be higher due to improved heat conduction properties. Obviously, the most favorable orientation represents a carefully weighted compromise between such conflicting requirements.

The properties and quality of a reinforced plastic heat shield composite are significantly affected by processing variables and fabrication techniques. There have been instances where high quality material which is readily available in lab specimen size could not be reproduced on full scale production hardware. There is still considerable controversy regarding the influence of fabrication pressure and temperature. The effect of post curing is also important. In certain cases a long post-cure time at moderate temperature can be beneficial, in that residual solvents and waters of reaction are removed. This benefit can be counteracted by increased porosity, unpredictable shrinkage and increased brittleness of the material. Manufacturing defects such as voids, cracks, resin rich or resin starved regions may adversely affect ablation performance.

B. *Material Systems*

The historical development of reinforced plastic composites as thermal protection systems is shown in Table 5. The first generation of heat protection systems was the solid metallic heat sink in which the aerodynamic heating rate cannot exceed the rate of heat conduction into the material, or the surface will melt. Hence, the most effective heat sink materials are thermally conductive and have high heat capacities. Beryllium and copper are the most commonly used heat sink materials.

The increased heat absorption efficiencies associated with ablative mechanisms in reinforced plastic composites soon led to a decision to use them in re-entry vehicles. The second generation thermal protection materials, first introduced in the middle 1950's were composed of inorganic and organic fiber-reinforced composites. A typical early approach used melamine fiberglass sheets which were bonded together and hogged out to form a nose cone shape. This composite was successfully flown on the Jupiter C IRBM and on several ICBM flights. The phenolic resins, with their superior char forming ability, have been used with a variety of reinforcements. For example, phenolic-silica was flown on the RVX-1 vehicle. The identification of the

TABLE 5. HEAT SHIELD/STRUCTURE MATERIAL SYSTEMS

FIRST GENERATION • HEAT SINK	FOURTH GENERATION • PHENOLIC GRAPHITE • PHENOLIC CARBON
SECOND GENERATION • MELAMINE FIBERGLASS • PHENOLIC FIBERGLASS • PHENOLIC ASBESTOS • PHENOLIC SILICA • PHENOLIC NYLON	FUTURE GENERATION • BORON FILAMENTS • POLYIMIDES • DIPHENYL OXIDES • POLY BENZIMIDAZOLES • HIGH MODULUS GRAPHITE FILAMENTS • SILICON CARBIDE FILAMENTS • WHISKER REINFORCEMENTS
THIRD GENERATION • MODIFIED EPOXIES • SILICONES	

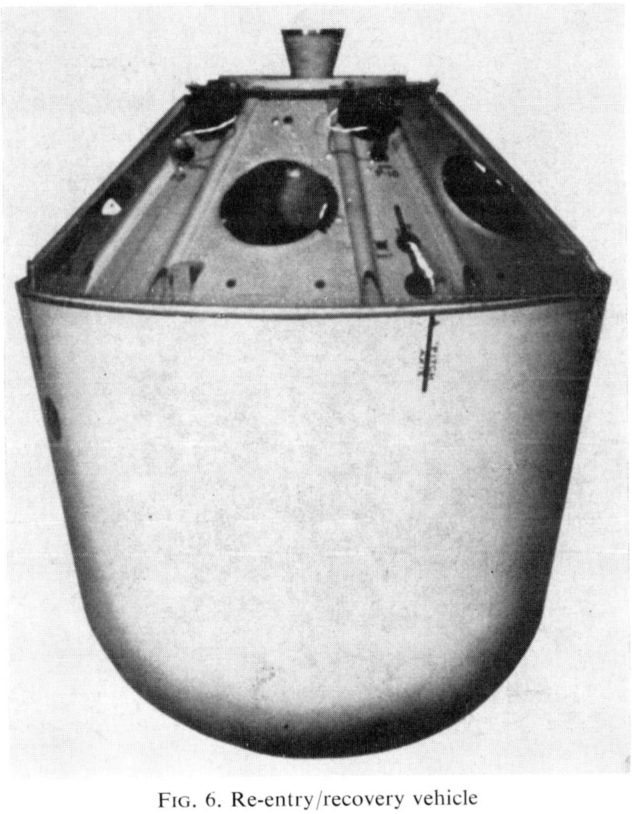

FIG. 6. Re-entry/recovery vehicle

capabilities of an all-organic thermal shield led to the development of phenolic-nylon composites, which were used on many second generation vehicles. This material has been used in the form of wound frustums and cylinders on ballistic re-entry vehicles, in a chopped molded form for the nose tips on re-entry vehicles, and as a parallel laminate on scientific space re-entry vehicles. The designs in each case utilized the mechanical characteristics of the nylon reinforcement to provide proper shield behavior in the ground and flight environment. For the scientific space vehicle shown in Fig. 6, a composite thermal shield/structural system was developed. This utilized a fiberglass-

FIG. 7. Operational re-entry vehicle

reinforced phenolic structural shell integrally cured with a phenolic-nylon thermal shield. The re-entry vehicle shown in Fig. 7 utilized phenolic-nylon heat shields throughout. The nose section was molded, in matched metal dies from 1/2 in. chopped squares of impregnated fabric and cured in an autoclave. The mismatch in thermal expansion between the heat shield and the metal substructure was accommodated by using a compressible, elastomeric bond which permitted relative motion between the heat shield and substructure.

Similar developments brought into operational usage phenolic-silica, and phenolic quartz composites. The TVX re-entry vehicle, Fig. 8, utilized tape-

FIG. 8. TVX re-entry vehicle

wound phenolic-silica thermal shields. These were machine wound from impregnated fabric and cured in an autoclave. The shields were bonded to the aluminum structure with a rigid epoxy resin. Because of the inorganic reinforcement, thermal expansion mismatches were minimized.

A further extension of the all-organic thermal shield technology led to the development of the third generation heat shield materials, a series of castable filled and unfilled epoxy resin systems which were developed, characterized by ground tests, and successfully flight tested. The need for low density thermal protection systems for satellite and space vehicles led to the development of silicone elastomer foam systems, both syntactic and chemically blown, which exhibit high strain capabilities over a wide temperature range.

The demonstrated success of the reinforced plastic composite heat shields on second and third generation vehicles has provided a sound basis for the design of more advanced re-entry vehicles. These vehicles are characterized by a streamlined design, higher ballistic parameter, higher entry speeds, longer range, increased accuracy, low detectability, and protection against enemy countermeasures. To meet these needs, a fourth generation of heat shield materials was developed, graphite-phenolic and carbon-phenolic composites. These provided a dual thermal protection and structural function in that the thermal shield could be used as a load bearing member of the system. These materials, yielding approximately 85% carbonaceous residue after decomposition, provide a degraded surface capable of reradiating at very high temperatures and providing heat absorption by a sublimation process. Thus, they combine many of the advantages of the graphites with the relative ease of fabrication of reinforced plastics. In conjunction with a metallic substructure, the stiffness and elongation characteristics of these fourth generation materials control its performance efficiency as a load bearing member. These characteristics are dependent upon the cure and reinforcement orientation within the composite heat shield and must be carefully considered in developing a shield system which meets the thermal, mechanical, and ablation requirements. The moderately high thermal conductivity of the carbon and graphite fibers results in greater thermal penetration into the heat shield during re-entry, which can result in increased thermal and mechanical erosion phenomena. Design techniques have been developed to minimize these effects. The GTV vehicle, Fig. 9, provided flight

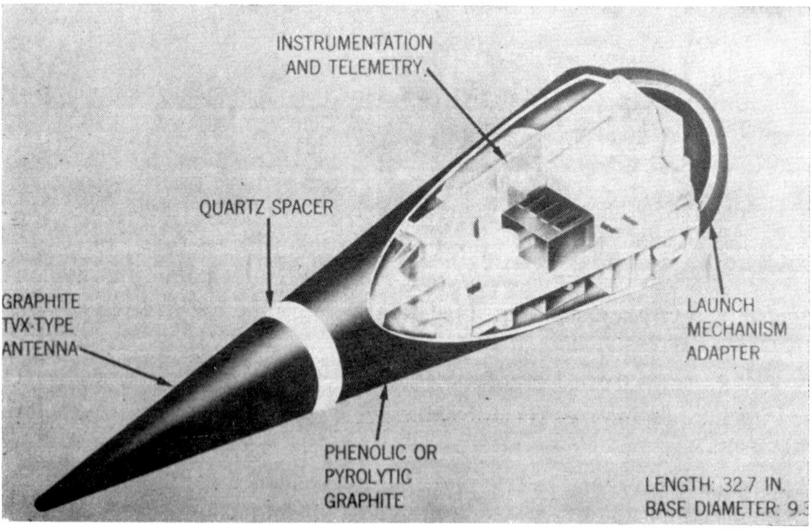

FIG. 9. GTV—graphite test vehicle

test data on graphite-phenolic and carbon-phenolic composites. The noses were machined from blanks which were high-pressure molded in matched metal dies from layers of impregnated cloth arranged in the desired orientation. The shields were tape wrapped by machine, cured under high pressure in hydroclave, and bonded to an aluminum substructure using a rigid epoxy resin adhesive.

For future generation material systems, high modulus boron and graphite filaments used in conjunction with high temperature (600–1000°F) resin

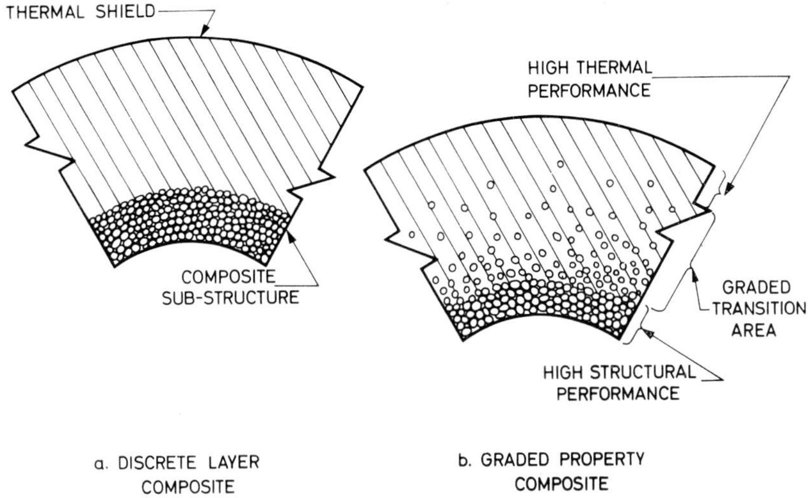

FIG. 10. Integrated shield/substructure concepts

systems such as polyamides, diphenyl oxides and polybenzimidazoles may make it possible to achieve an integrated shield/structure system such as is illustrated in Fig. 10. Integration of the reinforced plastic heat shield with a reinforced composite substructure having a similar resin matrix gives the designer additional opportunities to expand the capabilities and performance of re-entry vehicles. Due to the low temperature capability of bonds in conventional heat shield/substructure systems, it is necessary to increase the thickness of the heat shield in order to keep the bond line temperature at allowable levels. Thus, the designer must pay a weight penalty in both bond weight and heat shield insulation weight. The integrated heat shield/substructure eliminates the bond layer by using the same resin for both heat shield and substructure and curing these simultaneously. The temperature drop across the bond is eliminated, and the heat capacity of the reinforced composite structure is utilized. The required heat shield thickness can then be controlled by the decomposition temperature of the matrix of the structural composite and by its strength–temperature characteristics. The gradual change in material properties through the thickness will provide minimum thermal incompatibility problems between the heat shield reinforcements and the structural reinforcements. Other newly developed high performance filaments and whiskers, along with advanced fabrication techniques should make it possible for the designer to arrive at even more efficient composite designs, with reinforcement provided in all three axes of the composite.

C. *Thermostructural Optimization*

Designing with reinforced plastic composites is somewhat more complex than with conventional metallic engineering materials. The number of important variables is large and they are interrelated in a very complex manner. However, the design process has been greatly simplified in recent years by the utilization of parametric studies to determine the degree and magnitude of related design factors and by high-speed computers for obtaining design data.

Considerable effort has been spent on optimization and synthesis studies to evaluate and provide means of identifying optimum thermostructural systems and correlating these systems with vehicle system requirements. Digital computer techniques have been developed for this purpose. Synthesizing vehicle re-entry loads, heat shield performance data, bond properties and structural behavior (material and configuration), it is possible to arrive at minimum weight solutions. These techniques can evaluate monocoque, ring stiffened, waffle reinforced, honeycomb sandwich, corrugation and corrugated core sandwich substructural shells designed with a wide range of metals and reinforced plastics. The optimum solution will represent a trade-

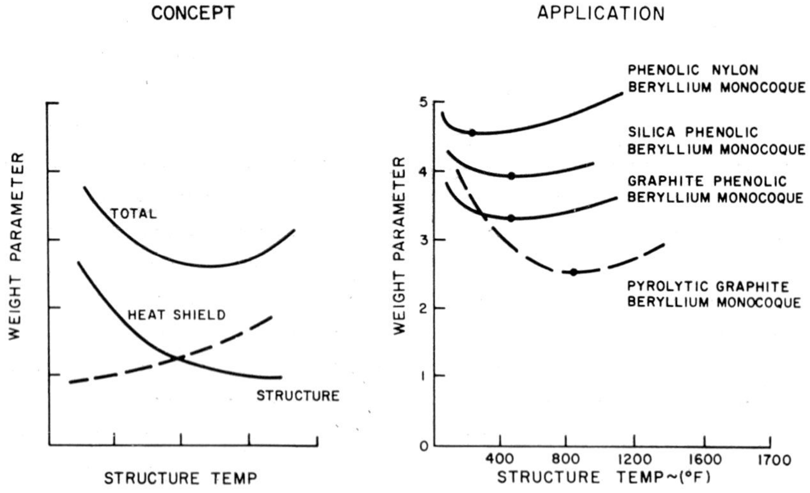

FIG. 11. Shield–structure optimization

off between the reduction in heat shield weight due to decreased insulation requirements as the allowable structural temperature is increased and the increase in structural weight due to decreased properties as the structural temperature increases. This concept is shown in Fig. 11 with the identification of a minimum value of the total shield/structure weight. The application shows a trade-off for four heat shield materials using a common substructure, beryllium monocoque, and indicates that graphite phenolic represents a minimum weight solution when limited to a 400°F back face temperature. On the other hand the pyrolytic graphite solution has greater promise if a high temperature bond can be used.

Re-entry vehicle thermostructural weights have improved significantly since the first introduction of reinforced plastic composites in 1956, as

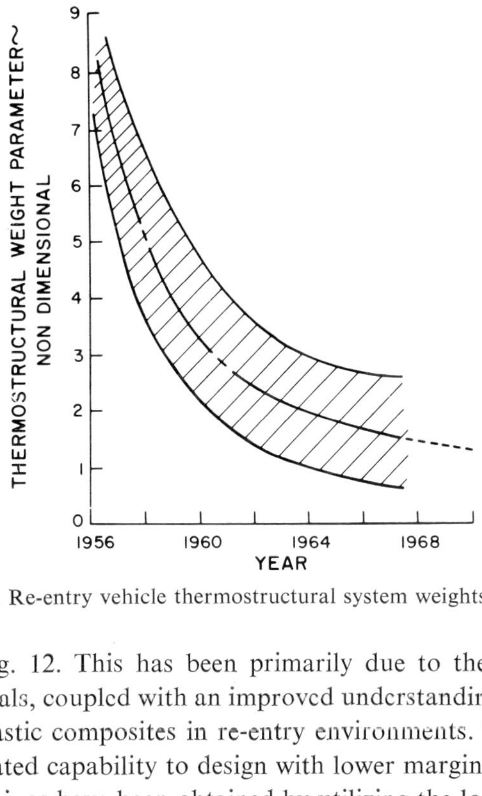

FIG. 12. Re-entry vehicle thermostructural system weights trends

illustrated in Fig. 12. This has been primarily due to the development of improved materials, coupled with an improved understanding of the behavior of reinforced plastic composites in re-entry environments. This has resulted in the demonstrated capability to design with lower margins of safety. Additional weight savings have been obtained by utilizing the load carrying capability of the heat shield.

In addition to optimization for thermostructural loads during re-entry, other thermal environments such as ground handling, transportation, and storage temperature extremes play an important role in obtaining an efficient design. This is particularly true in the case of orbiting re-entry vehicles which can undergo substantial circumferential temperature gradients as well as cyclic temperature changes (±250°F) during each orbit prior to re-entry. Thermal shields and substructures, which are chosen primarily on the basis of meeting re-entry requirements, must also satisfy these additional environments. Most heat shield materials have a coefficient of expansion that differs widely from that of the substructure they are protecting, as shown in Fig. 13.

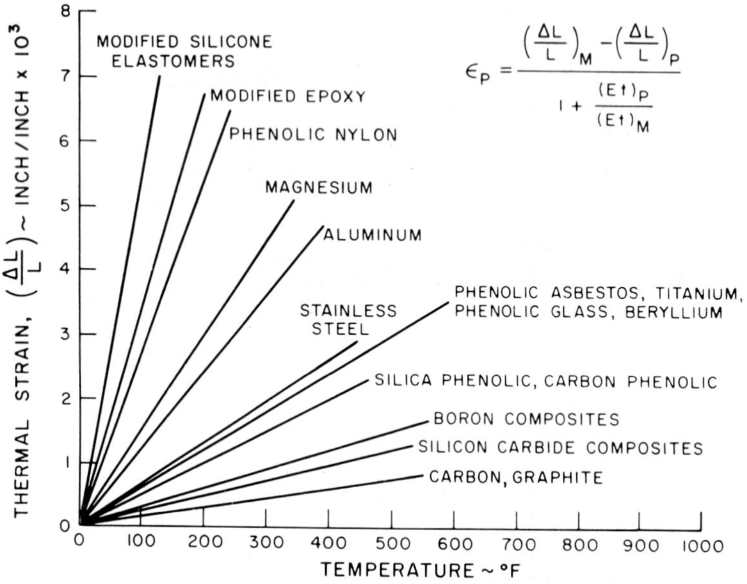

FIG. 13. Thermal expansion of typical heat shield substructure materials

There are several ways to compensate for this difference. The shield material may be modified in an attempt to match the expansion characteristics of the structure or obtain increased elongation characteristics. However, this approach often results in a weight penalty associated with decreased thermal performance during re-entry. In some cases the differential expansion characteristics can be compensated for by proper selection and design of the bonding system. This has resulted in the consideration and utilization of rigid and semirigid bonds as well as prefoamed and foamed in place compressible bonds. However, most nonrigid bond systems do not have high temperature capabilities and result in a weight penalty associated with increased heat shield insulation weight.

D. *Thermomechanical Design Aspects*

Thermostructural optimization trade-offs are affected to varying degrees by thermomechanical stresses in the degraded material. The principal thermal stress problems are associated with the steep temperature gradients through the thickness of a heat shield. Mechanical problems are the result of external stagnation pressures, shear forces at the solid–gas interface, and internal pressures created by the vaporization of the degraded materials. Depending on the type of heat protection material used, these stresses can result in limitations on heat shield design. Thermal stresses are important in the behavior of reinforced plastic composites such as carbon phenolic. Figure 14 shows a cross-section of a carbon phenolic nose tip after exposure in a rocket engine

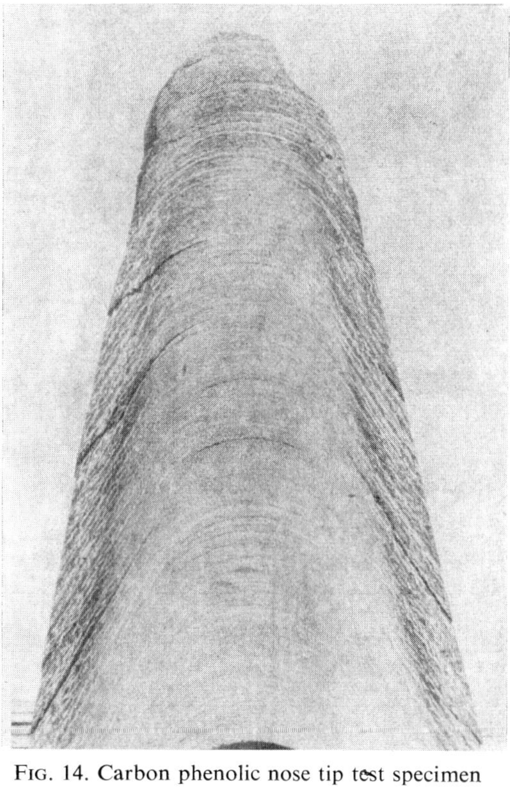

FIG. 14. Carbon phenolic nose tip test specimen

test facility. The nose is machined from blankets that have been molded from layers of impregnated cloth arranged in the orientation shown. The darkened area indicates the depth of degraded material. Interlaminar cracking in the degraded material is evident. The surface cracking in this laminated material is due to its orthotropic thermal expansions. The phenolic matrix has a much larger coefficient of thermal expansion and a lower thermal conductivity than the carbon fibers. Figure 15 shows specimens of phenolic silica and phenolic nylon which have been similarly exposed in a hyperthermal environment. Under the same exposure, plastic reinforced with inorganic fibers, such as silica, chars much deeper than the same plastic reinforced with organic fibers, such as nylon. However, the resistance to shear forces of the phenolic silica char layer is much greater than that of the phenolic nylon. What appears to be an inherent limitation of the organic reinforcement is its inability to anchor a surface char layer to the virgin substrate. Such materials generate gases of decomposition before the matrix material is sufficiently degraded and porous enough to prevent pressure build up in the

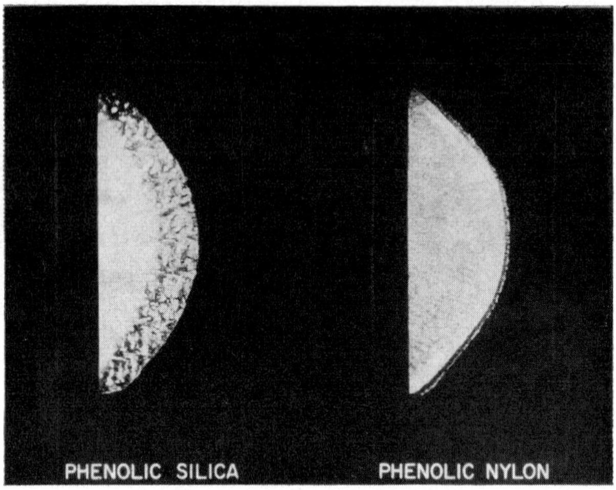

FIG. 15. Test specimens

composite. Thus, the char can pop off locally unless special provisions are made for venting.

Inorganic fibers such as silica perform much better in environments which involve high mechanical forces of pressure and shear. The fibers tend to remain essentially intact in the hot char layer, providing a degree of structural reinforcement. In the case of such inorganic fibers as silica, surface melting of the fibers can take place, resulting in the appearance of droplets or a film.

As the mechanical effects of the environments such as stagnation pressure increase, premature removal of surface material takes place with an attendant reduction in material efficiency. Most reinforced plastic composites begin to exhibit departures from thermochemical ablation theory at about 20 atm. This is illustrated in Fig. 16, where recession rate is plotted against stagnation pressure. The lower curve is based on thermochemical predictions. The cross hatched area between the lower and upper curves represents the departure from theory in erosion rate performance of typical plastic reinforced composites and shows the relative effects of thermochemical ablation compared to mechanical erosion as stagnation pressure increases. Depending on cloth orientation, the erosion rates may vary from thermochemical predictions by a factor of 3 at stagnation pressures of 100 atm. At these stagnation pressures the importance of cloth orientation becomes critical.

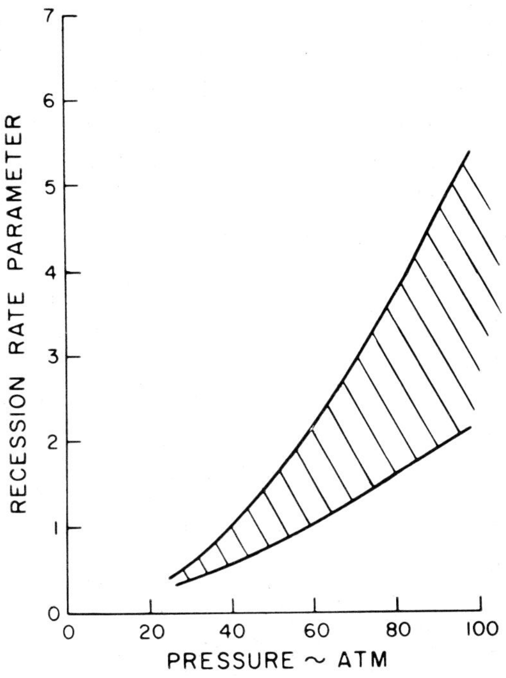

FIG. 16. Recession rate vs. pressure

E. *Thermal Stresses*

The thermal stress performance of plastic-reinforced composites during re-entry is complicated by the anisotropic characteristics of the composite, by the effects of temperature on properties, by the very steep gradients imposed on the material and by the limited amount of quantitative information available regarding the failure mechanics of these materials in re-entry environments. Theoretical prediction techniques have been developed based upon macroscopic continuum mechanics principles which require the use of high-speed digital computers. Utilization of these methods results in thermo structural phenomena that differ significantly from those previously encountered in other classes of thermal mechanical problems, such as those observed in heat transfer and heat power apparatus. The temperature distributions through the thickness of a reinforced phenolic tape wrapped heat shield as a function of original (prior to re-entry) thickness at eleven altitudes or re-entry times is shown in Fig. 17. At very high altitudes, greater than 100,000 ft, very little erosion takes place and surface temperatures rise to 3000°R. At lower altitudes, of the order of 100,000 ft, the denser atmosphere coupling with the re-entry velocity (20,000 ft/sec) results in increasing erosion and intensive heating of the material. Surface temperatures reach as high as

6800°R. During this phase of re-entry, the temperature profile is very steep with very little penetration in depth as the material erosion rate is in phase with the rate of heat conduction into the shield. As the vehicle continues to lower altitudes, its velocity has become significantly reduced and therefore the heat flux also is considerably reduced. The surface temperature drops

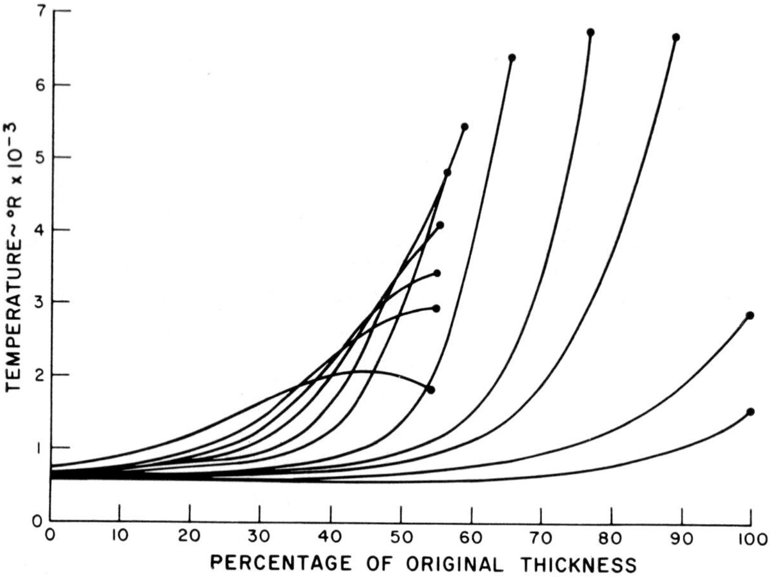

FIG. 17. Temperature profiles

below the critical surface erosion temperature and the rate of erosion approaches zero. Convective cooling of the surface begins to take place so that by the time ground impact takes place, the surface material temperature is less than the temperature of the interior material. It can be seen that material located within 30 to 50% of the original heat shield thickness undergoes a considerable heating and cooling cycle.

Thermal expansion is one of the material properties which significantly affects the thermal stresses developed in a heat shield during re-entry. Figure 18 shows the thermal expansion and thermal contraction characteristics of a typical tape wrapped reinforced phenolic heat shield as a function of temperature and wrap angle. As the material is heated from room temperature, it expands until chemical decomposition of the phenolic resin begins to take place. Then contraction takes place as the resin decomposes, until temperatures of 2000°F are reached, after which the entire composite starts expanding again as a carbonaceous material. Subsequent cooling from high temperature levels proceeds along a different path than followed during the heating cycle. The anisotropic and irreversible characteristics of the thermal expansion data

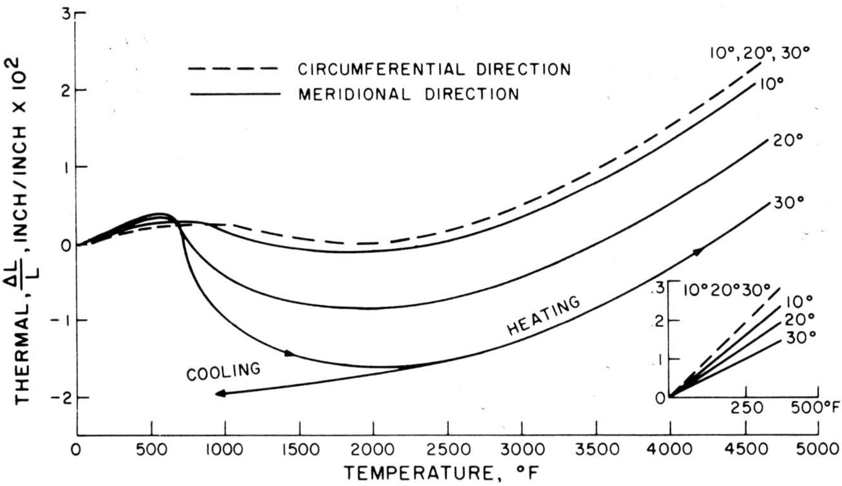

FIG. 18. Thermal expansion of a laminated plastic reinforced heat shield

in conjunction with similar characteristics for the moduli and Poisson ratios require the application of high speed digital computers to accurately assess the thermal stress fields in the composite. Figure 19 shows the thermal stress distributions during early (heating) and late (cooling) re-entry times in the axial and hoop directions. At the innermost fiber the hoop and axial stresses are always tensile. As one moves outward towards the heat shield surface, the stresses become compressive, which is expected for materials where expansion

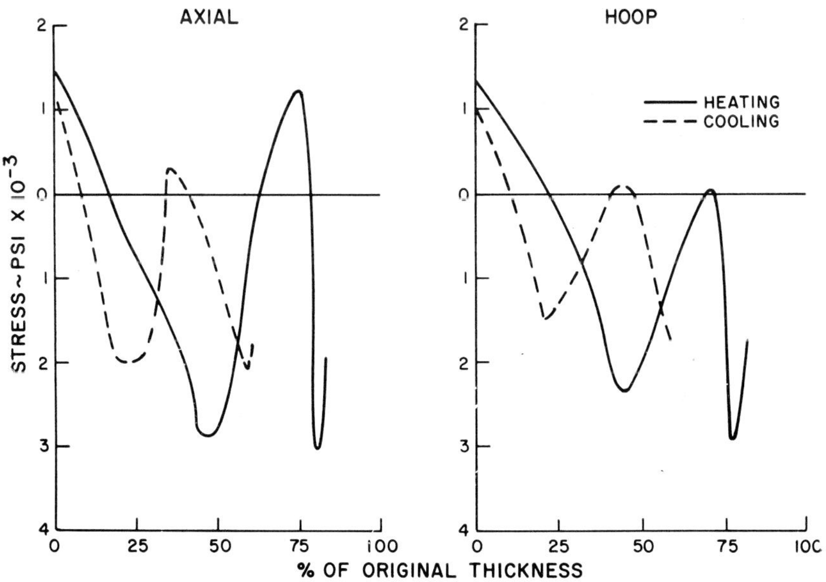

FIG. 19. Thermal stresses in laminated reinforced plastic heat shield

increases with temperature. The axial stress distribution during heating shows that a maximum compressive value is reached and then the stress becomes tensile very close to the surface. This tensile stress is due to the fact that the portion of the shield at which it occurs is in the temperature range which has experienced contraction according to its thermal expansion behavior. It can be noted that the hoop stress during heating also proceeds to become tensile after reaching its maximum compressive value; however, the magnitude of this tensile stress is much lower than that experienced in the axial direction. The difference is due to the fact that the material contracts much less in the hoop direction than it does in the axial direction. The stresses in both the axial and hoop directions at the outer surfaces are compressive because the outermost material is in a temperature range which corresponds to a large expansion, relative to the other portions of the shield. The stresses in the cooling phases of re-entry possess the same type of distribution through the thickness. Of major significance, however, is the fact that material in the heat shield subjected to one state of stress does not remain so throughout re-entry. For example, material at the 50% station will be subjected to compressive axial stresses during the heating phase which become tensile stresses during cooling. This cyclic change in stress direction can be of critical importance with respect to the propagation of cracks through the heat shield.

F. *Filament Wound Structures*

The advent of new filamentary materials (advanced glass compositions, boron, graphite, carbon, silicon carbide, and boron nitride) has resulted in the possibility of substantial weight and performance improvement by integrating the reinforced plastic composite heat shield with a filament wound substructure. The unique advantages of filament wound composites lie in the characteristics of the filaments; the many continuous, small diameter, high strength fibers imbedded in a matrix.

Since the fibers are unidirectional, structural efficiency is obtained by orienting the fibers to align them with the directions of applied loads. Realization of the potential of these materials requires an interactive design and end fabrication procedure that blends the four major phases of composites engineering: loads analysis, filament and matrix materials selection, composition and orientation of the component materials to give desired directional properties and control of fabrication to retain the required design properties. Such concept to end product integration is fundamental in the application of filament technology to re-entry vehicle structures.

The design logic flow chart in Fig. 20 illustrates the manner in which micromechanical analyses are fused together with conventional macromechanical procedures.

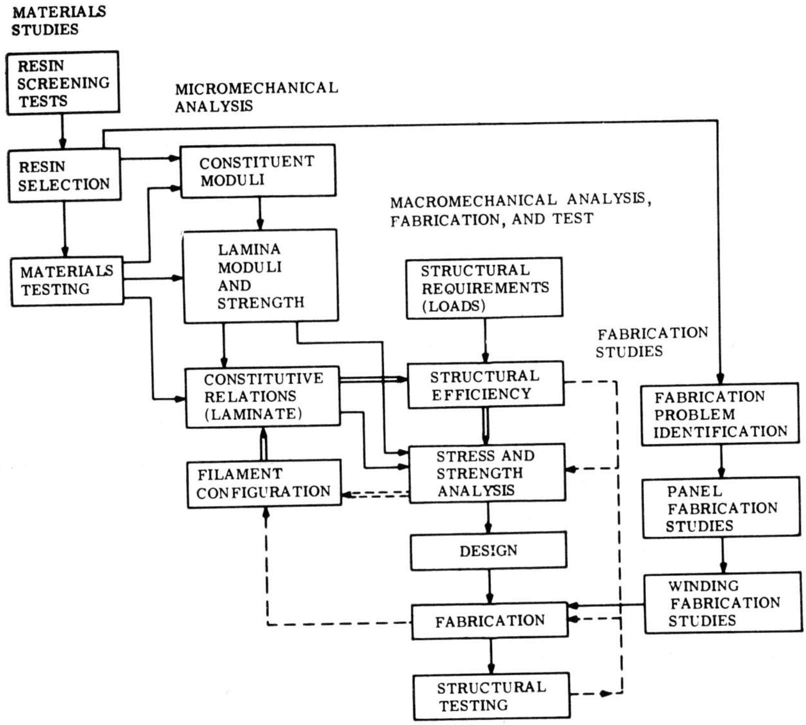

FIG. 20. Composite structures—design logic

In micromechanics, the true character of a composite material is studied, taking into account its heterogeneous nature (fibers embedded into a matrix). Based upon this, the constitutive equations are developed. These equations relate the macroscopic (lamina) stress and moduli to the constituent variables, such as the volumetric composition, geometric arrangement of the fiber, and constituent material properties. As a result of the micromechanics analysis the lamina (or thin sheet of the composite containing many parallel filaments in a representative volume element) is defined and subsequently treated as being macroscopically homogeneous. This analysis in conjunction with screening test data over the temperature range of concern, results in the initial resin-fiber selection.

Filament wound composite structures are normally fabricated by a lamination process in which the adjacent laminae are oriented in different fiber directions. The macromechanic analysis predicts the response of the structure in terms of these laminae orientations and properties and results in definition of the stress and strain level in each lamina in the composite. This is used in conjunction with a failure criteria, lamina strength allowables and design loads to predict the structural efficiency of various resin systems as well as the optimum lamina orientations for the critical loading conditions. In addition, the gross laminate effective moduli are used to determine the buckling capability of the structure.

With this information, it is obvious that the selection of an optimum material and lamina orientation in the composite is an iterative process. Testing is required to obtain the basic properties of the resin and fiber constituents to screen the minimal resin strength requirements, to screen the manufacturing process variables, to verify the micromechanical analysis, to establish the lamina strength allowables, to verify the strength of the composite and the failure criteria from macromechanical analysis, to establish the lamina strength allowables, to verify the strength of the composite and the failure criteria from macromechanical analysis and to validate the capability of the final hardware. With elevated temperature requirements, these tests must be performed at different temperatures. Fabrication using automated winding equipment results in fabrication limitations (such as minimum helix winding angles, maximum volume to prevent fiber variation of helix angle along length, etc.) that must be factored back into the filament configuration and structural efficiency of the specific design. Several iterations may be needed before the final design is obtained.

IV. SUMMARY

The thermal protection and ablative performance of reinforced plastic composites in very high temperature environments has greatly increased the importance of this type of engineering material. During the past decade re-entry heating problems have been successfully solved with the use of plastic reinforced composites. As mission requirements have become more and more severe, developments in reinforced plastics technology have been able to keep up with the demands.

In the future, aside from the higher re-entry velocities and the more severe aerothermodynamic environments, there will be other mission requirements superimposing added environments on the basic heat shield function. Refurbishable thermal protection systems for manned re-entry vehicles is an example. To meet these requirements more complex reinforced plastic composites will be developed which will require parallel advances in design, analysis, and fabrication techniques. Many new and different uses for reinforced plastic composites in high temperature environments will be found which will result in developments far beyond our present concepts.

ACKNOWLEDGEMENTS

The authors wish to express their appreciation for the assistance provided by Mr. L. H. Shenker and Mr. P. B. Cline of the GE-RESD Research and Engineering Department.

REFERENCES

1. Brunner, M. J. Phenolic Carbon Heat Protection System Requirements for Ballistic and Maneuvering Applications, General Electric Report No. 64SD252, May 1965.
2. Dow, N. F., Rosen, B. W. and Kingsbury, H. B. Evaluation of the Potential of Advanced Composite Materials for Aircraft Structures, Air Force Material Lab Report TR-66-144, Wright-Patterson Air Force Base, Ohio, May 1966.
3. Dow, N. F. and Rosen, B. W. Development of a Rational to Guide the Utilization of Boron Filament Reinforced Composite Materials for Aerospace Structures, General Electric Co., Valley Forge Pa., August 1964.
4. Gruntfest, J. J. and Shenker, L. H. Behavior of Reinforced Plastics at Very High Temperatures, *Modern Plastics*, **35**, June 1958, p. 155.
5. Gruntfest, J. J. and Shenker, L. H. Behavior of Reinforced Plastics at Very High Temperatures—Part 2, *Modern Plastics*, **36**, April 1959, p. 110.
6. Gruntfest, J. J. and Shenker, L. H. Ablation, *International Science and Technology*, July 1963, p. 48.
7. Hess, T. E., Bailey, J. C. and Most, H. Ballistic Re-entry Vehicle Thermo-Structural Optimization, G. E. Report No. 62 SD 199, General Electric Co., Phila., Pa., OT 1962.
8. Hurwicz, H. Thermal Protection Systems, TDR No. ML-TDR-64-82, AFML, Research & Technology Division, Wright-Patterson Air Force Base, Ohio, Jan. 1965.
9. Lafazan, S., and Walsh, W. The Charring Ablator Concept Application to Lifting Orbital and Super Orbital Entry, Aerospace Corporation Report ATN-63(3153-01)-1, Jan. 1963.
10. Scala, S. M. Sublimation in a Hypersonic Environment, General Electric Report No. R58SD 289, Phila., Pa., Nov. 1958.
11. Scala, S. M. The Thermal Degradation of Reinforced Plastics during Hypersonic Re-entry, General Electric Report No. R59SD401, Phila., Pa., July 1959.
12. Scala, E. Material Aspects of the Re-entry Problem, p. 245 in *High Temperature Structures and Materials*, Pergamon Press, N.Y., 1964.
13. Schmidt, D. L. Ablation, p. 815 in *Engineering Design For Plastics*, Reinhold, New York, 1964.
14. Schmidt, D. L. Research Trends in Ablative Plastics and Composites, AIAA 6th Structures & Materials Conference, Palm Springs, Cal., April 1965.
15. Schmidt, D. L. Ablative Polymers for Hypersonic Atmosphere Flight, Air Force Materials Lab Report TR-66-78, Wright-Patterson Air Force Base, Ohio, May 1966.
16. Shenker, L. H. Re-entry Considerations—Lecture Notes for Short Course on Design Factors for Materials Application in Spacecraft, June 1964.
17. Shenker, L. H., Dolan, C. M. and Edighoffer, H. Thermal Shield Bonding Systems for Re-entry Vehicles, General Electric Report No. 64SD256, June 1964.
18. Swann, R. T. and South, A. Theoretical Analysis of Effects of Ablation on Heat Transfer to an Arbitrary Axisymmetric Body, NACA TN-D-741, April 1960.
19. Warren, W. R. and Diaconis, N. S. The Performance of Ablation Material as Heat Protection for Re-entry Satellites, JAS paper 60 49, 1960.
20. Wood, R. M. and Tagliani, R. J. Heat Protection by Ablation, Douglas Aircraft Corporation Engineering Report No. 923, Jan. 1960.

APPLICATIONS OF COMPOSITE MATERIALS IN SPACE VEHICLE STRUCTURES

RICHARD R. HELDENFELS†

NASA Langley Research Center

ABSTRACT

The characteristics of filamentary composite materials are reviewed and compared with the structural design requirements of vehicles used in space operations. Numerous potential applications of filamentary composites in space vehicles are identified but few have been made or planned. Additional applications await solution of basic problems in analysis, design, fabrication, and reliability.

I. INTRODUCTION

Filamentary composite materials have become the subject of great attention in recent years[1-4] because lower structural mass or other advantages can accrue from their use in structural applications. Over one hundred million dollars have been expended in research and development on filaments and composites in just the last few years. Only about 10% of this effort, however, has been directed toward space vehicle structures and only a few applications have been made. Why have composites been used so little to date and what are the potential uses of these materials in space vehicles? Applications of composites in space vehicle structures will be reviewed in this paper to examine reasons for slow acceptance of composites and those situations in which future utilization appears most promising. Characteristics of composites will be considered first, followed by a discussion of the space vehicle design process and several examples of specific applications in which composites appear to have merit.

The term "composite material" can encompass many things, consequently, herein it will be limited to composites composed of long, small-diameter filaments uniformly oriented in an appropriate matrix. In addition, the term "space vehicle" will be used to describe those vehicles, used for operations in space, that can be categorized as launch, space, entry, and landing craft. Within these definitions, there are numerous filamentary composite materials and a large variety of vehicles and space missions that make generalization

† Chief, Structures Research Division.

difficult. Therefore, only a few representative structural applications and composite materials can be considered.

The International System of Units (SI) is used to express all physical quantities in this paper.[5]

II. CHARACTERISTICS OF FILAMENTARY COMPOSITES

The selection of materials for flight vehicle structures is based on numerous properties, parameters, and considerations. For example, the NASA Special Committee on Materials Research for Supersonic Transports[6] used 14 parameters to screen candidate materials. That list has been modified and extended herein to provide a basis for a generalized evaluation of filamentary composites for use in space vehicle structures (Fig. 1). Properties and factors that influence the selection of typical structural materials for representative space vehicle applications are listed under rating categories of Plus, Minus, and More Data Needed that indicate the relative merits of each characteristic of composites when compared to conventional materials. Some of these items can be established quantitatively, but others are qualitative or subjective. The

Plus	*More data needed*	*Minus*
Low density	Toughness	Reliability
High strength	Thermal stability	Maturity
High fabricated strength	Fatigue resistance	Design methodology
High stiffness	Corrosion resistance	Joints and connections
Low thermal stress	Impact resistance	Cost
Formability	Space environmental effects	
Tailored design	Damping	
	Producibility	
	Availability	

FIG. 1. An evaluation of filamentary composite characteristics important in space vehicle structures applications

list is not intended to be exhaustive and many of these ratings are debatable. Certainly, any such general evaluation is controversial and exceptions can be found for every item.

A detailed discussion of each item could fill as many papers, so a few general comments must suffice. Composites have numerous "pluses" that include unique features such as tailored design, but many other important factors are not good or are still in doubt. For example, much more data on the effects of the space environment on composite materials are needed before the place of these materials in space structures can be established. The greatest deficiencies are the result of the relative newness of this technology, limited availability of required design methods, and the cost and complexity of designing functional structures with materials which, in turn, are small-scale structures that require rigorous stress analysis. The relationship of these factors to space vehicle structural applications should become somewhat clearer after specific cases are discussed in the sections to follow.

The plus characteristics of filamentary composites that make them most interesting for use in structures are the high specific strength and stiffness available in filaments. The discussion that follows, therefore, will concentrate on space vehicle components in which strength and stiffness are important. Composites are highly interesting, however, for many space structures because of other unique or outstanding properties; a few such cases will be included, too. The typical composites in most examples cited in this paper will be existing filaments in an existing resin matrix; other matrices, such as metals, may provide additional advantages, but are not given detailed consideration herein because their future development is not expected to alter the conclusions substantially.

Figure 2 shows the ratios of strength and stiffness to density, on logarithmic scales, for a variety of filaments, composites, and metals. The figure also

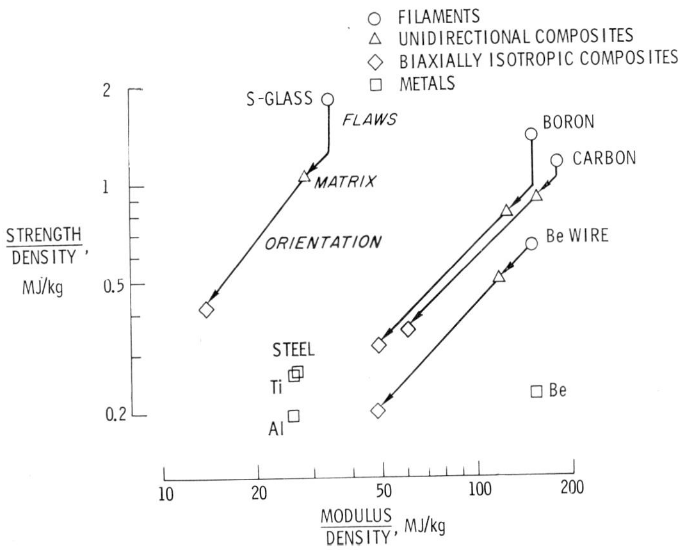

FIG. 2. Specific strength and stiffness of filaments, composites, and metals

shows the reduction in properties that results from flaws in the filament, from combining filaments with a resin matrix to produce a unidirection reinforcement, and from providing filament orientations that give isotropic properties in the plane of the laminate. These factors, present in practical structures, considerably reduce the outstanding characteristics of filaments, but many composites still exceed significantly the properties of conventional materials.

The use of composites to replace sheet metal, where isotropic properties are required, does not give particularly outstanding results, but the gains are substantial in those structural applications in which nearly unidirectional composites can be used. Since many structures require only small amounts of biaxiality, the ability to tailor the orientation of the filaments in each stress field provides a significant design advantage. On the basis of strength or stiffness, then, a composite may have greater material efficiency than most metals. But these two properties alone are not necessarily direct indicators of greater structural efficiency.

Figure 3 considers the structural efficiency of a simple component, a circular cylinder in axial compression.[7] Structural density is plotted as a function

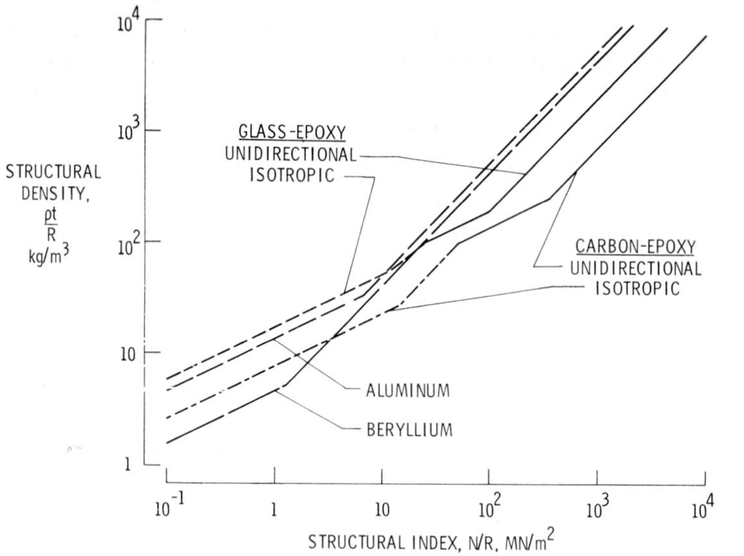

FIG. 3. Structural efficiency of materials for monocoque cylinders under axial compression

tion of the structural index for cylinders of four materials—two metals, aluminum and beryllium, and two composites, glass in epoxy and carbon in epoxy. Note that for each composite, two sets of curves are used; the lower sets are for isotropic laminates with filaments in three directions, 60° apart, whereas the upper sets are for unidirectional composites operating at the yield stress. Other orientations may be used to advantage in the intermediate region. Experiments are needed, however, to determine if composites can achieve these calculated efficiencies since they may experience failure modes not present in ductile metals. The benefits obtainable from composite construction vary significantly with the loading intensity. The calculations indicate that composites are not superior to beryllium metal in stiffness-critical applications but they exceed the metals when advantage can be taken of their high strength in unidirectional applications. Charts such as this provide comparison of the efficiency of materials in simple components under simple loading conditions in which only strength, stiffness, and density are important. In specific applications, strength and stiffness affect the structural mass in more complicated ways. In addition, many other material characteristics must be considered in the selection of practical structural materials.

III. SPACE VEHICLE DESIGN CONSIDERATIONS

To determine how materials are selected for space vehicle structures, it is enlightening to review the space vehicle design process outlined on Fig. 4. An iterative cycle of mission analysis, establishment of design criteria, and selection of appropriate structural configurations and materials is employed.[8] Each phase is influenced by the loads and environments that the spacecraft experiences during the mission. Mission analysis considers alternate ways in which the mission objectives might be accomplished to arrive at the optimum approach; it usually identifies one or more mission modes in which state-of-

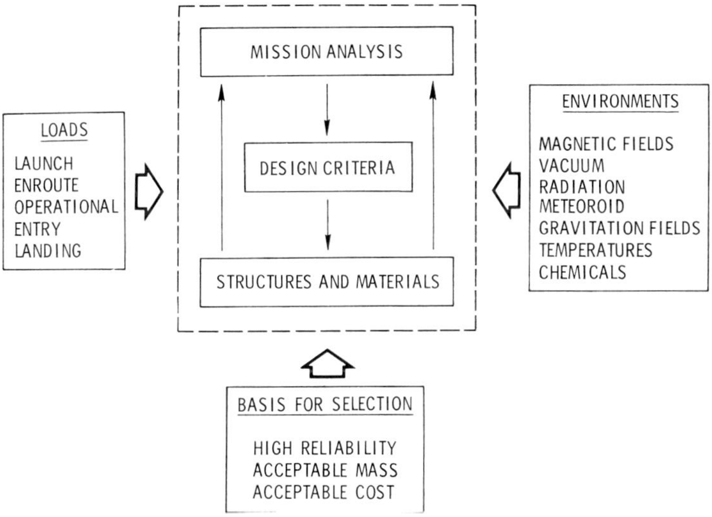

FIG. 4. The space vehicle design process

the-art structures and materials technology are acceptable. The selection of mission mode, structural configuration, and materials is guided by three general criteria. The primary consideration for the system is high reliability and maximum assurance that the mission will be performed, as designed, on the first flight. The approach selected must also have acceptable mass and cost; that is, they must be within the limits of the project. Because mass reduction is expensive, the achievement of the lowest cost and the lowest mass is not compatible; thus, the system with the lowest structural mass is not necessarily the optimum one. Tight schedules and the reliability needed to guarantee mission success generally eliminate consideration of materials which are still in the development phase. In addition, the limited data on effects of space environment on bulk and surface properties of composites, and the present lack of methods which reliably predict some types of composite failure have discouraged space vehicle applications.[9, 10] Therefore, the opportunities that composites offer for significant reduction in space vehicle structural mass are not likely to be realized until all pertinent composite characteristics are established with high confidence. Applications may be found, however, in which one of the unique characteristics of composites, other than strength or stiffness, provides essential capabilities not otherwise available. For example, composites may offer the best solution if a structure must be transparent to radio frequency radiation or if particular thermal expansion characteristics are required.[11, 12] The latter would take advantage of the designer's opportunity to tailor composites to the application. This opportunity to tailor materials to the application could be the greatest advantage of the composite if structural designers were more proficient in determining the optimum combination of all material properties required for each structure.

IV. SPACE VEHICLE APPLICATIONS

Some of the potential applications of filamentary composite materials in space vehicle structures are shown in Fig. 5. Four types of space vehicles are listed on the left and structural components of each vehicle type, in which com-

Vehicle	*Component*
Launch vehicles Solid, liquid, reusable	Propellant tanks Interstage structures Payload shrouds Rocket nozzles
Spacecraft Observatories, explorers manned stations	Pressure vessels Antennas Power arrays Meteoroid shields Trusses
Entry vehicles Ballistic, lifting earth, mars, venus	Thermal protection Aeroshells Deployable decelerators
Landing craft Landers, shelters, rovers earth, moon, mars	Energy absorbers Frameworks Cabin walls

FIG. 5. Potential applications of filamentary composite structures in space vehicles

posites may have application, are listed on the right. Every space vehicle structural element or component is a likely candidate for utilization of filamentary composites, but those listed here are representative possibilities for consideration in the following sections of this paper. Applications in current or planned vehicles, however, are much more limited. Many small structural parts of composite materials (particularly of glass in resin) have been used in all types of space vehicles,[(10)] but only launch vehicles have utilized major structural components of filamentary composites. Plans for future space vehicles do not include much more extensive utilization. Applications must be discussed, therefore, in terms of what may come to pass rather than of what has occurred.

A. *Launch Vehicles*

Figure 6 shows two types of future launch vehicles. The expendable type takes off vertically and is an extension of current launch vehicle technology. The reusable type, that has been the subject of much study, debate, and plan-

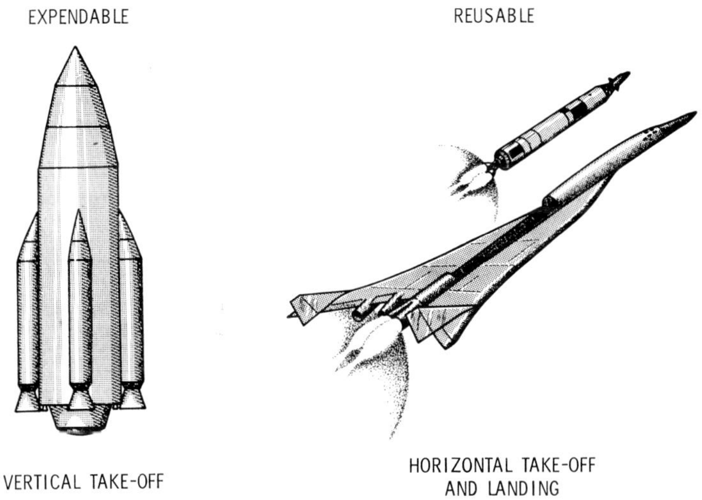

FIG. 6. Types of future space launch vehicles

ning, probably will have horizontal take-off and landing capabilities. A reusable vehicle could be one stage to orbit (aerospace plane) but the two-stage approach illustrated is more likely. The first stage is a hypersonic airplane with airbreathing propulsion while the second stage is a conventional rocket vehicle. The outlook for major applications of filamentary composites in the reusable, airplane-like vehicle is poor because most of the structural materials must have high efficiency at temperatures above 800° K. This capability is not characteristic of prospective filamentary composites.

The expendable launch vehicle illustrated utilizes a combination of solid and liquid propellants, the solids being clustered around the liquid core. Current space launch vehicles such as Titan III-C and the thrust-augmented Deltas are similar combinations while the Scout uses solid rockets exclusively and the other space launch vehicles use only liquid propellants. Solid rocket motor cases and nozzles are the outstanding example of major space vehicle structural components that now use filamentary composites.(13) Solid rockets for current space vehicles, however, use both metal and composite cases, with the latter more prevalent in the smaller sizes, but the prospects for increased future use are good.

Composites also have been used in the payload shrouds on at least two launch vehicles.(14) The best known is the Atlas-Agena shroud used to launch a Mariner spacecraft. This shroud, of phenolic-glass honeycomb sandwich construction, failed during launch and was replaced by a ring-stiffened magnesium shroud for launching the successful Mariner IV to Mars.(15) The Titan III used shrouds of fiberglass honeycomb sandwich construction on several successful flights, but because of a failure, a replacement of stiffened aluminum was designed.(14) In each case, the greater structural efficiency provided by the composite was lost because of design and environmental problems. The Mariner shroud delaminated due to the increased differential pressure in the unvented honeycomb cells resulting from the decreasing external pressure, increasing temperature, and thermal outgassing during launch. A vented honeycomb would not have failed, but this approach creates other problems. Venting to the interior, for example, may significantly contaminate spacecraft surfaces.

Analytical studies(16) have been made of the reductions in structural mass that might accrue from utilizing advanced structures and materials in future, liquid-propellant, launch vehicles. Some results are summarized in Fig. 7. The sketch at the left indicates the vehicle studied, a 9,140,000 kg, two-stage launch vehicle having a payload of 370,000 kg. A comprehensive computer analysis was made to determine the loads experienced and the structural mass required. The base line design utilized ring- and stringer-stiffened aluminum-alloy construction typical of present launch vehicles. Honeycomb sandwich provided the lowest mass of the structural configurations examined. Relative structural mass for these two types of structures of several metals and composites is listed. Note that converting to honeycomb construction in aluminum alloy results in a substantial reduction of structural mass. Titanum provides a slight additional reduction when used in sandwich construction and beryllium is even better. Beryllium, however, is about equally efficient in either the integrally stiffened or honeycomb sandwich configurations. The use of filamentary composites in honeycomb sandwich construction saves some additional mass compared to the aluminum sandwich, but this gain is only half as great as that obtained by changing the aluminum structure from stiffened skin to honeycomb sandwich. Moreover, this benefit from composites may not be fully attainable because of practical problems such as propellant compatibility.

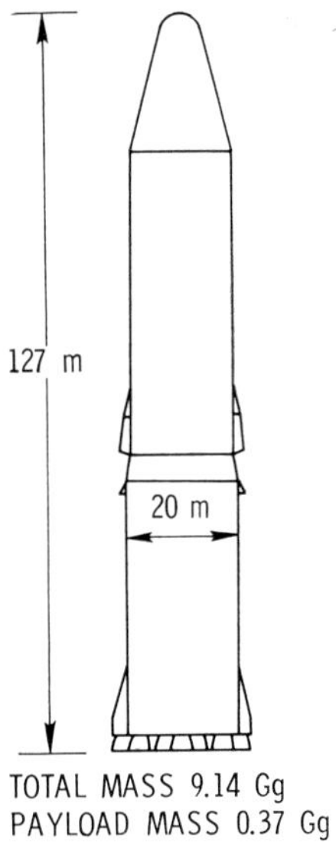

INTEGRALLY STIFFENED	
ALUMINUM	1.00
TITANIUM	1.04
BERYLLIUM	.55
HONEYCOMB SANDWICH	
ALUMINUM	0.66
TITANIUM	.61
BERYLLIUM	.53
GLASS IN EPOXY	.67
CARBON IN ALUMINUM	.47

FIG. 7. Relative structural mass of large, liquid-propellant launch vehicles

All three stages of the Saturn V launch vehicle use stiffened skins of aluminum instead of aluminum honeycomb sandwich with its lower mass but greater complexity and cost. Considering the opportunities to construct more efficient launch vehicle structures of materials that have been available for many years, the prospects appear dim that the designer of the next space launch vehicle will tackle the special problems of an advanced composite sandwich.

B. *Spacecraft*

Spacecraft have a wide variety of configurations; the one shown in Fig. 8 is the most complex expected in the near future. This manned laboratory in earth orbit may be used for scientific and engineering research in the NASA Apollo Applications program. The components and appendages of the vehicle offer numerous opportunities for effective use of filamentary composites. Potential application areas include meteoroid shields for the cabin;[17] the

FIG. 8. Future research laboratory in space

man-rated pressurized enclosure; the furniture and interior structure of the station; the beams, columns, and trusses used to support the various appendages; the communications antennas; the solar cell arrays that supply power; and the astronomical telescope. However, only two simple structural components, common to most spacecraft, will be discussed.

Spacecraft carry many pressure vessels for storing various fluids. For example, the Apollo space vehicle (Saturn V with Lunar Excursion, Command and Service Modules) contains six main propellant tanks, and 88 other pressure vessels that vary in volume from about 0.02 to over 4 m^3. This basic structural component offers an excellent opportunity for introducing composite materials technology into the space vehicles in the near future.[18] A precedent also exists since the solid rocket motor case, where filamentary composites have had the greatest use in aerospace vehicles, is primarily a pressure vessel.

Figure 9 shows how pressure vessel mass can be reduced by using the high strength of filamentary composites. Most of the small pressure vessels used in space vehicles are made of titanium because of its high specific strength. Problems have arisen, however, due to stress corrosion attack by various fluids. Both cylindrical and spherical shapes have been used to meet packaging constraints although cylinders have a higher mass per unit volume. If a filamentary composite is used, an isotensoid winding can provide the same specific mass in a wide range of shapes while significantly reducing the specific mass below that for the best metallic sphere. Many combinations of filament and matrix can be used in addition to the three listed, but the least mass will always be provided by the filament with the highest useful strength (glass). The benefits indicated here may not be achievable, however, because of practical problems such as chemical incompatibility between composite and fluid (which may be a problem for metals, also), minimum gage limitations, and porosity. Filaments in an epoxy matrix may result in a container so

MATERIAL	RELATIVE MASS CYLINDER L/D = 3		SPHERE
ALUMINUM	1.97		1.57
TITANIUM	1.25		1.00
GLASS IN EPOXY		0.49	
CARBON IN EPOXY		.56	
CARBON IN GLASS		.59	

FIG. 9. Relative mass of pressure vessels

porous to the fluid that a liner must be added.[19] Other matrices may solve such problems. Therefore, the combination of carbon filaments in a glass matrix is suggested. Methods are not now available for manufacturing this composite but it is an interesting possibility for future development. A glass surface containing the fluid could provide the utility of a soft drink bottle in a structure of the efficiency expected in aerospace applications.

Figure 10 shows a truss structure that has been designed for a space vehicle application in which the design loads arise from vibrations during launch. Design calculations have been made for a variety of materials, all in tubular form, and the relative mass of the resulting structures is shown by the bar graphs. The efficiency of all materials is reduced 20 to 30% by the mass of the joints, with the joint mass penalty increasing as the material efficiency increases. Welded joints can be used with aluminum but the tubes of all other materials are much more difficult to connect. Fairly sophisticated connections are required to attain the low joint mass given by these calculations. Again, the advanced composites are only slightly better than beryllium, but substantially better than aluminum or glass in epoxy. The problem of joining components of composite materials will occur in all types of structures and

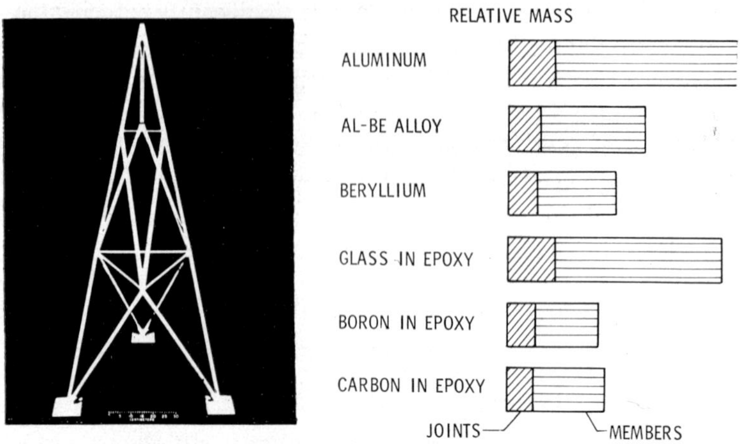

FIG. 10. Relative mass of truss structures

can have a major influence on selection of the best configuration and material. Composites will achieve more of their outstanding potential in those applications where major structural components can be fabricated in a single (filament or tape) winding operation.

C. *Entry Vehicles*

Ablation materials for thermal protection of entry vehicles are composites but not of the filamentary types used to obtain high strength and stiffness. Figure 11 shows a replaceable heat-shield panel for a lifting-body type,

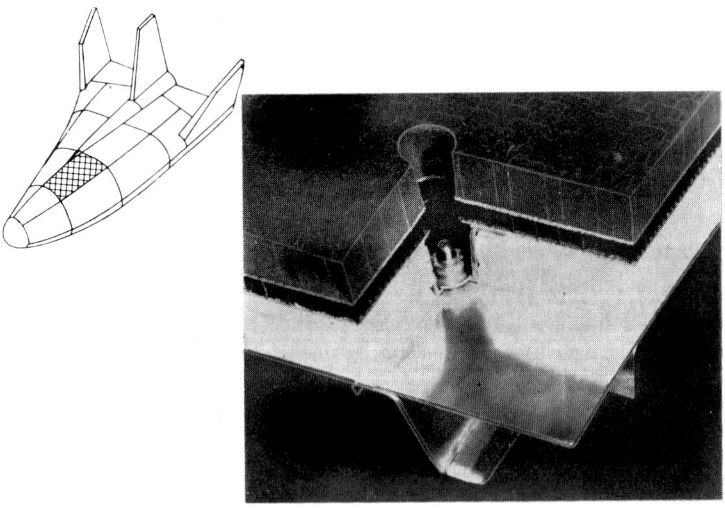

FIG. 11. Replaceable ablation heat shield for a lifting body entering from low earth orbit

earth-orbital entry vehicle to which a unique combination of properties of filamentary composites has made a significant contribution.[20] The ablation material is contained in a honeycomb and supported on a glass-phenolic honeycomb-sandwich panel firmly attached to the underlying vehicle structure at a number of points. Insulation fills the space between the panel and vehicle structure. The composite sandwich provided the best approach in this particular application because it had favorable fabrication and cost characteristics and a combination of load-carrying ability, low thermal stresses, and strain compatibility with the ablator that permitted the panel to be anchored at the support points. Use of this glass-plastic honeycomb core panel on fixed supports permits the use of replaceable panels of any convenient size, whereas the best metal sandwich panel required flexible supports that significantly limit panel size. These larger panels simplify the initial vehicle construction and reduce the time required to refurbish the thermal protection system after each flight.

Figure 12 shows the sequence of events in a representative entry and landing on Mars. Vehicles for landing payloads on Mars have many new and

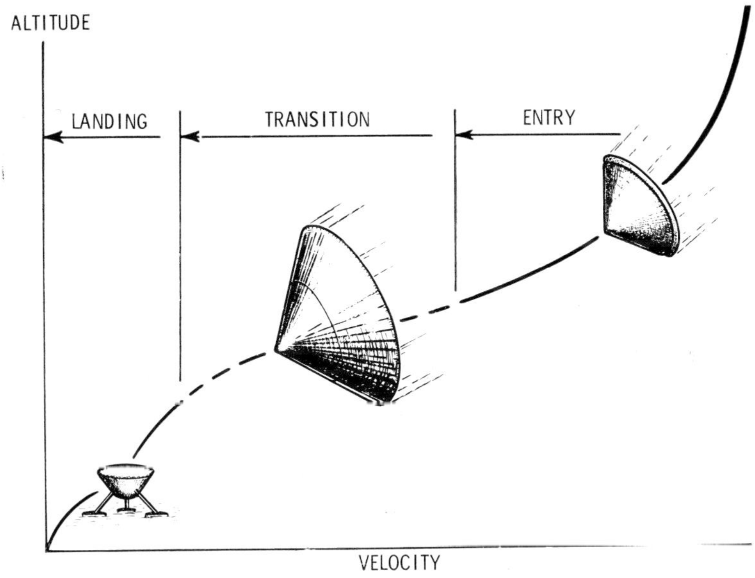

FIG. 12. Atmospheric entry and landing on Mars

difficult structural design problems.[21] Composites may be needed on the first mission, and may make a more substantial contribution to future landings of larger vehicles.

The entry starts with a low-density body that decelerates to supersonic speeds in the very thin atmosphere of Mars. Then, one or more variable-geometry decelerating devices, such as the inflatable afterbody illustrated, are deployed to slow the lander to the speed at which the retrorocket landing system can be activated. The requirement for very low-density vehicles places great emphasis on mass reduction if large payloads are to be landed by a single vehicle. Numerous alternate ways of performing such a mission are currently under investigation but three opportunities for use of advanced composites are evident, one in the structural shell of the entry body (the aeroshell), a second in the variable-geometry decelerator, and a third in the lander. Landing craft will be discussed briefly in the next section.

The inflatable decelerator requires a flexible, impermeable, high-strength membrane that can survive the impact loads and heating of supersonic deployment. These are characteristics that are best obtained by combining several materials in a composite. A laminate of appropriately oriented thin films of high-temperature plastic strengthened by a unidirectional layer of very small-diameter glass filaments may satisfy this requirement. Such membranes could be useful in many other variable-geometry or pressure-stabilized structures of space vehicles.

Figure 13 shows a 120° conical aeroshell for an unmanned Mars entry and the relative structural mass of several combinations of structures and

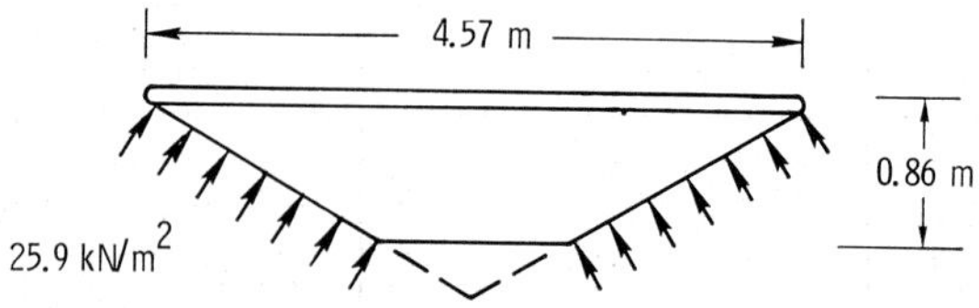

TOTAL MASS = 111 kg (7.3 kg/m²)

MATERIAL	RELATIVE WALL MASS	
	SANDWICH	RING-STIFFENED
ALUMINUM	1.00*	1.71
MAGNESIUM	1.02	1.32
BERYLLIUM	.57*	.63
GLASS IN EPOXY	1.25	1.92
CARBON IN EPOXY	.90*	.92

*MINIMUM GAGE LIMITATIONS

FIG. 13. Relative mass of low-density aeroshells

materials for the shell wall. The shell is designed by buckling under an external pressure. In addition to the shell wall, which constitutes 40% of the structural mass, large stiffening rings are required at each end and reinforcements are needed because of the discontinuity stresses introduced by the rings. The composites provide a reduction in wall mass, but the difference is not as great as anticipated. This is a very lightly loaded shell and minimum gage limitations affect the structural mass in certain cases; the composite could be used to greater advantage in the larger structure of a manned vehicle. The relative masses shown are for only the wall of the shell. End rings of composites could also be beneficial but the potential gains have not been determined yet. Structures of metals are more competitive, on a mass basis, in this case than in the others examined herein.

D. *Landing Craft*

The last type of space vehicle to be considered is that which lands on the surface of a planet, or other body, for subsequent operations there. The Lunar Excursion Module (LEM) of the Apollo spacecraft and the Martian

FIG. 14. Expandable lunar shelter

lander shown in Fig. 12 are examples. The structure and landing gear of such vehicles could utilize composites but one of the most likely applications is an expandable shelter. Figure 14 shows a structure designed to extend the lunar exploration time of the Apollo astronauts.[22] It is carried to the lunar surface in a small package on the LEM for subsequent deployment and use. The wall construction, Fig. 15, is a composite that uses combinations of fabrics, plastics, and metal fibers to perform the required functions. Elastic recovery of the flexible foam, compressed in the container, erects and maintains the shape of the shelter without internal pressurization. The foam also provides the primary meteoroid protection but since it is an excellent insulator, copper filaments are required for proper thermal balance between the inner and outer surfaces. Steel wires provide the load carrying capability for the internal pressure bladder and the fabric outer surface, which provides a meteoroid bumper and container, has a suitable thermal control coating. The filamentary or

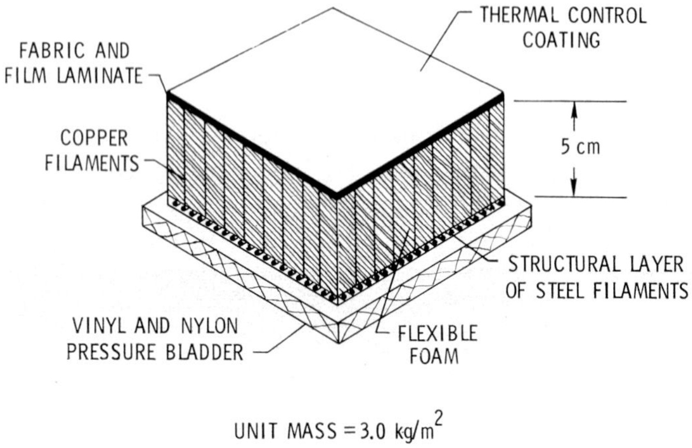

FIG. 15. Wall construction of lunar shelter in Fig. 14

matrix materials used in this shelter are state of the art and such structures would not necessarily require more advanced materials. However, it is an excellent example of how the unique characteristics of filaments provide capabilities that are not readily attainable with other material forms.

V. CONCLUDING REMARKS

A number of potential applications of filamentary composite materials in space vehicles structures have been reviewed and cases identified in which the use of composites instead of conventional materials could significantly reduce the structural mass. The calculated mass reductions, however, are not spectacular when beryllium is the competitor and stiffness is important. Other potential attributes such as high specific strength, foldability, formability into complex shapes, and the opportunity to tailor material properties to structural requirements appear to be of greater interest in space vehicle applications. Filamentary composites have had few such applications but their utilization is expected to increase.

Figure 16 lists several space vehicle structural components wherein composites should find increased or early use, probably in the order listed. Solid rockets already make extensive use of composites and aerodynamic decelerators require combinations of material properties best supplied by composites. Similarly, composites can meet the unique needs of replaceable heat shields and expandable space structures. Pressure vessels benefit from the high specific strength of filaments whereas certain beams, columns, and truss structures can capitalize on their unidirectional properties. Large aeroshells for planetary entry may be the space vehicle application in which minimum structural mass is most important; beryllium appears best but advanced composites may be competive when all factors are considered.

Solid rocket motor cases and nozzles
Aerodynamic decelerators
Replaceable heat shields
Expandable space structures
Pressure vessels
Beams, columns, and trusses
Large aeroshells

FIG. 16. Promising future applications of filamentary composites in space vehicle structures

The lack of widespread use of filamentary composite structures is primarily due to the relative newness of this technology and the limited design methods, low reliability and confidence, high cost, and low producibility that are characteristic of an immature engineering material. These are problems to be solved by research and development and do not constitute a physical barrier to extensive future applications.

REFERENCES

1. DIETZ, ALBERT G. H. Composite Materials. 1965 Edgar Marburg Lecture. American Society for Testing and Materials, 1965.
2. GRIMES, DAVID L. Trends and Applications of Structural Composite Materials. Advisory Group for Aerospace Research and Development Report 523, November 1965.
3. MAYER, NORMAN J. Structural Applications for Advanced Composite Materials. *Metals Engineering Quarterly*, **5**, No. 3, August 1965, pp. 44–49.
4. Anon. Report of the Ad Hoc Committee on Composites. Publication MAB-215-M. National Academy of Sciences. National Research Council. Washington, D.C., November 1965.
5. Anon. ASTM Metric Practice Guide. Prepared by Ad Hoc Committee on Metric Practice. American Society for Testing and Materials, Second Edition, December 1966.
6. RARING, RICHARD H., FREEMAN, J. W., SCHULTZ, J. W. and VOORHEES, H. R. Progress Report of the NASA Special Committee on Materials Research for Supersonic Transports. NASA TN D-1798, 1963.
7. DOW, NORRIS F. and ROSEN, B. WALTER. Structural Efficiency of Orthotropic Cylindrical Shells Subjected to Axial Compression. *AIAA Journal*, **4**, No. 3, March 1966, pp. 481–485.
8. OSGOOD, CARL C. *Spacecraft Structures*. Prentice-Hall, Inc., 1966.
9. GOETZEL, C. G., RITTENHOUSE, J. B. and SINGLETARY, J. B. Space Materials Handbook.

Air Force Materials Laboratory, Wright-Patterson Air Force Base, Ohio, ML-TDR-64-40, January 1965.

10. RITTENHOUSE, JOHN B. and SINGLETARY, JOHN B. Space Materials Handbook. Supplement 1 to the Second Edition, *Space Materials Experience*. NASA SP-3025, 1966.
11. TSAI, STEPHEN W. Strength Characteristics of Composite Materials. NASA CR-224, 1965.
12. GRESZCZUK, LONGIN B. Thermoelastic Considerations for Filamentary Structures, *Proceedings 20th Anniversary Technical Conference, SPI Reinforced Plastics Division, The Society of the Plastics Industry Inc., Feb. 2–4, 1965, Chicago, Illinois*, Section 5-C, pp. 1–10.
13. MORRIS, E. E., SANGER, M. J. and DARMS, F. J. Filament-Wound Structures. *Space/Aeronautics*, **47**, No. 2, February 1967, pp. 86–92.
14. RAYBURN, LOUIS C. and WESTBURG, PETER W. Shrouds for Space Payloads. *Space/Aeronautics*, **47**, No. 2, February 1967, pp. 66–73.
15. FREEMAN, R. F., BOUNDY, R. A. and HARRINGTON, R. Polymeric Applications for the Mariner Mars, 1964 Spacecraft. Jet Propulsion Laboratory Technical Report 32-1031. (Reprinted from *Rubber and Plastic Age*, July 1966.)
16. GOMERSALL, EDWARD W., BUCY, RAYMOND W., JR. and NISHIOKA, KENJI. A Survey of Advanced Structural Technologies Applicable to Future Large Launch Vehicles. Presented at SAE Space Technology Conference on Low Cost Orbital Transportation, Palo Alto, California, May 9–12, 1967.
17. STEPKA, FRANCES S. Projectile-Impact-Induced Fracture of Liquid-Filled, Filament-Reinforced Plastic or Aluminum Tanks. NASA TN D-3456, 1966.
18. GERARD, GEORGE. Aerospace Pressure Vessel Design Synthesis. NASA CR-287, 1965.
19. FRISCHMUTH, ROBERT W., JR. Investigation of Thin Films as Floating Liners for Fiber-Glass Cryogenic Propellant Tanks. NASA TN D-3205, 1966.
20. LAPORTE, A. H. Research on Refurbishable Thermostructural Panels for Manned Lifting Entry Vehicles. NASA CR-638, 1966.
21. ROBERTS, LEONARD. Entry Into Planetary Atmospheres. *Astronautics and Aeronautics*, **2**, No. 10, October 1964, pp. 22–29.
22. TYNAN, CHARLES I., JR. Lunar Stay Time Extension Module. Selected Papers on Environmental and Attitude Control of Manned Spacecraft by Langley Research Center Staff, NASA TM X-1325, 1966, pp. 67–77.

INTERNATIONAL FORUM

COMPOSITES RESEARCH IN ENGLAND

J. E. GORDON

I CANNOT give a full account of the work in England because, of course, it would take much too long. First, just a word about the sort of scale of this work. I suppose it is about an order of magnitude less than the composite program in this country. But, on the other hand, we spend very much less in proportion on hardware. We cannot afford, I suppose, just to make anything that takes our fancy, the way you fortunate people seem to be able to. So we spend perhaps more time at the bench, and less in the workshop. Also for the same sort of reasons, we do concentrate rather more on fewer projects. If you take those projects that we are actually working on, I would say the effort is probably roughly comparable to any similar project in this country.

Our objectives are perhaps a little different from yours. Again for rather similar reasons, we are looking not only for materials for aircraft and space, etc., but also for commercial materials, in the long run. You have got to remember that in England a great deal of industrial research is done in government establishments. For this reason, as far as I know, there is no work on boron. Although the properties may be excellent, it is not likely to be a cheap commercial fiber. The four fibers being pursued are glass, carbon, silicon nitride and silicon carbide. Glass, of course, is a bit old hat now. But incidentally we still do export an awful lot of glass (reinforced plastic) boats to this country. As a matter of interest, asbestos fibers, which you can regard virtually as a poor man's whisker, is awfully good stuff and costs only about 10 cents a pound. I don't know really why it is being so neglected. The artificial fibers, of course, are carbon and whiskers. I cannot say a great deal about carbon, but briefly, they were developed primarily at R.A.E.,† and now two industrial firms are producing these fibers commercially. These fibers are produced in various grades, and have specific gravities varying from 1.8 to 1.9, tensile strength varying from 200,000 to 350,000 psi, and Young's moduli ranging from 25 to 60×10^6 psi. There is still quite a bit of potential for further improvement. We are very much conscious of the cost aspect as I have been saying. Very roughly, the order of cost which is being practically realized and considered at the moment is between $140 and $200, which is subject to significant reduction as the production volume increases.

† R.A.E. = Royal Aircraft Establishment.

At E.R.D.E.,† we have stopped making gunpowder, and we do in fact make silicon nitride and silicon carbide whiskers. Now the properties for these two whiskers are as follows: silicon nitride, Si_3N_4, has a specific gravity of about 3.2, and like most whiskers, exhibits a mean tensile strength of about 2% of the Young's modulus. We find that the strength is remarkably consistent, statistically, and is somewhere about 800,000 psi. It's Young's modulus is rather poor, being about 55×10^6 psi. The other whisker which we are making is silicon carbide, which has about the same specific gravity of 3.2. It has excellent strength, and again is about 2% of the Young's modulus, which is equivalent to 1,400,000 psi. The Young's modulus is somewhere between 68 to 70×10^6 psi, according to our measurements. In comparing the specific properties (tensile strength or Young's modulus divided by specific gravity) of various fibers you can see that the specific modulus of silicon carbide whiskers pretty well keeps up with the Joneses of boron and beryllium; they are, perhaps, a little behind carbon.When you come to specific strength there is no argument about it at all; the silicon carbon whiskers are two or three times as strong as any continuous fiber that I know of. We have adopted a rather different approach with regard to whiskers, since our first priority is on scaling up the growth of whiskers, rather than concentrating on experiments with whisker composite materials first.

I can tell you something about how to make silicon nitride whiskers, since we are not making them now, except to fulfill occasional orders from America. In growing these whiskers from the vapor phase the main problem is the low vapor pressure of silicon. The vapor pressure of elemental silicon, at the usual growth temperature, is far too low to get any transport of elemental species. Thus you have to work with a transportable species, which in this case is silicon monoxide. To facilitate the formation and transport of silicon monoxide some hydrogen is added to the nitrogen atmosphere in the furnace. A by-product of the reaction is water. If there is any appreciable amount of water present, the formation of silicon monoxide will be suppressed, so carbon is also used in the furnace. Hypothetically, some free silicon is formed and will react with nitrogen to form silicon nitride. But of course from a very elemental chemistry point of view, $Si + CO + H_2$ will also yield $SiC + H_2O$, so you might perhaps get silicon carbide. If you want to know whether any of these reactions are going to go at all, the thermodynamics of the various reactions can be programmed into a computer. The results indicate that silicon carbide will be produced. However, only a few percent of silicon carbide is produced; the rest of it is silicon nitride. The reason for this is purely kinetic, and slow as the silicon nitride reaction is it is still far faster than making silicon carbide by this route, and so you end up with silicon nitride. Now we think the equipment which you use for this process is im-

† E.R.D.E. = Explosives Research Development Establishment.

portant because it governs the whole economics and practicability. We use a type of furnace, which we call the "bran tub" because of its shape. However, it does work and is a fairly cheap, and a very flexible type of large experimental furnace. It is a large steel vessel, made by ship's boiler makers, and is insulated with ordinary ceramic bricks. It is heated electrically with a carbon resistance rod, and has all sorts of connections here and there. In the case of growing silicon nitride, we employ ceramic fuel tubes within the furnace. The tubes contain a stoichiometric mixture of silicon and silica. Argon is percolated through the tube when the furnace is at temperature, and serves as a carrier gas for the silicon monoxide vapor. Nitrogen gas containing a little hydrogen is introduced through another tube, and the various gases are allowed to react in the furnace chamber. Whiskers then begin to grow around the fuel tubes; and eventually fill the furnace cavity. We have two of these tubs shown in the slide. There is a baby tub of 70 l. and a larger 2000 l. tub (5 to 6 ft in diameter). This one has the top off, and that one is sealed for a run. It is really a simple thing to operate. Although they are not fully automatic, they are run entirely by unskilled labor. I do not know how many whiskers are grown in the tub normally, perhaps 10^{12} or something of that order. So it is all very well that whiskers are grown, but the really important question is whether they can be grown economically. We have given a lot of thought to the economics of these furnaces and the growth process. Now these furnaces that I just showed you cost about 0.2 cents per liter per hour to run, and this includes all the capital costs, interest, depreciation, engineering costs, etc. The power cost is only about 10% of that, incidentally. I suppose that you can consider the costs of the whiskers as being made up of the furnace operating costs and the raw material costs. Unfortunately, silicon nitride whiskers grow at an exceedingly slow rate. All our research seems to indicate that we cannot speed it up very much, because the rate controlling process cannot be easily altered. The growth rate is about 120th of a gram of whiskers per liter per hour. This is very slow, and leads to a furnace cost, per pound of whisker, of about $120 per pound. Furthermore, we put gases (and expensive gases like argon) through in comparatively large quantities. The material costs are also high, being about $120 per pound. When you add the material cost to the size operating cost we get a figure in the region of $240 a pound. Thus, one of the reasons why we have abandoned silicon nitride whiskers, is that we can't speed the reaction up to an appreciable extent, and furthermore it has not got all that good mechanical properties. We are now making silicon carbide in the same furnace slightly altered. The main advantage is that silicon carbide whiskers grow so much faster; growth rates in the tub are about one tenth of a gram per liter per hour, and on small scale experiments are about an order of a magnitude higher than that. Thus, if you take a tenth of a gram per liter per hour as compared with a 120th, that brings your furnace costs down to about $10 per pound, and straight away that is very important indeed. Also, the raw materials are much cheaper and you end up with a probable cost in the general region of $15 to $30 dollars a pound. This process yields high-grade whiskers at what we think is quite a promising low cost. But of course, there is a long gap between these experiments and getting commercial quantities of whiskers from the manufacturers.

The whiskers come out of the furnace in a tangled up mess, like raw cotton, and one of the objections which many people have raised to us is that you cannot handle them easily. But this, I think, is not really true. We disperse these whiskers mechanically in a liquid and put them through a series of hydrocyclones and special sieves which removes the debris and classifies the whiskers according to size. The classified whiskers are then mixed with ammonium alginate, which is an extract of seaweed used by the cooks in making meringues in restaurants. The whiskers are fairly dilute and, in suspension in this alginate, are squirted out through a suitably shaped orifice at quite a high speed into an acid coagulating bath. This process aligns the whiskers very accurately and "freezes" them in that position in a sort of noodle or macaroni like thread which is wound onto a drum. The whole thing runs very fast, and impresses the most visitors. We have now a million feet of it, or something of that kind. The next figure† shows an optical micrograph of a bit of our whisker thread. It isn't awfully well resolved, but I think you can see the fibers are parallel. This is just an off cut from a mass of this material, the result is a dry whisker thread which is a few mils thick and can be handled just like any other continuous filament. When we want to make an ordinary resin composite out of it, we burn out the alginate, substitute an epoxy resin or whatever is required, and mold it up in the usual way.

The whisker growths in the furnace cover a fairly wide spectrum of sizes, and you sort these; the long ones go into the high-grade epoxy type materials, the middle-size ones tend to go into a metal, because you can exploit the use of short fiber more effectively in metals because of their high shear strengths. We are just beginning to extrude whiskers in SAP‡ as a continuous rod. This seems to be perhaps quite a promising process. There are still whiskers remaining after the classification—say whiskers under 50 microns long or something of that kind—and those we find can go quite well into nylon, which does in fact improve the properties very substantially indeed.

† Figure not shown here.
‡ SAP, Sintered aluminum powder.

FRENCH DEVELOPMENTS IN THE FIELD OF THERMAL-RESISTANT AND MECHANICAL COMPOSITES

JACQUES DUFLOS

Directeur Commercial,
Pierre Genin & Cie, S.A., France

THE programs we have had under study in France, though limited, have urged our industry toward more elaborate research in the field of advanced structural materials. Our means have been modest but the results have not always been negligible.

We have certainly gone into all the possible parameters of the various problems less than you did, our budgets being limited. We very often had to choose quickly such solutions that were both rapid and efficient.

The composite materials and in particular the glass/resin composites had already been studied and rather widely used in the aircraft industry. 1500 kg of RP.—mainly polyester-glass—have been utilized in the "Caravelle". The bodies of new Sud-Aviation helicopters are, for their larger parts, made of RP.

When it became necessary to reconvert some of the laboratories and plants for the composites to be used in rockets, it was easy enough for the men to adapt themselves. Most of the time the persons who had been interested in the development of polyester/glass laminates are now creating phenolic/silica or phenolic/carbon laminates necessary for the missiles.

The problems we thus had to solve were problems connected with the protection of the rockets when leaving the atmosphere, their protection during re-entry, problems concerning thermomechanical structures and improved mechanical structures.

I will give you a short description of the principal fields we have studied in each case and I will insist more particularly on some realizations our firm has dealt with, more closely.

I. THERMAL PROBLEMS

The thermal problems occur mostly in the following cases: when leaving the atmosphere, during re-entry, and for the inside and outside protection of tanks.

To start with, it was essential to obtain such materials having thermal properties. Consequently, we have developed pure fused quartz yarns; we have manufactured and used fabrics with a high percentage of silica, phenolic resins with a high coking rate, we have used silicone elastomers and created laminates capable of mechanically withstanding temperatures up to 300–400°C.

A pure fused quartz yarn—manufactured in France by Quartz and Silice—is made by progressive drawing out in gas flam of silica rods obtained by melting rock crystal.

One of the main problems in this type of manufacturing has been, above all, the choice of the sizing to be used. Together with the yarn manufacturer, we have developed a sizing having some compatibility with phenolic resins, as it contains an amino-silane coupling agent, A 1100, produced in France under license.

We have noticed that it is not essential to eliminate this sizing when making a laminate with silicone rubber; it is advisable to do so, however, when making a laminate with normal silicone resins. The fabrics developed for the reinforcement of silicone rubber are loose gauze fabrics, weighing 160 g/m^2 (4.6 oz/yd^2).

Pure fused quartz has also been used together with silicone rubber for a rocket protective nose where radio-electric problems occurred and they are now currently used for reinforcing the same products in the rear protections against the flame back-flashes around the nozzles.

Laminates made of quartz fabrics and phenolic resins have also been used in some heat shields for ballistic missiles. But, whenever the layers of the laminate were normal to the walls, the optical conductivity of the I.R. by the pure fused quartz itself did not allow their development. It was then better to use fabrics with a high percentage of silica obtained by chemical and thermal leaching of a glass fabric, because of the numerous internal refractions due to the porous structure of the material.

At the "Europlastic 1966" exhibition in Paris we displayed a missile nose-cone (Fig. 1) which had reached an altitude of 1200 km during a test launching. This element was 2.50 m high, and had a base diameter of 1.20 m.

During re-entry, the gases surrounding the head of this element reached a temperature of 4500°C. The surface temperature reached 1800°C. The surface melting of the silica was visible. The energy absorbed by the head and the surrounding atmosphere at the moment of re-entry was 35×10^8 joules.

The insulating material was 8 mm thick. Structures having a diameter of 2 m and, at the moment, a maximum height of about 4 m are manufactured in the same manner.

It might be interesting for you to know that we have developed a new form for these materials based on leached glass.

FIG. 1. Space capsule after re-entry

As you know, the fabrics made in large width have up to the present time been preimpregnated generally with phenolic resin, before being put in bias and cut into tapes. We have thought of manufacturing directly a bias tape by using a braiding process. This patented process allows the direct manufacturing of tapes in the desired width, the yarns of which are at an angle of 45° with regard to the direction of the tape (Fig. 2). Longitudinal yarns are included into the selvedge for an easier manufacturing and to avoid any stretch of the tape during the leaching and impregnating processes. The positioning for winding is easier, the working time is reduced by half, the final product is more homogeneous and, because of the perfect alignment of the tapes due to the longitudinal yarns, the inside surface can be kept rough from the mould, without any tooling or finishing. The external longitudinal strands have obviously to be cut to allow a good opening of yarns.

To obtain this kind of tape, we had to modify somehow the usual braiding machines. This process may be used with glass yarns (subsequently leached or not) as well as with carbon and graphite strands or yarns made of synthetic fibers. As a matter of fact, whenever the enthalpy became essential, we did choose such reinforcement combinations using both quartz and polyamide yarns.

The various possible combinations include either mixed yarns or fabrics made of both types, the one in the warp and the other in the weft. The percentage of the two types may vary according to the degree of enthalpy wanted.

In the same field of thermal problems, we must also quote the realization, for which a patent has been applied, of new types of carbon and graphite fabrics.

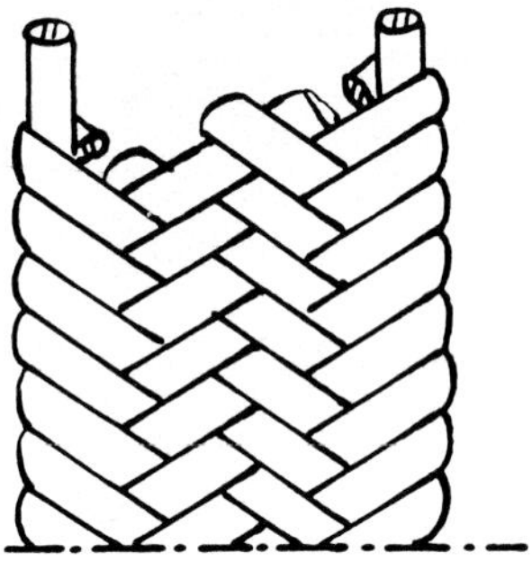

FIG. 2. Sleeved tape with longitudinal threads

In France, we have a perfectly round rayon yarn (patented by Mr. Drische) due to the processes developed by the Cie des Textiles Artificiels. This type of rayon has been called "B.X." Meryl. The carbonization of this very regular fiber gives a product which is itself well aligned and the mechanical strength of which is of interest. We have thus been able to double the mechanical properties as compared to the carbon and graphite fabrics previously produced in our country: a 260 g/m² fabric (7.4 oz/m²) for instance, has a strength of 20 to 30 kg/5 cm (22 to 33 lb/in). Its handling is more easy. We must, however, say that we have certainly to improve the present process of carbonization, to better the properties, and to obtain with this type of yarn such materials that we may use in the filament-winding process.

The general properties of these fabrics are the following:

Carbon fabric

Style T.C.M. 909	
Weight/m²	260 g (7.4 oz/yd²)
Specific gravity	1.42
Active surface	2 m²/g

Resistivity	0.9/ohm/cm²/cm
Carbon percentage	96%
Ashes	0.2%
Volatiles	5%
pH	7.4

Graphite fabric

Style T.G.M. 285	
Weight/m²	230 g (6.7 oz/yd²)
Specific gravity	1.5
Active surface	4 m²/g
Resistivity	0.9 ohm/cm²/cm
Carbon percentage	99%
Ashes	0.16%
Volatiles	$<1\%$
pH	6.1

In the case of the nozzles, three types of materials have been selected in France, according to the temperature and pressure conditions demanded in each case.

A. A laminate named "Ortho-strasil"—normal lamination with regard to the wall of the nozzle—is made with a leached glass fabric called Thermoproof 604/1100.

B. A laminate named "Para-strasil"-lamination presenting a lower angle than 90° to the nozzle wall made with a Thermoproof 604/1100 fabric.

C. A laminate with a material called "Astrakan" by one of the manufacturers because of its aspect before moulding, made of chopped carbon fibres preimpregnated with phenolic resin with a low percentage of nitrogen.

Concerning the phenolic resins used with these various reinforcements, the O.N.E.R.A. (National Organization for Aeronautical Research and Studies) and the Société Plastugil have developed high temperature resistant materials or high coking point materials.

The O.N.E.R.A. resin No. 1, which has a specific gravity of 1.225, is an ammoniacal resol obtained with formaldehyde phenol molar proportions of 2.5. The loss of mechanical properties after exposure 1.000 hours at 250°C for a laminate is only 10%. The Phenora 3 resin is an improvement which gives 20% better results in ablation than Phenora 1. The improving of the thermostability of these resins is under study through introducing mineral oxides into ammoniacal resols in the ratio of formaldehyde/phenol 1.3 or 2.5 and through condensing phenol and formol with various oxides or metal salts used as condensation catalysts.

These tests have shown that if a resin is heated in vacuum up to 400°C before use, it gives a high thermostability both in inert and oxidizing atmosphere and that the resins made from aluminium nitrate and those treated by the acid-butyl phosphorus, can be of great interest.

The Institut Français du Pétrole (the French Petroleum Institute) has been given the task to study the thermostable resins of the benzimidazole type. They have tried to develop a fusible and very soluble low polymerization degree polymer.

The Institut Français du Pétrole has used a new process and obtained a product which leads directly to the solution necessary for the impregnation and makes the control of the polymerization easy. The polymer thus obtained has a molecular weight of 420 to 450.

We have closely studied the best possible ways to impregnate the resins produced according to these principles. The two problems consist in trying to find a process which could not contaminate the operators on the one hand, and on the other hand in obtaining such materials which would be the easiest to mould, in sheets and in shaped parts as well. We have therefore developed two equipments, the one as usual by dipping the fabric into a batch, the other one for a static impregnation in a vacuum oven.

In the former case, the equipment is important: a large part of the material risks contact with the resins which means that a fairly long cleaning is necessary; a thorough ventilation is also required. In the latter case the running may be fairly clean, the transferring of the resin being made simply under the effect of the vacuum in the impregnation box. The resin is never in the open air and is never handled by the operator. Unfortunately the output of such an equipment is limited and its cost is high.

From the mechanical point of view we have observed no notable and interesting differences between the two processes.

A. *Vacuum Process*

The sheets made with a glass fabric equivalent to 1581 have given a cold flexural strength range of about 53 hbars (76,000 psi). At the temperature of 300°C, these characteristics keep in a range of 43 hbars (61,000 psi).

B. *Impregnation Process*

Cold flexural strength : 53 hbars (76,000 psi); at 300°C : 45 hbars (64,000 psi). We have therefore decided to develop an industrial method for impregnating by dipping and the best results we have obtained are as follows: cold flexural strength, 74 hbars (106,000 psi); at 300°C : 60 hbars (85,000 psi).

The moulding conditions for prepreg containing more than 30% of resin and 11% of volatiles are the following: the mould is heated to 125°C, the plates are closed without pressure for 15 min to allow the entrapped bubbles to escape. A pressure of 14 bars (20 psi) is then applied, the temperature brought up to 370°C in 35 min, and kept at this level for 3 hours; after cooling down to less than 100°C, the part is taken out of the mould. The amount of resin remaining on the sheets must be of at least 20% if the best mechanical strength and a good shear strength (more than 3 hbars—4000 psi) are required.

We have thus made a quick survey of the data from the main studies in the field of thermal resistant materials. I will now give you a brief account of the research work made in France on the mechanical structures.

II. MECHANICAL STRUCTURES

For the development of filament wound structures, it was necessary to have a better knowledge of the optimum possibilities of glass yarns and of the best conditions in which to use them.

O.N.E.R.A. has been asked to undertake a certain number of basic studies and has done systematic tests both on yarns and on resin-impregnated single plies. It has thus been possible to point out the optimum elastic mechanical properties of the resins and the best resin glass yarn modulus ratio. The tests have shown that the most favourable relative elongation of a resin for obtaining a better strength of a laminate is about 5.5 to 6% in the case of a glass yarn reinforcement under tension.

The O.N.E.R.A. tests have been performed on an equipment which allowed the individual pre-stressing of each of the glass yarns used in the making of the test specimen (Figs. 3 and 4).

On the other hand, the Société du Verre Textile has been given the task by the Technical Department of the Air Ministry to investigate the optimum possibilities of E glass. They have next been asked to develop a glass which would allow the making of filaments with a better breaking strength and possibly a better Young's modulus and to study the ageing or the fatigue under constant load of these materials after impregnation and curing.

The first conclusion to which the Société du Verre Textile has come is that each glass composition produces filaments having their own mechanical properties which can be evaluated by measuring the tensile strength of virgin filaments taken directly from under the bushing. It seemed that these characteristics could not be improved by the forming conditions: diameter of the holes, glass temperature, glass level in the bushing, drawing-out speed and even diameter of the yarn, it going from 3 to 50 microns.

All these tests have, indeed, been made in standard atmosphere, that is to

FIG. 3

FIG. 4

FIGS. 3. and 4. Device for making monolayer of glass fiber under individual load and impregnation

say 20°C and 65% R.H. It has been found that E-glass has an average tensile strength of 340 hbars ±1.5% (490,000 psi). As soon as such a yarn has been handled, this property drops down to 250 hbars (360,000 psi) ±12.5%. After twisting and plying there remain no more than 140 hbars (200,000 psi).

These data have brought the S.V.T. to think of impregnating the glass yarns immediately after forming from the cakes and by using 400 filament yarns.

A sizing had next to be developed which would allow an excellent wetting with the epoxy resins and a resin system chosen to provide a good stress transfer, an easy handling, and a sufficiently long shelf life with curing conditions that would be acceptable by the customers.

The adopted sizing called P9 is an epoxy resin sizing. The data obtained on test pressure vessels reproducing the conditions of industrial use have given a total instantaneous stress of 90 to 95 hbars (130,000 to 135,000 psi) which corresponded to a unidirectional stress of about 220 hbars (315,000 psi) on the filaments.

This kind of yarn has been widely used in the tanks of the French rockets and particularly in the booster of the third stage of the "Diamant" rocket which brought to orbit a first wholly French satellite.

This tank was 1200 mm long and had a diameter of 650 mm. Later on, the Société du Verre Textile has been endeavouring to develop compositions for glasses having a still higher mechanical strength, by studying diverse combinations with certain types of metals. Twenty glasses with three constituents containing Si_2O_3 and MgO have been investigated, 13 with SiO_2, Al_2O_3 and other metallic oxides; 91 with four constituents containing essentially SiO_2, Al_2O_3, MgO and various metallic oxides LiO_2, CaO, CeO_2, ThO_2, Ta_2O_5, etc., and lastly 42 glasses with five constituents.

The result of these investigations has been a composition melting around 1600°C called "R" glass and containing 59% SiO_2, 24.5% Al_2O_3, 9% CaO, 6% MgO and 1.5% miscellaneous.

The tensile strength of this glass is 440 hbars (630,000 psi) on virgin filaments instead of 340 (490,000 psi) for the E-glass, 380 (540,000) measured on the forming cake against 250 (360,000) for the E-glass and the practical strength usable on pressure vessels is 165 (236,000) against 140 (200,000). The specific gravity is 2.58, the coefficient of linear expansion 4.10.

With a 17 to 20% improvement in strength as compared to the E-glass, the R-glass provides a greater reliability after ageing under stress. The temperature resistance is better and so is the ageing under zero load in air or in water. For example, an R-glass pressure vessel under a constant stress of 10 bars (140 psi) for 9 months has, at the end of this period, a strength of 133 hbars (190,000 psi) measured on the filaments against 88 (126,000) for the same made of E-glass.

The diagrams give the temperature resistance and the ageing under zero load in air and water (Figs. 5 and 6).

To use these impregnated yarns the firms which manufactured the parts had to create special equipment and to develope a special technology.

Sud-Aviation has more particularly studied this problem. One example of this equipment is for making isotensoid geodesic windings.

The principal use is the Diamant booster with its two polar closures, a 12% one in front, a 39% one in the rear. The winding angle problem has been studied very closely by means of a coordinate reducing device consisting of a revolving plate on which comes a plaster model of each of the end-domes,

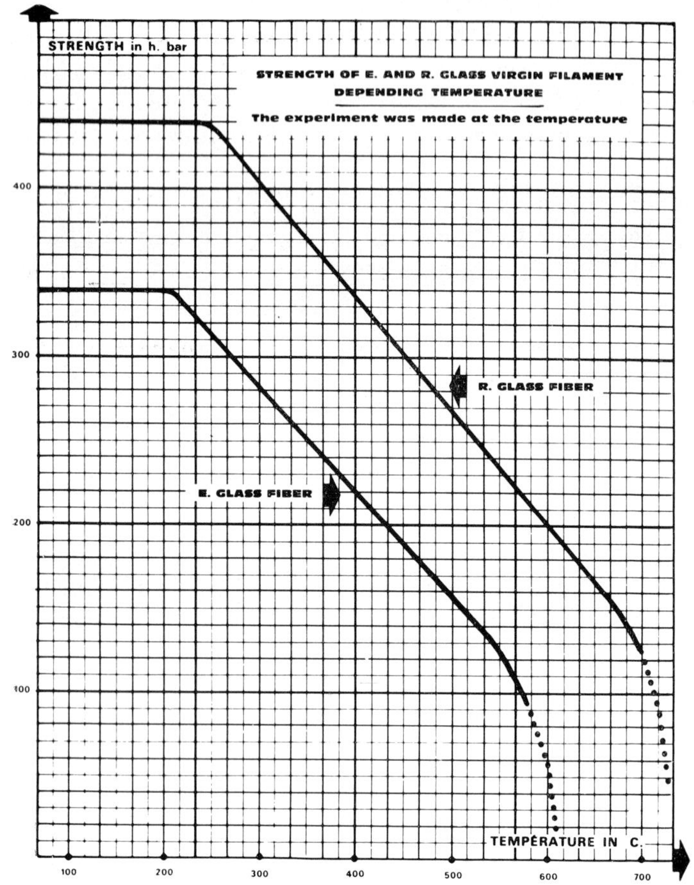

FIG. 5. Strength vs. temperature of filaments

the equilibrium of the yarns being studied by simply observing the path they tend to adopt themselves.

After determination of a diagram of the natural equilibrium, the nearest practical diagram has to be studied taking into account the requirements of winding and the compatibility of the angles for the two end-domes. The test is repeated with a second yarn which follows the chosen path. The numerical functions of the winding remained to be determined from the results. The angle of compromise which was found for the yarns was 14°. Owing to various technological reasons and to the great differences in the closures, the longitudinal winding has been slightly increased as compared to the circumferential layers which gave a mass increase of 6% as compared to the theory. For a total breaking stress of 90 hbars (130,000 psi), the circumferential layers are submitted to a load of 121 hbars (175,000 psi). It must be noted

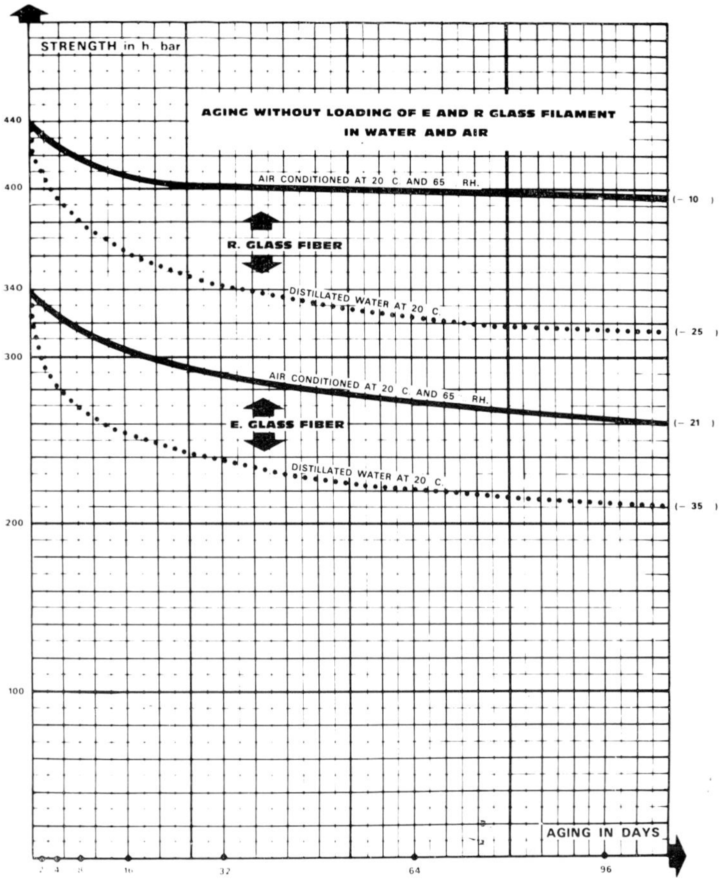

FIG. 6. Strength vs. aging of filaments

that, in the case of isotensoid geodesic movement, the crossing yarns give an alveolar pattern resulting in a star design around each of the ends. Sud-Aviation did find out that the number of points of these stars must be as limited as possible—3 to 6 seems to be a convenient figure. "Diamant" had 5.

For some boosters, particularly in the case of large ones, whenever the nozzle is wound in one operation together with the booster, close-fitting winding has been done with so-called planar trajectories. This type of winding takes much time and the winding of the end-domes is very long. Perhaps the equipment is not working in very economical conditions, but the results obtained seem to be interesting. The closure ratio cannot, however, be superior to 25%.

The excellent results obtained on the occasion of real and test launchings have shown the high reliability of the pressure vessels made according to these processes. This is why a number of designs have been made to extend this technique to large boosters. The largest ones now existing in France are 6 m long and 2 m in diameter.

As you may see, our realizations in France are not spectacular, but our engineers now have good experience with the problems of improving composite materials and the action of the very high temperatures on the heat protection.

These realizations are very small in comparison to yours. We merely hope to have been able, nevertheless, to have advanced somehow in Europe the knowledge of the composite materials which is a new and fascinating scientific field.

REFERENCES

Fibres de verre et stratifiés haute résistance. Conférence E. Grœne, Europlastique, 18 May 1962.

Elaboration et propriétés des fibres de verre. Conférence J. Münch, C.N.E.S., 7/11 February 1966.

Fils de verre à haute résistance. Conférence E. Grœne, Europlastique, 23/24 May 1966.

Etude comparative des propriétés du verre "E" et du verre "R". C. Janin and J. C. Breton.

COMPOSITES RESEARCH IN GERMANY

U. HÜTTER

BEFORE I start I wish to express my thanks for being invited here and to say that I consider it an honor as well as a pleasure to have the opportunity to speak at this meeting. The research work on composite material, as understood in modern terms, started in the Federal Republic of Germany in the early 50's. But even before World War II plastics reinforced by fibrous materials were under investigation.

One example of relatively early efforts is the development of a wind turbine. Work on the turbine blade configuration also provided the ground work for laboratory development of airplane wing structures.

The wind power plants were designed in 1956 by a team at the Allgaier Werke in Uhlingen, Germany. They have a rotor diameter of approximately 34 m, or in British units, 111 ft. The hub root diameter is about 30 in and the whole air-foil is made of a fiberglass reinforced composite.

The rotor has a loop flange attachment to the hub which proved to be one of the most reliable and fatigue resistant attachments yet devised. The fiber strands run along the attachment bolts, and any one of the bolts is surrounded by strands which go in infinite lengths from the tip to the root.

Experience gained with that rotor influenced the style and production of some helicopter rotor blades in the following years, for example, the model rotor for the high-speed helicopter system. Some shrouded propellers with blades of fiberglass reinforced plastic were designed for a circumferential speed of more than 350 m/sec. The composite blade design proved to be superior on a weight basis than even a beryllium blade. Strain measurements during flight were made via straingages embedded in the loop flange and proved the soundness of the design.

A recent application is the four-bladed hingeless rotor for the Boelkow 105 which was tested in flight at the end of April this year. These rotors are made in the same manner, having similar loop flange attachments.

During the middle 50's some experimental high performance gliders came out of production. The ring stiffeners and shell frames were built up of balsa-wood stiffened fiber reinforced plastics. Due to the fact that these shells were at least in some ranges low loaded structures, fabric covering was used. In 1957 the prototype Phoenix designed by Eduard Naegele and built by

Akademische Fliegergruppe Stuttgart made her maiden flights. Ninety units of the sister-plane Phoebus have been built already. Another glider plane, the H30,† has been built by Eugen Haenle who was my right hand during the construction of the big rotor of the wind-turbine. The H30 was given a wing featuring the concentrated flanges made of roving reinforced material.

To obtain the necessary material data for design, which are needed for such products, we started development of new types of test specimens. This effort brought to light the unreliability of standard specimens.

These engineering efforts were accompanied by research work in the field of glass-fiber reinforced plastics to clear up certain problems recognized during the development phase. This research work covered topics such as the relationship between fiber diameter and ultimate strength, high temperature effects, stability of resins, etc. Technological investigations have been undertaken by the following research institutions.

1. The Studiengesellschaft Windkraft EV in Stuttgart which financed the big wind-turbine.
2. The Akademische Fliegergruppe of the Technical Universities, Braunschweig, Darmstadt and Stuttgart.
3. The Deutsche Forschungsanstalt für Luft- und Raumfahrt in Stuttgart Institut für Flugzeugbau, T. H. Stuttgart, Institut für Kunststoffverarbeitung in Aachen, etc.

Since 1959, the Arbeitsgemeinschaft für Verstärkte Kunststoffe im Gesamtverband der Kunststoffverarbeitenden Industrie, which means, "study group on reinforced plastic within the Association of Plastics-User Industries", and similar reinforced plastics divisions in the states have helped to spread research results on reinforced plastics at their annual sessions. Out of the multitude of industrial corporations engaged in fiberglass plastics research and production only a few which significantly contributed to the work will be mentioned.

1. Glassfluegel Haenle in Schlatstadt.
2. Boelkow Entwicklungen in Ottobrunn near München.
3. Leichtflug Union in Bonn.
4. Vereinigte Flugtechnische Werke in Bremen.

In this short review some essential examples will be given out of the larger number of problems dealt with. When utilizing fibers as a reinforcement the question arose whether or not the increase of fiber strength with decrease in fiber diameter would be a like fact as it is for single fiber measurements. A statistical analysis showed that from purely theoretical considerations there exists no significant influence of fiber diameter upon ultimate strength of

† H stands for my brother, Wolfgang Hütter, who designed the glider.

unidirectional composites. An experimental survey of samples with different fiber diameters showed that experiment and analysis lie very close together.

In the same report there is some information about the results of a study dealing with the compressive failure mechanism of a roving reinforced material. Compression specimens were loaded until secondary failure occurred. Longitudinal photomicrographs show the typical failure modes of fibers buckling in the plasticized matrix environments. This observation, made in 1962, explained the compression failure. The micro-structure models have been used not only to investigate the compression failure, but also to get an insight into yielding of the matrix. These studies have been supplemented by the interpretations of photo-elastic observations to find the stress distribution in the gaps between adjacent single fibers.

Besides limits for the strength and stiffness values, per unit weight, there often are environmental conditions confining the utilization of composites. For instance one issue is the high temperature tolerance of composites. It is known that the dimensional changes are not uniform for a change in the constant of elasticity with temperature. If a resin is tested without fibers, a reliable value for the temperature limit can be gained by shear test of the laminated plate. Associated with this problem is the radiation heating of plastic surfaces by solar radiation. To illustrate: one very hot day after we had finished the blade of the wind turbine, we observed that the blade had taken on a corrugated shape. The blades had been stored in the sun lounge near the workship for but two hours! It was so bad that we had to reheat the blade in the mold and later protect it by some pigment. With an environmental temperature of 35°C, even for pigmented resin, (for instance those treated with titanium oxide) plastic composites show a surface temperature of 46°. Untreated surfaces are heated up to 70° to 80°C depending on the degree of coloring. This illustrates the importance of pigment protection to composite surfaces of aircraft. Let me mention here something else. An unpigmented surface of a composite structure of fiber glass and resin gives a very poor radar echo, but is very sensitive to heating. A pigmented surface gives good heat protection but also a good radar echo. This may not be desirable for radar signature reduction of aircraft for example.

To turn to another research example: from experiences with test rotors, it was known that rotors of VTOL aircraft and air screws are endangered by rain erosion. A research program tuned to this program did not cover only composite rotors but, in comparison, also rotors made of other materials. Relatively small test specimens were tested at the Dornier works by mounting them on a rotating arm having peripheral speeds up to 470 m/sec and moving them through mists of a given size droplet distribution and population. As a measure of erosion the loss of weight was plotted against time, drop size and mist population. In addition to such tests, other specimens have been impinged by a fast water jet with a jet velocity of more than 2000 m/sec. Rain erosion occurs in any material, but it proved to be most pronounced in glass. Better resistance to erosion is shown by some plastics, the best of which, for our sample choice, was polyurethane. Better yet were most ceramics, which finally were surpassed by hardened metals. Small stones and sand impinging on aircraft structures is a similar annoying environmental condition for VTOL aircraft. A series of different plastic coatings were investigated which are able to prevent damage because of the high local energy dissipation.

Since about 1960 investigation of the influence of several atmospheric and other environmental conditions on aging has been conducted with the co-operation of several research institutes. These research programs are tailored to improve the ability to resist radiation, that is achieve a certain color stability in open air. I would like to include here the research on flame resistance of composites which is most interesting to the building industry.

Research on long time behavior of composite materials has been in progress since 1961. Fatigue strength tests have been run to 10^8 cycles, but no asymptotic endurance limits were confirmed, neither for fabric laminated nor for strand laminated composites. The notch sensitivity of strand reinforced laminates was less than that of normal homogeneous materials. The small specimen fatigue-tests were also supplemented by several destruction tests for structural parts of rotors. Static tests and fatigue tests were done with actual blades using load correction factors derived from actual service conditions. Such tests confirmed the strength values predicted in the design process. Finally, performance tests were necessary to pass the airborne service requirements. Including the Boelkow Phoebus and the Hütter–Haenle Libelle, both high performance airplanes which are in production now, 130 gliders of 16 different types so far have been built from fiberglass reinforced plastics in Germany. All of them have already received the full airborne service certificates. The great number is due to the fact that the various academic student groups design and bring out their own aircraft, an old tradition in Germany.

In the research field of optimization of low loaded shells for wings and fuselage structures, progress has been made in connection with the design of the four-seat commercial aircraft entirely constructed of fiber glass fabric-reinforced composites. Scores of sandwich shells and thin-walled, corrugated or similar shaped structures have been investigated. Sample structures for the structural tests have been experimentally evaluated in a combined bending and torsional load rig, which is specially designed to permit application of any desired load combination.

As far as high-load density structures are concerned, my paper tomorrow will give sufficient detailed information so I can now omit mentioning this most important topic.

As a representative of missile structures, the pressure case of the third stag of the ELDO rocket, may be mentioned. I am a little reluctant to show you e picture of it because we have seen such an excellent film by Mr. Duflos. Tha casing is produced by a filament winding procedure and features almost the same dimension, but not the same perfection I think as Mr. Duflos demonstrated. To facilitate manufacture, a special machine for polar winding has been designed and built.

It may be mentioned that a group of students of the AGRR runs a small research program, producing rocket nozzles, using refrasil fibers in a resin matrix, in such a way that the fibers point in a radial direction at a small angle to the nozzle's axis to obtain better ablation resistance. The design features a special filament winding process for production.

So far no reference has been made to the application of composites in civil engineering, although fiberglass reinforced plastic road vehicles, boat hulls, etc., have already been in use for some time.

Special attention has been paid throughout all the research to the development of test specimens, which really represent the load conditions of actual structural parts. As to shells, the tube specimens give the most reliable results. At least this was found to be the case as far as our own experience is concerned. We observed a spread of less than 3% with such specimens specially produced by a well controlled filament winding process. The only disadvantage is it is a more expensive manufacturing process. From the beginning of research into composite structures the load induction problem was carefully investigated. Extensive tests with plane and slant angle lap joints as well as loop flanges have been made. By a rational normalization it was possible to compare different designs with respect to like construction quality. We defined a parameter which we call the "Anschlussgüte" or "connection quality". This is something like the additional length of the flange which would result if I placed all the additional material needed for load transfer into an extension of the flange having equal cross-section. If we now normalize, for example by the diameter of the flange, we get a representative figure of merit. This normalization gives the possibility to compare any possible connections one can imagine.

To record the deformations in the load induction zone, a method has been developed to place strain gages submerged between the load carrying laminate and roving layers during manufacture. The results show that even the strand loops points to the application of advanced composites load inducing techniques. After embedding the strain gages in the part, they can be located again by X-ray. This embedding of strain gages in the material enables us to observe the stresses even inside of the most complicated parts in the flying airplane.

The demand for higher working stresses in practical applications lead to technological investigations for better uniformity, manufacturing control methods, etc. An example for improvement of the measures of impregnating fabrics with resin when producing members with large surfaces, should be cited. After scanning many methods we found that injection measures were superior to many conventional processes. The technologies applied when impregnating the roving have a marked influence on the strength values obtained. Parallelism of fibers in the laminates, voids in the matrix, and inclusion of foreign bodies, extrusion velocities, etc., are other influencing parameters. In view of the experience gained some impregnation measures have been suggested which in themselves improve the strength of laminates by some 13% and more while assuring very small scatter in the strength values and moduli.

The research work on composites is closely coordinated in Germany, in fundamental investigations as well as in fields of application. There is good cooperation between independent research institutes and industry. The already mentioned AG Verstärkte Kunststoffe meets at annual meetings where many papers are presented. Moreover, the AG Verstärkte Kunststoffe advances the coordination of the research work in the different institutes by a committee comprised of the heads of the research institutes which are involved in the composite research.

I may remark that the problems solved in the last decade concerning load induction, as well as highly loaded shell structures, were treated in such a manner that the change in working with advanced composites can be done without fundamental changes in the design methods or design philosophy. Some research on advanced composites worth mentioning is only just beginning; for example, adaption and investigation of production methods for monocrystilline whiskers, the handling of whiskers, and the orientation of whiskers in the matrix by hydrodynamic boundary effects. Production of advanced fibers, such as boron fibers is just now starting in Germany. The work ahead certainly looks exciting and I believe we are entering an era of ever-widening applications.

SESSION III

MICROMECHANICS

THEORY OF COMPOSITE MATERIALS†

ZVI HASHIN

Towne School,
University of Pennsylvania,
Philadelphia, Pennsylvania

1. INTRODUCTION

We are witnessing these days a veritable "composite materials explosion". The impetus for this development is largely due to the search for lightweight materials which can withstand the severe stresses and temperatures encountered in aerospace environments. The urgency of requirements has led to a situation where technology and scientific theory are being created simultaneously. While only a few years ago the theory of composites was an esoteric subject which occupied the minds of only a handful of scientists, it has now grown into a foremost research activity producing an ever-increasing amount of reports and papers.

Scientific activity proceeding at such sudden pace cannot be devoid of undesirable phenomena. There is much in present work on theory of composites which is of passing or dubious value, in spite of the undoubtedly good intentions of everybody concerned. In the writer's opinion too little emphasis is put on fundamental principles and on reasonable mathematical rigor. Moreover, available results of fundamental importance are all too often submerged under a flood of approximations and speculations, based on unjustifiable assumptions.

The time has come to ask the questions: Is this state of affairs a necessary characteristic of the theory of composites? Or can this subject be developed as a true scientific theory as has been done in other now well-established fields of applied mechanics and engineering science? This writer believes that the first question should be answered in the negative and the second in the affirmative. It is the purpose of this survey to show that a framework for a sound theory of composites already exists and that it can provide results of great engineering importance. It is of utmost *practical* importance to devote a much greater part of present efforts to this type of research activity for, as somebody has said, "There is nothing more practical than a good theory".

† Supported by the U.S. Army Research Office under contract DA-31-124-ARO-D-194 with the University of Pennsylvania.

2. GENERAL CONSIDERATIONS

A composite material may be defined as a solid heterogeneous medium. A heterogeneous medium is a mixture of several media of different physical properties with well-defined interfaces. It is assumed that the different materials form regions which are large enough to be regarded as continua. All regions that are filled by material of the same physical properties are called a *phase* and the totality of phase interfaces is called the *phase geometry*. It is further assumed that the phases are homogeneous, i.e. their physical properties are space constant. In that case the heterogeneous medium is a nonhomogeneous medium with piecewise constant properties.

The fundamental problems are:

(a) Prediction of macroscopic physical properties.

(b) Analysis of the internal fields in the heterogeneous medium.

This survey is exclusively concerned with problem (a). Problem (b) is of such extreme difficulty that only a few isolated results for very special cases are available.

The physical properties of a heterogeneous medium obviously depend upon the physical properties of the phases and the phase geometry. While phase physical properties are relatively easy to classify, this is not the case for phase geometry. The phase geometry of real materials is of a highly irregular nature. To measure it in detail for any specimen is a hopeless task. Furthermore, not much would be gained by such knowledge since a different specimen would have a different phase geometry. We are here confronted with a very complex statistical situation. Statistical theory of phase geometry is an important subject in its own right. The interested reader is referred to ref. 1. Here we shall only mention a few elementary aspects.

We first of all define a *multiphase material* as an arbitrary and general mixture of any number of phases. The most important case is that of a *two-phase material*. In the general multiphase material there is no geometrical bias of one phase over another or if there is, we are not aware of it because of ignorance. An important case of bias is a *suspension*. Consider for simplicity a two-phase suspension. Here one phase is a matrix while the other is in the form of inclusions which are embedded in the matrix. See Figs. 1, 2 for typical cases of general two-phase materials and suspensions. Suspensions may be further idealized by assuming particles of spherical, ellipsoidal or cylindrical shapes. It should, however, be borne in mind that these shapes occur only rarely in real materials and they are assumed for the sole purpose of mathematical treatment of an otherwise insoluble problem.

An interesting kind of heterogeneous medium is a polycrystalline aggregate. This may be regarded as a multiphase material with an infinite number of phases, each phase corresponding to the orientation parameters of the principal crystallographic axes of crystals. This subject will not be discussed in the present survey. The interested reader is referred to refs. 2, 3.

A fundamental assumption in the theory of heterogeneous media is statistical homogeneity. The precise definition of this term has to be given in terms of probability functions.[1] Here we shall give an intuitive explanation. Consider a large heterogeneous specimen which has certain macroscopic physical properties. If all large enough (relative to phase region size) subregions have the same physical properties as the whole specimen, the specimen is statistically homogeneous. Since the physical properties are functions of the statistics of the

FIG. 1. General two-phase material

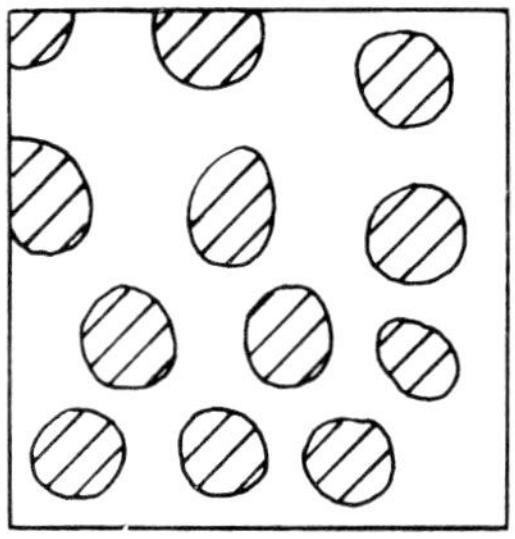

FIG. 2. Two-phase suspension

phase geometry, the assumption of statistical homogeneity means essentially that large enough subregions are statistically identical to the whole specimen.

Another important assumption, independent of statistical homogeneity, is that of *statistical isotropy*. This assumption means essentially that macroscopic properties are the same in all directions or to put it more precisely, are not affected by the orientation of the coordinate system to which the physical constants are referred. In what follows statistical homogeneity will always be assumed. In many cases statistical isotropy is an appropriate assumption; however, in some very important cases such as fiber reinforced materials, the macroscopic properties are predominantly anisotropic.

It is now necessary to define the concept of macroscopic physical properties. Discussion will be limited to the case when the phase constitutive relations are linear. We shall be concerned with three important cases.

Case (a). Time-independent Vector Fields

The phase constitutive relations are expressed in terms of vectors D_i and H_i related by

$$D_i(\mathbf{x}) = \alpha_{ij}(\mathbf{x})\, H_j(\mathbf{x}),\dagger \tag{2.1}$$

† Here and henceforth a subscript has the range 1, 2, 3, of three-dimensional cartesian coordinates, a repeated subscript denotes summation over 1, 2, 3 and a comma before a subscript indicates partial differentiation with respect to the variable identified by the subscript.

where

$$H_i = -\phi_{,i}, \tag{2.2}$$

and

$$D_{i,i} = 0. \tag{2.3}$$

The constitutive relations describe dielectric behavior, magnetostatics and thermal and electric conduction. Thus the problems of determination of macroscopic dielectric constant, magnetic permittivity and thermal and electric conductivity are mathematically identical. Hence for conciseness all these problems will be referred to as the dielectric problem.

In classical dielectric theory the constant dielectric tensor α_{ij} relates homogeneous electric displacement D_i to homogeneous field intensity H_i. If the fields are not homogeneous then they are assumed to be homogeneous within a differential element, and it is then said that (2.1) applies at a "point". In a heterogeneous body of statistically homogeneous phase geometry a relation resembling (2.1), only applies for statistically homogeneous D_i and H_i fields. Such fields are produced by homogeneous boundary conditions —i.e. boundary conditions that would produce homogeneous fields in a homogeneous body—applied to the surface of a very large heterogeneous body. Such boundary conditions are

$$\phi(S) = -H_i^\circ x_i \qquad \text{(a)}$$

or

$$D_n(S) = D_i^\circ n_i. \qquad \text{(b)} \tag{2.4}$$

Here n_i are the components of the outward normal to S, D_n is the normal component of the vector D_i and H_i° and D_i° are constants. Statistical field homogeneity implies that local averages over large enough subregions (representative volume element, abreviated RVE) are the same as body averages.

The effective dielectric tensor α_{ij}^* is defined by

$$\bar{D}_i = \alpha_{ij}^* \bar{H}_j, \tag{2.5}$$

where overbars denote volume averages. It is seen that the RVE replaces the differential element of classical continua and that statistical homogeneity replaces homogeneity. However, extension of a relation of type (2.5) to statistically nonhomogeneous fields is problematic. It appears that in that case multipolar theories may be needed.

For a statistically isotropic body

$$\alpha_{ij}^* = \alpha^* \delta_{ij} \tag{2.6}$$

where δ_{ij} is the Kronecker delta. Then

$$\bar{D}_i = \alpha^* \bar{H}_i. \tag{2.7}$$

For boundary condition (2.4a), it can be rigorously shown that

$$\bar{H}_i = H_i^\circ, \tag{2.8}$$

while for boundary condition (2.4b),

$$\bar{D}_i = D_i^\circ. \tag{2.9}$$

Thus in each case determination of effective dielectric properties amounts to calculation of an average field. Because of (2.4), (2.8) and (2.9) the definition (2.5) may also be interpreted as a simple consequence of the linearity of the problem.

A useful alternative definition of effective dielectric properties stems from the theorem:

$$U = \tfrac{1}{2}\int_V D(\mathbf{x})\, H_i(\mathbf{x})\, dV = \tfrac{1}{2}\bar{D}_i\bar{H}_i V, \tag{2.10}$$

if either one of (2.4) is prescribed over the whole boundary S. This rigorously valid formula is remarkable since the deviations from the averages $\bar{D}_i$ and $\bar{H}_i$ do not contribute to U. Fixing attention, for simplicity, on a statistically isotropic body we have for case (2.4a), by use of (2.7),

$$U^H = \tfrac{1}{2}\alpha^*\bar{H}_i\bar{H}_i V = \tfrac{1}{2}\alpha^* H_i^\circ H_i^\circ V \tag{2.11}$$

and for case (2.4b)

$$U^D = \frac{1}{2\alpha^*}\bar{D}_i\bar{D}_i V = \frac{1}{2\alpha^*} D_i^\circ D_i^\circ V, \tag{2.12}$$

which provide alternative definitions for α^* in terms of U.

Case (b). Time-independent Tensor Fields

When the phases are linear elastic the local stress–strain relation is:

$$\sigma_{ij}(\mathbf{x}) = C_{ijkl}(\mathbf{x})\,\varepsilon_{kl}(\mathbf{x}), \tag{2.13}$$

where

$$\varepsilon_{ij} = \tfrac{1}{2}(u_{i,j}+u_{j,i}), \tag{2.14}$$

and in the absence of body forces,

$$\sigma_{ij,j} = 0. \tag{2.15}$$

Homogeneous boundary conditions, which produce homogeneous stress and strain fields in homogeneous elastic bodies, are:

$$u_i(S) = \varepsilon_{ij}^\circ x_j \qquad \text{(a)}$$

$$\qquad\qquad\qquad\qquad (2.16)$$

$$T_i(S) = \sigma_{ij}^\circ n_j, \qquad \text{(b)}$$

where $\overset{\circ}{\varepsilon}_{ij}$ and $\overset{\circ}{\sigma}_{ij}$ are constant. For either one of (2.16) prescribed on S, the stress and strain fields are statistically homogeneous (provided that the phase geometry is statistically homogeneous). The effective elastic moduli C^*_{ijkl} are then defined by

$$\bar{\sigma}_{ij} = C^*_{ijkl}\bar{\varepsilon}_{kl}. \tag{2.17}$$

For statistical isotropy this becomes

$$\bar{\sigma}_{ij} = \lambda^*\bar{\varepsilon}_{kk}\delta_{ij}+2\mu^*\bar{\varepsilon}_{ij}, \tag{2.18}$$

where λ^* is an effective Lamé modulus and μ^* is the effective shear modulus. It should be borne in mind that relationships of type (2.17)–(2.18) strictly apply only for statistically homogeneous fields of stress and strain in statistically homogeneous bodies. For statistically nonhomogeneous fields such as torsion or bending the situation is more complicated and not clear at this time. Probably multipolar theories such as couple stress theory are needed to construct the proper stress–strain relations.[3]

A useful equivalent form of (2.18) is obtained by splitting $\bar{\sigma}_{ij}$ and $\bar{\varepsilon}_{ij}$ into isotropic and deviatoric parts. Then,

$$\bar{\sigma} = 3K^*\bar{\varepsilon} \qquad \text{(a)}$$

$$\bar{s}_{ij} = 2\mu^*\bar{e}_{ij}, \qquad \text{(b)} \tag{2.19}$$

where

$$\bar{\sigma} = \tfrac{1}{3}\bar{\sigma}_{kk}, \quad \bar{s}_{ij} = \bar{\sigma}_{ij}-\bar{\sigma}\delta_{ij} \tag{2.20}$$

$$\bar{\varepsilon} = \tfrac{1}{3}\bar{\varepsilon}_{kk}, \quad \bar{e}_{ij} = \bar{\varepsilon}_{ij}-\bar{\varepsilon}\delta_{ij} \tag{2.21}$$

$$K^* = \lambda^*+\tfrac{2}{3}\mu^*. \tag{2.22}$$

Here K^* is the effective bulk modulus. The effective Young's modulus E^* and Poisson's ratio ν^* are given by the usual relations:

$$E^* = \frac{9K^*\mu^*}{3K^*+\mu^*} \tag{2.23}$$

$$\nu^* = \frac{3K^*-2\mu^*}{2(3K^*+\mu^*)}. \tag{2.24}$$

For boundary condition (2.16a)

$$\bar{\varepsilon}_{ij} = \overset{\circ}{\varepsilon}_{ij} \tag{2.25}$$

while for boundary condition (2.16b)

$$\bar{\sigma}_{ij} = \overset{\circ}{\sigma}_{ij}, \tag{2.26}$$

both results being rigorous. Thus, as in the dielectric problem, determination of effective elastic properties amounts to calculation of an average field for a given boundary condition.

In analogy to the results (2.11)–(2.12), effective elastic moduli can be defined in terms of the strain energy stored in the elastic heterogeneous body. We have the rigorously valid relation

$$U = \tfrac{1}{2}\int_V \sigma_{ij}(\mathbf{x})\,\varepsilon_{ij}(\mathbf{x})\,dV = \tfrac{1}{2}\bar{\sigma}_{ij}\bar{\varepsilon}_{ij}V \tag{2.27}$$

if either one of (2.16) is prescribed on the boundary surface. By statistical homogeneity the strain energy W per unit RVE is given by

$$W = \tfrac{1}{2}\bar{\sigma}_{ij}\bar{\varepsilon}_{ij}\,. \tag{2.28}$$

Inserting (2.17) into (2.28) we obtain

$$W^{\varepsilon} = \tfrac{1}{2}C^{*}_{ijkl}\bar{\varepsilon}_{ij}\bar{\varepsilon}_{kl} \tag{2.29}$$

or

$$W^{\sigma} = \tfrac{1}{2}S^{*}_{ijkl}\bar{\sigma}_{ij}\bar{\sigma}_{kl}, \tag{2.30}$$

where S^{*}_{ijkl} is the inverse of C^{*}_{ijkl} and is called the effective compliances tensor.

When (2.16a) is prescribed the average strains in (2.29) are $\overset{\circ}{\varepsilon}_{ij}$ and when (2.16b) is prescribed the average stresses in (2.30) are $\overset{\circ}{\sigma}_{ij}$. Thus in each case the effective elastic moduli are defined in terms of strain energy.

For statistically isotropic bodies (2.29)–(2.30) greatly simplify. It is again convenient to use isotropic and deviatoric parts of the strain and stress tensors according to (2.20)–(2.21). Then

$$W^{\varepsilon} = \tfrac{1}{2}(9K^{*}\bar{\varepsilon}^{2}+2\mu^{*}\bar{e}_{ij}\bar{e}_{ij}) \tag{2.31}$$

$$W^{\sigma} = \frac{1}{2}\left(\frac{\bar{\sigma}^{2}}{K^{*}}+\frac{\bar{s}_{ij}\bar{s}_{ij}}{2\mu^{*}}\right). \tag{2.32}$$

Case (c). Time-dependent Tensor Fields

The case of linearly related time-dependent tensor fields arises in the linear theory of viscoelasticity. If the phases of a heterogeneous body are linearly viscoelastic we have local relationships between stress and strain in the form

$$\sigma_{ij}(\mathbf{x},\,t) = \int_0^t C_{ijkl}(\mathbf{x},\,t-\tau)\,\frac{\partial\varepsilon_{kl}(\mathbf{x},\tau)}{\partial\tau}\,d\tau \qquad \text{(a)}$$

(2.33)

$$\varepsilon_{ij}(\mathbf{x},\,t) = \int_0^t S_{ijkl}(\mathbf{x},\,t-\tau)\,\frac{\partial\sigma_{kl}(\mathbf{x},\tau)}{\partial\tau}\,d\tau \qquad \text{(b)}$$

where C_{ijkl} is the relaxation moduli tensor and S_{ijkl} is the creep compliances tensor. The time dependent strains ε_{ij} are related to the time-dependent displacements by (2.14) and in the absence of body forces and the usual neglect of inertia forces, (2.15) is also valid. It is here assumed that initial elastic responses may be incorporated in (2.33) by allowing the appearance of time delta functions in the integrands.

Homogeneous boundary conditions for viscoelastic bodies have the form:

$$u_i(S, t) = \tilde{\varepsilon}_{ij}(t)\, x_j \qquad \text{(a)}$$

or (2.34)

$$T_i(S, t) = \tilde{\sigma}_{ij}(t)\, n_j, \qquad \text{(b)}$$

where $\tilde{\varepsilon}_{ij}$ and $\tilde{\sigma}_{ij}$ are space constant and time-dependent. For either one of these boundary conditions applied to a statistically homogeneous heterogeneous viscoelastic body the stress and strain fields are spatially statistically homogeneous. The average strains for (2.34a) are rigorously

$$\bar{\varepsilon}_{ij}(t) = \tilde{\varepsilon}_{ij}(t), \tag{2.35}$$

and the average stresses for (2.34a) are rigorously

$$\bar{\sigma}_{ij}(t) = \tilde{\sigma}_{ij}(t) \tag{2.36}$$

The effective stress–strain relation then has the form

$$\bar{\sigma}_{ij}(t) = \int_0^t C^*_{ijkl}(t-\tau)\,\frac{d\bar{\varepsilon}_{kl}(\tau)}{d\tau}\,d\tau \qquad \text{(a)}$$

(2.37)

$$\bar{\varepsilon}_{ij}(t) = \int_0^t S^*_{ijkl}(t-\tau)\,\frac{d\bar{\sigma}_{kl}(\tau)}{d\tau}\,d\tau. \qquad \text{(b)}$$

Here C^*_{ijkl} is the effective relaxation moduli tensor and S^*_{ijkl} is the effective creep compliances tensor. If the body is statistically isotropic, (2.37) are greatly simplified. For that case it is again convenient to split the stresses and strains into isotropic and deviatoric parts, (2.20)–(2.21). Then

$$\bar{\sigma}(t) = 3\int_0^t K^*(t-\tau)\,\frac{d\bar{\varepsilon}(\tau)}{d\tau}\,d\tau \qquad \text{(a)}$$

$$\bar{s}_{ij}(t) = 2\int_0^t \mu^*(t-\tau)\,\frac{d\bar{e}_{ij}(\tau)}{d\tau}\,d\tau \qquad \text{(b)}$$

(2.38)

$$\bar{\varepsilon}(t) = \frac{1}{3}\int_0^t I^*(t-\tau)\,\frac{d\bar{\sigma}(\tau)}{d\tau}\,d\tau \qquad \text{(c)}$$

$$\mathrm{e}_{ij}(t) = \frac{1}{2}\int_0^t J^*(t-\tau)\,\frac{d\bar{s}_{ij}(\tau)}{d\tau}\,d\tau. \qquad \text{(d)}$$

Here K^* is effective bulk relaxation modulus, μ^* the effective shear relaxation modulus, I^* the effective bulk creep compliance and J^* the effective shear creep compliance.

The physical interpretation of the effective relaxation moduli and creep compliances is more clearly brought out if the time dependence in (2.34) is taken in the form of a step function, i.e.

$$\tilde{\varepsilon}_{ij}(t) = \overset{\circ}{\varepsilon}_{ij}H(t) \qquad \text{(a)}$$

$$\tilde{\sigma}_{ij}(t) = \overset{\circ}{\sigma}_{ij}H(t), \qquad \text{(b)} \qquad (2.39)$$

where $\overset{\circ}{\varepsilon}_{ij}$ and $\overset{\circ}{\sigma}_{ij}$ are space and time-independent. In that event for (2.34a) prescribed

$$\bar{\sigma}_{ij}(t) = C^*_{ijkl}(t)\,\overset{\circ}{\varepsilon}_{kl} \qquad (2.40)$$

and for (2.34b) prescribed

$$\bar{\varepsilon}_{ij}(t) = S^*_{ijkl}(t)\,\overset{\circ}{\sigma}_{kl}. \qquad (2.41)$$

For statistically isotropic bodies the simplification of (2.40)–(2.41) is completely analogous to the simplification of (2.37) to the form (2.38).

It is now clear which problems have to be solved in order to calculate effective relaxation moduli and creep compliances: For boundary conditions of type (2.34a) with (2.39a) the average stresses have to be found or for boundary conditions (2.34b) with (2.30b) the average strains have to be calculated. If this can be done, then (2.40)–(2.41) serve to identify the effective relaxation moduli and creep compliances. For further discussion see ref. 4.

It is quite easy to generalize (2.27) to time-dependent fields (see ref. 5) but the results are not useful for alternative definition of effective relaxation moduli and creep compliances, as has been done for the dielectric and elastic cases.

It should be noted that time-dependent vector fields can be discussed in a completely analogous fashion. Linearly related time-dependent vector fields are encountered in the theory of dielectric losses. This subject will not be discussed here. The interested reader is referred to refs. 6, 7, 55.

The preceding discussion demonstrates a remarkable unity in the definition of effective physical properties of linear heterogeneous media. It is seen that in one definition an average field must be calculated for homogeneous boundary condition on the surface of a heterogeneous body. In general, calculation of the average field requires knowledge of the detailed field within the heterogeneous body. Since such a field must satisfy continuity conditions at the phase interfaces, the problem is in general intractable and can be solved only for a few simple cases. This method of solution will henceforth be referred to as the averaging method.

Fortunately, the alternative definitions of effective physical constants in terms of energy integrals put at our disposal the powerful method of variational bounding. Since extremum principles can be used to provide lower and upper bounds for (2.10) and (2.27) and since these quantities can be defined in terms of effective physical constants, (2.11)–(2.12), (2.29)–(2.32), the extremum principles can be directly used to construct bounds for effective physical constants.

In what follows methods of attack and results in theory of composite materials will be discussed for different types of physical behavior and different types of composites. Several surveys of the subject are already available in the literature. A general survey of theory of mechanical behavior is given in ref. 2, elasticity in refs. 3, 8, 9, dielectrics in ref. 10, statistical approaches in refs. 11, 12 and plasticity and failure in ref. 13.

3. ELASTIC PROPERTIES OF STATISTICALLY ISOTROPIC TWO-PHASE MATERIALS

Due to the enormous mathematical difficulties involved, direct rigorous calculation of effective elastic moduli of two-phase materials by the averaging method can be performed only in very few cases. The simplest case is that of a dilute suspension of spherical particles. This may be described by a matrix with elastic moduli K_1, μ_1, in which are imbedded spherical particles which may be of different diameters, with elastic moduli K_2, μ_2. The fractional volume of the particles is

$$v_2 = c. \tag{3.1}$$

The expression dilute suspension is defined by the assumption that $c \ll 1$. The calculation has been performed independently by Dewey,[14] Eshelby[15] and Hashin[16] by use of energy methods which are equivalent to the averaging method. The results are:

$$K^* = K_1+(K_2-K_1)\frac{3K_1+4\mu_1}{3K_2+4\mu_1}c \tag{3.2}$$

$$\mu^* = \mu_1+(\mu_2-\mu_1)\frac{5(3K_1+4\mu_1)}{9K_1+8\mu_1+6(K_1+2\mu_1)\mu_2/\mu_1}c. \tag{3.3}$$

Of interest is the special case of rigid particles which is obtained from (3.2)–(3.3) by setting

$$\frac{K_2}{K_1} \to \infty \qquad \frac{\mu_2}{\mu_1} \to \infty \tag{3.4}$$

(see Hashin[17]). Another interesting case is that of spherical holes, i.e. a porous solid. This is obtained from (3.2)–(3.3) by setting

$$\frac{K_2}{K_1} \to 0 \qquad \frac{\mu_2}{\mu_1} \to 0 \tag{3.5}$$

(see Mackenzie[18]).

It should be borne in mind that (3.2)–(3.3) are only of little practical interest since in most two phase materials the phase fractional volumes are of the same order. Expressions (3.2)–(3.3) agree with experiment for $c = 1$–2%. They quickly become much smaller than experimental values when c increases. (For particles stiffer than the matrix.)

It is seen that (3.2)–(3.3) depend only on phase elastic moduli and phase fractional volumes. This is due to the fact that because of the dilute concentration it does not matter where the spheres are placed, as long as it can be assumed that their mutual interaction can be neglected. However, the effective elastic moduli of a two-phase elastic spheres suspension of arbitrary phase fractional volumes will be a function of c *and* all the details of spheres distribution. If we have at our disposal different suspension specimens with same phase elastic moduli and same c, their K^* and μ^* will in general be different. But if c is reduced in all specimens to 1–2% the K^* and μ^* of all specimens will be given by (3.2)–(3.3).

Eshelby[15] has also analyzed the case when the particles are randomly oriented ellipsoids. Because of mathematical complexity the results could only be stated implicitly.

We now turn to the case of two-phase spheres suspensions of arbitrary phase fractional volumes. The only rigorous closed result available here is for a very special arrangement of particles which has been named composite spheres assemblage. This arrangement is described by filling out a volume of arbitrary shape with composite spheres. Each sphere consists of a spherical particle of radius a_n of material 2 and a concentric spherical shell with radii a_n, b_n of material 1. In each sphere

$$\left(\frac{a_n}{b_n}\right)^3 = c. \tag{3.6}$$

The external radii b_n of the composite spheres may be given any values. The filling out process is started with relatively large composite spheres. Smaller and smaller spheres are placed in the remaining spaces. No spheres should overlap. In the limit the entire volume can be filled out with spheres and then becomes a composite spheres assemblage (see Fig. 3). Hashin[19] has calculated the effective bulk modulus of this composite spheres assemblage which is given by

$$K^* = K_1+(K_2-K_1)\frac{(3K_1+4\mu_1)\,c}{3K_2+4\mu_1+3(K_1-K_2)\,c}\,. \tag{3.7}$$

It is seen that for very small c, (3.7) agrees with (3.2) as it should. It should be noted that the only geometrical information entering into (3.7) is c. This

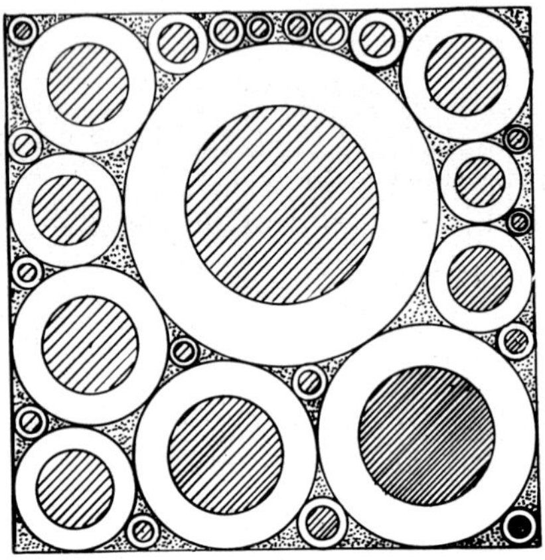

FIG. 3. Composite spheres assemblage

surprising state of affairs must be due to the very special distribution of spherical particles implied by the model described. However, the actual distribution of spheres is not known.

Again the special cases of rigid particles and cavities are obtained by simplifying (3.7) by use of (3.4) or (3.5), respectively. The last case is of particular interest since it describes some special porous solid. The result for the porous solid is

$$K^* = K_1\left[1-\frac{(3K_1+4\mu_1)\,c}{3K_1c+4\mu_1}\right]. \tag{3.8}$$

Recently Walsh, Brace and England[20] have given experimental results for the bulk modulus of a porous solid as a function of porosity which agree very well with (3.8). Apparently these authors were not aware of the result (3.8) in its present interpretation. However, the expression (1) in their paper, which was given by MacKenzie[18] with a different connotation is identical to (3.8).

It is also of interest to note that Hill[18] has shown that (3.7) applies for another case. It is the rigorous expression for the bulk modulus of a two-phase material of arbitrary phase geometry when the phase bulk moduli are different K_2, K_1, and the shear moduli are equal and are both given by μ_1.

Another interesting investigation is due to Molyneux and Beran.[21] They consider the problem of a random elastic medium with small variation in properties. This means that at each point in the body, the variations of K and μ are small in comparison to their averages. For this case they were able to give expressions for K^* and μ^* in terms of averages and correlations of K and μ. The two-phase material with small variations in properties from phase to phase is a special case of their more general treatment.

The results discussed so far appear to be the only exact expressions derived to date for effective elastic moduli of statistically homogeneous and isotropic two-phase materials. It is seen that all of them hold only in some special situations such as: small fractional volume of one phase, special spherical phase geometry and small fluctuations of elastic moduli.

Consider a statistically homogeneous and isotropic two-phase material of irregular phase geometry. The elastic moduli of the first phase are K_1, μ_1 and those of the second phase are K_2, μ_2. The simplest geometrical information we can think of are the volume fractions of the phases, v_1 and v_2, where, of course,

$$v_1+v_2 = 1. \tag{3.9}$$

As has been emphasized again and again such rudimentary information is completely insufficient for the determination of the effective elastic moduli. But we can pose the question differently: What can we know about the effective elastic moduli if the only geometrical information available are the phase volume fractions? The answer to this question is that we can bound the effective elastic moduli in terms of this simple information.

Bounding methods are based on two fortunate circumstances: (a) The strain or stress energy stored can be expressed in terms of effective elastic moduli and average strains or stresses, eqns. (2.29)–(2.32), (b) The strain or stress energy can be bounded by use of the extremum principles of the theory of elasticity. The reader who is interested in general aspects of such bounding methods is referred to refs. 2, 3, 19. Here we shall only discuss results.

The first bounds for the effective elastic moduli of a two-phase material of irregular geometry were given by Paul.[22] These very simple results are:

$$\frac{1}{\dfrac{v_1}{K_1}+\dfrac{v_2}{K_2}} \leqslant K^* \leqslant K_1v_1+K_2v_2 \tag{3.10}$$

$$\frac{1}{\dfrac{v_1}{\mu_1}+\dfrac{v_2}{\mu_1}} \leqslant \mu^* \leqslant \mu_1v_1+\mu_2v_2. \tag{3.11}$$

An upper bound for the Young's modulus E^* is obtained by substituting the right sides of (3.10)–(3.11) for K^* and μ^* into (2.23). Similarly a lower bound for E^* is obtained by substitution of the left sides of (3.10)–(3.11) in (2.23).

The right sides of (3.10)–(3.11) are sometimes called the law of mixtures. This is an unfortunate choice of words since (3.10)–(3.11) are good bounds only when the difference between phase moduli is very small. For appreciable differences, experimental values of K^* and μ^* are generally very much below the weighted averages in the right sides of (3.10)–(3.11).

The derivation of the bounds (3.10)–(3.11) is based on the classical elasticity extremum principles of minimum potential and minimum complementary energy. Substantially improved bounds in terms of phase moduli and phase volume fractions have been derived by Hashin and Shtrikman[23] on the basis of some new elasticity extremum principles which were derived by them.[24] These bounds have the following form

$$K^*_{(-)} = K_1 + \frac{v_2}{\dfrac{1}{K_2-K_1} + \dfrac{3v_1}{3K_1+\mu_1}} \tag{3.12}$$

$$K^*_{(+)} = K_2 + \frac{v_1}{\dfrac{1}{K_1-K_2} + \dfrac{3v_2}{3K_2+4\mu_2}} \tag{3.13}$$

$$K^*_{(-)} \leqslant K^* \leqslant K^*_{(+)} \tag{3.14}$$

$$\mu^*_{(-)} = \mu_1 + \frac{v_2}{\dfrac{1}{\mu_2-\mu_1} + \dfrac{6(K_1+2\mu_1)\,v_1}{5\mu_1(3K_1+4\mu_1)}} \tag{3.15}$$

$$\mu^*_{(+)} = \mu_2 + \frac{v_1}{\dfrac{1}{\mu_1-\mu_2} + \dfrac{6(K_2+2\mu_2)\,v_2}{5\mu_2(3K_2+4\mu_2)}} \tag{3.16}$$

$$\mu^*_{(-)} \leqslant \mu^* \leqslant \mu^*_{(+)} \tag{3.17}$$

$$K_2 > K_1, \qquad \mu_2 > \mu_1.\dagger \tag{3.18}$$

While the original derivation of the bounds involved some operations whose mathematical legality may be questioned, subsequent rederivations, Hill,[8] ((3.12)–(3.13) only), Hashin,[25] Walpole[26] have established the mathematical rigor of the results.

It is of considerable importance to note that (3.12) is identical with the composite spheres assemblage result (3.7). This is established after putting $v_2 = c$, $v_1 = 1-c$ in (3.12). Similarly (3.13) may be identified with another

† This restriction has recently been removed by Walpole.[26]

composite spheres assemblage result, in the case when the spherical particles are of material 1 with fractional volume v_1 and the shells of material 2, with fractional volumes v_2. This identification of the bounds with exact results provides the very important information that the bounds (3.12)–(3.13) are *best possible* in terms of the geometrical information used (i.e. volume fractions).[23] In other words, to obtain better bounds for K^* more geometrical information is needed.

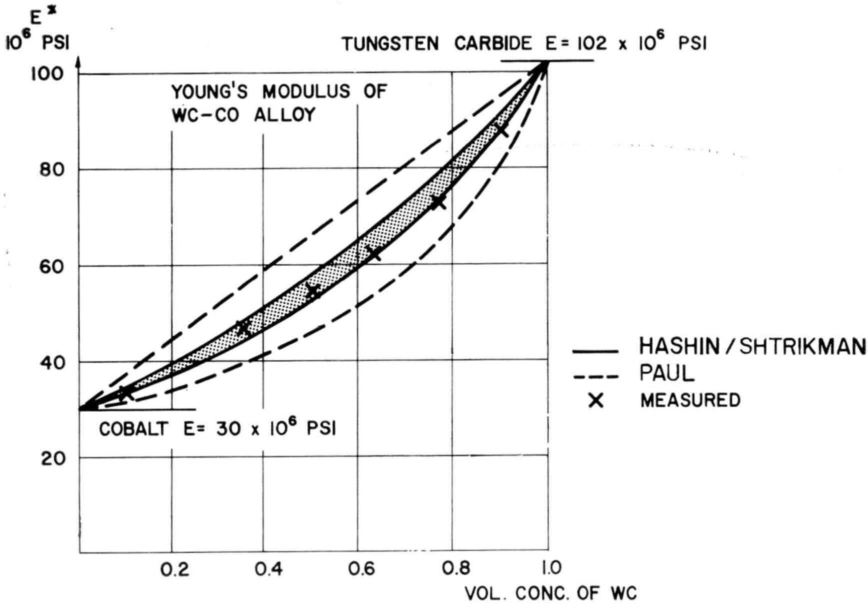

FIG. 4. Theoretical bounds and experimental results for Young's modulus of two-phase alloy

To obtain bounds for the Youngs modulus E^* the upper bounds (3.13,) (3.16) are substituted into (2.8) to obtain $E^*_{(+)}$ and the lower bounds (3.12,) (3.15) give $E^*_{(-)}$.

The bounds are reasonably close when the ratios between elastic moduli of the stiffer phase to those of the more compliant phase are not too large. For an example see Fig. 4. The bounds mostly agree quite well with experimental data. For verifications see, for example, Fig. 4 based on Nishimatsu and Gurland,[27] Hasselmann and Fulrath,[28] Umekawa and Sherby.[29]

In the extreme cases of rigid reinforcement or of a porous solid the bounds become impractical. In the first case the upper bounds become infinite while in the second case the lower bounds vanish. This is physically plausible since the two-phase materials treated are very general. In particular they cover suspensions as special cases. Thus for a rigid phase the bounds must also apply for the situation when we have a rigid matrix. But then obviously the effective elastic moduli are infinite. Similar considerations hold for the porous case.

Interpretation of the bounds (3.12)–(3.13), (3.15)–(3.16) in the literature is not always free of misconceptions. Thus it was stated that the lower bounds are based on the assumption that the more compliant phase is a matrix and the stiffer phase is in the form of particles and vice versa for the upper bounds. This is incorrect and nothing of the kind is involved in the derivation. All that can be said is that, intuitively, for two materials with same volume fractions and same phase moduli, where in the first the stiffer phase is a matrix while in the second the more compliant phase is a matrix, the actual effective moduli of the first material will be closer to the upper bound while those of the second will be closer to the lower bound.

Finally it is mentioned that the results (3.10)–(3.11) and (3.12)–(3.17) are easily generalized to *n*-phase materials (see refs. 22, 23).

Bounding methods can also be applied when the phase geometry is a very special one. An example for this is the bounding of shear modulus μ^* of a composite spheres assemblage (see Hashin[19]). The composite spheres assemblage is a not unreasonable model for a suspension or a particulate composite. The general bounds (3.15)–(3.16) have an important bearing on the shear modulus bounds for a composite spheres assemblage. Since this has never been discussed in detail in the literature it is worthwhile to give a brief discussion here.

Let the effective shear modulus of the composite spheres assemblage be denoted μ_C^*. Then

$$\mu_{C(-)}^* \leqslant \mu_C^* \leqslant \mu_{C(+)}^*, \tag{3.19}$$

where the bounds in (3.19) are the ones given in ref. 19. Now it can be shown that if the particles are stiffer than the matrix

$$\mu_{C(-)}^* \leqslant \mu_{(-)}^*, \tag{3.20}$$

and when the matrix is stiffer than the particles

$$\mu_{(+)}^* \leqslant \mu_{C(+)}^*, \tag{3.21}$$

where $\mu_{(-)}^*$ and $\mu_{(+)}^*$ are the bounds given by (3.15)–(3.16). Since the bounds (3.15)–(3.16) are valid for all phase geometries they also apply for the composite spheres assemblage. Thus improved bounds for μ_C^* can be written as follows

$$\mu_{(-)}^* \leqslant \mu_C^* \leqslant \mu_{C(+)}^* \quad \text{when} \quad K_2 > K_1, \quad \mu_2 > \mu_1 \tag{3.22}$$

$$\mu_{C(-)}^* \leqslant \mu_C^* \leqslant \mu_{(+)}^* \quad \text{when} \quad K_2 < K_1, \quad \mu_2 < \mu_1 . \tag{3.23}$$

Here the subscript 2 always refers to the spherical particles.

The improved bounds (3.22)–(3.23) are mostly in good agreement with experimental results for particulate composites and they are quite close together even for particles which are much stiffer than the matrix.

The bounds (3.22)–(3.23) will here be given explicitly for the special cases of rigid particles and spherical voids only. For rigid particles

$$1+\frac{15(1-\nu)}{2(4-5\nu)}\,\frac{c}{1-c} \leqslant \frac{\mu_C^*}{\mu} \leqslant 1+\frac{\dfrac{15(1-\nu)}{2(4-5\nu)}\,c}{1-c\left[1+\dfrac{63}{4(7-10\nu)(4-5\nu)}\,\phi(c)\right]}. \tag{3.24}$$

For a porous material

$$1-\frac{\dfrac{15(1-\nu)}{7-5\nu}\,c}{1+c\left[\dfrac{2(4-5\nu)}{7-5\nu}-\dfrac{126}{(7+5)(7-5\nu)}\,\phi(c)\right]} \leqslant \frac{\mu_C^*}{\mu} \leqslant 1-\frac{15(1-\nu)c}{7-5\nu+2(4-5\nu)c} \tag{3.25}$$

where

$$\phi(c) = \frac{(1-c^{2/3})^2}{1-c^{7/3}}, \tag{3.26}$$

and μ and ν are the shear modulus and Poisson's ratio of the matrix respectively.

Care should be taken with the idealization to rigid particles. Calculations have shown that even for particle to matrix shear moduli ratio of value 30, the rigid particle approximation is too idealized and that it is necessary to use the bounds for elastic particles as given in ref. 19.

We have seen that for two-phase materials of irregular geometry the information contained in the specification of the volume fractions only, seems to have been fully exploited. The question may now be asked: what do we do in order to improve the bounds? We obviously need to have more geometrical information about the phase geometry and the question then arises what kind of information should be used? Examples for additional information are not difficult to find. One possibility is to generalize the volume fraction type of information in a probabilistic sense. The volume fraction of a phase may be interpreted as the probability that a point which is thrown at random into the two-phase material falls within that phase. Instead we could throw a rigid rod into the material (Buffon needle) which is defined by two points, its extremities, which are a distance r apart. We can speak of the probability $P_{11}(r)$ that both points are in phase one, or the probability $P_{12}(r)$ that one point is in phase one and the second in phase two, etc. We can similarly define three-point and four-point probabilities and in principle n-point probabilities. Specification of statistical geometry in terms of such probabilities is certainly a logical approach. It is, however, not at all clear that for the purpose of study of effective physical properties this is the most advantageous characterization. It is very easy to think of different kinds of geometrical information such as the moment of inertia of particles, mean particle distance, root mean square of particle diameters, etc.

It is also very important to bear in mind that from a practical point of view it is not very helpful to exploit additional geometrical information if this information cannot be controlled in the process of manufacture of a two-phase material. Thus the volume fractions are easily controlled by simply specifying phase quantities. On the other hand, how to control a two-point probability is anyone's guess.

Recently Beran and Molyneux[30] were able to derive bounds on the bulk modulus K^* of heterogeneous media in terms of additional statistical information. The bounds are expressed in terms of three-point probability functions. While very little is known about such probability functions for heterogeneous media there is one special case in which a three-point probability function can be expressed in terms of the phase volume fraction. This case is described as a symmetric heterogeneous medium. For a two-phase material the two-phases must have equal volume fractions and, to use a very rough expression, they must be geometrically indistinguishable from the statistical point of view. For the symmetric two-phase medium the bounds obtained are better than (3.12)–(3.13). It should be borne in mind that this does not contradict the statement that the bounds (3.12)–(3.13) are best possible in terms of volume fractions, since to improve the bounds additional information, i.e. statistical symmetry, has been exploited.

Miller[31] has considered two-phase materials whose statistical geometry is that of symmetric or asymmetric cell materials. In that case the three-point probability functions entering into the bulk modulus bounds derived in ref. 30 can be determined in terms of one geometrical parameter only for the symmetric cell material and in terms of two geometrical parameters for the asymmetric cell material.

To the best of the writer's knowledge the discussion up to this point has dealt with all the theoretically sound results for two-phase isotropic elastic materials which have been published. The practician who intends to use semi-empirical formulae and rough approximations in order to estimate elastic properties of a two-phase material would do well to first convince himself that none of the results discussed above applies to the situation with which he is confronted.

We now turn to discussion of some approximate methods. The most primitive kind of approximation is to fit some mathematical expression to agree more or less with experimental results. In view of the complexity of the present problem this is a hopeless undertaking. Those who devise such empirical formulae mostly assume that only phase-elastic moduli and volume fractions have to be accounted for. The result is then an empirical expression which fits one set of data but will hardly ever fit another set of data, obtained by another experimentalist.

In other approaches expressions are constructed which contain an empirical parameter which is to be determined by experiment. The question must then be asked: Why not determine the effective moduli experimentally to begin with?

We shall here restrict ourselves to approximations which do not require experimental information. A well-known scheme of approximation is the method known as the "self-consistent scheme" (henceforth abbreviated as SCS). It is quite easy to show that if a two-phase material is subjected to statistically homogeneous fields of stress and strain, knowledge of the average strain or stress in one phase (i.e. $\bar{\varepsilon}_{ij}^{(1)} = \frac{1}{V_1} \int_{V_1} \varepsilon_{ij}\, dV$) will enable one to exactly calculate the effective elastic moduli. To compute such an average precisely is, of course, an intractable problem. The SCS proceeds by estimating it.

Consider a two-phase suspension. A typical particle is assumed to be of spherical shape and to be imbedded directly in a *homogeneous* material whose elastic moduli are the K^* and μ^* of the suspension. The last assumption is the crucial one and can, of course, not be correct since it is only on a relatively large scale of magnitude that a heterogeneous material can be replaced by an equivalent homogeneous one. A heterogeneous element which is very large in comparison to the particles has elastic moduli K^* and μ^* but all that can be said about the immediate neighborhood of a particle is that the the elastic properties *at points* erratically assume the phase moduli values. Bearing all this in mind one may nevertheless proceed with the calculation. The results are then given in the following form:

$$K^* = K_1 + (K_2 - K_1)\frac{3K^* + 4\mu^*}{3K_2 + 4\mu^*}\, c \tag{3.27}$$

$$\mu^* = \mu_1 + (\mu_2 - \mu_1)\frac{5(3K^* + 4\mu^*)c}{9K^* + 8\mu^* + 6(K^* + 2\mu^*)\mu_2/\mu^*}. \tag{3.28}$$

Equations (3.19)–(3.20) are two simultaneous equations in which the unknowns are K^* and μ^*. They have in general to be solved numerically. These results have been given independently by Budiansky[32] and Hill.[33] It is interesting to note that (3.27)–(3.28) bear a formal resemblance to the dilute suspension results (3.2)–(3.3). This is due to an inherent connection between the methods used to arrive at these results.

The approximate results (3.27)–(3.28) are related in an interesting way to the bounds (3.12)–(3.17). To see this relationship let K^*_{SCS} and μ^*_{SCS} as predicted by (3.27)–(3.28) be plotted as a function of the fractional volume $v_2 = c$ and let the bounds (3.12)–(3.17) be plotted in a similar way. The curves are shown schematically in Fig. 5. It has been shown by Hill[33] that always

$$K^*_{(-)} \leqslant K^*_{SCS} \leqslant K^*_{(+)} \tag{3.29}$$

$$\mu^*_{(-)} \leqslant \mu^*_{SCS} \leqslant \mu^*_{(+)} \tag{3.30}$$

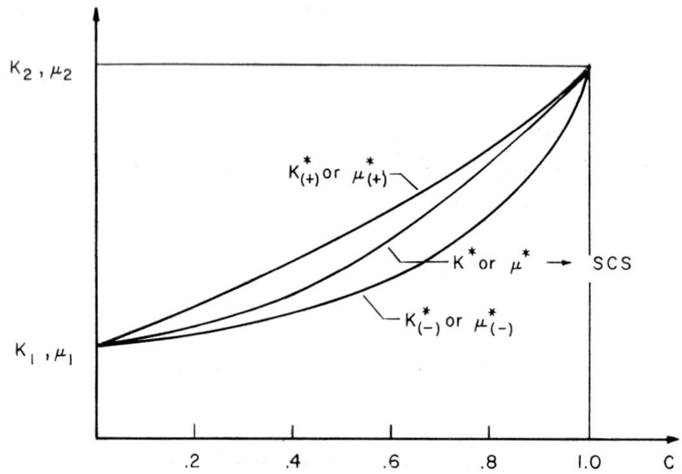

FIG. 5. Self-consistent scheme prediction and bounds

where the left sides of (3.29)–(3.30) are given by (3.12) and (3.15), respectively, and the right sides are given by (3.13) and (3.16), respectively. Thus K^*_{SCS} and μ^*_{SCS} do not violate the bounds (3.12)–(3.17). If they did violate the bounds they would have to be rejected since the bounds are exact. Furthermore, K^*_{SCS} is tangent to $K^*_{(-)}$ at $c \to 0$ and tangent to $K^*_{(+)}$ at $c \to 1$. Similarly μ^*_{SCS} is tangent to $\mu^*_{(-)}$ at $c \to 0$ and tangent to $\mu^*_{(+)}$ at $c \to 1$ (see Fig. 5). This is due to the fact that at $c \to 0$, $K^*_{(-)}$ and K^*_{SCS} become identical with the dilute suspension results (3.2), and $\mu^*_{(-)}$ and μ^*_{SCS} become identical with (3.3). To reach a similar conclusion for $c \to 1$ we can consider a dilute suspension in which spherical particles of volume fraction $1-c$ are phase one, while the matrix is phase two. The results for this case are simply found by interchanging 1 by 2 in (3.2)–(3.3) and replacing c by $1-c$. Then at $c \to 1$ $K^*_{(+)}$, and K^*_{SCS} and $\mu^*_{(+)}$, μ^*_{SCS} become identical with these dilute suspension results.

On the basis of the preceding discussion we can obtain some idea regarding the fit of the SCS results to experimental data. Since the measured values of K^* and μ^* should always lie between the bounds (3.12)–(3.14) and (3.15)–(3.17), respectively, and since the SCS results are always between those bounds and moreover are tangent to the lower bounds at $c \to 0$ and tangent to the upper bounds at $c \to 1$, it is to be expected that whenever the bounds are not far apart the SCS results should give a good fit with experimental data. However, when the bounds are far apart, considerable scatter in the measured K^* and μ^* must be expected and then the SCS results cannot be expected to provide a good approximation. Indeed these speculations are verified by some experiments.

It is of course very easy to construct by a pure curve fitting procedure a curve which always lies between the bounds and is tangent to the lower bounds at $c \to 0$ and tangent to the upper bounds at $c \to 1$. The SCS results

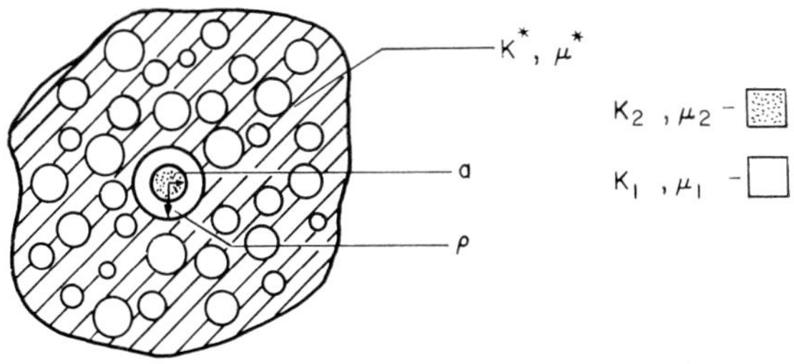

FIG. 6. Modified self-consistent scheme

also form such a curve. The question must then be asked: is there an inherent advantage in the SCS curve over any other such curve, due to the approximate physical scheme on which its derivation is based? At the present time this remains an open question.

If we now recall the basic imbedding assumption involved in the SCS, we note a certain arbitrariness. Specifically, it has been assumed that a spherical particle is directly imbedded in a homogeneous material with moduli K^* and μ^*. But the imbedding may be done differently. It may be assumed that the particle of radius a, say, is imbedded in a concentric matrix shell of radius ϱ, and from there on starts the homogeneous material with unknown moduli K^* and μ^* (Fig. 6). Such a procedure would even seem more reasonable from a physical point of view, however, the calculation of μ^* is now much more difficult and to the writer's knowledge has not yet been performed for an arbitrary value of ϱ. In principle the calculation would yield results for K^* and μ^* which depend on c and ϱ. These results would generalize the previous results in which $\varrho = a$. But there is no reason whatsoever to believe that $\varrho = a$ is a particularly good choice and we must conclude that ϱ is of the nature of an unknown parameter. This is a serious drawback to the self-consistent scheme.

A choice which immediately comes to mind is

$$\left(\frac{a}{\varrho}\right)^3 = c. \qquad (3.31)$$

This means that in the imbedded composite sphere (Fig. 6) the ratio of particle to matrix material is the same as in the entire composite. The SCS analysis on the basis of (3.31) has apparently been attempted by Kerner.(34) The writer has unfortunately not been able to understand this paper, but very interestingly the results obtained by Kerner are precisely the lower bounds (3.12) and (3.15). It seems that for any intermediate choice of ϱ, i.e.

$$a < \varrho < \frac{a}{c^{1/3}}, \qquad (3.32)$$

the SCS would yield a result between those given by (3.19)–(3.20) and the lower bounds. The writer has proved this for dielectrics but not for elasticity. If this is true it would mean that the whole region between the bounds can be covered by SCS results with different choices of ϱ, and this would unfortunately also imply that the SCS is of little value for two-phase materials.

The SCS in Kerner's(34) version, i.e. on the basis of the assumption (3.31), has also been treated by Van der Poel.(35) Van der Poel has obtained the same result as Kerner for K^*, that is to say the lower bound (3.12). However, his result for μ^* is different and much higher than (3.15), which is Kerner's result. This curious state of affairs is due to a fundamental error in Van der Poel's paper. He has used a wrong elasticity solution for a concentric composite sphere problem arising in the theory (eqs. (6), (7) and (8) in ref. 35). Unfortunately the error has been compounded by several authors who have used Van der Poel's μ^* for comparison with experiment and have in some cases found good agreement.

We close the discussion of the SCS by mentioning the work of Wu(36) who has used the SCS to estimate the effective elastic moduli of a two-phase suspension with randomly oriented ellipsoidal particles. The form of the SCS used is the one first discussed in which the particle is assumed to be directly imbedded in homogeneous material with moduli K^* and μ^*.

4. ELASTIC PROPERTIES OF FIBER-REINFORCED MATERIALS

The major part of present research activity in composites is devoted to fiber-reinforced material. The simplest kind of fiber-reinforced material consists of an elastic matrix, here labelled 1, in which are imbedded parallel fibers made of a material which is labelled 2. It is assumed that fibers and matrix are linearly elastic and isotropic.

As a convenient representative specimen of such a material we may choose a cylinder whose generators are parallel to the fibers and whose cross-section includes a very large number of fiber cross-sections (Fig. 7). It may be assumed that the fibers extend throughout the entire length of the cylinder. This condition is usually not fulfilled in practice for there always are some broken fibers. However, usually the broken fibers are long enough so that with

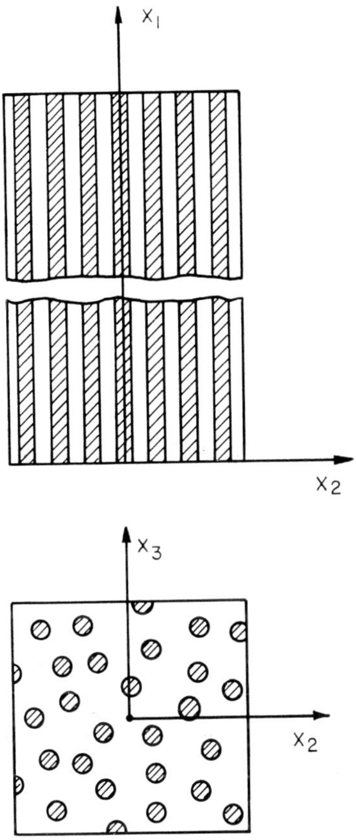

FIG. 7. Fiber-reinforced material

respect to elastic properties, the assumption of continuous fibers only involves the neglect of completely insignificant end effects.

It is assumed that fiber cross-sections are randomly located in the transverse plane cross-section of the cylindrical specimen. The fiber-reinforced material may then be assumed to be statistically homogeneous. Obviously it is not statistically isotropic but it may be assumed to be transversely isotropic, in the statistical sense, with respect to any axis which is parallel to the fibers. We choose a Cartesian coordinate system $x_1x_2x_3$ in which x_1 is in fiber direction and consequently x_2x_3 are in the transverse plane, normal to the fibers (Fig. 7). The stress–strain law (2.1), relating average stresses and strains, then assumes in view of transverse isotropy the following form (see Hashin and Rosen):[37]

$$\begin{aligned}
\bar{\sigma}_{11} &= C^*_{11}\bar{\varepsilon}_{11}+C^*_{12}\bar{\varepsilon}_{22}+C^*_{12}\bar{\varepsilon}_{33} && \text{(a)}\\
\bar{\sigma}_{22} &= C^*_{12}\bar{\varepsilon}_{11}+C^*_{22}\bar{\varepsilon}_{22}+C^*_{23}\bar{\varepsilon}_{33} && \text{(b)}\\
\bar{\sigma}_{33} &= C^*_{12}\bar{\varepsilon}_{11}+C^*_{23}\bar{\varepsilon}_{22}+C^*_{22}\bar{\varepsilon}_{33} && \text{(c)}\\
& && \qquad (4.1)\\
\bar{\sigma}_{12} &= 2C^*_{44}\bar{\varepsilon}_{12} && \text{(d)}\\
\bar{\sigma}_{23} &= (C^*_{22}-C^*_{23})\,\bar{\varepsilon}_{23} && \text{(e)}\\
\bar{\sigma}_{31} &= 2C^*_{44}\bar{\varepsilon}_{31}. && \text{(f)}
\end{aligned}$$

It is seen that the material has five independent effective elastic moduli. Thus the situation is much more complicated than for a statistically isotropic material.

From the physical point of view it is better to choose a different set of effective elastic moduli, which are defined in terms of the moduli appearing in (4.1). These are:

$$k^*_t = \tfrac{1}{2}(C^*_{22}+C^*_{23}) \tag{4.2}$$

$$\mu^*_t = \tfrac{1}{2}(C^*_{22}+C^*_{23}) \tag{4.3}$$

$$\mu^*_a = C^*_{44} \tag{4.4}$$

$$E^*_a = C^*_{11}-\frac{2C^{*2}_{12}}{C^*_{22}+C^*_{23}} \tag{4.5}$$

$$\nu^*_a = \frac{C^*_{12}}{C^*_{22}+C^*_{23}}\,. \tag{4.6}$$

The physical significance of these moduli is as follows: k^*_t is a plane strain bulk modulus. Thus if axial deformation of the cylindrical specimen is prevented and the only non-vanishing average strains are

$$\bar{\varepsilon}_{22} = \bar{\varepsilon}_3 = \bar{\varepsilon}, \tag{4.7}$$

then it follows from (5.1) that the only non-vanishing average stresses are $\bar{\sigma}_{11}$ and

$$\bar{\sigma}_{22} = \bar{\sigma}_{33} = \bar{\sigma}. \tag{4.8}$$

Then we have the relation

$$\bar{\sigma}^\circ = 2k^*_t\bar{\varepsilon}. \tag{4.9}$$

The modulus (4.3) is a transverse shear modulus for a shearing action in the transverse x_2x_3 plane, as is easily seen from (4.1e). Similarly (4.4) is an axial shear modulus governing shearing action in x_1x_2 and x_1x_3 planes. This follows from (4.1d, f).

The modulus (4.5) is an axial Young's modulus. Thus if the only non-vanishing average stress is $\bar{\sigma}_{11}$ then the axial average strain $\bar{\varepsilon}_{11}$ is given by

$$\bar{\varepsilon}_{11} = \frac{\bar{\sigma}_{11}}{E_a^*}. \tag{4.10}$$

In this case, however, there will also be average transverse strains $\bar{\varepsilon}_{22}$, $\bar{\varepsilon}_{33}$ which are given by

$$\bar{\varepsilon}_{22} = \bar{\varepsilon}_{33} = -\nu_a^* \bar{\varepsilon}_{11}. \tag{4.11}$$

This defines the axial Poisson's ratio (5.6).

The basic loadings which involve the above discussed moduli are shown in Fig. 8.

The theory of elastic properties of fiber-reinforced materials is strongly related to the theory developed for statistically isotropic heterogeneous elastic media. The composite spheres assemblage model is replaced by a composite cylinder assemblage model. Each composite cylinder consists of a

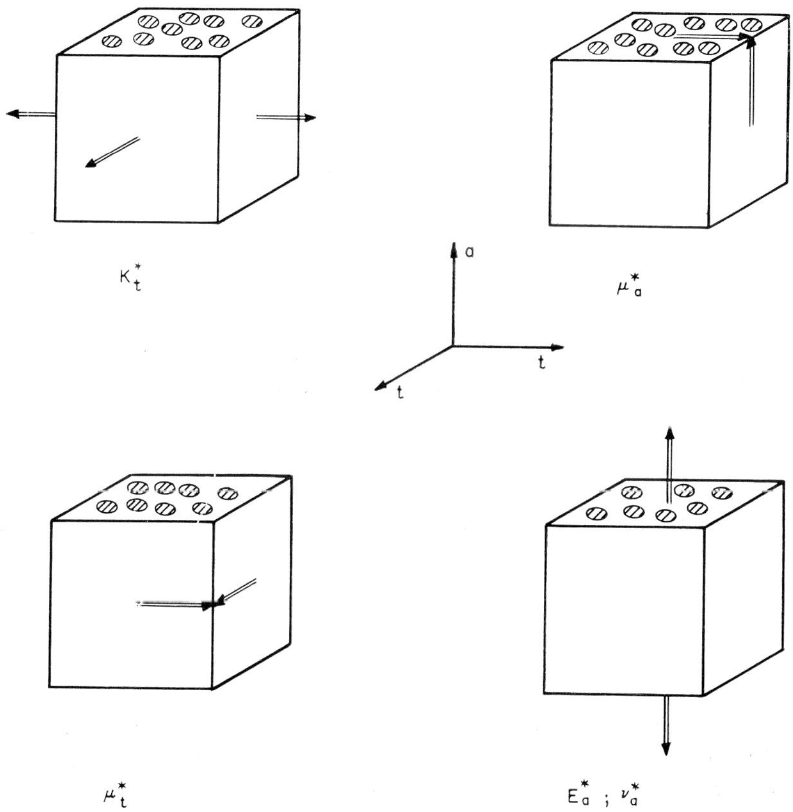

FIG. 8. Basic loadings for definition of elastic moduli of fiber-reinforced material

central circular fiber and a concentric matrix shell. For a detailed description of the model see ref. 37. Also, Fig. 3 may be imagined to represent a cut through a composite cylinder assemblage. The elastic properties of a composite cylinder assemblage have been analyzed in ref. 37 for hollow as well as solid circular fibers. Here we shall only give results for solid fibers. For this purpose it is convenient to define the plane strain bulk modulus k given by

$$k = K+\frac{1}{3}\mu, \tag{4.12}$$

in terms of the three-dimensional bulk modulus K and the shear modulus μ. The results are then:

$$k_t^* = k_1+\frac{v_2}{\dfrac{1}{k_1-k_2}+\dfrac{v_1}{k_1+\mu_1}} \tag{4.13}$$

$$\mu_a^* = \mu_1+\frac{v_2}{\dfrac{1}{\mu_2-\mu_1}+\dfrac{v_1}{2\mu_1}} \tag{4.14}$$

$$E_a^* = E_1v_1+E_2v_2+\frac{4v_1v_2(\nu_1-\nu_2)^2}{v_1/k_2+v_2/k_1+1/\mu_1} \tag{4.15}$$

$$\nu_a^* = \nu_1v_1+\nu_2v_2+\frac{v_1v_2(\nu_2-\nu_1)(1/k_1-1/k_2)}{v_1/k_2+v_2/k_1+1/\mu_1}\,. \tag{4.16}$$

Unfortunately no closed expression for μ_t^* is available. This modulus could only be bounded from above and below (see ref. 37). It is important to note that the lower bound given in ref. 37 has been superseded by a very simple better lower bound as will be explained below. The upper bound for the case of solid fibers is expressed in terms of a parameter which is one of the six unknowns of a linear system of equations. Bounds on other important derived moduli follow easily in terms of (4.12)–(4.16) and the μ_t^* bounds.[37]

It should be noted that (4.13)–(4.16) have been given in different notation in ref. 37. There were unfortunately also some misprints which were corrected in *J. Appl. Mech.*, **32,** 219 (1965). In particular (4.15)–(4.16) were given as very complex expressions. The much simpler equivalent forms (4.15)–(4.16) were later given by Hill.[38] The interested reader will find numerical plotted examples for (4.13)–(4.14) in ref. 37. Comparison with experimental results, which is quite good in many cases, may be found in ref. 39 and also in ref. 22.

It is of some interest to note that in analogy with dilute spheres suspensions it is possible to have a dilute fiber suspension. This is simply the case when v_2 is very small compared to unity and consequently v_1 is very close to unity.

Such results are easily obtained from (4.13)–(4.16) by setting $v_1 = 1$ everywhere.

As in the case of three-dimensional two-phase materials it is possible to construct bounds for fiber-reinforced materials in which the fibers have irregular shapes. Again, as in the three-dimensional case, we may consider the following problem: given a statistically homogeneous and transversely isotropic material consisting of two elastic phases. The phase geometry is cylindrical, i.e. every transverse cut of a cylindrical specimen has precisely the same phase geometry. The only information available is the elastic moduli and the volume fractions of the phases. The problem is to construct bounds in terms of this information. This problem has been treated by Hill[38] and Hashin.[25] The lower and upper bounds obtained are:

$$k^*_{t(-)} = k_1 + \frac{v_2}{\dfrac{1}{k_2-k_1} + \dfrac{v_1}{k_1+\mu_1}} \tag{4.17}$$

$$k^*_{t(+)} = k_2 + \frac{v_1}{\dfrac{1}{k_1-k_2} + \dfrac{v_1}{k_2+\mu_2}} \tag{4.18}$$

$$\mu^*_{a(-)} = \mu_1 + \frac{v_2}{\dfrac{1}{\mu_2-\mu_1} + \dfrac{v_1}{2\mu_1}} \tag{4.19}$$

$$\mu^*_{a(+)} = \mu_2 + \frac{v_1}{\dfrac{1}{\mu_1-\mu_2} + \dfrac{v_2}{2\mu_2}} \tag{4.20}$$

$$E^*_{a(-)} = E_1 v_1 + E_2 v_2 + \frac{4 v_1 v_2 (\nu_1 - \nu_2)^2}{v_1/k_2 + v_2/k_1 + 1/\mu_1} \tag{4.21}$$

$$E^*_{a(+)} = E_1 v_1 + E_2 v_2 + \frac{4 v_1 v_2 (\nu_1 - \nu_2)^2}{v_1/k_2 + v_2/k_1 + 1/\mu_2} \tag{4.22}$$

$$\nu^*_{a(-)} = \nu_1 v_1 + \nu_2 v_2 + \frac{v_1 v_2 (\nu_2 - \nu_1)(1/k_1 - 1/k_2)}{v_1/k_2 + v_2/k_1 + 1/\mu_1} \tag{4.23}$$

$$\nu^*_{a(+)} = \nu_1 v_1 + \nu_2 v_2 + \frac{v_1 v_2 (\nu_2 - \nu_1)(1/k_1 - 1/k_2)}{v_1/k_2 + v_2/k_1 + 1/\mu_2} \tag{4.24}$$

$$\mu^*_{t(-)} = \mu_1 + \frac{v_2}{\dfrac{1}{\mu_2-\mu_1} + \dfrac{(k_1+2\mu_1)v_1}{2\mu_1(k_1+\mu_1)}} \tag{4.25}$$

$$\mu^*_{t(+)} = \mu_2 + \frac{v_1}{\dfrac{1}{\mu_1-\mu_2} + \dfrac{(k_2+2\mu_2)v_2}{2\mu_2(k_2+\mu_2)}}. \tag{4.26}$$

The bounds (4.21)–(4.24) have been derived in ref. 38, the bounds (4.19)–(4.20) and (4.25)–(4.26) in ref. 25 and the bounds (4.17)–(4.18) have been given in both papers. In all bounds it is assumed that

$$\mu_2 > \mu_1, \qquad k_2 > k_1 \tag{4.27}$$

It is interesting to note that (4.17), (4.19), (4.21) and (4.23) are identical to (4.13)–(4.16) respectively, which are composite cylinder assemblage results. Similarly (4.18), (4.20), (4.22) and (4.24) may be identified with composite cylinder assemblage results in which phase 1 is a matrix and phase 2 is in the form of fibers. Thus the lower and upper bounds can be identified with exact solutions. Therefore these bounds are best possible in terms of the information used and further improvement of bounds can only be carried out on the basis of additional geometrical information. This has been shown in ref. 25. No such conclusion can at present be made for the bounds (4.25)–(4.26). It is, however, believed that they are also best possible.

It is of interest to note that the lower bound (4.25) is always above the lower bound for μ_t^* given by Hashin and Rosen[37] for the composite cylinder assemblage. Since (4.25) is based on a very general geometry which includes as a special case the composite cylinder assemblage geometry, it supersedes the μ_t^* lower bound given in ref. 37.

As for the numerical margins between the various bounds, these depend primarily on the ratios μ_2/μ_1 and k_2/k_1, i.e. on the relative stiffness of the phases. The margin between the bound pairs (4.17)–(4.18), (4.19)–(4.20) and (4.25)–(4.26) is related to these ratios in the same numerical order of magnitude as the margins between the bounds (3.12)–(3.13) and (3.15)–(3.16). In many practical applications the relative stiffness ratios of fibers and matrix are quite high (of the order of 25–50). Then the margin between the three bound pairs mentioned becomes quite appreciable and consequently no good numerical estimate is provided by these bounds. This also shows that appreciable scatter in values of the three elastic moduli K_t^*, μ_a^* and μ_t^* is to be expected when only volume fractions are controlled.

On the other hand the bounds (4.21)–(4.22) and (4.23)–(4.24) provide good estimates even when the fibers are much stiffer than the matrix. Indeed in (4.21)–(4.22) the last term can usually be neglected as numerically insignificant and with less accuracy the same can be done in (4.23)–(4.24). Thus we obtain the very important approximations

$$E_a^* \approx E_1 v_1 + E_2 v_2 \tag{4.28}$$

$$\nu_a^* \sim \nu_1 v_1 + \nu_2 v_2 \tag{4.29}$$

for arbitrary fiber geometry. Note that (4.28)–(4.29) become rigorous equalities when the Poisson's ratios of fibers and matrix are equal.

As an illustrative example Fig. 9 shows bounds for the transverse Young's modulus E_t^* (normal to the fibers) for irregular phase geometry in terms of the simplest information and also bounds based on the composite cylinder assemblage model, together with experimental results. The material consists of an epoxy matrix and boron fibers. It is seen that the margin between the irregular phase geometry bounds is very wide, that the composite cylinder assemblage bounds are very close together and that the experimental results are in good agreement with the latter bounds.

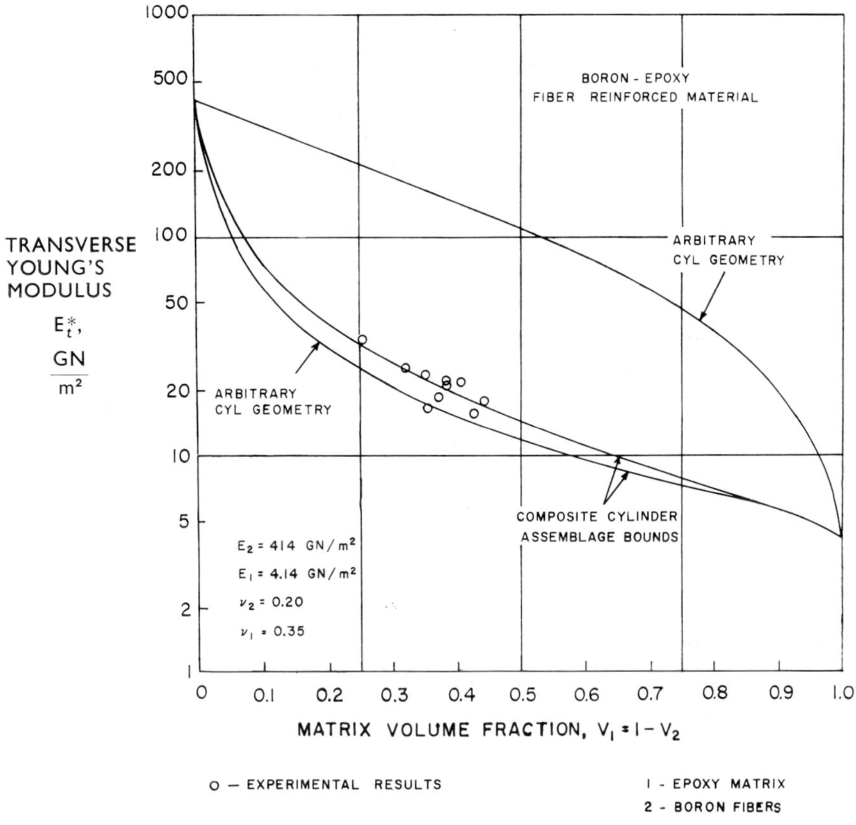

FIG. 9. Theoretical bounds and experimental results for transverse Young's modulus of fiber-reinforced material

The problem of elastic behavior of uniaxially fiber-reinforced materials has been the subject of many other papers. These may be divided into two classes: computer solutions for periodic square and hexagonal arrays of equal circular fibers and approximate treatments. It is an amazing fact that the hexagonal array computer solutions are extremely close to the results given in ref. 37 on the basis of the composite cylinder assemblage model, though the geometries involved are quite different. For comparison see, for example, Chen and Cheng.[40]

The various approximate treatments given in the literature are based on assumptions whose validity is dubious and thus a great deal of confusion has been created. In the writer's opinion the least objectionable approximation is the self-consistent scheme which has been discussed before in section 3. Self-consistent scheme results for fiber-reinforced materials will here be briefly outlined.

The method is entirely analogous to the one used for statistically isotropic composites. Hill[41] has given results on the basis of the assumption that a fiber is directly imbedded in a hypothetical transversely isotropic and homogeneous material whose elastic properties are the unknowns of the problem. All the results found lie between the general bounds (4.17)–(4.26). Moreover, they are tangent to the appropriate lower bounds for vanishing fiber volume fractions and tangent to the upper bounds for vanishing matrix volume fractions.

Recently Hermans[42] has used the modified SCS in which a composite cylinder is imbedded in the hypothetical equivalent material. The volume fractions of fiber and matrix in the composite cylinder are those of the entire composite. The results very interestingly coincide with the lower bounds (4.17), (4.19), (4.21), (4.23) and (4.25) and therefore also with the composite cylinder assemblage results except for μ_t^*. It is not difficult to show that in all cases where closed composite cylinder assemblage results are available this coincidence is to be expected. This coincidence was not noticed in ref. 42. The reader will notice the striking analogy with the relation between SCS and composite spheres assemblage results and bounds for statistically isotropic two phase materials, as discussed in section 3.

Of considerable technical importance are biaxially fiber-reinforced materials. Such materials consist of layers of uniaxially reinforced materials in which the reinforcement angle alternates from layer to layer. If the layers are many and of equal thickness the composite is orthotropic and has nine effective elastic moduli. Prediction of the four in-plane effective elastic moduli has been given in ref. 43 on the basis of the assumption that layers are homogeneous transversely isotropic with known elastic properties. A more refined analysis on the basis of the composite cylinder assemblage model for each layer has been given in ref. 44.

5. CONDUCTIVITY, DIELECTRIC CONSTANT AND MAGNETIC PERMITTIVITY OF COMPOSITES

As has been explained in section 2, the title problems are all mathematically analogous and are therefore all called for reasons of convenience—the dielectric problem.

Development of the theory of dielectric properties of composites has followed the same lines as the theory of elastic composites. Indeed almost all the results given in sections 3 and 4 have their dielectric counterparts. It will therefore be sufficient to briefly list the available results for the dielectric problem. It is also to be noted that the mathematics involved in the dielectric problem is much simpler than in the elastic case.

The first result ever given is the dielectric constant of a dilute suspension of spheres, Maxwell[45]†

$$\alpha^* = \alpha_1 \left[1 + \frac{\alpha_2 - \alpha_1}{\alpha_2 + 2\alpha_1} v_2 \right] \tag{5.1}$$

where α_2, α_1 are the dielectric constants of spherical particles and matrix respectively and v_2 is the volume fraction of particles. This result was generalized to ellipsoidal particles in ref. 46.

There followed a great deal of approximation of dubious value. For discussion see, for example, Boettcher.[47]

The fundamental difficulty of the problem was not understood until 1955 when W. F. Brown, Jr.[48] showed that the dielectric constant of a composite is dependent upon the entire statistics of the phase geometry and is therefore in itself a statistical quantity. This paper may be regarded as a milestone in the theory of composites for it showed the futility of attempts to find expressions for effective physical constants of composites in terms of the volume fractions or other simple geometrical parameters. For related investigation see Prager.[49]

Bounds for α^* of statistically isotropic two phase materials were first given by Wiener.[50] These are given by

$$\frac{1}{\dfrac{v_1}{\alpha_1} + \dfrac{v_2}{\alpha_2}} < \alpha^* < \alpha_1 v_1 + \alpha_2 v_2 \tag{5.2}$$

and are the analogue of the bounds (3.10)–(3.11).

† Actually Maxwell treated the case of a regular cubical array of spheres.

The best possible bounds in terms of volume fractions and phase dielectric constants were derived by Hashin and Shtrikman.[51] These are

$$\alpha^*_{(-)} = \alpha_1 \frac{v_2}{\dfrac{1}{\alpha_2-\alpha_1}+\dfrac{v_1}{3\alpha_1}} \tag{5.3}$$

$$\alpha^*_{(+)} = \alpha_2 + \frac{v_1}{\dfrac{1}{\alpha_1-\alpha_2}+\dfrac{v_2}{3\alpha_2}} \tag{5.4}$$

$$\alpha^*_{(-)} < \alpha^* < \alpha^*_{(+)}, \quad \alpha_2 > \alpha_1. \tag{5.5}$$

These bounds are in very good agreement with experimental results. Figure 10 gives an example for the bounds (5.3)–(5.4). Note the considerable scatter in the experimental results (circles). It is of interest to note that DeLoor[52] on the basis of extensive experimental investigation, alone, speculated that (5.3)–(5.4) are bounds for α^* before this fact was established theoretically. It was also shown in ref. 57 that (5.3) is an exact result for α^* of a composite spheres assemblage if the spherical cores are labelled 2 and the shells are labelled 1. Similarly (5.4) is an exact result for another composite spheres assemblage.

Beran and Molyneux[53] analyzed statistically the case of random dielectrics with small spatial variations of the dielectric constant. Later Beran[54] and Brown[10] gave bounds for α^* in terms of statistical information. Miller[31] exploited the results derived in ref. 54 for the special geometrical situation described as symmetric and asymmetric cell materials.

A very interesting recent development is due to Schulgasser[55] who generalized the theory given in ref. 51 for dielectrics with losses and was able to give bounds for the loss factor and the dielectric constant of such composites.

Regarding the dielectric constant of fiber-reinforced materials, the situation is much simpler than for elastic moduli of such materials. A transversely isotropic fiber-reinforced material has just the two principal dielectric constants α^*_a, in fiber directions and α^*_t, normal to the fibers. A most elementary calculation shows that

$$\alpha^*_a = \alpha_1 v_1 + \alpha_2 v_2, \tag{5.6}$$

rigorously, for arbitrary fiber shapes. The α^*_t situation is much more complicated. For a composite cylinder assemblage it was shown by Hashin,[56] that

$$\alpha^*_t = \alpha_1 + \frac{v_2}{\dfrac{1}{\alpha_2-\alpha_1}+\dfrac{v_1}{2\alpha_1}} \tag{5.7}$$

where 2 refers to fibers and 1 refers to matrix. Furthermore, it is not difficu show that the best possible bounds in terms of phase vol ume fraction and properties for α_t^* are given by

$$\alpha_{t(-)}^* = \alpha_1 + \frac{v_2}{\dfrac{1}{\alpha_2-\alpha_1}+\dfrac{v_1}{2\alpha_1}} \qquad (5.8)$$

$$\alpha_{t(+)}^* = \alpha_2 + \frac{v_1}{\dfrac{1}{\alpha_1-\alpha_2}+\dfrac{v_2}{2\alpha_2}}. \qquad (5.9)$$

The results (5.6)–(5.7) are here given for the first time. Note the complete analogy of (5.7)–(5.9) with (4.14), (4.19)–(4.20). This is no coincidence; it may be shown that the underlying mathematics in both cases is exactly the same.

The self-consistent scheme approximation is readily applicable to the dielectric problem. Results for the case when a typical spherical particle is thought to be directly imbedded in the equivalent material have been given by Landauer.[57] This result has the form

$$v_1 \frac{\alpha_1-\alpha^*}{\alpha_1+2\alpha^*} + v_2 \frac{\alpha_2-\alpha^*}{\alpha_2+2\alpha^*} = 0. \qquad (5.10)$$

In analogy to the elastic case it may be shown that (5.10) is tangent to the lower bound (5.3) for $v_2 \to 0$, tangent to the upper bound (5.4) for $v_2 \to 1$

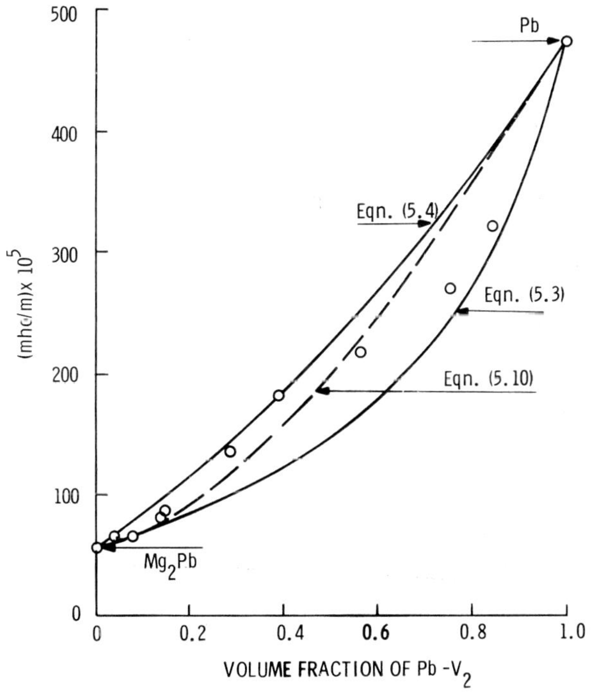

FIG. 10. Theoretical and experimental results for electrical conductivity of Mg_2Pb–Pb mixture. Experimental results given by Stepanow (see ref. 57)

and lies between the bounds for all values of v_2. An example is shown in Fig. 10 where the intermediate curve is based on (5.10). It is clearly seen that the SCS curve does not agree very well with the (scattered) experimental results. Again, as in the elastic case, the writer believes that any agreement of (5.10) with experimental results is due to the fact that it is always between the bounds (5.3)–(5.4) and therefore good agreement is to be expected only when these bounds become close. But then the approximation (5.10) is not needed.

The modified SCS used by Kerner for the elastic problem (see section 3) has also been applied by him[58] for the dielectric problem. The result is the lower bound (5.3) and again the writer was unable to understand his reasoning.

Recently the writer has applied the SCS for composite sphere embedding where ϱ obeys (3.24). It has been shown that for any such ϱ the results always lie between (5.3) and (5.10). Since it is not at all clear what the value of ϱ should be, this analysis shows that the SCS really cannot provide more information than the bounds (5.3)–(5.4).

6. VISCOELASTIC COMPOSITES

Early work on viscoelastic properties of composites has been devoted exclusively to dilute spheres suspension. Fröhlich and Sack[59] analyzed the properties of a Newtonian viscous fluid containing a small amount of elastic spheres. Oldroyd[60] treated the dilute spheres suspension problem for any viscoelastic properties of constituents and gave specific results for an elastic solid containing a small amount of spherical holes, filled with viscous fluid.

The general definition of effective relaxation moduli and creep compliances of viscoelastic composites has been first given in ref. 4 and has been discussed in section 2 in this paper. At the present time the principal tool for quantitative determination of such effective properties is a correspondence principle which relates effective mechanical properties of elastic and viscoelastic heterogeneous media. This principle has been established by Hashin[4] for statistically isotropic viscoelastic composites, but as stated there it applies equally well, with completely analogous derivation for anisotropic viscoelastic composites.

Consider two statistically homogeneous composite specimens *whose phase geometry is entirely identical*. The first specimen consists of elastic phases and the second specimen consists of viscoelastic phases (or elastic and viscoelastic phases). Let it be assumed that the effective elastic moduli C^{*E}_{ijkl} of the elastic composite specimen are known explicitly. Thus

$$C^{*E}_{ijkl} = C^{*E}_{ijkl} \text{ (phase moduli, phase geometry).} \tag{6.1}$$

Define the Laplace transforms of the effective relaxation moduli $C^*_{ijkl}(t)$ [see (2.37a), (2.40)] of the viscoelastic composite specimen by

$$\hat{C}^*_{ijkl}(p) = \int_0^\infty C^*_{ijkl}(t)e^{-pt}\,dt. \tag{6.2}$$

The correspondence principle then asserts that $\hat{C}^*_{ijkl}(p)$ are found by replacement of the phase elastic moduli in (6.1) by certain Laplace transforms of relaxation moduli of the viscoelastic phases of the viscoelastic composite specimen. Thus if (6.1) are known explicitly in some case, the Laplace transforms (6.2) can be determined and consequently the actual values of $C^*_{ijkl}(t)$ follow by inversion of the Laplace transformation.

The method is immediately applicable to composite spheres and cylinder assemblages as models for statistically isotropic viscoelastic composites and fiber-reinforced materials, respectively. These applications have been given in refs. 4, 61. Here we shall only quote some of the simplest results obtained in these investigations. Particularly simple results are obtained for viscoelastic composites consisting of a viscoelastic matrix and rigid particles. If such a statistically isotropic composite is described by a composite spheres assemblage then the effective bulk relaxation modulus [see (2.38a)] is given by

$$K^*(t) = [K_m(t)+\frac{4}{3}\,G_m(t)\,c]\frac{1}{1-c} \tag{6.3}$$

where $K_m(t)$ and $G_m(t)$ are bulk and shear relaxation moduli respectively, of the matrix and c is the volume fraction of particles. Note that (6.3) makes it possible to directly exploit measured viscoelastic matrix properties.

No similar result is available for the effective shear relaxation modulus $\mu^*(t)$ of a composite spheres assemblage. On the other hand there is the following useful result for $\mu^*(t)$ of an incompressible rigid particle viscoelastic composite for any phase geometry:

$$\frac{\mu^*(t)}{\mu_m(t)} = \frac{\mu^{*E}}{\mu_m^E}. \tag{6.4}$$

Here $\mu_m(t)$ is the shear relaxation modulus of the matrix, μ^{*E} and μ_m^E are the effective elastic shear modulus and elastic matrix shear modulus, respectively, of an elastic rigid particle composite specimen, having the same phase geometry as the viscoelastic composite specimen. In particular (6.4) shows the somewhat surprising result that the ratio between effective shear relaxation modulus and matrix relaxation modulus is not a function of time. Experimental findings of Bodner and Lifshitz[(62)] are in agreement with (6.4). Further results for effective viscoelastic properties of isotropic viscoelastic composites can be found in ref. 4.

The correspondence principle for elastic and viscoelastic composites has been exploited in ref. 61 for analysis of viscoelastic properties of fiber-reinforced materials, on the basis of the composite cylinder assemblage model. Again only a few simple results are quoted here and for further details the reader is referred to ref. 61.

In the case of rigid fibers and viscoelastic matrix, which is a good approximation for epoxy matrix and glass or boron fibers, there is a closed expression for $k_T^*(t)$ i.e. the plane strain bulk relaxation modulus,

$$k_T^*(t) = K_m(t)+\frac{1}{3}\mu_m(t)+\left[K_m(t)+\frac{4}{3}\mu_m(t)\right]\frac{c}{1-c}. \tag{6.5}$$

Here $K_m(t)$ and $\mu_m(t)$ are the bulk and shear relaxation moduli, respectively, of the isotropic matrix and c is the volume fraction of fibers.

For the axial shear relaxation modulus $\mu_a^*(t)$ there is the very simple result

$$\mu_a^*(t) = \frac{1+c}{1-c}\mu_m(t) \tag{6.6}$$

and the axial shear creep compliance $g_a^*(t)$ is given by

$$g_a^*(t) = \frac{1-c}{1+c}g_m(t) \tag{6.7}$$

where $g_m(t)$ is the shear creep compliance of the matrix. Expressions (6.5)–(6.7) make it possible to directly use measured matrix viscoelastic properties.

It has also been shown in ref. 61 that to a very good approximation the viscoelastic effect for uniaxial stress in fiber direction may be neglected.

For investigation of vibrations of viscoelastic bodies the so-called complex moduli of a viscoelastic material are needed. The reader who is not familiar with this topic is referred to Ferry[63] and Nowacki.[64] There is a well-known relation between the static relaxation moduli and the complex moduli in homogeneous viscoelastic materials.[64] It is not difficult to show that a similar relationship relates effective relaxation moduli and effective complex moduli in composites if the frequencies of vibrations are not too high, or more precisely the wavelength of the propagation of disturbance is large in comparison to the size of inhomogeneities. Let the effective complex moduli be denoted as

$$\mathcal{C}_{ijkl}^*(\iota\omega) = \mathcal{C}_{ijkl}^{*R}(\omega)+\iota\mathcal{C}_{ijkl}^{*I}(\omega) \tag{6.8}$$

where ω is the circular frequency and $\iota = \sqrt{-1}$. The following relationship can then be proved

$$\mathcal{C}_{ijkl}^*(\iota\omega) = \iota\omega\hat{C}_{ijkl}^*(\iota\omega).\dagger \tag{6.9}$$

The quantities $\hat{C}_{ijkl}^*(\iota\omega)$ in the right side of (6.9) are the Laplace transforms

† Result (6.9) can be stated in simpler form: the effective complex moduli are obtained by replacement of phase elastic moduli by phase complex moduli in expressions for effective elastic moduli of a composite with identical phase geometry.

(6.2) (supposed to be known by the correspondence principle) in which the transform variable p has been replaced by $\iota\omega$.

Some simp e applications of (6.9) will now be mentioned. For an isotropic composite consisting of rigid particles and incompressible viscoelastic matrix, the complex shear modulus is

$$\tilde{\mu}^* = \mu^{*R} + \iota\mu^{*I} \tag{6.10}$$

and the loss angle δ^* is given by

$$\tan \delta^* = \frac{\mu_C^{*R}}{\mu^{*I}}. \tag{6.11}$$

It may be shown from (6.9) and (6.4) that $\tan \delta^* = \tan \delta_m$ where δ_m is the loss angle of the matrix. Thus in the case discussed the presence of rigid particles does not affect the loss angle. This has been experimentally verified by Bodner and Lifshitz.[62]

In a fiber-reinforced material with rigid fibers described by the composite cylinder assemblage the axial complex shear modulus $\tilde{\mu}_a^*$ is simply given by

$$\tilde{\mu}_a^* = \frac{1+c}{1-c}\,\tilde{\mu}_m \tag{6.12}$$

where $\tilde{\mu}_m$ is the matrix complex shear modulus. The complex Young's modulus in fiber direction, for elastic fiber and viscoelastic matrix is given to a high degree of accuracy by

$$\tilde{E}_a^* = cE_f + (1-c)E_m^R + \iota(1-c)\,E_m^I \tag{6.13}$$

where E_f is the Young's modulus of the fibers and E_m^R and E_m^I are the real and imaginary parts, respectively, of the complex Young's modulus of the matrix.

The discussion and results concerning the complex moduli of viscoelastic composites have here been given for the first time. A detailed exposition will be published in the literature.

The reader will have noted that unlike the elastic and dielectric cases nothing has been said so far about bounding of effective viscoelastic properties of composites. The only contributions on this subject which are known to the writer have been made recently by Schulgasser[55] and the interested reader is referred to this work.

7. PLASTICITY OF COMPOSITES

The theory of plasticity is by an order of magnitude more difficult than the linear theories of dielectrics, elasticity and viscoelasticity. The same is true for problems of plastic behavior of composites.

Two general problems are of great interest:

(a) Prediction of elasto-plastic stress–strain relations of a composite.

(b) Prediction of failure load of a composite in the case when at least one phase is ideally plastic.

Problem (a) is much more difficult than problem (b). The only rigorous investigation, to the writer's knowledge, is due to Chu and Hashin who considered the behavior of a suspension consisting of elastic particles and elastoplastic matrix under hydrostatic stress, on the basis of the composite spheres assemblage model.[65, 66] It was found that even though it was assumed that the matrix behaves elastically under isotropic stress, the composite exhibits macroscopic plasticity under average hydrostatic stress (volume plasticity). The existence of such an effect is easy to understand since the presence of inclusions produces local shear stresses although the boundary loading is purely hydrostatic. Thus the macroscopic dilatational stress–strain relation may be described by a tangent bulk modulus. Let $\bar{\sigma}$ be the average hydrostatic stress and $\bar{\varepsilon}$ the average hydrostatic strain. Then

$$\bar{\sigma} = 3K_t^*(\bar{\sigma})\,\bar{\varepsilon} \tag{7.1}$$

where K_t^* is the tangent bulk modulus. Calculations showed that K_t^* varies little with $\bar{\sigma}$ if the inclusions are elastic. However, in the case of voids (porous medium) the variation of K_t^* with $\bar{\sigma}$ is appreciable resulting in a strongly nonlinear relation between $\bar{\sigma}$ and $\bar{\varepsilon}$ and thus an appreciable volumetric plastic effect is produced.

It would be of great interest to consider a similar problem for shear, but unfortunately the mathematical difficulties are so great that there seems to be little hope for any kind of solution.

Problem (b) is within the framework of the theory of limit analysis. According to this theory the limiting load of a body is reached, or plastic failure occurs, when the deformation of the body can increase without increase in load. It is quite clear, therefore, that a composite must have at least one ideally plastic (non-strain-hardening) phase in order that limit analysis methods be at all applicable. Limit analysis proceeds by construction of statically admissible stress fields and kinematically admissible velocity fields which are used to construct lower and upper bounds on the limiting load. The quality of the bounds obtained depends upon successful construction of such admissible fields, which generally requires considerable ingenuity on the part of the investigator.

The first limit analysis result for composites seems to be due to Drucker.[67] He showed that the limiting load of a composite consisting of ideally plastic matrix and rigid particles in simple tension is equal to the matrix limiting load, provided that the phase geometry is such that a plane at 45° to the simple tension direction can be passed entirely through the matrix, without passing through any particles. Similar conclusions were reached by Hashin[61] for fiber reinforced materials consisting of an ideally plastic matrix and rigid fibers. Shu and Rosen[68] exploited the limit analysis theorems to bound limiting loads for fiber-reinforced materials described by the composite cylinder assemblage model for various cases of loadings.

Drucker[13] computed the limiting load of a two-dimensional composite consisting of rigid equal hexagonal "fibers", regularly arranged and imbedded in ideally plastic matrix for the limiting case of very small matrix volume fraction.

Hashin (unpublished) used limit analysis methods to bound the yield surface of multiphase materials consisting of n ideally plastic phases which obey Mises yield conditions. If the yield condition of the rth phase is

$$s_{ij}s_{ij} = 2k_r^2 \tag{7.2}$$

then the yield surface of the composite in average stress space is bounded by two cylindrical surfaces of radii

$$k^*_{(-)} = k_1 \tag{7.3}$$

$$k^*_{(+)} = \sum_{r=1}^{r=n} k_r v_r \tag{7.4}$$

where k_1 is the smallest k_r and v_r is the volume fraction of the rth phase. The lower bound is not very interesting since it merely expresses the fact that the composite plastic strength is not less than that of the weakest phase. The upper bound is of more interest. As special cases of (7.4) it follows that the plastic strength of a composite in simple tension or pure shear is bounded from above by the volume fraction weighted averages of the simple tension or pure shear strengths, respectively, of the phases. Of particular interest is the case of a porous medium. If the plastic strength in simple tension is σ_0, the plastic strength of the porous medium σ_0^* and c the volume fraction of voids, then the theory leading to (7.4) shows that

$$\sigma^*_0 \leq \sigma(1-c). \tag{7.5}$$

On the basis of the usual experience with limit analysis results it is believed that the upper bounds obtained will be much closer to the limiting load than the trivial lower bounds.

For further discussion of plasticity of composites the papers by Drucker[13, 69] and Hill[71] are recommended. A related and very important subject is plasticity of polycrystalline aggregates, see for example the surveys ref. 2, 70.

8. CONCLUSIONS

The material which has been covered in the present survey may be regarded as a hopeful beginning of a theory of composites. Very much additional work which is based on firm theoretical foundation is needed. To discuss all the problems which merit further investigation would be an endless task. Only a few important broad areas of further research will be mentioned.

There is great need for additional development of statistical theories, for statistically homogeneous as well as for statistically nonhomogeneous fields. It seems that for this purpose theories of differential equations with stochastic coefficients are needed. Unfortunately not very much is known about this difficult subject.

Present knowledge in prediction of strength of composites is pitifully small. Yet this is one of the most important aspects of the subject.

Wave propagation and dynamic effects in composites is another very important subject which is in urgent need of theoretical development.

REFERENCES

1. Frisch, H. L. Statistics of Random Media, *Trans. Soc. Rheology*, **9,** 293 (1965).
2. Hashin, Z. Theory of Mechanical Behavior of Heterogeneous Media, *Appl. Mech. Rev.* **17,** 1 (1964).
3. Hashin, Z. Elasticity of Random Media, *Trans. Soc. Rheology*, **9,** 381 (1965).
4. Hashin, Z. Viscoelastic Behavior of Heterogeneous Media, *J. Appl. Mech.* **32,** 630 (1965).
5. Hashin, Z. Viscoelastic Behavior of Heterogeneous Media, Office of Naval Research, Contract Nonr 551 (42), Appendix B, June 1964.
6. Smyth, C. P. *Dielectric Behavior and Structure*, McGraw-Hill, New York, 1955.
7. Scaife, B. K. P. Dispersion and Fluctuations in Dielectrics, *Progress in Dielectrics*, **5,** 145 (1964).
8. Hill, R. Elastic Properties of Reinforced Solids: Some Theoretical Principles, *J. Mech. Phys. Solids*, **11,** 357 (1963).
9. Hashin, Z. Elasticity of Ceramic Systems, *Ceramic Microstructures: Their Analysis, Significance and Production*, R. M. Fulrath, and J. A. Pask, Eds. Proceedings of the 3rd Berkeley International Mechanics Conference, June 13–16, 1966.
10. Brown, W. F. Dielectric Constants, Permeabilities and Conductivities of Random Media, *Trans. Soc. Rheology*, **9,** 357 (1965).
11. Beran, M. J. Statistical Continuum Theories, *Trans. Soc. Rheology*, **9,** 339 (1965).
12. Beran, M. J. Statistical Theory of Heterogeneous Media, *Proceedings of the International Conference on the Mechanics of Composite Materials, May 8–10, 1967.*
13. Drucker, D. C. Engineering and Continuum Aspects of High Strength Materials, in *High Strength Materials*, Proc. 2nd Berkeley Int. Materials Conference, p. 795. V. F. Zackay, Ed., John Wiley, 1965.
14. Dewey, J. M. The Elastic Constants of Materials Loaded with Non-rigid Fillers, *J. Appl. Phys.* **18,** 578 (1947).
15. Eshelby, J. D. The Determination of the Elastic Field of an ellipsoidal Inclusion and Related Problems, *Proc. Roy. Soc., London (A)*, **241,** 376 (1957).
16. Hashin, Z. The Moduli of an Elastic Solid, containing Spherical Particles of Another Elastic Material, Proc. IUTAM Symp. on Nonhomogeneity in Elasticity and Plasticity, Warsaw, Poland, Pergamon Press, New York, 1959, p. 463.
17. Hashin, Z. The Moduli of an Elastic Solid Reinforced by Rigid Particles, *Bull. Res. Council Israel*, **5C,** 46 (1955).
18. Mackenzie, J. K. The Elastic Constants of a Solid containing Spherical Holes, *Proc. Phys. Soc. (B)*, **63,** 2 (1950).
19. Hashin, Z. The Elastic Moduli of Heterogeneous Materials, *J. Appl. Mech.*, **29E,** 143 (1962).

20. WALSH, J. B., BRACE, N. F. and ENGLAND, A. W. Effect of Porosity on Compressibility of Glass, *J. Am. Ceramic Soc.* **48,** 605 (1965).
21. MOLYNEUX, J. and BERAN, M. J. Statistical Properties of the Stress and Strain Fields in a Medium with Small Random Variations in Elastic Coefficients, *J. Math. Mech.* **14,** 337 (1965).
22. PAUL, B. Prediction of Elastic Constants of Multiphase Materials, *Trans. AIME,* **218,** 36 (1960).
23. HASHIN, Z. and SHTRIKMAN, S. A Variational Approach to the Theory of the Elastic Behaviour of Multiphase Materials, *J. Mech. Phys. Solids,* **11,** 127 (1963).
24. HASHIN, Z. and SHTRIKMAN, S. On Some Variational Principles in Anisotropic and Nonhomogeneous Elasticity, *J. Mech. Phys. Solids,* **10,** 335 (1962).
25. HASHIN, Z. On Elastic Behaviour of Fibre Reinforced Materials of Arbitrary Transverse Phase Geometry, *J. Mech. Phys. Solids,* **13,** 119 (1965).
26. WALPOLE, L. J. On Bounds for the Overall Elastic Moduli of Inhomogeneous Systems, *J. Mech. Phys. Solids,* **14,** 151 (1966).
27. NISHIMATSU, C. and GURLAND, J. Experimental Survey of the Deformation of the Hard-ductile Two-phase Alloy System WC-CO, *Trans. ASM,* **52,** 469 (1960).
28. HASSELMAN, D. P. H. and FULRATH, R. M. Strength and Elastic Properties of Two-phase Materials, *Ceramic Microstructures: Their Analysis, Significance and Production,* R. M. Fulrath and J. A. Pask, Eds., Proceedings of the 3rd Berkeley International Mechanics Conference, June 13–16, 1966.
29. UMEKAWA, S. and SHERBY, O. D. Young's Modulus of Eutectic Alloy Systems with Special Emphasis on the Silver-Lead System, *J. Mech. Phys. Solids,* **14,** 65 (1966).
30. BERAN, M. J. and MOLYNEUX, J. Use of Classical Variational Principles to Determine Bounds for the Effective Bulk Modulus in Heterogeneous Media, *Quart. Appl. Math.* **24,** 108 (1966).
31. MILLER, M. N. Bounds for Effective Dielectric Constant and Bulk Modulus of Heterogeneous Materials, in Terms of Statistical Information, Ph.D. Dissertation, Towne School, University of Pennsylvania, 1967.
32. BUDIANSKY, B. On the Elastic Moduli of Some Heterogeneous Materials, *J. Mech. Phys. Solids,* **13,** 223 (1965).
33. HILL, R. A Self-consistent Mechanics of Composite Materials, *J. Mech. Phys. Solids,* **13,** 213 (1965).
34. KERNER, E. H. The Elastic and Thermoelastic Properties of Composite Media, *Proc. Phys. Soc. (B),* **69,** 808 (1956).
35. VAN DER POEL, C. On the Rheology of Concentrated Dispersions, *Rheologica Acta,* **1,** 198 (1958).
36. WU, T. T. The Effect of Inclusion Shape on the Elastic Moduli of a Two-phase Material, *Int. J. Solids, Structures,* **2,** 1 (1966).
37. HASHIN, Z. and ROSEN, B. W. The Elastic Moduli of Fiber Reinforced Materials, *J. Appl. Mech.* **31,** 223 (1964).
38. HILL, R. Theory of Mechanical Properties of Fibre-strengthened Materials. I. Elastic Behaviour, *J. Mech. Phys. Solids,* **12,** 199 (1964).
39. DOW, N. F., ROSEN, B. N. and HASHIN, Z. Study of Mechanics of Filamentary Composites, NASA CR-492, June 1966.
40. CHEN, C. H. and CHENG, S. Mechanical Properties of Fiber Reinforced Composites, *J. Composite Mats.* **1,** 30 (1967).
41. HILL, R. Theory of Mechanical Properties of Fibre-strengthened Materials. III. Self-consistent Model, *J. Mech. Phys. Solids,* **13,** 189 (1965).
42. HERMANS, J. J. The Elastic Properties of Fiber Reinforced Materials when the Fibers are Aligned, *Proc. Koninkl. Nederl. Akad. Wetenschappen,* **B70,** 1 (1967).
43. DOW, N. F. and ROSEN, B. W. Evaluations of Filament Reinforced Composites for Aero-space Structural Applications, NASA CR-207, April 1965.
44. HASHIN, Z., BRULL, M. A., CHU, T. Y. and ZUDANS, Z. Determination of the Effective Elastic Properties for Biaxially Fiber Reinforced Materials, Contract DA 44-177-AMC

351 (T), USAAML, TR. 67-41, Final Report No. F-B2455, Franklin Institute Research Labs., Philadelphia, Penn., 1966.
45. MAXWELL, J. C. *Treatise on Electricity and Magnetism*, 3rd ed, Vol. 1, p. 440. Oxford University Press, 1904.
46. BURGER, H. C. Das Leitvermögen verdünnter mischkristallfreier Legierungen, *Phys. Z.* **20,** 73 (1919).
47. BOETTCHER, C. J. F. *Theory of Electric Polarisation*, Elsevier, Houston, 1952.
48. BROWN, W. F. JR., Solid Mixture Permittivities, *J. Chem. Phys.* **23,** 1514 (1955).
49. PRAGER, S. Diffusion in Inhomogeneous Media, *J. Chem. Phys.* **33,** 122 (1960).
50. WIENER, O. *Abhandl. math.–phys. Kl. Königl. Sachsischen Gesell.* **32,** 509 (1912).
51. HASHIN, Z. and SHTRIKMAN, S. A Variational Approach to the Theory of the Effective Magnetic Permeability of Multiphase Materials, *J. Appl. Phys.* **33,** 3125 (1962).
52. DELOOR, G. P. Dielectric Properties of Heterogeneous Mixtures, Thesis, University of Leiden, Leiden, 1956.
53. BERAN, M. J. and MOLYNEUX, J. Statistical Properties of the Electric Field in a Medium with Small Random Variations in Permittivity, *Nuovo Cimento*, **30,** 1046 (1963).
54. BERAN, M. J. Use of the Variational Approach to Determine Bounds for the Effective Permittivity in Random Media, *Nuovo Cimento*, **38,** 771 (1965).
55. SCHULGASSER, K. Macroscopic Behavior of Heterogeneous Hereditary Media, Dissertation, University of Pennsylvania, 1967.
56. HASHIN, Z. in Hollow Glass Fiber Reinforced Laminates, Contract NOw-63-0674-c, Final Report, MSVD, General Electric Co., Philadelphia, Penn. 1964.
57. LANDAUER, R. The Electrical Resistance of Binary Metallic Mixtures, *J. Appl. Phys.* **23,** 779 (1952).
58. KERNER, E. H. The Electrical Conductivity of Composite Media, *Proc. Phys. Soc. (B)*, **69,** 802 (1956).
59. FRÖHLICH, H. and SACK, R. Theory of the Rheological Properties of Dispersions, *Proc. Roy. Soc.* **A185,** 415 (1946).
60. OLDROYD, J. G. The Effect of Small Viscous Inclusions on the Mechanical Properties of an Elastic Solid, *Deformation and Flow of Solids*, p. 304. Grammel, Ed. Springer Verl. Berlin, 1956.
61. HASHIN, Z. Viscoelastic Fiber Reinforced Materials, *AIAA* J. **4,** 1411 (1966).
62. BODNER, S. R. and LIFSHITZ, J. Experimental Investigations on the Dynamic Strength of Composites, Proceedings of the International Conference on the Mechanics of Composite Materials, May 8–10, 1967.
63. FERRY, J. D. *Viscoelastic Properties of Polymers*, J. Wiley, New York, 1961.
64. NOWACKI, N. *Dynamics of Elastic Systems*, J. Wiley, New York, 1963.
65. CHU, T. Y. and HASHIN, Z. Plastic Behavior of Composites and Porous Media under Hydrostatic Pressure, *Proc. 5th U.S. Nat. Congress Appl. Mechanics, Minneapolis, Minn., 545, ASME, New York, 1966.*
66. CHU, T. Y. Plastic Behavior of Composites and Porous Media under Isotropic Stress, Dissertation, Towne School, Univ. of Pennsylvania, 1967.
67. DRUCKER, D. C. On Minimum Weight Design and Strength of Nonhomogeneous Plastic Bodies, in *Nonhomogeneity in Elasticity and Plasticity*, p. 139, W. Olszak, Ed., Pergamon Press, New York, 1959.
68. SHU, L. S. and ROSEN, B. W. Strength of Fiber Reinforced Composites by Limit Analysis Methods. *J. Composite Mats.* **1** 366, 1967.
69. DRUCKER, D. C. The Continuum Theory of Plasticity on the Macroscale and the Microscale, *J. Materials*, **1,** 873 (1966).
70. DORN, J. E. and MOTE, J. D. On the Plastic Behavior of Polycrystalline Aggregates, in *Materials Science Research*, Vol. I, p. 12, H. H. Stadelmaier and W. W. Austin, Eds., Plenum Press, New York, 1963.
71. HILL, R. The Essential Structure of Constitutive Laws for Metal Composites and Polycrystals, *J. Mech.-Phys. Solids* **15,** 79 (1967).

STATISTICAL THEORY OF HETEROGENEOUS MEDIA†

MARK J. BERAN

Towne School,
University of Pennsylvania,
Philadelphia, Pennsylvania

ABSTRACT

In this paper we shall first outline what we consider to be the statistical solution to a problem in which the material medium is described from a stochastic point of view. Subsequent to this we shall discuss the hierarchy of moment equations that result from averaging the basic governing equation. Here we indicate the solution for small perturbations in the elastic constants when the material is statistically homogeneous and isotropic and the average stress field is uniform. Effective constants will next be defined. Using variational principles and the statistical perturbation solutions previously obtained, expressions for bounds will be determined which contain higher order statistical information. The bulk modulus will be considered as an example. Finally, simple algebraic formulas will be obtained for the bounds by using a cell model for materials. The cell model was applied by Miller.[1]

I. INTRODUCTION

The purpose of this paper is to discuss the properties of heterogeneous media from an explicitly statistical point of view. The equations governing the displacement and shear in a linear elastic medium with spatially variable elastic constants may be written in the form

$$\mu\nabla^2 u_i+(\lambda+\mu)\frac{\partial\theta}{\partial x_i}+\frac{\partial\lambda}{\partial x_i}\,\theta+2\,\frac{\partial\mu}{\partial x_j}\,e_{ij}=0 \tag{1}$$

$$\tau_{ij}=\lambda\theta\delta_{ij}+2\mu e_{ij} \tag{2}$$

where

$u_i=$ displacement vector,

$\tau_{ij}=$ stress tensor,

$e_{ij}=\dfrac{1}{2}\left(\dfrac{\partial u_i}{\partial x_j}+\dfrac{\partial u_j}{\partial x_i}\right)=$ strain tensor,

$\theta=e_{ii}$,

$\mu=$ shear modulus,

$\lambda=k-\frac{2}{3}\mu$; $k=$ bulk modulus.

† This work was supported by the Army Research Office, Durham, N.C. Contract DA-31-124-ARO-D-340.

μ and λ are functions of position. We consider in this paper those materials for which μ and λ may be conveniently described from a statistical point of view.

We suppose that u_i is specified over some surface, S, and the problem we face is to find u_i in the interior region bounded by S.† Since we have chosen here to describe μ and λ from a statistical point of view we will also describe u_i in this fashion.

As we know from statistical mechanics and turbulence theory, it is useful to introduce the concept of an ensemble to enable us to precisely define the statistical properties of μ, λ and u_i. We shall discuss the concept of an ensemble in the next section and define the statistical properties of μ, λ and u_i that are necessary for a statistical solution to the problem. We shall point out that to obtain even so simple a quantity as the mean square value of u_i we must know an extraordinary amount of statistical information about μ and λ if we wish to obtain a solution without making approximations. To complete this section we shall discuss the notion of a hierarchy of moment equations and the use of perturbation theory to terminate the hierarchy.

Since the complete statistical solution to a problem is so difficult we are forced to settle for much less information in most problems. In particular, to partially circumvent the difficulty of dealing with elastic constants that are functions of position, many investigators replace eqs. (1) and (2) by the following equations

$$\mu^*\nabla^2\bar{u}_i+(\lambda^*+\mu^*)\frac{\partial\bar{\theta}}{\partial x_i}=0 \tag{3}$$

$$\overline{\tau_{ij}}=\lambda^*\bar{\theta}\delta_{ij}+2\mu^*\overline{e_{ij}} \tag{4}$$

where the bar indicates a mean value and μ^* and λ^* are effective constants. In this manner the detailed variations of μ and λ are smoothed over and the problem is now to determine μ^* and λ^*.

When the mean quantities like $\bar{u}_i$ vary slowly over distances in which λ and μ vary rapidly the above procedure is justified. When the boundary conditions are such that the average stress and strain fields are constants the procedure is exact. As we might expect, however, the effective constants μ^* and λ^* depend on all the statistical information needed to specify the μ and λ variations. Thus replacement of eqs. (1) and (2) by eqs. (3) and (4) has been a reformulation of the problem, not a solution.

It turns out, however, that while to determine μ^* and λ^* exactly requires considerable statistical information about the μ and λ fields, it is often possible to bound the values of μ^* and λ^* rather well with a minimum of statistical information about the μ and λ fields. This is done using variational principles. Professor Hashin in a previous paper has discussed what has been accom-

† For simplicity of presentation we do not consider boundary conditions on the tractions

plished for two-phase materials when the only information available is the volume fractions of the two phases. In section III we will show, as an example, how the bounds on the bulk modulus ($k^* = \lambda^* + \frac{2}{3}\mu^*$) may be improved if more statistical information than volume fractions is available. Although the mathematical development is involved the final results are simple algebraic formulas for a certain class of materials.

The author is aware that this symposium is intended for people of widely varying background many of whom are not particularly interested in the mathematical developments presented in the next two sections. Section IV, where final results are presented, has thus been prepared so that it may be read almost independently of the rest of the paper. It is only necessary that the reader be familiar with the concept of effective constants.

II. STATISTICAL FORMULATION

We suppose that we wish to find the properties of u_i within a material bounded by a surface S if u_i is specified on S, and μ and λ may be most conveniently treated from a statistical point of view. To do this we consider an ensemble of M material samples that are "macroscopically identical" but differ in the detailed variation of μ and λ. M is chosen to be a very large number. By macroscopically identical we mean that the surface geometry of all the samples is the same and that u_i is the same function on the surface of all the samples. In addition, the samples are prepared in the same overall manner. For example, if we consider an inclusion-matrix system we demand the volume fraction of inclusions to be the same in all samples and that the mixing procedure be the same. We suppose, however, that the preparation procedure does not determine the detailed variation of μ and λ.

The only information available to us is now statistical information. For example, to find the probability of μ having a value between μ and $\mu + d\mu$ at some point $\mathbf{x}_1$ in a sample we have not investigated (we denote this function by $P_1(\mu(\mathbf{x}_1))\, d\mu(\mathbf{x}_1)$), we must have previously measured the value of μ at $\mathbf{x}_1$ in large number of similarly made samples and determined the fraction of samples with a μ value between μ and $\mu + d\mu$.† We note that the mean value of $\mu(\mathbf{x}_1)$, $\overline{\mu(\mathbf{x}_1)}$, is found from the relation (μ is taken to be positive)

$$\overline{\mu(\mathbf{x}_1)} = \int_0^\infty \mu(\mathbf{x}_1)\, P_{1\mu}(\mathbf{x}_1))\, d\mu\, (\mathbf{x}_1). \tag{5}$$

Proceeding in a similar manner we may find

$$P_{2\mu}(\mu(\mathbf{x}_1),\ \mu(\mathbf{x}_2))\, d\mu(\mathbf{x}_1)\, d\mu(\mathbf{x}_2)$$

† We assume this fraction is well determined as the number of samples, M, approaches infinity.

defined as the probability that $\mu(\mathbf{x}_1)$ has a value between $\mu(\mathbf{x}_1)$ and $\mu(\mathbf{x}_1)+d\mu(\mathbf{x}_1)$ at point $\mathbf{x}_1$ and between $\mu(\mathbf{x}_2)$ and $\mu(\mathbf{x}_2)+d\mu(\mathbf{x}_2)$ at point $\mathbf{x}_2$. From P_2 we may calculate the correlation between $\mu(\mathbf{x}_1)$ and $\mu(\mathbf{x}_2)$ from the expression

$$\overline{\mu(\mathbf{x}_1)\,\mu(\mathbf{x}_2)} = \int_0^\infty\int_0^\infty \mu(\mathbf{x}_1)\,\mu(\mathbf{x}_2)\,P_{2\mu}(\mu(\mathbf{x}_1),\,\mu(\mathbf{x}_2))\,d\mu(\mathbf{x}_1)\,d\mu(\mathbf{x}_2). \tag{6a}$$

In general, to completely specify the statistical properties of the samples we must have determined a joint probability density of the form

$$P_{n\mu\lambda} \equiv P_n(\mu(\mathbf{x}_1),\,\ldots,\,\mu(\mathbf{x}_j),\,\lambda(\mathbf{x}_{j+1}),\,\ldots,\,\lambda(\mathbf{x}_n))$$

in the limit as $n \to \infty$. Any information less than P_n is an incomplete statistical description of a heterogeneous elastic medium.

Since μ and λ are specified statistically we can only specify the solution u_i statistically. A complete statistical description of u_i is given by the function

$$P_{nu_i\mu\lambda} \equiv P_n(u_i(\mathbf{x}_1),\,\ldots,\,u_k(\mathbf{x}_j),\,\mu(\mathbf{x}_{j+1}),\,\ldots,\,\mu(\mathbf{x}_{j+l}),\,\lambda(_{j+l+1}),\,\ldots,\,\lambda(\mathbf{x}_n))$$

in the limit $n \to \infty$. Thus from a conceptual point of view a statistical solution to the problem is the determination of $P_{nu_i\mu\lambda}$ from $P_{n\mu\lambda}$ as $n \to \infty$.

In the limit $n \to \infty$ (if we are careful of our limiting procedure) a functional equation governing $P_{nu_i\mu\lambda}$ may be derived using eq. (1). A boundary condition for the solution of this equation is $P_{n\mu\lambda}$. We note that P_{nu_i} may be obtained from $P_{nu_i\mu\lambda}$ by integration.

It is clear from the definition of $P_{n\mu\lambda}$ that the amount of information contained in this function is considerable and that practically it is not useful to consider functions like $P_{n\mu\lambda}$ and $P_{nu_i\mu\lambda}$ ($\lim n \to \infty$). Rather we would prefer to deal with quantities like

$$\bar{\mu},\ \bar{\lambda},\ \overline{\mu^2},\ \overline{\lambda^2},\ \overline{u_i},\ \overline{u_iu_j},\ \overline{u_i\mu},\ \overline{u_i\lambda}.$$

In fact it would be most desirable if, for example, we were able to obtain a solution for $\overline{u_iu_j}$ in terms of $\overline{\mu^2}$ and $\overline{\lambda^2}$. Unfortunately, it may be shown that this is not possible. In fact, if no approximations are made it can be shown that $\overline{u_iu_j}$ depends upon the entire function $P_{n\mu\lambda}$ ($n \to \infty$).

To see the difficulty that arises it is convenient to look at the hierarchy of equations governing moments of the form

$$\overline{u_i(\mathbf{x}_1)\,u_j(\mathbf{x}_2)\,\ldots\,u_k(\mathbf{x}_l)}.$$

Any set of equations we find for this moment, if l is finite, will be indeterminate if no approximations are made. We shall now briefly outline a derivation of this hierarchy for μ and λ fields that are statistically homogeneous and isotropic and for which the average stress and strain fields are uniform. Subsequent to this we shall show how to terminate the hierarchy if the fluctuations in μ and λ are small.

For convenience let us write

$$\begin{aligned}\mu &= \bar{\mu}+\mu' \\ \lambda &= \bar{\lambda}+\lambda' \\ u_i &= \overline{u_i}+u_i' \\ e_{ij} &= \overline{e_{ij}}+e_{ij}' \\ \tau_{ij} &= \overline{\tau_{ij}}+\tau_{ij}'\end{aligned}$$

where $$\overline{\mu'} = \overline{\lambda'} = \overline{u_i'} = \overline{e_{ij}'} = \overline{\tau_{ij}'} = 0.$$

We assume here that the μ' and λ' fields are statistically homogeneous and isotropic and that $\overline{e_{ij}}$ and $\overline{\tau_{ij}}$ are independent of position.

In terms of these quantities eqs. (1) and (2) yield, after some manipulation,

$$\bar{\mu}\nabla^2\bar{u}_i+(\bar{\lambda}+\bar{\mu})\frac{\partial\bar{\theta}}{\partial x_i} = 0 \tag{6b}$$

$$\bar{\mu}\nabla^2 u_i'+(\bar{\lambda}+\bar{\mu})\frac{\partial\theta'}{\partial x_i} = -\frac{\partial}{\partial x_j}(\lambda'\bar{\theta}\delta_{ij}+2\mu'\overline{e_{ij}}+\lambda'\theta'\delta_{ij}+2\mu'e_{ij}') \tag{7}$$

$$\overline{\tau_{ij}} = \bar{\lambda}\bar{\theta}\delta_{ij}+\overline{\lambda'\theta'}\delta_{ij}+2\bar{\mu}\bar{e}_{ij}+\overline{2\mu'e_{ij}'}. \tag{8}$$

Because of the fact that we have assumed that $\overline{e_{ij}}$ and $\overline{\tau_{ij}}$ are independent of position it is possible in this special case to find $\overline{u_i}$ from eq. (6b); $\overline{u_i}$ is simply a linear function of $\mathbf{x}$. If $\overline{e_{ij}}$ and $\overline{\tau_{ij}}$ were functions of position this would not be possible. We see, however, that even here $\overline{\tau_{ij}}$ cannot be found from eq. (8) since $\overline{\lambda'\theta'}$ and $\overline{\mu'e_{ij}'}$ are unknown. To find these functions we require additional equations. Additional equations may be found by multiplying eq. (7) successively by $\mu'(\mathbf{x}')$ and $\lambda'(\mathbf{x}')$ and averaging. We find then

$$\bar{\mu}\nabla^2\overline{u_i'(\mathbf{x})\mu'(\mathbf{x}')}+(\bar{\lambda}+\bar{\mu})\frac{\partial}{\partial x_i}\overline{\theta'(\mathbf{x})\mu'(\mathbf{x}')} \tag{9a}$$

$$= -\frac{\partial}{\partial x_j}\overline{\lambda'(\mathbf{x})\mu'(\mathbf{x}')}\bar{\theta}\delta_{ij}-2\frac{\partial}{\partial x_j}\overline{\mu'(\mathbf{x})\mu'(\mathbf{x}')}\;\overline{e_{ij}}$$

$$-\frac{\partial}{\partial x_j}\overline{\lambda'(\mathbf{x})\theta'(\mathbf{x}')\mu'(\mathbf{x}')}\delta_{ij}-2\frac{\partial}{\partial x_j}\overline{\mu'(\mathbf{x})e_{ij}'(\mathbf{x})\mu'(\mathbf{x}')}$$

$$\bar{\mu}\nabla^2\overline{u_i'(\mathbf{x})\lambda'(\mathbf{x}')}+(\bar{\lambda}+\bar{\mu})\frac{\partial}{\partial x_i}\overline{\theta'(\mathbf{x})\,\lambda'(\mathbf{x}')} \tag{9b}$$

$$=\frac{\partial}{\partial x_j}\overline{\lambda'(\mathbf{x})\,\lambda'(\mathbf{x}')}\bar{\theta}\delta_{ij}-2\frac{\partial}{\partial x_j}\overline{\mu'(\mathbf{x})\,\lambda'(\mathbf{x}')}\,\overline{e_{ij}}$$

$$-\frac{\partial}{\partial x_j}\overline{\lambda'(\mathbf{x})\theta'(\mathbf{x})\lambda'(\mathbf{x}')}\,\delta_{ij}-2\frac{\partial}{\partial x_j}\overline{\mu'(\mathbf{x})\,e_{ij}'(\mathbf{x})\,\lambda'(\mathbf{x}')}.$$

These equations are also indeterminate since they contain terms like $\overline{\lambda'(\mathbf{x})\theta'(\mathbf{x})\mu'(\mathbf{x}')}$. To find equations for quantities like $\overline{\lambda'(\mathbf{x})\theta'(\mathbf{x})\mu'(\mathbf{x}')}$ we multiply eq. (7) by quantities like $\lambda'(\mathbf{x}')\mu'(\mathbf{x}'')$ and average. The resulting equations are still indeterminate since these last equations contain quantities like

$$\overline{\lambda'(\mathbf{x})\theta'(\mathbf{x})\lambda'(\mathbf{x}')\mu(\mathbf{x}'')}.$$

Continuing in this manner we generate a hierarchy of equations governing the moments of a single u_i' and various combinations of μ' and λ'. To find a hierarchy of equations containing equations governing all moments of the form

$$\overline{u_i'(\mathbf{x}_1)\,\ldots\,u_j'(\mathbf{x}_l)\mu'(\mathbf{x}_{l+1})\,\ldots\,\mu'(\mathbf{x}_{l+p})\lambda'(\mathbf{x}_{l+p+1})\,\ldots\,\lambda'(\mathbf{x}_{k+p+s})},$$

it is only necessary to multiply eq. (7) by all combinations of terms like

$$u_i'(\mathbf{x}_1)\,\ldots\,u_j'(\mathbf{x}_k)\mu'(\mathbf{x}_{k+1})\,\ldots\,\mu'(\mathbf{x}_{k+q})\lambda'(\mathbf{x}_{k+q+1})\,\ldots\,\lambda'(\mathbf{x}_{k+q+r})$$

and average.

The resulting hierarchy of equations governing the moments in the limit l, p and $s\to\infty$ is considered to represent a solution to the statistical problem. Although any finite set of the hierarchy is indeterminate, the infinite set is considered determinate and equivalent to the equations governing $P_{nu_i\mu\lambda}$ (lim. $n\to\infty$).

To make the hierarchy usable it must be terminated at a low order. In most cases it is extremely difficult to accomplish this, but if the fluctuations μ' and λ' are small compared to $\bar{\mu}$ and $\bar{\lambda}$ a simple perturbation analysis allows this to be done. To lowest order a perturbation analysis implies that in eqs. (9a) and (9b) the terms like $\overline{\lambda'(\mathbf{x})\theta'(\mathbf{x})\mu'(\mathbf{x}')}$ may be neglected when compared to terms like $\overline{\lambda'(\mathbf{x})\mu'(\mathbf{x}')}\,\bar{\theta}$. If this is done consistently, a determinate set of equations result and we may solve for quantities like $\overline{u_i'(\mathbf{x})\mu(\mathbf{x}')}$ in terms of $\overline{\mu'(\mathbf{x})\mu'(\mathbf{x}')}$, $\overline{\lambda'(\mathbf{x})\mu'(\mathbf{x})}$ and $\overline{\lambda'(\mathbf{x})\lambda'(\mathbf{x}')}$. The details of this procedure are given in Molyneux and Beran.[2]

III. EFFECTIVE CONSTANTS

When the statistics of λ and μ are homogeneous and isotropic and the average stress and strain fields in a medium are uniform, we may consider eq. (4)

$$\overline{\tau_{ij}} = \lambda^* \bar{\theta} \delta_{ij} + 2\mu^* \bar{e}_{ij} \tag{4}$$

as an equation defining the effective elastic constants λ^* and μ^*. If we compare this to eq. (8)

$$\overline{\tau_{ij}} = \bar{\lambda}\bar{\theta}\delta_{ij} + \overline{\lambda'\theta'}\delta_{ij} + 2\bar{\mu}\bar{e}_{ij} + 2\overline{\mu' e_{ij}} \tag{8}$$

we see that effective constants may be defined in terms of the correlation functions $\overline{\lambda'\theta'}$ and $\overline{\mu' e'_{ij}}$.

Comparing eqs. (4) and (8) we may write for example,

$$\mu^* = \bar{\mu}(1 + \overline{\mu' e'_{ij}}/\bar{\mu}\overline{e_{ij}}); \quad \overline{e_{ij}} \neq 0 \text{ (no sum on } i, j, i \neq j). \tag{10}$$

In the case of small fluctuations λ' and μ', λ^* and μ^* may be found explicitly. When the fluctuations λ' and μ' are large, however, we do not, in general, know how to terminate the hierarchy and hence cannot find $\overline{\lambda'\theta'}$ and $\overline{\mu' e'_{ij}}$. As Professor Hashin[3] has pointed out it is possible, nevertheless, to bound λ^* and μ^*. We shall now show how this may be done using the statistical approach we have outlined above. All the results we obtain are consistent with the Hashin, Shtrikman[3] bounds.

To simplify the discussion we shall now consider bounds only on the effective bulk modulus. We first note that there is an alternate method of defining effective constants that is equivalent to eq. (4). The energy density of an elastic body is

$$w_E = \frac{1}{2} k\theta^2 + 2\mu\pi_{ij}\pi_{ij} \tag{11}$$

where

$$\pi_{ij} = e_{ij} - \tfrac{1}{3}\theta\delta_{ij}.$$

For a heterogeneous medium that is statistically homogeneous and isotropic and subject to a uniform stress and strain fields, we may write

$$\bar{w}_E = \frac{1}{2} k^* \bar{\theta}^2 + 2\mu^* \overline{\pi_{ij}}\,\overline{\pi_{ij}}. \tag{12}$$

It can be shown that, since in this case $\overline{e'_{ij}\tau'_{ij}} = 0$, the definitions given in eqs. (4) and (12) are equivalent (we note $k^* = \lambda^* + \frac{2}{3}\mu^*$).

The expression for w_E given by eq. (11) and an alternate expression in terms of τ_{ij} may be used to bound k^*. We will outline a method to obtain the upper bound on k^*. The lower bound calculation proceeds similarly. The details of the method used to obtain both bounds are given in Beran and Molyneux.[4]

From the principle of minimum potential energy we find that subject to the subsidiary condition,

$$e_{ij} = \frac{1}{2}\left(\frac{\partial u_i}{\partial x_j}+\frac{\partial u_j}{\partial x_i}\right),$$

$$\tilde{w}_E = \frac{1}{V}\int\limits_V w_E \, d\mathbf{x}$$

is a minimum and stationary for strains satisfying the condition

$$\frac{\partial}{\partial x_j}(k\theta\delta_{ij}+2\mu\pi_{ij}) = \frac{\partial \tau_{ij}}{\partial x_j} = 0.$$

V is the volume of the body and we must assume that $\delta u_i = 0$ on S. For any admissible function, i.e., a function satisfying the subsidiary condition and the condition $\delta u_i = 0$, we have

$$\tilde{w}_E \leqslant \frac{1}{2V}\int (k\theta_A^2+2\mu\pi_{ijA}\pi_{ijA})\, d\mathbf{x}. \tag{13}$$

Here
$$e_{ijA} = \frac{1}{2}\left(\frac{\partial}{\partial x_j}u_{iA}+\frac{\partial}{\partial x_i}u_{jA}\right).$$

Since we are considering statistically homogeneous media subjected to uniform stress and strain fields we assume an ergodic type hypothesis:

$$\tilde{w}_E = \bar{w}_E. \tag{14}$$

In addition we replace the condition $\delta u_i = 0$ on S by the condition $\bar{u}_{iA} = \bar{u}_i$. Then we find

$$\tfrac{1}{2}k^*\bar{\theta}^2+2\mu^*\bar{\pi}_{ij}\bar{\pi}_{ij} \leqslant \overline{\tfrac{1}{2}k\theta_A^2+2\mu\pi_{ijA}\pi_{ijA}}. \tag{15}$$

To isolate the bulk modulus we choose a field of the form

$$u_i = -\beta x_i.$$

This yields

$$k^* \leqslant \frac{1}{\bar{\theta}^2}\,\overline{k\theta_A^2+2\mu\pi_{ijA}\pi_{ijA}}. \tag{16}$$

An upper bound on k^* is found by choosing a suitable admissible field e_{ijA}.

If e_{ijA} is chosen to be a constant we find after some manipulation that

$$k^* \leqslant \bar{k}. \tag{17}$$

(See Paul[5] for this type of calculation.) Determination of $\bar{k}$ requires a knowledge of only the volume fractions of the two phases. To determine a bound on k^* in terms of higher-order statistical information we use an admissible solution with more information. This may be done by choosing e_{ijA} to be the first-order perturbation solution obtained from eq. (7) (actually, this solution with an adjustable multiplicative constant). If this solution is substituted in eq. (16) we need only calculate quantities of the form

$$\overline{\theta'^2}, \quad \overline{k'\theta'}, \quad \overline{\pi'_{ij}\pi'_{ij}}.$$

Those quantities may be obtained by solving equations like eqs. (9a) and (9b), neglecting third-order terms. The details of the calculations are very lengthy and we list only the result. We find

$$k^* \leqslant \bar{k} - \frac{\overline{k'^2}}{\bar{\lambda}+2\bar{\mu}+\dfrac{\overline{\mu' k'^2}}{\overline{k'^2}}+\dfrac{2J}{\overline{k'^2}}} \tag{18}$$

where

$$J = \frac{1}{16\pi^2}\int_\xi\int_\eta \frac{\partial^2}{\partial\xi_i\,\partial\xi_j}\,\frac{\partial^2}{\partial\eta_i\,\partial\eta_j}\,\overline{(\mu'(0)k'(\xi)k'(\eta))}\,\frac{\mathrm{d}\xi\,\mathrm{d}\eta}{|\xi|\,|\eta|}\,. \tag{19}$$

A lower bound may be found from similar calculations using the principle of complementary energy. We find here

$$\left\{\overline{\frac{1}{k}} - \frac{\left(\overline{\dfrac{k'}{k}}\right)^2}{\overline{\dfrac{k'^2}{k}}-\dfrac{3}{8}\,\overline{\dfrac{k'^2}{\mu}}+\dfrac{9}{8}J'}\right\}^{-1} \leqslant k^* \tag{20}$$

where

$$J' = \frac{1}{16\pi^2}\iint_{\xi\,\eta} \frac{\partial^2}{\partial\xi_i\,\partial\xi_j}\,\frac{\partial^2}{\partial\eta_i\,\partial\eta_j}\,\overline{\frac{k'(\xi)\,k'(\eta)}{\mu(0)}}\,\frac{d\xi\,d\eta}{|\xi|\,|\eta|}\,. \tag{21}$$

Equations (18) and (19) determine the bounds in terms of the two correlation functions

$$\overline{\mu'(0)\,k'(\xi)\,k'(\eta)}$$

and

$$\overline{\frac{k'(\xi)\,k'(\eta)}{\mu(0)}}\,.$$

Presently, at the University of Pennsylvania, experiments are being conducted by Corson to evaluate these functions for Pb–Al and Pb–Fe mixtures. From the measured functions J and J' the bounds on k^* will be calculated and a comparison will be made to see if the experimentally determined values of k^* lie within the bounds.

An alternate approach has been used by Miller.[1] He considers a class of two-phase materials that may conveniently be described in terms of a cell model. His work is outlined in the next section.

IV. EVALUATION OF THE BOUNDS ON k^* USING A PARTICULAR CELL MODEL

To evaluate eqs. (18) and (19), Miller assumes that a two-phase material may be described by a system of cells like those pictured in Fig. 1. The cross-hatched cells have properties λ_1, μ_1, k_1 and occupy the volume fraction v_1; the plain cells have properties λ_2, μ_2, k_2 and occupy the volume fraction v_2.

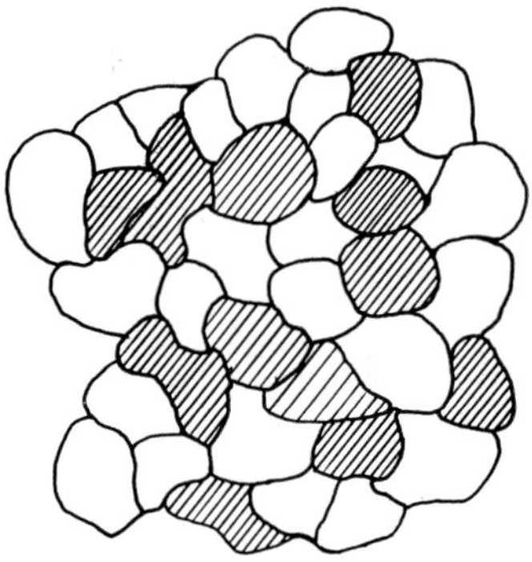

FIG. 1. Cellular material

The geometrical character of the two-cell types may be quite different, but Miller makes the very important assumption that the probability of the material property of a cell having a particular value is statistically independent of the material property of an adjacent cell. With this assumption he is able to show that

$$\overline{\mu'(0)\,k'(\xi)\,k'(\eta)} = \mu_1' k_1'^2 \phi \left[g_1 - \frac{\phi^2}{(1-\phi)^2}\, g_2 \right] \tag{22}$$

$$\frac{\overline{k'(\xi)\,k'(\eta)}}{\mu(0)} = \frac{k_1'^2}{\mu_1}\,\frac{\phi}{(1-\phi)}\,\{(\beta-1)\,[\phi^2 g_2-(1-\phi)^2 g_1] + (\beta+\phi(1-\beta))\,[(1-\phi)f_1^{11}+\phi f_2^{11}]\} \tag{23}$$

where

$$\mu' = \mu_1 - \bar{\mu},$$
$$k_1' = k_1 - \bar{k},$$
$$\beta = \mu_1/\mu_2,$$

ϕ = volume fraction of material "one",

g_j = the conditional probability that three points (coordinates 0, ξ, η) are in the same cell of material property "j" given that one of the points is in a cell with material property "j",

f_j^{11} = the conditional probability that two points (coordinates ξ, η) are in the same cell of material property "j" given that one of the points is in a cell with material property "j".

Equations (22) and (23) may be substituted into eqs. (18). We find

$$\frac{k^*}{(k_1k_2)^{1/2}} \leqslant \frac{1}{\alpha^{1/2}}\left[1-\phi(\alpha-1)- \frac{\phi(1-\phi)\alpha^2}{\alpha-(\alpha-1)\,\phi+2\gamma\left[1-\frac{4}{3}\phi+\frac{\beta}{3}(4\phi-1)+3(\beta-1)\,(G_1(1-\phi)^2-G_2\phi^2)\right]}\right] \tag{24}$$

$$\frac{k^*}{(k_1k_2)^{1/2}} \geqslant \frac{\alpha^{1/2}}{\phi(1-\alpha)+\alpha-\dfrac{\phi(1-\phi)\,(\alpha-1)^2}{1+\phi(\alpha-1)+\dfrac{3\alpha}{8\beta\gamma}[3(\beta-1)(3[G_2\phi^2-G_1(1-\phi)^2])-\frac{4}{3}\phi+3[\beta-1])^2]}} \tag{25}$$

where

$$\alpha = \frac{k_1}{k_2},$$

$$\gamma = \frac{\mu_2}{k_2}.$$

G_1 and G_2 are two numbers which represent the geometry of the cells. Miller has shown that if the upper bound is to be greater than zero and the lower bound to be finite

$$\tfrac{1}{9} \leqslant G_1 \leqslant \tfrac{1}{3},$$
$$\tfrac{1}{9} \leqslant G_2 \leqslant \tfrac{1}{3}.$$

For these values the bounds on k^* always fall within the bounds derived by

Hashin and Shtrikman.[3] A plot of the bounds on k^* is given in Fig. 2 for $\alpha = 10, \beta = 10, \gamma = 1.0$. Many other plots may be found in Miller's thesis.

For materials that satisfy Miller's independence hypothesis G_1 and G_2 may be easily found. For spherical cells, $G = \frac{1}{9}$; for plate-like cells, $G = \frac{1}{3}$. The reader is cautioned to note, however, that materials composed of inclusions in a matrix probably do not satisfy the independence hypothesis for intermediate values of ϕ.

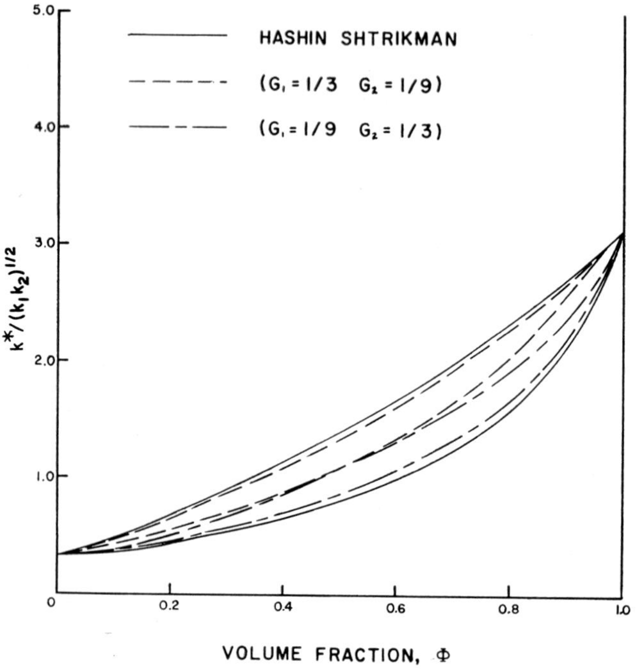

FIG. 2. Bounds on effective bulk modulus asymmetric cell material for $\alpha = \beta = 10$, $\gamma = 1.0$

REFERENCES

1. MILLER, M. Ph.D. Thesis, Towne School, Univ. of Penna. (1967).
2. MOLYNEUX, J. and BERAN, M. *J. Math. Mech.* **14**, 337 (1965).
3. HASHIN, Z. and SHTRIKMAN, S. *J. Mech. Phys. Solids* **11**, 127 (1963).
4. BERAN, M. and MOLYNEUX, J. *Quart. J. Applied Math.* **24**, 107 (1966).
5. PAUL, B. *Trans. Met. Soc. AIME* **218**, 36 (1960).

STRESS–STRAIN RELATIONS FOR COMPOSITE MATERIALS IN SHELLS OF ARBITRARY GEOMETRY

MAURICE A. BRULL†

University of Pennsylvania,
Philadelphia, Pennsylvania

I. INTRODUCTION

The design of shell structures constructed of fiber-reinforced materials requires that rational methods be available to calculate the stiffness properties of the shell. The objective of this paper is to present simple methods of calculating these stiffness properties, starting from bounds on elastic moduli derived by Hashin and Rosen[(1)] for unidirectionally reinforced materials.

The development is based on the use of a layered shell model as well as the usual assumptions underlying the classical theory of shells.[(2, 3)] In particular, it will be shown that a coupling between membrane and flexural variables appears in the constitutive relations. For homogeneous shells, it is well known[(2, 3)] that such a coupling exists only if higher order geometric terms are retained in the constitutive relations. For composite shells, however, this effect is a direct result of material behavior and is present in plates as well as in shells of any order of "thinness".

II. GENERAL REMARKS

In the remainder of this paper, we will make use of a tensor notation in which all quantities will be surface tensors referred to base vectors $\mathbf{a}^\alpha$ or $\mathbf{a}_\alpha$ ($\alpha = 1, 2$) in the shell reference surface and (where necessary) to a unit vector $\mathbf{a}^3$ or $\mathbf{a}_3$ normal to the reference surface. All tensor transformations are within the surface and the metric is that associated with the first fundamental form. Greek indices take on the values 1, 2, and the summation convention applies throughout.

Based upon the Kirchhoff kinematic hypothesis, the displacement vector for any point of the shell is taken in the form[(3)]

$$\mathbf{u} = (v_\alpha + z\beta_\alpha)\,\mathbf{a}^\alpha + w\mathbf{a}^3, \tag{1}$$

† Professor and Chairman, Division of Engineering Mechanics.

where $\mathbf{u}$ = displacement vector,
v_α = covariant components of the membrane displacement,
β_α = covariant components of the change of inclination of the normal to the shell reference surface,
z = coordinate in the direction of the normal to the shell reference surface,
w = normal displacement of the reference surface.

From the neglect of transverse shear deformation, we obtain the following relations between the rotation β_α and the displacements:

$$\beta_\alpha = -w_{,\alpha} - b_\alpha^\nu v_\nu, \tag{2}$$

where b_α^ν (or $b_{\alpha\beta}$) are the coefficients of the second fundamental form of the reference surface and are therefore related to the principal radii of curvature; the comma denotes partial differentiation.

With these assumptions it can easily be shown that the covariant components of strain may be expressed in the form[3]

$$\varepsilon_{\alpha\beta} = p_{\alpha\beta} + zq_{\alpha\beta}, \tag{3}$$

where $\varepsilon_{\alpha\beta}$ are the covariant components of strain.

The tensors $p_{\alpha\beta}$ and $q_{\alpha\beta}$ are the natural strain measures of shell theory, which can easily be expressed in terms of the displacements; the result is:

$$p_{\alpha\beta} = \tfrac{1}{2}[v_{\alpha/\beta} + v_{\beta/\alpha} - (b_{\alpha\beta} - b_{\beta\alpha})\, w], \tag{4}$$

$$q_{\alpha\beta} = \tfrac{1}{2}[\beta_{\alpha/\beta} + \beta_{\beta/\alpha} - b_\alpha^\nu v_{\nu/\beta} - b_\beta^\nu v_{\nu/\alpha} + (b_\beta^\nu b_{\nu\beta} + b_\beta^\nu b_{\nu\alpha})\, w]. \tag{5}$$

In eqs. (4) and (5) the notation $v_{\alpha/\beta}$ denotes the β covariant derivative of v_α with respect to the surface metric $a_{\alpha\beta}$. With these definitions for strain measures, it can be shown that strain energy density (per unit reference area) for the shell may be expressed in the form

$$W = \tfrac{1}{2}[D_0^{\nu\lambda\varepsilon\varkappa} p_{\nu\lambda} p_{\varepsilon\varkappa} + 2D_1^{\varepsilon\varkappa\nu\lambda} q_{\nu\lambda} p_{\varepsilon\varkappa} + D_2^{\nu\lambda\varepsilon\varkappa} q_{\nu\lambda} q_{\varepsilon\varkappa}], \tag{6}$$

where the quantities $D_n^{\nu\lambda\varepsilon\varkappa}$ $(n = 0, 1, 2)$ are the various shell stiffness tensors defined below. The constitutive relations must, for consistency with the laws of thermodynamics, take the form

$$N^{\alpha\beta} = \frac{\partial w}{\partial p_{\alpha\beta}}, \qquad M^{\alpha\beta} = \frac{\partial w}{\partial q_{\alpha\beta}}, \tag{7}$$

where $N^{\alpha\beta}$ and $M^{\alpha\beta}$ are suitably defined stress resultant and stress couple tensors (both symmetric). From eqs. (6) and (7), we may write the constitutive relations as

$$N^{\alpha\beta} = D_0^{\alpha\beta\nu\lambda} p_{\nu\lambda} + D_1^{\alpha\beta\nu\lambda} q_{\nu\lambda}, \tag{8}$$

$$M^{\alpha\beta} = D_1^{\nu\lambda\alpha\beta} p_{\nu\lambda} + D_2^{\alpha\beta\nu\lambda} q_{\nu\lambda}. \tag{9}$$

It should be noted that (due to the symmetry of the strain measures, stress resultants and stress couples) the tensors $D_n^{\alpha\beta\gamma\delta}$ satisfy the symmetry relations

$$D_n^{\alpha\beta\gamma\delta} = D_n^{\beta\alpha\gamma\delta} = D_n^{\alpha\beta\delta\gamma}. \tag{10}$$

In addition, in order for the strain energy function to be positive definite, we must have

$$D_0^{\alpha\beta\gamma\delta} = D_0^{\gamma\delta\alpha\beta}, \quad D_2^{\alpha\beta\gamma\delta} = D_2^{\gamma\delta\alpha\beta}. \tag{11}$$

It is interesting to note that the tensor $D_1^{\alpha\beta\gamma\delta}$ need not exhibit this last type of symmetry (with respect to pairs of indices). The form of the shell stiffness tensors may easily be derived by integration of the constitutive relations of linear elasticity across the shell thickness.[3] The result of this integration is outlined below:

$$D_0^{\nu\lambda\varepsilon\varkappa} = E_0^{\nu\lambda\varepsilon\varkappa} + \tfrac{1}{2}\,[b_\alpha^\varkappa(B_1^{\nu\lambda\varepsilon\alpha} + B_1^{\lambda\nu\varepsilon\alpha}) + b_\alpha^\varepsilon(B_1^{\nu\lambda\varkappa\alpha} + B_1^{\lambda\nu\varkappa\alpha}) + b_\alpha^\lambda(B_1^{\nu\alpha\varepsilon\varkappa} + B_1^{\nu\alpha\varkappa\varepsilon}) + b_\alpha^\nu(B_1^{\lambda\alpha\varepsilon\varkappa} + B_1^{\lambda\alpha\varkappa\varepsilon})], \tag{12}$$

$$D_1^{\varepsilon\varkappa\nu\lambda} = \tfrac{1}{2}\,[B_1^{\varepsilon\varkappa\nu\lambda} + B_1^{\varepsilon\varkappa\lambda\nu} + b_\gamma^\varepsilon(B_2^{\varkappa\gamma\nu\lambda} + B_2^{\varkappa\gamma\lambda\nu})], \tag{13}$$

$$D_2^{\nu\lambda\varepsilon\varkappa} = \tfrac{1}{4}\,[B_2^{\nu\lambda\varepsilon\varkappa} + B_2^{\nu\lambda\varkappa\varepsilon} + B_2^{\lambda\nu\varepsilon\varkappa} + B_2^{\lambda\nu\varkappa\varepsilon}]. \tag{14}$$

In eqs. (12) to (14) the various quantities are defined as

$$E_0^{\nu\lambda\varepsilon\varkappa} = \int_h \mu \bar{A}^{\nu\lambda\varepsilon\varkappa}\, dz, \tag{15}$$

$$B_n^{\nu\lambda\varepsilon\varkappa} = \int_h \mu\bar{\mu}_\pi^\lambda \bar{\mu}_\varrho^\varkappa \bar{A}^{\nu\pi\varepsilon\varrho} z^n\, dz \qquad (n = 0,\ 1,\ 2), \tag{16}$$

where the integrations are carried out over the shell thickness. In these equations, we have introduced the shifters μ_β^α, defined as

$$\mu_\beta^\alpha = \delta_\beta^\alpha - x^3 b_\mu^\alpha, \tag{17}$$

while the "inverse shifters" are defined by the equation

$$\mu_\beta^\alpha \bar{\mu}_\gamma^\beta = \delta_\gamma^\alpha, \tag{18}$$

and we have used the notation

$$\mu = |\mu_\beta^\alpha|. \tag{19}$$

In eqs. (15) and (16) the quantities $\bar{A}^{\nu\lambda\varepsilon\varkappa}$ are the shifted components of the elasticity tensor for the material; the term "shifted" indicating that these components are associated with a system of base vectors consisting of the reference surface base vectors and the unit vector normal to the middle surface.

It should be noted that the integrations, indicated in eqs. (15) and (16), can be carried out exactly even though the shifters $\bar{\mu}^{\alpha}_{\beta}$ are functions of the thickness coordinate z. For many cases of practical interest, however, it is permissible to neglect terms of order h/R in comparison to unity where h denotes the shell thickness and R the least radius of curvature. The result of such a procedure is a first order shell theory which is in wide use. In the remainder of the present work we will restrict our consideration to such a first order theory. Introducing the approximation $h/R \ll 1$ ($\mu^{\alpha}_{\beta} = \delta^{\alpha}_{\beta}$) in eqs. (15) and (16) leads to the following results for a single-layer homogeneous shell:

$$
\begin{aligned}
D_0^{\alpha\beta\gamma\delta} &= \bar{A}^{\alpha\beta\gamma\delta} h, \\
D_1^{\alpha\beta\gamma\delta} &= 0, \\
D_2^{\alpha\beta\gamma\delta} &= \bar{A}^{\alpha\beta\gamma\delta} \frac{h^3}{12} .
\end{aligned}
\tag{20}
$$

Thus, the constitutive relations are now reduced to the form

$$
\begin{aligned}
N^{\alpha\beta} &= D_0^{\alpha\beta\gamma\delta} p_{\gamma\delta} = \bar{A}^{\alpha\beta\gamma\delta} h p_{\gamma\delta}, \\
M^{\alpha\beta} &= D_0^{\alpha\beta\gamma\delta} p_{\gamma\delta} = \bar{A}^{\alpha\beta\gamma\delta} \frac{h^3}{12} q_{\gamma\delta}.
\end{aligned}
\tag{21}
$$

III. STIFFNESS TENSOR FOR FIBER-REINFORCED SHELLS

In this section we consider shells constructed of a fiber-reinforced material and apply the general theory outlined in the previous section to the calculation of the stiffness tensors for such shells. We restrict ourselves to the first order shell theory, for which the assumptions $h/R \ll 1$ or $\mu^{\alpha}_{\beta} \approx \delta^{\alpha}_{\beta}$ hold. With these assumptions, it is clear that eqs. (12) through (16) reduce to the following relations:

$$D_0^{\nu\lambda\varepsilon\varkappa} = \int_h \bar{A}^{\nu\lambda\varepsilon\varkappa} \, dz, \tag{22}$$

$$D_1^{\nu\lambda\varepsilon\varkappa} = \int_h \bar{A}^{\nu\lambda\varepsilon\varkappa} z \, dz, \tag{23}$$

$$D_2^{\nu\lambda\varepsilon\varkappa} = \int_h \bar{A}^{\nu\lambda\varepsilon\varkappa} z^2 \, dz. \tag{24}$$

It should be noted that these expressions apply directly to plates, and that the elastic constants $\bar{A}^{\nu\lambda\varepsilon\varkappa}$ are not modified by the shifting operation, since for a plate $\delta^{\alpha}_{\beta} = \mu^{\alpha}_{\beta}$.

In order to apply these results to fiber-reinforced shell structures, it is convenient to introduce a layered shell model. We thus consider the structure to be composed of m layers, each of which is unidirectionally reinforced and has fibers oriented at an angle θ_i from the coordinate directions. The elastic constants $\bar{A}^{\nu\lambda\varepsilon\varkappa}$ are then clearly functions of the normal coordinate z. If we denote by h_i the thickness of the ith layer and by z_i the distance of the middle surface of this layer from the reference surface, eqs. (22) through (24) become:

$$D_0^{\nu\lambda\varepsilon\varkappa} = \sum_{i=1}^{m} \bar{A}_{(i)}^{\nu\lambda\varepsilon\varkappa} h_i, \tag{25}$$

$$D_1^{\nu\lambda\varepsilon\varkappa} = \sum_{i=1}^{m} \bar{A}_{(i)}^{\nu\lambda\varepsilon\varkappa} z_i h_i \tag{26}$$

$$D_2^{\nu\lambda\varepsilon\varkappa} = \sum_{i=1}^{m} \bar{A}_{(i)}^{\nu\lambda\varepsilon\varkappa} \left[z_i^2 h_i + \frac{h_i^3}{12} \right]. \tag{27}$$

In these equations $\bar{A}_{(i)}^{\nu\lambda\varepsilon\varkappa}$ denotes the shifted components of the elasticity tensor for the ith layer referred to shell coordinates.

We now consider the ith layer of the shell which consists of a unidirectionally reinforced material. Such a material is transversely isotropic with the axis of isotropy parallel to the fiber direction. For such a material, there are five nonvanishing distinct components of the elasticity tensor referred to the natural axes. Among these constants, the ones entering into the shell stiffnesses are $\tilde{A}^{1111}$, $\tilde{A}^{2222}$, $\tilde{A}^{1122} = \tilde{A}^{2211}$, $\tilde{A}^{1212} = \tilde{A}^{1221} = \tilde{A}^{2121} = \tilde{A}^{2112}$; when the notation $\tilde{A}^{\alpha\beta\gamma\delta}$ denotes components of the elasticity tensor referred to material axes (i.e., one axis along the fibers and any two axes in the plane perpendicular to the fiber direction). It is assumed that the fiber direction is tangent to the shell reference surface. It has been shown by Hashin and Rosen[(1)] that four of the elastic constants for such a unidirectionally reinforced material may be calculated exactly, while the fifth may be bounded. In order to make use of these results in eqs. (25) to (27), we must first transform the elasticity tensor to the shell coordinates.

The transformation law for a surface tensor of rank 4 under a rotation of coordinates is written in the form

$$\bar{A}^{\nu\lambda\varepsilon\varkappa} = \alpha_\gamma^\nu \alpha_\eta^\lambda \alpha_\pi^\varepsilon \alpha_\varrho^\varkappa \tilde{A}^{\gamma\eta\pi\varrho}, \tag{28}$$

where, for the special transformation under consideration, the coefficients α_γ^ν are the direction cosines of the local tangents to the shell coordinate lines with respect to the local material axes. If the fibers of the ith shell layer are oriented at an angle θ_i to the local x^1 direction in the shell surface coordinates, the coefficients α_γ^ν are given by

$$\alpha_1^1 = \cos\theta_i, \quad \alpha_2^1 = -\sin\theta_i, \quad \alpha_1^2 = \sin\theta_i, \quad \alpha_2^2 = \cos\theta_i.$$

Making use of these coefficients in eqs. (28), we obtain the components of the elasticity tensor for the ith layer, referred to shell coordinates; the results are given in the matrix equation shown on p. 260.

$$
\begin{bmatrix} \bar{A}^{1111} \\ \bar{A}^{2222} \\ \bar{A}^{1122} \\ \bar{A}^{1212} \\ \bar{A}^{1211} \\ \bar{A}^{1222} \end{bmatrix}
=
\begin{bmatrix}
(\cos^4 \theta_i) & (\sin^4 \theta_i) & (4 \sin^2 \theta_i \cos^2 \theta_i) & 2 \sin^2 \theta_i \cos^2 \theta_i \\
(\sin^4 \theta_i) & (\cos^4 \theta_i) & (4 \sin^2 \theta_i \cos^2 \theta_i) & 2 \sin^2 \theta_i \cos^2 \theta_i \\
(\sin^2 \theta_i \cos^2 \theta_i) & (\sin^2 \theta_i \cos^2 \theta_i) & (-4 \sin^2 \theta_i \cos^2 \theta_i) & \sin^4 \theta_i + \cos^4 \theta_i \\
(\sin^2 \theta_i \cos^2 \theta_i) & (\sin^2 \theta_i \cos^2 \theta_i) & (\sin^2 \theta_i - \cos^2 \theta_i)^2 & -2 \sin^2 \theta_i \cos^2 \theta_i \\
(\sin \theta_i \cos^3 \theta_i) & (-\sin^3 \theta_i \cos \theta_i) & [2 \sin \theta_i \cos \theta_i (\sin^2 \theta_i - \cos^2 \theta_i)] & [\sin \theta_i \cos \theta_i (\sin^2 \theta_i - \cos^2 \theta_i)] \\
(\sin^3 \theta_i \cos \theta_i) & (-\sin \theta_i \cos^3 \theta_i) & [2 \sin \theta_i \cos \theta_i (\cos^2 \theta_i - \sin^2 \theta_i)] & [\sin \theta_i \cos \theta_i (\cos^2 \theta_i - \sin^2 \theta_i)]
\end{bmatrix}
\begin{bmatrix} \tilde{A}^{1111} \\ \tilde{A}^{2222} \\ \tilde{A}^{1212} \\ \tilde{A}^{1122} \end{bmatrix}
\tag{29}
$$

It is of interest to note that even though the fiber-reinforced material is orthotropic, the constants $\bar{A}^{1211} = \bar{A}^{1112} = \bar{A}^{2111} = \bar{A}^{1121}$ and $\bar{A}^{1222} = = \bar{A}^{2212} = \bar{A}^{2122} = \bar{A}^{2221}$ do not vanish when the elasticity tensor is referred to shell coordinates. Thus, a single-layer shell will be orthotropic only if the fiber direction is aligned with one of the shell coordinate directions. This character is, in fact, intuitively obvious, and a simple example would be a fiber-reinforced cylinder. It is clear that if the angle between the fiber direction and the generator direction does not vanish (this corresponds to a helical fiber wrapping) the shell will twist under internal pressure, a behavior characteristic of an anisotropic shell rather than an orthotropic one. This special type of anisotropy (often called skew- or rotated-orthotropy) is well known and Kingsbury[(4)] has obtained solutions for cylindrical shells exhibiting this behavior and has also indicated a method of obtaining solutions for general shells of revolution.

We note that components $\bar{A}^{1111}$, $\bar{A}^{1122}$, $\bar{A}^{2222}$ and $\bar{A}^{1212}$ of the elasticity tensor depend only on even powers of sin θ_i and are therefore even functions of θ_i. On the other hand, the constants $\bar{A}^{1211}$ and $\bar{A}^{1222}$ vary as odd powers of sin θ_i and are therefore odd functions of θ_i. It will be of interest to determine the effects of variations of θ_i across the shell thickness on these elastic constants. For this purpose we consider a shell composed of an even number of layers of equal thickness, and we choose the shell middle surface as a reference surface. We first consider a construction in which the fiber angle θ is an odd function of the thickness coordinate z. It is clear from our previous discussion that, for this case the moduli $\bar{A}^{1111}$, $\bar{A}^{1122}$, $\bar{A}^{2222}$ and $\bar{A}^{1212}$ will be even functions of z, while the moduli $\bar{A}^{1211}$ and $\bar{A}^{1222}$ will be odd functions of z. Using these results in eqs. (25), (26) and (27) (or (22) through (24)) we can determine the effect of this type of construction on the components of the shell stiffness tensors $D_n^{\nu\lambda\varepsilon\varkappa}$ ($n = 0, 1, 2$). Thus, we see that for θ odd in z, the nonzero components of $D_0^{\nu\lambda\varepsilon\varkappa}$ and $D_2^{\nu\lambda\varepsilon\varkappa}$ are

$$D_n^{1111},\ D_n^{1122},\ D_n^{2222},\ D_n^{1212} \qquad (n = 0, 2)$$

while the components D_n^{1211}, D_n^{1222} ($n = 0, 2$) vanish. However, the situation is different for $D_1^{\nu\lambda\varepsilon\varkappa}$. It is clear from eq. (26) or (23) that the components D_1^{1111}, D_1^{1122}, D_1^{2222} and D_1^{1212} vanish while the components D_1^{1211} and D_1^{1222} are nonzero. Thus, for this type of construction, the tensors $D_0^{\nu\lambda\varepsilon\varkappa}$ and $D_2^{\nu\lambda\varepsilon\varkappa}$ have the nonzero components corresponding to an orthotropic shell, but $D_1^{\nu\lambda\varepsilon\varkappa}$ exhibits anisotropic behavior and furthermore provides a coupling between the "membrane" and "flexural actions" of the shell.

As a second type of construction we again consider an even number of layers, but we let the fiber angle θ be an even function of the thickness coordinate z. It is then clear that all the elastic moduli $\bar{A}^{\nu\lambda\varepsilon\varkappa}$ are now even functions of z. From eq. (26) or (23) we may obviously write $D_1^{\nu\lambda\varepsilon\varkappa} = 0$ for all values of the superscripts.

It is also clear that *all* the components of the tensors $D_0^{\nu\lambda\varepsilon\varkappa}$ and $D_2^{\nu\lambda\varepsilon\varkappa}$ are nonvanishing, particularly D_0^{1211}, D_0^{1222}, D_2^{1211} and D_2^{1222}. Thus, with the even distribution of angles θ_i, the coupling between membrane and flexural actions disapppears, but the anisotropic behavior remains. This anisotropic character now appears directly in relations of the type $M^{\alpha\beta} = D_2^{\alpha\beta\gamma\delta} q_{\gamma\delta}$ or $N^{\alpha\beta} = D_0^{\alpha\beta\gamma\delta} p_{\gamma\delta}$ rather than through coupling terms. It is, in fact, likely that this even distribution of θ_i will lead to more strongly anisotropic behavior than the previously considered odd distribution of θ_i.

Finally, if we consider a general distribution of θ with z for the same type of shell (i.e., even number of layers), it is obviously possible to decompose a general distribution $\theta(z)$ in even and odd functions of z. Clearly then such a general distribution of $\theta(z)$ will give rise to both types of anisotropy.

We have shown, therefore, that every fiber reinforced shell is anisotropic unless all fibers are oriented along the middle surface coordinates (i.e., lines of curvature). It may be possible to show in specific cases that an odd distribution of θ_i, in which the anisotropy appears through the coupling term $D_1^{\nu\lambda\varepsilon\varkappa}$, gives rise to an *approximately* orthotropic shell. In this connection, it is of interest to note that Hashin *et al.*[5] obtained bounds on the elastic moduli of biaxially reinforced materials and proved that such materials would behave as orthotropic in the direction of the bisectors of the fiber direction. It is possible, in principle, to calculate shell stiffness tensors by introducing the moduli for biaxially reinforced materials in eqs. (25) to (27), and these would then exhibit orthotropic rather than anisotropic behavior. This apparent contradiction is due to the fact that the difference in the z coordinates of adjacent layers with fiber angles θ_i and $-\theta_i$ is effectively neglected if the moduli for biaxially reinforced materials are used. In fact, it is reasonable to expect that a shell composed of many pairs of layers having fiber angles θ_i and $-\theta_i$ with respect to the generator will be very nearly isotropic.

IV. CONCLUDING REMARKS

We have presented a simple method for the calculation of the stiffness tensors for shells composed of layers of unidirectionally reinforced materials. It was shown in particular that the constitutive relations thus obtained exhibit a coupling between membrane and flexural actions even for a first order shell theory. For homogeneous shells, such a coupling appears only when higher order shell theories are considered. In addition, it was established that the layered shell will exhibit anisotropic behavior for all possible distributions of the fiber angle θ with the thickness coordinate. If the fiber angle is an odd function of the thickness coordinate, the anisotropy appears only in the coupling terms between stress resultants and flexural strain measures and between stress couples and membrane strain measures. Thus, it is likely that certain types of construction can lead to a nearly orthotropic shell. An example would be a shell constructed of pairs of layers with fiber angles θ and $-\theta$ with respect to the generator.

The shell stiffness tensors defined by eqs. (22) through (24) are based on the effective properties obtained by Hashin and Rosen[1] for unidirectionally reinforced materials. It will be remembered that for such a transversely isotropic material, four of the five moduli were calculated exactly while only bounds could be obtained for the transverse shear modulus (i.e., shear modulus in the plane perpendicular to fiber direction). Equation (29) shows that the moduli $\bar{A}^{\alpha\beta\lambda\nu}$ entering into the calculation of the shell stiffness tensors depend on only four of the five moduli, namely $\tilde{A}^{1111}$, $\tilde{A}^{2222}$, $\tilde{A}^{1212}$, $\tilde{A}^{1122}$. The work of Hashin and Rosen[1] shows that these four moduli are independent of the transverse shear modulus so that the shell stiffness tensors calculated by the present method are exact within the frame-work of classical shell theory. This is a very fortunate occurrence with regard to the use of the present method of calculating shell stiffnesses, but it also serves to point out one of the inherent limitations in the present approach. It is clear that because of the neglect of transverse shear deformation, the transverse shear modulus for each layer drops out. Thus, we expect that this type of layered shell theory will not predict very accurately the interlaminar shear stresses which are often the controlling factor in structural failures of fiber-reinforced (or other layered) shells.

In fact, it is in order to close this paper with a few remarks concerning what we can expect to obtain analytically by the use of a combination of effective elastic moduli and a layered shell model. Clearly, the shell stiffness tensors calculated by the method presented will yield reasonably accurate values for the gross displacements and rotations. Similarly, we expect that the stress resultants and stress-couples, although less accurate than the displacements, will be determined to a suitable degree of approximation. The situation is, however, different for the calculation of stresses. It is well known that even for a homogeneous plate or shell, the classical expressions for stresses are really assumptions justified only by the fact that they either agree with three-dimensional elasticity for certain special cases, or can be shown to be leading terms of asymptotic solutions to the elasticity equations. For a shell constructed of heterogeneous layers, this no longer holds true, and the classical expressions for stresses can, at most, be considered to yield averages over areas which must include a suitably large number of fibers and layers. Furthermore, because of the heterogeneous character of the material the stresses are likely to vary over wide ranges, and the average information is clearly insufficient to calculate or even estimate peak stresses. Thus, it is doubtful that this type of analysis can be used to estimate the loads at which a shell will start to yield.

It is possible, however, that a useful improvement can be made in this area. If gross displacements are calculated on the basis of effective moduli and layered shell theory, it may be possible to investigate the magnitude of the local stresses by analyzing idealized models representing the microstructure features (such as composite cylinders) under the known displacements. This will be the subject of a future paper.

REFERENCES

1. Hashin, Z. and Rosen, B. W. The Elastic Moduli of Fiber Reinforced Materials, *Journal of Applied Mechanics*, **31**, 223 (1964).
2. Naghdi, P. M. Foundations of Elastic Shell Theory, *Progress in Solid Mechanics*, **4**, 1–90 (1963).
3. Zudans, Z. A. New Formulation and Evaluation of Elastic Shell Theory, Ph.D. dissertation, University of Pennsylvania, 1966.
4. Kingsbury, H. B. Stresses and Deformations in Anisotropic Shells, Ph.D. dissertation, University of Pennsylvania, 1964.
5. Hashin, Z., Brull, M. A., Chu, T. Y. and Zudans, Z. Determination of Effective Elastic Properties for Biaxially Fiber Reinforced Materials, Report No. FB2455, Franklin Institute Research Laboratories, 1966.

STRAINS AND STRESSES IN MATRICES WITH INSERTS

A. J. DURELLI†, V. J. PARKS‡, H. C. FENG††, and F. CHIANG‡‡

The Catholic University of America, Washington, D.C.

ABSTRACT

This paper is a contribution to the experimental stress analysis of composite structures made of a matrix reinforced with fibers. The analysis is conducted in oversize transparent models in which the matrix is cast around inserts of different rigidities. The manufacturing of the models is described. Two- and three-dimensional photoelastic methods are used for the analysis. Stresses and strains along the interfaces, and near the fiber ends, for different fiber end configurations, are studied in detail. Bonded and unbonded interfaces are analyzed. Two loading conditions are considered: restrained shrinkage and mechanically applied loading. Several of the difficulties encountered in the analysis are pointed out. The methods developed will have application in the fields of fiber-reinforced plastic structures, whisker-reinforced metals, auto tires, ceramics and reinforced concrete, among others.

I. INTRODUCTION

The determination of stresses, strains and displacements in composite structures is particularly difficult, both theoretically and experimentally. There is, however, a great technical interest in determinations of this kind for applications in the fields of fiber-reinforced plastics, whisker-reinforced metals, reinforced ceramics, reinforced concrete, etc. In recent years these fields have become extremely important, in particular in the manufacturing of rockets, and the stress analysis of composite materials has been given a great deal of attention by many investigators. Surveys of the present status of the mathematical analysis of stresses and strains in inserts and matrices are given by Sternberg(1) and Wilson.(2) Among the most recent contributions the work of Schuerch(3) should be mentioned. Contributions to the experimental analysis have been made by the author and associates(4,5,6) and by Schuster and Scala(7) among others. Considerations of the stresses present at the interface of matrix and inserts in some simple cases have been made by Ebert

† Professor, Civil Engineering and Mechanics Department.
‡ Associate Professor, Civil Engineering and Mechanics Department.
†† Research Associate, Civil Engineering and Mechanics Department.
‡‡ Post-Doctoral Fellow, Civil Engineering and Mechanics Department, now Assistant Professor, State University of New York (Stonybrook).

and Gadd.[8] In this paper an attempt will be made to approach the subject systematically, from an experimental standpoint, and to describe in detail three-dimensional techniques that permit the solution of problems that had not been solved previously. Two-dimensional techniques that permit the easier solution of some problems will also be described.

Some investigators have dealt with the overall behavior of the composite structure and determined stresses and strains using "effective" moduli concepts. The research reported in this paper deals with stresses and strains in matrices, with inclusions or inserts, at the scale of the dimensions of these inclusions. The solutions have been obtained for linear elastic materials within the framework of the classic theory of elasticity.

II. SPECIFICATION OF STRESS AND STRAIN CONDITIONS AT INTERFACE OF MATRICES AND INSERTS

In this section the boundary conditions of the problem will be defined. Simplifications that can be obtained for the particular case of incompressible materials and rigid inserts will be described.

A. *Three-dimensional Bodies*

Certain stress and strain conditions at the interface follow from the specification of perfect bond between matrix and insert (Fig. 1).

Obviously the laws of equilibrium require that:

$$\begin{aligned} \sigma_n)_m &= \sigma_n)_i \\ \tau_{nx})_m &= \tau_{nx})_i \\ \tau_{ny})_m &= \tau_{ny})_i \end{aligned} \qquad (1)$$

where σ and τ are the normal and shear stresses, respectively, m and i stand for matrix and insert, respectively; and n refers to the direction normal to the boundary.

Geometric considerations on bonded surfaces require that:

$$\begin{aligned} \varepsilon_x)_m - \varepsilon_x)_i &= \alpha_m - \alpha_i \\ \varepsilon_y)_m - \varepsilon_y)_i &= \alpha_m - \alpha_i \end{aligned} \qquad (2)$$

where α is the uniform isotropic free shrinkage and ε the direct (or "normal") strain.

If there is no shrinkage, and only mechanical loads are applied then

$$\begin{aligned} \varepsilon_x)_m &= \varepsilon_x)_i \\ \varepsilon_y)_m &= \varepsilon_y)_i. \end{aligned}$$

For either shrinkage or mechanical loading

$$
\begin{aligned}
\gamma_{nx})_m &= \frac{G_i}{G_m}\gamma_{nx})_i \\
\gamma_{nx})_m &= \frac{G_i}{G_m}\gamma_{ny})_i
\end{aligned}
\tag{3}
$$

where G is the shear modulus and γ the shear strain.

Figure 1 shows two special material requirements that simplify the bonding conditions on the interface. These requirements, which occur in a number of cases, are that the matrix is incompressible, the insert is rigid, or in some cases both.

	SHRINKAGE LOADING (WITH OR WITHOUT MECH. LOADING)	MECHANICAL LOADING (NO SHRINKAGE) $\alpha_m - \alpha_i = 0$
STRESSES	$\sigma_n)_m = \sigma_n)_i$ $\tau_{nx})_m = \tau_{nx})_i$ $\tau_{ny})_m = \tau_{ny})_i$	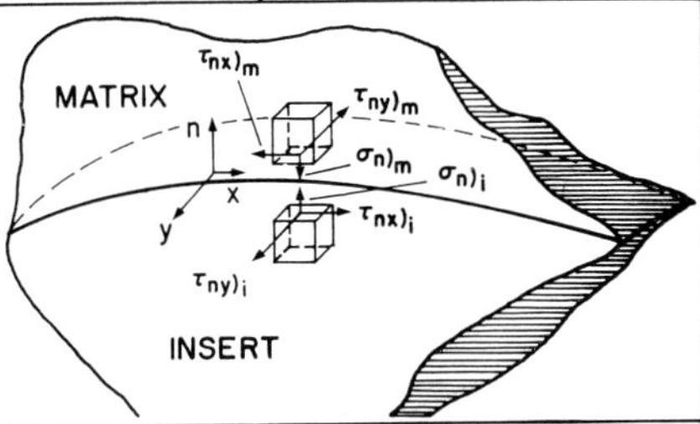
STRAINS	$\epsilon_x)_m - \epsilon_x)_i = \epsilon_y)_m - \epsilon_y)_i = \alpha_m - \alpha_i$ $\gamma_{nx})_m = (G_i/G_m)\,\gamma_{nx})_i$ $\gamma_{ny})_m = (G_i/G_m)\,\gamma_{ny})_i$	$\epsilon_x)_m = \epsilon_x)_i$ $\epsilon_y)_m = \epsilon_y)_i$ $\gamma_{nx})_m = (G_i/G_m)\,\gamma_{nx})_i$ $\gamma_{ny})_m = (G_i/G_m)\,\gamma_{ny})_i$
STRAINS (IF MATRIX IS INCOMPRESSIBLE) $\epsilon_x)_m + \epsilon_y)_m + \epsilon_n)_m = 0$	$\epsilon_x)_m - \epsilon_x)_i = \epsilon_y)_m - \epsilon_y)_i = \alpha_m - \alpha_i$ $\gamma_{nx})_m = (G_i/G_m)\,\gamma_{nx})_i$ $\gamma_{ny})_m = (G_i/G_m)\,\gamma_{ny})_i$ $\epsilon_x)_m + \epsilon_y)_m = -\epsilon_n)_m$	$\epsilon_x)_m = \epsilon_x)_i$ $\epsilon_y)_m = \epsilon_y)_i$ $\gamma_{nx})_m = (G_i/G_m)\,\gamma_{nx})_i$ $\gamma_{ny})_m = (G_i/G_m)\,\gamma_{ny})_i$ $\epsilon_x)_m + \epsilon_y)_m = -\epsilon_n)_m$
STRAINS (IF INSERT IS RIGID) $\epsilon)_i = 0$	$\epsilon_x)_m = \epsilon_y)_m = \alpha_m - \alpha_i$ (SHEAR EQUATIONS BREAK DOWN)	$\epsilon_x)_m = \epsilon_y)_m = 0$ (SHEAR EQUATIONS BREAK DOWN)
STRAINS (IF MATRIX IS INCOMPRESSIBLE & INSERT IS RIGID) $\epsilon_x)_m + \epsilon_y)_m + \epsilon_n)_m = 0$ $\epsilon)_i = 0$	$\epsilon_x)_m = \epsilon_y)_m = \alpha_m - \alpha_i$ $\epsilon_n)_m = -2(\alpha_m - \alpha_i)$ (SHEAR EQUATIONS BREAK DOWN) WHEN BOUNDARY IS PRINCIPAL $\gamma_{max} = 3(\alpha_m - \alpha_i)$	$\epsilon_x)_m = \epsilon_y)_m = \epsilon_n)_m = 0$ (SHEAR EQUATIONS BREAK DOWN) WHEN BOUNDARY IS PRINCIPAL $\gamma_{max} = 0$

FIG. 1. Stress and strain conditions at the interface between a matrix and an insert

1. INCOMPRESSIBLE MATRIX. If the matrix is incompressible then the sum of the strains at any point in the matrix will be zero

$$\varepsilon_x)_m + \varepsilon_y)_m + \varepsilon_n)_m = 0 \tag{4}$$

and the matrix strain in the direction normal to the boundary is then specified by

$$\varepsilon_n)_m = -[\varepsilon_x)_m + \varepsilon_y)_m]. \tag{5}$$

2. RIGID INSERT. If the insert is rigid then there is no strain on the insert $\varepsilon)_i = \gamma)_i = 0$. (This does not mean that the insert does not shrink, only that no stress producing strains are produced. Therefore $\alpha_i \neq 0$.) It is required that the elastic constants E and G approach infinity so that although strains are zero, stresses can still be transmitted by the insert to satisfy equilibrium. In Fig. 1 the note that the shear strain equation breaks down refers to the indeterminacy of an infinite G multiplied by a zero $\gamma)_i$.

Dropping the $\varepsilon)_i$ terms from the direct strain equation leads to

$$\varepsilon_x)_m = \varepsilon_y)_m = \varepsilon_t)_m = \alpha_m - \alpha_i \qquad \text{for shrinking loading}$$

and (6)

$$\varepsilon_x)_m = \varepsilon_y)_m = \varepsilon_t)_m = 0 \qquad \text{for mechanical loading}$$

where ε_t is introduced to represent any strain tangential to the interface.

3. INCOMPRESSIBLE MATRIX AND RIGID INSERT. When both these conditions of an incompressible matrix and a rigid insert occur together then the direct strain relation is further reduced to

$$\left.\begin{aligned} \varepsilon_t)_m &= \varepsilon_x)_m = \varepsilon_y)_m = \alpha_m - \alpha_i \\ \varepsilon_n)_m &= -2(\alpha_m - \alpha_i) \\ \varepsilon_t)_m - \varepsilon_n)_m &= 3(\alpha_m - \alpha_i) \end{aligned}\right\} \text{ for shrinkage} \tag{7}$$

and

$$\varepsilon_t)_m = \varepsilon_x)_m = \varepsilon_y)_m = \varepsilon_n)_m = 0 \quad \text{for mechanical loading.}$$

A further interface relation that results in the case of an incompressible matrix about a rigid insert is that, if the interface is perpendicular to a plane of symmetry, on planes perpendicular to the interface and the plane of symmetry

$$\gamma_{\max})_m = \varepsilon_t)_m - \varepsilon_t)_m = 3(\alpha_m - \alpha_i) \qquad \text{for shrinkage} \tag{8}$$

and $$\gamma_{\max})_m = \varepsilon_t)_m - \varepsilon_n)_m = 0 \qquad \text{for mechanical loading.}$$

These simplifications are useful in many experimental stress analysis applications because the material used for two-dimensional photoelasticity is in general a urethane rubber and the material used for three-dimensional photoelasticity is a frozen epoxy. Both the urethane rubber and the frozen epoxy have a value of Poisson's ratio close to 0.5, which corresponds to the condition of incompressibility. Both have low E values, approximately 500 psi for the rubber and 2000 psi for the frozen deformations in the epoxy, so that inserts of hard plastic or metal appear rigid.

Values of $\varepsilon_t - \varepsilon_n$ and γ_{max}, obtained by photoelastic means from an incompressible material along the interface of a rigid insert, can be compared with these deduced expressions to calibrate the material or to justify the assumptions of incompressibility and rigidity.

B. *Plates*

Consider the case of a thin flat plate (usually referred to as a two-dimensional plane stress problem) which has an inclusion with a bonded interface (Fig. 2). The state of stress and strain at this boundary is three-dimensional. The bonded interface is in the same condition as the one described above for the three-dimensional bodies (Fig. 1). Thus in the case of a rigid insert and Poisson's ratio equal to 0.5, if there is shrinkage the maximum shear strain on the matrix will be $3(\alpha_m - \alpha_i)$ at points on the interface at which the interface coincides with a principal direction. If there is no shrinkage and the strains are associated with applied loads the maximum shear stress at the same points will be zero, regardless of the amount of the mechanical load.

If the thin plate is a model of a prototype which is also thin, the out-of-plane restraint (or pinching) is part of the problem, and must be analyzed.

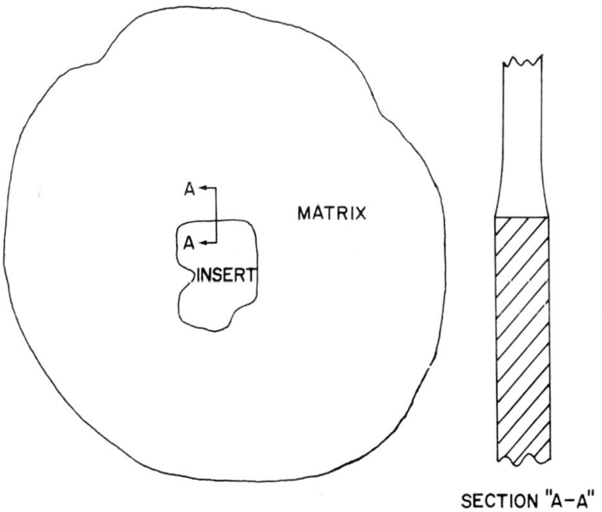

FIG. 2. Insert in a matrix plate showing pinching at interface

It may be necessary to use the three-dimensional freezing techniques in order to solve the problem completely. However, in many cases the thin flat plate is used to simulate the cross-section of a three-dimensional body. This follows the accepted relation between plane-stress and plane-strain. The distribution of in-plane stresses in a plane-stress solution is proportional to the distribution of in-plane stresses in a plane strain solution where the geometry and loading of the transverse cross-section of the plane-strain body is the same as the geometry and loading of the plane-stress body.

In a thin plate at the bonded boundary of a rigid insert the stresses and strains normal to the boundary can be written directly from Hooke's law. For shrinkage: $\varepsilon_t = \alpha_m - \alpha_i$, or if we define $\alpha = \alpha_m - \alpha_i$, $\varepsilon_t = \alpha$

$$\varepsilon_n = \frac{-2\nu\alpha + (\sigma_n/E)(1+\nu)(1-2\nu)}{1-\nu}.$$

For mechanical loading $\varepsilon_t = 0$ and

$$\varepsilon_n = \frac{(\sigma_n/E)(1+\nu)(1-2\nu)}{1-\nu}. \tag{9}$$

Compare these relations with those resulting from corresponding bonded boundary conditions restricted to plane stress.

For shrinkage $\varepsilon_t)_{\text{in-plane}} = \alpha$, and $\sigma_t)_{\text{out-plane}} = 0$

$$\varepsilon_n = -\nu\alpha + (\sigma_n/E)(1-\nu^2). \tag{10}$$

For mechanical loading $\varepsilon_t)_{\text{in-plane}} = 0$, and $\sigma_t)_{\text{out-plane}} = 0$

$$\varepsilon_n = (\sigma_n/E)(1-\nu^2).$$

Curves of these four relations are plotted for $\nu = 0.5$, 0.4 and 0.3 in Fig. 3. It is clear that the pinching effect deviates mostly from the non-pinched condition when $\nu = 0.5$.

In the case of the mechanical loading since the in-plane tangential strain $\varepsilon_t)_{\text{in-plane}}$ is zero, with or without pinching the curves for the mechanical loading also represent $(\varepsilon_n - \varepsilon_t)$. If the boundary is a principal plane, this term gives the maximum principal shear $\gamma_{\max}$.

$$\gamma_{\max} = \varepsilon_n - \varepsilon_t$$

and the curves of mechanical loading in Fig. 3 correspond to the observed isochromatic fringes where the boundary is principal.

In the case of shrinkage, $\varepsilon_t)_{\text{in-plane}} = \alpha$, and curves for $\varepsilon_n - \varepsilon_t$ are obtained by shifting the ε_n curves down one α unit. The shifted curves are shown dotted for the shrinkage loading. Again, if the boundary is a principal direction, $\gamma_{\max} = \varepsilon_n - \varepsilon_t$ and the dotted curves represent the values obtained from photoelastic observation at the boundary. ($\gamma_{\max}$ may take either the positive or negative signs at a point.)

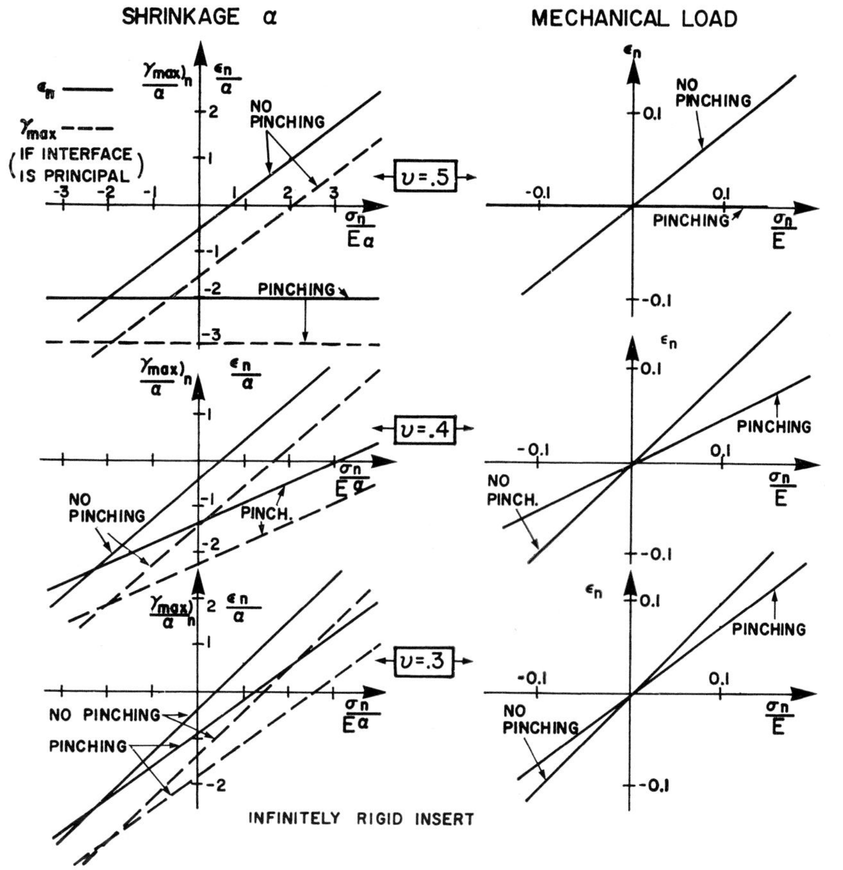

FIG. 3. Comparison of normal strains and stresses at a bonded boundary of a plate with and without pinching

At the boundary, if the tangential direction to the boundary is principal, for photoelastic analysis in the plane it is not necessary that the normal direction n be a true principal direction. In the case of pinching, it can be shown that although a principal direction tangent to the interface does not vary between the two faces of the plate, the normal direction will not be a true principal through the thickness. However, the direction n is a "secondary" principal throughout the thickness, that is, it is the in-plane perpendicular to the boundary which is a true principal direction. As such the observed fringes in the plane do represent an average of $\varepsilon_n - \varepsilon_t$ and γ_{max} values in that plane.

In order to simulate the plane strain condition with a thin plate, it might be reasoned that advantage should be taken of the fact that the plane strain condition already exists on the interface. However, the remainder of the plate is in plane stress, or in a transition zone from plane strain to plane stress, and it is felt that even at the interface the distribution may not be the same as if the whole plate is subjected to plane strain. Thus, an attempt was made to subject the plate to plane stress only. In order to obtain a plane stress condition at the bonded interface, and still retain the in-plane boundary condition that the tangential strain be equal to α_m—α_i, a knife-edged two-dimensional insert can be used.

C. *Visualization of Restrained Shrinkage Stresses and Strains at an Interface*

If a matrix shrinks while bonded to a rigid insert, because of the rigid restraint applied to the matrix, there can be no motion of the points of the matrix at the interface. However, stresses are set up in the matrix, and Hooke's law is used to associate strains with these stresses.

Assume that a matrix has no inserts but cavities of the same size as the size of the inserts and assume that it has been allowed to shrink freely. The linear dimension of the cavities will decrease by an amount equal to the free shrinkage of the matrix. If α_m were 0.01, the length, height and width of a cavity would decrease 1%. Assume that the inserts also shrink, this shrinkage is termed α_i. Now if the inserts are forced into this cavity the interference between the insert and the cavity will be characterized by the value $\alpha_m - \alpha_i$. If each point on the wall of the cavity is bonded to the point on the wall of the insert to which it corresponded before the shrinkage, then every fiber on the matrix interface will be subjected to the strain $\alpha_m - \alpha_i$. In general, it can be said that every fiber in the matrix has an initial (or unrestrained) length

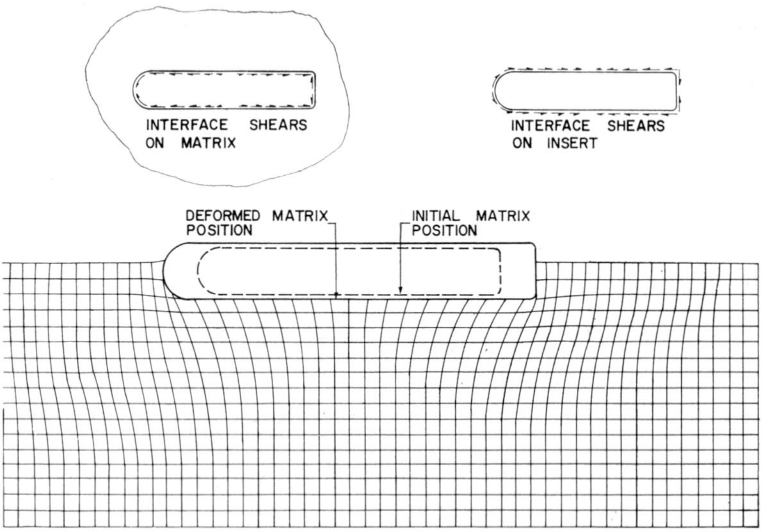

FIG. 4. Visualization of restrained shrinkage stresses at an interface

which is the length it would have had if there were no inserts and the matrix shrunk freely.

Figure 4 shows an exaggerated shrinkage ($\alpha = 20\%$) of a matrix with a square grid to demonstrate this concept. Far from the insert the matrix grid is shrunk 20%, but at the insert the grid is restrained. A 20% reduced insert is shown dotted to show where the cavity boundary of the grid would have been if it had not been restrained. The dotted line can then be considered the initial or undeformed position of the interface and the solid line the final or deformed position. The associated shear-stress boundary conditions on the matrix and insert are added to the figure to further picture the situation at the interface.

III. METHODS TO BE USED

A. *Experimental Approach*

Three-dimensional photoelasticity is the common experimental approach to the solution of problems in which stresses and strains are required within a three-dimensional body. Among the three-dimensional photoelasticity techniques, the so-called "freezing" method is the most precise and most commonly used. This method seems to be the most promising to approach, in a general way, the solution of the problem of the determination of stresses and strains in composite materials made of a matrix with reinforcing inclusions or inserts. However, some serious difficulties have been found, and will be discussed in the following section.

Two-dimensional photoelasticity can be used to solve two-dimensional simulations of the three-dimensional composite materials. The simplicity of two-dimensional photoelasticity makes this method well adapted to solve the problem of stress and strain distribution in models representing either the meridian or the transverse cross-sections of composite materials. Here also some difficulties have been encountered which did not permit, in many cases, an accurate evaluation of the data.

In what follows a short introduction to the photoelastic method will be presented and comments will be made as to the extent to which the above-mentioned difficulties have been overcome.

B. *The Photoelastic Method*

Photoelasticity is an experimental stress analysis method that takes advantage of the property of double refraction exhibited by certain isotropic transparent materials when subjected to stress or strain. This optical phenomenon, when the material is viewed in a field of polarized monochromatic light, manifests itself in the form of interference fringes or alternate dark and light bands. These fringes, referred to as isochromatics are ordered according to the number of darkness–brightness cycles that occur at any given point as the load is increased from zero to its final value.

It has been established that the fringe order at a given point is proportional to the thickness of the material in the direction of light propagation and to the difference of the principal stresses acting on planes containing the axis of propagation. The stress in the direction of the light propagation has no effect on the observed birefringence. Stated in equation form, this stress-optic law is expressed as follows:[9]

$$\frac{\sigma_1 - \sigma_2}{2} = \tau_{max} = \frac{nf}{t} \tag{11}$$

where n = fringe order,
t = model thickness,
σ_1, σ_2 = principal stresses,
τ_{max} = maximum shear stress,
f = material fringe constant.

To obtain stresses with this method of analysis, typically, a model is fabricated from a transparent plastic known to possess the required photoelastic properties. The model is machined geometrically similar to the prototype and loaded, also similarly, to the prototype loading.

In the case of inserts a somewhat different procedure is required. The photoelastic material is cast around the insert, and, if desired, allowed to bond to the insert. If a shrinkage load is desired advantage is taken of the curing shrinkage of the matrix (this includes a large proportion of ordinary thermal shrinkage). Mechanical loading may be superposed on the shrinkage load, or special precautions may be taken to avoid any shrinkage loading and then stresses due to mechanical loading are obtained separately.

1. Two-dimensional tests. In two-dimensional problems, the loaded model is examined in a field of polarized light. The fringe pattern gives the stress distribution, and direct visual observation can ordinarily be used to locate regions of high and low stress.

Essentially, two patterns can be obtained. If the polariscope is set to produce maximum darkness in the background outside the model (dark field), then the fringes are ordered $n = 0, 1, 2, 3$, etc. If the polariscope is arranged to produce maximum light in the field outside the model (light field), then the fringes have the values of the intermediate orders $n = 1/2$, $1\,1/2$, $2\,1/2$, $3\,1/2$, etc. By the simple process of counting fringes and multiplying their order by a calibration constant, the maximum shear stress distribution can be determined throughout the body of the model. The model stress distribution can then be converted by the use of appropriate scaling laws to the stress distribution in the prototype. When the stress field is uniaxial, as on free boundaries, one of the principal stresses is zero and the other can be determined directly from the photoelastic data. At interior points of the model, additional information is required for determining the principal stresses. The evaluation then becomes appreciably more complicated. A number of methods are available, however, to solve the two-dimensional problem completely.

2. THREE-DIMENSIONAL TESTS. For the photoelasticity solution of three-dimensional problems, a somewhat more involved technique is necessary, since the observation of the loaded model in a field of polarized light does not result in a fringe pattern which can readily be interpreted. To overcome this difficulty, a stress pattern is "locked-in" or "frozen" in the model, and thin slices are removed from the model wherever the stress distribution is required. By viewing the slices in polarized light, a fringe pattern is obtained which corresponds to the stress distribution in the three-dimensional model at the time of stress freezing.

The fringe pattern is frozen into the model by utilizing certain plastics which behave as if they were composed of two "phases", with different rates of change in the physical properties of each phase under temperature variation. When a model made of such material is heated to a certain "critical" temperature, one component or phase becomes soft, while the other is still relatively hard and perfectly elastic. When loaded in this condition and slowly cooled under load, the softened phase rehardens, locking-in or "freezing" the deformation and the corresponding fringe pattern permanently into all parts of the model. The load can then be removed and slices taken from the model. The removal of slices does not distort or change the fringe pattern in any manner if the machining operation is carefully performed.

The fringe patterns from the slices can be interpreted in much the same manner as patterns obtained from a two-dimensional model. Shear stress distributions throughout the slice and boundary stress distributions along free boundaries and along surfaces subjected to uniform pressure-type loadings can be determined directly from the patterns. At interior points of the slice, the problem of determining the principal stresses is considerably more difficult. In general, the experimental techniques available for providing the additional information required to solve completely a two-dimensional problem, are difficult to apply to the three-dimensional case. As a result, a method which utilizes photoelastic data and an integration procedure based on the equations of equilibrium has been developed. The procedure is cumbersome and subject to considerable inaccuracies unless it is applied with extreme care and skill. A complete knowledge of interior stresses is not always required, and a knowledge of boundary stress distributions is sufficient for many design problems. In this program principal stresses at interior points have been determined only along selected lines, the main purpose being to illustrate the application of the method. Shears acting in planes parallel to interfaces have been determined in many instances since they are likely the most significant stress components in composite material problems.

When stresses have been separated in this program the method used was the graphical integration of the equations of equilibrium.[10] These equations in finite form are of the following type:

$$\sigma_x = (\sigma_x)_0 - \sum_{x0}^{x} \Delta\tau_{yx}\frac{\Delta x}{\Delta y} - \sum_{x0}^{x} \Delta\tau_{zx}\frac{\Delta x}{\Delta z} \tag{12}$$

where $(\sigma_x)_0$ is the stress at some point along the line of integration, at which the stress is independently known. If this point is taken at a free boundary perpendicular to x, $(\sigma_x)_0 = 0$. The shear stress along the Cartesian direction can be computed from the maximum shear (isochromatics) and the isoclinic parameter.

At axes of rotational symmetry two of the normal stresses are equal, and starting at a point at a free boundary the principal stress, say σ_3 is given by:

$$\sigma_3 = -2\sum_{x_0}^{x} \Delta\tau_{yx}\frac{\Delta x}{\Delta y} \tag{13}$$

and

$$\sigma_1 = \sigma_2 = \sigma_3 + 2\tau_{max}. \tag{14}$$

The principal strains can be computed from the principal stresses using Hooke's law:

$$\varepsilon_2 = \varepsilon_1 = \frac{1}{E}[\sigma_1 - \nu(\sigma_1 + \sigma_3)] \tag{15}$$

$$\varepsilon_3 = \frac{1}{E}[\sigma_3 - \nu(2\sigma_1)]. \tag{16}$$

For materials exhibiting a Poisson's ratio equal to 0.5

$$\varepsilon_1 + \varepsilon_2 + \varepsilon_3 = 0$$

and using photoelastic data eqs. (15) and (16) become:

$$\varepsilon_1 = \frac{0.5}{E}(\sigma_1 - \sigma_3) = \frac{1}{E}\tau_{max} = \frac{2}{3}\frac{nf_\varepsilon}{t} \tag{17}$$

$$\varepsilon_3 = \frac{1}{E}(\sigma_1 - \sigma_3) = -\frac{2}{E}\tau_{max} = -\frac{4}{3}\frac{nf_\varepsilon}{t}. \tag{18}$$

C. *State of the Art and Development of New Techniques*

1. THREE-DIMENSIONAL TESTS. The commonly used three-dimensional photoelastic material is an epoxy resin mixed with a hardener. The following formula is known as Hysol 4290:

100 parts of Hysol R9-2053 (resin)
40 parts of Hysol H5-3401 (hardener) (19)
Curing temperature: 132° C.

A number of other epoxies[11] are also cured at about 132° C or above this temperature. A typical coefficient of thermal expansion of these materials is about: 85×10^{-6} per °C. The free shrinkage in the process of curing is about 1.75%.

When a plastic of this kind is cast on a material like steel or aluminum which has a much lower coefficient of expansion, and bonds to the metal in the process of curing, the plastic fails as it cools to room temperature. The restrained shrinkage of the plastic is larger than the maximum allowable strain and frequently the material breaks. Another epoxy formulation has been recommended recently by Sampson.[12] It requires much lower curing temperature. The formula is:

50 parts Bakelite ERL 2774 (resin)
50 parts Bakelite ERL 2795 (resin) (20)
25 parts Bakelite ZZL 0803 (hardener)
Curing temperature about 75° C.

The coefficient of expansion of this formula is also somewhat smaller, about 70×10^{-6} per °C. Since the drop from the "locking-in" temperature to room temperature is much smaller, the shrinkage is also appreciably smaller. The free shrinkage when curing is about 0.9%. With this formula several restrained shrinkage problems have been studied without difficulties.[13, 14]

Another serious difficulty encountered, when the freezing technique is used to stress analyze composite materials, develops in the process of slicing the model. Sawing steel or aluminum inserts raises the temperature of the epoxy matrix sufficiently to anneal several fringes of the frozen photoelastic pattern. In the development of the program reported in this paper it has been found that the epoxy prepared using formula (19) can be used very effectively for the manufacturing of the inserts, when the epoxy prepared using formula (20) is used for the matrix. No difficulties are encountered in either bonding or slicing.

The two previously reviewed difficulties can be overcome with the new techniques described above. There is, however, another difficulty that limits appreciably the usefulness of the three-dimensional photoelasticity method when applied to the solution of composite materials stress problems. It had not been possible so far to conduct a test to determine stresses produced by applied loads, without having at the same time the stresses produced by the restrained shrinkage. This difficulty is inherent in the nature of the freezing method. To freeze the birefringence associated with the applied load it is necessary to heat the specimen, to load it, and then to decrease the temperature. When this is done, the stresses associated with restrained shrinkage are also frozen in. Unless a material be developed that can freeze birefringence without "grabbing" simultaneously the inserts, it is necessary to conduct two tests, one giving the state of stress associated with restrained shrinkage and applied loading together, and the other giving the state of stress associated with shrinkage alone, and subtract the results of one from the results of the other. The procedure is lengthy but straightforward and is described in detail below. In an attempt to obtain results using one test only, epoxy has been cast around Plexiglas inserts. The technique is promising but has not been completely developed as yet.

2. Two-dimensional tests. Two-dimensional tests of composite material have a serious drawback. The photoelastic response at the interface is appreciably influenced by the restraint the insert produces on the matrix, perpendicularly to the plane. This "pinching" effect is presented in section II B and has been studied in detail elsewhere[15] and the use of a knife shaped boundary at the interface has been proposed to eliminate this effect. Several problems have been solved using this technique.[16, 17] The urethane rubber Solithane 113 is commonly used for the matrix manufacturing when the stress distribution problems include the effect of restrained shrinkage.

In two-dimensional tests as in three-dimensional tests the difficulty associated with shrinkage of the matrix around the inserts, in the process of curing, had to be overcome if the results of mechanical loading were to be obtained using one test only. Rubbers have been developed that permit the independent analysis of the two loading conditions. In particular, the urethane rubber Scotchcast 221 RTC has been found suitable using the standard techniques of casting.

3. Subtracting-out fringes due to shrinkage in a composite photoelastic model mechanically loaded. If a composite body is subjected to shrinkage and to a mechanical load, both conditions will produce stress and strains in the body. Since shrinkage is difficult to avoid in the three-dimensional photoelastic "freezing" of a model with inserts, consideration is given to some way of "subtracting-out" the shrinkage response.

Recall that isochromatics are associated with maximum shears and isoclinics with their directions. The shrinkage in the photoelastic model produces an optical response, a three-dimensional isochromatic and isoclinic fringe pattern throughout the model as the model is cooled from the critical temperature. If a load is applied a second isochromatic and isoclinic fringe pattern is added to this shrinkage pattern, essentially superposed with the shrinkage pattern. If identical models with identical inserts are cast together, and if one is loaded mechanically and the other is left unloaded then, in principle, the isochromatics of the unloaded model with isochromatics of the loaded model can be used to obtain the isochromatics associated only with the mechanical loading. Note, however, that these fringes are an optical effect, birefringence, represented by a second order tensor. To subtract one isochromatic from another it is necessary to subtract tensorially which means that the isoclinics also have to be used. That is, corresponding tensor components must be subtracted. The fringes are specified by a fringe order n obtained from the isochromatic pattern and a direction from a given orthogonal coordinate system, θ, which is obtained from the isoclinic pattern. The fringe components in the directions of the coordinate system can be computed as

$$n \cos 2\theta$$

$$n \sin 2\theta\,.$$

These values can be added or subtracted arithmetically in two equi-shaped bodies for components in the same directions.

If these components in the model subjected to restrained shrinkage are subtracted from the components in the model subjected to restrained shrinkage and to applied load, the resultants $(n \cos 2\theta)_r$ and $(n \sin 2\theta)_r$ are the fringe components (subscript r for resultant) corresponding to the applied load alone. The fringe order and angle associated with the applied load can be computed as

$$n_r = \sqrt{(n \cos 2\theta)_r^2 + (n \sin 2\theta)_r^2}$$

and

$$\theta_r = \arcsin \frac{(n \sin 2\theta)_r}{n_r}\,.$$

A material fringe value associating stresses with fringes can then be applied to obtain stresses in the usual way. This fringe value is obtained by subjecting a model of the same photoelastic material to a mechanical load which produces a known stress distribution; the load is applied over the same time and temperature range as the composite model.

It might be noted that the material fringe value used to compute stresses due to mechanical load need not be the same as one associated with the fringes due to shrinkage. The stress so obtained can be written in the usual way

$$\tau_{max} = \frac{\sigma_1 - \sigma_2}{2} = n_r \frac{f_\sigma}{t}$$

$$\tau_{xy} = n_r \frac{f_\sigma}{t} \sin 2\theta_r$$

$$\frac{\sigma_x - \sigma_y}{2} = n_r \frac{f_\sigma}{t} \cos 2\theta_r$$

where σ_1 and σ_2 are principal stresses,
- τ_{max} is the maximum principal shear stress,
- σ_x, σ_y and τ_{xy} are the normal and shear stresses in the $x-y$ direction,
- n_r is the resultant fringe order (corresponding to applied load only),
- θ_r is the resultant angle representing the angle between $x-y$ axes and the directions of σ_1 and σ_2,
- f_σ is the material fringe value obtained from the mechanical load calibration,
- t is the thickness of the slices from which the fringe orders were taken

D. *Curing an Epoxy Around an Insert*

If an epoxy is cured in a mold, has no inserts, and a parting compound is used on the walls of the mold, the epoxy shrinks freely to a final shape, geometrically similar to the mold shape, but smaller. In general, temperature is required to cure the epoxies used in experimental stress analysis. This is also the case in many industrial applications. Some particular aspects of the phenomenon of curing at high temperature require special comments.

The free shrinkage is the relative change in linear dimensions after accounting for the contraction of the mold (the total thermal and curing contraction of the epoxy). If a steel thick ring with an inside diameter D_R is used as a mold for an epoxy disk the free shrinkage α_{fe} of the epoxy made using formula (20) can be computed as:

$$\alpha_{fe} - \alpha_{fs} = \frac{D_R - D_D}{D_D}$$

where α_{fs} is the total thermal contraction of the steel and D_D is the diameter of the epoxy disk. Using this formula the drop in temperature is about 50° C. The free shrinkage of the epoxy to be computed is about

$$\alpha_{fe} = 0.0097.$$

This free shrinkage is, in general, larger than the one associated with the freezing phenomenon because, in general, the "locking-in" takes place when the material is in the rubbery state, at a temperature higher than room temperature. The free shrinkage can be thought of as made of two parts. First, the shrinkage that takes place when the epoxy is brought from the curing temperature down to the freezing temperature. This shrinkage is frozen and cannot be recovered unless the material is annealed. The second part of the shrinkage is that which takes place from the freezing temperature down to room temperature. This second part of the shrinkage of the epoxy in the glassy state is much lower than the first one corresponding to the rubbery state. For the epoxy manufactured using formula (20) the coefficient of shrinkage in the glassy state, is about $70 \times 10^{-6}/°C$.

It is important to point out that the frozen patterns are associated with the shrinkage that takes place before the epoxy reaches the glassy state. It is this frozen shrinkage to which α_m in the three-dimensional tests refers. The associated α_i will be the shrinkage that takes place in the insert before the epoxy reaches the glassy state.

It should also be pointed out that, if, in the process of curing the epoxy bonds to other materials that restrain its free shrinkage (either inserts, or material bonded on the outside surface), these materials should be completely separated from the epoxy to relieve the glassy state shrinkage. If a slice is cut from the specimen and contains part of an insert, the epoxy slice birefringence will be the result of the frozen pattern produced by the total restraint on the specimen, plus an elastic pattern produced by the partial elastic restraint on the slice. The piece of insert must be removed to have the frozen pattern alone.

E. *Application of Analysis to Compressible Matrices*

Since the available photoelastic model materials are incompressible, ($\nu = 0.5$) the application of the analysis to compressible matrices ($\nu < 0.5$) may be questioned. Earlier two-dimensional studies[4, 5] with harder plastics, that were compressible, did not show a marked difference from the studies with rubber presented here.

In the case of three-dimensional analysis, it is generally accepted that the difference in Poisson's ratio between prototype and model is usually not critical in first boundary value problems (problems in which all the boundary loading is specified in terms of stresses or forces). However, bonded interfaces specify mixed or second boundary value problems. In mixed or second boundary value problems where some or all of the boundary conditions fix strains and displacements, the influence of incompressibility is not as well understood, and requires more study.

IV. THREE-DIMENSIONAL ANALYSIS

A. *Matrix Shrinking About an Embedded Sphere*

The case of the rigid sphere embedded in a large shrunk matrix is the first problem solved with the new techniques. The problem has been solved theoretically and the correlation with the experimental results will permit an evaluation of the method and of its precision. There is symmetry of geometry

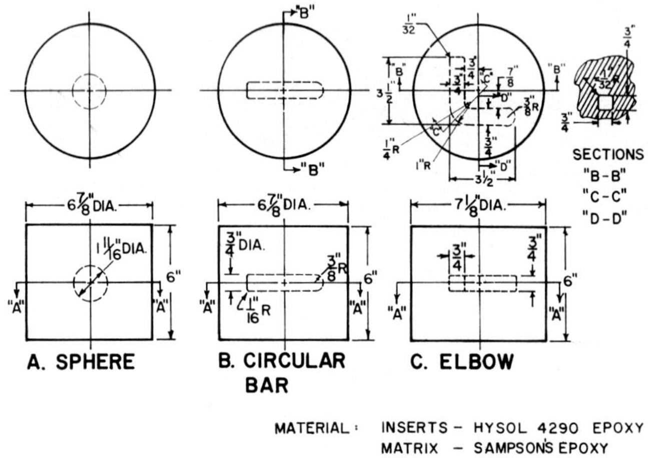

FIG. 5. Geometry of three models of matrices cast about inserts

and stresses about the center and any plane slice through the center will exhibit concentric isochromatics. The dimensions of the model are shown in Fig. 5(A).

The theoretical solution for the case of the uniformity applied pressure p_i can be found in:[18]

$$\sigma_r = -p_i\left(\frac{a}{r}\right)^3$$
$$\sigma_\theta = \frac{1}{2}p_i\left(\frac{a}{r}\right)^3. \tag{21}$$

If the sphere is rigid then in eq. (2) the strains in the insert are zero and the boundary strain of the matrix is

$$\varepsilon_t)_m = \alpha_m - \alpha_i .$$

Dropping the m subscript here and calling

$$\alpha = \alpha_m - \alpha_i$$

$\varepsilon_\theta = \alpha$ at the boundary $a = r$, where α is the difference between the matrix and the insert free shrinkage. Using Hooke's law

$$p_i = \frac{2\alpha E}{1+\nu}. \tag{22}$$

Substituting the value of the pressure in eqs. (21) for the value in terms of the shrinkage α:

$$\sigma_r = -\frac{2\alpha E}{1+\nu}\left(\frac{a}{r}\right)^3 \tag{23}$$

$$\sigma_\theta = \frac{\alpha E}{1+\nu}\left(\frac{a}{r}\right)^3 \tag{24}$$

and by Hooke's law:

$$\frac{\varepsilon_r}{\alpha} = -2\left(\frac{r}{a}\right)^{-3} \tag{25}$$

$$\frac{\varepsilon_\theta}{\alpha} = \left(\frac{r}{a}\right)^{-3} \tag{26}$$

$$\frac{\gamma_{\max}}{\alpha} = 3\left(\frac{r}{a}\right)^{-3}. \tag{27}$$

These values of strain, normalized in respect to the shrinkage are shown in Fig. 6. The curve corresponding to $\gamma_{\max}$ should correlate with the photoelastic isochromatic patterns along the radial line.

B. *Matrix Shrinking About an Embedded Circular Bar*

The embedded sphere presents a rather trivial case of stress distribution. Its study permitted, however, to evaluate the new method. A more interesting case is the embedded rod or circular bar with different end configurations. One of these ends in a hemisphere, the other a flat connected by means of small fillets (Fig. 5(B)). The isochromatics of a meridian slice are shown in Figs. 7 and 8. The field of isoclinics is shown in Fig. 9.

Frequently the most important strains are the ones present at the interface. The analysis will be limited to the determination of the shear strains tangential to the interface in the meridian plane and, to illustrate the general procedure of stress separation, the principal stresses along the axis of symmetry.

Figure 10 shows the maximum shear strains $\gamma_{\max}$ and the direction of the principal stresses at the interface obtained from the patterns in Figs. 7, 8 and 9.

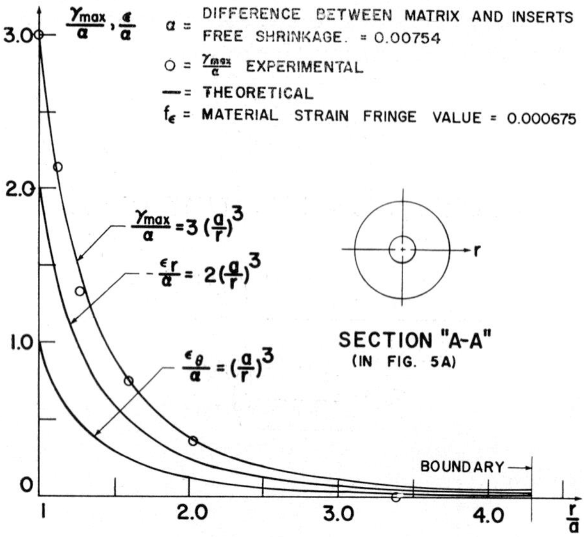

FIG. 6. Strains in a matrix shrunk around a spherical insert

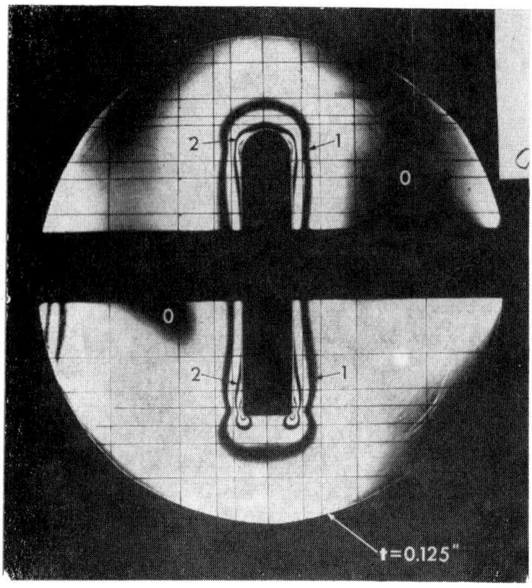

FIG. 7. Isochromatics in the meridian plane of a matrix shrunk around a circular bar with a flat and a round end (dark field)

Figure 11 shows the maximum shear strains tangential to the interface and the difference of normal strains obtained from the values in Fig. 10 using the expressions

$$\frac{\gamma_{nt})_m}{\alpha} = \frac{\gamma_{max}}{\alpha}\sin 2\theta$$

$$\frac{\varepsilon_t - \varepsilon_n}{\alpha} = \frac{\gamma_{max}}{\alpha}\cos 2\theta. \tag{28}$$

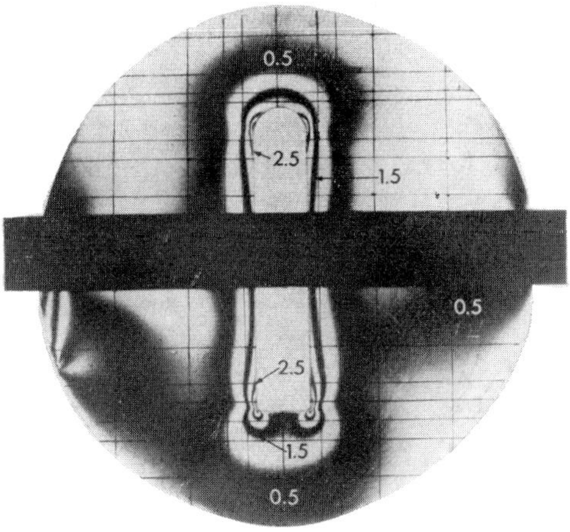

FIG. 8. Isochromatics in the meridian plane of a matrix shrunk around a circular bar with a flat and a round end (light field)

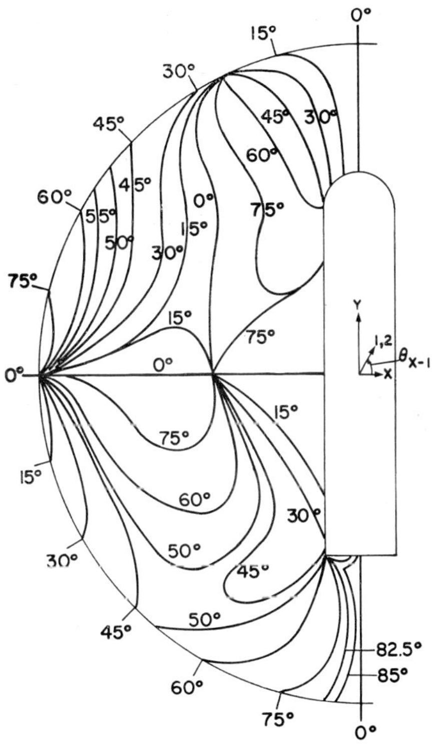

FIG. 9. Isoclinics in the meridian plane of a matrix shrunk around a circular bar with a flat and a round end

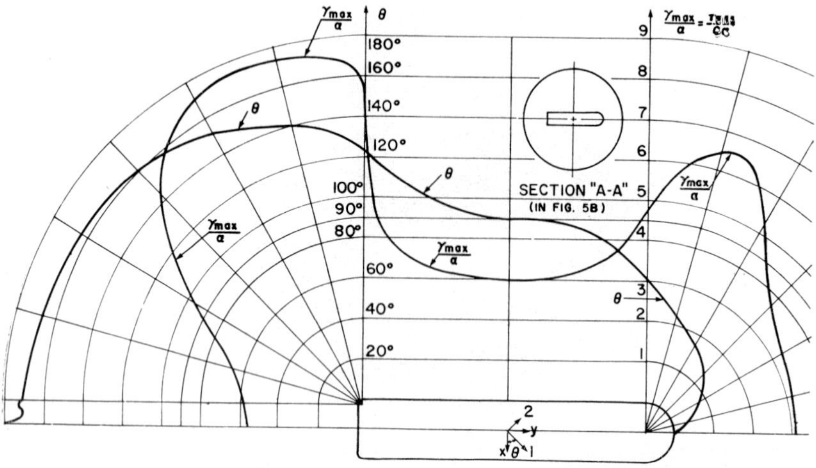

FIG. 10. Maximum shear strain and isoclinics at the interface between a matrix and an embedded circular bar with a flat and a round end (restrained shrinkage)

As pointed out previously, the normal strains at the interface for the case of incompressible matrix material and rigid insert are given by:

$$\varepsilon_t = \alpha$$
$$\varepsilon_n = -2\alpha.$$

The values of $(\varepsilon_t - \varepsilon_n)/\alpha$ given in Fig. 11 can be compared with

$$\frac{\varepsilon_t - \varepsilon_n}{\alpha} = 3. \tag{29}$$

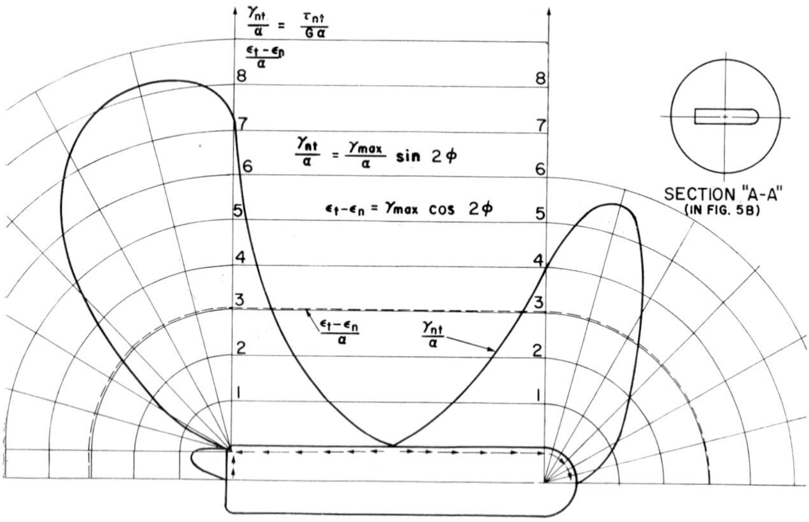

FIG. 11. Strains in the meridian plane, at the interface between a matrix and an embedded circular bar, with a flat and a round end (restrained shrinkage)

Recalling[19] that:

$$\gamma_{xy} = \frac{\tau_{xy}}{G}$$

$$\gamma_{yz} = \frac{\tau_{yz}}{G} \qquad (30$$

$$\gamma_{zx} = \frac{\tau_{zx}}{G}$$

where G is the modulus of rigidity

$$G = \frac{E}{2(1+\nu)};$$

all the values given for shear strains can easily be converted into shear stresses.

To separate the principal stresses the conventional method of integration of the equations of equilibrium can be used as shown in eq. (13). Principal strains can be computed directly from the isochromatic fringes using eqs. (17) and (18). Values so obtained for the normalized principal stresses and strains are shown in Fig. 12.

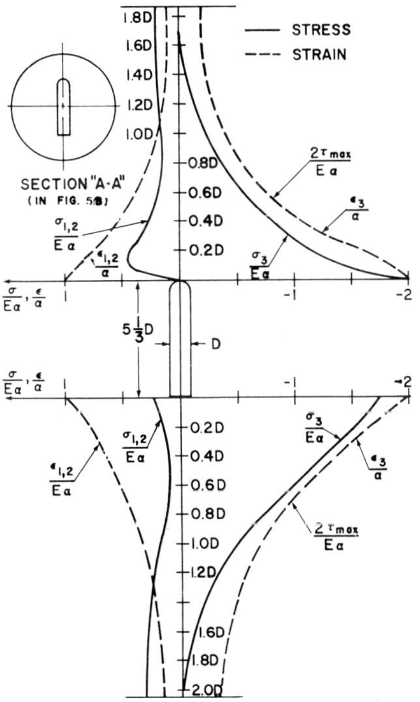

FIG. 12. Principal stresses and strains at the axis of symmetry of a matrix shrunk around a circular bar with a flat and a round end

C. *Matrix Shrinking About an Embedded Elbow of Square Cross-section, with a Flat and a Circular End*

This problem will illustrate a general application of the method of the determination of strains in matrices shrunk about inserts. There is only one plane of symmetry in this problem, section *A–A* as shown in Fig. 5(C). However, the plane perpendicular to section *A–A* and at 45° to the two arms of the elbow is practically a plane of symmetry since the two arms are equal with the exception of the localized zone at their ends. The transverse cross-sections *B–B* and *D–D* in Fig. 5(C) will also exhibit stress distributions which are practically symmetric.

Dark field and light field isochromatics in a transverse slice containing the plane of symmetry are shown in Figs. 13 and 14. (The dark bands in these photographs indicate the position of the longitudinal slices that had been removed before the cutting of the transverse slice.) Isoclinics and isostatics for the principal stresses in that same plane are shown in Figs. 15 and 16. The isostatic field permits a good visualization of the direction and relative value of principal stresses. It should be recalled that family 1 isostatics is everywhere parallel to the largest principal stress and family 2 always parallel to the smallest principal stress and that by definition $\sigma_1 > \sigma_2$.

Maximum shear strains at the interface in the plane of symmetry are plotted in Fig. 17. Shear strains in the plane of symmetry, at the interface, and in a direction tangential to the interface, are shown in Fig. 18. As already noticed

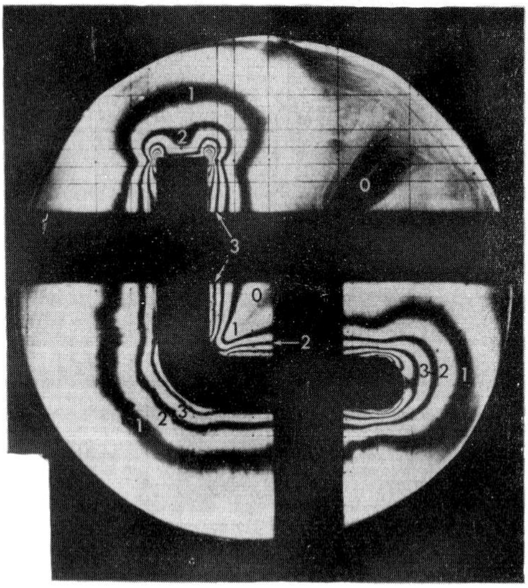

FIG. 13. Isochromatics in the plane of symmetry of a matrix shrunk around an elbow of square cross-section with a flat and a circular end (dark field)

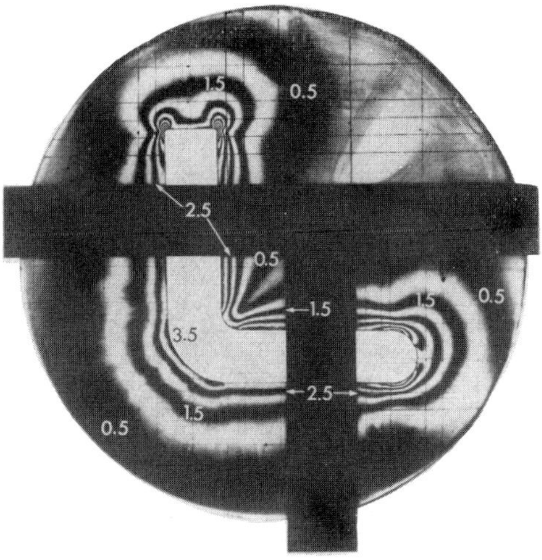

FIG. 14. Isochromatics in the plane of symmetry of a matrix shrunk around an elbow of square cross-section with a flat and a circular end (light field)

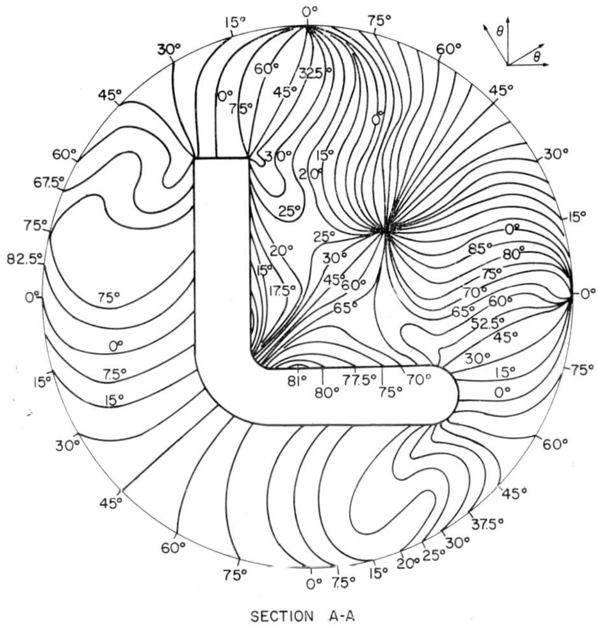

FIG. 15. Isoclinics in the plane of symmetry of a matrix shrunk around an elbow of square cross-section with a flat and a circular end

in Fig. 11, these shears concentrate near the ends of the bars and decrease rapidly for points in the central region. The corner of the elbow shows no concentration of shears. On the other hand, even at the smooth circular end of the arm, the concentration is high.

Equation (8) requires that the insert be rigid. In that case $\varepsilon_t - \varepsilon_n = 3\alpha$. Computed values of $\varepsilon_t - \varepsilon_n$ for the elbow are represented in Fig. 19. The elbow insert was apparently bent which decreased the inside 90° angle. This is suggested by the values of $\varepsilon_t - \varepsilon_n$ above 3α on the outer edge of the corner of the elbow and the values less than 3α on the inner edges. The elbow is apparently subjected to a bending which produced tensile strains on the outer edge and compressive strains on the inner edge. In the matrix this increases the ε_t

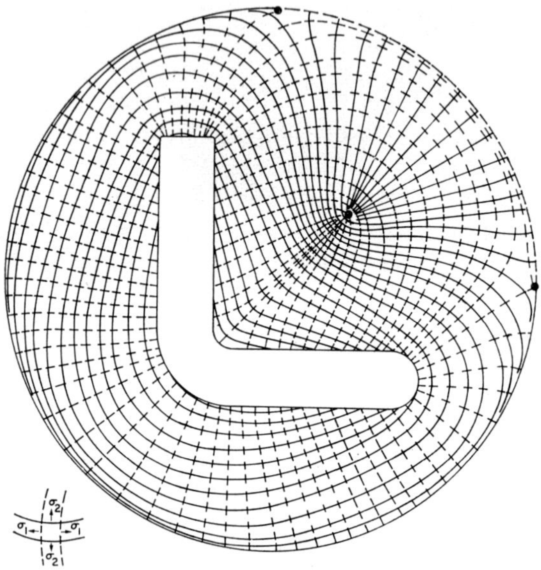

——— σ_1 FAMILY
.... σ_2 FAMILY

FIG. 16. Isostatics in the plane of symmetry of a matrix shrunk around an elbow of square cross-section with a flat and a circular end

strain at the outer edge and reduces the ε_t at the inner edge. The high values of $\varepsilon_t - \varepsilon_n$ at the 1/32 radius corners could not be explained satisfactorily. Large strains may be present in the insert at these corners.

Strains analysis in a transverse cross-section (as section *B–B* in Fig. 5(C)) is also of interest. The isochromatics are shown in Fig. 20. The values of maximum shears and directions of principal strains at the interface are plotted in Fig. 21. The evaluated shear and linear strains along the interface are represented in Fig. 22. There again the difference between the linear strains departs appreciably from 3α at some zones of the interface.

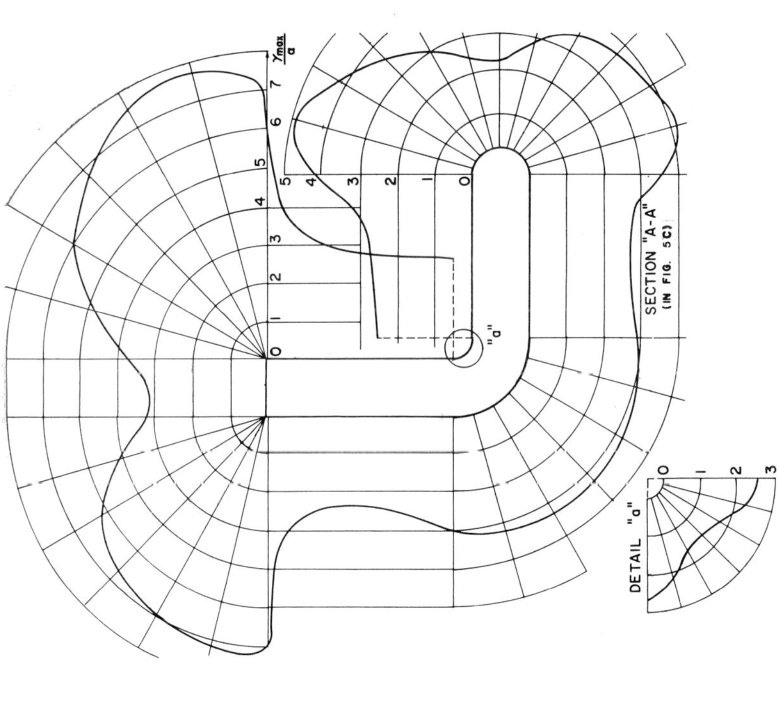

FIG. 17. Maximum shear strain in the plane of symmetry at the interface between a matrix and an embedded elbow with a flat and a circular end (restrained shrinkage)

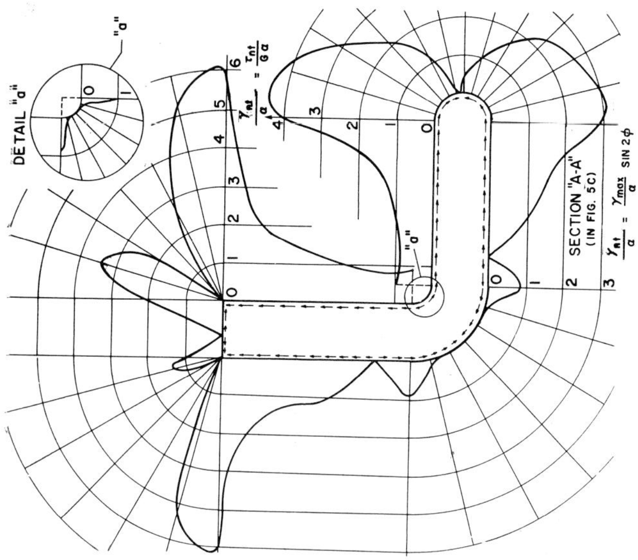

FIG. 18. Shear strains in the plane of symmetry and tangential to the interface between a matrix and an embedded elbow with a flat and a circular end (restrained shrinkage)

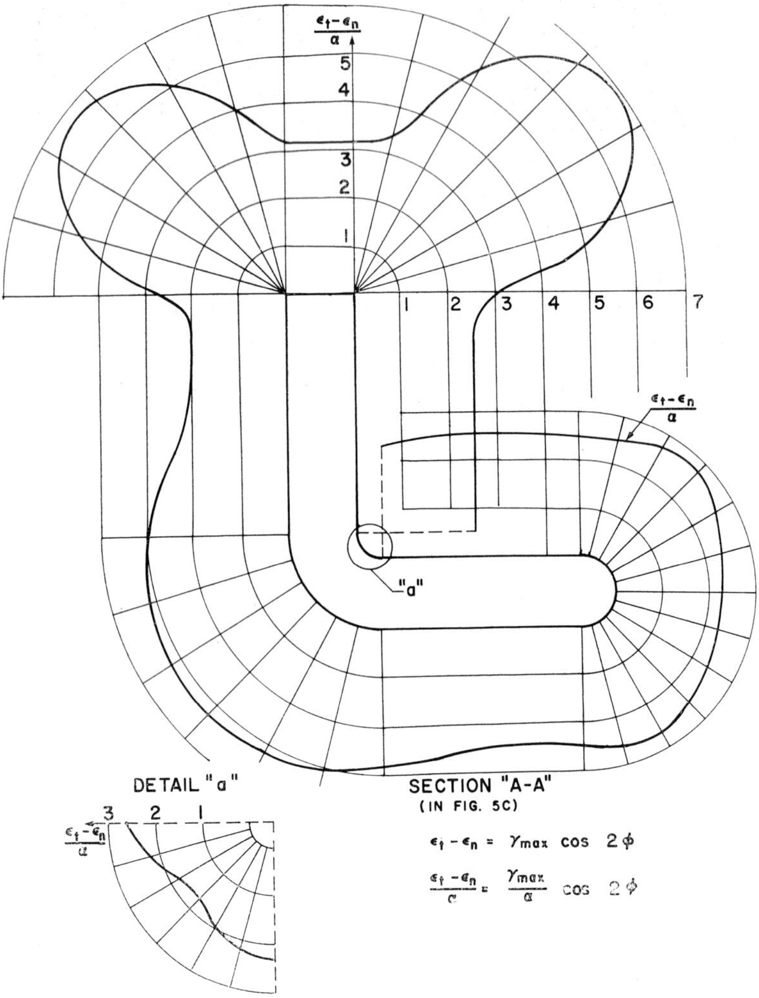

FIG. 19. Difference between normal strains in the plane of symmetry at the interface between a matrix and an embedded elbow with a flat and a circular end (restrained shrinkage)

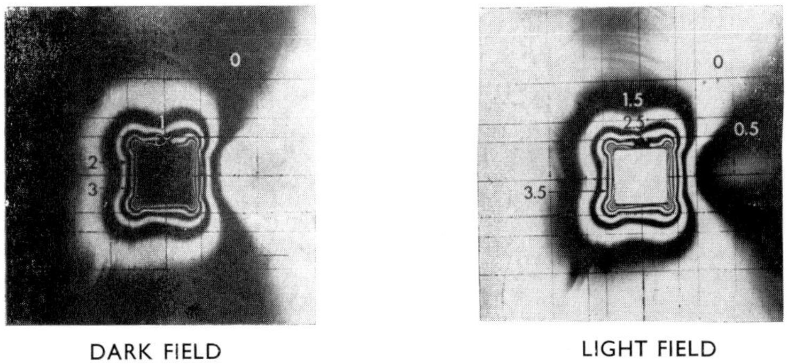

FIG. 20. Isochromatics in a transverse cross-section of a matrix shrunk around an elbow with a flat and a circular end

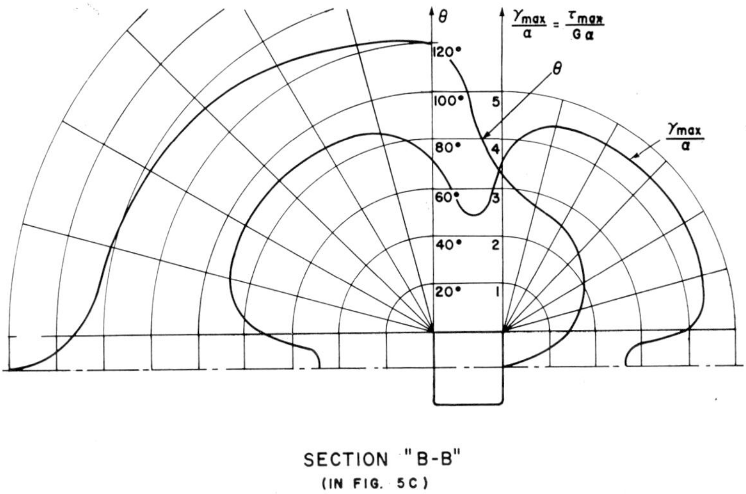

FIG. 21. Maximum shear strains and isoclinics in a transverse cross-section at the interface between a matrix and an embedded elbow with a flat and a circular end (restrained shrinkage)

D. *Matrix Shrinking About Two Embedded Circular Bars*

Of particular interest may be the analysis of stresses at the interface of two bars separated a short distance by the shrinking matrix. Two cases will be considered. In the first case the bars' ends are semi-spherical, in the second case they are flat with rounded fillets (Fig. 23). The two bars can be aligned or make an angle. The material used for the bars was Hysol 4290 epoxy (eq. 19), and the one used for the matrix was Sampson's epoxy (eq. 20).

Figure 24 shows the light field isochromatics in the meridian plane of the aligned bars. A thin layer of insert can be seen at the interface. It remained

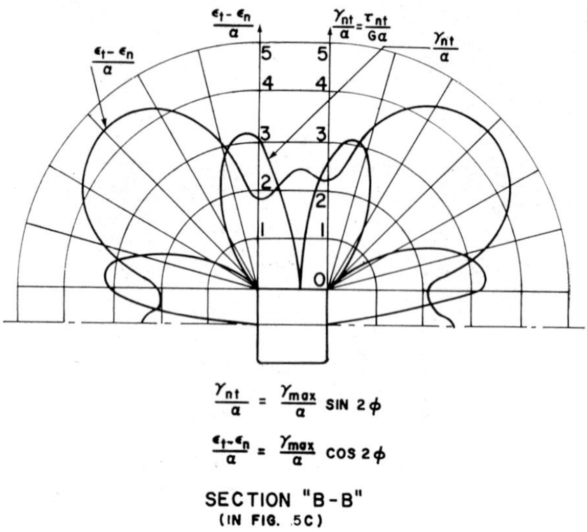

FIG. 22. Strains in a transverse cross-section at the interface between a matrix and an embedded elbow with a flat and a circular end (restrained shrinkage)

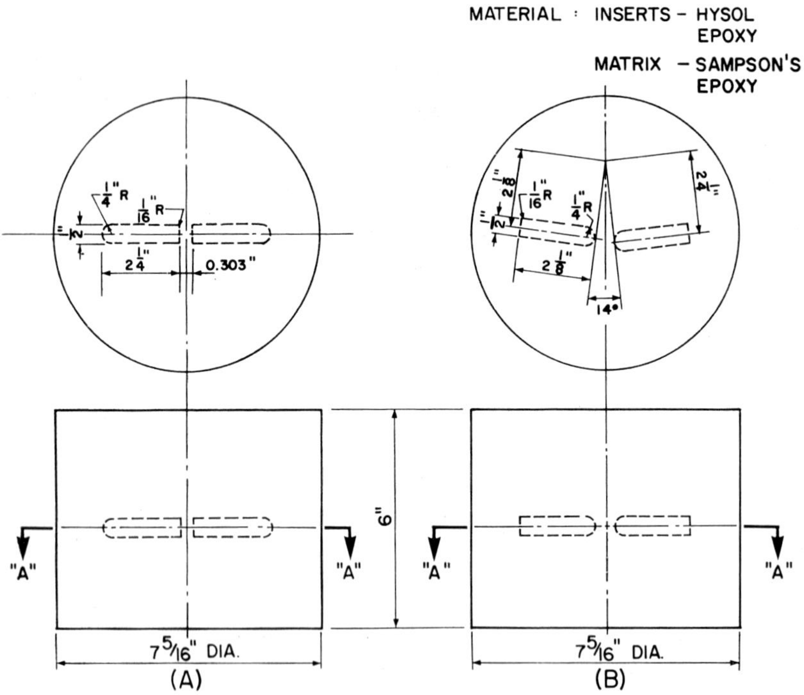

FIG. 23. Geometry of two models of matrices cast about circular bars

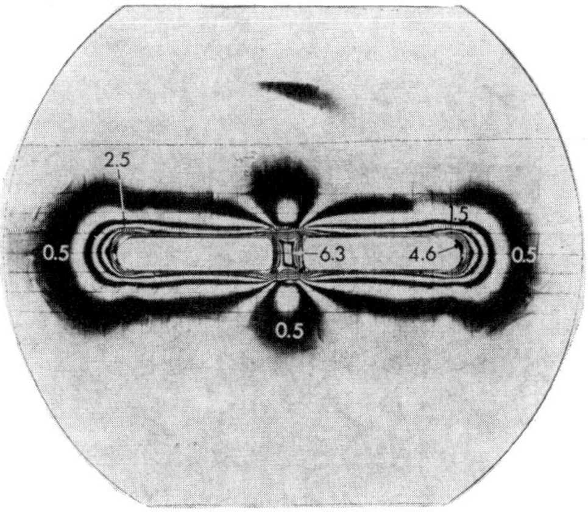

FIG. 24. Isochromatics in the meridian plane of a matrix shrunk around two circular bars, aligned and with flat ends a short distance from each other (light field)

after most of the insert had been removed. As explained in section III D, the insert has to be removed to conduct a correct analysis of the matrix. In this case the small amount of insert le influences the stresses in the matrix only in a negligible amount and it was considered better technique to leave the thin layer rather than remove some of the matrix in the process of removing the insert.

The maximum shear strains and the isoclinics at the interface are shown in Fig. 25. The result of evaluating the shear strains tangential to the interface

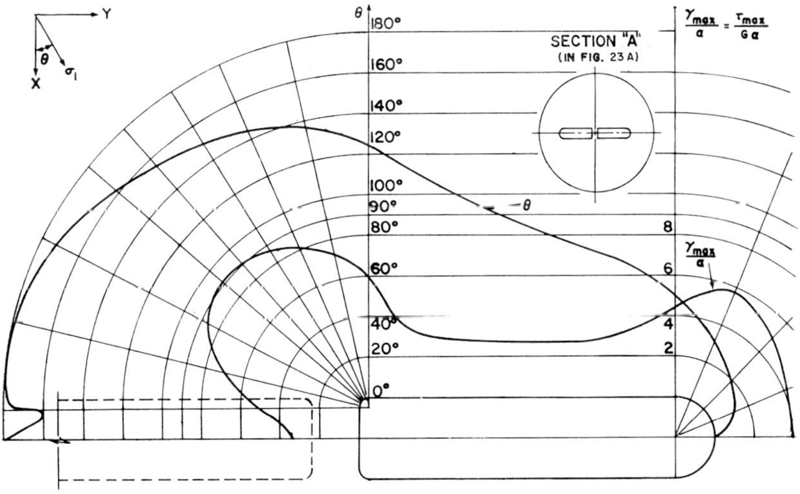

FIG. 25. Maximum shear strains and isoclinics at the interface between a matrix and two aligned circular bars in a matrix separated a short distance by the matrix (restrained shrinkage)

is reported in Fig. 26. Here again the difference between the normal strains near the corners is larger than predicted by eq. (8). The principal strains along the axis of symmetry in between the two bars are shown in Fig. 27.

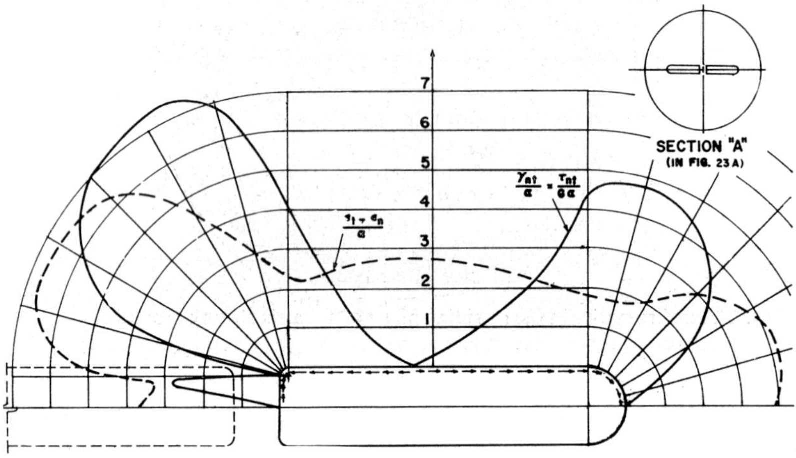

FIG. 26. Tangential shear strains along the interface of the aligned circular bars separated a short distance by the shrinking matrix

The dark field isochromatics for the case of the near aligned bars, facing each other at the semi-spherical end, is shown in Fig. 28. The maximum shear strains and the tangential strain at the interface are shown in Figs. 29 and 30, respectively.

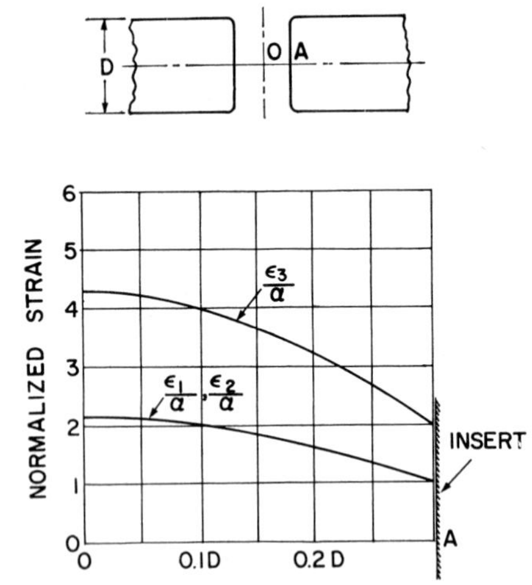

FIG. 27. Principal strains along the axis of symmetry between two circular bars, aligned, and with flat ends a short distance from each other

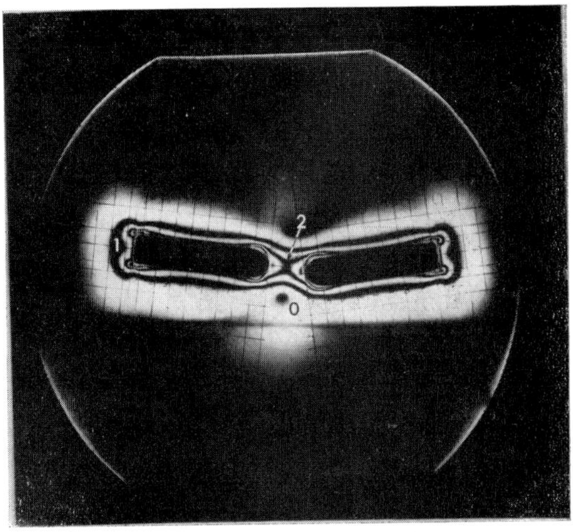

FIG. 28. Isochromatics in the meridian plane of a matrix shrunk around two circular bars, with the circular ends a short distance from each other (dark field)

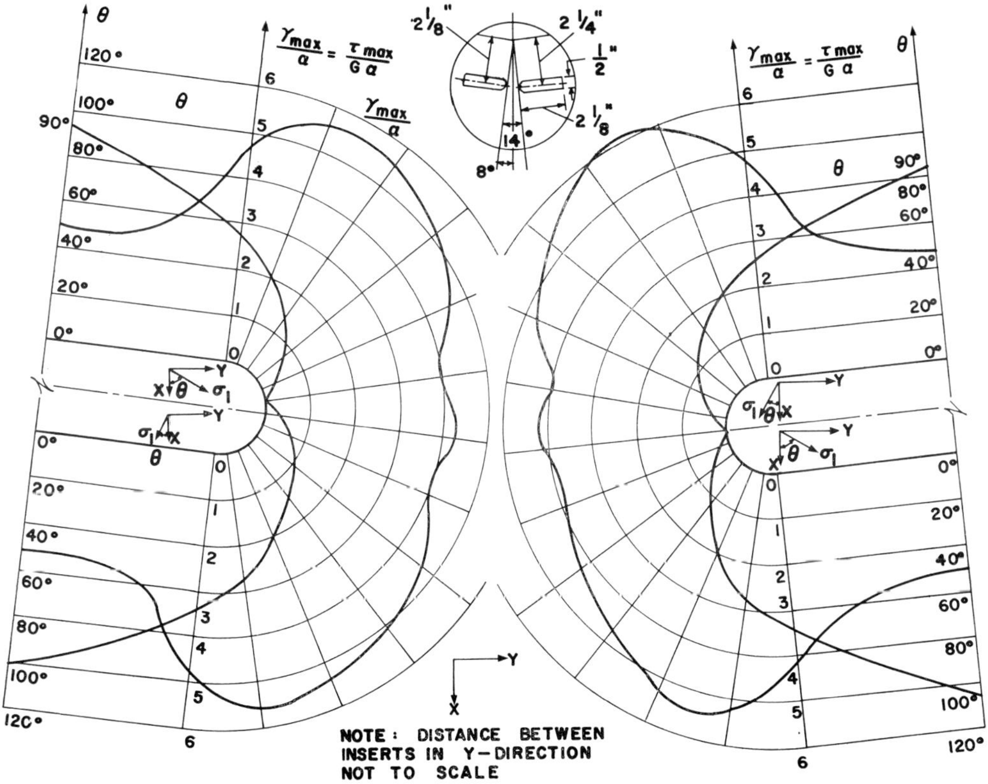

FIG. 29. Maximum shear strains and isoclinics at interfaces between a matrix and two non-aligned circular bars separated a short distance by the matrix

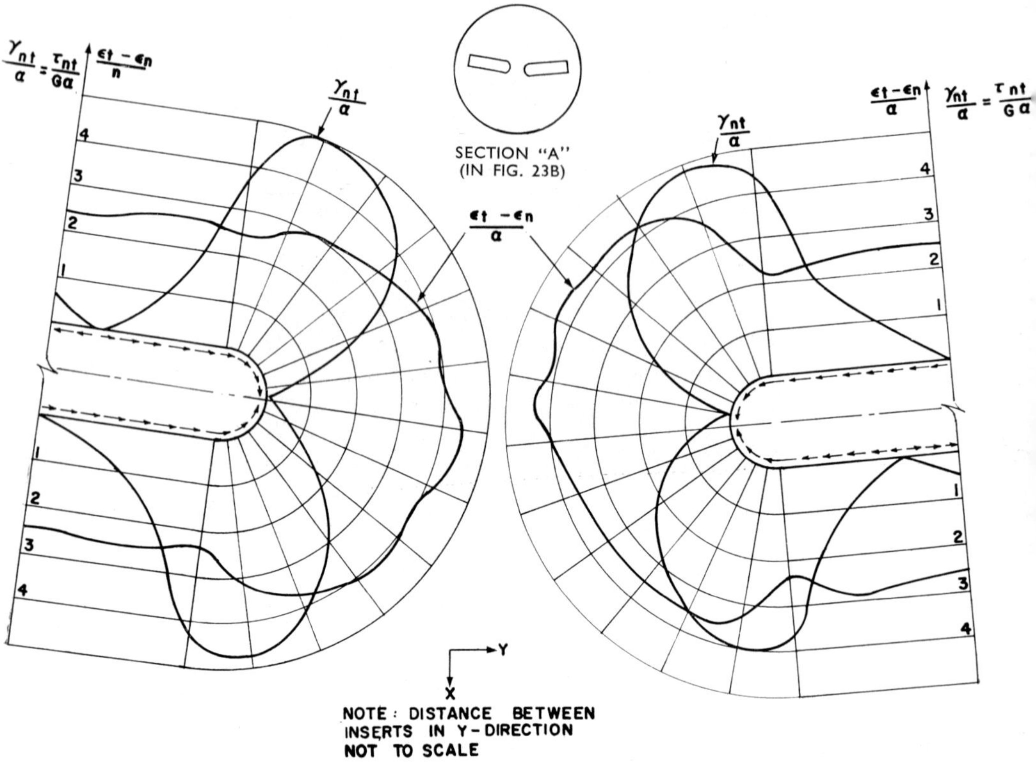

FIG. 30. Tangential shear strains along the interfaces of the non-aligned circular bars separated a short distance by the shrinking matrix

E. *Two Aligned Bars Eccentric in a Shrinking Rectangular Beam Subjected to a Concentrated Load*

In the previous sections the matrix was relatively large with respect to the size of the bars and for practical purposes could be considered as infinite. A problem of great interest in practice and which will be studied in this section is the distribution of stresses in a reinforced beam, as the case of the reinforced concrete beam. The reinforcing elements (usually steel bars) are parallel to the long sides of the beam, and are located near the boundary subjected to tensile stresses. In general, the beam is subjected to two main loading conditions: restrained shrinkage and bending loads. The determination of stress distributions associated with restrained shrinkage follows the same approach illustrated in the previous sections. The determination of the stress distribution associated with applied loading presents some particular features which deserve special consideration.

The geometry of the beams analyzed is shown in Fig. 31. As for the tests described in the previous sections, the matrix was made of Sampson's epoxy

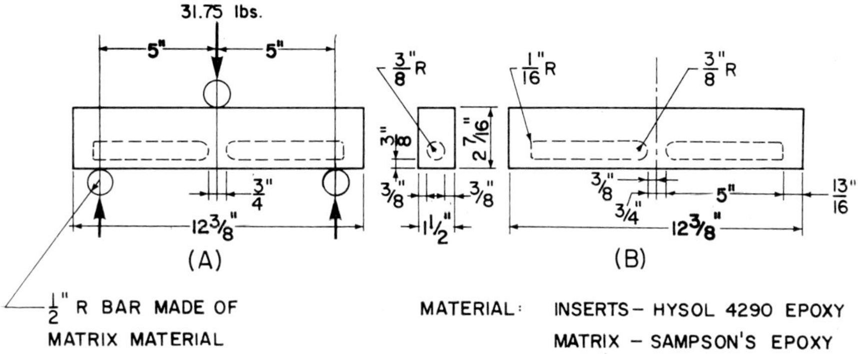

FIG. 31. Geometry and loading conditions of two composite beams: (a) Mechanical loading and restrained shrinkage. (b) Restrained shrinkage

and the inserts using Hysol's epoxy. The slicing plans are shown in Fig. 32. Photographs of the beams, before slicing, when put in the polariscope are shown in Fig. 33.

These patterns are difficult to interpret since they correspond to an integral effect along the path of light, an effect which varies in an unknown manner from point to point.

The isochromatics associated with the vertical meridian slice of the beam subjected to restrained shrinkage only are shown in Fig. 34(a). The value of the maximum shear strains and the direction of the principal strains, for points along the interface are shown in Fig. 35. The shear stresses, in the plane tangent to the interface and in the longitudinal direction of the bar, are shown in Fig. 36.

The other beam, shown in Fig. 31(A), has been subjected to a concentrated load applied to the middle point of its top boundary, and supported on two supports at the bottom. The birefringence associated with the applied loading was "frozen" in the specimen by subjecting the beam to the heating cycle described in section III C 1 when under load. This procedure locks-in in the

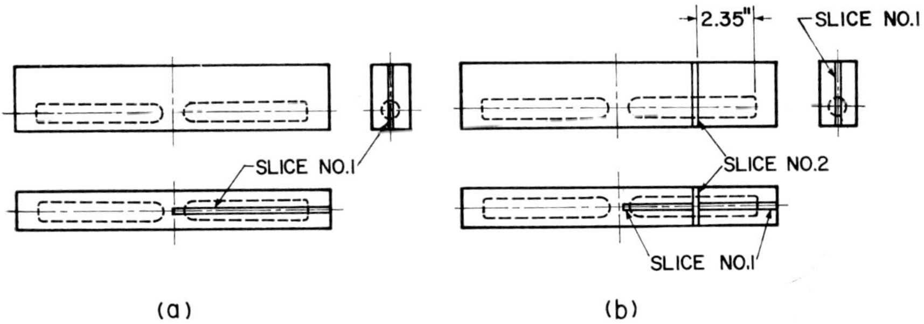

FIG. 32. Slicing plans of two composite beams

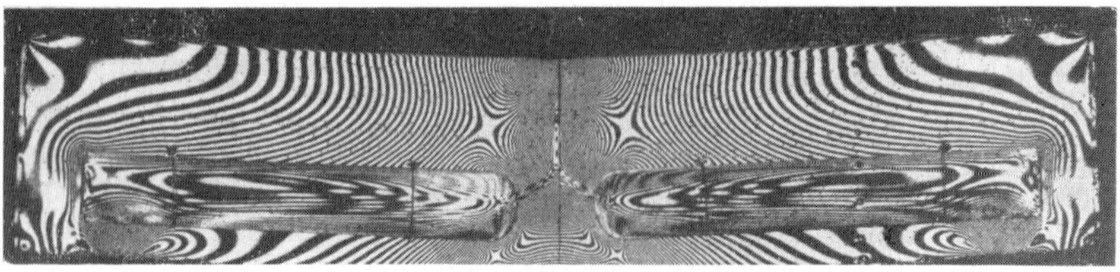

(a) MECHANICAL LOADING AND RESTRAINED SHRINKAGE

(b) RESTRAINED SHRINKAGE

FIG. 33. Photographs of two composite beams in the polariscope, before slicing

beam, at the same time, the birefringence associated with restrained shrinkage, and the one associated with applied loading. Since heat is necessary to lock-in the effect, and since restrained shrinkage is associated with the heating of the specimen, both effects cannot be physically separated.

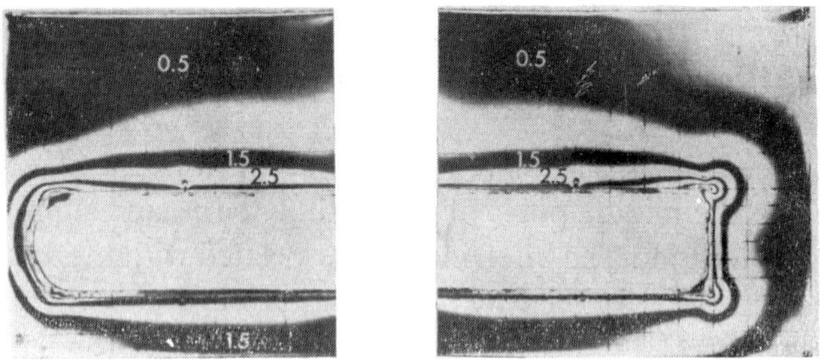

(a) RESTRAINED SHRINKAGE ONLY

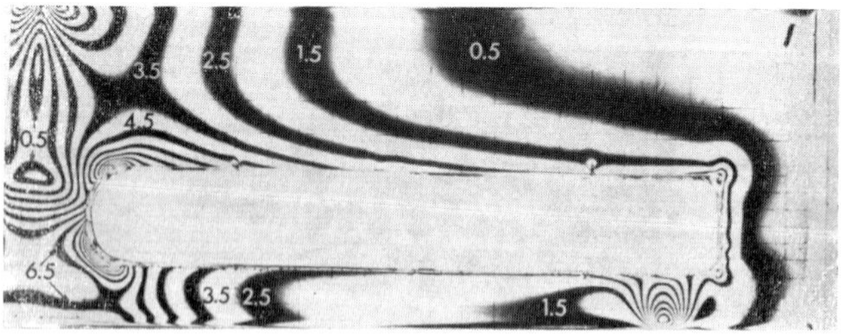

(b) MECHANICAL LOADING PLUS RESTRAINED SHRINKAGE

FIG. 34. Isochromatics of two composite beams reinforced with two circular bars

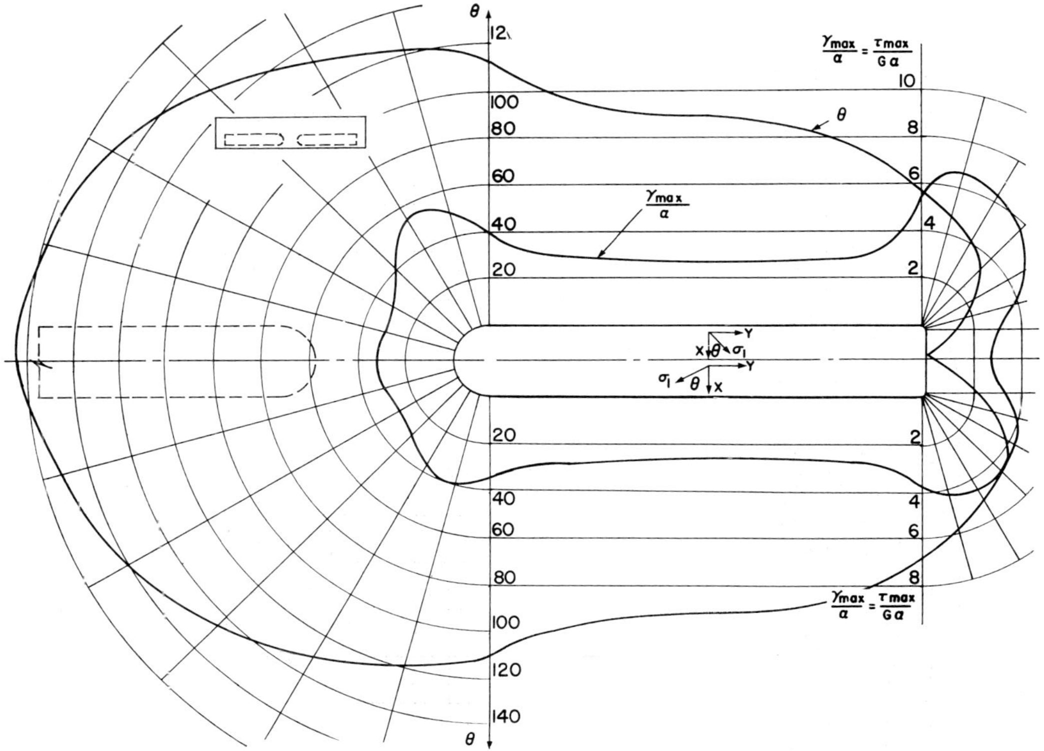

FIG. 35. Maximum shear strains and isoclinics at the interface of the insert in the composite beam subjected to restrained shrinkage

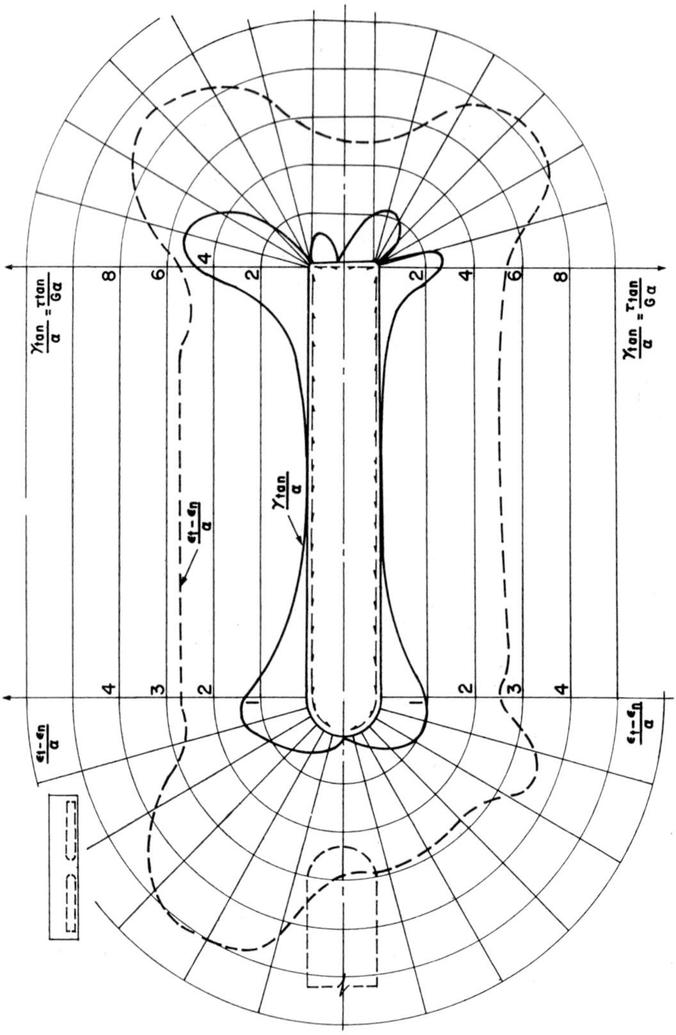

FIG. 36. Tangential shear strains at the interface of the insert in the composite beam subjected to restrained shrinkage

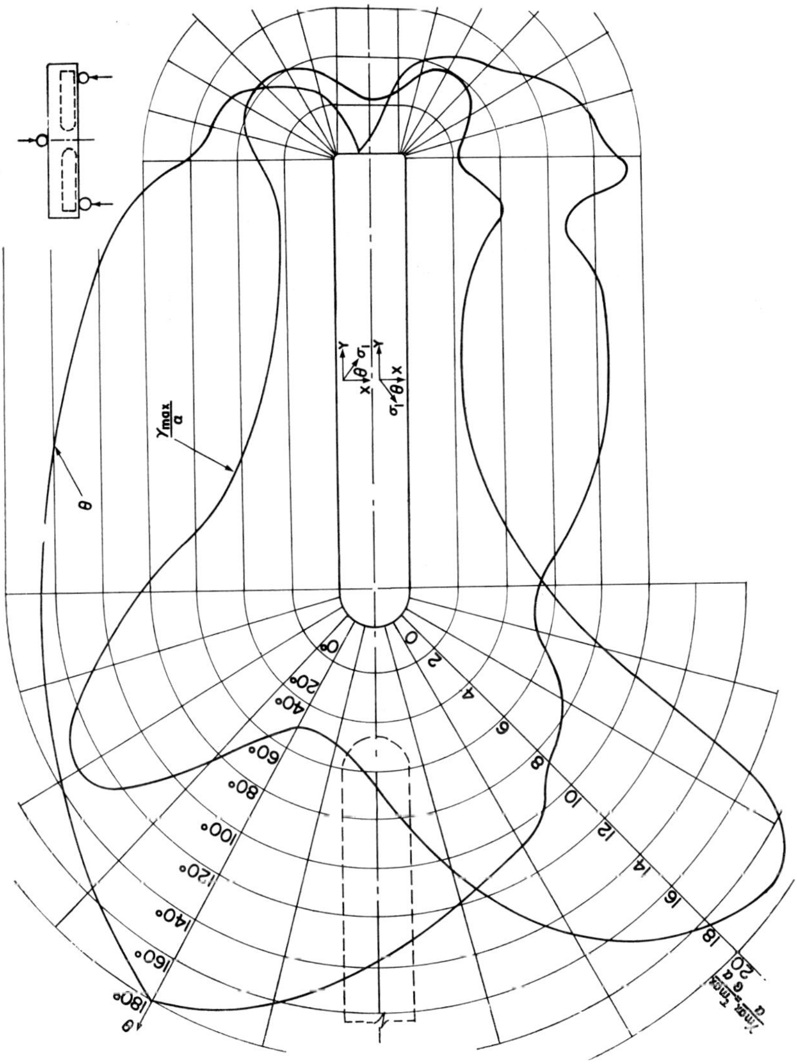

FIG. 37. Maximum shear strains and isoclinics at the interface of the insert in the composite beam subjected to mechanical loading and restrained shrinkage

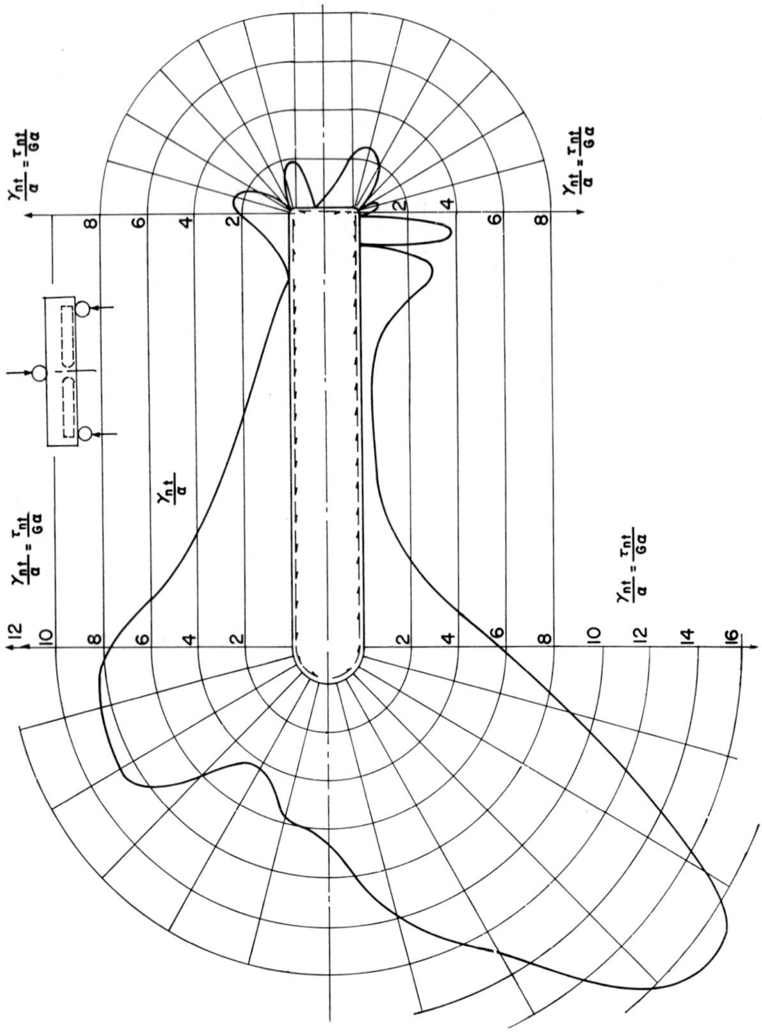

FIG. 38. Tangential shear strains at the interface of the insert in the composite beam subjected to mechanical loading and restrained shrinkage

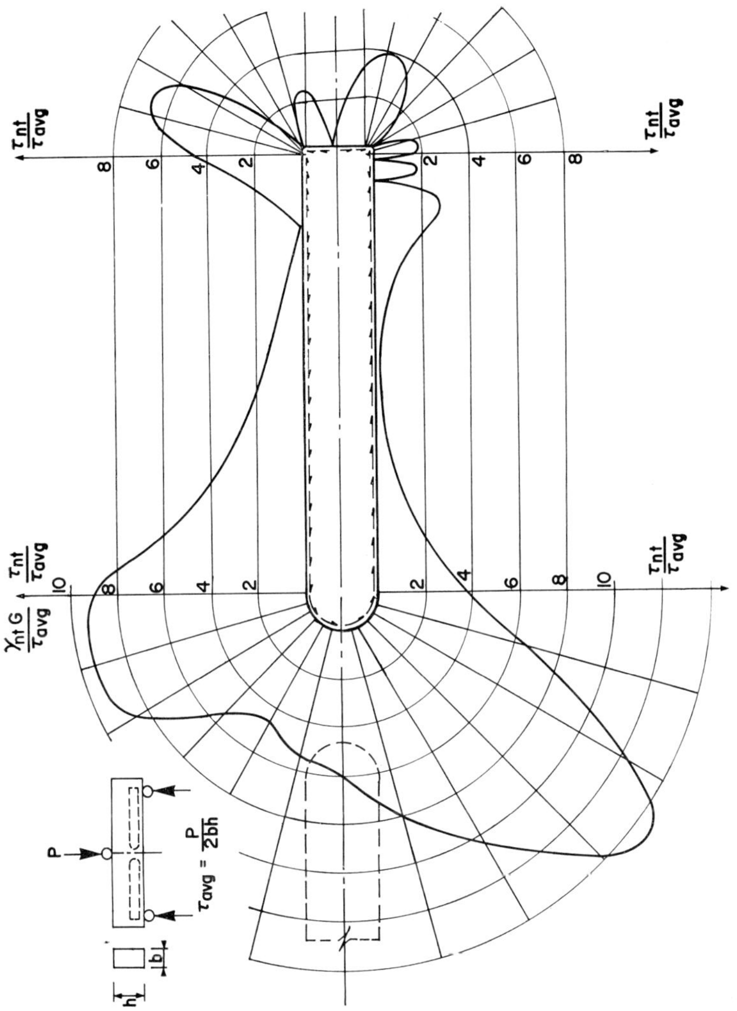

FIG. 39. Tangential shear stress at the interface of the insert in the composite beam subjected to mechanical loading

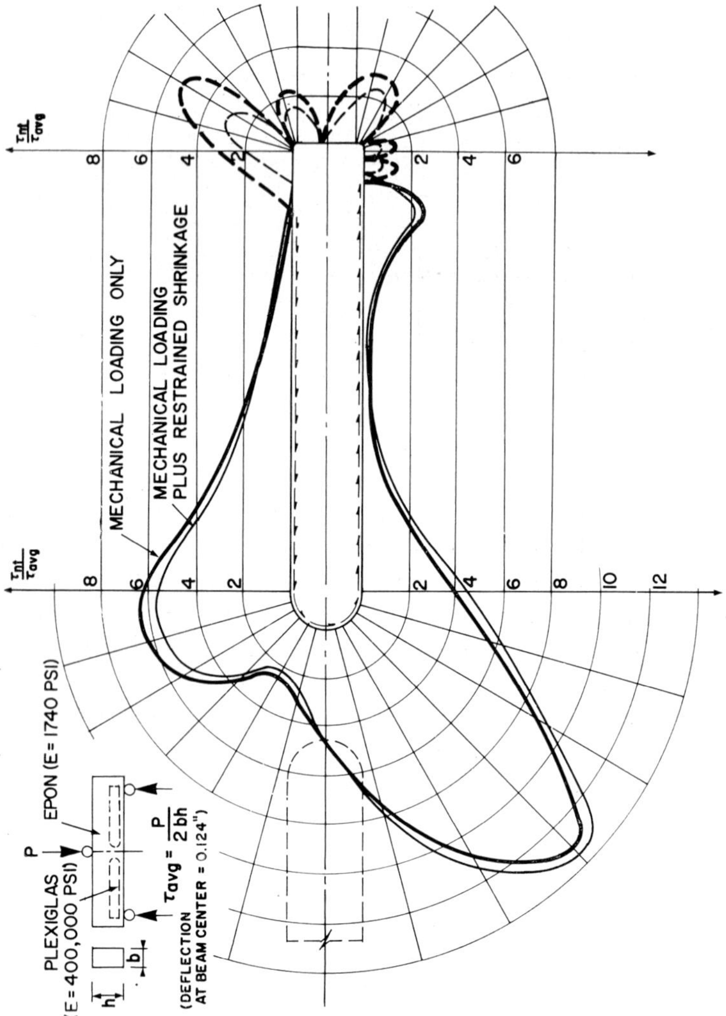

FIG. 40. Tangential shear strains at the interface of the insert in the composite beam subjected to mechanical loading with negligible restrained shrinkage

The isochromatics in the slice cut vertically through the middle of the beam are shown in Fig. 34(b). The values of the maximum shear strains and of the direction of the principal strain at points along the interface are shown in Fig. 37. The evaluation of this data gives the value of the shear strains tangential to the interface. They are given as Fig. 38.

To determine the strains associated with applied load, the shear strains tangential to the interface, and associated with restrained shrinkage (Fig. 36) have to be subtracted from the shear strains tangential to interface and associated with both the applied loading and restrained shrinkage as pointed out in section III C 3 (Fig. 38). The result of this operation is shown in Fig. 39.

In all these figures the scale of the ordinates is given both in terms of shear strains and in terms of shear stresses since as shown in eqs. (30) one can be directly converted into the other.

To avoid the time consuming operation of subtracting the optical response associated with shrinkage from the total response associated with shrinkage and mechanical loading in order to obtain the response associated only with mechanical loading, an attempt has been made to freeze an epoxy around a rigid insert, without introducing shrinkage stresses, or introducing only a negligible amount of them. At present best results have been obtained using an Epon epoxy resin cured for several days at room temperature and two days at 120° F. The insert used is Plexiglas. The same rectangular beam with eccentric bars studied above, was manufactured using Epon and Plexiglas. The result of the analysis is shown in Fig. 40. It can be seen in the figure that stresses associated with restrained shrinkage are very small in comparison to stresses associated with applied load. The plot of the shear stresses along the interface is not exactly the same as the one obtained using the subtraction method explained above. The differences are due mainly to a slight difference in the geometry of the two beams and the fact that the first beam was subjected to an appreciably larger deflection (0.329 in. against 0.124 in.).

F. *Matrix Shrinking About Embedded Multiple Circular Bars*

The last two sections emphasized analyses of the matrix around inserts lying end to end. Another commonly found condition is the case of inserts lying side by side, parallel to each other in arrays or bundles. To illustrate this type of multiple insert some stresses of interest have been determined in matrices shrinking about two, three and four circular bars, parallel to each other, and with one rounded end. The other end of all the bars is flat, and flush with the free surface of the cast matrix. Geometric details are shown in Fig. 41. Representative isochromatics are shown in Figs. 42 through 46.

Maximum shear strains at points in the axial plane of the matrix are shown in Fig. 47. The results from two transverse slices are included, one through

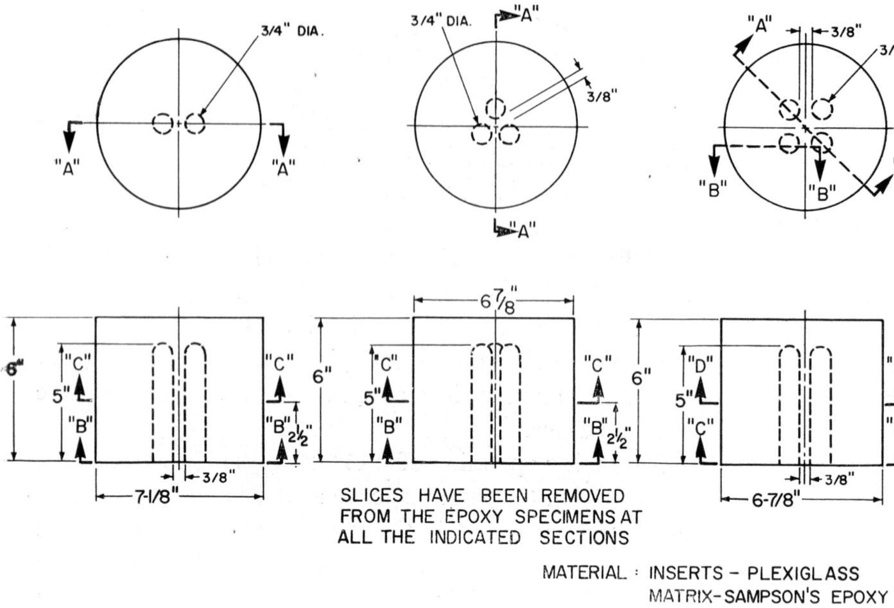

FIG. 41. Geometry of three models of matrices cast about multiple inserts

the middle of the cylindrical matrix and the other at its face surface (section *C–C* and *B–B* in Fig. 41). For comparison purposes, Lamé's solutions for plane stress and plane strain have also been plotted. It is noticeable that the middle slice is approximately in a state of plane stress, although there is about a 20% discrepancy at the interface.

The results obtained from a two-dimensional test using Solithane rubber and bonded steel inserts (reported below) have been added to Fig. 41. At the inter-

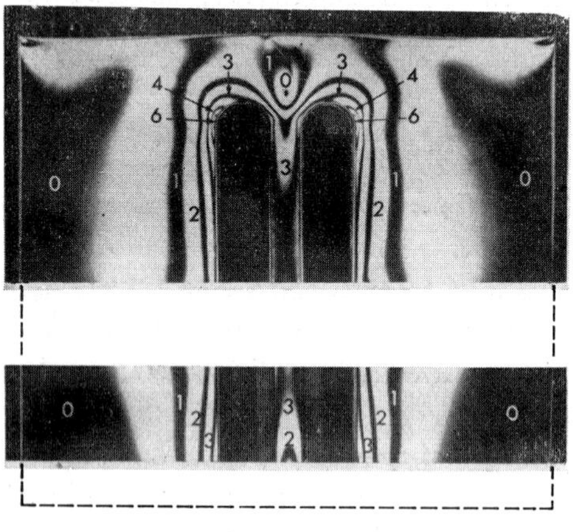

FIG. 42. Isochromatics in a meridian slice of a matrix shrunk about two circular bars

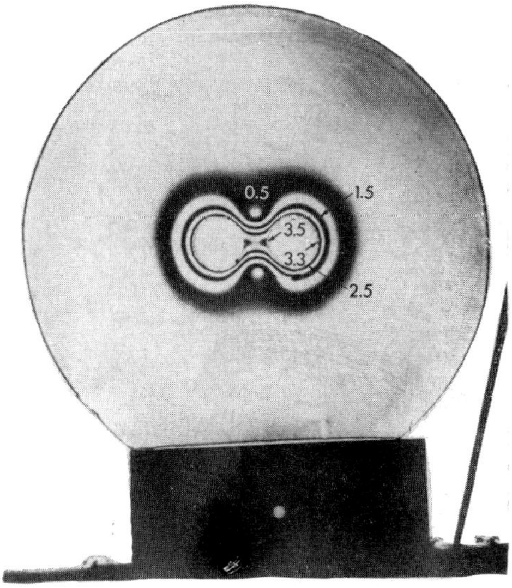

FIG. 43. Isochromatics in a transverse cross-section (middle slice) of a matrix shrunk about two circular bars

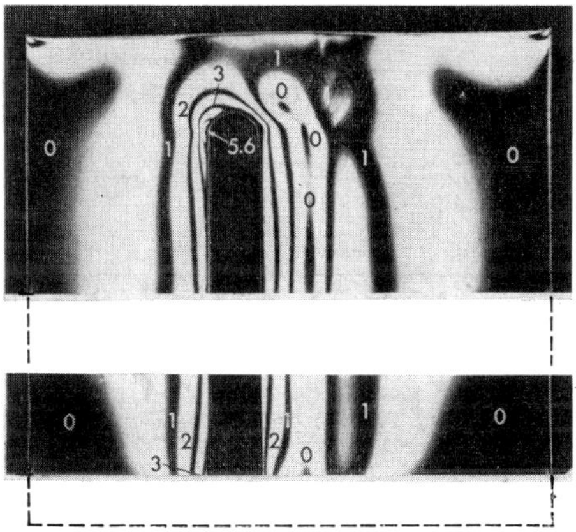

FIG. 44. Isochromatics in a longitudinal slice of a matrix shrunk about three circular bars

face they coincide with the plane strain solution and they approach the plane stress solution as points are further removed from the interface.

Comparison, between maximum shear strains obtained from transverse middle and face surface slices for points in between inserts, for the stress cases of two, three and four inserts are shown in Figs. 48, 49, and 50. Again results obtained from two-dimensional "pinched" tests are added for comparison. Figure 51 shows a meridian plane analysis of an interface for the case of two inserts.

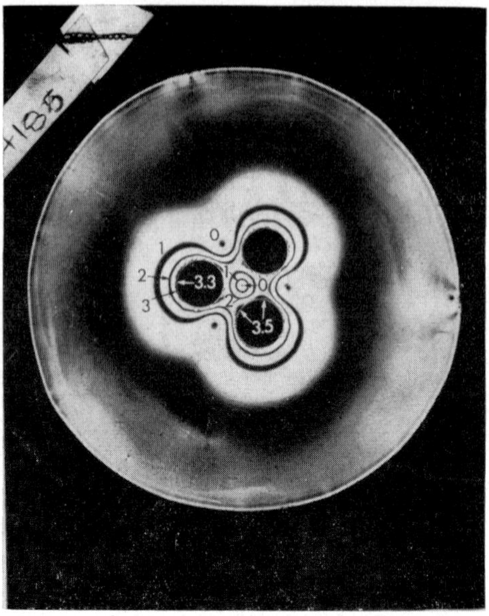

FIG. 45. Isochromatics in a transverse cross-section (middle slice) of a matrix shrunk about three inserts

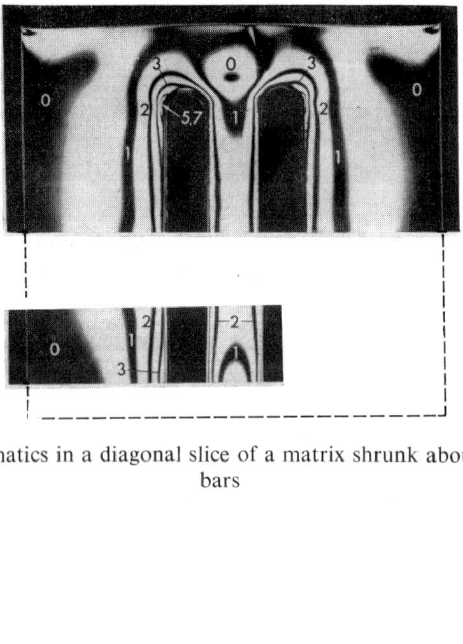

FIG. 46. Isochromatics in a diagonal slice of a matrix shrunk about four circular bars

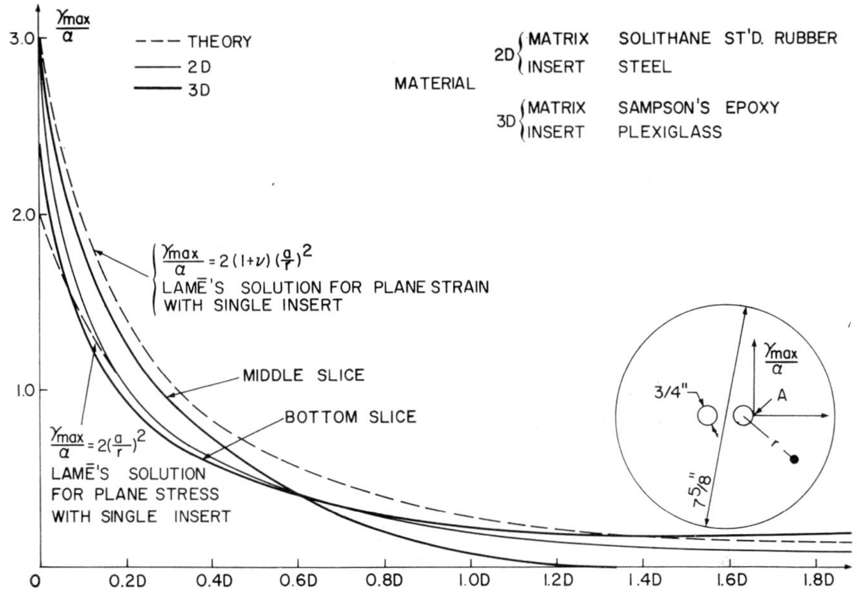

FIG. 47. Maximum shear strains at points in the axial plane of a matrix cast about two circular inserts, and shrunk

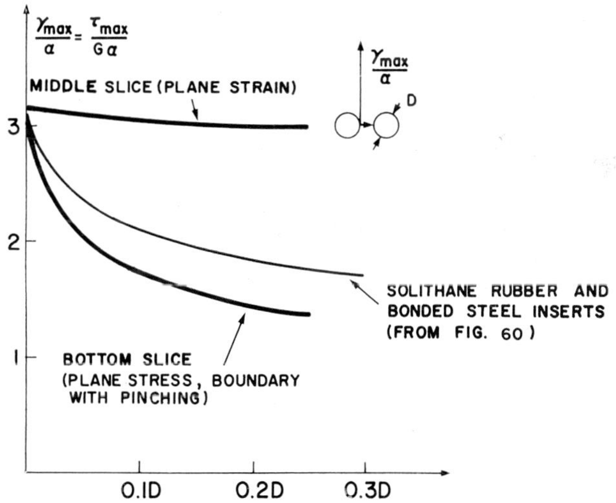

FIG. 48. Maximum shear strains in between two circular bars embedded in a shrinking matrix (transverse cross-sections)

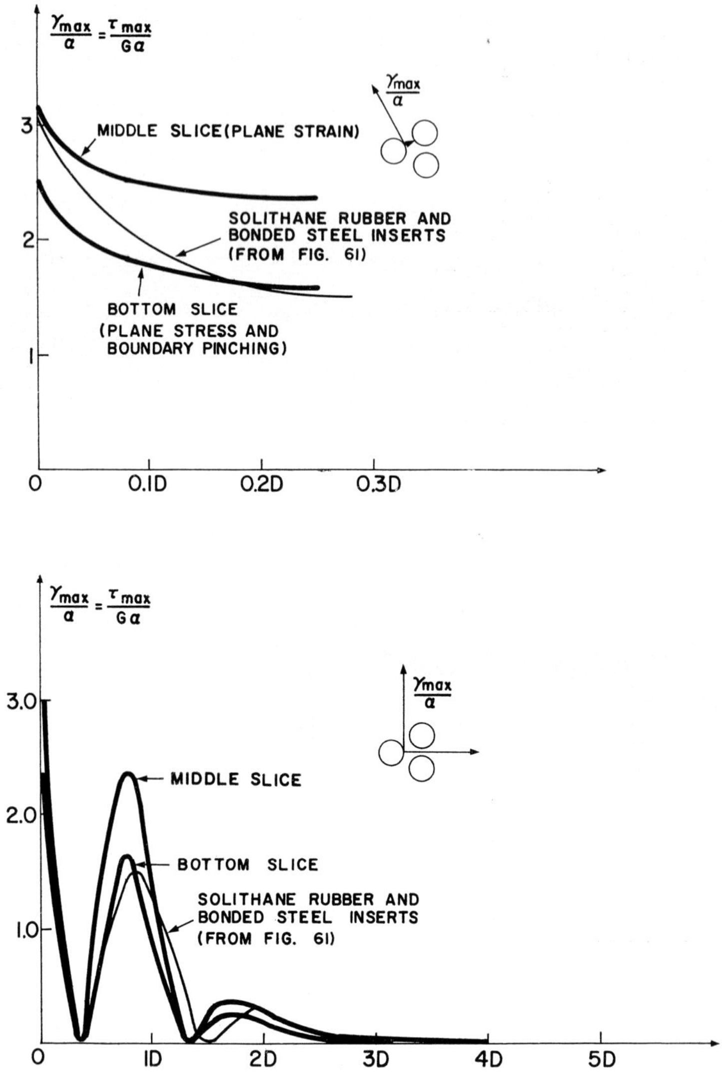

FIG. 49. Maximum shear strains in a shrinking matrix with three embedded circular bars (transverse cross-sections)

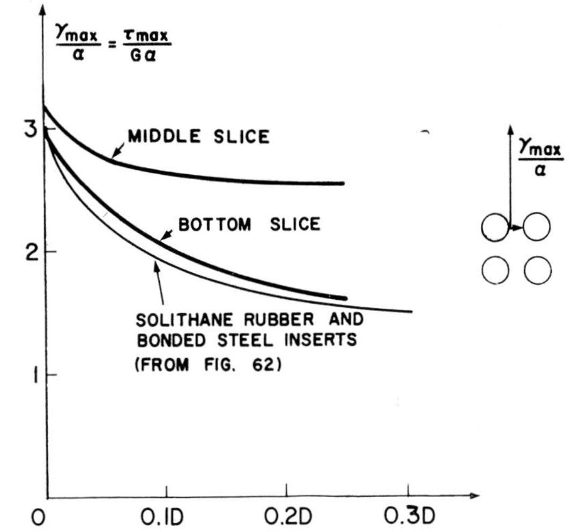

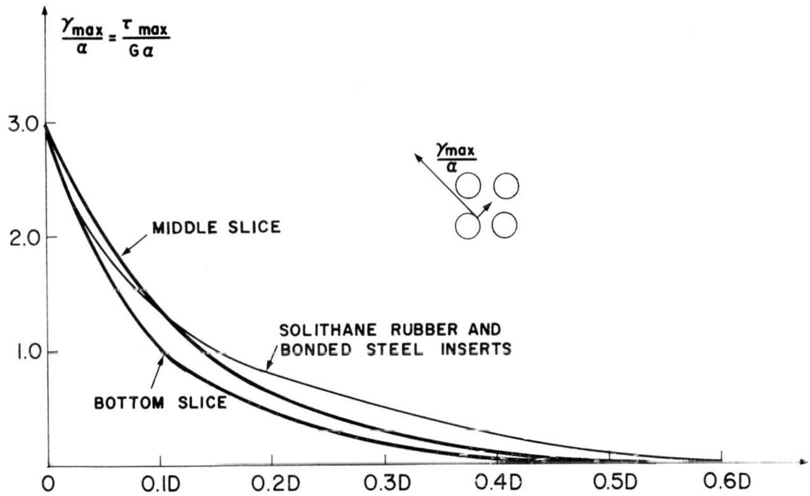

FIG. 50. Maximum shear strains in a shrinking matrix with four embedded circular bars (transverse cross-sections)

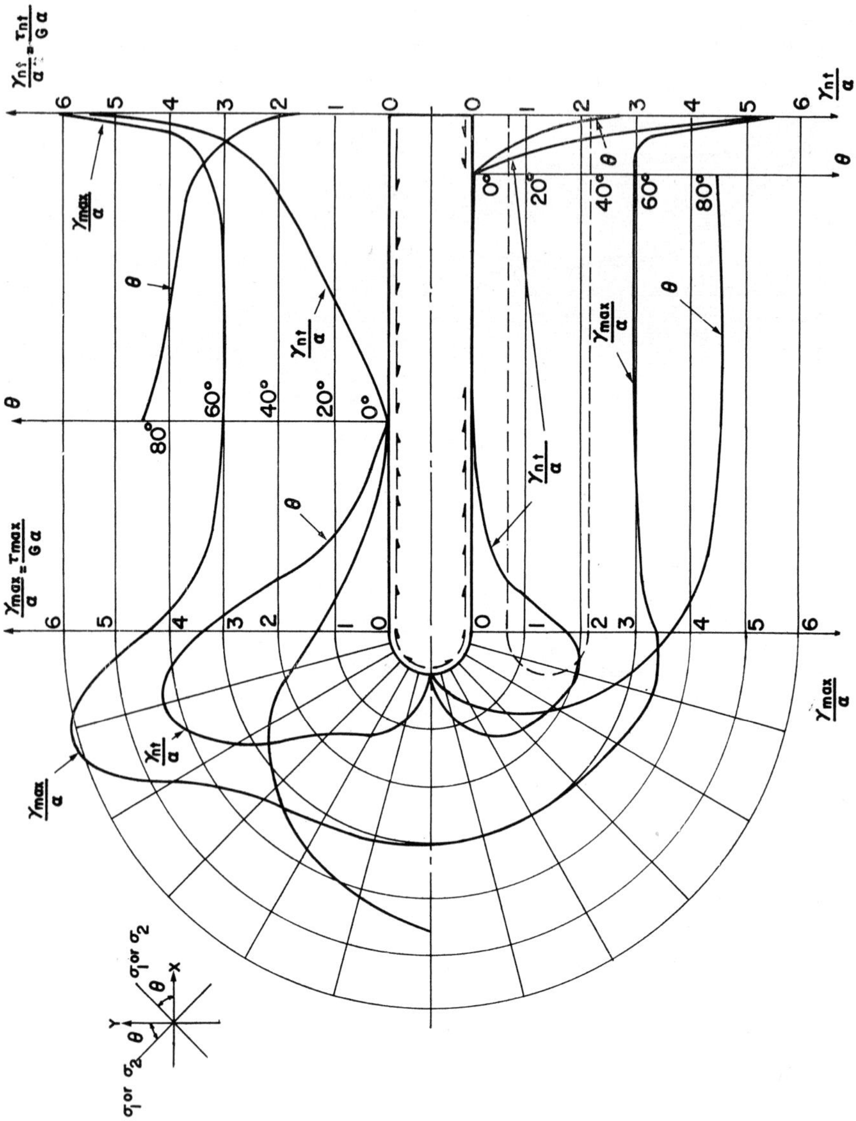

FIG. 51. Maximum shear strains, isoclinics and tangential shear strains along the interface of one of two circular bars embedded in a shrinking matrix

V. TWO-DIMENSIONAL ANALYSIS (BONDED INSERTS)

A. *Circular Insert Bonded to an Annular Matrix Diametrically Loaded*

This test illustrates the method of determining the distribution of stresses in two-dimensional composite bodies subjected to mechanical loadings. The matrix material used is the urethane rubber Scotchcast described in section IV B which shrinks only a negligible amount in the process of curing and permits the determination of the stresses associated with mechanically applied loadings in one test without the need of the subtraction of shrinkage stresses. The specimen is a ring the inside boundary of which is bonded to a disk which fits the inner diameter of the ring (Fig. 52). The loading is diametral compression.

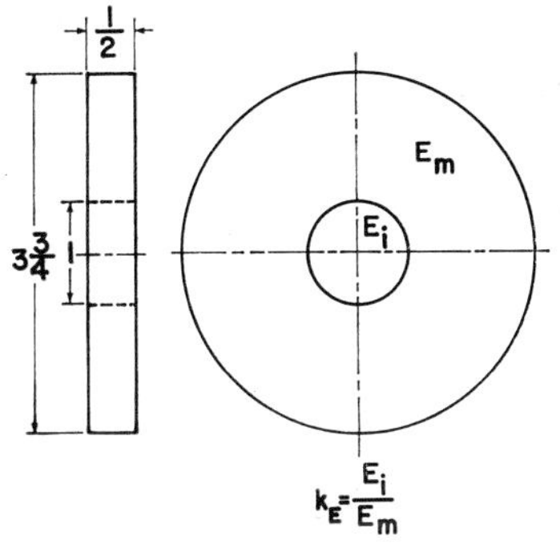

FIG. 52. Geometry of ring, with insert, to be loaded under diameter compression

The parameter used to compare the results systematically is:

$$k_E = \frac{E_i}{E_m},$$

the ratio between the moduli of elasticity of the insert and the matrix. A plain disk ($D' = 3.75$ in., $t = 0.5$ in.) of matrix material corresponds to $k_E = 1$ and an empty circular ring (O.D. = 3.75 in.; I.D. = 1 in.; $t = 0.5$ in.) of matrix material corresponds to $k_E = 0$. Intermediate values were obtained using inserts of another urethane rubber (Hysol, $k_E = 0.9$) and a silicone rubber ($k_E = 0.74$). An epoxy insert gave results corresponding to the practically infinitely rigid case ($k_E = 720$).

Isochromatics for the case $k_E = 0.9$ are shown in Figs. 53 and 54. The normalized maximum shear stresses along the horizontal and vertical axis of the circular specimens are shown in Figs. 55 and 56, respectively. The fringe value of the matrix material was obtained from the fringe order at the center of the

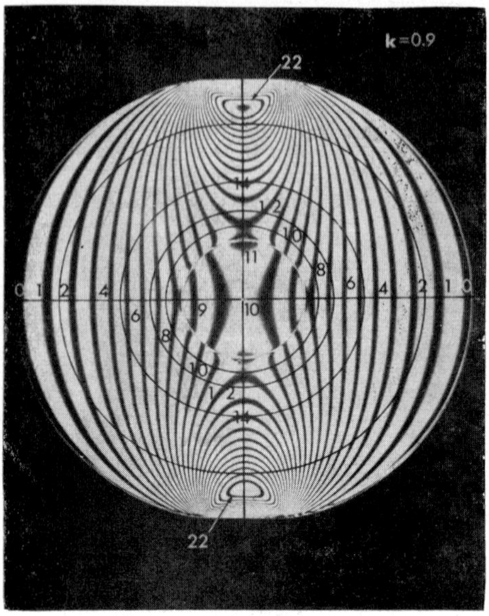

FIG. 53. Dark field isochromatics in a ring, filled with an insert, subjected to diametral load

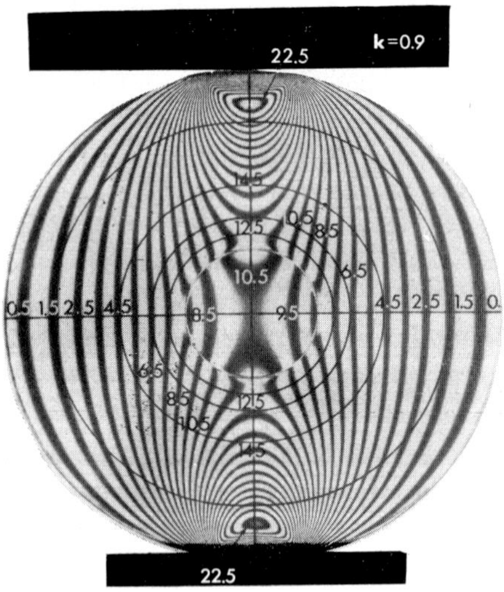

FIG. 54. Light field isochromatics in a ring, filled with an insert, subjected to diametral load

plain disk by:

$$f_\sigma = \frac{4P'}{\pi D' n'}$$

where P' is the load applied to the disk, D' the diameter of the disk and n' the fringe order at center, produced by the load P'.

The normalization was obtained from:

$$k_\tau = \frac{\tau_{\max}}{\sigma_a}$$

where

$$\sigma_a = \frac{P}{tD},$$

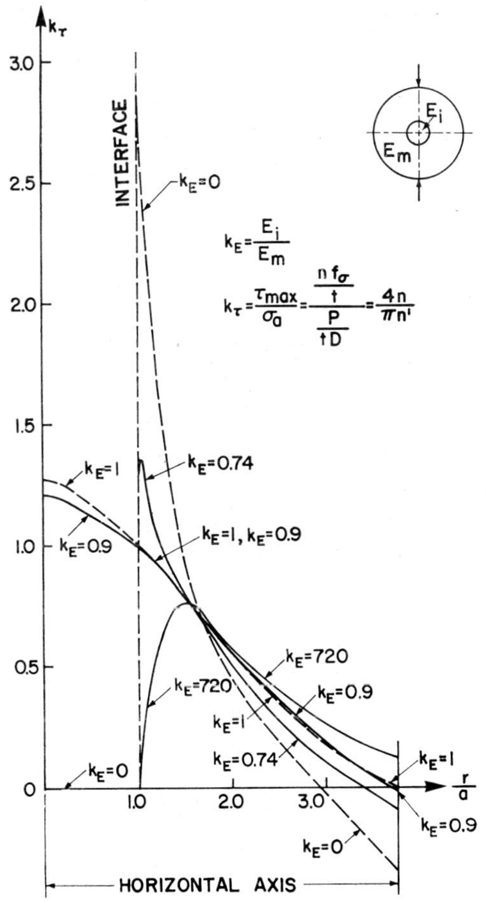

FIG. 55. Normalized maximum shear stresses at the horizontal axis of a ring with an insert

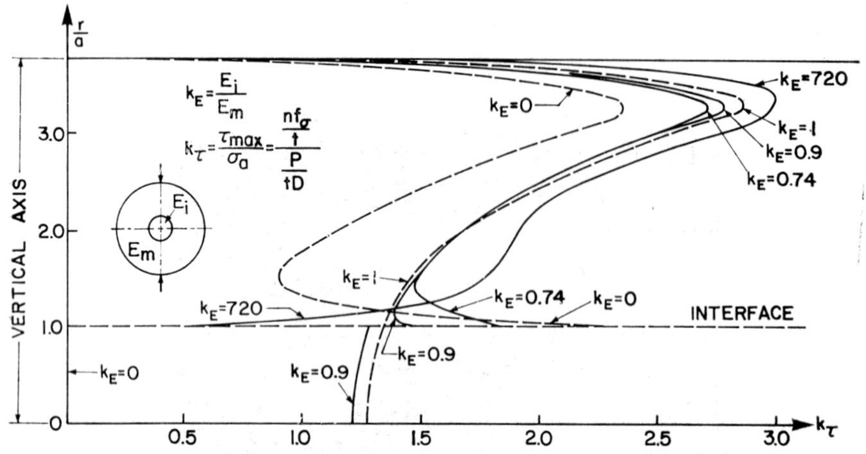

FIG. 56. Normalized maximum shear stresses at the vertical axis of a ring with an insert

and P is the load applied to the specimen and t and D are the thickness and outer diameter of the specimen. Since

$$\tau_{max} = \frac{nf_\sigma}{t},$$

if $P' = P$ and $D' = D$, then the normalized maximum shear can be expressed as a ratio of fringe orders

$$k_\tau = \frac{4}{\pi} \frac{n}{n'}.$$

In Figs. 55 and 56 the influence of "pinching" near the interface is noticeable for the case of $k_E = 0.9$, $k_E = 0.74$ and $k_E = 720$. It is obvious from this analysis that the problem solved exhibits a three-dimensional restrain at the interface and the solution obtained corresponds to the two-dimensional problem only at certain distances away from the interface.

B. *Circular Inserts Bonded in Plates Subjected to Restrained Shrinkage*

Using the urethane rubber Solithane 113 (or a similar product known as Hysol 4485) it is easy to manufacture flat matrices by casting the resin around any kind and number of inserts. Usually the inserts are made of steel and they are infinitely rigid with respect to the matrix. Examples of maximum shear strain distributions are shown in Figs. 57, 58 and 59. In these figures some selected isoclinics have also been recorded. Distribution of maximum shears at points along selected lines in the matrix located in between the inserts are shown in Figs. 60, 61 and 62.

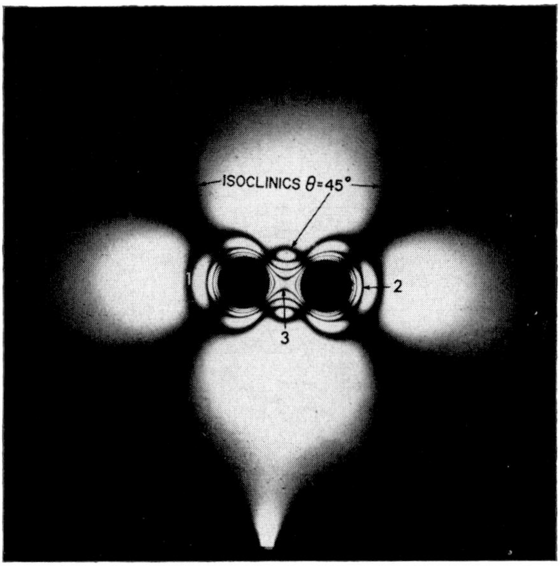

FIG. 57. Isoclinics ($\theta = 45°$) and isochromatics around two rigid inserts in a plate subjected to restrained shrinkage

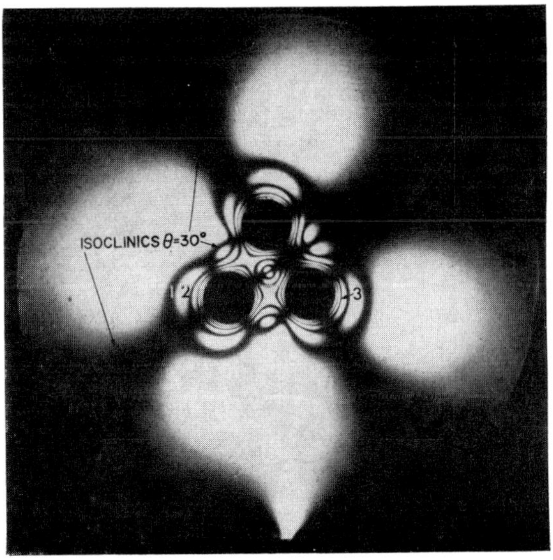

FIG. 58. Isoclinics ($\theta = 30°$) and isochromatics around three rigid inserts in a plate subjected to restrained shrinkage

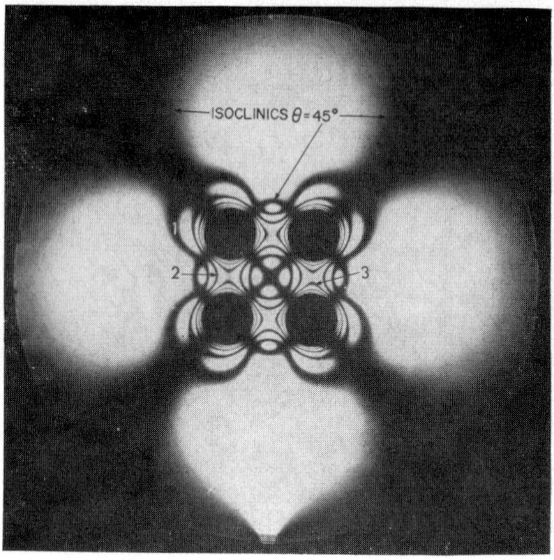

FIG. 59. Isoclinics ($\theta = 45°$) and isochromatics around four rigid inserts in a plate subjected to restrained shrinkage

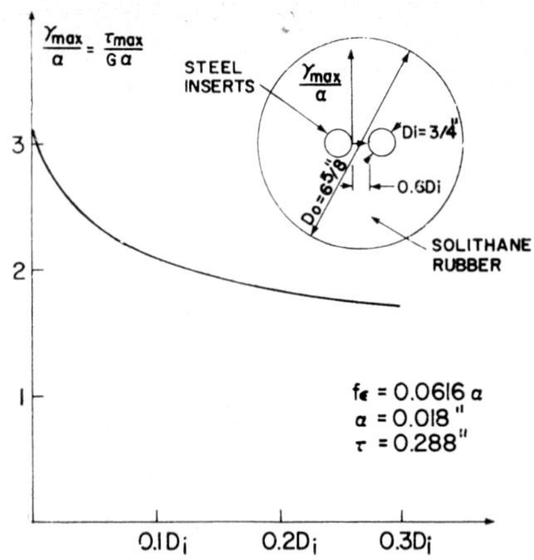

FIG. 60. Maximum shear strains in between two circular inserts embedded in a shrinkage matrix

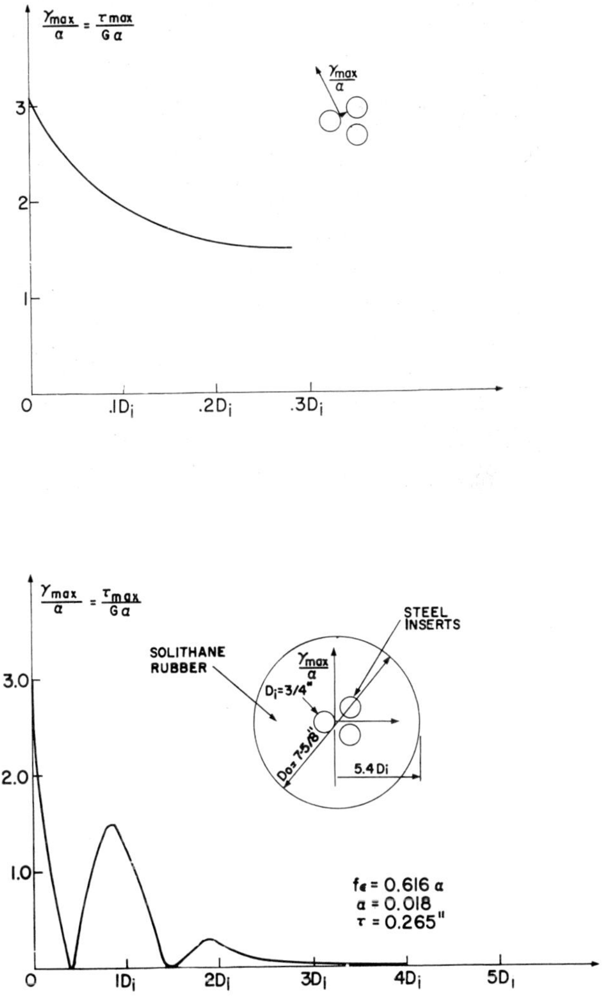

FIG. 61. Maximum shear strains in between three circular inserts embedded in a shrinking matrix

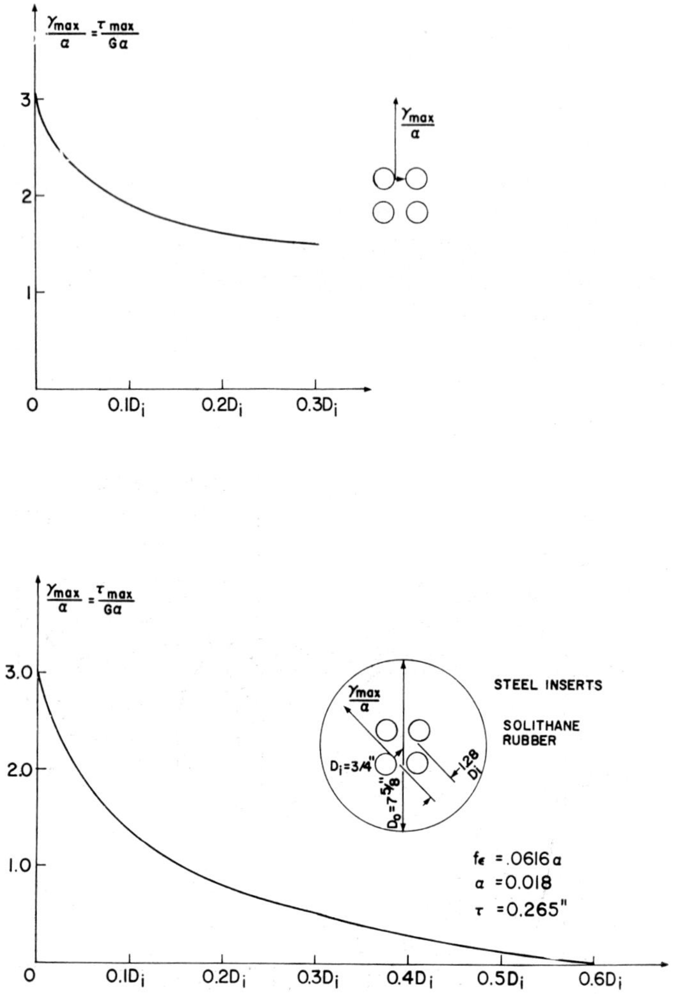

FIG. 62. Maximum shear strains in between four circular inserts embedded in a shrinking matrix

VI. TWO-DIMENSIONAL ANALYSIS (NO-PINCHING)

A. *Elimination of the "Pinching" Effect at the Interface*

The results of the test reported in the previous section correspond to a problem which is not a plane stress nor a plane strain problem, but a problem that approaches the plane stress situation at certain distances from the plug, and has the three-dimensional characteristics near the interface which in other papers have been called "pinching"[15]. To insure the conditions of plane stress at the boundary the knife edge mentioned in sections II B and IV B has been proposed. The geometry of this knife edge and isochromatics from a three-dimensional test[20] conducted to study its characteristics are shown in Fig. 63. Many problems have been solved using this technique and the results seem to be, in general, satisfactory. The plane stress or strain situation is well reproduced when the boundary is straight[21] or has large radii (Fig. 64). When the radius of the boundary is small the simulation is not as good. The design illustrated in Fig. 65 is being studied as an improvement.

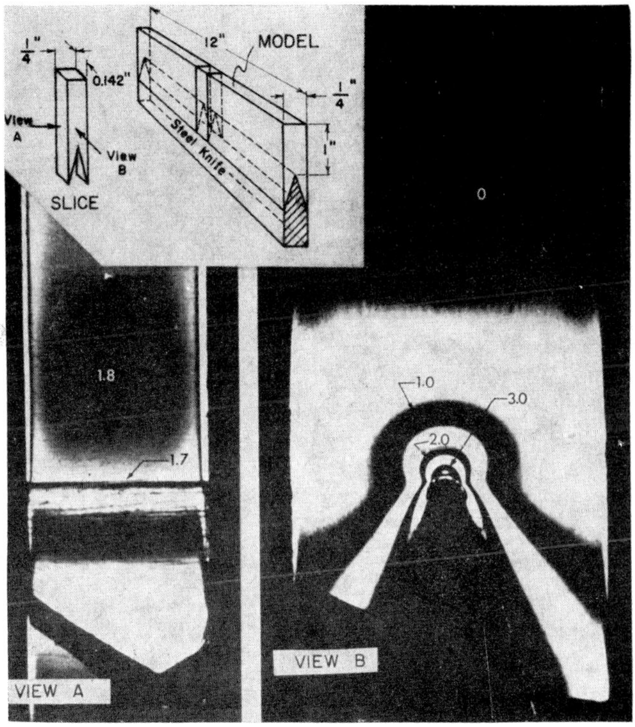

FIG. 63. Two isochromatic patterns of a slice from the center of an epoxy strip bonded to a knife edge

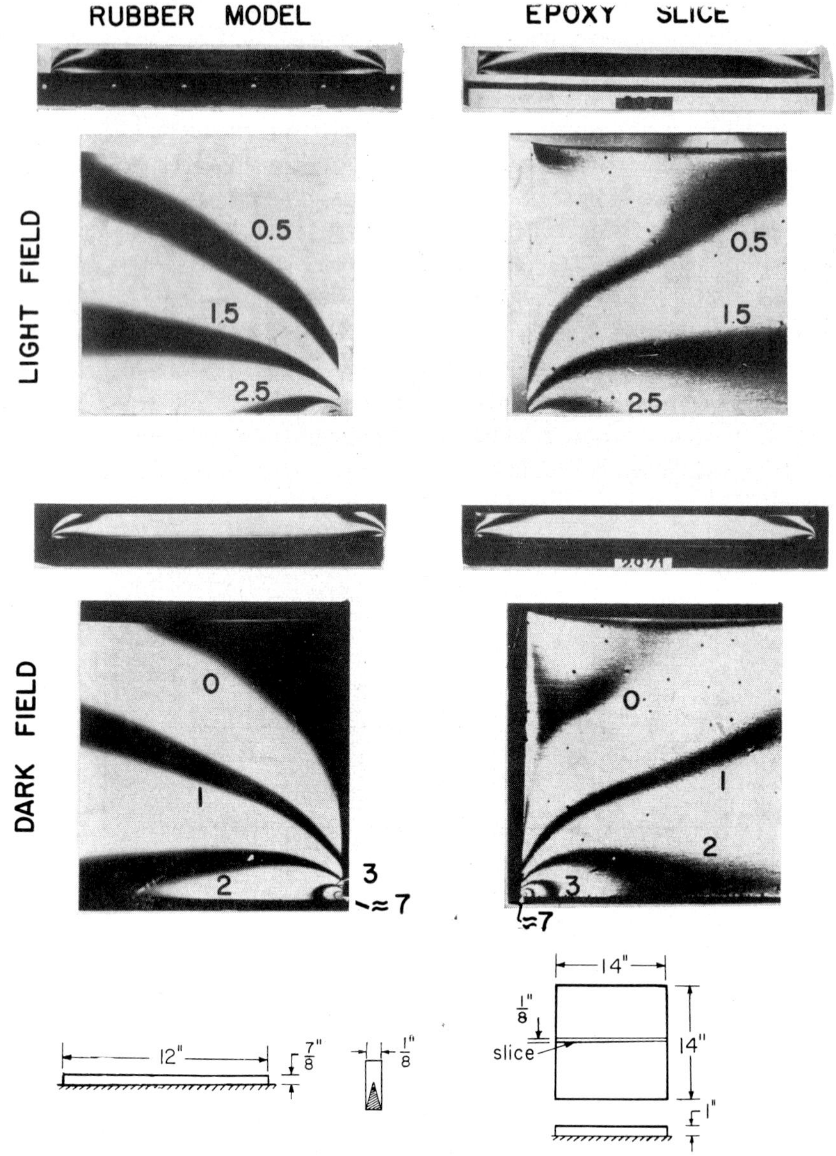

FIG. 64. Isochromatic patterns: *Left side*—7/8 in. × 12 in. × 1/8 in. thick rubber model bonded to a steel strip with a knife and shrunk. *Right side*—1 in. × 14 in. × × 1/8 in. thick frozen central slice cut from an epoxy model which was cast, bonded and shrunk on a steel slab

When the matrix is made of rubber it is possible to eliminate the pinching effect by removing the plug, oil it lightly if necessary, and replace it in the hole. This method eliminates most of the pinching, but may not give the desired boundary conditions since the non-bonded interface will not permit tensile stresses to develop normal to the interface and shear stresses tangential to the interface may also be eliminated. When stresses normal to the interface

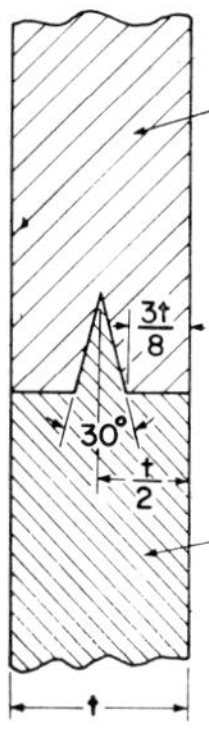

FIG. 65. New type of knife edge being studied to obtain plane stress condition at interfaces

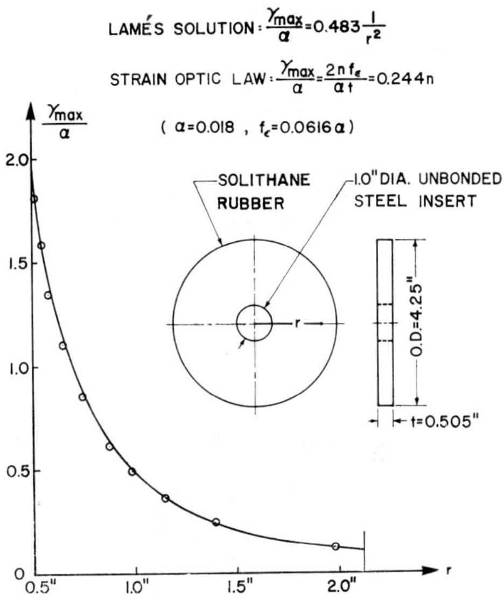

FIG. 66. Photoelastic calibration of a matrix shrinking about a circular insert

are compressive and principal the method is acceptable and can be used effectively for calibration. A round insert in the center of a circular shrinking matrix will satisfy Lamé's solution (Fig. 66).

B. *Rectangular Plate Subjected to Biaxial Restrained Shrinkage*

This problem has some interest as a help in the visualization of the stress distribution in a matrix of finite length between two fibers parallel to each other and interrupted at both ends by voids or cracks, and restrained from shrinkage by the two fibers. The two fibers may keep their relative positions or move toward each other.

1. BOUNDARY CONDITIONS. The problem as indicated in Fig. 67 is a mixed boundary value problem of elasticity. The short edges of the plate are free. The long edges are subjected to applied displacements. The strain applied in

the longitudinal direction along the long edges is constant and equal to α. The displacement of one long edge with respect to the other can be controlled by varying the rigidity of the frame. The analysis reported here has an average strain value between the two bonded edges of 0.59α.

The edges of the restraining bars have the shape of the knife described in section VI A.

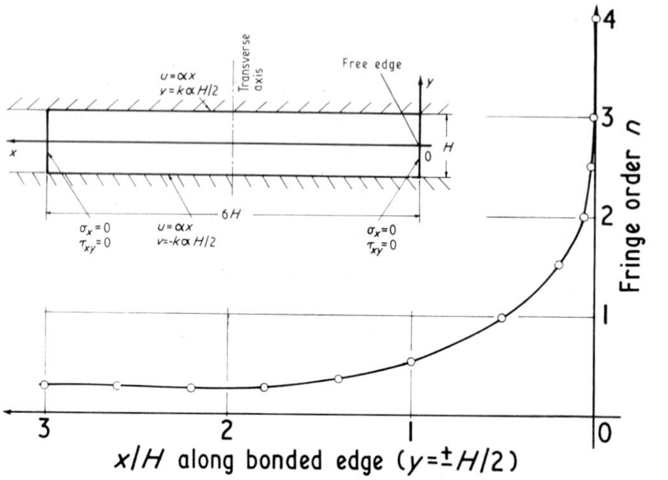

FIG. 67. Isochromatics along the two bonded edges of a rectangular plate subjected to biaxial restrained shrinkage

2. ANALYSIS. The normal stresses along the longitudinal and transverse axes were separated by using the shear difference method. These results were checked by using the oblique incidence method, and the agreement was found to be good. These stresses were initially obtained in terms of fringes. Subsequent to the calibration shown below, they were normalized and are reported below.

The evaluation of stresses along the bonded boundary presented some difficulties because neither the shear difference method nor the oblique incidence method can be applied easily. Hence, it was decided to evaluate the strains along the bonded edge from the fringe order and the applied boundary conditions. From Mohr's circle of strains and the strain-optic law of photoelasticity, the following formulae can be derived:

$$
\begin{aligned}
\varepsilon_1 &= \varepsilon_n - nF_\varepsilon \cos 2\theta + nF_\varepsilon \\
\varepsilon_2 &= \varepsilon_n - nF_\varepsilon \cos 2\theta - nF_\varepsilon \\
\varepsilon_t &= \varepsilon_n - 2n F_\varepsilon \cos 2\theta \\
\gamma_{\max} &= 2n F_\varepsilon \text{ (strain-optic law)} \\
\gamma_{nt} &= 2n F_\varepsilon \sin 2\theta
\end{aligned}
\tag{31}
$$

where ε_1, ε_2, ε_n, ε_t, and γ_{nt} are the principal and boundary components of strain, respectively, n is the fringe order in the isochromatic pattern (Fig. 67), and θ is the angle between the bonded boundary and the principal direction of the strain ε_1 obtained from the isoclinic pattern (Fig. 68). The maximum strain $\gamma_{max} = 2nF_\varepsilon$ is the diameter of Mohr's circle and F_ε is the model strain fringe value.

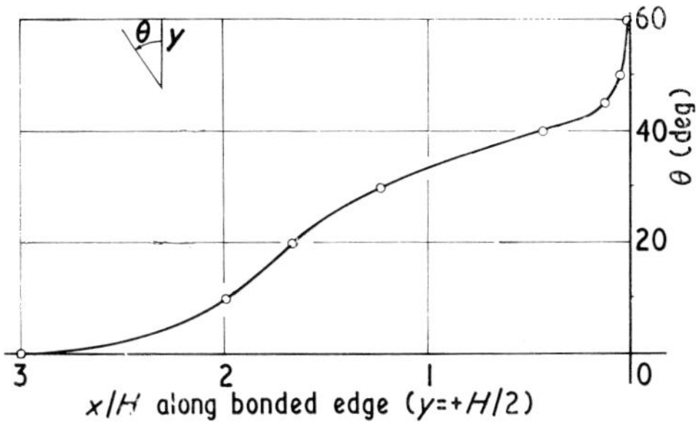

FIG. 68. Isoclinics along the two bonded edges of a rectangular plate subjected to biaxial restrained shrinkage

F_ε is obtained from the value of the stresses at the intersection of the transverse axis and the bonded edge as follows. These stresses in terms of fringes are:

$$\frac{\sigma_x}{F_\sigma} = 2.92; \quad \frac{\sigma_y}{F_\sigma} = 2.38 \tag{32}$$

where F_σ is the model stress fringe value. At all points along the bonded interface $\varepsilon_x = \alpha$ and from Hooke's law for $\nu = 0.48$,

$$\frac{\varepsilon_x}{F_\sigma} = \frac{1}{E}\left(\frac{\sigma_x}{F_\sigma} - \nu\frac{\sigma_y}{F_\sigma}\right) = \frac{1.78}{E}$$

or (33)

$$\alpha = 1.78\frac{F_\sigma}{E}.$$

The two model fringe constants, F_σ and F_ε, are related by the formula

$$F_\varepsilon = \frac{F_\sigma}{E}(1+\nu). \tag{34}$$

Substituting eq. (33) into (34) gives a usable expression of F_ε

$$F_\varepsilon = \frac{\alpha}{1.78}(1+\nu) = \frac{\alpha}{1.20}. \tag{35}$$

Substituting this expression into eq. (31) and dividing by α, normalized values of ε_1, ε_2, and ε_y along the bonded edge in terms of photoelastic values are obtained:

$$\frac{\varepsilon_1}{\alpha} = 1-\frac{n}{1.20}\cos 2\,\theta+\frac{n}{1.20}$$

$$\frac{\varepsilon_2}{\alpha} = 1-\frac{n}{1.20}\cos 2\,\theta-\frac{n}{1.20} \qquad (36)$$

$$\frac{\varepsilon_t}{\alpha} = \frac{\varepsilon_y}{\alpha} = 1-\frac{2n}{1.20}\cos 2\,\theta.$$

From these equations (36) the normal and the principal stresses were determined by application of Hooke's law. Figures 69 and 70 show the principal and the Cartesian strains along the bonded edge.

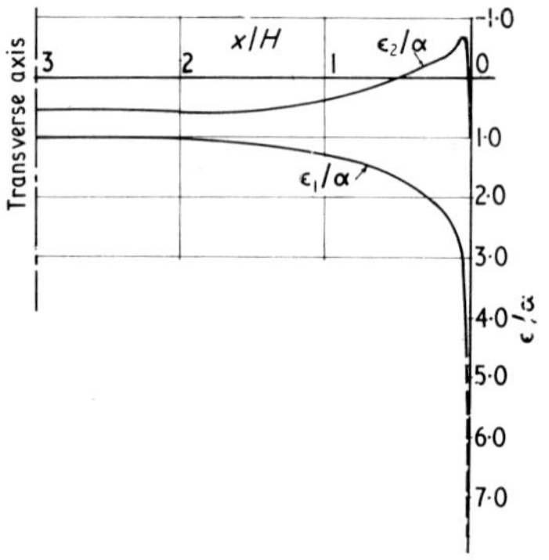

FIG. 69. Principal strains along the bonded edges ($y = \pm H/2$) of a rectangular plate subjected to biaxial restrained shrinkage

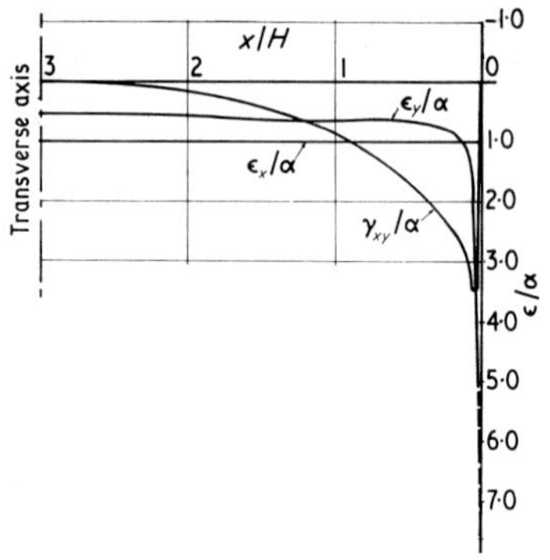

FIG. 70. Cartesian components of strain along the bonded edges ($y=\pm H/2$) of a rectangular plate subjected to biaxial restrained shrinkage

Figures 71 and 72 show the principal and Cartesian stresses along the bonded edge.

Returning now to the analysis along the longitudinal and transverse axes, it is noted that the shear difference and oblique incidence analysis gave σ_1 and σ_2 in terms of fringes (as in eq. 32). But from eq. (33) it can be seen that the actual values of the normalized stresses in eq. (32) are:

$$\frac{\sigma_x}{E\alpha} = \frac{2.92}{1.78} = 1.64$$

$$\frac{\sigma_y}{E\alpha} = \frac{2.38}{1.78} = 1.34. \qquad (37)$$

This same scaling was applied to all the curves on the longitudinal and the transverse axes and to the free boundary.

Figure 73 shows the principal stresses on the longitudinal axis. Figure 74 gives the strains on the transverse axis and on the free boundary. Figure 75 gives the stresses for the same two lines.

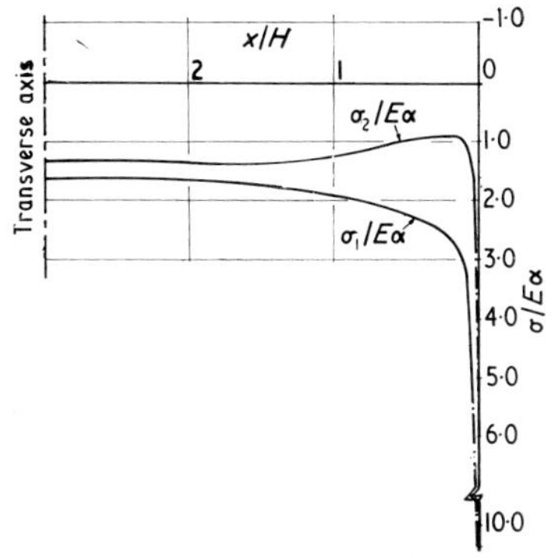

FIG. 71. Principal stresses along the bonded edges ($y = \pm H/2$) of a rectangular plate subjected to biaxial restrained shrinkage

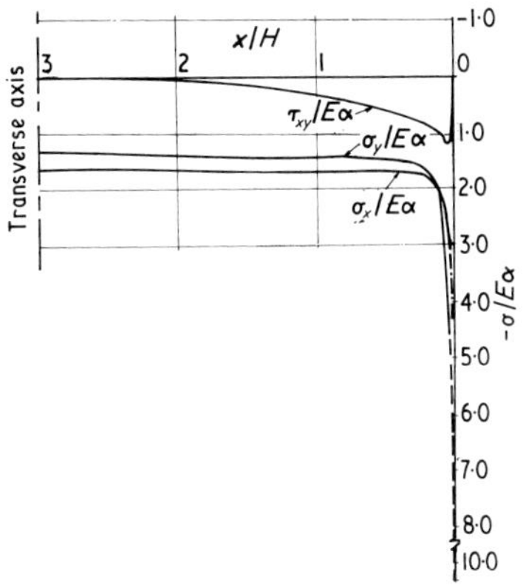

FIG. 72. Cartesian components of stress along the bonded edges ($y = \pm H/2$) of a rectangular plate subjected to biaxial restrained shrinkage

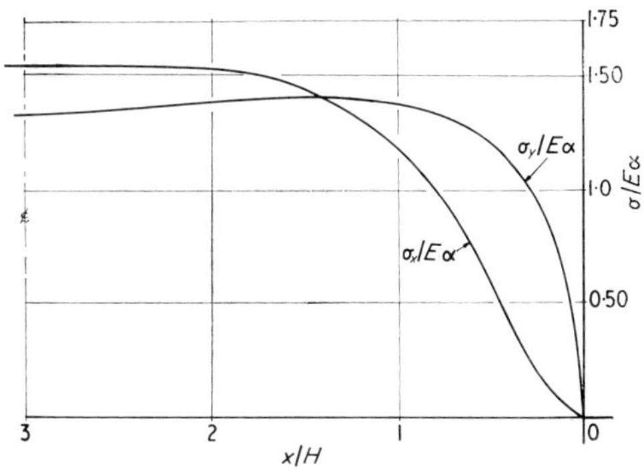

FIG. 73. Cartesian components of stress along the horizontal axis of a rectangular plate subjected to biaxial restrained shrinkage

The values obtained from calibration of the ring and disk for reference are:

Free shrinkage	$\alpha = 0.807\%$
Young's modulus	$E = 565$ psi
Material strain fringe constant	$f_\varepsilon = 0.0010 \dfrac{\text{in./in.}}{\text{fringe}}.$

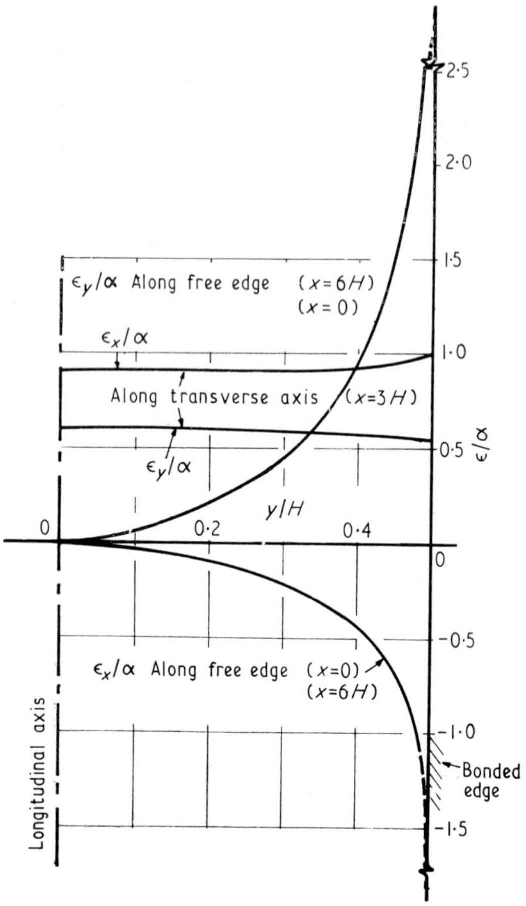

FIG. 74. Normal strain along the transverse axis and the two free edges of a rectangular plate subjected to biaxial restrained shrinkage

C. Two-dimensional Shrinkage Stresses Around an Insert the Geometry of Which is the Meridian Cross-section of a Round Bar

In general, two-dimensional photoelasticity tests are easier to conduct, faster and less expensive than three-dimensional photoelasticity tests. In some cases the three-dimensional problem can be simulated with a two-dimensional model and the results obtained may be useful in stress analysis and design. This may be particularly true when the influence of changes in parameters is being studied. A parametric study can be conducted using two-dimensional photoelasticity which would be often prohibitive using three-dimensional photoelasticity. The results may not be quantitatively the same, but the

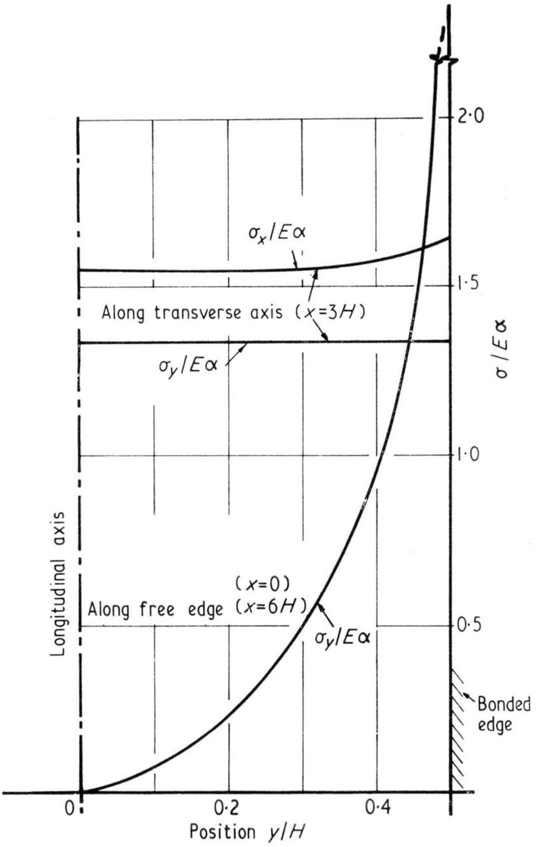

FIG. 75. Normal stresses along the transverse axis and the two free edges of a rectangular plate subjected to biaxial restrained shrinkage

trends will likely be the same. In other words several alternative shapes will grade stress-wise in the same order in two and three dimensions. Several studies have been conducted to justify this approach and to take advantage of it.[22]

In composite materials two simulations would be particularly useful, the one corresponding to a cross-section transverse to the longitudinal axis of the inserts, and the other corresponding to a meridian cross-section through an insert. Examples of the transverse cross-section simulations will be shown later. Here an illustration is given of the restrained shrinkage stresses around a two-dimensional model with the same geometry as the geometry of the meridian cross-section of the circular bar described in section IV B. (With the exception of the radii of the fillets at the flat end.) The insert was made of aluminum with knife edge, and the matrix was made of urethane rubber.

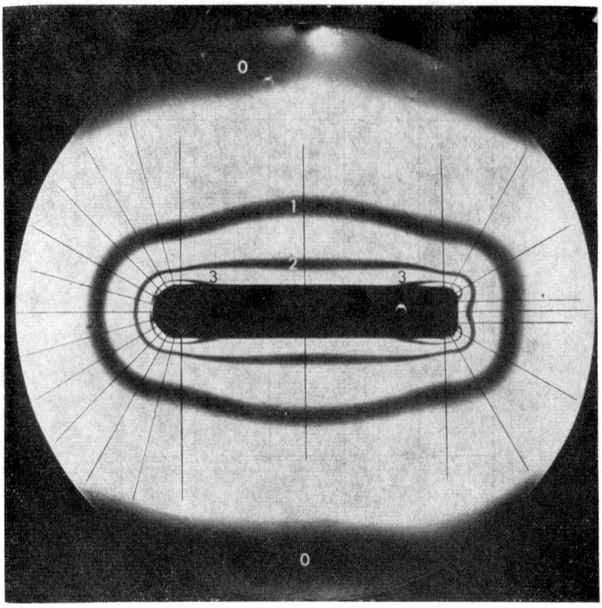

FIG. 76. Isochromatics around a flat bar in a shrinking matrix

Figure 76 shows the isochromatics associated with the restrained shrinkage stresses. Figure 77 shows the corresponding isoclinics. The maximum shear strains and the direction of the principal stresses at points on the interface are represented in Fig. 78. Since the solution of the strains at the interface of

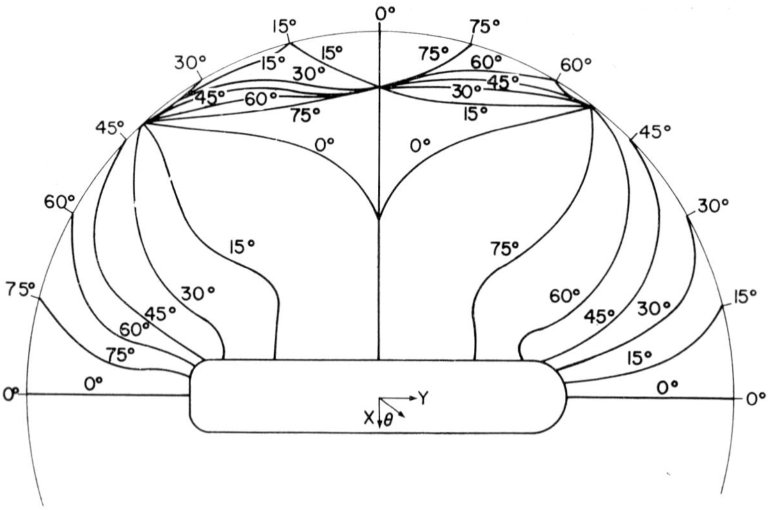

FIG. 77. Isoclinics around a flat bar in a shrinking matrix

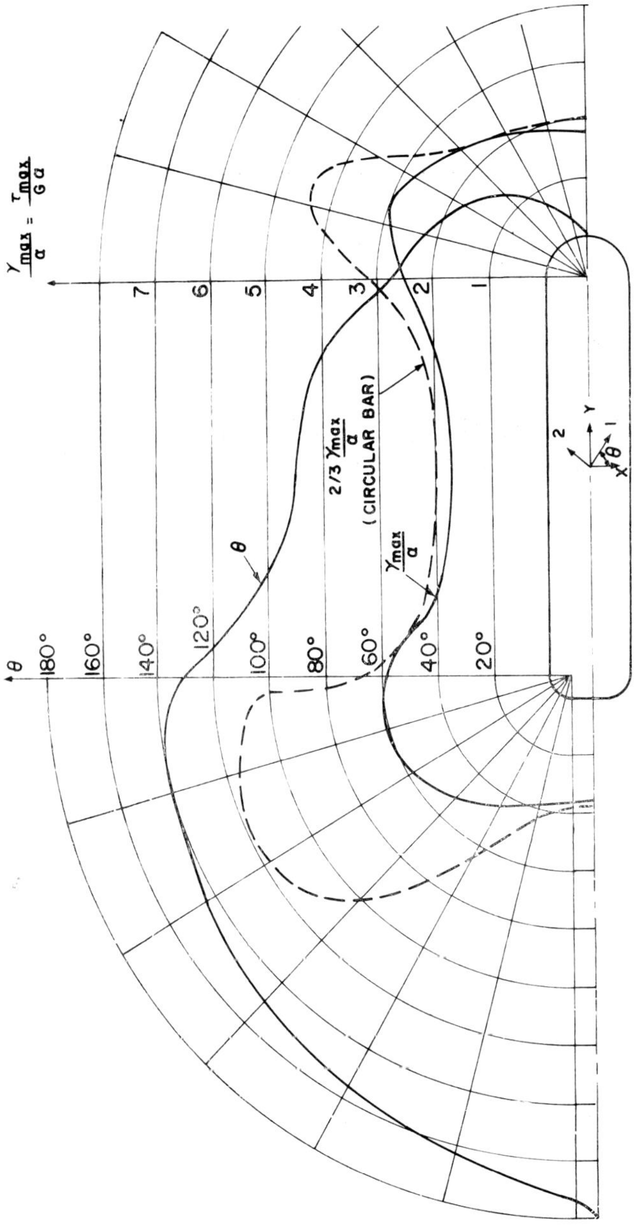

FIG. 78. Maximum shears and directions of principal strains at interface of a flat bar in a shrinking matrix

an embedded sphere is

$$\frac{\alpha_{\max}}{\alpha} = 3$$

and the solution of the strains at the interface of the embedded circle is

$$\frac{\alpha_{\max}}{\alpha} = 2$$

the experimentally obtained three-dimensional solution (Fig. 10) has also been represented in Fig. 78 multiplied by $\frac{2}{3}$. At the round end the comparison of the two curves is acceptable. At the other end no comparison should be made since the radii of the fillets are different in the two- and the three-dimensional models.

VII. CONCLUSIONS

In the development of this paper some simple facts have been pointed out. These may be particularly useful to the analyst and to the designer and will be listed here for convenience.

A. *On Elastic Restrictions*

The linear strain in the matrix in a direction normal to the interface, when the insert is infinitely rigid, the matrix is practically incompressible and the composite is subjected to restrained shrinkage, is equal to minus twice the difference in contraction of the matrix and the insert:

$$\varepsilon_n)_m = -2(\alpha_m - \alpha_i). \tag{38}$$

The difference between the linear strains perpendicular and tangential to the interface in the matrix, when the insert is infinitely rigid and the matrix incompressible and the composite is subjected to restrained shrinkage, is equal to three times the difference in contraction of the matrix and the insert:

$$\varepsilon_t)_m - \varepsilon_n)_m = 3(\alpha_m - \alpha_i). \tag{39}$$

In plates with inserts, with bonded interface, when the matrix is incompressible and the insert infinitely rigid and the composite is subjected to restrained shrinkage, the maximum shear strain at matrix points at the interface at which the interface coincides with a principal direction, is equal to three times the difference in contraction of the matrix and the insert:

$$\gamma_{\max} = 3(\alpha_m - \alpha_i).$$

In plates with inserts, with bonded interface, when the matrix is incompressible and the insert infinitely rigid and the composite is subjected to applied loads, the maximum shear strain at matrix points at the interface at which the interface coincides with a principal direction is equal to zero:

$$\gamma_{\max} = 0.$$

B. *On Methods*

It is possible to solve three-dimensional stress analysis problems of composite materials casting a special kind of epoxy (formula 20) about inserts made of conventional epoxy (formula 19). Stresses and strains associated with restrained shrinkage and with applied loads can be analyzed using the three-dimensional photoelasticity methods.

Casting a particular kind of epoxy (Epon) around inserts made of Plexiglas produces small shrinkage stresses. Stresses associated with applied load can be "frozen-in". The method seems promising to solve problems of stresses and strains associated with mechanical loading of composites without having to subtract the effect of restrained shrinkage.

Strains around inserts with bonded interface, in plates subjected to restrained shrinkage can be determined by casting a urethane rubber (as Solithane 113) around the inserts. At principal sections of the interface an approximation to the plane strain solution is obtained.

Strains around inserts with bonded interface in plates subjected to applied loads can be determined directly by casting a non-shrinking urethane rubber (as Scotchcast) around the inserts. At principal sections of the interface an approximation to the plane strain solution is obtained.

Strains around inserts with unbonded interface can be determined by casting a urethane rubber (as Solithane) around the inserts. The inserts are then removed, oiled and replaced. The method will give the plane stress solution, provided no tensile stresses act normal to the interface.

The "pinching" effect at the plate interface can be eliminated by using a properly designed knife edge.

ACKNOWLEDGEMENTS

The research work reported in this paper was sponsored by the Office of Naval Research. The authors would like to acknowledge the financial support received from this office and, in particular, would like to thank Dr. H. Liebowitz and Mr. J. M. Crowley for their encouragement and understanding. Most of the specimens were cast by Mr. H. Miller. Mrs. Doreen Smith was in charge of the typing of the paper and Mr. J. McDonough of the drawings.

REFERENCES

1. STERNBERG, E. Three-Dimensional Stress Concentrations in the Theory of Elasticity, *Appl. Mech. Rev.* **11,** 1–4 (1958).
2. WILSON, H. B. Mathematical Studies of Composite Materials, Rohm and Haas Co., Report S-42, Contract DA-01-021-ORD-11878, December 1963.
3. SCHUERCH, H. A Contribution to the Micromechanics of Composite Materials Stresses and Failure Mechanics Induced by Inclusions, NASA CR-582, September 1966.
4. DANIEL, I. M., and DURELLI, A. J. Photoelastic Investigation of Residual Stresses in Glass-Plastic Components, *16th Conf. of Reinforced Plastics, Div. of Soc. of Plastics Industry*, Sect. 19-A, pp. 1–8, February 1961.
5. DANIEL, I. M. and DURELLI, A. J. Shrinkage Stresses Around Rigid Inclusions, *Exp. Mech.*, **2,** No. 8, August 1962, pp. 240–244.
6. DANIEL, I. M. Photoelastic Studies of Mechanics of Composites, in I.I.T.R.I. Report M-6132: Application of Advanced Fibrous Reinforced Composite Materials, Contract AF 33 (615)-3323, July 1966.
7. SCHUSTER, D. M. and SCALA, E. The Mechanical Interaction of Sapphire Whiskers with a Birefringent Matrix, *Trans. Metal. Soc. of AIME*, **230,** 1635–1640 (1964).
8. EBERT, L. J. and GADD, J. D. A Mathematical Model for the Mechanical Behavior of Interfaces in Composite Materials, Chap. 5 in *Fiber Composite Materials*, Seminar of A.S.M., October 17, 1964.
9. DURELLI, A. J. and RILEY, W. F. *Introduction to Photomechanics*, Prentice-Hall, 1965, p. 52.
10. Ibid., p. 189.
11. Ibid., p. 146.
12. SAMPSON, R. A Three-Dimensional Photoelastic Method for Analysis of Differential Contraction Stresses, *Exp. Mech.*, **3,** No. 10, October 1963, pp. 225–235.
13. DURELLI, A. J., PARKS, V. J. and DEL RIO, C. J. Stresses in Square Slabs with Different Edge Geometries when Bonded on One Face to a Rigid Plate and Shrunk, *Exp. Mech.*, **7**, No. 11, November 1967, pp. 481–484.
14. DURELLI, A. J., PARKS, V. J. and DEL RIO, C. J. Stresses in a Square Slab Bonded on One Face to a Rigid Plate and Shrunk, *Acta Mechanica*, **3,** No. 4, 1967, pp. 352–359.
15. PARKS, V. J. and DURELLI, A. J. Photoelastic Analysis of Plates Subjected to Restrained Shrinkage, *Journ. Appl. Mech.*, **32,** No. 3, 1965, pp. 504–510.
16. DURELLI, A. J. and PARKS, V. J. Photoelastic Analysis of Stresses in Two- and Three-Dimensional Bodies Partially Restrained from Shrinkage, *Proc. of the Fifth U.S. Natl. Conf. App. Mech.*, p. 264, 1966.
17. PARKS, V. J. and DURELLI, A. J. Stress Distribution at Variously Shaped Corners in Plates, Bonded on Two Long Edges, and Subjected to Restrained Shrinkage, Mech. Behavior Working Group, *Interagency Ch. Prop. Bull. 119*, No. 1, pp. 366–392, October 1966.
18. TIMOSHENKO, S. and GOODIER, J. N. *Theory of Elasticity*, 2nd ed., p. 359. McGraw-Hill, 1951.
19. DURELLI, A. J., PHILLIPS, E. and TSAO, C. *Introduction to the Theoretical and Experimental Analysis of Stress and Strain*, p. 87. McGraw-Hill, 1958.
20. DURELLI, A. J. and PARKS, V. J. Experimental Stress Analysis of Loaded Boundaries in Two-Dimensional Second Boundary Value Problems, *Exp. Mech.*, **7**, No. 9, September 1967, pp. 381–385.
21. DURELLI, A. J. and PARKS, V. J. Restrained Shrinkage Stresses in the Neighborhood of Corners of Long Strips with Different End Configurations, *Bull. Am. Ceramic Society*, **46**, No. 6, June 1967. pp. 582–586.
22. DURELLI, A. J. Experimental Strain and Stress Analysis of Solid Propellant Rocket Motors, in *Mechanics and Chemistry of Solid Propellants*, p. 404, edited by Eringen, Liebowitz, Koh, and Crowley, Pergamon Press, 1967.

WAVE PROPAGATION IN LAMINATED AND FIBER-REINFORCED COMPOSITES

GEORGE HERRMANN† and JAN D. ACHENBACH‡

Northwestern University, Evanston, Illinois

ABSTRACT

A discussion is presented of several dynamic theories for laminated and fiber-reinforced composites advanced recently by the authors. These theories all employ the concept of "effective stiffnesses" and thus include a description of essential properties of dynamic behavior which the commonly used "effective moduli" theories do not contain.

The features of the newly developed theories are exhibited by the study of propagation of plane free harmonic waves in an unbounded composite solid. Dispersion curves are displayed for both transverse and longitudinal motions, and for a laminated medium, comparisons are made with curves obtained by using the equations of the theory of elasticity for all layers.

LIST OF SYMBOLS

a	radius of fiber
a	elastic constants
b	inertia constants
c	phase velocity
h	thickness of reinforcing sheet
i	$=(-1)^{1/2}$
k	wave number
r	radius of gyration
t	time
u_1, u_2	displacement components
u_{01}, u_{02}	displacement at centerline
v	volume
x_1, x_2	Cartesian coordinates
$\bar{x}_2, \bar{x}_3$	local Cartesian coordinates
A	cross-sectional area of fiber
C_f	compressional stiffness
C_{44}	effective modulus
D_f	flexural stiffness

† Professor, Department of Civil Engineering, The Technological Institute.

‡ Associate Professor, Department of Civil Engineering, The Technological Institute.

E_f	Young's modulus
G_f	shear stiffness
H	thickness of matrix layer
I_f	moment of inertia
S	area of element
S_f	$= C_f/h$
T	kinetic energy density
V	strain energy density
β	$= c/(\mu_m/\varrho_m)^{1/2}$, dimensionless phase velocity
γ	$= \mu_f/\mu_m$
δ	$= 2(1-\nu)/(1-2\nu)$
ε_{ij}	components of strain tensor
η	density of reinforcing elements
$\bar{\eta}$	$= h/H$
θ	$= \varrho_f/\varrho_m$
$\varkappa_i$	correction factor
λ	Lamé elastic constant
μ	Lamé elastic constant (shear modulus)
ν	Poisson's ratio
ξ	dimensionless wave number
ϱ	mass density
ϱ_c	$= \eta\varrho_f + (1-\eta)\,\varrho_m$
φ_{2i}	function describing deformation
ψ_{2i}	function describing deformation

Superscripts

f	pertaining to reinforcing elements
k	pertaining to kth element
m	pertaining to matrix
bar	pertaining to fictitious matrix

Subscripts

f	pertaining to reinforcing elements
m	pertaining to matrix

I. INTRODUCTION

If one wishes to describe the propagation of time-harmonic waves in a composite material and to limit attention to wave lengths which are large compared to the smallest characteristic length of the reinforcing fibers or laminates, it is meaningful to attempt the construction of a dynamical continuum theory for this heterogeneous body, rather than formulate boundary value problems for each element separately, together with appropriate interface conditions, as may be necessary for short wave lengths.

In constructing such a dynamical theory, a mere application of the well-known "effective modulus" theories for composites, established for static analysis, has to be viewed with suspicion. Indeed, if one considers a uniformly laminated or unidirectionally fibrous composite, one would be inclined to predict that the periodic structure of the medium should entail dispersion of free harmonic waves. This expectation has been confirmed for layered composites by solving appropriate boundary value problems of the theory of elasticity, as discussed in the body of this paper.

The authors propose to employ a conceptually entirely different approach in constructing continuum models for the dynamic analysis of laminated or fiber-reinforced composites. Instead of introducing a representative homogeneous medium by means of "effective moduli", the elastic and the geometric properties of one or both constituents are combined into "effective stiffnesses". Depending upon certain required supplementary kinematical constraints, different theories are evolved. All of these, however, contain at least some dispersive properties.

It is the purpose of the present contribution to review the various dynamic theories of composites recently established by the authors, to discuss their implications as regards the propagation of free harmonic waves and to note the relation of these theories to the theories of elasticity with couple stresses and with microstructure.

To remain within reasonable confines of manuscript length, only displacement equations of motion for plane strain and some of their solutions will be discussed here, omitting presentation of explicit constitutive relations, boundary conditions, uniqueness, etc.

II. THE LAMINATED COMPOSITE—EFFECTIVE SINGLE STIFFNESS THEORY

In deriving the displacement equations of motion of an elastic body, it is convenient to employ Hamilton's principle

$$\delta \int_{t_0}^{t_1} \int_{v} (T-V)\, dv\, dt = 0 \tag{1}$$

where v is the volume and t is the time.

The kinetic energy density T in plane strain is defined by

$$T = (\varrho/2)\,[(\partial u_1/\partial t)^2 + (\partial u_2/\partial t)^2] \tag{2}$$

where ϱ is the mass density.

(x_2^{fk}), now have become continuously varying functions of x_2. To indicate this change we omit the superscript k, i.e., we write u_{01}^f, u_{02}^f and ψ_{21}. By eqs. (12) and (13) we also have introduced energy densities T_f and V_f, whose functional dependence on u_{01}^f etc. is the same as that of T_f^k and V_f^k on u_{01}^{fk} etc.

As regards the kinetic energy T_m and the strain energy density V_m of the matrix, they essentially remain in the forms eqs. (2), (3), except that moduli $\bar{\lambda}$ and $\bar{\mu}$ are now introduced which are not the moduli λ_m and μ_m of the matrix, but are constants which must be computed from the limit cases at long wave lengths. For greater detail see ref. 1.

The total kinetic and strain energy densities are then given by

$$T = (\eta/h)T_f + (1-\eta)T_m \tag{15}$$

$$V = (\eta/h)V_f + (1-\eta)V_m. \tag{16}$$

The energy densities eqs. (15) and (16) involve both u_{0i}^f and u_i^m. We now define, however, these two field variables as being one and the same, henceforth referred to as u_i. In this manner we secure interaction of the reinforcing sheets and the matrix, essentially through continuity of the displacements at the centerlines of the reinforcing sheets.

Upon application of Hamilton's principle, eq. (1), the densities eqs. (15) and (16) lead to the following equations of motion

$$(1-\eta)\left[(\bar{\lambda}+2\bar{\mu})\frac{\partial^2 u_1}{\partial x_1^2}+\bar{\mu}\frac{\partial^2 u_1}{\partial x_2^2}+(\bar{\lambda}+\bar{\mu})\frac{\partial^2 u_2}{\partial x_1 \partial x_2}\right]+\eta S_f\frac{\partial^2 u_1}{\partial x_1^2} = \varrho_c\frac{\partial^2 u_1}{\partial t^2} \tag{17}$$

$$(1-\eta)\left[(\bar{\lambda}+2\bar{\mu})\frac{\partial^2 u_2}{\partial x_2^2}+\bar{\mu}\frac{\partial^2 u_2}{\partial x_1^2}+(\bar{\lambda}+\bar{\mu})\frac{\partial^2 u_1}{\partial x_1 \partial x_2}\right]$$
$$+\eta\mu_f\left(\frac{\partial^2 u_2}{\partial x_1^2}+\frac{\partial \psi_{21}}{\partial x_1}\right) = \varrho_c\frac{\partial^2 u_2}{\partial t^2} \tag{18}$$

$$D_f\frac{\partial^2 \psi_{21}}{\partial x_1^2}-G_f\left(\psi_{21}+\frac{\partial u_2}{\partial x_1}\right) = I_f\frac{\partial^2 \psi_{21}}{\partial t^2}. \tag{19}$$

In these equations ϱ_c is the effective mass density defined by

$$\varrho_c = \eta\varrho_f+(1-\eta)\,\varrho_m \tag{20}$$

and I_f is the moment of inertia $(1/12)\varrho_f h^3$.

Plane waves propagating in the x_1 direction (the direction of lamination), which imply $\partial/\partial x_2 = 0$, separate the first equation of motion (governing longitudinal motions) from the remaining two equations (governing transverse motions).

If, for the transverse system, it is assumed

$$u_2 = \mathrm{A} \exp [ik(x_1 - ct)] \tag{21}$$

$$\psi_{21} = \mathrm{B} \exp [ik(x_1 - ct)] \tag{22}$$

one obtains the dispersion relation

$$[(1-\eta)\bar{\mu}+\eta\mu_f - c^2\varrho_c]\,[D_f k^2 + G_f - I_f k^2 c^2] - \eta\mu_f G_f = 0 \tag{23}$$

where k is the wave number and c is the phase velocity. For vanishing wave number the phase velocity becomes

$$c^2 = (1-\eta)\,\bar{\mu}/\varrho_c \tag{24}$$

and it has been shown that one has to choose[1]

$$\bar{\mu} = \frac{\mu_f \mu_m}{[\eta\mu_m + (1-\eta)\mu_f]\,(1-\eta)} \tag{25}$$

so as to achieve agreement with the exact solution as discussed below.

The phase velocities of the lower mode which represent an essentially transverse motion are computed from eq. (23) and are plotted in Fig. 2 by interrupted lines for three ratios of $\gamma = \mu_f/\mu_m$. The ordinate has been chosen as the dimensionless phase velocity $\beta = c/(\mu_m/\varrho_m)^{1/2}$, while the abscissa is the dimensionless wave number $\xi = kh$. It is noted that already for small wave numbers the dispersion curves depart from the constant phase velocity given by the effective modulus theory. Inspection of the equations of motion, eqs. (17), (18), (19), reveals that neither longitudinal waves in the x_1 direction nor waves in the x_2 direction are dispersive.

III. THE LAMINATED COMPOSITE—EXACT TREATMENT

To test the accuracy of the proposed effective single stiffness theory, it is now required to compare the approximate dispersion equation (23) with an exact treatment.

For a stratified medium consisting of alternating plane parallel layers of two homogeneous isotropic materials an exact dispersion equation can be obtained by employing the equations of the theory of elasticity for the layers. Considering motions for which the deformations of all layers of thickness h are the same and for which, moreover, the deformations of all layers of thickness H are the same, we can restrict our attention to one pair of neighboring layers and obtain two dispersion equations as four by four determinants, governing essentially transverse and essentially longitudinal motions. Exact dispersion curves have been obtained by calculating numerically the roots of the transcendental equations, and they have been plotted in

Figs. 2 and 3. The lowest antisymmetric mode at small wave numbers is essentially a transverse motion. The limit at $k = 0$ (infinite wave length) yields the same phase velocity as obtained by the effective modulus theory.[2] A detailed derivation and discussion of the exact dispersion equations is contained in ref. 3.

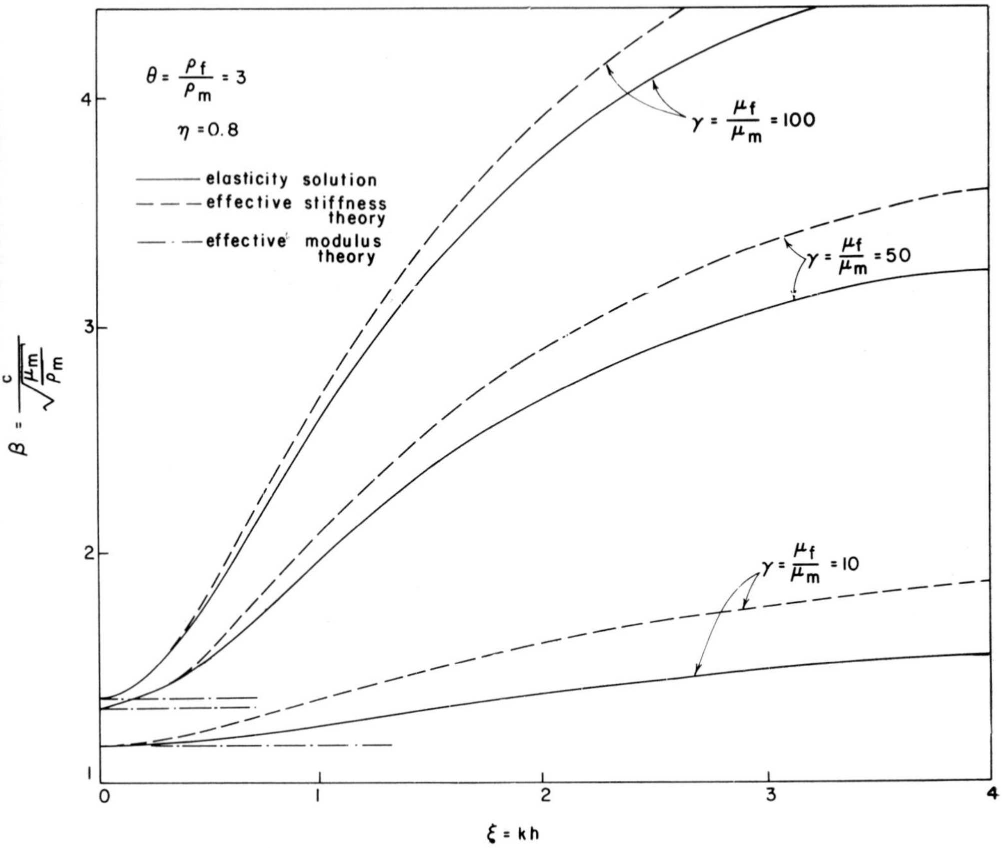

FIG. 2. Dispersion curves according to effective single stiffness theory

It is seen from Fig. 2 that rather accurate dispersive behavior is predicted for small ξ by the approximate theory discussed in the previous section and that the accuracy, for given ξ, improves as γ increases.

As already mentioned, the longitudinal motion as described by eq. (17) does not predict any dispersion at all. On the other hand, the phase velocity of the second lowest (the symmetric) mode, calculated "exactly" shows a steep decrease at small values of the wave number, as seen from Fig. 3. To gain some insight into this very pronounced dispersion behavior, one can consider disp rsion curves for a free reinforcing layer and a clamped matrix layer, which should be a good approximation for large γ, and compare the results with the ones obtained for the layered system. The comparison, as shown in Fig. 4, suggests that the phase velocity of the essentially longitudinal (the symmetric) motion decreases rapidly because most of the motion is

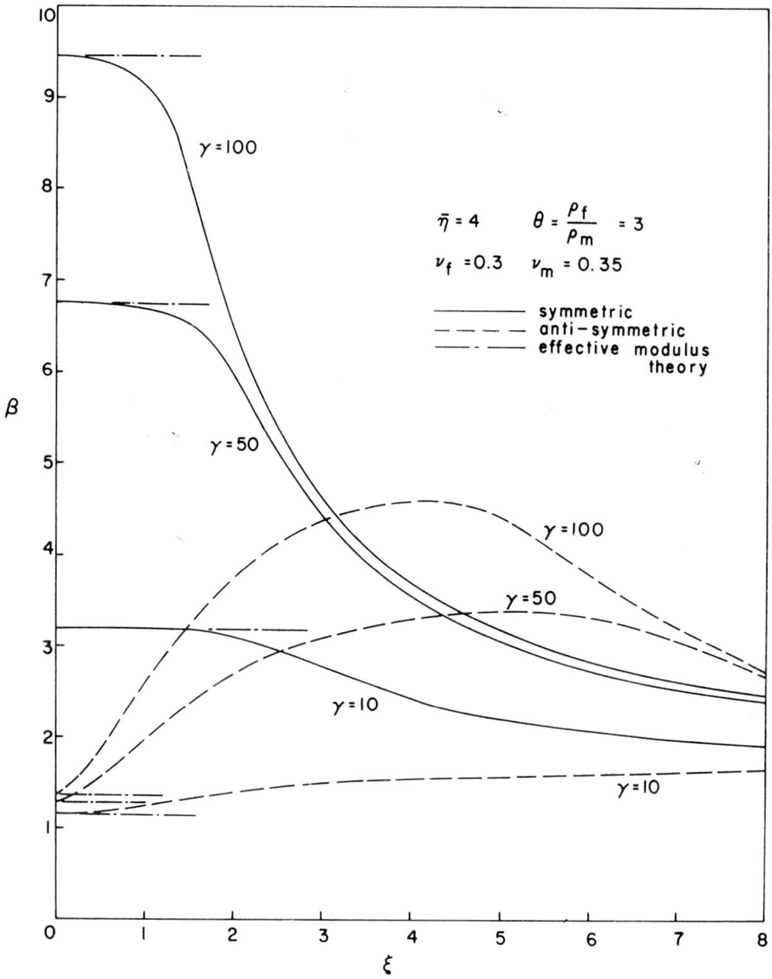

FIG. 3. Lowest symmetric and antisymmetric modes

transferred from a distribution over both layers at zero wave number to the low modulus matrix layers only, where the latter behave essentially as clamped layers in the symmetric shear mode. Similarly, for the lowest, essentially transverse (the antisymmetric) mode of the layered medium, at increasing wave numbers, the motion is concentrated in the (stiff) fiber layers which behave as free plates in the antisymmetric shear mode.

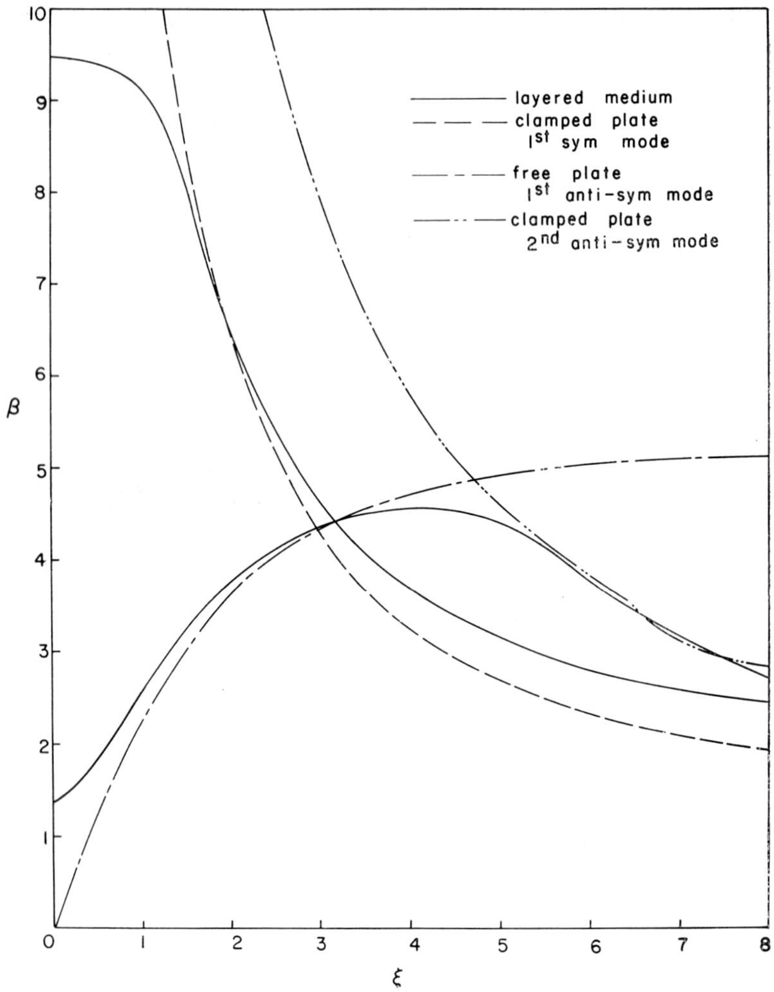

FIG. 4. Comparison of dispersion curves for $\gamma = 100$ with curves for free and clamped layers

IV. THE LAMINATED COMPOSITE—EFFECTIVE DOUBLE STIFFNESS THEORIES

The insight gained into the dispersion of longitudinal motion leads to the conclusion that the effective single stiffness theory which provided good results for the transverse waves cannot possibly be adequate for longitudinal waves since this theory does not account for the dispersive properties of the matrix. The authors were thus motivated to construct a more accurate theory,[(4)] in which the displacements of both the reinforcing sheets and the matrix layers are represented as linear expansions about the midplanes of the layers. In this manner it becomes possible to introduce effective stiffnesses for both the matrix layers and the reinforcing layers. For perfect bond between the layers the displacement at the interface of each pair of laminates must be continuous which results in a relation between the gross displacement gradient in the direction normal to the layering and a weighted sum of the functions which describe the local deformations of two adjacent layers.

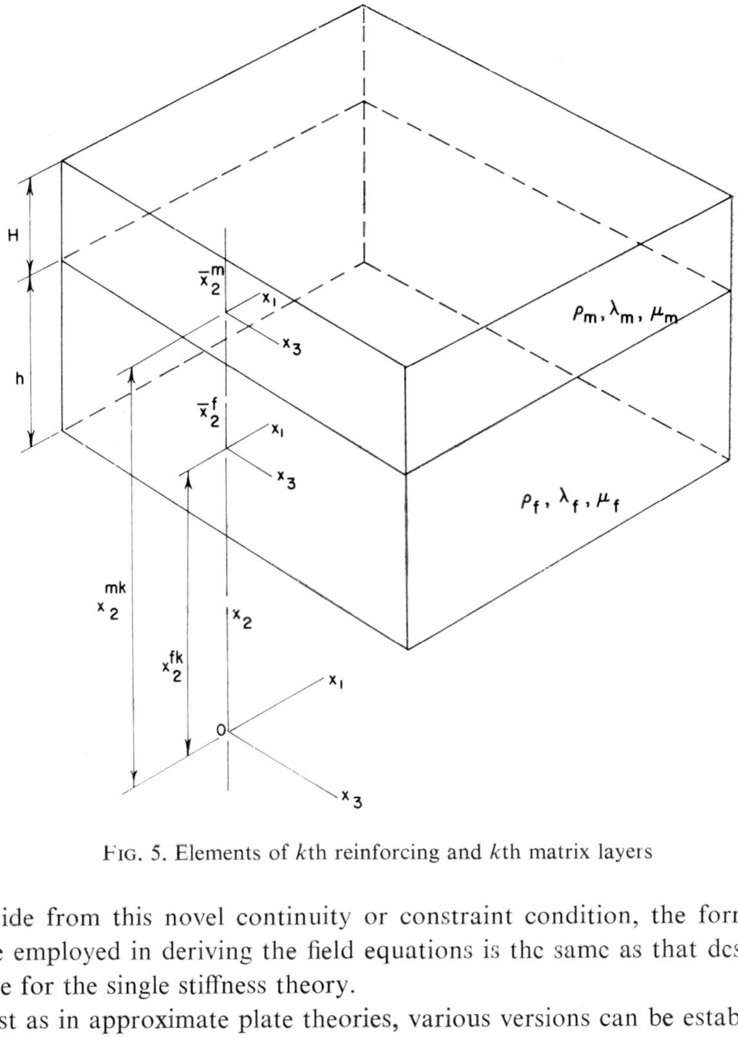

FIG. 5. Elements of kth reinforcing and kth matrix layers

Aside from this novel continuity or constraint condition, the formalism to be employed in deriving the field equations is the same as that described above for the single stiffness theory.

Just as in approximate plate theories, various versions can be established, depending upon the number of terms retained in the expressions of the displacements of the kth reinforcing layer and the kth matrix layer. A first-order theory is obtained if one considers the displacements in the form (see Fig. 5)

$$u_i^{fk} = u_{0i}^{fk}(x_1, x_2^{fk}, t) + \bar{x}_2^f \psi_{2i}^{fk}(x_1, x_2^{fk}, t) \tag{26}$$

for the reinforcing sheets, and

$$u_i^{mk} = u_{0i}^{mk}(x_1, x_2^{mk}, t) + \bar{x}_2^m \psi_{2i}^{mk}(x_1, x_2^{mk}, t) \tag{27}$$

for the matrix layers. Since the displacements are continuous at the interface of the kth fiber and the kth matrix layer, the relation results

$$u_{0i}^{mk}(x_1, x_2^{mk}) - u_{0i}^{fk}(x_1, x_2^{fk}) = \tfrac{1}{2} h\psi_{2i}^{fk}(x_1, x_2^{fk}) + \tfrac{1}{2} H\psi_{2i}^{mk}(x_1, x_2^{mk}) \tag{28}$$

which must hold at all times t.

Displacement equations of motion are established by computing first the strain energies and the kinetic energies in elements of surface area unity of the kth pair of reinforcing and matrix layers. The total kinetic and strain energies stored in a volume of the laminate containing n pairs of fiber and matrix layers are

$$T = \sum^n (T^{fk} + T^{mk}) \tag{29}$$

$$V = \sum^n (V^{fk} + V^{mk}). \tag{30}$$

As already discussed, the basic premise of effective stiffness theories is that a smoothing operation may be employed to replace the summation over n discrete points x_2^{fk} and x_2^{mk} by an integration over x_2. Thus we approximate

$$\sum^n (T^{fk} + T^{mk}) \simeq \int \frac{1}{h+H} (T^f + T^m)\, dx_2 \tag{31}$$

$$\sum^n (V^{fk} + V^{mk}) \simeq \int \frac{1}{h+H} (V^f + V^m)\, dx_2. \tag{32}$$

By means of this smoothing operation we have, in fact, replaced the layered medium by a homogeneous continuum. The field variables u_{0i}^{fk}, u_{0i}^{mk}, ψ_{2i}^{fk} and ψ_{2i}^{mk} which were so far defined only at certain discrete lines, namely x_2^{fk} and x_2^{mk}, have now become continuous functions also of x_2 and we delete henceforth the superscript k.

The kinetic and strain energy densities are expressions in terms of u_{0i}^f, u_{0i}^m, ψ_{2i}^f, ψ_{2i}^m and their spatial and temporal derivatives. The displacement u_{0i}^f and u_{0i}^m should be considered as representing the same field variable, namely the "macro-displacement" u_{0i}. By contrast, ψ_{2i}^f and ψ_{2i}^m represent "micro-deformations" of the fiber and matrix, respectively, and are thus not identical but related, see eq. (28), through

$$u_{0i}(x_1, x_2^{mk}) - u_{0i}(x_1, x_2^{fk}) = \tfrac{1}{2} h\psi_{2i}^f(x_1, x_2^{fk}) + \tfrac{1}{2} H\psi_{2i}^m(x_1, x_2^{mk}). \tag{33}$$

Noting that

$$x_2^{mk} = x_1^{fk} + \tfrac{1}{2}(h+H) \tag{34}$$

we assume that the thicknesses of the layers are sufficiently small such as to permit the replacement of the difference relation (33) by the differential relation

$$(1+\bar{\eta})\frac{\partial u_{0i}}{\partial x_2} = \bar{\eta}\psi_{2i}^{f}+\psi_{2i}^{m} \tag{35}$$

where $\bar{\eta} = h/H$.

This differential relation permits us to eliminate one of the microrotations from the set of field variables. We choose to eliminate ψ_{2i}^{m} and therefore only four field variables u_{01}, u_{02}, ψ_{21}^{f}, ψ_{22}^{f} are left to describe the deformation. For these, the simpler notation u_1, u_2, ψ_{21}, ψ_{22} will be used in the sequel.

Now application of Hamilton's principle, eq. (1), leads directly to the displacement equations of motion which are rather lengthy. They are not reproduced here but may be found in ref. 4. For plane waves propagating in the x_1 direction (i.e., $\partial/\partial x_2 = 0$) they take on the simpler form

$$a_2\frac{\partial^2 u_1}{\partial x_2^2}+a_7\frac{\partial \psi_{22}}{\partial x_1} = b_3\frac{\partial^2 u_1}{\partial t^2} \tag{36}$$

$$a_{10}\frac{\partial^2 u_2}{\partial x_1^2}+a_{12}\frac{\partial \psi_{21}}{\partial x_1} = b_3\frac{\partial^2 u_2}{\partial t^2} \tag{37}$$

$$a_{12}\frac{\partial u_2}{\partial x_1}+a_{14}\frac{\partial^2 \psi_{21}}{\partial x_1^2}+a_{15}\psi_{21} = b_4\frac{\partial^2 \psi_{21}}{\partial t^2} \tag{38}$$

$$a_7\frac{\partial u_1}{\partial x_1}+a_{16}\frac{\partial^2 \psi_{22}}{\partial x_1^2}+a_{17}\psi_{22} = b_4\frac{\partial^2 \psi_{22}}{\partial t^2}. \tag{39}$$

The coefficients a_i and b_i stand for

$$a_2 = \frac{\bar{\eta}(\lambda_f+2\mu_f)+(\lambda_m+2\mu_m)}{1+\bar{\eta}} \tag{40}$$

$$a_7 = \varkappa_1\bar{\eta}\frac{\lambda_f-\lambda_m}{1+\bar{\eta}} \tag{41}$$

$$a_{10} = \varkappa_2^2\frac{\bar{\eta}\mu_f+\mu_m}{1+\bar{\eta}} \tag{42}$$

$$a_{12} = \varkappa_2^2\bar{\eta}\frac{\mu_f-\mu_m}{1+\bar{\eta}} \tag{43}$$

$$a_{14} = -\frac{H^2}{12}\bar{\eta}^2\frac{\bar{\eta}(\lambda_f+2\mu_f)+(\lambda_m+2\mu_m)}{1+\bar{\eta}} \tag{44}$$

$$a_{15} = \varkappa_2^2\bar{\eta}\frac{\mu_f+\bar{\eta}\mu_m}{1+\bar{\eta}} \tag{45}$$

$$a_{16} = -\frac{H^2}{12}\bar{\eta}^2\frac{\bar{\eta}(\varkappa_3^f)^2\mu_f+(\varkappa_3^m)^2\mu_m}{1+\bar{\eta}} \tag{46}$$

$$a_{17} = \varkappa_1^2 \bar{\eta} \frac{\lambda_f + 2\mu_f + \bar{\eta}(\lambda_m + 2\mu_m)}{1+\bar{\eta}} \tag{47}$$

$$b_3 = \frac{\bar{\eta}\varrho_f + \varrho_m}{1+\bar{\eta}} \tag{48}$$

$$b_4 = -\frac{H^2}{12}\bar{\eta}^2 \frac{\bar{\eta}\varrho_f + \varrho_m}{1+\bar{\eta}}. \tag{49}$$

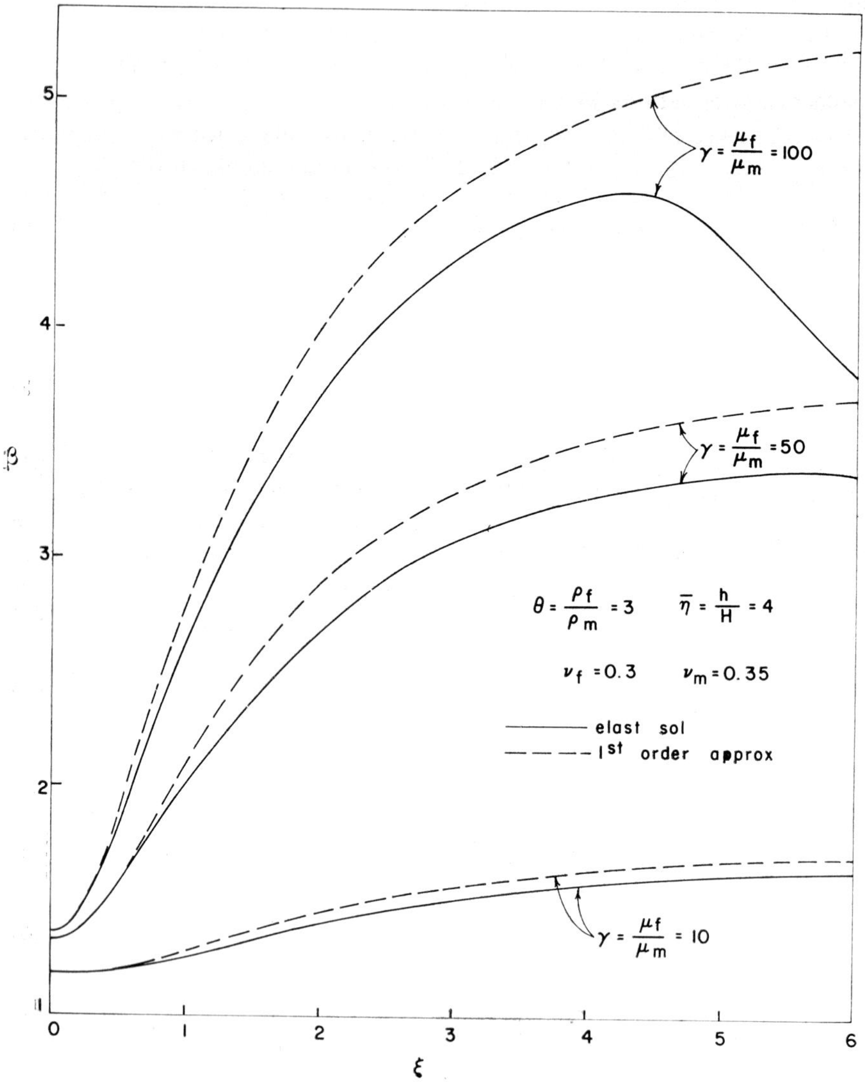

FIG. 6. Transverse mode according to effective double stiffness theory

The correction coefficients $\varkappa_i$ are discussed in detail in ref. 4.

It is noticed that for plane waves the system of equations again separates into a transverse and a longitudinal system, eqs. (37), (38) and eqs. (36), (39), respectively.

Let us consider first transverse motions defined by solutions of the form

$$u_2 = Ae^{ik(x_1-ct)} \tag{50}$$

$$\psi_{21} = Be^{ik(x_1-ct)} \tag{51}$$

which lead to the dispersion relation

$$\begin{aligned}(1+\theta\bar{\eta})^2\xi^2\beta^4 - [\varkappa_2^2(1+\theta\bar{\eta})\,(1+\bar{\eta}\gamma)\xi^2 + (1+\theta\bar{\eta})\,(\bar{\eta}\gamma\delta_f+\delta_m)\xi^2 \\ +12\,\varkappa_2^2\bar{\eta}(1+\theta\bar{\eta})\,(\gamma+\bar{\eta})]\beta^2 \\ +[\varkappa_2^2(1+\bar{\eta}\gamma)\,(\bar{\eta}\gamma\delta_f+\delta_m)\xi^2 + 12\varkappa_2^4\bar{\eta}\gamma(1+\bar{\eta})^2] = 0.\end{aligned} \tag{52}$$

Here $\theta = \varrho_f/\varrho_m$. The constants δ_f, δ_m depend on the Poisson's ratios of the reinforcing and matrix layers and are

$$\delta_f = \frac{2(1-\nu_f)}{1-2\nu_f} \tag{53}$$

$$\delta_m = \frac{2(1-\nu_m)}{1-2\nu_m}. \tag{54}$$

The lowest transverse mode described by the dispersion relation eq. (52) has been evaluated numerically for a set of parameters, and has been plotted in Fig. 6. It is worth pointing out that at vanishing wave number the phase velocity calculated from eq. (52) is the same as that obtained from an analysis employing the equations of elasticity for all layers, together with appropriate boundary conditions.[3] For small γ an improved dispersion curve is obtained as compared to the one in Fig. 2 based on the single-stiffness theory. It should be noted further that for given ξ the accuracy of the approximation becomes less satisfactory with increasing $\gamma = \mu_f/\mu_m$.

For longitudinal motion solutions similar to eqs. (50) and (51) are considered, leading to the dispersion relation

$$\begin{aligned}(1+\theta\bar{\eta})^2\xi^2\beta^4 - \{(1+\theta\bar{\eta})[1+(\varkappa_3^f)^2\bar{\eta}\gamma+\bar{\eta}\gamma\delta_f+\delta_m]\xi^2 \\ +12\,\varkappa_1^2\bar{\eta}(1+\theta\bar{\eta})\,(\gamma\delta_f+\bar{\eta}\delta_m)\}\beta^2 \\ +\{(\bar{\eta}\gamma\delta_f+\delta_m)[1+(\varkappa_3^f)^2\bar{\eta}\gamma]\xi^2 + 12\,\varkappa_1^2\bar{\eta}(\gamma\delta_f+\bar{\eta}\delta_m)\,(\bar{\eta}\gamma\delta_f+\delta_m) \\ -12\,\varkappa_1^2\bar{\eta}^2(\gamma\varepsilon_f-\varepsilon_m)^2\} = 0\end{aligned} \tag{55}$$

where $\varepsilon_f = 2\nu_f/(1-2\nu_f)$ and $\varepsilon_m = 2\nu_m/(1-2\nu_m)$. The lower mode described by this relation is plotted in Figs. 7 and 8 by interrupted lines. It is noted that the accuracy of the proposed first order approximation is satisfactory for certain ranges (see Fig. 8) but is rather poor for the set of parameters used in Fig. 7, and further improvement appears to be required.

It is instructive to plot the numerical results also as curves of dimensionless frequency versus dimensionless wave number, as shown in Figs. 9 and 10. The dispersion manifests itself now as departure from a linear relation, while the slope measures the group velocity.

The theory discussed above may be extended by including higher order terms in the expressions of the displacements of the kth reinforcing and the kth matrix layers, so as to contain the symmetric shear mode in the matrix. For

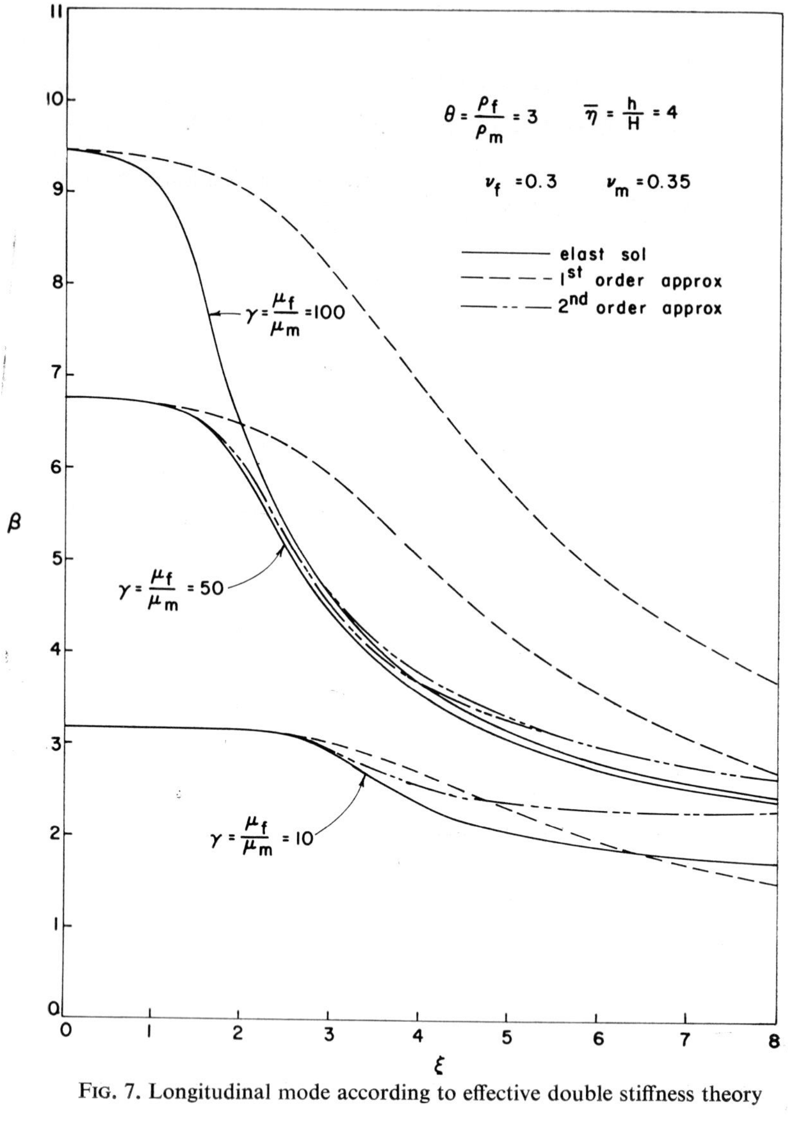

FIG. 7. Longitudinal mode according to effective double stiffness theory

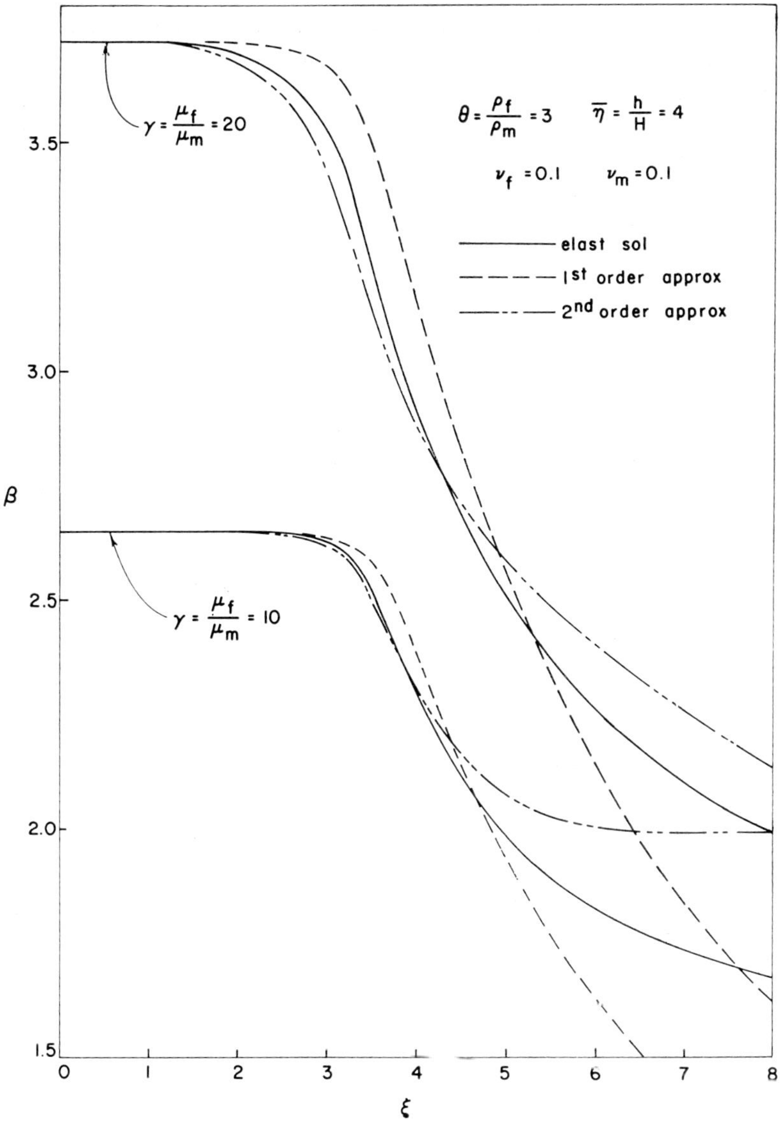

FIG. 8. Longitudinal mode according to effective double stiffness theory

example, for the kth reinforcing layer we may write

$$u_i^{fk} = u_{0i}^{fk}(x_1, x_2^{fk}, t) + \bar{x}_2{}^f \psi_{2i}^{fk}(x_1, x_2^{fk}, t) + \left[\frac{4(\bar{x}_2^f)^2}{h^2} - 1\right] \varphi_{2i}^{fk}(x_1, x_2^{fk}, t). \tag{56}$$

The resulting field equations are rather cumbersome and will not be repro-

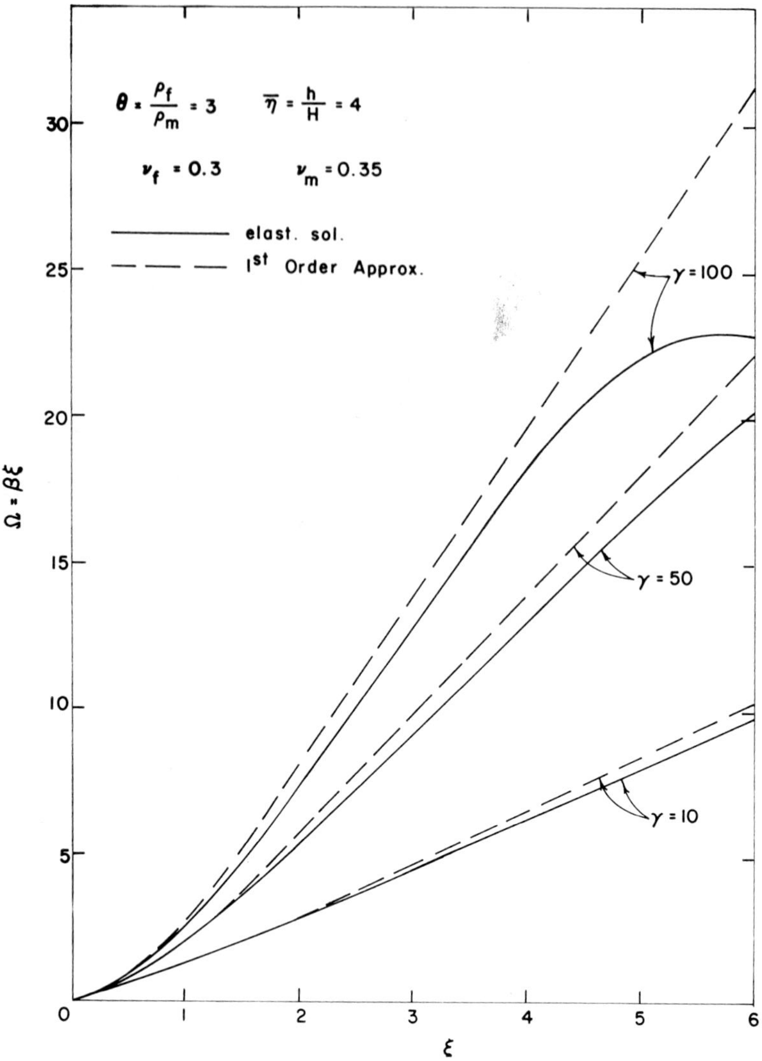

FIG. 9. Frequency vs. wave number for transverse mode

duced here. Since the representation of only the longitudinal motion needs to be improved it will be sufficient to consider expressions in the form

$$u_1^{fk} = u_{01}^{fk}(x_1, x_2^{fk}, t) + \left[\frac{4(\bar{x}_2^f)^2}{h^2} - 1\right] \varphi_{21}^{fk}(x_1, x_2^{fk}, t) \tag{57}$$

$$u_2^{fk} = \bar{x}_2^f \psi_{22}^{fk}(x_1, x_2^{fk}, t) \tag{58}$$

with analogous expressions for the matrix layers. Calculations of dispersion curves based on equations derived from displacement distributions given by

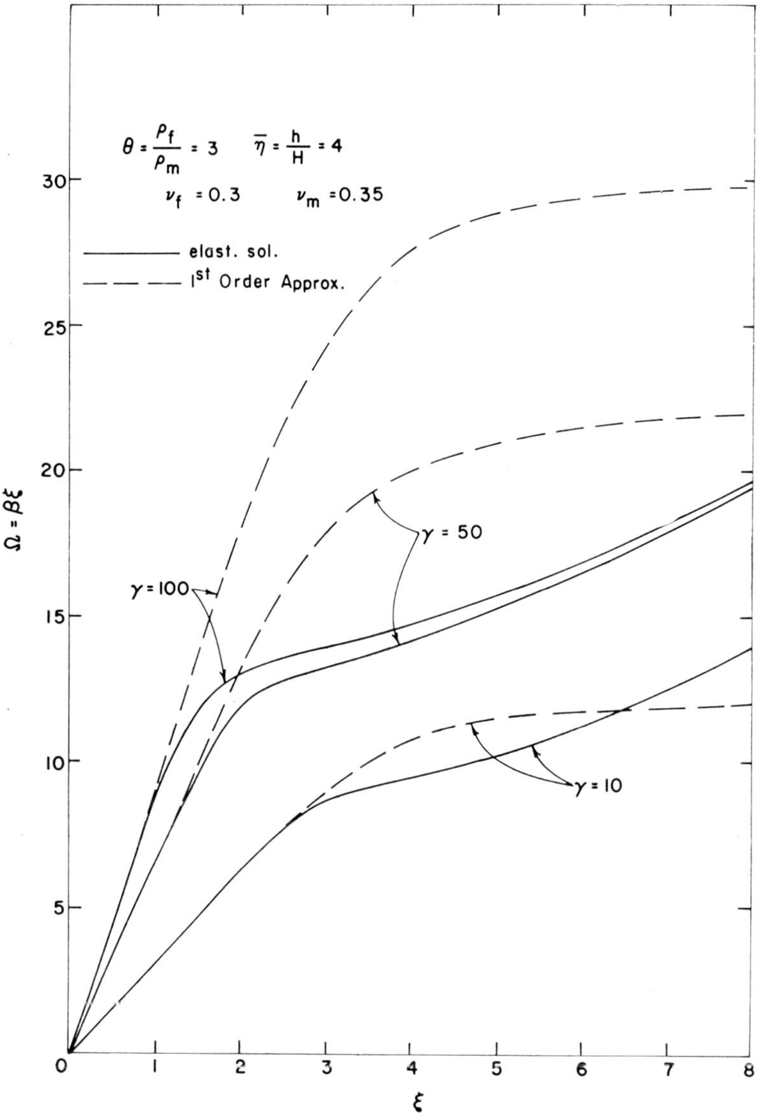

FIG. 10. Frequency vs. wave number for longitudinal mode

eqs. (57), (58) have been carried out, after appropriate correction factors and continuity conditions had been introduced, as discussed in detail by the authors in ref. 4. These second order approximation curves given in Figs. 7 and 8 indicate that satisfactory accuracy has now been achieved, particularly for large $\gamma = \mu_f/\mu_m$.

In a similar manner propagation of plane waves in the x_2 direction can be considered. The waves are dispersive,[4] but no further discussion of this motion is included here.

V. THE FIBER-REINFORCED COMPOSITE

The considerations which have lead the authors to construct effective stiffness theories for laminated composites may also be used to describe fiber-reinforced composites.[5] In the following only a summary is presented.

Let a cross-sectional area S of the composite contain n fibers, each of circular cross-sectional area A. Confining attention to transverse waves in the x_2 direction propagating in the x_1 direction, the governing equations are derived as

$$[C_{44}+\eta\varkappa\mu_f]\frac{\partial^2 u_2}{\partial x_1^2}-\eta\varkappa\mu_f\frac{\partial\psi_{21}}{\partial x_1}=\varrho_c\frac{\partial^2 u_2}{\partial t^2} \tag{59}$$

$$E_f r^2\frac{\partial^2\psi_{21}}{\partial x_1^2}+\varkappa\mu_f\left(\frac{\partial u_2}{\partial x_1}-\psi_{21}\right)=\varrho_f r^2\frac{\partial^2\psi_{21}}{\partial t^2} \tag{60}$$

where ϱ_c is defined by eq. (20) and η is the fiber density

$$\eta = nA/S \tag{61}$$

and

$$C_{44}=\mu_m\frac{\gamma(1+\eta)+(1-\eta)}{\gamma(1-\eta)+(1+\eta)}\,. \tag{62}$$

E_f is Young's modulus of the fiber material, $\varkappa$ is a correction coefficient whose value can be taken the same as in the Timoshenko beam theory,[6] and r^2 is the radius of gyration of the fiber cross section. An element of cross sectional area S is sketched in Fig. 11.

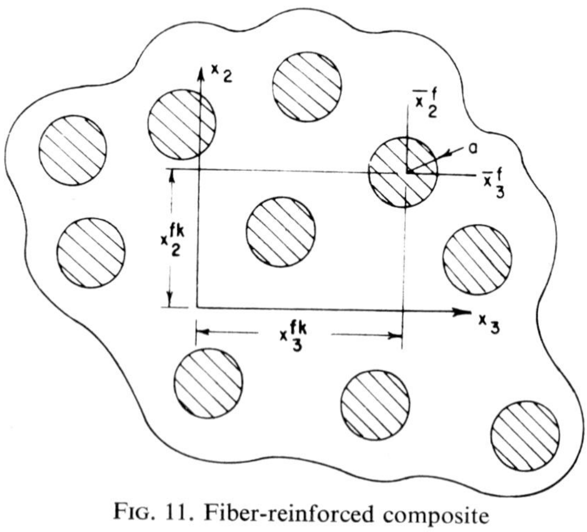

FIG. 11. Fiber-reinforced composite

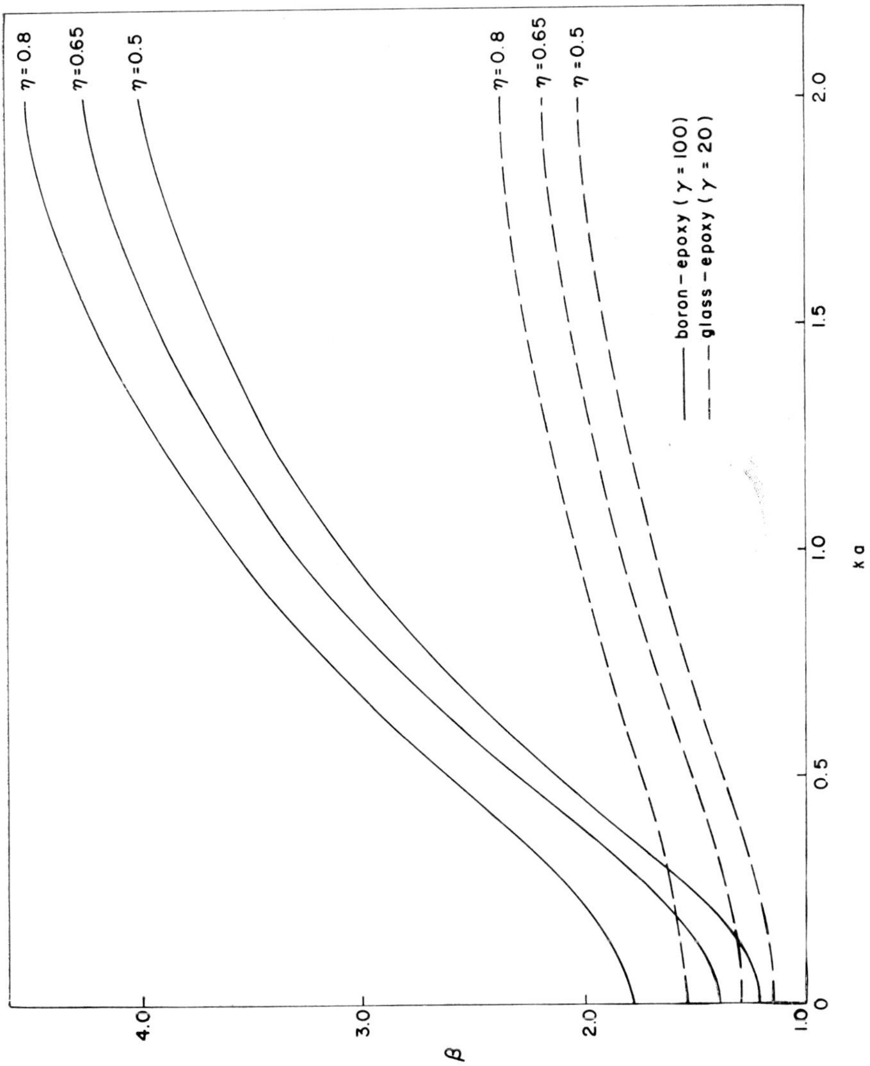

FIG. 12. Phase velocity for transverse waves

For time-harmonic waves of the usual form one obtains the dispersion relation

$$[C_{44}+\eta\varkappa\mu_f-c^2\varrho_c]\,[E_f r^2 k^2+\varkappa\mu_f-\varrho_f r^2 k^2 c^2]-\eta\varkappa^2\mu_f^2 = 0. \tag{63}$$

For various values of the fiber density η, dispersion curves are shown in Fig. 12. The numerical values for the parameters employed are those representative for boron-epoxy and glass-epoxy composites.[7] The ratio of the shear moduli γ was chosen as 100 for a boron-epoxy and 20 for a glass-epoxy composite. For all computations we use $\nu_f = 0.20$ and $\theta \equiv \varrho_f/\varrho_m = 3$.

It is noted that the dispersion curves depart at small values of the dimensionless wave number $\xi = ka$, where a is the radius of the fiber cross section, from the horizontal lines representing the phase velocities according to the effective modulus theory. The dispersion is particularly notable for the boron-epoxy composite. It must be borne in mind, however, that for a boron filament small values of the wave number ξ correspond to quite short wave lengths.

VI. CONCLUDING REMARKS

Various distinct effective stiffness theories have been discussed in the preceding sections, valid for both laminated and fibrous composites. For uniformly laminated composites the approximate theories of differing complexity could be tested by comparing solutions obtained on the basis of such theories with solutions obtained on the basis of the theory of elasticity.

The question could be raised as to what is the most "efficient" theory, i.e., a theory containing the least possible number of dependent field variables, yet predicting with reasonable accuracy dispersion of both transverse and longitudinal waves. For the range of parameters considered and for a large ratio of μ_f/μ_m it appears that the first antisymmetric shear mode of the stiffening layers must be included, as well as the symmetric shear and thickness-stretch modes in the matrix. Thus the displacement should be of the form, for plane strain

$$u_1^{fk} = u_{01}^{fk}+\bar{x}_2^f\psi_{22}^{fk} \tag{64}$$

$$u_2^{fk} = u_{02}^{fk} \tag{65}$$

$$u_1^{mk} = u_{01}^{mk}+\bar{x}_2^m\frac{\partial u_{02}^{mk}}{\partial x_1}+[4(\bar{x}_2^m)^2/H^2-1]\varphi^{mk} \tag{66}$$

$$u_2^{mk} = u_{02}^{mk}+\bar{x}_2^m\psi_{22}^{mk}. \tag{67}$$

Because of continuity requirements, only four equations of motion governing four field variables will result.

It is worth noting that the various dynamical continuum theories for laminated and fiber-reinforced composites advanced by the authors, as discussed in this paper, bear close resemblance to the theories of linear elasticity with couple stresses[8, 9, 10] and with microstructure,[11] sometimes referred to as generalized Cosserat continua. The motivation to consider such continua is dictated by the desire to describe various phenomena on the scale of the atomic lattice (or molecular distances) which ordinary continuum mechanics is not able to incorporate. Such more general continuum theories entail the introduction of additional material constants which so far have not been determined directly for any material. If one interprets these theories now as pertinent for a composite material, the nonclassical material constants are simply functions of the geometric layout and of the classical constants of the two constituent homogeneous materials. The authors have presented the details of this juxtaposition in a separate paper.[12]

ACKNOWLEDGMENTS

This work was supported by the Office of Naval Research under Contract ONR Nonr. 1228(34) with Northwestern University. The assistance of C. T. Sun, graduate student in the Department of Civil Engineering, in many phases of the work presented here is gratefully acknowledged.

REFERENCES

1. HERRMANN, G. and ACHENBACH, J. D. On Dynamic Theories of Fiber-Reinforced Composites, Northwestern University Structural Mechanics Laboratory Technical Report No. 67-2 (1967); also in *Proc. AIAA/ASME 8th Structures, Structural Dynamics & Materials Conf., AIAA, New York*, p. 112 (1967).
2. POSTMA, G. W. Wave Propagation in a Stratified Medium, *Geophysics*, **20**, 780 (1955).
3. SUN, C. T., ACHENBACH, J. D. and HERRMANN, G. Time-Harmonic Waves in a Stratified Medium Propagating in the Direction of the Layering, *J. Appl. Mech.*, **35**, pp. 408–411 (1968).
4. SUN, C. T., ACHENBACH, J. D. and HERRMANN, G. Effective Stiffness Theory of Laminated Media, Northwestern University Structural Mechanics Laboratory Technical Report No. 67–4 (1967).
5. ACHENBACH, J. D. and HERRMANN, G. Dispersion of Free Harmonic Waves in Fiber-Reinforced Composites, *AIAA J.*, **6**, No. 10, pp. 1832–1836 (1968); also Northwestern University Structural Mechanics Laboratory Technical Report No. 67-3 (1967).
6. MINDLIN, R. D. and DERESIEWICZ, H. Timoshenko's Shear Coefficient for Flexural Vibrations of Beams, *Proc. 2nd U.S. National Congress of Appl. Mech., ASME, New York*, p. 175 (1955).
7. WHITNEY, J. M. and RILEY, M. B. Elastic Properties of Fiber Reinforced Composite Materials, *AIAA J.*, **4**, No. 9, p. 1537 (1966).
8. MINDLIN, R. D. and TIERSTEN, H. F. Effects of Couple-stresses in Linear Elasticity, *Arch. Rat. Mech. Anal.*, **11**, 415 (1962).
9. TOUPIN, R. A. Elastic Materials with Couple-stresses, *Arch. Rat. Mech. Anal.* **11**, 385 (1962).

10. SCHAEFER, H. Das Cosserat-Kontinuum. To be published in *ZAMM.*
11. MINDLIN, R. D. Micro-structure in Linear Elasticity, *Arch. Rat. Mech. Anal.* **16,** 51 (1964).
12. HERRMANN, G. and ACHENBACH, J. D. Applications of Theories of Generalized Cosserat Continua to the Dynamics of Composite Materials. *Proc. IUTAM Symposium on the Generalized Cosserat Continuum Theory of Dislocations with Applications,* Springer-Verlag, Berlin, Heidelberg, New York, 69–79 (1968).

BASIC THERMOELASTIC PROBLEMS IN FIBER-REINFORCED MATERIALS†

R. B. TESTA‡ and B. A. BOLEY††

Columbia University, New York

ABSTRACT

This paper presents some basic considerations on the thermoelastic behavior of fiber-reinforced materials, with applications to two fundamental problems. In the first, bounds are constructed for the stress at any point in a beam and for the maximum stress in every phase of the composite. These apply for an arbitrary thermal load on a beam of arbitrary cross-section with fibers which are nearly axial. A simple algorithm is given to calculate the bounds on the maximum stress under certain conditions. The second problem considers the buckling of reinforcing fibers under thermal loading. Numerical results and explicit approximate solutions are given.

I. INTRODUCTION

Thermal loadings are encountered in various applications of composite materials. One application where the response to changes in temperature is of primary interest is the use of composites in heat shields for atmospheric re-entry. In addition, the formation process itself, whether effected at high temperatures (metal–metal composites) or accompanied by an exothermic reaction (glass–epoxy systems), often introduces thermal loadings which produce residual stresses in the composite.

The problems which arise in the study of thermal effects in composite materials are twofold; it is necessary to determine both the thermal properties (conductivity, specific heat, etc.) and the stresses and deformation under thermal load. Little attention has been paid to these in the literature on composite materials where studies of the metallurgical aspects and of the response to mechanical loading predominate (see for example, the bibliographics in refs. 1 and 2). The available data on thermal properties, which are not the main concern in the present study, has been summarized by Lowenstein.(2) Scala(1) and Lowenstein(2) have outlined some limited results for the stresses induced by temperature when the materials in a composite have different coefficients of thermal expansion. In addition, there have been studies of several related problems such as the layered beam in ref. 3.

† Supported in part by the Office of Naval Research.

‡ Associate Professor of Civil Engineering.

†† Professor of Civil Engineering, presently Cornell University, Ithaca, N. Y.

The response of composite materials when subjected to thermal loading has not received attention comparable to that for mechanical loading. Certain basic considerations are presented here in the form of two fundamental problems, as an initial effort towards development of a more comprehensive theory for the analysis of thermal stresses and deformations in composite materials. The first problem, which is presented in section II, considers a fiber-reinforced beam of arbitrary cross-section under an arbitrary thermal load; an expression for the axial stress is derived and bounds are established for the stress. Section III comprises the second problem in which the stability under thermal loading is considered for fibers in a matrix. In both problems, the materials are taken to be elastic, with material properties not affected by temperature, and the various phases of the composite are assumed to be perfectly bonded at the interfaces.

The thermoelastic response of a composite beam of rectangular cross section was considered in an earlier study.[4] There the material properties were taken to be discontinuous functions of position and a method of successive approximation[5] was employed to determine the stresses under arbitrary thermal loads. It was shown that the first approximation gives stresses identical to those obtained from the elementary beam theory in which plane sections are assumed to remain plane. The second approximation gave a correction term (or alternatively, an error estimate on the use of the elementary theory) for the case of reinforcing fibers at some angle to the axis of the beam. The square of the slope of the fibers appeared as a coefficient in this correction term, while successively higher powers of the slope occurred in the corrections for higher order approximations. The validity of the elementary beam theory was thus established (within an error of the order of 10%) for composite beams with nearly axial (up to 15° inclination) reinforcing fibers. This result is applied in section IIA to the derivation for the axial stress in a fiber-reinforced beam of arbitrary cross-section.

Knowledge of the maximum thermal stress in each material phase is a prerequisite for design of composites in any application in which temperature plays a role. It is shown in ref. 4 that bounds on the axial stress can be given explicitly for the case of fiber-reinforced beams of rectangular section under two-dimensional thermal loadings. Moreover, the bounds depend on the extreme values of αET in the section and not on the details of the distribution of temperature and materials. In practice, the conditions of symmetry in geometry and temperature in ref. 4 are not always satisfied and unsymmetrical bending results. Bounds on the stress at any point in a fiber-reinforced beam of arbitrary cross-section are established in section IIB for arbitrary temperature distributions. In section IIC bounds on the maximum stress in each material in the section are constructed, and an algorithm for the calculation of the bounds in certain composite materials is presented in the form of a beam-column analogy in section IID. Section IIE gives a summary of the procedure in applying the bounds. Besides their use in design, the bounds will serve also to estimate errors in determining stresses when there is some uncertainty about the temperature. In addition, uncertainties in the distribution of material also are taken into account in evaluating the bounds for the maximum stress.

Because of differences in coefficients of thermal expansion of the material phases in a composite, the internal stability of the composite under thermal loading becomes significant when the reinforcement is in the form of fibers. If the fibers buckle under an applied thermal load, then the resistance of the composite member to a superimposed mechanical load will be seriously impaired. Thus, it is especially necessary to ensure that the fibers in a composite do not buckle under the temperature changes occurring in the formation process. To observe the essential character of the problem for the stability of reinforcing fibers, a simple geometry is selected and a uniform temperature change is considered. In sections IIIA and B the problem is formulated by considering the buckled configuration as an incremental deformation superimposed on an initial equilibrium deformation in which the fibers remain straight.[6] Transcendental equations for the critical temperature and buckling mode are obtained from the formal solution in section IIIC. Results of the numerical solution of these equations are presented in section IIID in the form of graphs of the critical temperature and wavelength for various values of the parameters. Approximations applicable to certain composites are introduced, and explicit expressions for the critical values are obtained.

II. BOUNDS ON THERMAL STRESSES

A. *Formulation*

A beam of arbitrary cross-section which is uniform over the length of the beam (Fig. 1) is subjected to an arbitrarily distributed temperature increase. The beam is assumed to be free of external loads which, if present, would

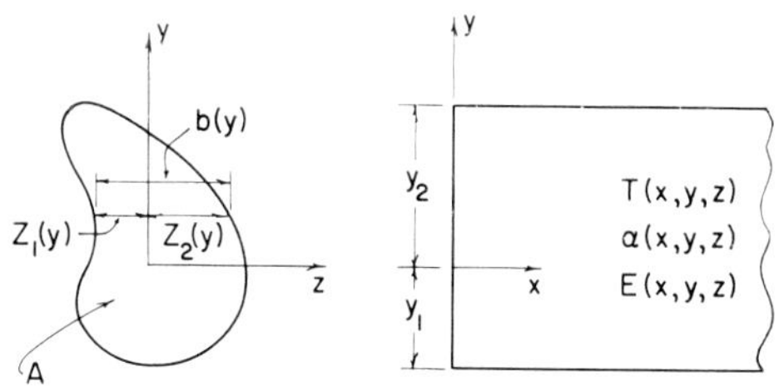

FIG. 1. Beam of arbitrary cross-section

give stresses which must be superimposed on the thermal stresses considered here. The material is a composite with reinforcing fibers embedded in and rigidly bonded to a matrix. The material properties are assumed to be unaffected by the temperature. If the fibers do not deviate greatly from the axial direction, then on the basis of the analysis for composite beams of rectangular section[4] together with the discussion in ref. 7 for homogeneous beams of arbitrary section, the axial stress is predicted with good accuracy by the elementary beam theory.

Axes are chosen at the centroid of the section in Fig. 1, so that

$$\int_A y\,dA = \int_A z\,dA = 0. \tag{1}$$

In addition, the distribution of materials in the section is assumed to be such that

$$\int_A Ey\,dA = \int_A Ez\,dA = 0. \tag{2}$$

However, the analysis can be extended to the case where eq. (2) does not hold. In eqs. (1) and (2), A is the area of the cross-section, $E(x, y, z)$ is Young's modulus and $\alpha(x, y, z)$ the coefficient of linear thermal expansion. In order to simplify the resulting expressions, and without loss of generality, the axes are also chosen to satisfy the condition

$$\int_A \alpha ETz\,dA \int_A Ey^2\,dA = \int_A \alpha ETy\,dA \int_A Eyz\,dA \tag{3}$$

under the temperature distribution† $T(x, y, z)$. For certain loadings, eq. (3) is the condition that z be the neutral axis in the section.[8]

The assumption of elementary beam theory that sections initially plane remain plane after deformation requires that the axial strain at points in a cross-section be linear in y and z,[8]

$$\varepsilon = f_0(x)+yf_1(x)+zf_2(x) \tag{4}$$

which in the elementary theory gives an axial stress

$$\sigma = E(\varepsilon-\alpha T). \tag{5}$$

The functions f_0, f_1, f_2 are evaluated by the condition that, in the absence of mechanical loading, the following stress resultants vanish:

$$\int_A \sigma\,dA = \int_A y\sigma\,dA = \int_A z\sigma\,dA = 0. \tag{6}$$

† A dependence of the temperature on time might also be indicated, but since inertia effects are neglected in the quasistatic theory employed here, time would appear only as a parameter.

After substituting eqs. (4) and (5) into eq. (6) and simplifying by eqs. (2) and (3) the axial stress in a composite beam of arbitrary section is given by

$$\sigma = -\alpha ET + \frac{P_T}{A\bar{E}} E + \frac{M_T}{\overline{EI}} Ey \tag{7}$$

where
$$\begin{aligned} P_T &= \int_A \alpha ET\, dA = \text{thermal load}, \\ M_T &= \int_A \alpha ETy\, dA = \text{thermal moment}, \\ \bar{E} &= \frac{1}{A}\int_A E\, dA = \text{effective modulus},^{(4)} \\ \overline{EI} &= \int_A Ey^2\, dA = \text{effective stiffness}.^{(4)} \end{aligned} \tag{8}$$

Equation (7) is identical to the expression derived in refs. 4 and 5 as the first term of a series solution for the thermal stress in nonhomogeneous beams of rectangular cross-section. However, it must be emphasized that here eq. (7) applies to any section provided the axes are chosen to satisfy eq. (3).

Consider the stress at points in a specific cross-section. The value of the coordinate x is then fixed and it can be omitted from the equations. The axial stress may be rewritten in the form

$$\sigma(\eta, \zeta) = -\tau(\eta, \zeta) + \frac{r(\eta, \zeta)}{A'} \int_{-\eta_1}^{1} \int_{-\zeta_1}^{\zeta_2} \tau(\eta', \zeta') \left[1 + \frac{\eta\eta'}{\varrho'^2}\right] d\zeta'\, d\eta' \tag{9}$$

in terms of the dimensionless variables

$$\begin{aligned} \eta &= y/y_2 \qquad & \eta_1 &= y_1/y_2 \\ \zeta &= z/y_2 & \zeta_1 &= z_1/y_2 \qquad & \zeta_2 = z_2/y_2 \end{aligned} \tag{10}$$

and the quantities

$$\tau(\eta, \zeta) = \alpha(\eta, \zeta)\, E(\eta, \zeta) T(\eta, \zeta) \tag{11a}$$

$$r(\eta, \zeta) = E(\eta, \zeta)/\bar{E} = r_i \tag{11b}$$

$$\varrho'^2 = \frac{\varrho^2}{y_2^2} = \frac{\overline{EI}/A\bar{E}}{y_2^2} \tag{11c}$$

$$A' = A/y_2^2 = \int_{-\eta_1}^{1} \int_{-\zeta_1}^{\zeta_2} d\zeta'\, d\eta' = \int_{-\eta_1}^{1} \beta(\eta')\, d\eta'. \tag{11d}$$

Figure 2 shows the dimensionless variables and identifies two points in the

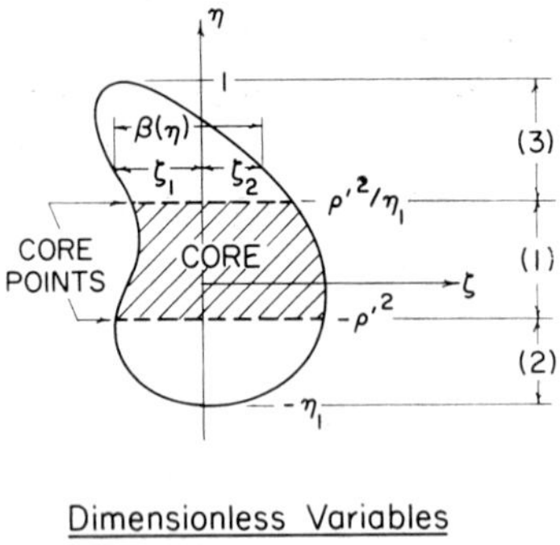

FIG. 2. Dimensionless variables

section which for homogeneous materials (where $\overline{EI}/\overline{AE} = I/A$) are the core points of the cross-section as discussed in refs. 9 and 10. The inequality

$$\varrho'^2 \leqslant \eta_1 \tag{12}$$

which is verified in Appendix A is employed in Fig. 2 and in the subsequent analysis. Note that $r(\eta, \zeta)$, eq. (11b), has only a discrete number of values, one for each of the materials in the composite, and these are denoted by r_i. However $\tau(\eta, \zeta)$, eq. (11a), may vary in an arbitrary manner.

B. *Maximum Stress at a Point*

Given the properties of the materials in the composite and their distribution in a cross-section of specified geometry, it is desired to establish bounds on the stress, eq. (9). At any point (η, ζ) in the section, the bounds are defined by

$$\sigma_m(\eta, \zeta) \leqslant \sigma(\eta, \zeta) \leqslant \sigma_M(\eta, \zeta) \tag{13}$$

and their values will depend on which material occupies the point in question. The extreme values of the functions σ_m and σ_M, namely

$$\begin{aligned} \Sigma_m &\leqslant \sigma_m(\eta, \zeta) \\ \Sigma_M &\geqslant \sigma_M(\eta, \zeta), \end{aligned} \tag{14}$$

give bounds on the maximum stress in the section. These must be determined for each of the materials in the composite.

In the cross section considered, let the extreme values of the function τ, eq. (11a), be given by

$$-\tau_m \leqslant \tau(\eta, \zeta) \leqslant \tau_M. \tag{15}$$

Note that the quantities τ_M and τ_m need not correspond to the values of

$\tau = \alpha ET$ at the points where the maximum and minimum temperatures occur. However, if the temperature does not vary rapidly in the section and if the materials are uniformly distributed these values will be associated with the temperature extremes in the section. When the quantities τ_M and τ_m are known, bounds on the stress, eq. (13), can be constructed according to the procedure in ref. 9 as follows:

A portion of the integrand in eq. (9) is designated by

$$f(\eta, \eta') = 1 + \frac{\eta\eta'}{\varrho'^2}. \tag{16}$$

The function τ in the integral, eq. (9), is replaced by τ_M wherever f is positive in the range of integration and by $-\tau_m$ where f is negative. The resulting expression is an upper bound for the value of the stress at any point (η, ζ). A lower bound is obtained by reversing the roles of τ_M and τ_m. In order to write more general expressions for these bounds, the function $\tau(\eta, \zeta)$ appearing as the first term on the right hand side of eq. (9) can also be replaced by its maximum or minimum value in the section. The resulting bounds are those defined in eq. (13), from which the overall bounds in eq. (14) can be determined. In the expressions thus obtained there will appear the parameter r_i identifying the various materials in the composite, thereby permitting evaluation of bounds on the stress in any material phase.

It is necessary then to determine the sign of the function f, eq. (16), for any value of η in the section and for all values of η' in the range of integration. Figure 3 is a section through $f = 0$ of a three-dimensional plot of f versus η and η', on which the ranges of η and η' are outlined by the dashed lines. It shows three ranges of the coordinate η, corresponding to the regions similarly labelled on the cross section in Fig. 2, in which the sign of $f(\eta, \eta')$ is classified. For any value of η in range (2), for example, f is positive for η' less than

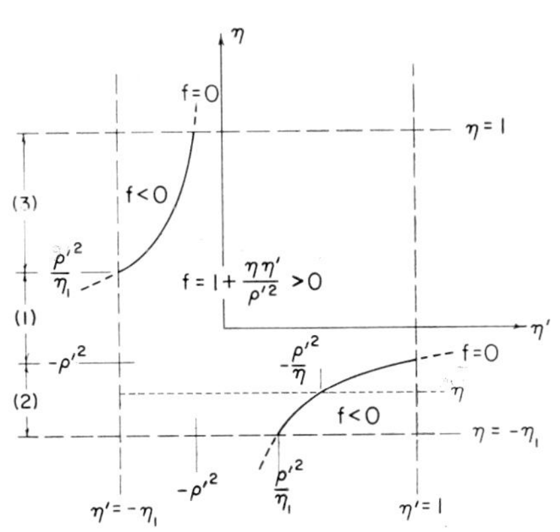

FIG. 3. Section through $f = 0$ for plot of f versus η

$-\varrho'^2/\eta$ and negative for η' larger than this value. This is illustrated in Fig. 3 where, for a particular value of η in range (2), the path of integration in eq. (9) is shown as a dotted line from $\eta' = -\eta_1$ to $\eta' = 1$.

The bounds now can be constructed according to the previously outlined procedure, with the three ranges identified in Fig. 3 considered individually.

1. RANGE (1), $-\varrho'^2 < \eta < \varrho'^2/\eta_1$ (FIG. 2). In this range which comprises points η in the core of the section the function f, eq. (16), is positive (Fig. 3), over the entire range of integration in eq. (9). Upper and lower bounds on the stress are obtained by substituting τ_M and τ_m, respectively, for τ in eq. (9). After carrying out the integration in ζ', the bounds are

$$\begin{aligned}
\sigma_M(\eta, \zeta) &= \tau_m + \frac{r_i \tau_M}{A'} \int_{-\eta_1}^{1} \beta(\eta') \left[1 + \frac{\eta\eta'}{\varrho'^2}\right] d\eta' \\
\sigma_m(\eta, \zeta) &= -\tau_M - \frac{r_i \tau_m}{A'} \int_{-\eta_1}^{1} \beta(\eta') \left[1 + \frac{\eta\eta'}{\varrho'^2}\right] d\eta'.
\end{aligned} \tag{17}$$

By means of eqs. (1) and (11d) the bounds reduce to

$$\begin{aligned}
\sigma_M(\eta, \zeta) &= \tau_m + r_i \tau_M = r_i(\tau_m + \tau_M) + \tau_m(1 - r_i) \\
\sigma_m(\eta, \zeta) &= -\tau_M - r_i \tau_m = r_i(\tau_m + \tau_M) - \tau_M(1 - r_i).
\end{aligned} \tag{18}$$

2. RANGE (2), $-\eta_1 \leqslant \eta \leqslant -\varrho'^2$ (FIG. 2). The range of integration in eq. (9) is separated into two parts, in each of which the sign of the function f is determined in Fig. 3. The upper bound is

$$\begin{aligned}
\sigma_M(\eta, \zeta) = \tau_m + \frac{r_i}{A'} \Bigg\{ \tau_M \int_{-\eta_1}^{-\varrho'^2/\eta} \beta(\eta') \left[1 + \frac{\eta\eta'}{\varrho'^2}\right] d\eta' \\
- \tau_m \int_{-\varrho'^2/\eta}^{1} \beta(\eta') \left[1 + \frac{\eta\eta'}{\varrho'^2}\right] d\eta' \Bigg\}
\end{aligned} \tag{19}$$

and a similar expression is obtained for $\sigma_m(\eta, \zeta)$. The bounds can be written in the form

$$\begin{aligned}
\sigma_M(\eta, \zeta) &= \frac{r_i}{A'} (\tau_m + \tau_M) \left[A' - A(\eta) - \frac{\eta}{\varrho'^2} Q(\eta)\right] + \tau_m(1 - r_i) \\
\sigma_m(\eta, \zeta) &= -\frac{r_i}{A'} (\tau_m + \tau_M) \left[A' - A(\eta) - \frac{\eta}{\varrho'^2} Q(\eta)\right] - \tau_M(1 - r_i)
\end{aligned} \tag{20}$$

where

$$A(\eta) = \int_{-\varrho'^2/\eta}^{1} \beta(\eta')\, d\eta' = A' - \int_{-\eta_1}^{\varrho'^2/\eta} \beta(\eta')\, d\eta'$$
$$Q(\eta) = \int_{-\varrho'^2/\eta}^{1} \eta'\beta(\eta')\, d\eta' = -\int_{-\eta_1}^{\varrho'^2 1} \eta'\beta(\eta')\, d\eta'. \tag{21}$$

The quantities defined in eqs. (21) can be interpreted as the area and moment of area of a portion of the cross-section. They will be identified subsequently for certain values of η.

3. RANGE (3), $\varrho'^2/\eta_1 \leqslant \eta \leqslant 1$ (Fig. 2). The bounds are derived as in the previous case with the result

$$\sigma_M(\eta, \zeta) = \frac{r_i}{A'}(\tau_m + \tau_M)\left[A(\eta) + \frac{\eta}{\varrho'^2} Q(\eta)\right] + \tau_m(1 - r_i)$$
$$\sigma_m(\eta, \zeta) = -\frac{r_i}{A'}(\tau_m + \tau_M)\left[A(\eta) + \frac{\eta}{\varrho'^2} Q(\eta)\right] - \tau_M(1 - r_i). \tag{22}$$

In order to evaluate the bounds at a point in the section, it is necessary to specify the shape of the cross-section, i.e., $\beta(\eta)$, so that the integrals in eq. (21) can be evaluated. For example, if the present results are evaluated for a rectangular section the bounds are those derived in ref. 4 (provided the axes specified by eqs. (1)–(3) are parallel to the edges of the section). Figure 4 shows these bounds as functions of location in the cross-section for the special case where $\varrho'^2 = 1/3$ which applies either to homogeneous materials or to a composite section for which $\overline{EI} = \bar{E}I$. This latter case is considered in a subsequent section.

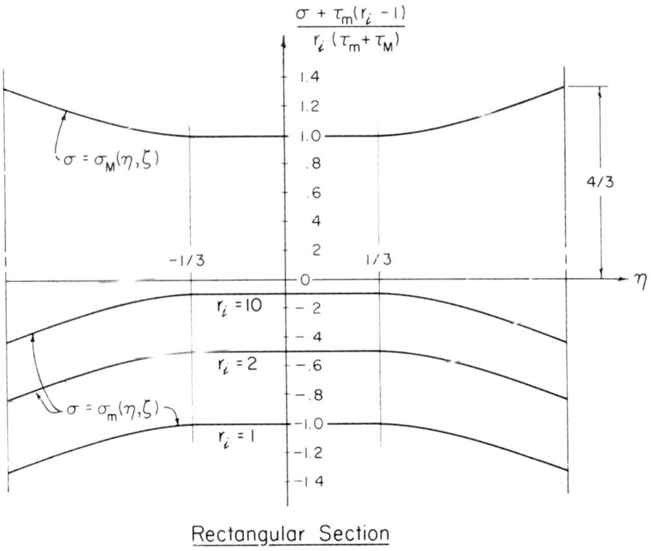

FIG. 4. Bounds for rectangular cross-section

It is observed in Fig. 4 and in eqs. (18), (20) and (22) that the upper and lower bounds on the stress need not be of equal magnitude, but that the following equality applies:

$$\sigma_M(\eta, \zeta) - \tau_m(1-r_i) = -\sigma_m(\eta, \zeta) - \tau_M(1-r_i). \tag{23}$$

For the case $r_i = 1$ (e.g., for a homogeneous material) eq. (23) gives $|\sigma_M(\eta, \zeta)| = |\sigma_m(\eta, \zeta)|$ which is the result obtained in ref. 9.

In a given composite, any one of the discrete values r_i, eq. (11b), specifies a material phase of the composite. The bounds when evaluated for that value of r_i (e.g., one of the curves in fig. 4) give upper and lower bounds on the stress at any point of the cross-section which may be occupied by that material. If the bounds are constructed for all values of r_i, the maximum stress which can occur at any point of the section is determined for any material phase of the composite. From these and some knowledge of the distribution of materials, the maximum possible stress in any material phase can be calculated if the extreme values of τ are given.

C. *Maximum Stress in a Section*

For a rectangular cross-section, Fig. 4 shows that the bounds are constant for all points within the core of the section and increase in magnitude monotonically from a core point to the adjacent extreme fiber. Therefore, the maximum value of a bound for the stress in any one material (i.e., using one value r_i) is the value of the bound at the point furthest from the centroid occupied by that material. For example, bounds on the maximum stress in a material (r_i) occupying a point at the extreme fibers of the rectangular section are

$$\frac{\Sigma_M - \tau_m(1-r_i)}{r_i(\tau_m + \tau_M)} = -\frac{\Sigma_m + \tau_M(1-r_i)}{r_i(\tau_m + \tau_M)} = 4/3. \tag{24}$$

For a section of arbitrary shape, a similar procedure gives bounds on the maximum stress in any material in the section. The values of the bounds are constant for all points in the core of the section, eq. (18), and there is a smooth transition to the values given by eqs. (20) and (22) for points outside the core. Moreover, the magnitude of each of the latter bounds increases monotonically from the core point to the adjacent extreme fiber, as indicated by the following calculation: In the range (3) (Fig. 2) the upper bound, eq. (20), when differentiated with respect to η gives

$$\frac{\partial \sigma_\eta}{\partial \eta} = \frac{r_i(\tau_m + \tau_M)\, Q(\eta)}{A' \varrho'^2} \geqslant 0. \tag{25}$$

Consequently, in the range of applicability (i.e., from the core point $\eta = \varrho'^2/\eta_1$ to the extreme fiber $\eta = 1$) the bound increases in magnitude with increasing values of η. To obtain the maximum value of the bound for any one material (r_i), the bound is evaluated at the largest value of η at which that material is found in the section.

A conservative bound for the maximum stress, eqs. (14), in a material phase is given by the values of eqs. (20) and (22) at the extreme fibers, using the appropriate value of r_i. In terms of the dimensional variables the values are

$$\begin{aligned}
\sigma_M(-y_1, z) &= r_i(\tau_m+\tau_M)\left[\frac{A_1}{A}-\frac{Q_1 y_1}{\overline{EI}/\overline{E}}\right]+\tau_m(1-r_i) \\
\sigma_m(-y_1, z) &= -r_i(\tau_m+\tau_M)\left[\frac{A_1}{A}-\frac{Q_1 y_1}{\overline{EI}/\overline{E}}\right]-\tau_M(1-r_i) \\
\sigma_M(y_2, z) &= r_i(\tau_m+\tau_M)\left[\frac{A_2}{A}+\frac{Q_2 y_2}{\overline{EI}/\overline{E}}\right]+\tau_m(1-r_i) \\
\sigma_m(y_2, z) &= -r_i(\tau_m+\tau_M)\left[\frac{A_2}{A}+\frac{Q_2 y_2}{\overline{EI}/\overline{E}}\right]-\tau_M(1-r_i)
\end{aligned} \tag{26}$$

from which the bounds in eq. (14) are given by

$$\begin{aligned}
\Sigma_M &= \max\,[\sigma_M(-y_1, z), \quad \sigma_M(y_2, z)] \\
\Sigma_m &= \min\,[\sigma_m(-y_1, z), \quad \sigma_m(y_2, z)].
\end{aligned} \tag{27}$$

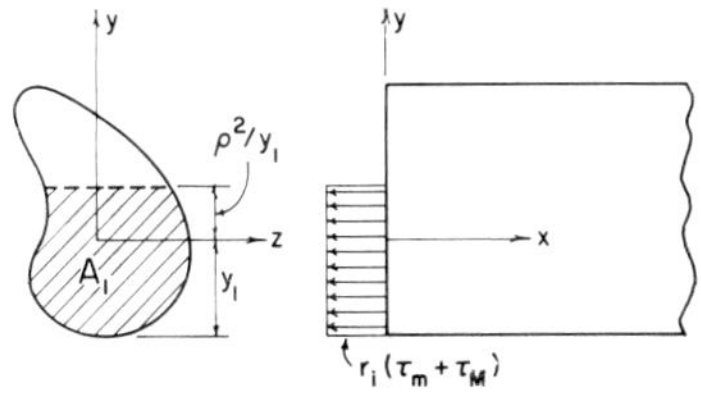

(a) Loading for bounds at $y = -y_1$

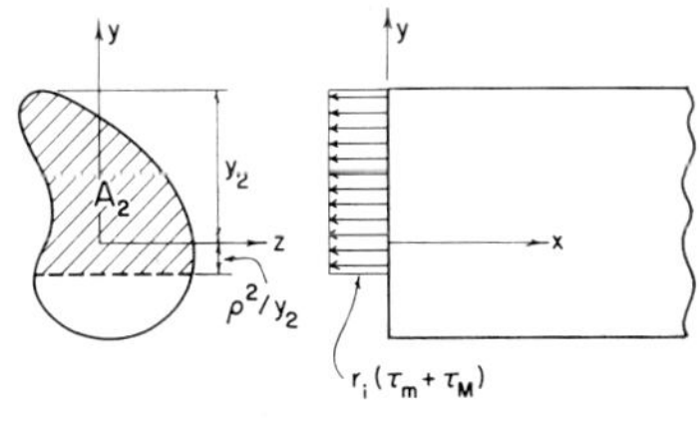

(b) Loading for bounds at $y = y_2$

Beam-Column Analogy $(\overline{EI} = \overline{E}I)$

FIG. 5. Beam column analogy

In eqs. (26), A_1 and A_2 are areas of the portions of the cross-section shown in Fig. 5, and Q_1, Q_2 are the first moments of these areas about the centroidal axis $y = 0$. (Note that Q_1 will be negative.)

D. *Beam-column Analogy*

For certain composite materials where the distribution of materials in the cross-section is such that

$$\overline{EI} = \bar{E}\, I \tag{28}$$

the bounds on the maximum stress, eqs. (27), can be evaluated quite easily by an analogy similar to the beam-column analogy developed in ref. 9. For an arbitrary distribution of materials the analogy in ref. 9 does not apply directly, because of the presence of the ratio $\overline{EI}/\bar{E}$ in eqs. (26). When eq. (28) holds,† this ratio reduces to I (the moment of inertia of the cross-section) and eqs. (26) and (27) can be rewritten in the form

$$\begin{aligned} \Sigma_M - \tau_m(1-r_i) &= \max\left\{r_i(\tau_m+\tau_M)\left[\frac{A_i}{A}+\frac{Q_i y_i}{I}\right]\right\} \\ \Sigma_m + \tau_m(1-r_i) &= \min\left\{-r_i(\tau_m+\tau_M)\left[\frac{A_i}{A}+\frac{Q_i y_i}{I}\right]\right\} \end{aligned} \tag{29}$$

where A_i, $y_i = A_1$, $-y_1$ or A_2, y_2 (Fig. 5) and Q_i is the first moment of the area A_i as defined previously. In addition, the quantity ϱ', eq. (11c) is then the dimensionless radius of gyration of the section and the core points in Fig. 2 coincide with those for a homogeneous section. Consequently the areas A_1 and A_2 for such a composite are identical to those defined in ref. 9.

The expression on the right-hand side of eq. (29) is the elementary beam-column formula for the stress at the fiber y_i in a homogeneous beam (with the same cross-section as the composite beam) subjected to a uniform axial loading $r_i(\tau_m+\tau_M)$ over the portion A_i of the section. Consequently, the following algorithm to calculate bounds on the maximum stress is established:

1. Apply a uniform axial load $r_i(\tau_m+\tau_M)$ to the area A_1 (Fig. 5(a)) and calculate the stress at the fiber $y = -y_1$ by the elementary beam-column formula as if the material were homogeneous.
2. Repeat step 1 with an axial load $-r_i(\tau_m+\tau_M)$ i.e., compression.
3. By the procedure of step 1, calculate the stress at the fiber $y = y_2$ when the loading $r_i(\tau_m+\tau_M)$ is applied to the area A_2 (Fig. 5(b)).
4. Repeat step 3 reversing the sign of the load.
5. The larger of the results of steps 1 and 3 gives the value $\Sigma_M - \tau_m(1-r_i)$, eq. (29).
6. The smaller result from steps 2 and 4 gives $\Sigma_m + \tau_m(1-r_i)$, eq. (29).

† However, a similar analogy could be developed.

As in the previous discussion of the bounds, this procedure will give the maximum stress which may occur in any one material phase if the appropriate value of r_i is selected.

The beam-column analogy for the evaluation of a bound on the maximum stress is similar to that given in ref. 9, except that here, the bound applies to the quantities on the left-hand side of eq. (29) and not to the stress itself. In ref. 9, values of the bound on the maximum stress are tabulated for homogeneous beams of various cross-sections. With the above modification in the bound, the tabulated values in ref. 9 can be applied to composite beams which satisfy eq. (28). In particular, the value 4/3 which is observed in Fig. 4 and eq. (24) also appears in the tabulation in ref. 9 for the rectangular section.

E. *Summary and Discussion*

In a composite beam of arbitrary cross-section, bounds on the axial stress resulting from an arbitrary thermal load are determined as follows:

1. Centroidal axes are selected which also satisfy eqs. (3).
2. If the extreme values of $\tau = \alpha ET$ in a cross-section are known, the upper and lower bounds for the stress at any point in the section are given explicitly by eqs. (18), (20) and (22). The bounds apply to a specific material at the point when the appropriate value of r_i, eq. (11b), is selected.
3. The maximum stress which can occur in any one material of the composite is determined by using the value of r_i for that material and evaluating the bounds at the point furthest from the axis $y = 0$ which is occupied by that material. For any material which occupies points at the extreme fibers, bounds on the maximum stress are given by eqs. (26) and (27). These also give a conservative bound on the maximum stress in any material, whatever its distribution in the section, and may be used when there is some uncertainty about the precise location of materials in the composite.
4. In a composite material satisfying eq. (28), bounds on the maximum stress are given by eq. (29) which can be evaluated by a simple beam-column analogy (Fig. 5). In this case, the core points and areas A_1 and A_2 are identical to the values in ref. 9 for homogeneous beams, and the results tabulated there, for various cross-sections can be applied to composite beams. However, the results in ref. 9 must be applied to the quantities on the left-hand side of eqs. (29) and not directly to the stress.

In view of the analysis in ref. 4, the foregoing analysis is valid only when the composite is a fiber-reinforced material with the fibers nearly axial. There is no restriction on the number of material phases, but their distribution in the composite has been assumed to satisfy eq. (2). A restriction on the selection of centroidal axes has been imposed by eq. (3). If this condition cannot be satisfied (for example with insufficient knowledge of the temperature distribution) the bounds constructed here must be evaluated for various orientations of the centroidal axes to determine their maximum values.

III. THERMAL BUCKLING OF FIBERS IN A MATRIX

The stability of reinforcing fibers embedded in a matrix is most conveniently formulated by means of the theory of incremental deformations.[11] An initial state of stress (without buckling of the fibers) is determined for the composite member under thermal loading. Buckling is then considered as a possible equilibrium incremental deformation superimposed on this initial state.

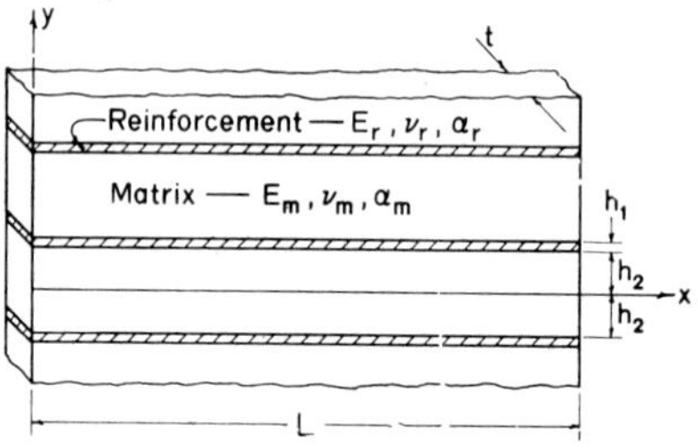

FIG. 6. Reinforced thin plate

The significant parameters governing the internal stability of a fiber reinforced material can be observed by considering the simple geometry of Fig. 6, in which a thin plate is reinforced by parallel and uniformly spaced fibers. The length of the plate (L) in the direction of the fibers is assumed to be much larger than the fiber spacing. The overall depth in the y direction is also taken to be large compared to the fiber spacing with the understanding that the results obtained here will be valid sufficiently far from a boundary. The member is subjected to a uniform temperature increase.

A. *Initial Deformation*

The stress in the composite member (Fig. 6) may be determined by the elementary theory[4] as long as the fibers remain straight. This is the initial stress and it is evaluated by eq. (7) which is identical in form to the expression given in ref. 4. Because of the uniform temperature and distribution of fibers,

$$M_T = 0 \tag{30}$$

and eq. (7) gives an initial axial stress

$$\sigma = E\left[\frac{P_T}{A\bar{E}} - \alpha T\right]. \tag{31}$$

Using subscripts m and r to denote properties of the matrix and reinforcement, respectively, eqs. (8) give

$$\frac{P_T}{A\bar{E}} = \frac{2h_2\alpha_m E_m + h_1\alpha_r E_r}{2h_2 E_m + h_1 E_r} T. \tag{32}$$

The stress, eq. (31), is a discontinuous function of y which has two possible values depending on the values of E and α. It is convenient to introduce these two values of the initial stress in the following dimensionless form:

$$\begin{aligned} p_m &= \frac{P_T}{A\bar{E}} - \alpha_m T = K_1 T \\ p_r &= \frac{P_T}{A\bar{E}} - \alpha_r T = K_2 T \end{aligned} \tag{33}$$

where

$$\begin{aligned} k_1 &= \frac{(\alpha_r - \alpha_m)r}{r + 2h_2/h_1} \\ k_2 &= \frac{\alpha_m - \alpha_r}{1 + rh_1/2h_2} \end{aligned} \tag{34}$$

and†

$$r = E_r/E_m. \tag{35}$$

The quantities p_m and p_r as shown in the inset of Fig. 9 represent the differences between the free expansions $\alpha_m T$, $\alpha_r T$ of the two materials and the actual strain in the composite, eq. (32), under the assumption of plane sections. Consequently, they are the portions of the thermal strains which produce stresses in the materials, and since the resultant of these stresses must vanish (for zero external load) the quantities in eqs. (33) are related by

$$p_r = -\frac{2h_2}{rh_1} p_m. \tag{36}$$

It is apparent from this interpretation that the fibers can buckle only for positive values of p_m—i.e., when there is tension in the matrix and compression in the fibers.

† The ratio r defined in eq. (35) should be distinguished from the quantities $r(\eta, \zeta)$ and r_i, eq. (11b), which were employed in section II.

B. *Incremental Deformation*

In the present theory, the equations for both the initial and incremental deformation are linear. Having included the temperature terms in the solution for the initial stress, the incremental deformation is determined under isothermal conditions and the equations in ref. 11 apply. In addition, since the geometry in Fig. 6 is identical to that considered in ref. 6 for buckling under mechanical loading, the formulation here follows closely that in ref. 6. The incremental deformation for the fibers is treated by classical beam theory, and for the matrix by the equations of generalized plane stress which are derived from the equations in ref. 11.

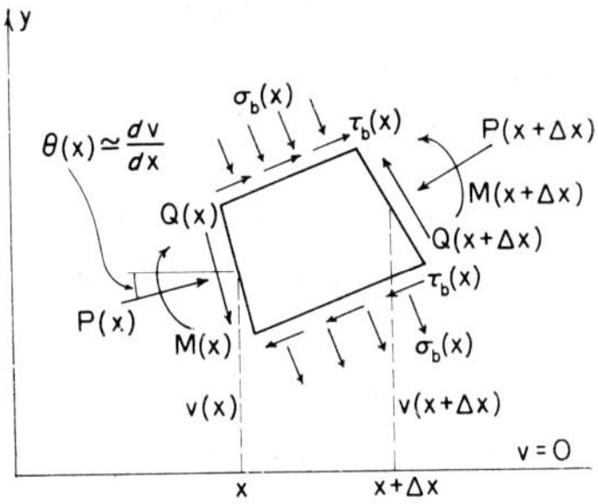

FIG. 7. Equilibrium conditions for fiber element

Let $v(x)$ be the deflection (in the y direction) of a fiber, $\sigma_b(x)$, $\tau_b(x)$ the stresses at the fiber-matrix interface (Fig. 7) and u_x, u_y the components of displacement in the matrix—all referring to the incremental deformation. Equilibrium of an element of fiber (Fig. 7), together with the moment curvature relation and the assumption of small deflections gives the equation

$$\frac{E_r h_1^3}{12}\frac{d^4 v}{dx^4} - E_r h_1 p_r \frac{d^2 v}{dx^2} - h_1 \frac{d\tau_b}{dx} + 2\sigma_b = 0 \tag{37}$$

in which the incremental axial stress has been neglected relative to the initial stress $E_r p_r$.[6] Equation (37) is valid when each reinforcing strip in Fig. 6 consists of a number of fibers placed side by side over the thickness t of the plate. If each strip is actually a thin plate, E_r in the first term of eq. (37) should be replaced by $E_r/(1-\nu_r^2)$.

The equations for the incremental deformation in the matrix under conditions of generalized plane stress are derived in ref. 6 by integrating over the thickness t the equilibrium equations in ref. 11 together with the isothermal stress-displacement relations for an isotropic elastic medium. With only one non-vanishing component of initial stress, $E_m p_m$, which is uniform throughout the matrix, these equations are

$$
\begin{aligned}
&[2-\nu_m(1+\nu_m)p_m]u_{x,\,xx}+(1-\nu_m)\,[1-(1+\nu_m)p_m]\,u_{x,yy}\\
&\quad+(1+\nu_m)\,[1+(1-2\nu_m)p_m]u_{y,\,xy}=0 \qquad (38)\\
&(1+\nu_m)\,[1-(1-\nu_m)p_m]u_{x,\,xy}+(1-\nu_m)\,[1+(1+\nu_m)p_m]u_{y,\,xx}+2u_{y,\,yy}=0.
\end{aligned}
$$

The matrix displacements u_x, u_y in eq. (38) are the values averaged over the thickness t.

In solving eqs. (37) and (38) for the incremental deformations, continuity conditions at the matrix–fiber interface must be satisfied because of the assumed bond between the phases in the composite. Consider the interface at $y = h_2$ and let the deflection v in eq. (37) pertain to the fiber adjoining this interface. Then continuity of the incremental displacements requires

$$
\begin{aligned}
v(x) &= u_y(x, h_2)\\
\frac{h_1}{2}\frac{dv(x)}{dx} &= u_x(x, h_2).
\end{aligned} \qquad (39)
$$

In addition, there must be continuity of the traction at the interface so that the stresses σ_b, τ_b in eq. (37) must be the incremental boundary stresses on the matrix. From the expressions in ref. 11 for the incremental stresses at a boundary, the corresponding expressions for the case of generalized plane stress are derived in ref. 6 by integrating over the thickness t giving the continuity conditions

$$
\begin{aligned}
\sigma_b(x) &= \frac{E_m}{1-\nu_m^2}\,[u_{y,\,y}+\nu_m u_{x,\,x}]_{y=h_2}\\
\tau_b(x) &= \frac{E_m}{2(1+\nu_m)}\,[1-(1+\nu_m)p_m]\,[u_{x,\,y}+u_{y,\,x}]_{y=h_2}\,.
\end{aligned} \qquad (40)
$$

Equations for the deflection of other fibers and continuity at other interfaces can also be written in this way. However, the additional information for other fibers and interfaces will be specified subsequently by the symmetry of the buckling modes which are considered.

C. *Solution*

The existence of a state of incremental deformation corresponding to sinusoidal modes of buckling of the reinforcement is investigated by means of a trial solution of eqs. (37) and (38) in the form

$$
\begin{aligned}
v(x) &= V\sin \pi x/\lambda\\
u_x(x, y) &= U_x(y)\cos \pi x/\lambda\\
u_y(x, y) &= U_y(y)\sin \pi x/\lambda
\end{aligned} \qquad (41)
$$

which must also satisfy eqs. (39) and (40) and certain symmetry conditions yet to be specified. The parameter λ introduced in eqs. (41) is the half wavelength of the buckling mode of the fiber. If the length L of the plate (Fig. 6) is an integral number of half wavelengths, the solution will be applicable for certain types of end conditions—e.g., for fiber ends pinned against displacement in the y direction. The validity of the present results for other end conditions is discussed subsequently.

The formal solution in the form of eqs. (41) now follows exactly as in ref. 6 and is only outlined here. Equations (41) are substituted into eqs. (38) giving solutions for U_x and U_y in terms of four undetermined constants:

$$
\begin{aligned}
U_x(y) &= A_1 \cosh \pi y/\lambda + A_2 \sinh \pi y/\lambda + A_3 \cosh \beta\pi y/\lambda + A_4 \sinh \beta\pi y/\lambda \\
U_y(y) &= A_1 \cosh \pi y/\lambda + A_2 \sinh \pi y/\lambda + K(A_3 \cosh \beta\pi y/\lambda + A_4 \sinh \beta\pi y/\lambda)
\end{aligned}
\tag{42}
$$

where

$$
\begin{aligned}
\beta &= \left[\frac{2+p_m(1+\nu_m)\,[2-\nu_m-p_m\nu_m(1+\nu_m)]}{2[1-(1+\nu_m)p_m]}\right]^{1/2} \\
K &= \frac{2-p_m(2-\nu_m+\nu_m^2)+p_m^2\nu_m(1-\nu_m^2)}{2\beta[1+(1-2\nu_m)p_m]}\,.
\end{aligned}
\tag{43}
$$

Equations (40) together with eqs. (41) and (42) give the values of σ_b and τ_b to be used in eq. (37) from which V in eq. (41) is determined, again in terms of the four constants in eqs. (42). There remains then to satisfy the conditions for continuity of displacements, eqs. (39), which can be achieved for two distinct configurations of the buckled fibers, namely, a symmetric and an antisymmetric mode (Fig. 8). These two cases are considered separately† by

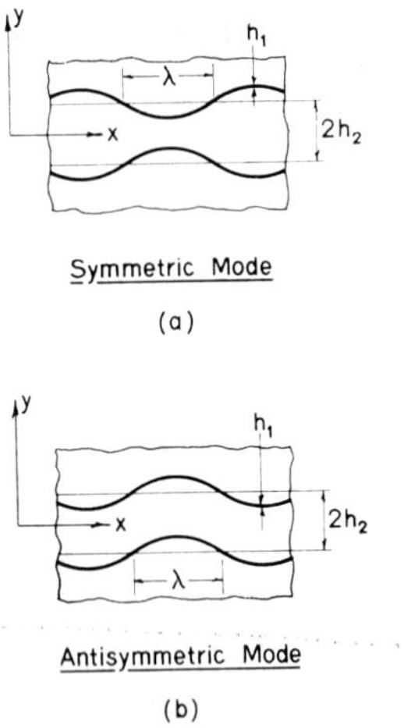

FIG. 8. Fiber buckling modes

† In ref. 6 it is shown that if the continuity conditions are written for the two interfaces $y = \pm h_2$, the resulting four equations in the four unknowns A_1, A_2, A_3, A_4 can be uncoupled to give separate solutions for the symmetric and antisymmetric modes.

taking $A_2 = A_4 = 0$ for the symmetric mode and $A_1 = A_3 = 0$ for the antisymmetric mode. Equations (39) then provide conditions for the two remaining constants in each case, in the form of homogeneous algebraic equations as follows:

1. Symmetric mode (Fig. 8(a))

$$\begin{aligned} a_1A_1+a_2A_3 &= 0 \\ a_3A_1+a_4A_3 &= 0 \end{aligned} \tag{44}$$

where

$$a_1 = -2\left\{rh_1\frac{\pi}{\lambda}(1+\nu_m)\left(p_r+\frac{\pi^2h_1^2}{12\lambda^2}\right)+\frac{\pi h_1}{\lambda}[1-p_m(1+\nu_m)]\right\}$$
$$\sinh\frac{\pi h_2}{\lambda}-4\cosh\frac{\pi h_2}{\lambda}$$

$$a_2 = -\left\{2\,rh_1\frac{\pi}{\lambda}(1+\nu_m)\left(p_r+\frac{\pi^2h_1^2}{12\lambda^2}\right)+\frac{\pi h_1}{\lambda}\right)(\beta+K)[1-p_m(1+\nu_m)\right\}$$
$$\sinh\frac{\beta\pi h_2}{\lambda}-\frac{4(\beta K-\nu_m)}{1-\nu_m}\cosh\frac{\beta\pi h_2}{\lambda} \tag{45}$$

$$a_3 = \cosh\frac{\pi h_2}{\lambda}-\frac{\pi h_1}{2\lambda}\sinh\frac{\pi h_2}{\lambda}$$

$$a_4 = \cosh\frac{\beta\pi h_2}{\lambda}-\frac{\pi h_1}{2\lambda}K\sinh\frac{\beta\pi h_2}{\lambda}.$$

A non-vanishing incremental deformation in the mode of Fig. 8(a) is possible only if the determinant of the coefficients in eq. (44) vanishes, namely,

$$a_1a_4-a_2a_3 = 0. \tag{46}$$

2. Antisymmetric mode (Fig. 8(b))

$$\begin{aligned} b_1A_2+b_2A_4 &= 0 \\ b_3A_2+b_4A_4 &- 0 \end{aligned} \tag{47}$$

where the coefficients b_1, b_2, b_3, b_4 are identical to the coefficients in eqs. (45) but with the functions sinh and cosh interchanged. Buckling may occur in this mode if the determinant of the coefficients vanishes, i.e.,

$$b_1b_4-b_2b_3 = 0. \tag{48}$$

The critical values of the initial deformation (p_m, p_r) and the corresponding buckling modes (h_2/λ) are given by eqs. (46) and (48) for symmetric and antisymmetric sinusoidal modes. Equation (36) can be used to eliminate p_r from these equations, but they remain highly intractable transcendental equations containing not only the quantities p_m and h_2/λ which are of immediate interest, but also three additional parameters—r, h_2/h_1 and ν_m. If the materials and geometry are specified, the latter parameters are fixed and eqs. (46) and (48) are transcendental equations relating p_m and h_2/λ, the roots of which can be found numerically. A continuous spectrum of roots will be obtained for fixed values of r, h_2/h_1 and ν_m, from which the smallest value of p_m is selected as the critical value and the corresponding value of h_2/λ gives the buckling mode.

D. *Results and Conclusions*

The roots of eqs. (46) and (48) and the critical values for the symmetric and antisymmetric modes were evaluated on an IBM 7094 computer for a matrix with $\nu_m = 0.3$, values of r, eq. (35), from 10 to 1000, and ratios

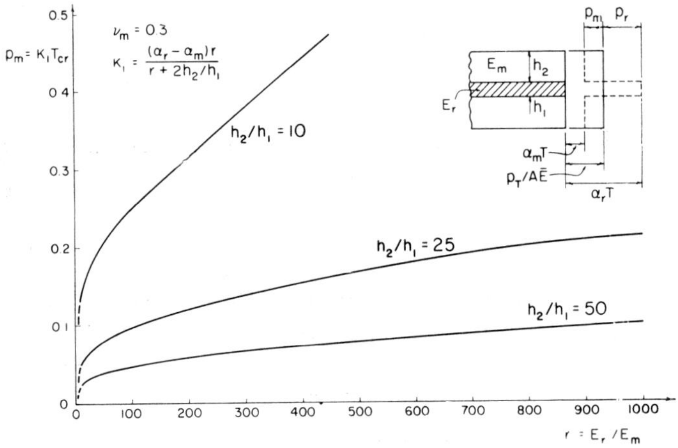

FIG. 9. Critical values of p_m as function of r

$h_2/h_1 = 10, 25, 50$ corresponding to volume fractions of reinforcement of approximately 5%, 2% and 1%. The critical values of p_m and p_r are shown in Figs. 9 and 10 as functions of r, and the corresponding values of h_2/λ are given in Fig. 11.

Since these results are based on elastic analysis, they are only applicable in ranges where neither p_m nor p_r exceed the elastic limits of the materials. It is apparent from Figs. 9 and 10 that for a given value of h_2/h_1, the present results are applicable over a range of values of r the upper and lower limits of which are determined, respectively, from Figs. 9 and 10 by the elastic limits of p_m and p_r. It is apparent, moreover, that the larger the volume fraction of reinforcement (smaller values of h_2/h_1) the smaller will be the range of

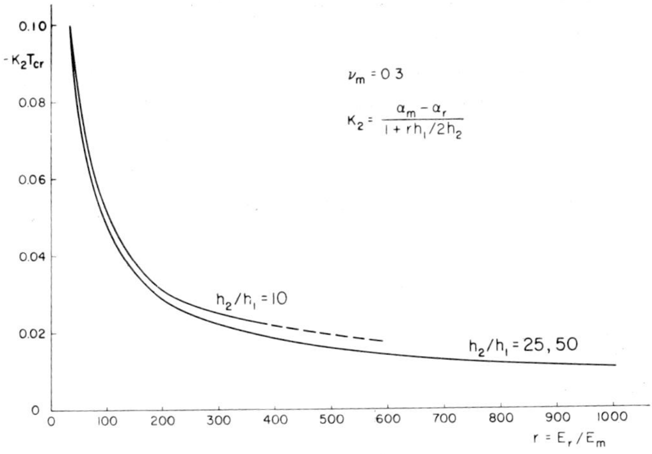

FIG. 10. Critical values of p_r as function of r

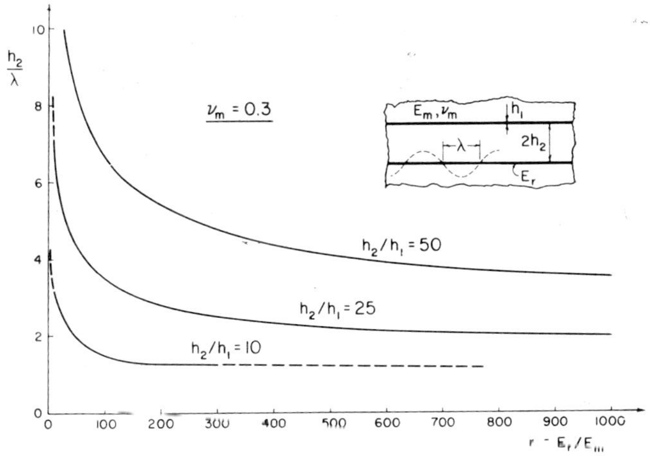

FIG. 11. Critical values of h_2/λ as function of r

validity of the elastic analysis, and it is for this reason that the relatively large values of h_2/h_1 have been considered in the numerical examples. In any subsequent work on the stability of reinforcing fibers, it appears desirable to include the inelastic behavior of matrix and fiber.

As a direct consequence of the large values of h_2/h_1 used in the numerical evaluation, the results for the symmetric and antisymmetric modes are the same, and only one set of curves appears in Figs. 9, 10 and 11. The buckling modes for these values of h_2/h_1 are short wavelength modes ($h_2/\lambda > 1$) for which it is shown in Appendix B that the same critical values of p_m are given by both eqs. (46) and (48). Consequently, in the buckling of fibers in a member for which h_2/h_1 is large, both symmetric and antisymmetric modes are equally likely to occur.

A second consequence of buckling in a mode of short wavelength is the applicability of the present results to members in which the end conditions are not those implied by the trial solution, eqs. (41), provided the length of the member is much larger than the fiber spacing. For if the buckling mode is such that $h_2/\lambda > 1$ and the member is long, $L \gg h_2$, then the actual end conditions are expected to be insignificant in determining the buckling modes at some distance from the ends.

On comparing the present results for buckling under thermal loading to the results in ref. 6, it is observed that the values of p_r are not significantly different from the critical values for buckling under mechanical load. Consequently, the presence of a tensile axial stress in the matrix in the present case has very little stabilizing effect on the fibers. However, this conclusion is not entirely unexpected in view of the result in ref. 6. There the matrix is in compression, and it is found that for the range of parameters of interest in fiber-reinforced materials, the matrix is effective only in providing lateral support for the fibers by virtue of its stiffness, while the possible instability of the matrix is not significant. This observation and its consequences in applying the theory of incremental deformations[11] to determine the stability of reinforcing fibers are discussed in greater detail in ref. 6.

The critical values of strain in the fiber (Fig. 10) are very insensitive to the value of h_2/h_1 for sufficiently large values of this parameter. It is noticed that a simple equation, namely,

$$-p_r = \frac{4.2}{r} + 0.006, \tag{49}$$

adequately reproduces the portion of the curve in Fig. 10 giving values of p_r which are likely to be in the elastic range, and may be useful in design. The corresponding values of p_m are given by eq. (36), and neither may exceed the elastic limit of the respective materials for the result to apply. The obvious disadvantages of such an empirical formula are its applicability for only the one value $\nu_m = 0.3$ used in the numerical calculation for the curve in Fig. 10, and no information is given concerning the buckling mode. Various approximations are considered in ref. 6 to obtain elementary expressions predicting the critical values of strain in the fiber. Since the state of initial stress in the matrix, be it tension or compression, appears to have no significant influence on these critical values of p_r for large values of h_2/h_1, the approximations in ref. 6 will apply to the present problem for the range of parameters considered. Consequently, the result

$$p_r = \frac{2(1-\nu_m)}{r(1+\nu_m)(3-\nu_m)} + \left[\frac{3}{r(1+\nu_m)(3-\nu_m)}\right]^{2/3} \tag{50}$$

which is derived in ref. 6 for large values of h_2/h_1 may be used to determine the critical values given in Fig. 10. While this expression is not as simple as the empirical formula, eq. (49), it shows a dependence on ν_m and reproduces the curve in Fig. 10 over a larger range of r. Moreover, the modes corresponding to the values of p_r in eq. (50) are also given in ref. 6 in the form

$$\frac{h_2}{\lambda} = \frac{2h_2}{\pi h_1}\left[\frac{3}{r(1+\nu_m)(3-\nu_m)}\right]^{1/3} \tag{51}$$

which applies to symmetric and antisymmetric modes.

APPENDIX A

The inequality in eq. (12) may be written in terms of the integrals in eqs. (8) in the form

$$\int_{-\eta_1}^{1}\int_{-\zeta_1}^{\zeta_2} E(\eta, \zeta)\,\eta^2 d\zeta\, d\eta \leqslant \eta_1 \int_{-\eta_1}^{1}\int_{-\zeta_1}^{\zeta_2} E(\eta, \zeta)\, d\zeta,\, d\eta. \tag{A1}$$

By means of the transformation $\theta = (\eta+\eta_1)/(1+\eta_1)$, eq. (A1) becomes

$$\int_{0}^{1}\int_{-\zeta_1}^{\zeta_2} E(\theta, \zeta)\,[(1+\eta_1)\theta-\eta_1]^2\, d\zeta\, d\theta \leqslant \eta_1 \int_{0}^{1}\int_{-\zeta_1}^{\zeta_2} E(\theta, \zeta)\, d\zeta\, d\theta. \tag{A2}$$

Equation (2) gives the condition

$$\int_{-\eta_1}^{1}\int_{-\zeta_1}^{\zeta_2} E(\eta, \zeta)\eta\, d\zeta\, d\eta = 0 \tag{A3}$$

which becomes, in terms of the variable θ,

$$(1+\eta_1)\int_{0}^{1}\int_{-\zeta_1}^{\zeta_2} \theta E(\theta, \zeta)\, d\zeta\, d\theta = \eta_1 \int_{0}^{1}\int_{-\zeta_1}^{\zeta_2} E(\theta, \zeta)\, d\zeta\, d\theta. \tag{A4}$$

Substitution of eq. (A4) into eq. (A2) gives

$$\int_{0}^{1} \theta^2 \int_{-\zeta_1}^{\zeta_2} E(\theta, \zeta)\, d\zeta\, d\theta \leqslant \int_{0}^{1} \theta \int_{-\zeta_1}^{\zeta_2} E(\theta, \zeta)\, d\zeta\, d\theta. \tag{A5}$$

In view of the inequalities

$$0 \leqslant \theta \leqslant 1 \tag{A6}$$

$$\int_{-\zeta_1}^{\zeta_2} E(\theta, \zeta)\, d\zeta > 0$$

the inequality (A5) and consequently eq. (12) are verified.

APPENDIX B

The transcendental equations for buckling in the symmetric and antisymmetric modes, eqs. (46) and (48) differ only by an interchange of the functions sinh and cosh of the same argument. If the arguments are sufficiently greater than one for the value h_2/λ corresponding to the smallest root p_m of the transcendental equation, then an interchange of the functions sinh and cosh will have no effect in the calculation of the root and eqs. (46) and (48) will give the same critical values of p_m and h_2/λ. That this is the case in the present results is shown below.

For the value $h_2/\lambda = 1$, the sinh and cosh of argument $\pi h_2/\lambda$ are approximately equal, within an error of less than 1%. This approximation also applies for the argument $\beta\pi h_2/\lambda$, since eq. (43) which can be rewritten in the form

$$\beta^2 = [1+(1+\nu_m)p_m]\left[1+\frac{(1-\nu_m/2)(1+\nu_m)p_m}{1-(1+\nu_m)p_m}\right] \tag{B1}$$

shows that

$$\beta \geqslant 1 \tag{B2}$$

provided

$$\begin{aligned} 0 \leqslant p_m &\leqslant \frac{1}{1+\nu_m} \\ 0 \leqslant \nu_m &\leqslant \frac{1}{2}. \end{aligned} \tag{B3}$$

Figure 11 shows that the buckling wavelength decreases as larger values of h_2/h_1 are considered. For $h_2/h_1 = 10$ it is found that $h_2/\lambda > 1$, and consequently for the values of h_2/h_1 considered in the numerical examples

$$\begin{aligned} \sinh \pi h_2/\lambda &\simeq \cosh \pi h_2/\lambda \\ \sin \beta\pi h_2/\lambda &\simeq \cosh \beta\pi h_2/\lambda. \end{aligned} \tag{B4}$$

Therefore, the critical values given by eqs. (46) and (48) are approximately equal and there is no appreciable difference between the values for the symmetric and antisymmetric modes for large h_2/h_1.

REFERENCES

1. Fiber Composite Materials, American Society for Metals, Metals Park, Ohio, 1965.
2. HOLLIDAY, L. (editor). *Composite Materials*, Elsevier Publishing Company, 1966.
3. ERINGEN, A. C. Thermal Stresses in a Multiple Layer Beam, *Proceedings of the First Midwestern Conference on solid Mechanics*, pp. 69–73 (1953).
4. BOLEY, B. A. and TESTA, R. B. Thermal Stresses in Composite Beams, to be published in *Int. Journal of Solids and Structures.*

5. BOLEY, B. A. The Determination of Temperature, Stresses, and Deflections in Two-Dimensional Thermoelastic Problems, *Journal of the Aeronautical Sciences*, **23**, No. 1, p. 67 (1956).
6. CHUNG, W. Y. and TESTA, R. B. The Elastic Stability of Fibers in a Composite Plate, *Journal of Composite Materials*, **3**, No. 1, p. 58 (1969).
7. BARREKETTE, E. S. Thermoelastic Stresses in Beams, *Journal of Applied Mechanics*, **27**, No. 3; *Transactions of the ASME*, Series E, **82**, 465 (1960).
8. BOLEY, B. A. and WEINER, J. H. *Theory of Thermal Stresses*, John Wiley and Sons, Inc. 1960.
9. BOLEY, B. A. Bounds on the Maximum Thermoelastic Stress and Deflection in a Beam and Plate, *Journal of Applied Mechanics*, **33**, No. 4, p. 881 (1966).
10. TIMOSHENKO, S. *Strength of Materials*, 3rd ed., Part I, p. 254, D. Van Nostrand Company, 1955.
11. BIOT, M. A. *Mechanics of Incremental Deformations*, John Wiley and Sons, Inc., 1965.

CONTINUUM THEORIES FOR COMPOSITE MATERIALS†

SEVERINO L. KOH

Purdue University, Lafayette, Indiana

I. INTRODUCTION

The applications of composite materials in engineering structures are too numerous to enumerate here. An indication of the extent of these applications and further projected usage is given in the several papers presented during the first day of this conference. One can only echo the enthusiasm that many have already expressed over these materials which have been called *the materials of the future*.

With the utilization of composite materials in applications that demand greater efficiency of design and allow less margin for engineering inexactitudes, a better understanding of composites is critically needed. Not only must one know the physical and mechanical properties of the elements that constitute the composite; more important, one must also be able to predict the *combined* behavior of these constituents. In numerous applications, the mechanical and/or the non-mechanical interaction of the constituents may result in a predominant effect that overrides the separate characteristics of the individual materials that make up the composite.

It is therefore little wonder that with the increased utilization of composite materials an accompanying upsurge of interest has been registered among theoreticians and experimentalists alike in proposing mathematical models to predict this *combined* behavior of the composite. The literature does not want for empirical laws characterizing these materials. Neither has there been a dearth of ideas on the theoretical analysis of composites.

Of the numerous theoretical methods advanced, the simpler approach considers only the effect due to one of the constituents and disregards altogether the contributions of the other elements of the composite. A point in case is that of a fiber-reinforced composite for which the behavior of the network of reinforcing cords is analyzed in lieu of calculations that would include the

† The work reported here was partly supported by NASA through a contract awarded to the General Technology Corporation, Lawrenceville, New Jersey. Esso Production Research Company, Houston, Texas also extended help in the preparation of the final copy of the paper.

effect of the matrix. This approach avoids the complicating interaction effects among the constituents and yields approximate solutions which may be adequate in certain applications.

Another approach is to consider all constituents but to treat each separately, i.e., each must satisfy the equations of motion (or equilibrium) and conditions of incompressibility (or continuity). With the use of classical methods, the problem under consideration is solved such that conditions of constraints are satisfied, e.g. conditions of continuity of displacement across interfaces between adjoining constituents. Methods have varied from the simplest averaging procedures such as those employed in elementary strength of materials (reducing the problem to one a second-year engineering student can tackle without difficulty) to the invocation of the most elegant of statistical concepts.†

On the other hand, there exists a large body of knowledge that somehow has not been tapped for application in the study of composites. Very little has been done to make use of the results of micromechanics as developed from the continuum viewpoint, notwithstanding the immediate application that seems apparent. This may be attributed to the fact that the more pertinent results applicable to the case under study have been developed only within the last few years, although the basic concepts may be traced back to as far as the turn of the century.‡ Then too, one cannot totally ignore the fact that the continuum mechanics approach has not found much favor among certain groups of engineers who criticize roundly the "rigid formalism" and the "intimidating mathematics" that seem to characterize papers by so-called "continuumists".

It is not the intent of this paper to expound on the merits and demerits of the continuum mechanics approach. The author will consider his purpose well served if he can demonstrate in this small contribution that physical significance may be given to certain concepts developed within the rigid formalism of continuum mechanics, to the end that the curiosity of workers in composites may be aroused to delve deeper into this wealth of knowledge available for ready application.

To make subsequent discussions more meaningful and for easy reference, a theory of simple microelastic solids is briefly presented in the next section.

† In a lecture presented at this conference, Hashin[(1)] reviews some of these theories. His survey extends beyond the elastic behavior of composites, covering as well theories of conductivity, dielectrics, viscoelasticity and plasticity.

‡ To review here the literature on continuum mechanics, even only that part of the field of obvious application to composite materials, would be a lengthy project and would clearly detract from the main purpose of this paper. Suffice it to cite several significant references which exemplify the variety of concepts in the field: E. and F. Cosserat,[(2)] whose pioneering contribution is now widely recognized; Toupin;[(3)] Eringen and Suhubi;[(4)] Mindlin;[(5)] and Green and Rivlin.[(6)] This list does not represent by any means an attempt at completeness.

The theory is largely based on that proposed by Eringen and Suhubi[4], but some results due to Mindlin[5] are also incorporated in this presentation. Only the elastic behavior of the composite is considered and such effects as due to viscosity and heat conduction are disregarded.

The application of the theory to various types of composites (particle composites, fiber-reinforced materials, etc.) is possible. However, for a simple illustration, only the case of laminates is considered in this paper. Specifically, the case of an elastic matrix reinforced by parallel sheets of a different elastic material is studied. A method of analysis is presented for the plane deformation of the reinforced medium.

II. THEORY OF SIMPLE MICROELASTIC SOLIDS

Let us consider a body which is a composite of several elastic materials. At some prescribed time t, which for reference we set equal to zero, the body has a configuration occupying volume V bounded by surface S. We take this to be the *reference configuration* and assume that the body is undeformed at this state.

With reference to a fixed rectangular Cartesian coordinate system, as shown in Fig. 1, we assume that at a point of the body defined by position vector $\mathbf{X}$

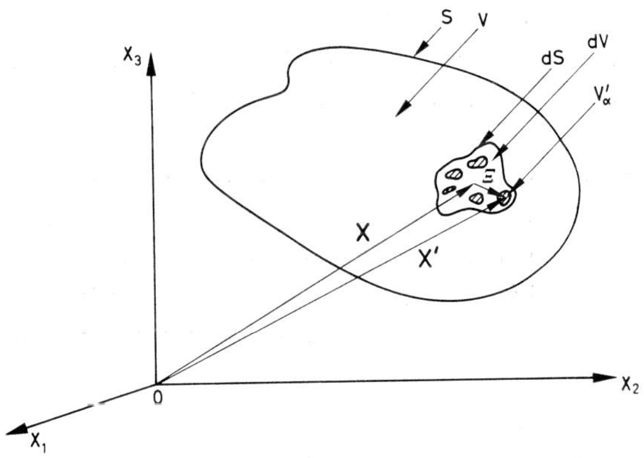

FIG. 1. Reference configuration ($t = 0$)

from the fixed origin O the volume element dV, called the *macrovolume*, contains sufficiently many particles of the different materials that constitute the composite. These particles are collected into cells, called *microvolumes*, each of which is treated as a continuum and designated by V'_α ($\alpha = 1, 2, \ldots$). In the case of a reinforced medium, we may consider the totality of particles of the matrix as constituting a cell and those of the reinforcing materials, the remaining cells, all compactly contained within the macroelement.

The material particles are identified by their position vectors $\mathbf{X}'$ from the fixed origin O. They may also be referred to the point $\mathbf{X}$ by vectors $\boldsymbol{\Xi}$. The total mass dM of the macroelement dV may be expressed in terms of an average mass density ϱ_0. This mass is, of course, the sum of the masses of the constituents in the macroelement. In other words,

$$dM = \varrho_0 \, dV = \sum_{\alpha=1,2,\dots} \int_{V'_\alpha} \varrho'_0 \, dV' = \int_{dV} \varrho'_0 \, dV' \tag{1}$$

where ϱ'_0 is the mass density at material point $\mathbf{X}'$, and the last integral is accomplished in the Lebesgue–Stieltjes sense.

We also take point $\mathbf{X}$ to be the mass center of the macroelement. Thus,

$$\mathbf{X}\varrho_0 \, dV = \sum_{\alpha=1,2,\dots} \int_{V'_\alpha} \mathbf{X}'\varrho'_0 \, dV' = \int_{dV} \mathbf{X}'\varrho'_0 \, dV'. \tag{2}$$

With motion, the body assumes at time $t>0$ a configuration which we will call the *deformed configuration*. The body now occupies the volume v bounded by the surface s. The macroelement dV is carried over to the spatial macrovolume dv with the mass center $\mathbf{X}$ of the macroelement in the reference configuration shifting to the spatial point $\mathbf{x}$ in the deformed configuration (Fig. 2).

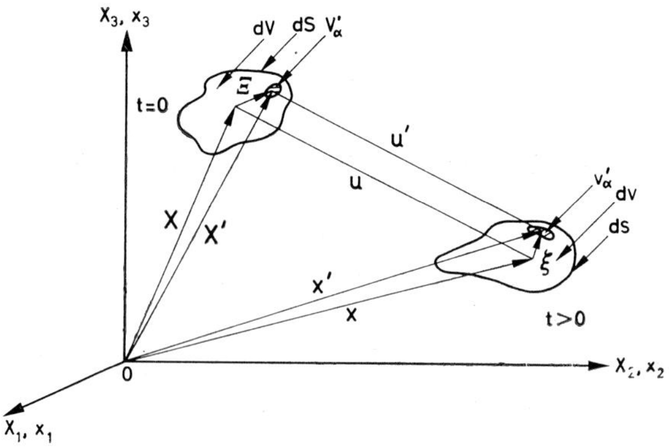

FIG. 2. Motion of microelement V'_α imbedded in macroelement dV

The material particle located at $\mathbf{X}'$ in the reference configuration now occupies the spatial point $\mathbf{x}'$, which with respect to point $\mathbf{x}$ is located by the position vector $\boldsymbol{\xi}$. Analogous definitions of mass densities for the deformed macrovolume dv may be made as those for the corresponding macrovolume dV in the reference configuration. Thus, in the deformed configuration we have

$$dm = \varrho \, dv = \sum_{\alpha=1,2,\dots} \int_{v_\alpha} \varrho' \, dv' = \int_{dv} \varrho' dv' \tag{3}$$

where ϱ and ϱ' are respectively the average mass density of the macroelement dv and the mass density of the continuum at point $\mathbf{x}'$.

The displacement of the material particle $\mathbf{X}'$ is given by

$$\mathbf{u}' = \mathbf{x}' - \mathbf{X}' \tag{4}$$

where $\mathbf{u}'$, called the *microdisplacement*, is distinguished from the *displacement* of the center of mass $\mathbf{X}$ given by the relation

$$\mathbf{u} = \mathbf{x} - \mathbf{X}. \tag{5}$$

With the motion and inverse motion of the body respectively prescribed by the relations

$$\mathbf{x}' = \mathbf{x}'(\mathbf{X}', t); \quad \mathbf{X}' = \mathbf{X}'(\mathbf{x}', t) \tag{6}$$

and with the assumption of an affine deformation of the macroelement specifically given by thc relation

$$\xi_k = \chi_{km} \Xi_m. \tag{7}$$

Eringen and Suhubi[4] showed that the motion carries the mass center $\mathbf{X}$ of dV to the center of mass $\mathbf{x}$ of the macroelement dv. In eq. (7), repeated indices m follow the summation convention over the range 1, 2, 3; and χ_{km} is the *microdeformation gradient* defined according to

$$\chi_{km} \equiv \left. \frac{\partial \xi_k}{\partial \Xi_m} \right|_{\Xi_m = 0} \cdot \tag{8}$$

Individually, the constituents that make up the composite are continuous media. (For the problem to be considered below, these constituents are taken to be elastic materials in the classical sense.) Therefore, the microelement in each macroelement may be treated as a continuum and at each point $\mathbf{x}'$ in this microvolume, the basic field equations of continuum mechanics are applicable[7]. In other words at each $\mathbf{x}'$ the following results based on local conservation axioms must be valid:

Conservation of mass

$$\frac{\partial \varrho'}{\partial t} + (\varrho' v_k')_{,k} = 0 \tag{9}$$

Balance of momentum

$$t'_{mk,m} + \varrho'(f'_k - \dot{v}'_k) = 0 \tag{10}$$

Balance of moment of momentum

$$t'_{mk} = t'_{km}, \tag{11}$$

where v_k' is the velocity, $\dot{v}_k' \equiv Dv_k'/Dt$ is the acceleration, t'_{mk} is the stress tensor, and f_k' is the body force, all quantities associated with the continuum instantaneously located at $\mathbf{x}'$ at time t. The comma notation $(\ )_{,k}$ used in eqs. (9) and (10) indicates partial differentiation with respect to $\mathbf{x}'_k$. Other field equations, e.g. that due to the local conservation of energy, are also valid. However, for the mechanical, non-heat-conducting case to be considered these other equations are no longer pertinent in the formulation of the problem.

For the analysis of the composite material, it is desirable to express the field equations in terms of the variables associated with and defined at the spatial point $\mathbf{x}$, the center of mass of the macroelement. To this end, we first apply an axiom of conservation of mass to the macroelement:

$$dM = dm; \quad \varrho_0\, dV = \varrho\, dv \tag{12}$$

which is also expressible in the form

$$\frac{D}{Dt}(\varrho\, dv) \equiv \frac{\partial \varrho}{\partial t} + (\varrho v_k)_{,k} = 0 \tag{13}$$

where $D(\ )/Dt$ denotes the material derivative of the expression enclosed in the parentheses and $v_k \equiv \dot{u}_k$ is the velocity of the mass center $\mathbf{x}$.

By integrating eq. (10) over the entire material volume $\mathcal{V}$ and with the application of the Green–Gauss theorem, we obtain

$$t_{mk,m} + \varrho(f_k - \dot{v}_k) = 0 \tag{14}$$

where the variables given are statistical averages definable in terms of tensor fields associated with the microelemental continua. Specifically, the "average" stress tensor t_{mk}, the "average" body force f_k, and the "average" inertia $\varrho\dot{v}_k$ are defined according to:

$$\begin{aligned} t_{mk}\, da_m &\equiv \int_{ds} t'_{mk}\, da'_m \\ \varrho f_k\, dv &\equiv \int_{dv} \varrho' f'_k\, dv' \\ \varrho \dot{v}_k\, dv &\equiv \int_{dv} \varrho' \dot{v}'_k\, dv' \end{aligned} \tag{15}$$

where integration over ds and dv is accomplished in the Lebesgue–Stieltjes sense.

If we take the first moment of eq. (10), i.e., take the outer product of each of the terms in eq. (10) with x'_n, and integrate over the material volume $\mathcal{V}$, we obtain the following equation which may be considered as the mathematical expression of the balance of first moment:

$$t_{km} - s_{km} + \lambda_{nmk,n} + \varrho(l_{mk} - \dot{\sigma}_{mk}) = 0. \tag{16}$$

The new variables introduced in eq. (16), namely the *microstress average* $s_{km}=s_{mk}$, the *first stress moment* λ_{nmk}, the *first body moment* l_{mk}, and the inertial spin $\dot{\sigma}_{mk}$ are defined respectively as follows:

$$s_{km}\,dv \equiv \int_{dv} t'_{km}\,dv'; \qquad \lambda_{nmk}\,da_n \equiv \int_{ds} t'_{nm}\,\xi_k\,da'_n;$$
$$\varrho l_{mk}\,dv \equiv \int_{dv} \varrho' f'_m\,\xi_k\,dv'; \qquad \varrho\dot{\sigma}_{mk}\,dv \equiv \int_{dv} \varrho'\,\dot{v}'_m\,\xi_k\,dv'. \tag{17}$$

By a simple calculation, Eringen and Suhubi showed that the inertial spin $\dot{\sigma}_{mk}$ may be expressed in the form

$$\dot{\sigma}_{mk} = I_{pq}\,\ddot{\chi}_{mp}\,\chi_{kq} = i_{pk}\,(\dot{\nu}_{mp}+\nu_{mq}\nu_{qp}) \tag{18}$$

where I_{pq} is akin to the moment of inertia tensor of the macroelement taken with respect to the mass center **X**:

$$I_{pq}\,\varrho_0\,dV \equiv \int_{dV} \Xi_p\,\Xi_q\varrho'_0\,dV'. \tag{19}$$

The *microinertia moment* i_{pk} is defined according to

$$i_{pk}\,\varrho\,dv \equiv \int_{dv} \xi_p\,\xi_k\,\varrho'\,dv' \tag{20}$$

and ν_{mq}, the *gyration tensor*, is given by

$$\nu_{mq} \equiv \dot{}_{mn}\,\chi^{-1}_{nq} \tag{21}$$

where χ^{-1}_{nq} is the inverse of the microdeformation gradient matrix χ_{nq}.

In addition to the above field equations, a suitable set of constitutive equations characterizing the composite medium is needed. Suhubi and Eringen[4] proposed a nonlinear constitutive theory for the microelastic medium which is expressible in the three sets of equations indicated below:

$$\begin{aligned} t_{ij} &= f_{ij}\,(e_{mn},\,\varepsilon_{mn},\,\gamma_{mnp}) \\ s_{ij} &= g_{ij}\,(e_{mn},\,\varepsilon_{mn},\,\gamma_{mnp}) \\ \lambda_{ijk} &= h_{ijk}\,(e_{mn},\,\varepsilon_{mn},\,\gamma_{mnp}) \end{aligned} \tag{22}$$

where the second order tensor functions f_{ij} and g_{ij}, and the third order tensor function h_{ijk} are functions of the variables e_{mn} *(strain)* ε_{mn} *(microstrain)*, and γ_{mnp} *(gradient of the microdeformation gradient)*. These tensors are defined according to

$$\begin{aligned} e_{mn} &\equiv \tfrac{1}{2}\,(\delta_{mn}-X_{p,\,m}\,X_{p,\,n}) \\ \varepsilon_{mn} &\equiv \delta_{mn}-X_{p,\,m}\,\chi^{-1}_{pn} \\ \gamma_{mnp} &\equiv X_{q,\,m}\,\chi^{-1}_{qn,\,p} \end{aligned} \tag{23}$$

where δ_{mn} is the Kronecker delta, and $X_{p,m} \equiv \partial X_p/\partial x_m$.

The explicit expressions for the nonlinear theory simply indicated in eqs. (22) are too lengthy to reproduce here. For most purposes, it is adequate to use the following linearized equations:

$$
\begin{aligned}
t_{ij} &= (\lambda+\tau)e_{mm}\,\delta_{ij}+2(\mu+\sigma)e_{ij}+\eta\varepsilon_{mm}\,\delta_{ij}+\nu\varepsilon_{j}+\varkappa\varepsilon_{ij} \\
s_{ij} &= (\lambda+2\tau)e_{mm}\,\delta_{ij}+2(\mu+2\sigma)e_{ij}+(2\eta-\tau)\varepsilon_{mm}\,\delta_{ij}+(\nu+\varkappa-\sigma)\,(\varepsilon_{ij}+\varepsilon_{ji}) \\
\lambda_{ijk} &= -\tau_1(\delta_{jk}\,\gamma_{imm}+\delta_{ik}\,\gamma_{mmj})-\tau_2(\delta_{jk}\,\gamma_{mim}+\delta_{ij}\,\gamma_{mmk}) \\
&\quad -\tau_3\,\delta_{jk}\,\gamma_{mmi}-\tau_4\,\delta_{ik}\,\gamma_{jmm}-\tau_5(\delta_{ik}\,\gamma_{mjm}+\delta_{ij}\,\gamma_{kmm}) \\
&\quad -\tau_6\,\delta_{ij}\gamma_{mkm}-\tau_7\gamma_{jki}-\tau_8(\gamma_{kij}+\gamma_{ijk}) \\
&\quad -\tau_9\gamma_{jik}-\tau_{10}\gamma_{kji}-\tau_{11}\gamma_{ikj}
\end{aligned}
\tag{24}
$$

where λ, τ, μ, σ, η, ν, $\varkappa$, τ_α ($\alpha = 1, 2, \ldots, 11$) are constitutive constants. Consistent with the linearization employed above, the kinematical variables e_{ij}, ε_{ij}, and γ_{ijk} are linearized to

$$
\begin{aligned}
e_{ij} &= \tfrac{1}{2}(u_{i,\,j}+u_{j,\,i}) \\
\varepsilon_{ij} &= u_{j,\,i}+\varphi_{ij} \\
\gamma_{ijk} &= -\varphi_{ij,\,k}
\end{aligned}
\tag{25}
$$

where u_i is the displacement and φ_{ij} is the *microdisplacement gradient* defined according to

$$\varphi_{ij} \equiv \delta_{ij}-\chi_{ij}^{-1}. \tag{26}$$

For the linear theory, the inertial spin also simplifies to

$$\dot{\sigma}_{jk} = i_{mk}\frac{\partial^2\varphi_{im}}{\partial t^2}. \tag{27}$$

To insure single-valuedness of u_i and φ_{ij}, compatibility conditions have to be satisfied. In the linearized theory, these conditions may be expressed as follows:[4]

$$
\begin{aligned}
e_{ikm}e_{jnp}(e_{kn})_{,\,mp} &= 0 \\
e_{imn}(\gamma_{jkn})_{,\,m} &= 0 \\
(e_{jk}-\varepsilon_{jk}-r_{jk})_{,\,i} &= \gamma_{kij}
\end{aligned}
\tag{28}
$$

where e_{ikm} is the permutation symbol and r_{jk} is the rotation tensor defined by $r_{jk} \equiv \frac{1}{2}(u_{j,\,k}-u_{k,\,j})$.

The boundary value problem is completely formulated with the specification of the boundary conditions. These may be given in terms of surface tractions and moments in the form

$$
\begin{aligned}
t_{km}n_k &= t^*_{(n)m} \\
\lambda_{kmp}n_k &= \lambda^*_{(n)mp}
\end{aligned}
\tag{29}
$$

where $\mathbf{t}^*_{(n)}$ and $\lambda^*_{(n)}$ are prescribed on the surface $\mathcal{S}$ of the solid.

It is possible to formulate the boundary value problem in terms of the displacement and the microdisplacement gradient. To this end, we express the constitutive equations in terms of u_i and φ_{ij}, and substitute the resulting equations into the equations of motion. The differential equations to be solved are then

$$\alpha_1 u_{k,ki}+\alpha_2 u_{i,kk}+\alpha_3\varphi_{kk,i}+\alpha_4\varphi_{ki,k}+\varrho_0 f_i = \varrho_0\frac{\partial^2 u_i}{\partial t^2}$$

$$\begin{aligned}&\beta_1(\varphi_{kk,ij}+\delta_{ij}\varphi_{km,mk})+\beta_2\delta_{ij}\varphi_{kk,mm}+\beta_3\varphi_{jk,ik}\\&\quad+\beta_4(\varphi_{ik,kj}+\varphi_{kj,ik})+\beta_5\varphi_{ki,kj}+\beta_6\varphi_{ji,kk}\\&\quad+\beta_7\varphi_{ij,kk}+\beta_8\varphi_{kk}\delta_{ij}+\beta_9\varphi_{ij}+\beta_{10}\varphi_{ji}\\&\quad+\beta_{11}u_{k,k}\delta_{ij}+\beta_{12}u_{i,j}+\beta_{13}u_{j,i}+\varrho_0 l_{ij} = \varrho_0 I_0\frac{\partial^2\varphi_{ij}}{\partial t^2}\end{aligned} \tag{30}$$

where $\alpha_\varkappa$ ($\varkappa = 1, 2, 3, 4$) and $\beta_\varkappa$ ($\varkappa = 1, 2, \ldots, 13$) are constants expressible in terms of the constitutive constants of the composite in equation (24).

The boundary conditions are then appropriately expressed in the form

$$u_i = u_i^*; \quad \varphi_{ij} = \varphi_{ij}^* \tag{31}$$

where u_i^* and φ_{ij}^* are prescribed on the bounding surface $\mathcal{S}$. In this class of boundary value problems, the set of 12 differential equations given in (30) are solved for the 12 unknowns u_i and φ_{ij} subject to the boundary conditions (31).

III. PLANE DEFORMATION OF A LAMINATE

To demonstrate the application of the theory of microelastic solids to the mechanical analysis of composites, we consider the case of an elastic material reinforced by parallel plane sheets of a different elastic material.

The problem of elastic materials reinforced by systems of cords has been examined by several investigators whose contributions to the problem have been summarized by Green and Adkins[(8)] and by Rivlin.[(9)] In these studies, the cords are considered to be inextensible but perfectly flexible with the ultimate analytical results evolving from the constraints imposed upon the composite by these reinforcing cords. Recently, Wozniak[(10)] presented a theory of fibrous materials based on the concepts of micromechanics. He considered the material to be composed entirely of a network of fibers and treated two cases: one with the microstructure perfectly rigid, and the other with the microstructures articulated. However, in both cases there is no consideration given to a reinforced matrix.

For our study we take into account the effects of both the matrix and the reinforcement. We confine our discussion to the case of an elastic matrix reinforced by uniformly spaced sheets, a small constant distance b separating any two consecutive sheets. In the microscopic scale we take the reinforcing sheet to be a fabric composed of two sets of elastic strings. The two families of strings are assumed to be initially straight, one family intersecting the other orthogonally. (This assumption simplifies the calculations in the analysis. However, no particular difficulty is anticipated if the strings were considered initially curved and non-orthogonal.) The plane deformation of the composite medium is examined with the displacement of the network of strings in the reinforcing sheet restricted to move in the plane of the sheet. The stability of the construction is not considered.

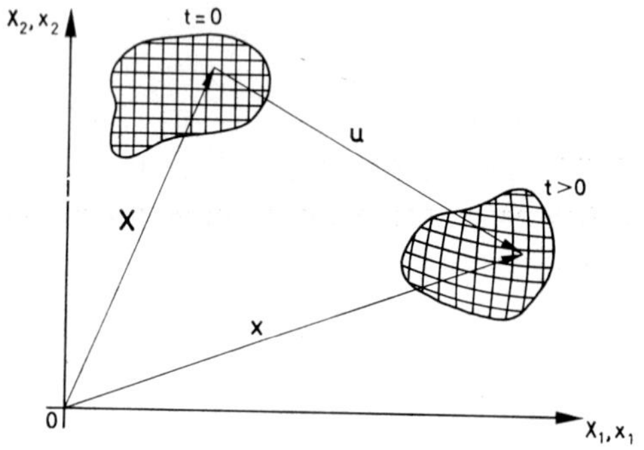

FIG. 3. Plane deformation of reinforced medium

Without loss in generality, we analyze a layer of the composite material of thickness b so as to include only one sheet reinforcement in this layer. The plane deformation of the reinforced medium is shown in Fig. 3. In the undeformed configuration, the strings belonging to one family are parallel, equally spaced, and perpendicular to the strings of the other set. In the deformed configuration, the strings are no longer straight nor is their distribution necessarily uniform.

However, in the deformed macroelement, such as dx_1dx_2 centered at (x_1, x_2) as shown in Fig. 4, for all practical purposes one may consider the strings to be straight. The strings of the same set are parallel to each other, but in general, they intersect the strings of the other set at a constant angle not necessarily orthogonal.

The macroelement is composed of sufficiently many microelements which are located from the mass center of the macroelement by position vectors defined in the $\xi_1 - \xi_2$ coordinate system. Let us consider the deformation of a microelement. Neglecting the rigid body displacement, we have in Fig. 5 an exploded view of this cell with its undeformed and deformed configurations superposed for comparison. Initially the microelement is rectangular with the mutually perpendicular strings $\Psi = \Psi_0$ and $\Phi = \Phi_0$ intersecting at the center of the microelement. Upon deformation, the microelement assumes the shape

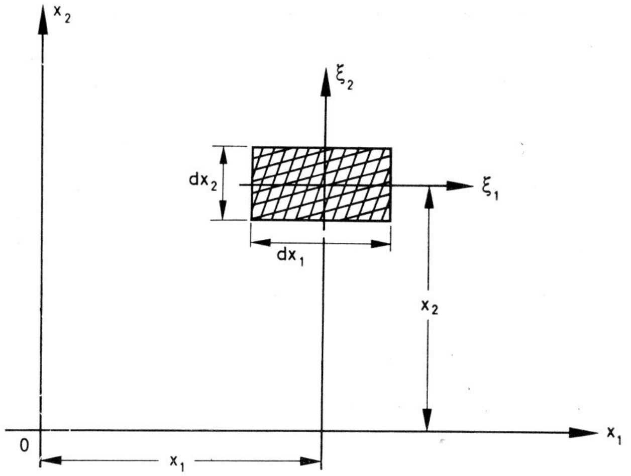

FIG. 4. Coordinate system in deformed macroelement

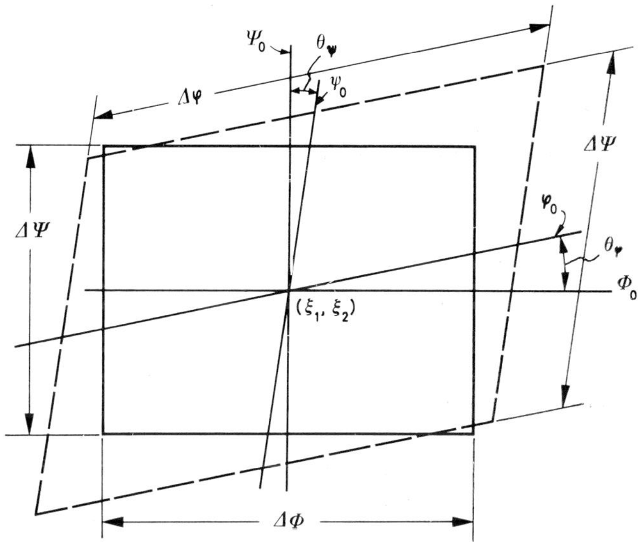

FIG. 5. Deformation of a cell

of the parallelogram and the strings are now deformed to positions $\psi = \psi_0$ and $\varphi = \varphi_0$, respectively.

The deformation of the strings may be considered to be the combination of an extension and a rotation: segment $\Delta\Psi$ has extended to $\Delta\psi$ and rotated by an angle θ_ψ with the vertical; while segment $\Delta\Phi$ has extended to $\Delta\varphi$ accompanied by a rotation θ_φ from the horizontal. Assuming that the elastic matrix in the microelement undergoes homogeneous deformation and that there is no relative displacement between contiguous parts of the composite material, the deformation and motion of the strings (assumed infinitesimal) may be used to calculate the normal strains and the shearing strains in the elastic matrix. A simple calculation shows that the strain components in the elastic matrix are

$$e'_{11} = \frac{T'_\varphi}{A_\varphi E_\varphi}\,;\quad e'_{22} = \frac{T'_\psi}{A_\psi E_\psi}\,;\quad e'_{12} = e'_{21} = \frac{1}{2}(\theta_\varphi+\theta_\psi)$$
$$e'_{33} = e'_{13} = e'_{31} = e'_{23} = e'_{32} = 0 \tag{32}$$

where T', A, E are respectively the tension, the cross-sectional area, and Young's modulus of the string identified by the subscript φ or ψ.

A significant assumption in the analysis is made regarding the variation of the string tensions in the microelement. Specifically, we assume that in a given macroelement the tension T'_φ, T'_ψ in the strings centered at the microelement located at (ξ_1, ξ_2) are related to the tensions T_φ, T_ψ in the strings at the mass center (x_1, x_2) of the macroelement according to a Taylor series expansion. In retaining the linear terms, we have the following relations:

$$T'_\varphi(x_1+\xi_1, x_2+\xi_2) = T_\varphi(x_1, x_2)+\xi_1\frac{\partial T_\varphi(x_1, x_2)}{\partial x_1}+\xi_2\frac{\partial T_\varphi(x_1, x_2)}{\partial x_2}$$
$$T'_\psi(x_1+\xi_1, x_2+\xi_2) = T_\psi(x_1, x_2)+\xi_1\frac{\partial T_\psi(x_1, x_2)}{\partial x_1}+\xi_2\frac{\partial T_\psi(x_1, x_2)}{\partial x_2}\,. \tag{33}$$

Imposing the condition that the tension along a given string segment in the macroelement is constant, we have

$$T'_\varphi = T_\varphi+\alpha_\varphi(\xi_1-\xi_2\cot\theta_\varphi)$$
$$T'_\psi = T_\psi+\alpha_\psi(\xi_1-\xi_2\tan\theta_\psi) \tag{34}$$

where θ_φ and θ_ψ are respectively the angles of rotation of the string segments $\Delta\varphi$ and $\Delta\psi$, and where for simplicity we introduced the notation

$$\alpha_\varphi \equiv \frac{\partial T_\varphi(x_1, x_2)}{\partial x_1}\,;\quad \alpha_\psi \equiv \frac{\partial T_\psi(x_1, x_2)}{\partial x_1}\,. \tag{35}$$

Using eqs. (34) in (32), we are able to express the strain in the matrix in terms of the new variables. Thus,

$$e'_{11} = \frac{1}{A_\varphi E_\varphi}[T_\varphi+\alpha_\varphi(\xi_1-\xi_2\cot\theta_\varphi)]$$
$$e'_{22} = \frac{1}{A_\psi E_\psi}[T_\psi+\alpha_\psi(\xi_1-\xi_2\tan\theta_\psi)] \tag{36}$$
$$e'_{12} = e'_{21} = \tfrac{1}{2}(\theta_\varphi+\theta_\psi)$$
$$e'_{33} = e'_{31} = e'_{32} = e'_{13} = e'_{23} = 0.$$

Since the matrix is taken to be a homogeneous, isotropic Hookean elastic solid, it is characterized by the constitutive equation

$$\tau'_{ij} = \lambda e'_{mm}\delta_{ij} + 2\mu e'_{ij} \tag{37}$$

where λ and μ are the Lamé constants of the matrix. Hence, by eqs. (36) and (37), the stress components in the matrix are given by

$$\begin{aligned}
\tau'_{11} &= \frac{\lambda+2\mu}{A_\varphi E_\varphi}[T_\varphi + \alpha_\varphi(\xi_1 - \xi_2 \cot\theta_\varphi)] + \frac{\lambda}{E_\psi A_\psi}[T_\psi + \alpha_\psi(\xi_1 - \xi_2 \tan\theta_\psi)] \\
\tau'_{22} &= \frac{\lambda+2\mu}{A_\psi E_\psi}[T_\psi + \alpha_\psi(\xi_1 - \xi_2 \tan\theta_\psi)] + \frac{\lambda+2\mu}{A_\varphi E_\varphi}[T_\varphi + \alpha_\varphi(\xi_1 - \xi_2 \cot\theta_\varphi)] \\
\tau'_{12} &= \tau'_{21} = \mu(\theta_\varphi + \theta_\psi) \\
\tau'_{33} &= \frac{\lambda}{A_\varphi E_\varphi}[T_\varphi + \alpha_\varphi(\xi_1 - \xi_2 \cot\theta_\varphi)] + \frac{\lambda}{A_\psi E_\psi}[T_\psi + \alpha_\psi(\xi_1 - \xi_2 \tan\theta_\psi)] \\
\tau'_{31} &= \tau'_{13} = \tau'_{23} = \tau'_{32} = 0.
\end{aligned} \tag{38}$$

To calculate the average stress t_{km}, the microstress average s_{km}, and the first stress moment λ_{kmn}, we use eqs. $(15)_1$, $(17)_1$ and $(17)_2$, respectively. However, before this can be done we have first to define some densities of distribution for the strings at each point (x_1, x_2). For our purpose, we let $d_\varphi(x_1, x_2)$ be the spacing between the φ-strings in a given macrovolume, and $d_\psi(x_1, x_2)$ be that between the ψ-strings in the same macrovolume. In addition, we define an effective thickness d of the reinforcing sheet sandwiched between the matrix which has an effective thickness $b' = b - d$. Figure 6 shows the macroelement for which the fundamental stress averages are calculated.

Although the restriction is not necessary, for simplicity the strings are taken to be of the same material ($E_\varphi = E_\psi \equiv E$) and of the same diameter

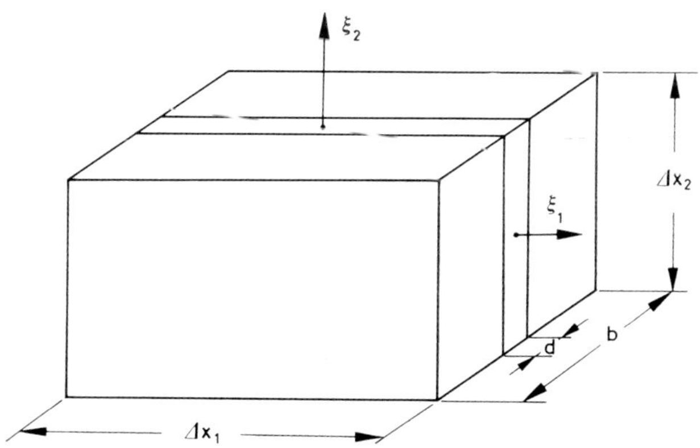

FIG. 6. View of macroelement

($A_\varphi = A_\psi \equiv A$). We denote the stiffness of the strings by $D \equiv A_\varphi E_\varphi = A_\psi E_\psi$. We also assume that the angles of rotation θ_φ and θ_ψ are small.

Without much difficulty, we calculate the average stress components to be

$$t_{11} = \left[(\lambda+2\mu)\left(T_\varphi+\alpha_\varphi\frac{l}{2}\right)+\lambda\left(T_\psi+\alpha_\psi\frac{l}{2}\right)\right]\frac{b'}{bD}+\left(T_\varphi+\alpha_\varphi\frac{l}{2}\right)\frac{1}{bd_\varphi} \quad (39)$$

$$t_{22} = \left[(\lambda+2\mu)\left(T_\psi-d_\psi\frac{l\theta_\psi}{2}\right)+\lambda\left(T_\varphi-\alpha_\varphi\frac{l}{2\theta_\varphi}\right)\right]\frac{b'}{bD}+\left(T_\psi-\alpha_\psi\frac{l\theta_\psi}{2}\right)\frac{1}{bd_\psi}$$

$$t_{33} = \frac{\lambda}{D}(T_\varphi+T_\psi)$$

$$t_{12} = \mu(\theta_\varphi+\theta_\psi)\frac{b'}{b}+\left(T_\varphi+\alpha_\varphi\frac{l}{2}\right)\frac{\theta_\varphi}{bd_\varphi}$$

$$t_{21} = \mu(\theta_\varphi+\theta_\psi)\frac{b'}{b}+\left(T_\psi-\alpha_\psi\theta_\psi\frac{l}{2}\right)\frac{\theta_\psi}{bd_\psi}$$

$$t_{13} = t_{31} = t_{23} = t_{32} = 0$$

where l denotes the characteristic length of the composite.

The microstress average s_{mk} is calculated with the use of eq. $(17)_1$, and the following components are obtained:

$$s_{11} = [(\lambda+2\mu)T_\varphi+\lambda T_\psi]\frac{b'}{bD}+\frac{T_\varphi}{bd_\varphi} \quad (40)$$

$$s_{22} = [(\lambda+2\mu)T_\psi+\lambda T_\varphi]\frac{b'}{bD}+\frac{T_\psi}{bd_\psi}$$

$$s_{33} = [T_\varphi+T_\psi]\frac{\lambda}{D}$$

$$s_{12} = s_{21} = \mu(\theta_\varphi+\theta_\psi)\frac{b'}{b}+\left(\frac{T_\psi\theta_\psi}{d_\psi}+\frac{T_\varphi\theta_\varphi}{d_\varphi}\right)\frac{1}{b}$$

$$s_{13} = s_{31} = s_{23} = s_{32} = 0.$$

Finally, the components of the third order tensor λ_{kmn}, the first stress moment, are calculated with the use of eq. $(17)_2$. To arrive at explicit results in the analysis, we make the further assumption that the strings are densely distributed. The components obtained are then given by

$$\lambda_{111} = \left[(\lambda+2\mu)\left(T_\varphi+\frac{\alpha_\varphi l}{2}\right)+\lambda\left(T_\psi+\alpha_\psi\frac{l}{2}\right)\right]\frac{b'}{b}\frac{l}{2D}+\left(T_\varphi+\alpha_\varphi\frac{l}{2}\right)\frac{d}{b}\frac{l}{2A}$$

$$\lambda_{121} = \mu(\theta_\varphi+\theta_\psi)\frac{b'}{b}\frac{l}{2}+\left(T_\varphi+\alpha_\varphi\frac{l}{2}\right)\frac{d}{b}\frac{l\theta_\varphi}{2A} \quad (41)$$

$$\lambda_{112} = -\left[(\lambda+2\mu)\frac{\alpha_\varphi}{\theta_\varphi}+\lambda\alpha_\psi\theta_\psi\right]\frac{b'}{b}\frac{l^2}{12D}-\frac{d}{b}\frac{l^2\alpha_\varphi\theta_\varphi}{12A}$$

$$\lambda_{122} = -\frac{d}{b}\frac{l^2\alpha_\varphi}{12A}$$

$$\lambda_{211} = \frac{d}{b}\frac{l^2\alpha_\psi\theta_\psi}{12A}$$

$$\lambda_{221} = [(\lambda+2\mu)\alpha_\psi+\lambda\alpha_\varphi]\frac{b'}{b}\frac{l^2}{12D}+\frac{d}{b}\frac{l^2\alpha_\psi}{12A}$$

$$\lambda_{212} = \mu(\theta_\varphi+\theta_\psi)\frac{b'}{b}\frac{l}{2}+\left(T_\psi-\alpha_\psi\theta_\psi\frac{l}{2}\right)\frac{d}{b}\frac{l\theta_\psi}{2A}$$

$$\lambda_{222} = \left[[\lambda+2\mu)\left(T_\psi-\alpha_\psi\frac{l}{2}\,\theta_\psi\right)+\lambda\left(T_\varphi-\alpha_\varphi\frac{l}{2\theta_\varphi}\right)\right]\frac{b'}{b}\frac{l}{2D}$$
$$+\left(T_\psi-\alpha_\psi\frac{l}{2}\,\theta_\psi\right)\frac{d}{b}\frac{l}{2A}.$$

All other components of λ_{kmn} are equal to zero.

It might be pointed out in passing that for the case of zero characteristic length l, all components of λ_{kmn} vanish, and the average stress t_{km} reduces to the microstress s_{km}. The apparent discrepancy in the shear stress components is not real. The last terms also vanish if one takes into account the fact that for this case d approaches zero and $b' \to b$. Thus, $t_{12} = t_{21} = s_{12} = s_{21} = \mu(\theta_\varphi+\theta_\psi)$. This indeed reduces the problem to the classical case of plane deformation of an isotropic Hookean solid.

The above stresses and stress moments must satisfy the field equations of microelastic solids as given in eqs. (14) and (16). For the static case of zero body forces f_k and zero body couples l_{mk}, these equations simplify to

$$t_{mk,\,m} = 0 \tag{42}$$
$$t_{km}-s_{km}+\lambda_{nmk,\,n} = 0.$$

Using eqs. (39), (40) and (41) in (42), we obtain six equations in the six unknowns T_φ, T_ψ, θ_φ, θ_ψ, d_φ, d_ψ. All other equations reduce to identities $0 = 0$. One will recall that the variables α_ψ and α_φ are related to the tensions T_φ and T_ψ by eqs. (35).

In summary, a boundary value problem on the plane deformation of the laminate will require the solution of the set of six differential equations derived by substituting (35), (39), (40) and (41) into (42). Boundary conditions have to be satisfied. These may be expressed directly in terms of the basic variables T_φ, T_ψ, d_φ, d_ψ, θ_φ and θ_ψ or derived from eqs. (33) for a statically equivalent system in terms of the basic variables.

IV. CONCLUSIONS

During the past decade or so, great advances have been registered in the field of continuum mechanics. However, the numerous significant results obtained have remained largely untouched by engineers. An attempt is made here to find physical meaning to some of the latest works in this rapidly advancing field of analysis.

The application of continuum theories of micromechanics in the study of the behavior of composites appears to be quite natural. The plane deformation problem discussed is but a simple illustration. The possibilities for further application in the engineering analysis of composite materials can only be restricted by the imaginativeness of the engineer. A challenge exists for him to explore the realm of the *continuumists*. He should not be intimidated by the mathematics involved nor the complexities of the notation so characteristic of the field.

Rather, he should grab the bull by the horns and show that the bull is not the ferocious monster it has been portrayed to be. Best of all, in this endeavor he can put the bull to good use.

REFERENCES

1. HASHIN, Z. *Proc. Intl. Conf. Mechanics of Composite Materials* (1967).
2. COSSERAT, E. and COSSERAT, F. *Théorie des corps déformables.* Paris, Hermann, 1909.
3. TOUPIN, R. A. *Arch. Rational Mech. Anal.*, **11**, 385 (1962).
4. ERINGEN, A. C. and SUHUBI, E. *Intl. J. Eng. Sci.* **2**, 189, 389 (1964).
5. MINDLIN, R. D. *Arch. Rational Mech. Anal.* **16**, 51 (1964).
6. GREEN, A. E. and RIVLIN, R. S. *Arch. Rational Mech. Anal.* **17**, 113 (1964).
7. ERINGEN, A. C. *Nonlinear Theory of Continuous Media*, New York, McGraw-Hill, 1962.
8. GREEN, A. E. and ADKINS, J. E. *Large Elastic Deformations*, Oxford, 1960.
9. RIVLIN, R. S. *Nonlinear Problems in Engineering*, p. 51, ed. Ames, Academic Press, 1964.
10. WOZNIAK, C. *Arch. Mech. Stos.*, **5**, 651; **6**, 777 (1965).

SESSION IV

STRUCTURAL MECHANICS

STRUCTURAL MECHANICS OF FIBER-REINFORCED COMPOSITES

JOHN M. HEDGEPETH† and DONALD F. HASKELL‡

Martin Marietta Corporation,
Baltimore Division,
Baltimore, Maryland

I. INTRODUCTION

The structural mechanics of fiber-reinforced, composite, structural materials is in a stage of rapid evolution. Prior to 1960 relatively few papers directly concerned with fiber-reinforced, composite, structural materials existed in the literature; since that time, hundreds have appeared. Much of the early analysis was concerned with definition of the potential of these composite materials. This work utilized available theory—both isotropic and classical anisotropic—and spurred the development of new theories directly applicable to advanced composites. Research has now progressed to the point that theoretical analyses of macroscopically uniform, composite materials and composite structures are fairly well developed. However, buckling predictions, accurate descriptions of creep and fatigue, and the behavior of joints and discontinuities are not as advanced. Throughout the whole field there appears to be a definite need for systematic experimentation for both development of new theory and verification of existing analyses.

Figures 1 and 2 illustrate some of the results of an attempt to survey the composites effort reported in this country. The total number of papers surveyed was 1187, and this number is termed the "total composite effort." The papers surveyed are classified according to their relative contribution to the total effort. Figure 1 illustrates the distribution of effort according to various work phases in the development of hardware. The largest effort has been in the development of a workable plan or method, i.e., the development of theoretical analyses and methods of analysis. The next largest is in design engineering of hardware, and, as is indicated, very little has been accomplished. This is the area into which increasing effort is being applied.

Thirty-six per cent of the "total effort" has been concerned with theoretical analysis (see Fig. 2). Of this work the smallest amount has been in character-

† Assistant Director, now with Astro Research Comp., Santa Barbara, Cal.
‡ Staff Engineer, now with Fairchild Hiller, Space and Electronic Systems Division, Germantown, Maryland.

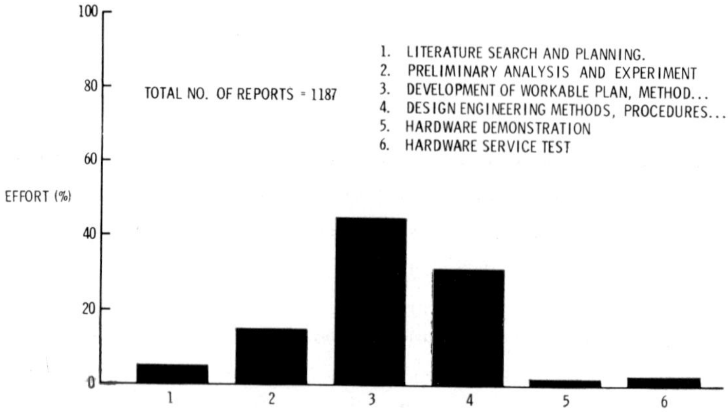

FIG. 1. Total composite effort—work phases

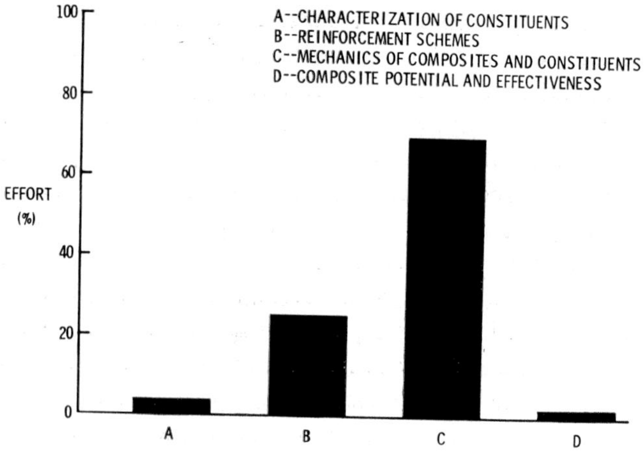

FIG. 2. Theoretical analysis (36% of total effort)

ization of constituents and composite potential and effectiveness. Considerably more has been done in reinforcement schemes. The preponderant amount of theoretical effort has been directed to the analysis of mechanics of composites and their constituents.

This survey certainly does not cover the entire field of structural mechanics of fiber-reinforced composites. Only certain topics relating to continuous, fiber-reinforced composites will be treated: macroscopically uniform composite materials, composite structures, buckling, creep, fatigue, and joints and discontinuities. Constitutive relations, theories of flaws and fracture, and residual and thermal stresses are considered from the viewpoint of macroscopically uniform composite materials. Cylinders, pressure vessels, plates, and beams are included in the discussion of bending and stretching of composite structures; the survey of buckling treats cylinders and plates.

II. UNIFORM COMPOSITE MATERIALS

A. *Constitutive Relations*

Methods for the prediction of the elastic moduli of unidirectionally fiber-reinforced composite materials are well advanced and techniques for the measurement of these moduli have been developed. Good agreement between theory and experiment is indicated for the various moduli, except the shear modulus.

Fiber-reinforced composites with transversely isotropic phases have been treated by Hill.[1] This author has shown that the macroscopic elastic moduli are connected by simple universal relations which are independent of the geometry at a given concentration of fibers. When the phases have equal transverse rigidities, exact values of the moduli are found. Otherwise, upper and lower bounds are obtained for arbitrary geometry.

A variational method has been utilized by Hashin and Rosen[2] to derive bounds and expressions for the effective elastic moduli of materials unidirectionally reinforced by hollow circular fibers. Exact results were obtained for hexagonal arrays of identical fibers and approximate results for randomly arrayed fibers of equal or unequal cross sections. Typical numerical results were presented for the moduli of a random array for both solid and hollow fibers. Unfortunately, the paper contained errors in two of the equations as well as a computer error in the typical solutions for the hollow fibers. These errors were uncovered and later corrected by the authors. Apparently, the errors in the equations did not significantly affect the numerical results; however, the computer error was reported to cause the curves for hollow fibers to be greatly in error. These errors were pointed out and corrected in a later report.[3] An important result of these studies was the conclusion that the upper and lower bounds are in agreement for four out of five elastic constants if an appropriate choice is made of the set of constants. Furthermore, the bounds for the fifth constant are reasonably close. An example is shown in Fig. 3. In addition, the indicated agreement with experimental results is apparent.

In an investigation of methods to enhance the transverse properties of fibrous composites, Dow extended an analysis, available in the literature, in order to develop a method for the determination of the elastic constants of a variety of configurations of three-dimensional fiber reinforcement. The approach is applied to the following cases: (1) unidirectional triangular filaments in the x-direction; (2) orthogonal elliptical filaments in the x- and y-directions, with minor axes of the ellipses oriented in the z-direction; (3) round filaments oriented in the x- and y-directions and at angles thereto; and (4) a case in which the effects of variations in the properties of filaments and

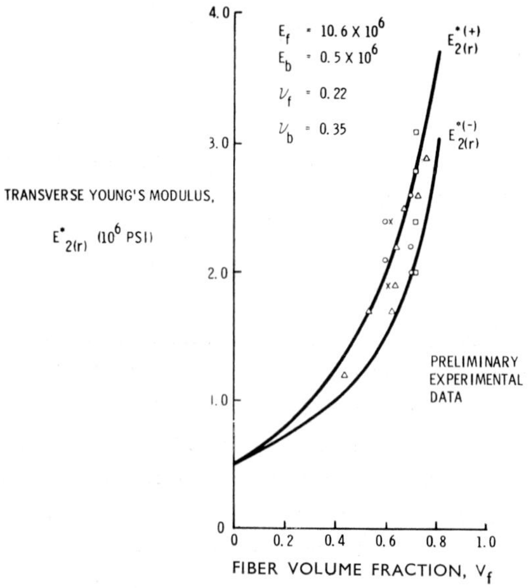

FIG. 3. Transverse Young's modulus—theory and experiment, ref. 3

matrix are considered for the orthogonal filaments of cases (2) and (3). The analytical approach relates to that followed by Dow, Libove, and Hubka in NACA TR 1195 (1954) in which the properties of integrally stiffened plates are presented. The basis for the extension of this integral stiffening analysis to the analysis of fiber composites is considered to be in the geometric similarity of the matrix material between fibers in a unidirectionally reinforced filamentary composite to the skin and integral webbing of the integrally stiffened plates.

Tsai[(4)] developed a small deformation theory for the prediction of the gross behavior of fiber reinforced composites from the characteristics of the constituent materials. His elastic constant definitions incorporated a measure of fiber contiguity. Experimental techniques to verify the theoretical predictions were presented. The theory included an evaluation of the relationship and contribution of the fiber and matrix properties, matrix content, composite density, and the geometric configuration. The constitutive equations were developed for unidirectional composites, cross-ply and angle-ply laminates, and pressure vessels.

An attempt was made by Pickett[(5)] to determine the elastic moduli for unidirectionally fiber-reinforced plastic material from a knowledge of the elastic moduli of the constituents. Two geometries were investigated: a regular hexagonal array of uniform circular fibers, and a tape with a single row of circular fibers. Computer programs were written to compute the composite elastic moduli and also the microstress in terms of microstrains at the fiber–

resin interface. Approximately periodic solutions of the elasticity equations were obtained for fiber and resin. The coefficients in the solution were determined by satisfying the boundary conditions by the method of least squares or by a combination of the Fourier Method with the Method of Least Squares. The computer program for all composite elastic constants and microstresses on the fiber–resin interface was successful in the case of the regular hexagonal array, whereas the program for the tape was not.

Point matching has been applied by Leissa(6) to determine the overall elastic constants of a unidirectionally fiber-reinforced composite on the basis of plain strain elasticity theory. Leissa presented several numerical examples of the method for fibers of circular, elliptical, and polygonal shape.

Classical theory of elasticity has been applied by Whitney and Riley(7) to simplified models of a unidirectionally fiber-reinforced composite unit cell to obtain approximate equations for longitudinal modulus, transverse modulus, and Poisson's ratio. An equation available from the literature was suggested for the shear modulus. Comparison between theory and experiment for longitudinal and transverse modulus is reasonable; however, "very poor agreement with theory is obtained in the case of shear modulus."

Simple closed form expressions have been given by Greszczuk(8) for the five principal elastic constants of unidirectional laminates: two moduli of elasticity, two Poisson's ratios, and an in-plane shear modulus. These expressions were based on a mathematical model consisting of a rectangular array of circular filaments imbedded in an elastic matrix.

Good agreement was found between theoretical and experimental results and even better agreement was obtained when the theory was corrected for the effect of voids. The magnitude and distribution of interfilament stresses was considered for a laminate subjected to tension transverse to the filament direction, and approximate expressions were also given for interfilament stress concentrations in laminates subjected to transverse tension and in-plane shear.

An analysis of the longitudinal shear loading of a unidirectional composite has been formulated by Adams and Doner.(9) Shear stiffness was determined for a doubly symmetric but otherwise arbitrary shape contained in an elastic matrix subjected to shear stress components in the directions of the fibers. Solution of the governing equations of elasticity was obtained by means of finite difference representation and a systematic over-relaxation procedure. Results are shown in Fig. 4. Evidently, changing the fiber geometry drastically changes the transverse shear stiffness. This strong dependence on shape also results in the wide disparity between upper and lower bounds for transverse shear modulus reported by Dow and Rosen(3) for so-called arbitrary phase geometry in which the fibers can be randomly located and are of random shape (see Fig. 5).

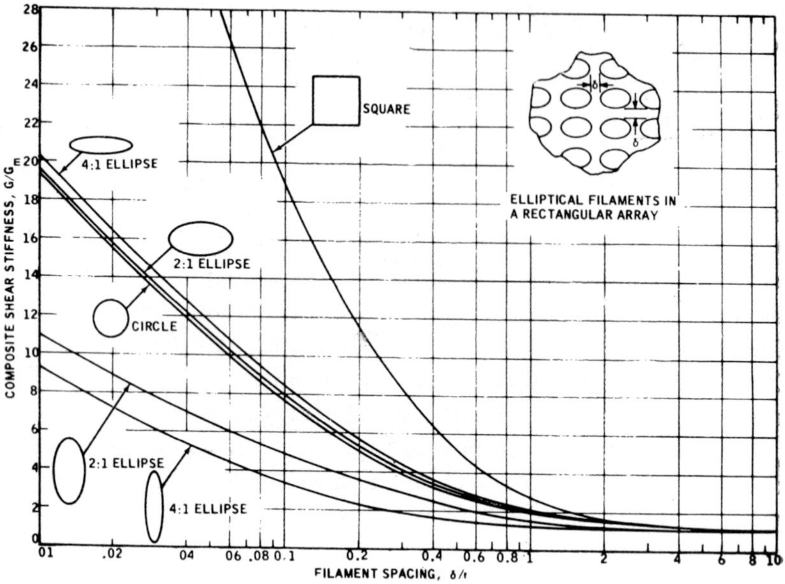

FIG. 4. Longitudinal shear loading of boron filaments in epoxy ($G_f/G_m = 120$), ref. 9

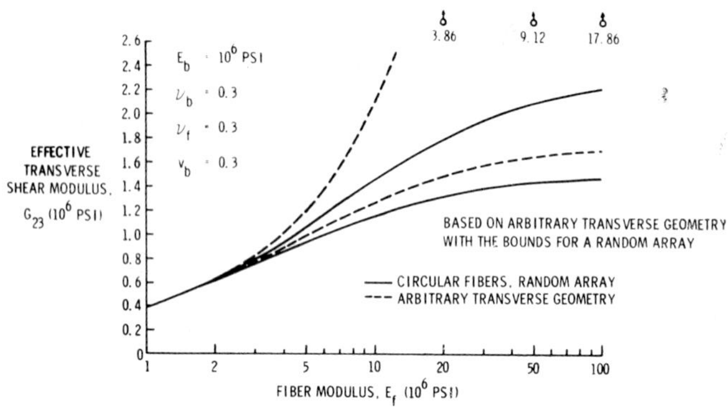

FIG. 5. Effective transverse shear modulus bounds, ref. 3

B. *Flaws and Fracture*

A considerable effort concerning development of theories of strength of fiber composites has been reported. These theories have been based on the mechanics of fabrics, continuum mechanics yield theories of strength, or flaw theories and micromechanics analyses.

The primary load-carrying element in a composite is the fiber. Kies[10] has discussed the statistics of failure of fibers. An example is shown in Fig. 6, wherein the probability distribution curve is seen to be bimodal. Fibrous materials apparently are subject to random flaws which produce weak points and cause early failures. A composite of these fibers would tend to exhibit numerous breaks at relatively low stresses but would continue to carry load. Rosen, Dow, and Hashin[11] have generated a statistical theory of tensile failure based on the concept of an "ineffective length." Failure only occurs

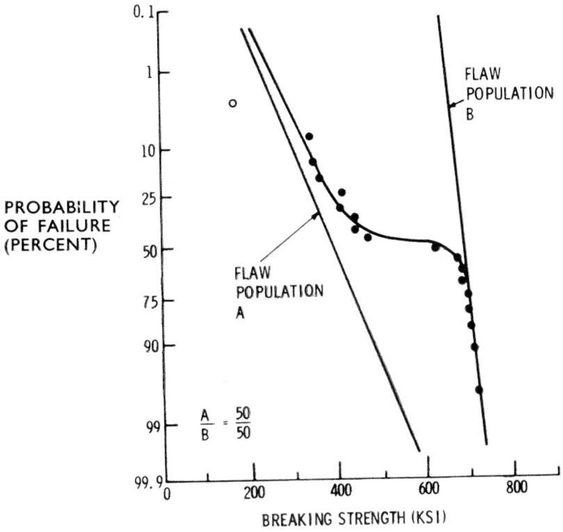

FIG. 6. Failure probability for S glass fibers (8 cm long), ref. 10

when enough fibers break so that the ineffective lengths accumulate sufficiently to weaken the cross-section (see Fig. 7).

Fabric mechanics has been highly developed[12] and in its simplest form, netting analysis, is principally applied to filament-wound vessels, such as cylindrical or spherical pressure vessels under uniform pressure loading. In this theory, the fiber is assumed to provide all of the strength and rigidity of the composite, with the role of the matrix essentially ignored. The method possesses the virtue of simplicity and is convenient for approximate results (it is known to be inaccurate in compression and flexure[12]); however, its use in sophisticated design tasks is no longer considered appropriate.[13]

A good review of available yield criteria for strength predictions was presented recently by Tsai.[12] As may be expected, these proposed theories derive from the various continuum mechanics yield conditions. They are based on strain energy, maximum principal stress, maximum shear stress, maximum stress, maximum strain, both stress and strain conditions, and conditions that account for polar strength properties.

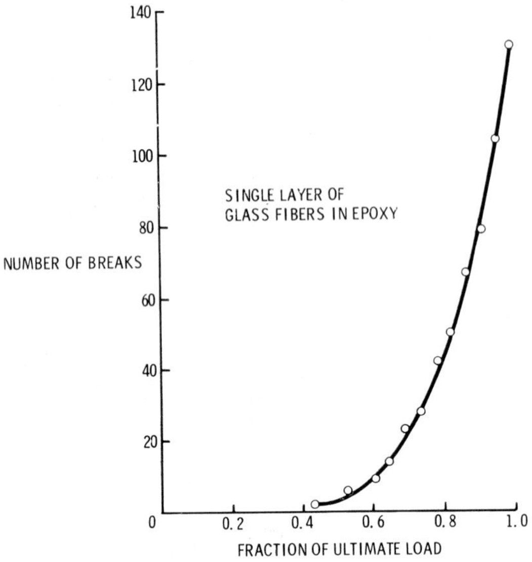

FIG. 7. Fiber breaks as function of applied load, ref. 11

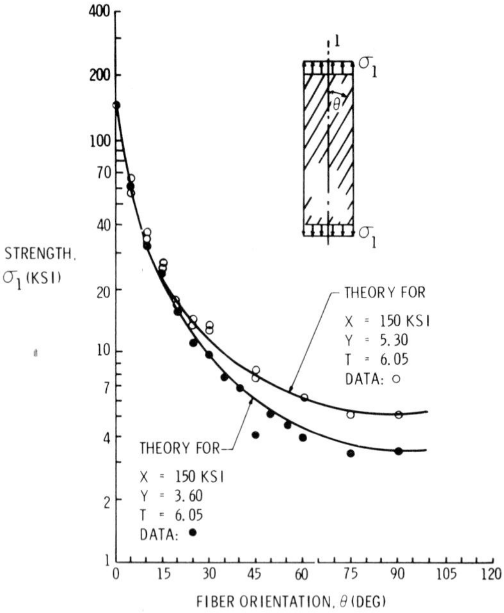

FIG. 8. Strength of unidirectional composite as a function of fiber orientation, ref. 17

Recently, Spencer[14] treated the tensile strength of a unidirectionally fiber-reinforced material in which the fiber was assumed to be rigid-brittle and the matrix was assumed to behave as a rigid-plastic material. Two failure modes were analyzed: matrix yield and fiber fracture. The matrix was assumed to satisfy Tresca's yield condition and a limit analysis theorem of Prager and Hodge was employed to obtain a lower bound to the yield-point load. Fair agreement was found between the predicted results and published experimental results.

A similar analysis has been reported by Wendt[15] for compressive strength of a unidirectionally reinforced laminate. The fibers were assumed in this treatment to be elastic-brittle and the matrix was treated as elastic-plastic. Yield was assumed to be governed by Mises' criterion and upper and lower bounds on the limit load were obtained by limit analysis. No comparison with experiment was presented.

Hill's anisotropic distortional energy criterion obtained by an extension of Mises' yield criterion was employed by Tsai[16] and Azzi and Tsai[17] in their analysis of the tensile strength of unidirectionally fiber-reinforced laminates. Good agreement was found between their predictions and experimental measurements, as illustrated in Fig. 8. They also applied their method of analysis to include combined loading and thermal stress.[18] Results of the theory were found to lie just outside the limits of experimental scatter for uniaxially loaded cross-ply laminate test data, although excellent agreement with angle-ply test data was demonstrated.

In a comparison of maximum stress, maximum strain, and maximum work strength theories with test results for unidirectionally glass-fiber-reinforced epoxy laminates subjected to tensile and compressive loading, Tsai[19] has concluded that, of the three theories, the maximum work criterion appears to present the best correlation with experiment. However, as he has pointed out, since "statistically reliable data for the verification of the theoretical yield condition are rare",[12] it is not certain at present which yield condition is best.

The designer is nevertheless faced with establishing multiaxial allowable stresses. Rogers and his group at Convair, Fort Worth, have done so as part of the design procedures for a boron-reinforced F-111 tail. Examples are shown in Figs. 9 and 10.

One of the earliest contributions to micromechanics and flaw theories of strength was a treatment by Hedgepeth[21] of the static and dynamic stress concentrations caused in a sheet of parallel filaments by one or more filaments being broken. Small deflection elasticity theory was used to study the stress distributions in a sheet of parallel filaments which were assumed to carry only normal loads and were imbedded in a matrix capable of only transmitting shear. Particular attention was paid to the dynamic overshoot resulting from a sudden break in the filaments. As the number of broken filaments increased from one to infinity, the dynamic response factor increased from 1.15 to 1.27. A somewhat lower dynamic response factor was found when a hole is suddenly caused in the filament sheet (see Fig. 11).

Hedgepeth's approach was later used by Schuerch[22] to analyze the effect of flaws on the nonductile failure of uniaxial fiber-matrix materials. In another

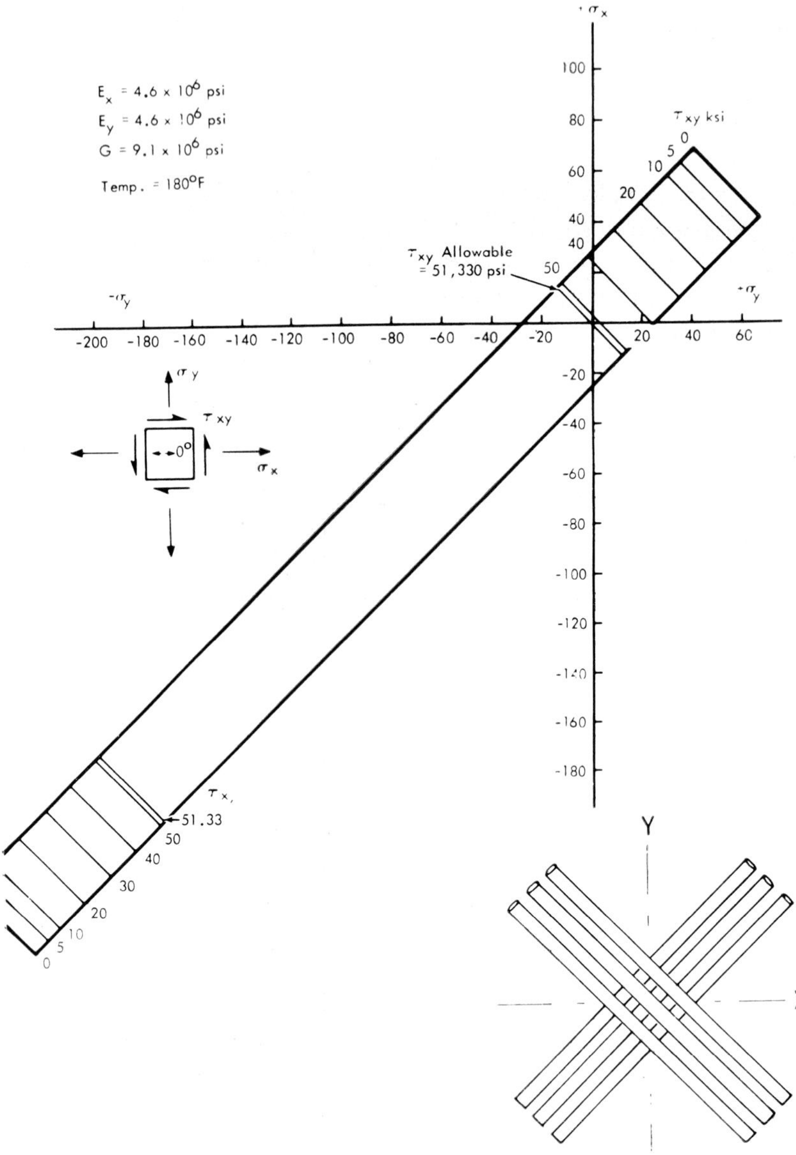

FIG. 9. Projected proportional limit data: 100%±45° laminate, ref. 20

study, Schuerch[23] analyzed the compressive strength and an instability failure mechanism of an idealized two-dimensional model of a unidirectional fiber-reinforced composite. Two cases of microstability failure were considered: an antisymmetric (or shear) buckling of the fibers and a symmetric fiber (transverse) buckling mode. Predicted strength values appear to correlate reasonably well with limited experimental results (see Fig. 12).

An investigation by Dow and Rosen[3] also indicated that the compressive strength of a fiber composite loaded in a direction parallel to the fibers is

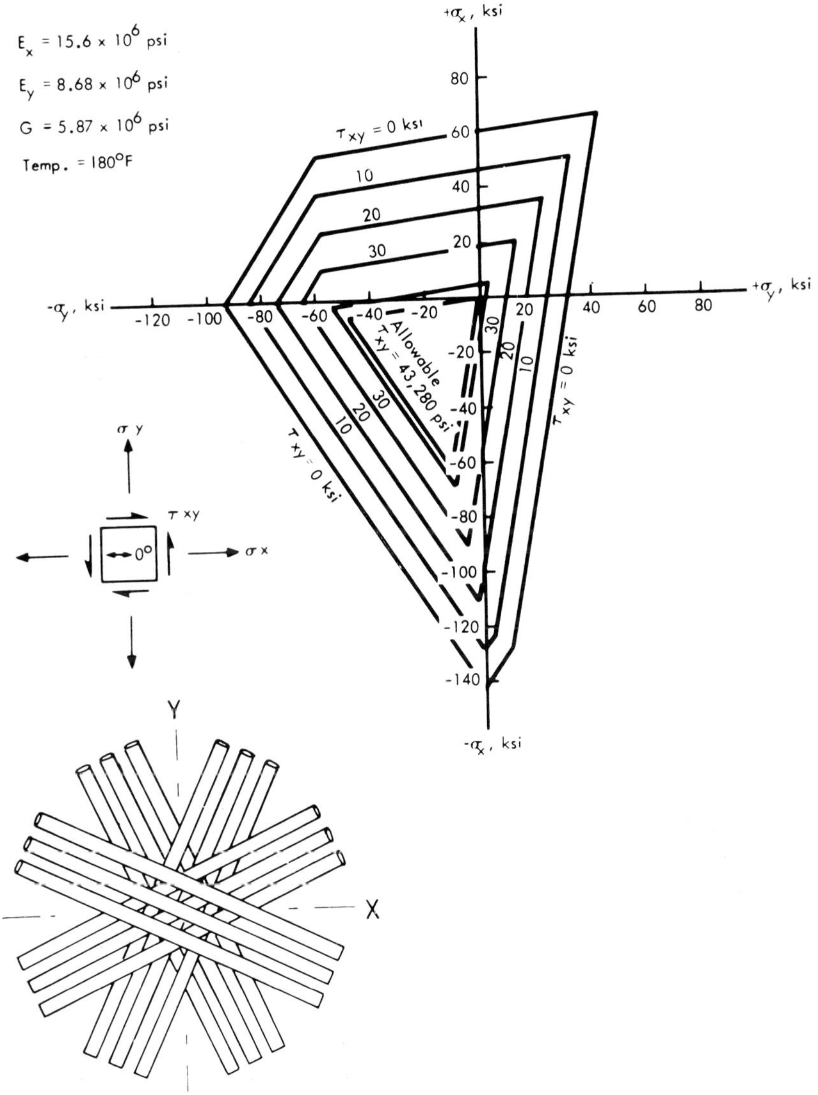

FIG. 10. Projected proportional limit data: 70%±25°, 30%±65° laminate, ref. 20

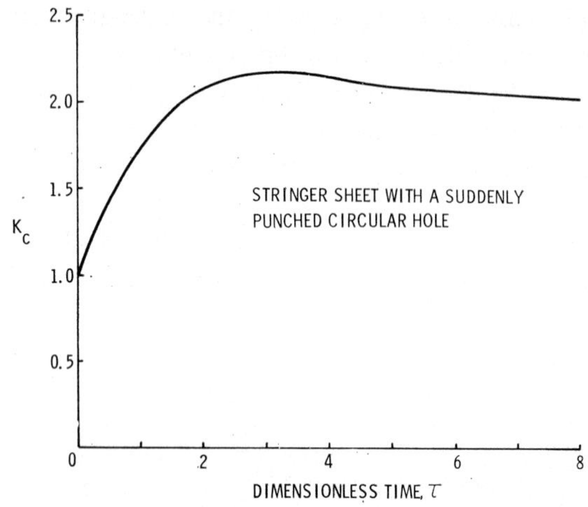

FIG. 11. Variation of stress-concentration factor, ref. 21

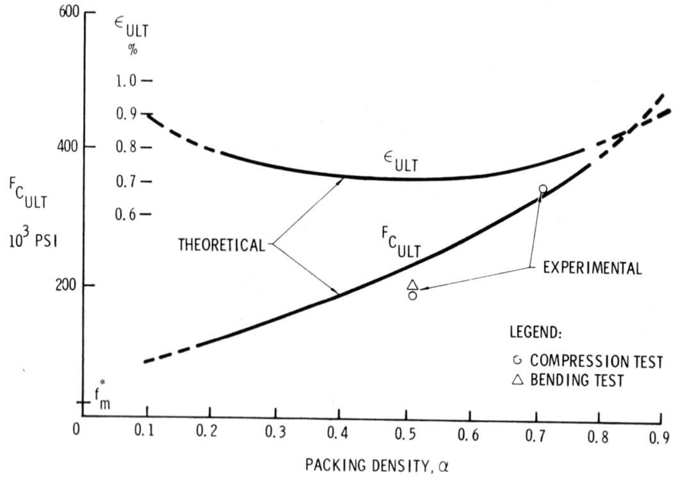

FIG. 12. B/Mg composite, compressive strengths, ref. 23

governed by an instability mode similar to the buckling of a column on an elastic foundation. Qualitative support for this failure mode was provided by experimental results for the compressive strength of hollow glass-fiber composites.

C. *Residual and Thermal Stresses*

Islinger *et al.*[24] have presented a limited analytical and experimental study of residual stresses in polyester specimens reinforced by a single layer of glass fibers located near one surface. Photoelasticity and curvature produced by resin shrinkage were used to study the characteristics of the residual stresses and to compute their magnitude. For the conditions studied, tensile stresses from 170 to 1930 psi were found in the polyester. Studies of resin shrinkage effects in single filaments and on fabric-based laminates by Stavsky and McGarry[25] revealed helical buckling of the encapsulated single filaments

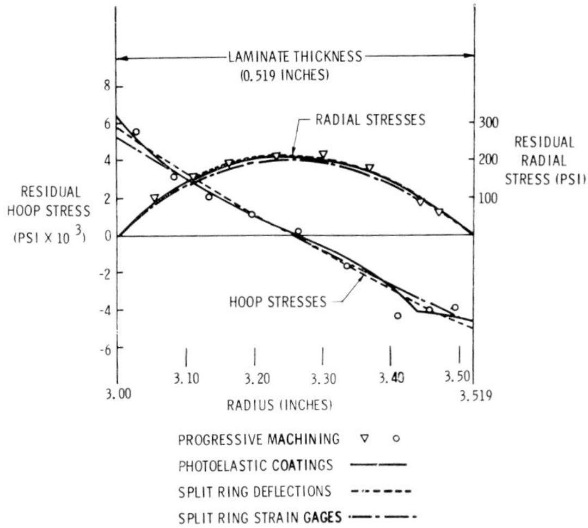

FIG. 13. Residual stresses in dispersed cylinders, ref. 27

produced by axial shrinkage of the resin. As post-cure temperature of the laminates was increased, extensive internal cracking occurred.

Residual stresses may be produced during the production of filament-wound cylinders. Residual stresses that arise during the winding of large thick cylinders were investigated by Myers and Fink.[26] The residual stresses in many 6 in. diameter models were measured and correlated with predictions from a previously developed analysis. Reasonable agreement was found between the predictions and measurements. Figure 13 illustrates residual stresses in thick glass filament-wound cylinders measured by Stone and Greszczuk.[27] The four methods used to measure these residual stresses are in close agreement. Tensile hoop stresses at the inside laminate surface and compressive hoop stresses at the outside were present in all the specimens analyzed. The residual stresses were small, being approximately 1% of the expected ultimate strength. Also, no apparent relation was found to exist between residual stress and interlaminar shear strength.

A method has been developed by Tsai and Azzi[18] for the analysis of laminated anisotropic materials subjected to combined states of stress resulting from applied loads and from mechanical and thermal interactions induced by the process of lamination. The behavior of the laminate is described by a piecewise linear stress–strain relation, and the theory is specialized to thermal contraction.

III. COMPOSITE STRUCTURES

In view of the wide use of anisotropic and laminar structural elements in modern structures, it is understandable that a great deal of work has been accomplished in understanding the behavior of composite structures. A large number of articles have been published in the general area of anisotropic and laminar plates and shells, and these, in many instances, may be directly applicable to the problems peculiar to fiber-reinforced composites. In fact, a number of these available theories have been applied, essentially unchanged or extended, to describe the characteristics of fiber-reinforced composite plates and shells.

A. *Cylinders*

A small deflection theory in which the electrostatic extension and flexure of thin laminated anisotropic shells is described has been formulated by Dong, Pister, and Taylor.[28] In their theory the shell wall may be composed of an arbitrary number of laminates, each of which may possess different thickness, orientation, and/or anisotropic elastic properties. The five equations of equilibrium of a shell element, together with the compatibility equation for in-plane strain components of the reference surface are simplified by the Donnell approximation, and the generalized Donnell equations are further specialized for orthotropic laminae of equal thickness. A procedure is discussed for determining stresses in an individual lamina as well as interlaminar shear stresses. Axisymmetric bending of laminated cylindrical shells is discussed in detail. Implications of the neglect of coupling between extensional and flexural deformation for this class of problems is pointed out.

Steyer[29] has presented an axisymmetric plane strain analysis for stresses and strains in a long, anisotropic, laminated, thick-walled, circular cylinder subjected to internal and external pressure. The analysis is applicable to a laminated fiber-wound shell with an arbitrary fiber axis orientation for each lamina. Laminae elastic properties may vary in the axial, radial, and tangential directions. Each lamina is assumed to be homogeneous, anisotropic, and in a state of plain strain. To facilitate determination of the equivalent three-dimensional elastic constants in terms of lamina matrix and filament properties, two crucial assumptions are employed. The fibers are assumed to be of square cross-section, and the opposite sides of a composite elemental cube are assumed to remain straight and parallel after deformation. Typical results obtained by use of this theory and the theory of Greszcuk[30] for glass fiber-reinforced materials are found to be in considerable disagreement. The major discrepancies in the results are traced by Steyer to his assumption that opposite sides of the composite elemental cube remain straight and parallel after deformation.

The stresses and small deformations of thick, orthotropic, circular cylindrical, homogeneous shells under arbitrary load and support conditions have been treated by Crouzet–Pascal *et al.*[31] In addition, they analyzed thick, circular, cylindrical, sandwich shells with orthotropic cores by incorporating the equations developed for the homogeneous shell into the formulation of the field equations and associated boundary conditions for the sandwich shell. In their most general form, the equations developed describe the stresses and deformations of a three-layered, circular, cylindrical shell. The core exhibited bending and stretching rigidity.

One important aspect of fiber-reinforced composite materials is the role of shear deformations. In an analytical determination of the failure stresses of two different ring-stiffened cylinder configurations, Abildskov[32] found widely differing results when shear deflection effects were neglected. However, when these effects were taken into consideration, close agreement was obtained. It must be emphasized that shear deformations should not be neglected in analysis, since fiber-reinforced composite structures apparently undergo much larger shear deformations than comparable metal structures.

B. *Pressure Vessels*

In a recent review Habip[33] surveyed several contributions in the field of filament-wound structures. These contributions were analytical studies of pressure vessels which included the following topics: a monotropic membrane theory for isotensoid pressure vessels; axisymmetric rotating and pressurized filamentary structures; nonrotating bottle-shaped and toroidal pressure vessels; isotensoid end closures; isotensoid torous, ovaloid, planar wound vessels; and geodesic-isotensoid design analysis for cones, spheres, ellipsoids, and cylinders.

In addition to geodesic-isotensoid pressure vessel investigations, other work has been reported on deformations and reliability of thick-walled filament-wound pressure vessels. Shaffer[34] analyzed a pressurized spiralloy circular cylinder. The cylinder was assumed to be sufficiently long so that, except near the ends, stress and strain may be assumed to be independent of the axial coordinate. The cylinder was symmetrically wound with a constant filament winding angle and was considered to be orthotropic with constant material properties. A generalized plane strain analysis was performed for the condition of a closed tube with internal pressure. Two sets of equations were obtained for the stress, strain, and radial displacement: one set being applicable for incompressible material and the other for compressible material. Computations were presented for the incompressible orthotropic cylinder and the results were compared with those for an isotropic incompressible cylinder (see Fig. 14). The difference between the radial stress distributions associated with each solution is so small that the plotted results appear to

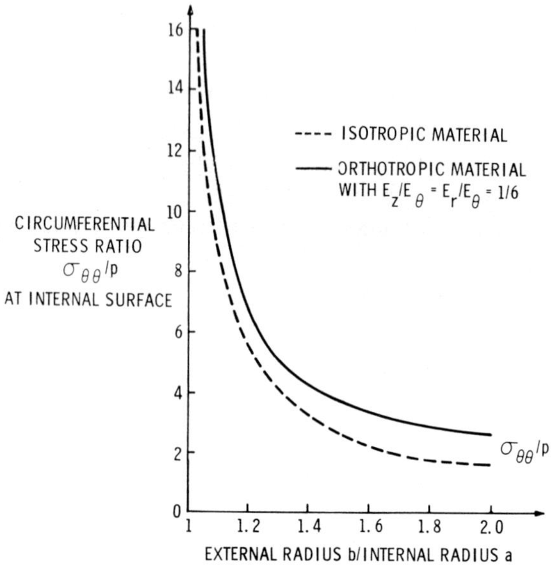

FIG. 14. Circumferential stress in incompressible orthotropic tubes, ref. 34

coincide. However, the difference between circumferential stresses associated with each solution increases with the ratio of external radius to internal radius.

Another analysis of a thick-walled cylinder has been reported by Steyer[29] for the case of a multilayered, filament-reinforced, anisotropic, elastic cylinder with an arbitrary orientation of the filament axis for each layer. He developed a method for analyzing the stresses and strains due either to internal or external radial pressure. In the analysis, the mechanical properties may vary from layer to layer in the axial, radial, and tangential direction.

If the anisotropic properties of the material are properly accounted for and a full-fledged shell analysis is performed, good results can be expected for pressure vessels. One comparison from Azzi and Tsai[35] is shown in Fig. 15. Note that an approximate netting analysis yields poor results.

In regard to reliability of filament-wound pressure vessels, Majerus and Ferriera[36] concluded, from a detailed literature survey and analysis, that "reliability increases as the resin shear modulus and ductility increases and as the lower fiber strengths increase. Also, the Poisson's ratio of the glass fibers and resin should be kept as small as possible for any particular Young's modulus and should be compatible to reduce the resin strain concentrations." By assuming that the fibers possessed a modified Weibull flaw distribution and that the resin was a bilinear elastic material, Majerus and Ferriera were able to study the influence of the fiber-to-matrix bond on reli-

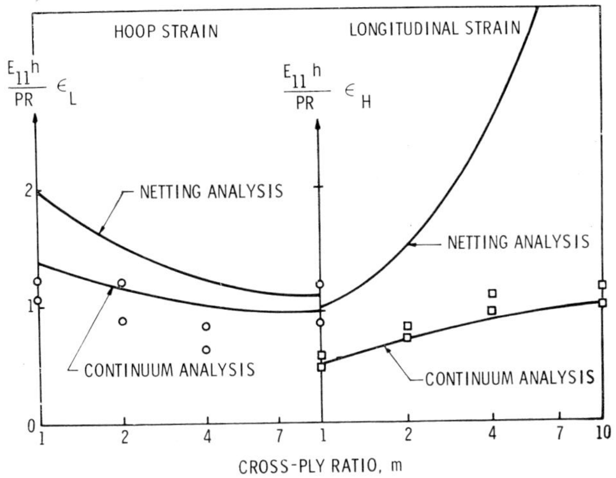

FIG. 15. Cylindrical pressure vessels—experiment and theory, ref. 35

ability. The "reliability load" was calculated for the cases of bonded and unbonded resin. The predicted bond reliability was found to be substantially higher than the predicted unbonded reliability, and the difference between the two cases (15% to 47%) was dependent upon the distribution of ultimate fiber strengths.

C. *Plates*

Bending and stretching of laminated anisotropic plates has been discussed by a number of investigators on the basis of classical linear plate theory. The general theory has been developed to a considerable extent for application to composite plates, and some experimental work has been undertaken.

Bending and stretching of plates consisting of two orthotropic sheets of equal thickness, which are laminated in such a way that the axes of elastic symmetry are a tequal and opposite angles to the x, y-plate axes, has been treated by Reissner and Stavsky.[37] A coupling phenomena between the in-plane stretching and transverse bending occurs for plates of this type. This coupling does not occur in the theory of homogeneous plates as pointed out by Reissner and Stavsky and has not been considered in earlier work for such plates. Explicit solutions for two specific plate problems were used to illustrate the general results of the paper. The results of Reissner and Stavsky's theory were compared to the results of theories which do not consider all the cross-coupling terms associated with the assumed kind of anisotropy and heterogeneity. In one of the problems solved, an infinite plate, with $E_{xx} = 10E_{yy} = E_{zz}$ and a total included angle between the axes of elastic symmetry of the two laminae equal to 60°, was subjected to a sinusoidal transverse load which varied along the x- and y-directions of the plane of the plate. The shear stresses at the midplane of the plate calculated both on the basis of no coupling and coupling between bending and in-plane stretching, differ by approximately 47% for these two cases. The higher stress is the result of including the cross-coupling effects.

This coupling of bending and stretching and the phenomena of cross-elasticity, which occurs in the stress-resultants and couples relations for the small bending and stretching of laminated aeolotropic plates, has also been described in a later report by Stavsky and McGarry.[25] One of the basic contributions of this report was to present a governing eight-order system of differential equations for the subject plate. This general eighth-order system was used to treat some simple boundary value problems for rectangular plates in cases where the bending-stretching coupling disappears and where both the bending-stretching and cross-elasticity effects vanish. This treatment "revealed that homogeneous plate theory can be used to analyze such special cases."

A second theory, in which transverse shear and normal stress deformations are taken into account, was presented in Stavsky and McGarry's report for the small bending of symmetrically laminated isotropic plates. A sixth-order system of differential equations with three boundary conditions was obtained by the added assumption of a constant Poisson's ratio. A tenth order system with five boundary conditions would result if Poisson's ratio were not assumed to be constant. It was considered by Stavsky and McGarry that "The basic features of the latter have no counterpart in the theory of homogeneous plates and shells."

In an experimental program, Azzi and Tsai[35] used glass fiber-reinforced, resin cross-ply and angle-ply laminated plates as test specimens to examine the effects of coupling between in-plane strain and curvature. This effect was shown to be significant in certain cases, depending on materials and geometric lamination parameters. Use of the generalized Hooke's law in analysis was found to be justified, since the properties of unidirectional composites tested transformed in accordance with the fourth-rank tensor. In addition, experiments for both cross-ply and angle-ply laminates confirmed that the "coupling between the in-plane strain and moment and between curvature and stress resultant is very strong" for an even number of laminae, being strongest for the case of two laminae. This coupling gives rise to internally induced normal stresses for cross-piled laminates and shear stresses for angle-ply laminates, which stresses are additive to the externally imposed axial or bending stresses.

The effect of lamina properties and orientation on the shear stresses in a biaxially stiffened matrix have been studied by Rosen, Dow and Hashin.[11] They determined shear stresses by using the methods previously developed by Rosen for determination of the elastic constants of laminates of orthotropic plates. The particular case analyzed considers the elastic axes-of-symmetry of the two laminae to be at equal and opposite angles to the loading direction. For the particular case considered, it was shown that an increase in the transverse to axial Young's modulus ratio decreases the ratio of shear stress on weak shear planes to the applied stress.

The theory developed by Dong, Pister, and Taylor,[28] which was primarily described for thin laminated anisotropic shells, may also be applied to stretching and bending of plates. This may be accomplished by allowing the radius of curvature of the shell to approach infinity thereby obtaining a system of equations for laminated plates. As in the shell theory, the plates may be composed of an arbitrary number of laminae, each of which may possess different thickness, orientation, and/or anisotropic properties.

D. *Beams*

The work that has been accomplished so far on the structural mechanics of fiber-reinforced composite beams has primarily been concerned with the design and test of box beams with fiber-composite faced sandwich for the covers and spars. Examples of this work include studies of helicopter main blades and tail rotors, horizontal and vertical airplane tail stabilizers, and development-type cantilever box beams.

Crichlow and Sorenson[38] reported the design and test of a cantilever, single-cell (two-spar) box beam with upper and lower surface, spars, and end-ribs of honeycomb sandwich construction. The faces were large diameter glass fiber-reinforced plastic laminates. To allow for material property variations, the design allowable stresses were reduced 75% of average coupon test data. The beam, design allowables, and applied stresses are shown in Fig. 16. During static bending test, the tension cover failed at stresses well below the design values. In the failure area, which was not a critical area according to analysis, the computed stresses were from 64 to 73% of the reduced design allowable strength of 60,000 psi. Failure was attributed to poor binding strength of the resin.

A program for the design of aircraft components (fuselage components and main wing and horizontal tail stabilizer box beams) has been described by Waddoups.[20] This program includes the following analysis programs: a micro and macroproperties program for lamina optimization, a laminated composite point stress analysis program for in-plane and out-of-plane loads, an orientation optimization program for membrane loads, minimum weight design techniques to account for anisotropy, fatigue analysis procedures, static stress analysis procedures, dynamic analysis procedures, and structural

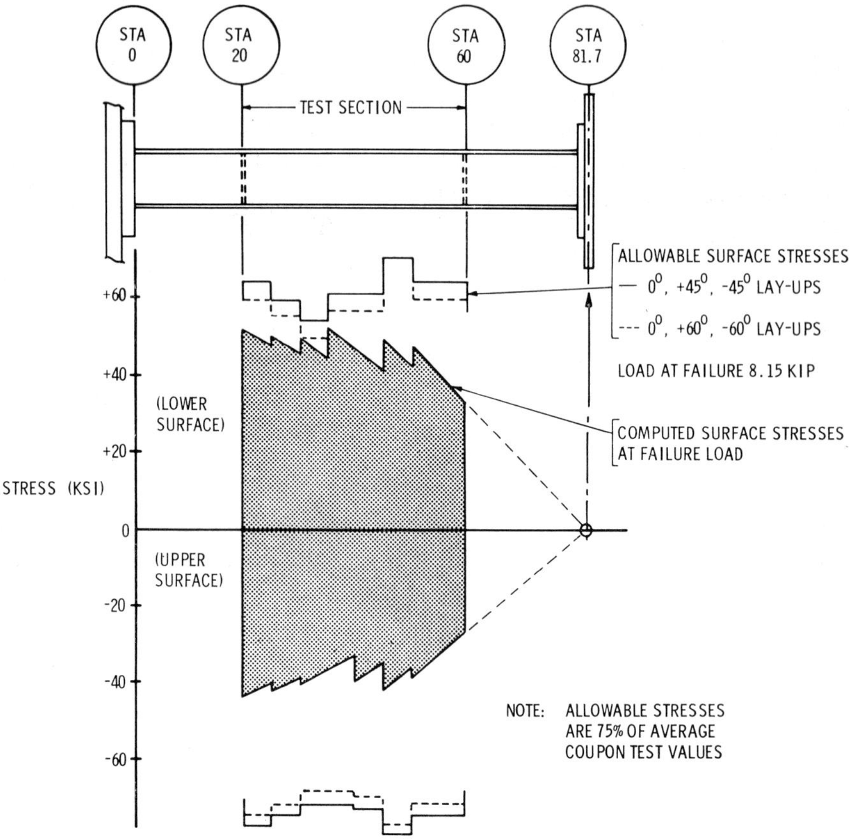

FIG. 1. ..llowable and computed stresses for box surfaces, ref. 38

optimization techniques. These programs are being developed on a continuing basis. The St. Venant yield criterion has been employed as the basis for optimum orientation of laminae. Preliminary design and analysis of a horizontal tail stabilizer main box beam was performed by simulating the test box by plate and beam elements and utilizing existing stiffness matrices. Only one half of the structure was simulated because of symmetry about the middle surface of the component. The substructure was simulated with beams and the facings of the beam sandwich covers with generally orthotropic quadrilateral plates and anisotropic triangular plates. Flangeless beam shear webs were used to simulate the core. In this analysis, it was assumed that the material was linear elastic to the proportional limit. Shear stress–strain data from test specimens showed a marked departure from this assumption.

IV. BUCKLING

Recently reported analyses that are directly concerned with fiber-reinforced composite cylinders include treatments of unstiffened, ring-stiffened, and sandwich circular cylindrical shells. Theories have been developed for hydrostatic, axial, and combined loading of unstiffened cylinders; and ring-stiffened cylinders have been treated for hydrostatic loading. Sandwich cylinders have been studied for the cases of axial and hydrostatic loading. Very little systematic experimental work has been performed to verify the analyses.

Some analytical work on the buckling of composite plates has been reported; however, this work has used available theories. To the authors' knowledge, no original theory directly applicable to fiber-reinforced composite plate buckling has been published.

A. *Cylinders*

Table 1 lists recent work on the buckling of cylindrical composite shells. The articles have been grouped according to the type of construction, either unstiffened, ring-stiffened, or sandwich shells.

An important general finding for axially compressed cylinders is that either anisotropy or heterogeneity through the thickness degrades the strength of the shell (see Figs. 17 and 18).

Buckling of fiber wound cylindrical shells due to stresses caused by windings has been investigated.[49, 50] Euler buckling appears to be possible if the fibers are not bonded well to the shell and can separate from the shell at the point of buckling. However, this Euler buckling may have little significance since a sensitive snap-through condition has been found to exist. Imperfections in the shell may increase the imminence of collapse for a particular winding pressure.

TABLE 1. STATUS OF CYLINDRICAL SHELL BUCKLING ANALYSES

Unstiffened and Ring-stiffened Shells

Reference number	Cylinder construction	Theory and/or experiment	Type of loading	Basis of theory or results	Comparison with experiment
39	Homogeneous, aelotropic—both ends hinged or both ends clamped—laminated shell wall—unstiffened	New theory	Combined axial, radial and torsional	Flugge's equilibrium equation, Donnell's method followed to establish boundary conditions for classical small deflection solution—anisotropic shell material	None
40	Unstiffened	Available theory and test	Hydrostatic	Small deflection theory for orthotropic cylinder shell buckling by Stein and Mayers	Collapse pressure consistently higher than predictions—due to greater end restraint in tests than in equation (hinged ends)
41	Unstiffened hollow glass fiber-reinforced thick-wall cylinder	Test	Hydrostatic	Stein and Mayers orthotropic shell buckling equation used by Jacobson(40)	Collapse pressure consistently higher than predictions
42 43	Unstiffened and ring-stiffened	Test	Hydrostatic	Test results appear to compare well with isotropic material buckling equation of Windenburg and Trilling—thin shell theory	Experiment only
44	Unstiffened with clamped ends	Available theory and test	Axial	Cheng and Ho's theory(39) used	Test loads 65% to 85% of predictions for shells with diameter to thickness ratios from 167 to 643

45	Unstiffened	Slight modification of available theory plus test	Axial	Classical Gerard and Becker solution modified by "orthotropic modification factors": Ey/Ex	Results considered satisfactory for design purposes
46	Heterogeneous cylinder—laminated unstiffened cylinder	Extension of available theory—Cheng and Ho type solution	Axial	Donnell type for orthotropic material. Heterogeneity deleterious effect on stability. Most efficient layer arrangement dependent upon material properties and degree of homogeneity	No experiments—only numerical examples
			Sandwich Shells		
32	Sandwich with composite faces and foam core	Analytical design study	Hydrostatic	Shear deformation in fiber composite structure must be included in analysis	None
47	Unstiffened cylinders and sandwich cylinder	Theory and test	Axial	Modification of large deflection theory of isotropic shells	Reasonably good agreement (plywood cylinders)
48	Sandwich-core rigid in shear	Structural efficiency analysis	Axial	Stein and Mayers equation used. Results: (a) need for adequate in-plane shear stiffness for efficient shells (b) isotropic lay-up is more efficient	No experiment—only numerical examples

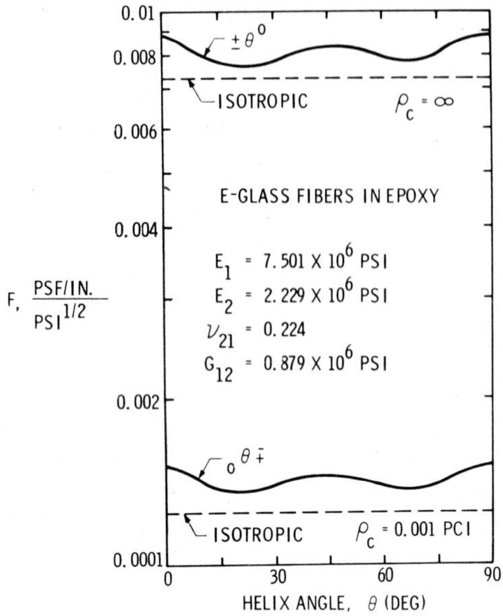

FIG. 17. Elastic structural efficiency of symmetric biaxial laminates, ref. 3

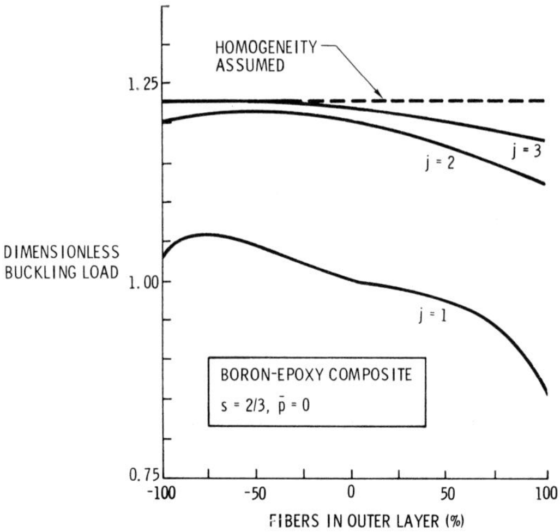

FIG. 18. Buckling load for multilayered cylinder, ref. 46

B. *Plates*

In regard to plate buckling, Rosen *et al.*(51, 11) have contributed a comparison between the stability of biaxially stiffened plates and isotropic plates as well as the stability of flat plates that contain oriented voids. The comparison between the stability of a simply supported, long orthotropic plate subjected to uniaxial in-plane compressive load and that of similarly loaded isotropic plates is presented in the form of the plate buckling stress ratio for orthotropic plates normalized with respect to an isotropic plate with equal longitudinal properties as a function of the lamina orientation angle. Figure 19 shows the effect of lamina properties on laminate buckling stress. Maximum stress is achieved for lamina orientation of 45°, and at 45° the critical stress is independent of the shear modulus. For a given shear modulus, the variation in stress associated with change in lamina orientation is small while the variation in weight is even smaller.

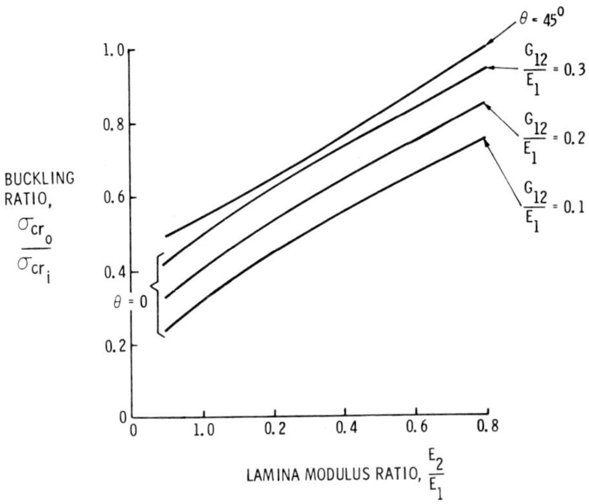

FIG. 19. Effect of lamina properties on laminate buckling stress, ref. 51

In a study of the effect of oriented voids on the stability of an in-plane, compressive loaded flat plate, Rosen, Dow and Hashin[11] concluded that the orientation of the voids is not important in the elastic range. They suggest that "if any criterion exists it is that the holes should, in general, be aligned in the direction of principal stress, to delay as long as possible the onset of plasticity".

V. CREEP

Very little research concerning creep behavior of fiber-reinforced composites has been reported. Systematic experimental creep investigations are lacking although some isolated creep tests have been performed on unstiffened and ring-stiffened glass fiber-reinforced plastic cylinders in which the cylinders were subjected to external hydrostatic pressure. In one test, 5% creep over an 18 hour period was measured with the shell loaded to 3000 psi or 80% of the collapse pressure.[52] In other tests reported by Hom, Buhl and Couch,[43] two unstiffened laminated cylindrical shells were subjected to long duration loading. One cylinder collapsed at a pressure of 3400 psi after a total loading time of 113 days applied in seven periods. The length of these periods ranged from 27 hours to 26 days. In contrast, the collapse load of an identical cylinder tested under short-term loading was 3550 psi. The second long-term loaded cylinder collapsed under a short-term load of 9025 psi after sustaining a pressure of 7315 psi for 50 days. The short-term collapse pressure of a duplicate cylinder was 8975 psi. Since only two creep collapse tests were performed, these results are not conclusive; however, they do give an indication that the collapse pressures of short- and long-term loaded cylinders are not too far removed from each other.

One attempt to theoretically evaluate creep in fiber-reinforced materials treated the composite as a combination of Hookean solid and a Newtonian fluid.(53) Later, Hashin employed a "composite cylinder assemblage model" and a correspondence principle for elastic and viscoelastic heterogeneous media to describe macroscopic viscoelastic properties of fiber-reinforced materials consisting of linear viscoelastic matrix and unidirectional elastic fibers. Explicit results in terms of general linear viscoelastic matrix properties were derived in some cases, and in other cases, analytical expressions for effective relaxation and creep functions were obtained for a viscoelastic spring-dashpot matrix model.

VI. FATIGUE

A good summary of the results generated over the previous four years on compressive and shear fatigue programs sponsored by the Bureau of Ships is given by Freund and Silvergleit.(54) They report uniaxial compressive, biaxial compressive and interlaminar shear cyclic fatigue data. All of the data reported showed a large degree of scatter. Biaxial compressive fatigue results contained strength variations of 114%, and the variation in uniaxial compressive fatigue strengths showed a variation of 110%. Lowest scatter was found for interlaminar shear fatigue and even in this case the strength varied 44%. These results represent the findings of a large number of tests—for example, the results of more than 200 tests were included in the biaxial compressive fatigue data. The majority of the unidirectional specimens were found to delaminate under cyclic loading with subsequent failure. Both the interlaminar shear and the compressive strength decreased with an increase in void content by approximately the same percentage. However, it was pointed out that shear failure becomes much more prominent, since the ultimate interlaminar shear strength is approximately 7 to 10% of the ultimate compressive strength. Shear failure was concluded to be the primary cause of compressive fatigue failure of glass fiber-reinforced plastic.

Freund and Silvergleit suggested that a "percentage of ultimate" concept be used to predict the approximate fatigue life of glass reinforced plastic material. In this "percentage of ultimate" concept the applied fatigue stress is expressed as a percentage of the true ultimate static stress. However, as a wise precaution, it is pointed out that "this predicting method is approximate since the true static stress is difficult to ascertain at present, except by destructive testing." Figure 20 shows the results of reducing the cyclic fatigue stress to a percentage of ultimate. These curves represent the lower boundary curve for each of the three types of cyclic fatigue tests.

Crichlow and Sorenson[38] reported conventional axial load tensile S–N fatigue tests as well as spectrum fatigue tests of large diameter (0.003–0.005 in

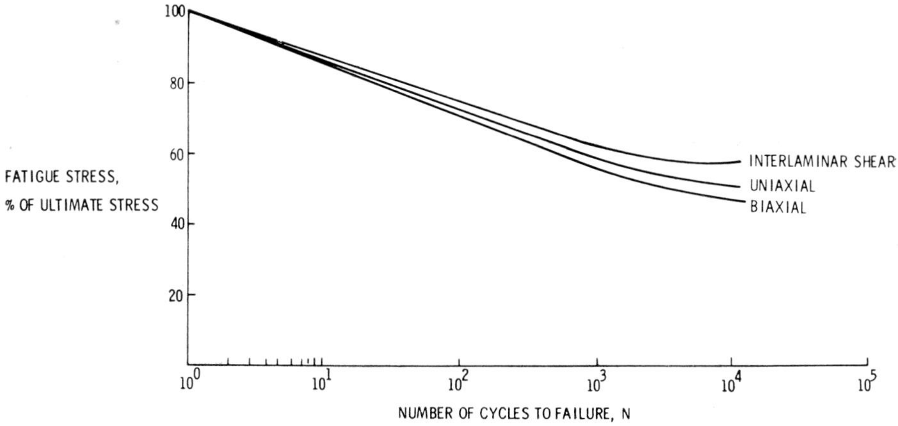

FIG. 20. Fatigue life, ref. 54

diameter) "S" glass fiber unidirectional and multidirectional reinforced laminates. A stress ratio of 0.1 and 1800 cycles per minute were employed in the S–N tests. Superior fatigue strength-density ratios were obtained for these laminates compared with 2024–T3 aluminum alloy sheet in the low stress–high cycle range and equal or inferior results for the high stress–low cycle range (see Fig. 21). Titanium 8AL-1Mo-1V duplex-annealed sheet, however, was superior to these laminates in all stress ranges.

For the spectrum fatigue tests, two loading sequences were employed—one a flight-by-flight spectrum consisting of 81 cycles per flight and the other a block spectrum for 1020 flights also containing 81 cycles per flight. It was found that the flight-by-flight results, as indicated in Fig. 22, are more conservative than the block loading, which is considered consistent with similar experiments in metals. Figure 22 also shows the number of flights between discovery of the fatigue crack and final failure or termination of the test. The indicated rate of crack growth was considered to be "sufficiently low to expect good performance from the standpoint of ample inspection intervals and good residual static strength in the presence of fatigue cracks or other damage."

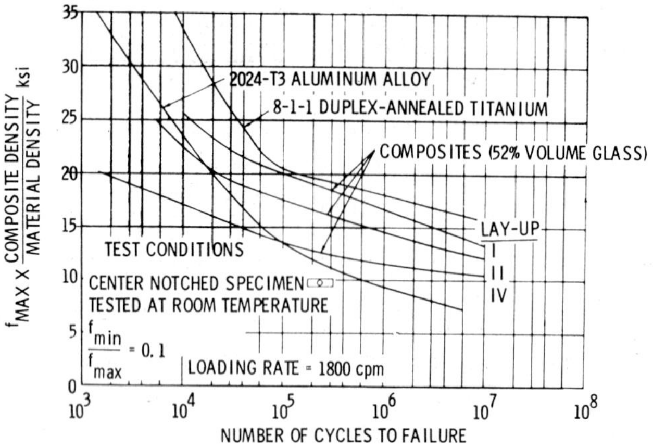

FIG. 21. S–N curves metals and large-diameter S glass epoxy composites, ref. 38

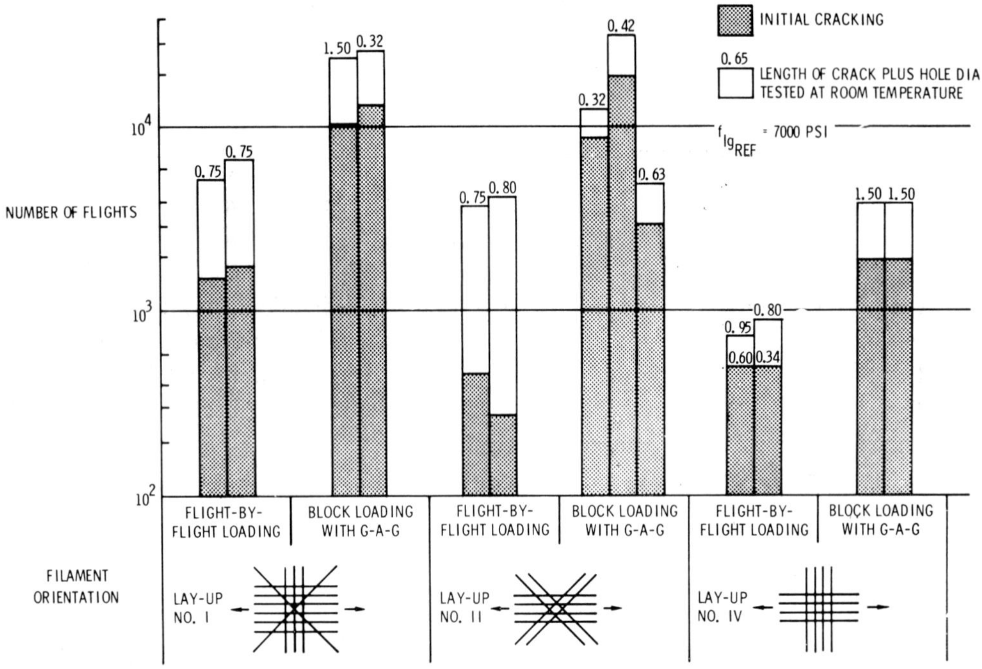

FIG. 22. Spectrum fatigue tests, ref. 38

VII. JOINTS AND DISCONTINUITIES

Love[55] has presented a method of analysis for the discontinuity stresses at the cylinder-to-dome juncture in filament-wound pressure vessels. In general, agreement between calculated and measured strains recorded during hydrotest of a vessel is good, particularly in the lower stress range. Most of the discrepancies found between predicted and measured strains was attributed to the neglect of resin shear strains and the assumption of linear elasticity throughout the entire applied stress range.

Moments at the edge of an adhesive lap joint between fiber-reinforced laminates have been measured by Gnapp(56) and found to be in good agreement with predictions from the Goland–Reissner expressions.

The joint and penetration problem in compression-loaded deep diving submersibles appears to be in hand. In a recent work reported by Meyers and Fink(26) it was stated that "the incorporation of penetrations and both closure-cylinder and transverse cylinder joints appears feasible if proper design procedures are employed and adequate sealing methods are utilized in test." The contribution of the various components of a complete vessel design toward the overall weight/displacement ratio for the basic cylinder design investigated by Meyers and Fink is shown in Table 2.

TABLE 2. PERCENTAGE OF OVERALL WEIGHT/DISPLACEMENT RATIO, REF. 26

	%
Basic shell	65.6
Frames	15.0
Cylinder cutout	8.5
Closure with cutout	3.1
Shell end build-ups	2.8
Spacers	1.9
Closure without cutout	1.6
Transverse joint	1.5

Of the total vessel weight/displacement ratio of 0.65 the basic ring-stiffened cylinder comprises approximately 80%. In comparison, the other major contributor is the cylinder cutout at 8.5% of the total weight/displacement ratio. These joints and penetrations do not appear to impose an excessive weight penalty. Later work by Abildskov(57) has indicated that "weight/displacement ratios as low as 0.56 are possible in sandwich-wall models having titanium closures which are capable of withstanding 17,778 psi (40,000 ft depth)". Abildskov states that the discontinuity stresses developed at the joint can be effectively predicted and controlled.

It appears that joint and penetration problems in compression-loaded structures, submersibles in particular, are being solved, or at least, solutions to these problems appear feasible. However, this does not seem to be the case in regard to joints for aerospace applications. For example, effective load introduction to tension panels fabricated from high strength- and high modulus-to-density ratio fiber-reinforced material is a major problem—particularly for practical applications. It appears that a lot more work will have to be accomplished in joint design and the coupling of joint design and fabrication efforts before effective usage of advanced composites in practical structures will be possible.

VIII. SUMMARY

The structural mechanics of fiber-reinforced composites is in a stage of rapid evolution. Theoretical analyses of macroscopically uniform composite materials and composite structures are fairly well developed. However, buckling predictions and accurate descriptions of creep and fatigue, and the behavior of joints and discontinuities are not as advanced. Accurate experimental data are needed to verify existing theories as well as to aid in the development of new ones.

Constitutive relations are available for the prediction of the elastic behavior of unidirectional fiber-reinforced composites. These relations have been found to provide predictions that correlate reasonably well with experiment.

The bulk of the effort in the area of theories of strength of continuous fiber-reinforced composites has primarily emphasized the tensile strength of unidirectional reinforced laminae with only little effort concerned with compressive strength. Practically nothing has been published on shear strength prediction, transverse strength, and strength under combined loading. Even so, the progress toward accurate prediction of tensile strength is encouraging. More experimental effort is needed to develop and verify strength theories.

Residual and thermal stresses in cylindrical fiber-reinforced shells have been treated both analytically and experimentally. The magnitude of these stresses has been found to be small in compression loaded cylinders and the test data correlate reasonably well with predictions.

The analysis of pressure vessels is well in hand and theories for the bending and stretching of cylinders (other than pressure loaded vessels) are available although there is not complete correlation with experiment. Available classical-linear theories for anisotropic and laminar plates have been applied to fiber-reinforced composite plates. These theories have not been fully substantiated by experiment. Box beams have been treated analytically and experimentally. The experiments have not been completely successful because of fabrication and joint design difficulties.

Theoretical analyses of creep and fatigue are apparently just beginning to be reported—the bulk of the work so far has been experimental. Systematic experiments are vitally needed for the development of creep theories of fiber-reinforced composites.

Working joints and penetrations have been developed for submersibles; however, much development work is needed for other structures.

REFERENCES

1. HILL, R. Theory of Mechanical Properties of Fiber Strengthened Materials: I, Elastic Behaviour, *J. Mech. Phys. Solids* **12** (1964), 199–212.
2. HASHIN, Z. and ROSEN, B. W. The Elastic Moduli of Fiber Reinforced Materials, *J. Appl. Mech.* **31F**, no. 2 (June 1964), 223–232.
3. DOW, N. F. and ROSEN, B. W. Structural Efficiency of Orthotropic Cylindrical Shells Subjected to Axial Compressions, *AIAA J.* **4**, 3 (March 1966), 481–485.
4. TSAI, S. W. Structural Behavior of Composite Materials, NASA CR-71 (1964).
5. PICKETT, G. Analytical Procedures for Predicting the Mechanical Properties of Fiber Reinforced Composites, AFML-TR-65-220 (June 1965).
6. LEISSA, A. W. Application of Point-Matching to Problems in Micromechanics, Conf. Fund. Aspects Fiber Reinf. Plas. Compos. (May 24–26, 1966), spons. by WPAFB.
7. WHITNEY, J. M. and RILEY, M. B. Elastic Properties of Fiber Reinf. Composite Materials, *AIAA. J.* **4**, no. 9 (Sept. 1966), 1537–1542.
8. GRESZCZUK, L. B. Theoretical and Experimental Studies on Properties and Behavior of Filamentary Composites, 21st Ann. Mtg. RPD, Soc. Plas. Ind. (Feb. 1966), sec. 8A.
9. ADAMS, D. F. and DONER, D. R. Longitudinal Shear Loading of a Unidirectional Composite, *J. Comp. Mtls.* **1** (Jan. 1967), 4–17.
10. KIES, J. A. The Strength of Glass Fibers and the Failure of Filament Wound Pressure Vessels, U.S. Nav. Res. Lab. Dept. 6034 (Feb. 1964), AD 439 217.
11. ROSEN, B. W., DOW, N. F. and HASHIN, Z. Mechanical Properties of Fibrous Composites, NASA CR-31 (1964).
12. TSAI, S. W. A Review of Strength Theories of Composite Materials, *Major Status Report – Advanced Composites Hardware* (Sept. 1966), spons. by WPAFB, 323–327.
13. MCGARRY, F. J. Composite Structural Materials, *Exptl. Mech.* (June 1966), 331–336.
14. SPENCER, A. J. M. A Theory of the Failure of Ductile Materials Reinforced by Elastic Fibers, *Internat. J. Mech. Sci.* **7**, 3 (March 1965), 197–209.
15. WENDT, F. W. Relationships Between Micromechanics and Structural Behavior of Fiber Reinforced Plastic Composites, Conf. Fund. Aspects Fiber Reinf. Plas., spons. by WPAFB, Dayton, Ohio (May 24–26, 1966).
16. TSAI, S. W. Strength Characteristics of Composite Materials, NASA CR-224 (April 1965).
17. AZZI, V. D. and TSAI, S. W. Anisotropic Strength of Composites, *Exptl. Mech.* **5** (1965), 283–288.
18. TSAI, S. W. and AZZI, V. D. Strength of Laminated Composite Materials, *AIAA. J.* **4**, 2 (Feb. 1966), 296–301.
19. TSAI, S. W. Deformation and Strength of Filamentary Structures, Conf. Fund. Aspects Fiber Reinf. Plas. Compos., spons. by WPAFB, Dayton, Ohio (May 24–26, 1966).
20. WADDOUPS, M. E. Design of Aircraft Components With Composite Materials, Gen. Dyn., Ft. Worth Div. (Sept. 21, 1966).
21. HEDGEPETH, J. M. Stress Concentrations in Filamentary Structures, NASA TND-882 (May 1961).
22. SCHUERCH, H. A Semi-Atomistic Model for Flaw-Induced Fracture in Non-Ductile Materials, Astro Res. Corp., ARC-R-200 (Oct. 1965), Contract NASw-652.
23. SCHUERCH, H. Prediction of Compressive Strength in Uniaxial Boron Fiber-Metal Matrix Composite Materials, *AIAA J.* **4**, 1 (Jan. 1966), 102–106.
24. ISLINGER, J. S., GUTFREUND, K., BRZUSKIEWICZ, J. and OLSON, O. H. Mechanism of Reinforcement of Fiber-Reinforced Structural Plastics and Composites, WADC TR 59–600 (June 1960).
25. STAVSKY, Y. and MCGARRY, F. J. Investigation of Mechanics of Reinforced Plastics, WADD TR-60-746 (June 1962).
26. MYERS, N. C. and FINK, B. Filament Wound Structural Model Studies for Deep Submergence Vehicles, *Naval Engineers J.* **77**, 2 (April 1965), 275–290.
27. STONE, F. E. and GRESZCZUK, L. B. Study of Residual Stresses in Thick Glass Filament

Reinforced Laminates, Douglas MSSD for Bu. Ships, Dept. Navy, Quart. Prog. Rpt. No. 2 (Dec. 30, 1964), AD 609–610.

28. DONG, S. B., PISTER, K. S. and TAYLOR, R. L. On the Theory of Laminated Anisotropic Shells and Plates, *J. Aerospace Sci.* **29** (Aug. 1962), 969–975.
29. STEYER, C. C. Analysis of a Thick-Walled Layered Anisotropic Elastic Cylinder Subjected to a Radial Pressure, *AIAA J. Spacecraft* **3**, no. 4 (April 1966), 464–471.
30. GRESZCZUK, L. B. Analysis of Filament Wound Pressure Vessels, Lecture 10, Course X414PQ, Univ. of Cal. at Los Angeles (July 12, 1963).
31. CROUZET-PASCAL, J., PIFKO, A. B., MAHONEY, J. B. and SALERNO, V. L. On the Deformation of Thick Composite, Orthotropic Circular Cylindrical Shells, Grumman Res. Dept. Rpt. RE-161, ONR, Contract Nonr-3465(00) FBM, Project NR-64-463, Tech. Rpt. No. 1 (Sept. 1962).
32. ABILDSKOV, D. and DAINES, J. Investigation of Advanced Design Concepts for Deep Submersibles, H. I. Thompson Fiberglass Co., First Quart. Rpt. (April 1964), Contract NObs-90180.
33. HABIP, L. M. A Review of Recent Work on Filamentary Structure, *Internat. J. Mech. Sci.* **7**, 4 (April 1965), 297–299.
34. SHAFFER, B. W. Generalized Plane Strain of a Pressurized Spiralloy Tube, Hercules Powder Co. (Aug. 1962), AD 282 756.
35. AZZI, V. D. and TSAI, S. W. Elastic Moduli of Laminated Anistotropic Composites, *Exptl. Mech.* **5**, (1965), 177–185.
36. MAJERUS, J. N. and FERRIERA, S. K. Fracture and Reliability of Filament Wound Chambers, *Proc. ASCE*, **91** EMI (*J. Engng. Mech. Div*) (Part I), (Feb. 1965), 107–136.
37. REISSNER, E. and STAVSKY, Y. Bending and Stretching of Certain Types of Heterogeneous Aeolotropic Elastic Plates, *J. Appl. Mech.* **28**, 3 (Sept. 1961), 402–408.
38. CRICHLOW, W. J. and SORENSON, V. S. Advancements in Monofilament Structural Composite Technology, *J. Aircraft*, **3**, no. 5 (Sept.–Oct. 1966), 431–435.
39. CHENG, S. and HO, B. P. C. Stability of Heterogeneous Aeolotropic Cylindrical Shells Under Combined Loading. *AIAA J.* **1**, 4 (April 1963), 829–898.
40. JACOBSON, H. R. Optimum Construction of Reinforced Plastic Cylinders Subjected to High External Pressures, MSSD, Douglas Aircraft Co., Inc. Quarterly Progress Rpt. No. 1, BuShips Project S-F103-05-03, Task 1025 Contract NObs 88425 (Mar. 63–Mar. 64), AD 433 652.
41. WRIGHT, F. C. Hollow Glass Filament Reinforced Plastics for Deep Submergence Vessels, 21st Ann. Mtg., RPD, Soc. Plas. Ind. (Feb. 1966), sec. 14B.
42. BUHL, J. E., JR., PULOS, J. G. and GRANER, W. R. Reinforced Plastics for Hydrospace Vehicles, David Taylor Model Basin, Rpt. 1524 (Aug. 1961), AD 264 812.
43. HOM, K., BUHL, J. E., JR. and COUCH, W. P. Hydrostatic Pressure Tests of Unstiffened and Ring-Stiffened Cylindrical Shells Fabricated of Glass-Filament Reinforced Plastics, David Taylor Model Basin Rpt. 1745 (Sept. 1963), AD 422 520.
44. TSAI, J., FELDMAN, A. and STANG, D. A. The Buckling Strength of Filament-Wound Cylinders Under Axial Compression, NASA CR-266 (July 1965).
45. WANG, S. C. Longitudinal Stability Problems of Filament Wound Shells, Soc. Plas. Ind. 21st Ann. Mtg. (Feb. 1966), sec. 14-E.
46. TSAI, J. Effect of Heterogeneity on the Stability of Composite Shells Under Axial Compression, *AIAA J.* **4**, 6 (June 1966), 1058–1062.
47. KUENZI, E. W. Buckling of Layered Orthotropic and Sandwich Cylindrical Shells in Compression, *Collected Papers on Instability of Shell Structures*, NASA TN D-1510 (1962), 323–330.
48. DOW, N. F. and ROSEN, B. W. Structural Efficiency of Orthotropic Cylindrical Shells Subjected to Axial Compression, *AIAA J.* **4**, 3 (March 1966), 481–485.
49. LANGHAAR, H. L., BORESI, A. P., MARCUS, L. and LOVE, G. Buckling of a Long Fiber-Wound Cylindrical Shell Due to Stresses Caused by Windings, *Trans. ASME* **32E** (*J. Appl. Mech.*), 1 (March 1965), 81–86.
50. MIKULAS, M. M. JR. and STEIN, M. Buckling of a Cylindrical Shell Loaded by a Pre-Tensioned Filament Winding, *AIAA J.* **3**, 3 (March 1965), 560–561.

51. ROSEN, B. W., KETLER, A. E. JR. and HASHIN, Z. Hollow Glass Fiber Reinforced Plastics, Gen. Elec. Co. for. Bur. Nav. Wpns., U.S. Navy Dept., Final Rpt., Contract No. NOw 61-0613-d (Nov. 1962), AD 400-237.
52. PULOS, J. G. and BUHL, J. E. JR. Hydrostatic Pressure Tests of an Unstiffened Cylindrical Shell of a Glass Fiber Reinforced Epoxy Resin, David Taylor Model Basin, Rpt. No. 1413 (April 1960), AD 430 296.
53. HASHIN, Z. Viscoelastic Fiber Reinforced Materials, *AIAA J.* **4**, 8 (August 1966), 1411–1417.
54. FREUND, J. F. and SILVERGLEIT, M. Fatigue Characteristics of Glass Filament Reinforced Plastic Material, Soc. Plas. Ind., 21st Ann. Mtg., RPD (Feb. 1966), sec. 17-B.
55. LOVE, GEORGE G. Structural Analysis of Orthotropic Shells-Minimizing Discontinuity Stresses in Filament Wound Pressure Vessels, AIAA Launch and Space Vehicle Shell Structures Conf,. Palm Springs, Calif. (April 1–3, 1963), paper no. 2919-63.
56. GNAPP, J. I. Measurements of Moments at Edges of Adhesive Lap Joints Between Glass Reinforced Plastic Laminates, Feltman Res. Labs., Army Contract No. 1C0244D14115 (Sept. 1965).
57. ABILDSKOV, D. Investigation of Closure and Closure Attachments for GRP Cylindrical Deep Submergence Vessels, 21st Ann. Mtg. of RPD, Soc. Plas. Ind. (Feb. 1966), sec. 17-A.

THE APPLICATION OF FINITE ELEMENT STIFFNESS MATRIX ANALYSIS FOR COMPOSITE STRUCTURES

M. E. WADDOUPS† and J. R. BLACKLOCK‡

General Dynamics Corporation, Fort Worth, Texas

I. NOMENCLATURE

γ	shear strain
δ	average element displacement
ε	strain
ε^0	laminate reference surface strain
ν	Poisson's ratio
σ	stress
$\bar{\sigma}$	average laminate stress
τ	shear stress
$\bar{\tau}$	average laminate shear stress
χ	curvature
$a, b, c, \ldots$	stress assumption arbitrary coefficients
f	flexibility matrix
h	lamina reference coordinates
k	lamina index (kth lamina)
u	displacement
x	reference coordinates
x, y, z	plate coordinates
A	inplane stiffness matrix
$\bar{A}$	inplane average laminate stiffness matrix
B	stiffness coupling matrix
D	flexural stiffness matrix
E	Young's modulus
G	shear modulus
H	equilibrium matrix
M	stress couples
N	stress resultants

† Senior Structures Engineer—Structural Materials and Process Development.
‡ Senior Structures Engineer—Advanced Structural Dynamics.

P	force coordinates
Q	inplane compliance matrix
S	compliance matrix—element supported stiffness matrix
$\bar{S}$	element unsupported stiffness matrix
U	matrix of stress assumption independent coefficients
V	boundary traction matrix

II. INTRODUCTION

The process of analyzing a composite material as a quasi-homogeneous material has been limited in the past by the difficulty of accounting for anisotropy in complex structures. This is evident when we review the number of solutions for the stress distribution in such simple elements as plates and bars for a material which is anisotropic. Although the solutions are not particularly difficult, the amount of cataloging completed remains quite marginal. The purpose of this research has been to close this problem area by creating a finite stiffness matrix element for use in the solution of plane problems. A major constraint on the development was that the elements were to be embedded into operational General Dynamics matrix methods programs, and therefore the mid-side force deflection coordinates were required for the plate elements.

Of special interest is that the work was done on an airplane component design program.[†] This provided an ideal situation for this research because the finite element work could be checked against experimental data. The task was essentially three-phased:

(1) Material characterization.
(2) Element formulation and checkout.
(3) Structural analysis and test.

Each area of the investigation was tempered by the overall goals of the major contract, the development of efficient composite structure.

III. BASIC STRESS–STRAIN RELATIONS

The modeling of the composite material begins with the determination of the elastic properties of the mono-layer. Because the basic boron-epoxy mono-layer consists of a uniform array of parallel unidirectional filaments, there exists on the macroscopic scale three natural planes of symmetry. The

† This work was carried out under the Air Force contracts AF33(615)-3323, "Application of Advanced Fibrous Reinforced Composite Materials" and AF33(615)-5257, "Structural Airframe Application of Advanced Composite Materials".

thin mono-layers are assumed to be quasi-homogeneous and in a state of plane stress. The resulting stress-strain relations may be written in the form[1, 2]

$$\varepsilon_i = S_{ij}\sigma_j \qquad i, j = 1, 2, 6. \tag{1}$$

If the reference axes are coincident with lamina symmetry axes, the S_{ij} are related to the engineering elastic constants as follows:

$$S_{11} = \frac{1}{E_{11}}$$

$$S_{12} = -\frac{\nu_{12}}{E_{11}} = -\frac{\nu_{21}}{E_{22}}$$

$$S_{22} = \frac{1}{E_{22}}$$

$$S_{66} = \frac{1}{G_{12}}$$

$$S_{16} = S_{26} = 0.$$

The engineering elastic constants for the analysis were obtained by static test.[3,4] The typical stress strain curves for the boron-epoxy mono-layer are shown in Fig. 1.

Lamination may be accounted for by first inverting eqs. (1):

$$\sigma_i = Q_{ij}\varepsilon_j \qquad i, j = 1, 2, 6. \tag{2}$$

Again, if the reference axes are coincident with the lamina symmetry axes, the Q_{ij} are related to the engineering moduli as follows:

$$Q_{11} = \frac{E_{11}}{1-\nu_{12}\nu_{21}}$$

$$Q_{12} = \frac{\nu_{12}E_{22}}{1-\nu_{12}\nu_{21}} = \frac{\nu_{21}E_{11}}{1-\nu_{12}\nu_{21}}$$

$$Q_{22} = \frac{E_{22}}{1-\nu_{12}\nu_{21}}$$

$$Q_{66} = G_{12}$$

$$Q_{16} = Q_{26} = 0.$$

The appropriate transformations for the Q_{ij} in the cases for which the reference axes and lamina symmetry axes are not coincident are found in ref. 1.

It is assumed that for thin laminated plates the Kirchoff–Love hypothesis is valid and the strain displacement relations are:

$$\varepsilon_i = \varepsilon_i^0 + Z\chi_i \qquad i = 1, 2, 6 \tag{3}$$

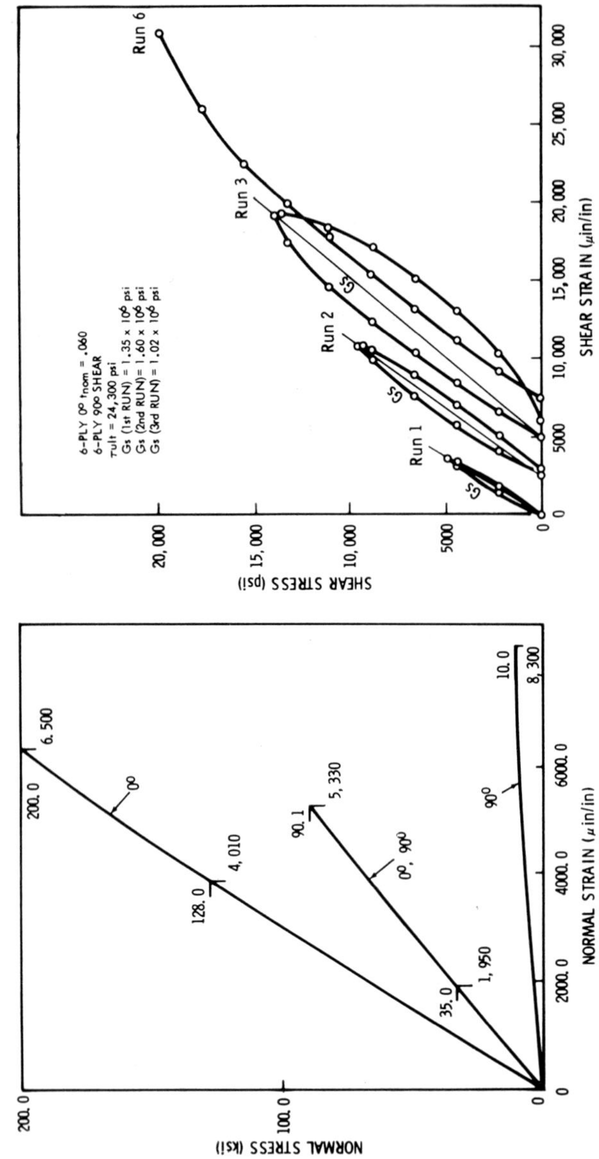

FIG. 1. Primary stress–strain

where the inplane strains and curvature changes are related to the displacements as follows:

$$\varepsilon_1^0 = \frac{\partial u_1}{\partial x_1} \qquad \chi_1 = \frac{\partial^2 u_3}{\partial x_1^2}$$

$$\varepsilon_2^0 = \frac{\partial u_2}{\partial x_2} \qquad \chi_2 = \frac{\partial^2 u_3}{\partial x_2^2}$$

$$\varepsilon_6^0 = \frac{\partial u_1}{\partial x_2} + \frac{\partial u_2}{\partial x_1} \qquad \chi_6 = \frac{2\partial^2 u_3}{\partial x_1 \partial x_2}.$$

Substitution of eq. (3) into eq. (2) and integrating across the thickness of the plate gives the equation for the stress resultants and stress couples:

$$(N_i, M_i) = \int_{-h/2}^{h/2} \sigma_i(1, z)\, dz. \tag{4}$$

Integration of eq. (4), assuming that the lamina are quasi-homogeneous, gives:

$$\begin{aligned} N_i &= A_{ij}\varepsilon_j^0 + B_{ij}\chi_j \qquad i, j = 1, 2, 6 \\ M_i &= B_{ij}\varepsilon_j^0 + D_{ij}\chi_j \end{aligned} \tag{5}$$

where

$$\begin{aligned} A_{ij} &= \sum_{k=1}^{n} Q_{ij}^{(k)}(h_{k+1} - h_k) \\ B_{ij} &= \sum_{k=1}^{n} Q_{ij}^{(k)}(h_{k+1}^2 - h_k^2) \\ D_{ij} &= \sum_{k=1}^{n} Q_{ij}^{(k)}(h_{k+1}^3 - h_k^3). \end{aligned} \tag{6}$$

Under the condition of pairwise symmetric lamination about the laminate geometric middle surface the B_{ij} reduce the zero. It is important to note that this uncouples the plane and bending problem and allows for independent solutions. As shown by Lekhnitski[5] the plane problem may be cast into a more familiar form by the reduction:

$$\bar{\sigma}_i = N_i/h = (A_{ij}/h)\varepsilon_j^0 \qquad i, j = 1, 2, 6 \tag{7}$$

and let:

$$A_{ij}/h = \bar{A}_{ij} \qquad i, j = 1, 2, 6.$$

IV. FINITE ELEMENT FORMULATION AND CHECKOUT

For clarity in this portion of the development the following reference coordinate notation changes will be made:

$$(1, 2, 6) = (x, y)$$

$$\bar{\sigma}_1 = \bar{\sigma}_x \qquad \varepsilon_1^0 = \varepsilon_x$$

$$\bar{\sigma}_2 = \bar{\sigma}_y \qquad \varepsilon_2^0 = \varepsilon_y$$

$$\bar{\sigma}_6 = \bar{\tau}_{xy} \qquad \varepsilon_6^0 = \gamma_{xy}.$$

Two elements will be considered:

(1) Linear stress assumption triangle.
(2) Incomplete quadratic stress assumption quadrilateral.

For either element the equations of equilibrium

$$\frac{\partial \bar{\sigma}_x}{\partial x} + \frac{\partial \bar{\tau}_{xy}}{\partial y} = 0$$

$$\frac{\partial \bar{\sigma}_y}{\partial y} + \frac{\partial \bar{\tau}_{xy}}{\partial x} = 0 \qquad (8)$$

and compatibility

$$\frac{\partial^2 \varepsilon_x}{\partial y^2} - \frac{\partial^2 \gamma_{xy}}{\partial x \partial y} + \frac{\partial^2 \varepsilon_y}{\partial x^2} = 0 \qquad (9)$$

must be satisfied. The elements will be of the stress assumption–average displacement type, and it is convenient to write eq. (9) in terms of the average stresses

$$\bar{A}_{11} \frac{\partial^2 \bar{\sigma}_x}{\partial y^2} + \bar{A}_{12} \frac{\partial^2 \bar{\sigma}_y}{\partial y^2} + \bar{A}_{16} \frac{\partial^2 \bar{\tau}_{xy}}{\partial y^2}$$

$$+ \bar{A}_{12} \frac{\partial^2 \bar{\sigma}_x}{\partial x^2} + \bar{A}_{22} \frac{\partial^2 \bar{\sigma}_y}{\partial x^2} + \bar{A}_{26} \frac{\partial^2 \bar{\tau}_{xy}}{\partial x^2} \qquad (10)$$

$$+ \bar{A}_{16} \frac{\partial^2 \bar{\sigma}_x}{\partial x \partial y} + \bar{A}_{26} \frac{\partial^2 \bar{\sigma}_y}{\partial x \partial y} + \bar{A}_{66} \frac{\partial^2 \bar{\tau}_{xy}}{\partial x \partial y} = 0.$$

The stress assumption for the triangular plate, together with the force–displacement coordinates, are shown in Fig. 2. The stress assumption identically satisfies both equations, (8) (equilibrium) and (10) (compatibility). The stress assumption and the force displacement coordinates for the quadrilateral plate are shown in Fig. 3. The stress assumption satisfies equilibrium equation (8), but for the compatibility equation (10) to be satisfied the shear coupling terms,

$$\bar{A}_{16} = \bar{A}_{26} = 0. \qquad (11)$$

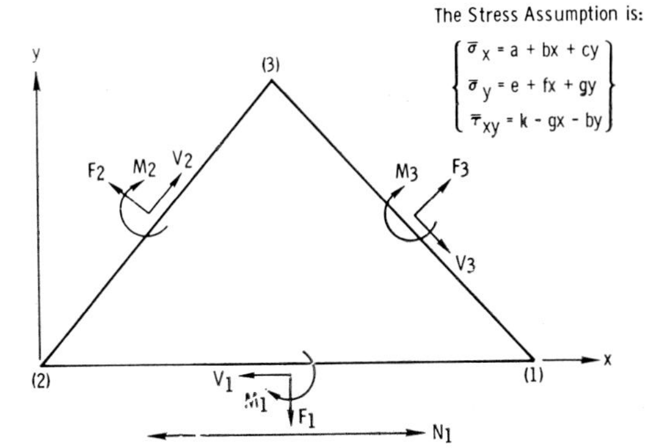

FIG. 2. Anisotropic triangular element with a linear stress assumption

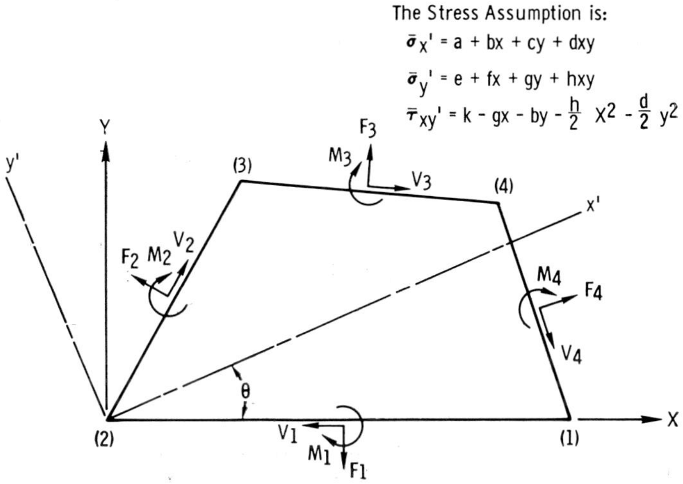

FIG. 3. Orthotropic quadrilateral plate with a quadratic stress assumption

This completely defines the limits of applicability for both elements:

(1) Triangular element—plane, anisotropic laminates with middle surface symmetric orientations.

(2) Quadrilateral element—plane, generally orthotropic laminates with middle surface symmetric orientations.

The restriction of having the laminate symmetry axes be coincident with the element reference axes (Fig. 3) for the quadrilateral element was removed by writing the stress assumption for the (x', y') axes at an angle θ to the element reference axes.

The formulation of the elements may be completed with the linear theory proposed by Best.[6] The same technique was used for both the triangular and quadrilateral elements.

The following definitions are made:

$$k = \text{col}\,(a, b, c, d \ldots)$$
$$\sigma = \text{col}\,(\sigma_x, \sigma_y, \tau_{xy}) \tag{12}$$
$$\varepsilon = \text{col}\,(\varepsilon_x, \varepsilon_y, \gamma_{xy})$$

where k is a vector of arbitrary constants and σ and ε are the stress and strain components respectively for the plane problem. The strain is related to the stress by the equation (in contracted matrix notation):

$$\varepsilon = Q\sigma \qquad (Q_{(3x3)}) \tag{13}$$

and the stresses are related to the vector k by the independent variables of the stress assumption in the form:

$$\sigma = Uk \qquad (U_{(3xm)}) \tag{14}$$

where m is the number of force–displacement coordinates for the supported element.

The forces may be determined by evaluation of the line integrals of eq. (14) on each unsupported side of the element and are expressed in the form

$$P = Vk \qquad (V_{(mxm)}) \tag{15}$$

where the P_i are the forces for the supported structure. Now the stresses and strains may be written in terms of the forces in the form:

$$k = V^{-1}P \tag{16}$$
$$\sigma = UV^{-1}P \tag{17}$$
$$\varepsilon = \bar{A}^{-1}UV^{-1}P. \tag{18}$$

The flexibilities may be determined by evaluation of the integral:

$$f = \iiint_V \varepsilon^T \cdot \sigma \, dv \tag{19}$$

where ε is the column strain vector at point P due to a unit load at coordinate i. σ is the column stress vector due to a unit load at coordinate j. This results in an integral of the form (some subscripts are dropped for clarity):

$$f = e_i^T V^{-1T} \left[\iiint_v U^T \bar{A}^{-1} U \, dv \right] V^{-1} e_j \tag{20}$$

where $e_i^{\prime}$ and e_j are column vectors with 1 in the ith and jth places respectively and zeros elsewhere.

The supported stiffness matrix may be written:

$$P = S\delta \tag{21}$$

where

$$S = f^{-1} \tag{22}$$

and S is the average displacement at each point P.

The unsupported stiffness matrix may be formed by equilibrium of the forces P_m and the remaining three forces. The unsupported stiffness matrix may be written:

$$\bar{S} = HSH^T \qquad (H_{(m+3,\, m)}).$$

The triangular plate element has been checked out on the element level only. This implies that the element satisfied equilibrium and the stress assumption is consistent. The quadrilateral plate was checked on the element level and then several trials were made on a sample problem. The problem selected was a cantilevered anisotropic beam.(5) A summary of the geometry and results are shown in Figs. 4 and 5. The force displacement results are acceptable but as the angle between the element reference axes and the laminate symmetry axes approaches 45 degrees the stress results become ill conditioned. Near 45 degrees for a right rectangular plate the maximum coupling of the stress assumption coefficients occurs and this is a probable cause of the problem. The 40 degree checkout problem did reinforce confidence in the force displacement relations accuracy even in light of the stress difficulties.

V. STRUCTURAL ANALYSIS AND TEST

The structural analysis of the two-third-scale F-111 horizontal tail was accomplished with the aid of several digital computer programs. The basis for these programs was the matrix deflection method of analysis. They were used to determine the internal force distribution, the elastic deflected shape and the stresses in the boron composite cover skins.

The matrix deflection method of analysis can be divided into four basic steps:

(1) The idealization of the physical structure into a system of small elements or substructures.
(2) The generation of stiffness matrix models of each element of the substructures.
(3) The addition of the element stiffness matrices into a master stiffness matrix, application of boundary conditions and, finally, solution of the resulting set of simultaneous equations for a deflected shape of the structure.
(4) Computation of the element forces and stresses by applying the deflections obtained.

The idealization is made after one has determined what finite stiffness elements will be available. Since we were to use the linear stress triangular and the quadratic stress quadrilateral plate elements; their capability to represent a gradient in the stress field across their boundaries was used as a

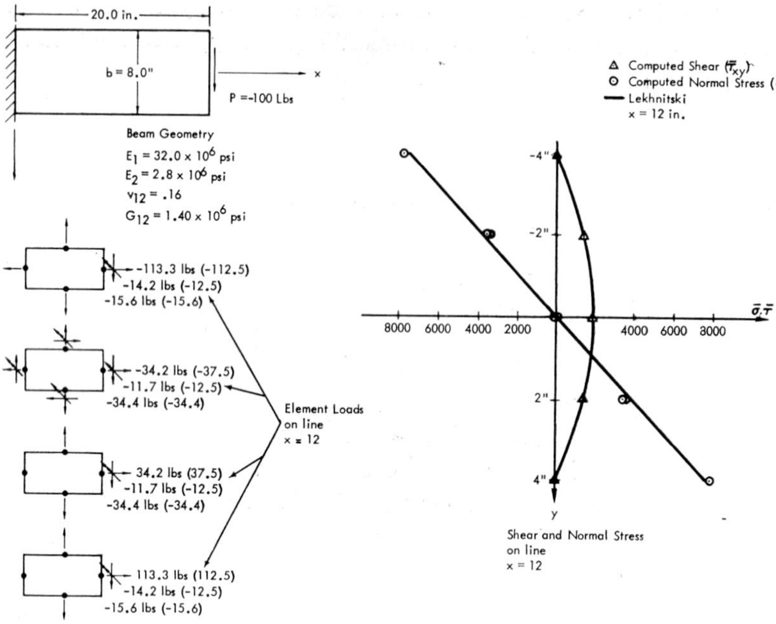

FIG. 4. Orthotropic beam

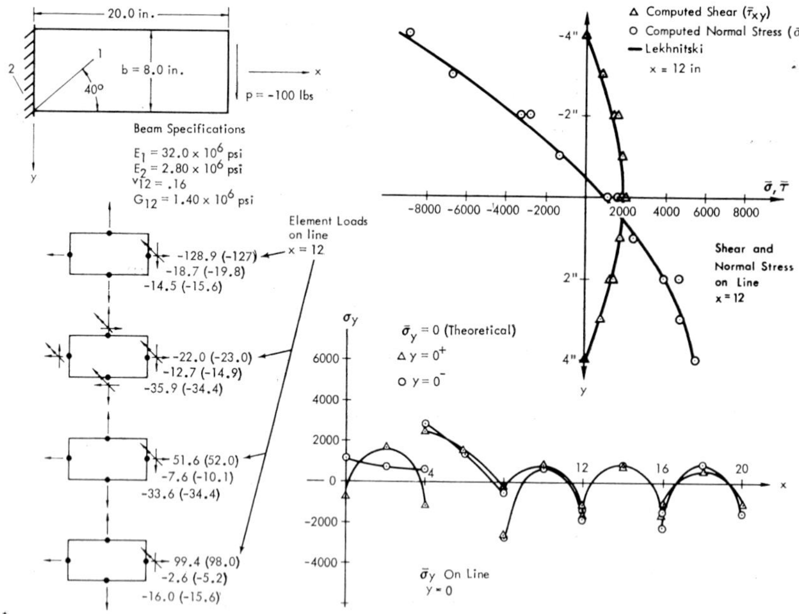

FIG. 5. Generally orthotropic beam

guide to determine how fine the idealization grid should be. In this manner we arrived at the grid system of 49 surface plates. The element stiffnesses were determined from Fig. 6 and the lamina orientation.

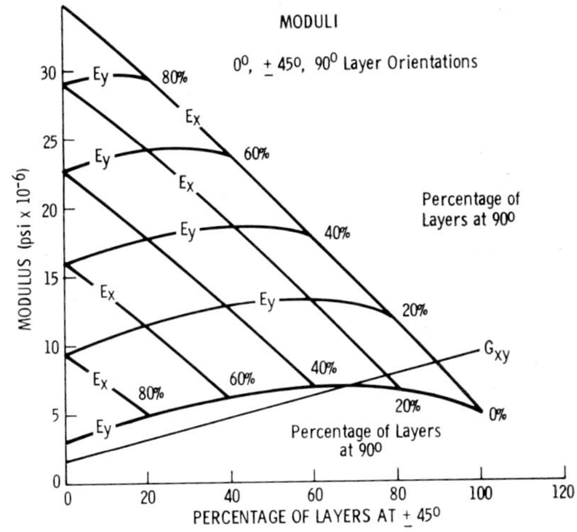

FIG. 6. Computed mechanical properties

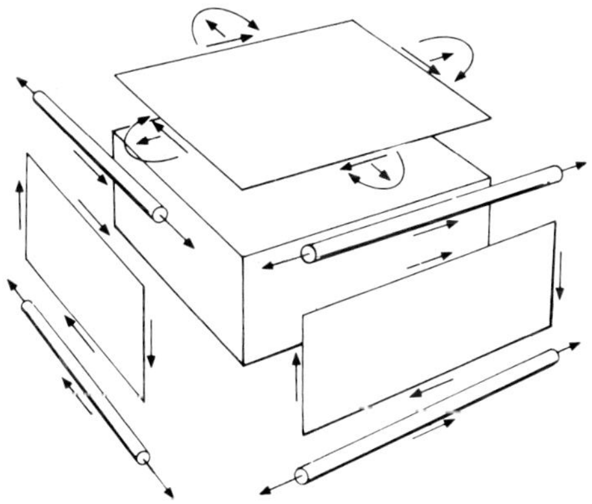

FIG. 7. Orthotropic plate cell

The stiffness matrix for the horizontal tail was generated using the finite element idealization indicated in Fig. 7. This idealization composed of cover skins, shear webs and bulkhead flanges represented only that half of the tail above its center plane. Appropriate boundary conditions were then imposed in order to model the full tail for the antisymmetric loading which was to be applied.

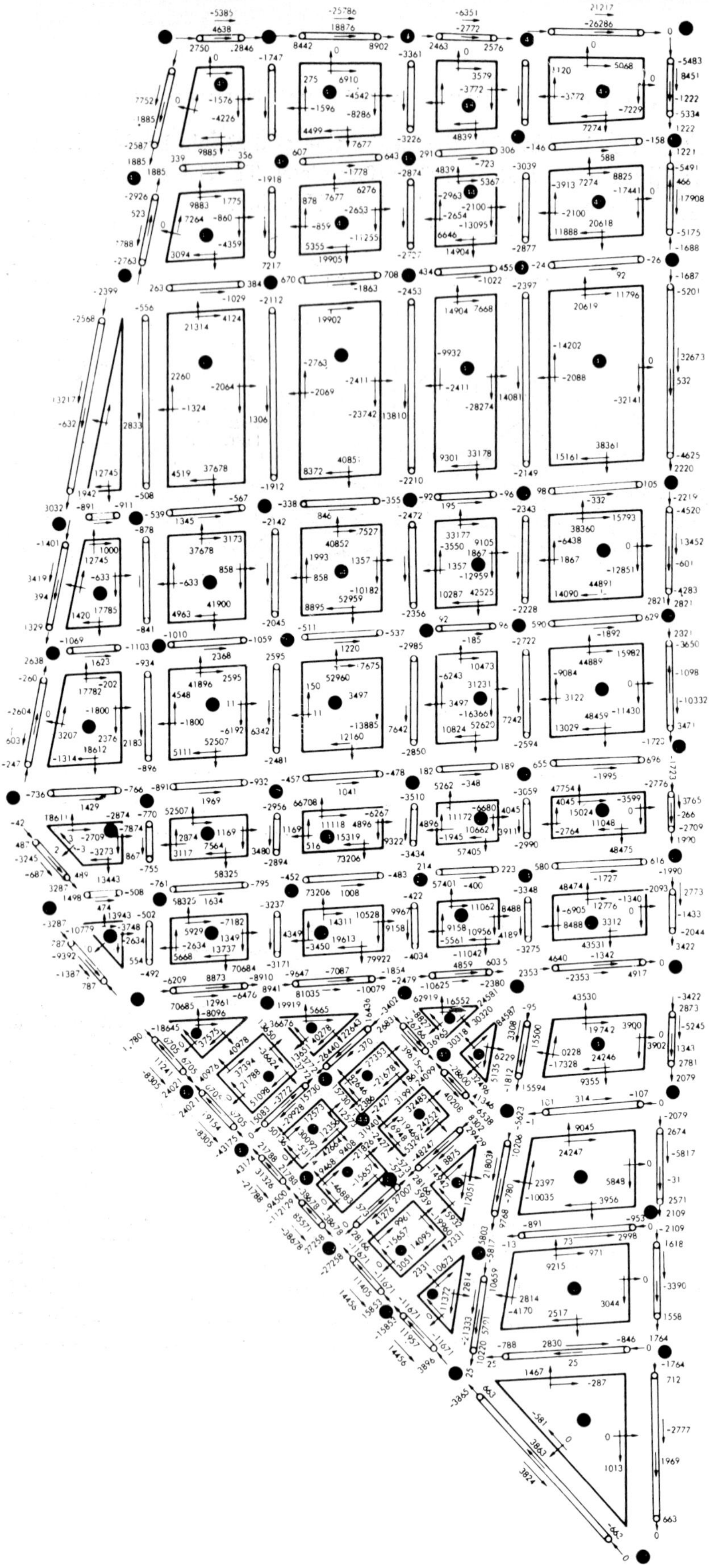

FIG. 8. Computer simulation

FIG. 9. Pre-test of horizontal tail

The resulting stiffness matrix was a size 479×479 symmetric matrix. The static test loads were applied at the tip spar. The solution to these 479 equations was obtained by the block Gauss–Seidel iteration method with a Aitken–Thiele Delta Process accelerator.(7) The solution required 891 successive iterations to obtain convergence. The term "convergence" here means that the ratio of the maximum deflection change to its total value during the 891st iteration was less than 1.0×10^{-6}. This accuracy produced forces which were typically accurate to five significant figures.

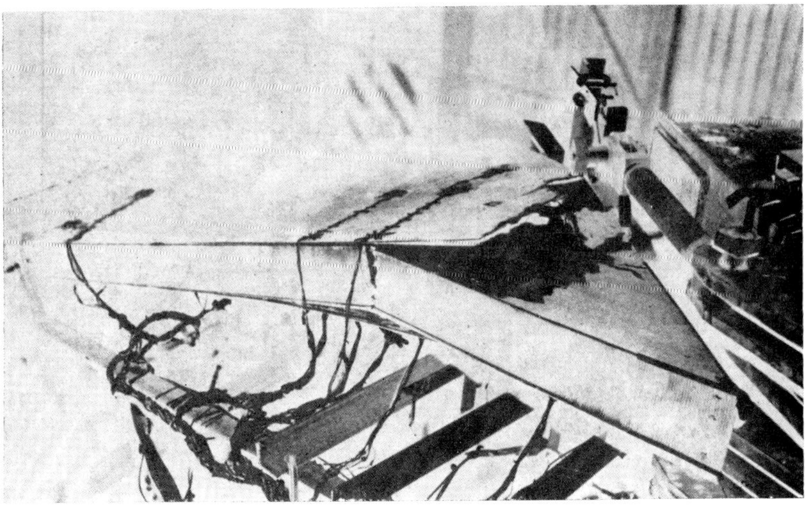

FIG. 10. Horizontal tail after test

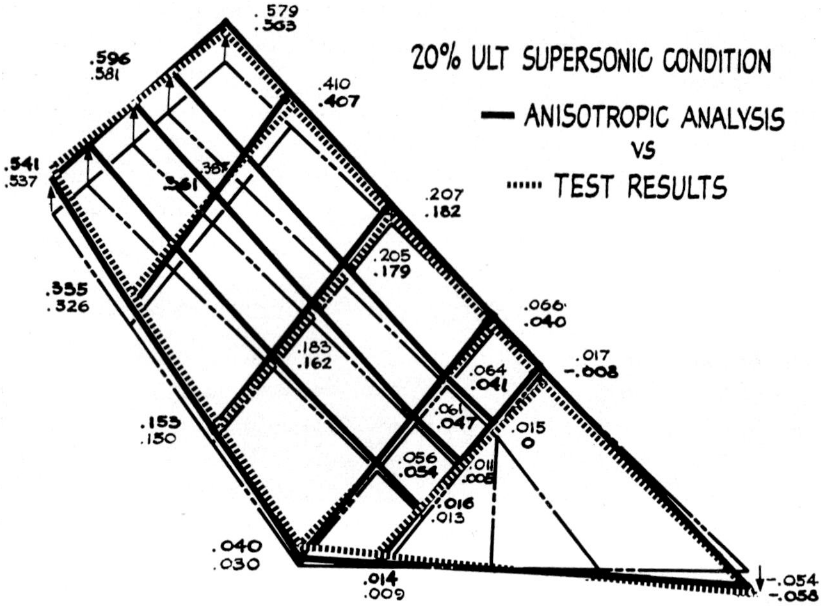

FIG. 11. Deflected shape

The final results were verified by static test. Figures 9 and 10 are photographs of the part in the test fixture. The results of the matrix method portion of the structural analysis are presented in Figs. 8, 11 and 12. You will notice that the deflected shape obtained analytically matches closely that which was obtained in the laboratory test. The deflection disparity, near the forward rib, is attributed to the coarse grid system of plates which was used in that

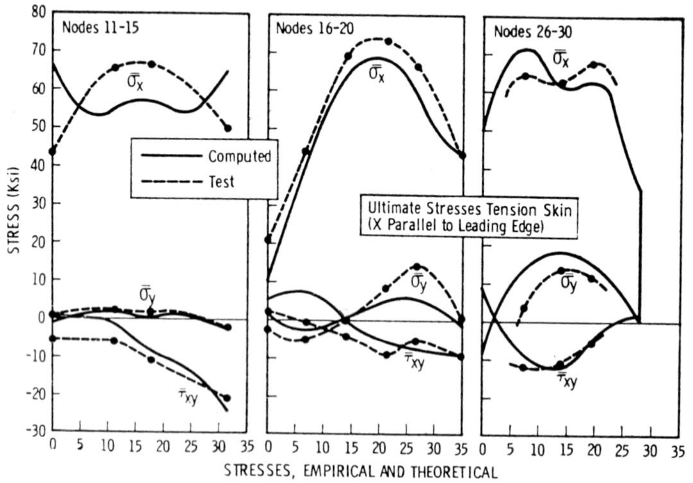

FIG. 12. Empirical and theoretical stresses

area. This is supported by a look at the stress curve. It can be noted that over the area of a single plate the strain gage data shows a state of stress which the 2nd degree complex quadrilateral plate could not match. However, experience has shown that even when the stress data contains discrepancies, accurate forces may be obtained.

The internal force distribution shown in Fig. 8 was used to calculate the stresses in the joints and fittings and to size the understructure in general.

VI. CONCLUSIONS

The finite element matrix technique is an adequate tool for the closing of solutions for complex anisotropic structures. The correlation with experimental results for this study verified both the element development and the material characterization phases of the program.

One distinct problem area emerged during the program. The stress coefficients for the quadrilateral plate were found to become ill conditioned when the principal axis reference angle approached 45 degrees. This problem can be avoided, however, by a judicious selection of the reference side for each plate. In addition, it was determined that even when the stress coefficients became ill conditioned the stiffness coefficients and hence the force deflection remained stable.

REFERENCES

1. TSAI, S. W. Structural Behavior of Composite Materials, NASA CR-71 (July 1964).
2. TSAI, S. W. Mechanics of Composite Materials, Part II – Theoretical Aspects, AFML-TR-66-149 (November 1966).
3. WADDOUPS, M. E. and SHOCKEY, P. D. Strength and Modulus Determination of Composite Materials with Sandwich Beam Tests, General Dynamics Fort Worth Division (September 1966).
4. ROGERS, C. W. *et al.* Application of Advanced Fibrous Reinforced Composite Materials, Final Report Contract No. AF33(615)-3323 (September 1966).
5. LEKHNITSKI, S. G. *Anisotropic Plates*, University Microfilms, Inc. 1963.
6. BEST, G. C. A Formula for Certain Types of Stiffness Matrices of Structural Elements, *AIAA J.* **1**, No. 1 (1963), pp. 212–213.
7. BODEWIG, E. *Matrix Calculus*, North-Holland Publishing Company—Amsterdam, 1956.

FIBER-REINFORCED PLASTICS FOR HYDROSPACE APPLICATIONS

KENNETH HOM[†]

ABSTRACT

Construction of a deep-diving submersible with adequate payload requires use of materials with exceptionally high strength-to-weight ratios. By virtue of their high strength and low density, fiber-reinforced plastics show promise for this application. This paper highlights some of the significant results and problems related to use of glass-reinforced plastics as a material for pressure hull structures.

I. INTRODUCTION

Although overshadowed by the advance of and interest in aerospace technology, increasing attention is being focused on exploration and exploitation of the ocean depths. Today, man has only a limited capability with which to accomplish these desires, not only for scientific purposes but also for commercial and military reasons. With the nuclear-powered submarines of today, he can move at will in a horizontal plane, but he is restricted to limited depth. With the presently available bathyscaphs *Trieste* and *Archimede*, he can probe to great depth, but his horizontal movement is extremely limited. A prime requisite for achieving the capability of "going deep" is the development of long-endurance, highly-maneuverable, deep-diving hydrospace vehicles.

To obtain such vehicles, there is a need to develop lightweight, highly reliable pressure hulls from structural materials possessing a high strength-to-density ratio. This is not to imply that high strength and low density are the only consideration, but they are basic factors in selection and evaluation of materials for pressure hull studies. Those materials that are currently being given most consideration include the family of high-strength metallics such as steel, aluminum, and titanium alloys, and some of the high-strength nonmetallics such as glass, ceramics, and fiber-reinforced plastics. To place these materials in their proper perspective, the common factor relating the strength-to-weight characteristics for hulls of various materials to a geometric configuration for a specified design depth is the ratio of the weight of the pressure hull to the weight of sea water displaced by the submerged hull.

[†] Head, Structures Evaluation Branch, Submarine Structures Division, Naval Ship Research and Development Center, Washington, D.C.

This factor is referred to as the weight–displacement ratio (W/D). Figure 1, taken from ref. 1, shows the depth potential of near-perfect spheres superimposed on the familiar distribution curve of the ocean depths. The portion of the bars above the depth–distribution curve corresponds to hulls having a 0.5 W/D ratio, and the portion beneath the curve shows the greater depth attainable by heavier hulls with 0.7 W/D ratio. The ratios of 0.5 and 0.7 may appear to be arbitrary, but small sized vehicles can normally be designed with W/D ratios of 0.5 or less, whereas vehicle displacements become quite large as W/D ratios approach 0.7. The use of these values permits a meaningful

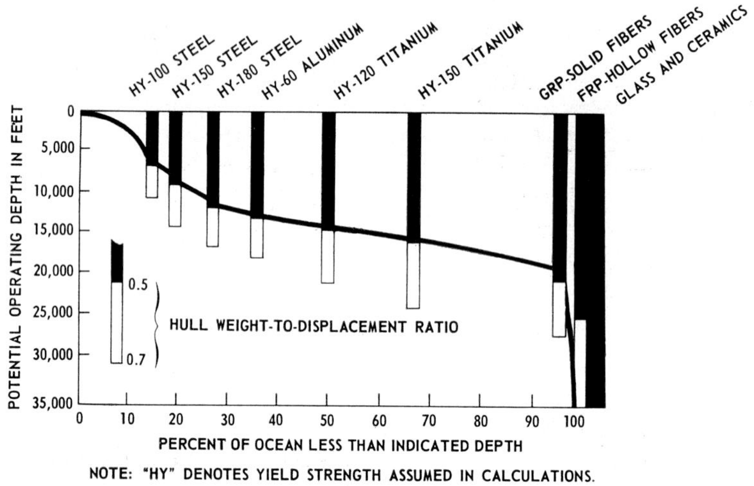

FIG. 1. Operating depth potential of spherical pressure hulls

comparison of the depth potential of various hull materials. An examination of Fig. 1 reveals that of all the metallic pressure hull materials taken into consideration, the very best results would permit operations to about a depth of 20,000 ft only at the expense of increased displacement. It is significant that the nonmetallic materials shown in Fig. 1 would permit operations to 20,000 ft and more with minimum displacement vehicles.

Development of suitable lightweight floats for deep operations would permit considerable increase in the W/D ratio of the main pressure hull. Such a combination would permit the depth potential to be greater than that shown in Fig. 1. To provide a frame of reference showing the need for lightweight floats to be used with metallic pressure hulls, information was taken from a study conducted by the Naval Ship Engineering Center to develop the curves shown in Fig. 2, showing the effect of the pressure hull and the flotation system on total weight of a vehicle designed for an operating depth of 20,000 ft. The curves of Fig. 2 are cited for a pressure hull with a volume of 382 ft^3, which corresponds to a 9-ft diameter sphere, and a maximum vehicle speed of 5 knots with an endurance of 30 hr at 3 knots. These curves represent a potential search vehicle to be developed under the Deep Submergence Systems Project. A family of curves is given which represents *W*/*D* ratios of pressure hulls fabricated from various hull materials under consideration. Cross-hatched zones are superimposed on the figure to indicate applicable regions of pressure hull materials for HY-140 to HY-220 steel; HY-105 to HY-140

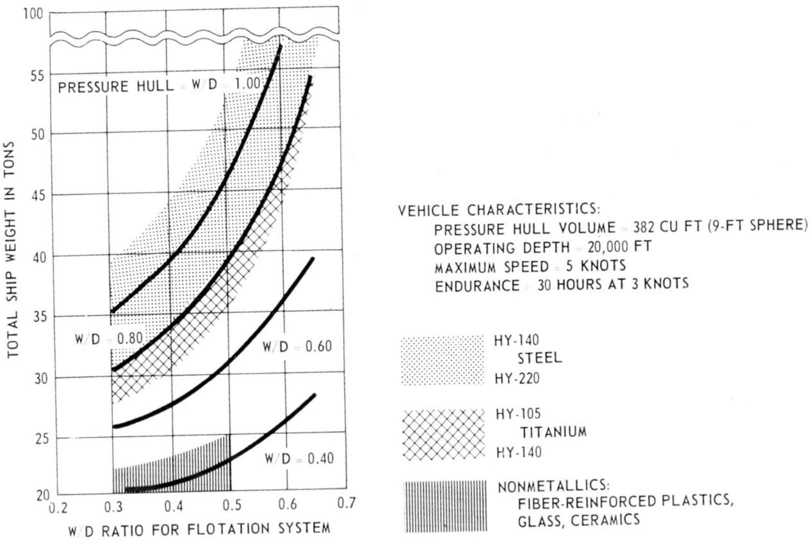

FIG. 2. Effect of pressure hull and flotation system on vehicle weight

titanium; and the nonmetallics—fiber-reinforced plastics, glass, and ceramics. Present state of the art would dictate an HY-140 steel pressure hull and from 42 to 44 lb/ft^3 of syntactic foam as a flotation system. This would result in vehicle weight of approximately 100 tons. Use of HY-110 titanium with improved foam of 37 lb/ft^3 would result in vehicle weights of about 45 tons. If consideration is given to nonmetallic spheres as floats, vehicle weight can be reduced further. Floats made of materials such as glass, ceramics, and fiber-reinforced plastics offer potential *W*/*D* ratios from 0.3 to 0.4. An HY-120 titanium pressure hull with a 0.4 *W*/*D* float would have a vehicle weight of approximately 33 tons. Further substantial reduction in vehicle weight would require using nonmetallic pressure hulls. For example, a vehicle with the aforementioned performance and a nonmetallic pressure hull of 0.5 *W*/*D* ratio would weigh 27 tons, while a 0.4 ratio would result in a weight of approximately 22 tons.

It is apparent from Figs. 1 and 2 that to achieve depths of 20,000 ft and deeper with lightweight, mobile vehicles, the designer is confronted with the need to use nonmetallic materials whether they be for the flotation system or the main pressure hull.

It is for these reasons that the United States Navy is interested in developing nonmetallic materials for deep submergence application having properties of high strength and low density. Unfortunately, these materials also offer formidable problems in applications to pressure hull structures. This paper highlights some of the results and problems related to one of the nonmetallic materials—glass-reinforced plastics.

II. STRUCTURAL BEHAVIOR OF FILAMENT-WOUND GRP CYLINDERS

Test specimens of thick-walled, unstiffened cylinders with the ends restrained by metallic end-closure plates to prevent premature instability failure have developed compressive strengths from 150 to 170 ksi and an effective modulus of 6×10^6 psi when subjected to hydrostatic pressure loading.[2, 3] These cylinders were filament wound from S–HTS glass roving preimpregnated with an epoxy resin. Similar strength levels have also been realized with flat-specimen tests of 2 : 1 orthogonal laminates under unidirectional compressive loading.[4]

In realistic pressure hull structures, a stiffening system has to be incorporated into a cylindrically shaped hull to prevent premature instability failure and thereby to assist in utilizing the high-strength properties available in the material. When GRP cylinders were stiffened by ring frames, lower strength levels were observed as a result of localized bending and shearing stresses induced by the ring frames. Shear failure occurred in the cylindrical shell at the toe of the frame. The importance of considering the shear sensitivity of GRP material can be illustrated by the test results of two identical models in which one of the models incorporated generous fillets at the toe of the frame. Figure 3 shows the configuration of these two models. By including the fillets in the structure, the weight-displacement ratio was increased from 0.524 to 0.544. The model without fillets failed at a pressure of 12,200 psi, whereas the model with fillets failed at a pressure of 16,100 psi. The higher collapse pressure of the latter model more than offset the increase in weight of 4% that resulted from incorporating fillets into the hull structure. However, effective strength levels of only 115 to 120 ksi were developed. Failures were directly related to the interlaminar shear strength of the material.[5]

The best strength-weight performances obtained to date for ring-stiffened cylinders are shown in Fig. 4 along with the results of sandwich cylinders and hollow-glass, unstiffened cylinders.

A concept offering high instability resistance and reducing shear and bending is the sandwich cylinder with a uniform core of syntactic foam. Tests have demonstrated that shells of this type can develop stress levels of 160 ksi

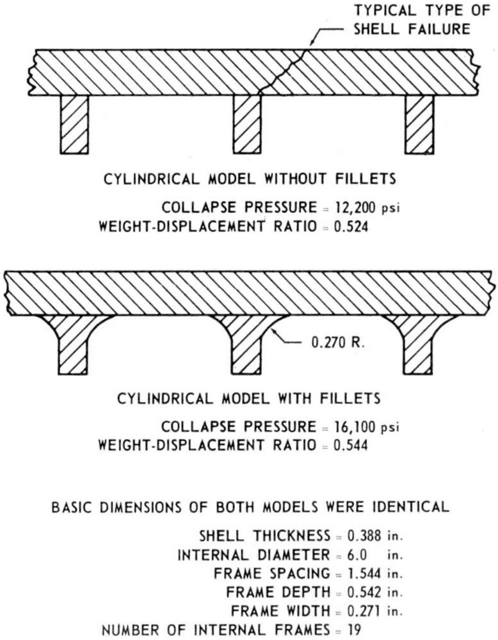

FIG. 3. Strength performance to two ring-stiffened cylinders

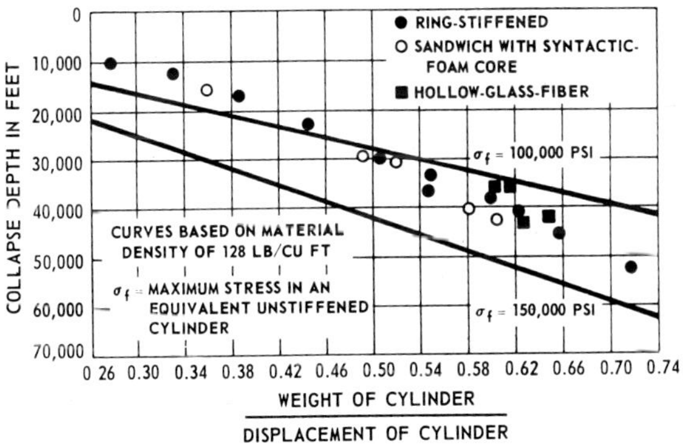

FIG. 4. Test results for glass-reinforced plastic cylinders

in the GRP facings. However, only a marginal increase in static-strength performance over the ring-stiffened cylinders was achieved because of the relatively low strength-to-weight characteristics of present foams. The results of these tests are shown also in Fig. 4.

The simplest method of reducing shearing and bending stresses is to use an unstiffened cylinder. Development of lower density, hollow fiber, glass-reinforced plastics has made this concept practical for hulls having a collapse depth of or exceeding 35,000 ft, provided that the overall length of the cylinder is four diameters or less. The increase in thickness of the cylinder afforded by the lighter weight, hollow fibers more than offsets the loss of stability due to lower elastic modulus of the composite material. Also, because of the lower rigidity of the hollow fibers, better compatibility exists between the glass reinforcement and the resin binder. To date, little or no static strength-weight advantage has been found for the hollow glass cylinder over the ring-stiffened cylinder. However, improved cyclic performance and simpler fabrication procedures are anticipated.

Still another potential method for alleviating the stiffener problem in cylindrical hulls is to use fibers with a higher modulus than glass fibers. Higher-modulus composites would inherently give rise to hull structures with higher resistance against the instability modes of failure. Lighter frames would be required. Thus, shear and bending stresses would be reduced, and higher structural strength could be obtained. Higher-modulus fibers would also lead to more efficient utilization of composite materials for sandwich hull structures. The structure would be more stable and thus less core material would be required. Also, due to the substantial difference between moduli of the facing and core materials, less load would be transmitted to the low-strength core.

Calculations indicate that a cylinder with a W/D ratio of 0.44 and made of carbon-filament-reinforced plastics with an effective composite modulus of 15×10^6 psi would not require a stiffening system. With this W/D ratio an unstiffened cylinder of semi-infinite length would have an elastic buckling depth exceeding 70,000 ft. At a depth of 40,000 ft a composite stress of 100 ksi would be developed. Present problems with high-modulus, carbon-fiber composites have been the inability of laminates to take high-compressive stresses. It is recognized that these fibers have only recently been available in the commercial market, and development of improved matrix-reinforcement, bonding efficiency, and fabrication techniques may be expected. It is projected that in the period from 1970 to 1980 glass-fiber reinforcement will probably be replaced by superior materials offering large and sudden improvements over the upper limit of properties in GRP composites. The 25×10^6 psi modulus fibers, which are now available, are expected to be the forerunner of carbon reinforcements having even higher properties. Modulus of 50×10^6 psi and tensile strengths of 400 ksi for future carbon fibers have been forecasted. Densities of composites made from these fibers would only be 85 to 95 lb/ft^3—less than 75% of the weight of GRP. These materials cannot be overlooked as future potential materials for deep-submergence application.

III. STRUCTURAL DETAILS

Studies with filament-wound cylinders have been extended to investigation of design details that arise in realistic pressure hulls such as closures, openings, and joints. Much of this work was conducted by H. I. Thompson Fiber Glass Company under the sponsorship of Naval Ship Engineering Center (formerly the Bureau of Ships).[6] Figure 5 is a drawing representing the type of structure that was investigated. The entire pressure hull was designed to obtain a collapse depth of 30,000 ft and to sustain 10,000 excursions to a depth of 15,000 ft without loss in overall strength. The basic ring-stiffened cylinder shown in Fig. 5 had a *W*/*D* ratio of 52%. It represented the lightest weight hull obtained that satisfied the strength requirements. The stiffening rings were relatively lightweight and in conjunction with the shell provided adequate resistance to premature failure due to overall instability. The ring frames utilized only 19% of the material in the region representing efficient hull design. In the region of the cylindrical hull opening, a thicker shell, larger frames, and greater frame spacing were used. The latter geometry, representing a *W*/*D* ratio of 62% and less efficiency, was selected to provide sufficient space for an opening in the shell without interfering with the adjacent frames. The shell was of S-glass filaments wound to a 4C:2L fiber distribution; the frames were of the same type of material but utilized a 9C:L fiber distribution. The end closures consisted of S-HTS glass cloth layups with a composite strength of 60,000 psi and were fabricated by a vacuum-bag molding process. Closures were attached to the cylindrical hull by adhesive-bonded lap joints. A disconnectable transverse joint was incorporated in the cylindrical hull to provide a means of access for equipment and machinery. The openings into the pressure hull, both in the cylindrical section and the closure, were reinforced by 17-4 PH stainless steel fittings designed to carry in-plane shell loads about the opening in both compression and bending. Light flanges were provided to locally support the cut-fiber ends of the shell and thereby to assist in the transfer of high-compressive bearing loads into the fitting.

One model such as that shown on Fig. 5 was tested to failure under static loading. The static model collapsed at a pressure of 12,000 psi. Two other models were subjected to 10,000 cycles to a pressure of 6700 psi and then were tested to failure under static loading, resulting in collapse pressures of 10,700 and 11,800 psi. In addition to these models, there are other similar models that are presently being subjected to 2 cycles per day for an intended period of two years to determine the response of the structure under long-term, low-cycle fatigue. It appears that the incorporation of design details is feasible if proper design procedures are employed and adequate sealing methods are utilized in the test. However, weight penalties are imposed on the overall pressure hull due to the addition of structural details. It was mentioned previously that the basic ring-stiffened cylinder had a W/D ratio of 0.52; the overall pressure hull, however, had a W/D ratio of 0.65. It is apparent that in order to achieve efficient pressure hulls, attention must be given to new concepts and approaches to obtain lightweight, watertight closures, joints, and penetrations.

Studies have revealed the importance of maintaining the integrity of a structure subjected to high-pressure water environment.[5] The use of a watertight metallic jacket is a possible method of overcoming this problem. This type of construction would permit cylinders or rings made of fiber-reinforced plastics to be placed side by side, not physically joined together, and would provide the major resistance to hydrostatic loading while the jacket would ensure watertight integrity, provide longitudinal strength, and prevent contact of sea water with the composite material. This type of construction would

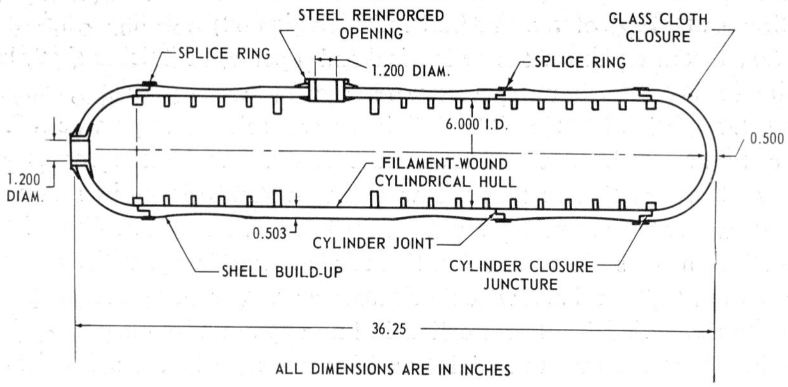

FIG. 5. Pressure hull of glass-reinforced plastic showing structural details investigated

permit the manufacturer to produce the primary elements in shorter, easier-to-fabricate sections. Construction would become a matter of merely placing the rings together and surrounding them with the metal jacket. Studies have been conducted along these lines.[7] Hydrostatic tests with 1-ft diameter models have shown that a titanium jacket provides a feasible method of protecting GRP laminates. Studies to date have been limited to models having the basic configuration of a ring-stiffened cylinder. An effective stress of 100 ksi was developed in these structures. Again, the primary mode of failure was due to the lack of shear strength in GRP materials. At one-half of the design collapse depth, the jacket was stressed to the yield strength. If it were possible to utilize the potential strength of GRP, the jacket would be required to undergo an extreme stress range due to the threefold difference in elastic modulus between GRP and titanium. This could cause a fatigue problem. A high-modulus carbon-fiber composite with a titanium jacket appears particularly attractive since it would permit stressing the primary structural elements to an ultimate composite stress without overstressing the metallic jacket. The modulus of the carbon-fiber elements would be compatible with titanium. Likewise, the shear problem would greatly be reduced as to structural details, such as metallic reinforced openings and closures.

IV. STRUCTURAL TESTS OF DOUBLE CURVED SHELLS

Although the circular cylinder stands out as one of the most practical and convenient shapes for filament-wound pressure hull structures, it might not, however, be the shape that offers the highest resistance to shear type of failure. Doubly curved shells such as spheroids and spheres would offer improved

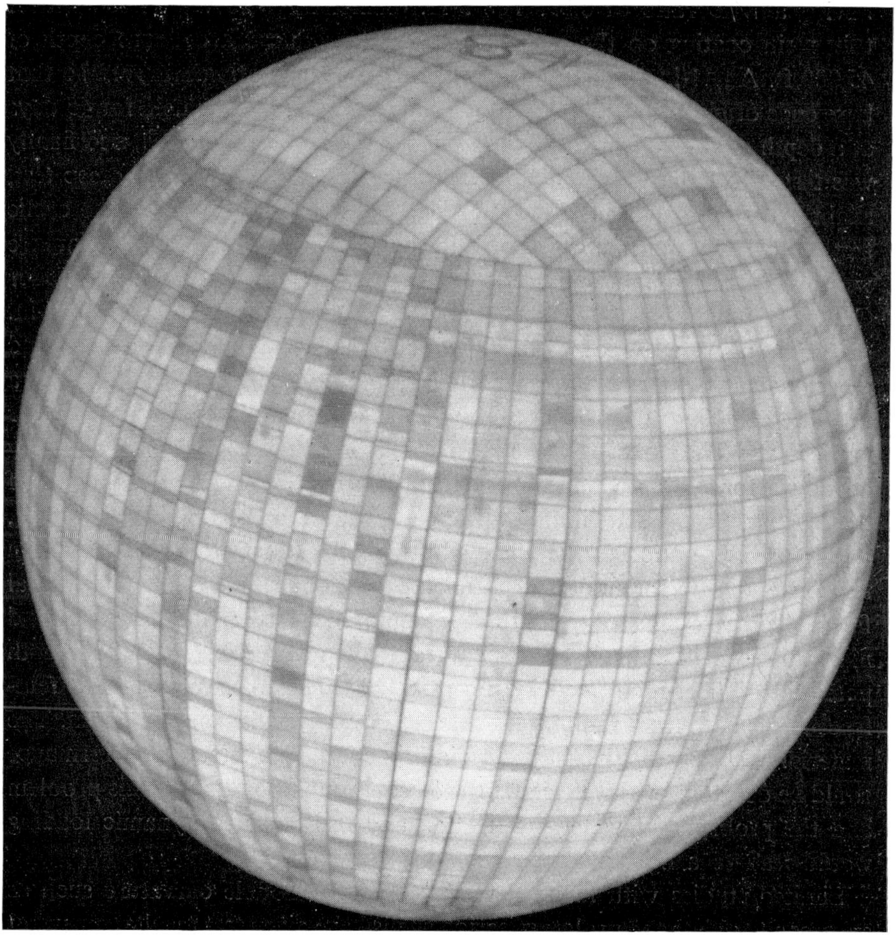

FIG. 6. Eleven-inch diameter radial fiber sphere

stability characteristics and associated improvements in shear levels. Also, these are desirable shapes for small-sized vehicles. Extremely encouraging results have been observed in tests of GRP spheres fabricated by the United States Rubber Company in which a unique fabrication technique is used to achieve radial orientation of each fiber. The spheres are of a mosaic block construction very similar to the method used to build igloos. Small blocks cut from cured, unidirectional GRP plates are placed inside a mold with the fibers oriented in the radial direction and are then impregnated with resin and cured to form spherical segments. These segments are bonded together to form a complete sphere. The observed strength-weight characteristics of these shells have been more than double those obtained for GRP cylinders.[(8)] Figure 6 shows a photograph of one of the 11-in. diameter spheres prior to testing. Recent tests were conducted with spheres 3 in. and 11 in. in diameter having a W/D ratio of 0.39. The 3-in. diameter sphere was tested with a thin resin coating on the outer surface and produced a collapse depth of 45,000 ft. An 11-in. diameter sphere with the resin coating removed so that bare ends of glass fibers were exposed to water pressure was held for 45 days at a depth of 26,000 ft, then it was cycled 5000 times to 22,500 ft, and finally was tested to failure. The collapse depth was 56,000 ft. After the creep test, slight extrusions of resin along the bond lines were noted. After the cyclic test, an original hairline crack of 2 in. grew only 3/4 in. No other damage was observed. The fabrication methods employed by United States Rubber Company can realistically be projected to larger diameter spheres. A 32-in. diameter sphere was recently proof-tested at the Naval Ship Research and Development Center to a pressure of 10,000 psi and then was subjected to a pressure of 9000 psi for 30 hr. This sphere is presently being scheduled for cyclic tests. The concept of radial-fiber-GRP spheres appears to be very attractive for use as a flotation system. The high-strength performance of these spheres can be attributed to lack of shear and bending actions. Also, due to the orientation of the fibers and the type of loading, the fibers are in tension and therefore not susceptible to local buckling of the individual fibers. These results serve to further demonstrate that hull concepts and material disposition play a very important role in realizing the full strength potential of fiber-reinforced plastic materials. The introduction of a reinforced opening into a radial fiber sphere is a very challenging problem. The material is shear sensitive in the radial direction. The strength performance could be defeated by the effects of the discontinuity. Solution to this problem and the problems of water penetration and resistance to dynamic loading would lead to a lightweight primary pressure hull.

Limited studies with filament-wound shells of double curvature such as spheres and oblate and prolate spheriod shells have given test results somewhat less than desired. To filament wind these structures, a helical winding pattern is generally used. In addition to being highly anisotropic and thus less tractable analytically, shells with crossed-over fibers have lower glass compaction, more resin-enriched regions, prone to glass failure at the crossover points, and have higher void content. New and unique methods of fabrication for doubly curved shells without crossed-over fibers are needed. The use of a "shingle method" of fabrication appears to be very attractive, and work in this direction has recently been initiated. The method is akin to a "building-block" procedure in which preimpregnated, pressure-compacted tiles of preoriented glass fibers are placed over a mold to form the pressure hull. The method offers the possibility of molding in lightweight GRP reinforcements around openings as well as a conceivable approach to fabricate doubly curved shells and nested spheres.

V. CONCLUSIONS

The full strength potential of glass-reinforced plastics has been demonstrated with spheres fabricated with fibers oriented in the radial direction. Although similar strength levels have not yet been realized with filament-wound structures, there is considerable potential in use of these materials for deep-submergence application. Since the dominant mechanism of failure of laminated-GRP structures can be attributed to shear, there is a continuing need to develop concepts and structural details which would lend themselves to these materials. It is recognized that composite materials are still in early stages of development for submarine hull application. With the development of improved matrices, reinforcements, matrix-reinforcement bonding, and fabrication techniques, markedly superior strength-weight characteristics can be expected for future pressure hull structures.

REFERENCES

1. KRENZKE, M. A. Structural Aspects of Hydrospace Vehicles, AIAA Paper No. 65-575 presented at the AIAA Propulsion Joint Specialist Conference, Colorado Springs, Colorado (14–18 June 1965).
2. MYERS, N. C. and FINK, B. Filament-Wound Structural Model Studies for Deep Submergence Vessels, *Naval Engineers Journal* (April 1965).
3. HOM, K. *et al.* Elastic Material Constants of Filament-Wound Cylinders Fabricated from E-HTS/E787 and S-HTS/E787 Prepreg Rovings, David Taylor Model Basin Report 1823 (February 1966).
4. Mechanical Properties of Orthogonal Filament-Wound Plastic Laminates, U.S. Naval Applied Science Laboratory Project 6189, Technical Memorandum 6 (17 June 1964).
5. HOM, K. and COUCH, W. P. Investigation of Filament-Reinforced Plastics for Deep-Submergence Application, David Taylor Model Basin Report 1824 (November 1966).

6. Myers, N. C. *et al.* Investigation of Structural Problems with Filament-Wound Deep Submersibles, H. I. Thompson Fiber Glass Company, Final Report, Bureau of Ships Project S-F013-05-03, Task 1025, Contract Nobs 88351 (January 1964).
7. Proffitt, J. L. Hydrostatic Pressure Tests of Glass-Reinforced Plastic Cylinders with a Titanium Jacket, David Taylor Model Basin Report 1851 (January 1965).
8. Couch, W. P. Hydrostatic, Creep, and Cyclic Tests of Radially Oriented Glass-Fiber Reinforced Plastic Spheres, David Taylor Model Basin Report 2089 (September 1965).

PROBLEMS IN THE DESIGN OF JOINTS AND ATTACHMENTS

K. R. BERG

Whittaker Research and
Development Division
San Diego, California 92123

I. INTRODUCTION

A primary consideration for the design of structures using composite materials is the stiffness and strength evaluation of these composite materials in sheet form or in filament-wound cylindrical, spherical, or structural shapes. However, an equally important parallel requirement for the complete design of practical structures is the development of attachment methods, joint designs, and the problem of load introduction in composite structures.

Due to the filament matrix characteristics in composite materials, the attachment by means of bolts, screws, and rivets is not as simple as for isotropic materials. Typically, the basis for this is the weakness in planar shear, bearing, transverse tension, and interlaminar shear. This is especially true for advanced high-modulus composites in polymeric matrix systems where modulus ratios and strength ratios may be as high as 120 : 1 between the filament and the matrix.

Due to the continuum load transfer associated with adhesive bonding, this method of attachment for joints is favored in composite materials. However, even in adhesive-bonded structures, due to anisotropic properties, care must be exercised to minimize stress concentration problems. Consideration must be given to special design features to overcome certain composite weaknesses.

II. MECHANICAL JOINTS

A. *Attachment by Means of Bolts, Screws, and Rivets*

1. PINS AND BOLTS. A logical carry-over from present aircraft construction is the assembly of joints in composite materials through the use of bolts, screws, and rivets. Composite materials do have some capability to withstand loads introduced by this type of mechanical attachment; however, unlike isotropic materials, the design of the material itself strongly influences the allowable

load transfer which this type of joint permits. For example, the simplest orientation, unidirectional filaments, with its low inherent inplane shear strength results in a shear tear-out mode of failure. An improvement for this mode of failure would be the use of $\pm 45^\circ$ filament orientation, but this results in typically low net tension capability. The use of a so-called isotropic filament orientation, 0°, 60°, 120°, combines improved shear strength and tension strength, but at the same time reduces considerably the efficiency of the material. Figure 1 shows a typical failure for a near unidirectional laminate loaded by pin bearing. Figures 2, 3, and 4 show representative failure for $\pm 45^\circ$,

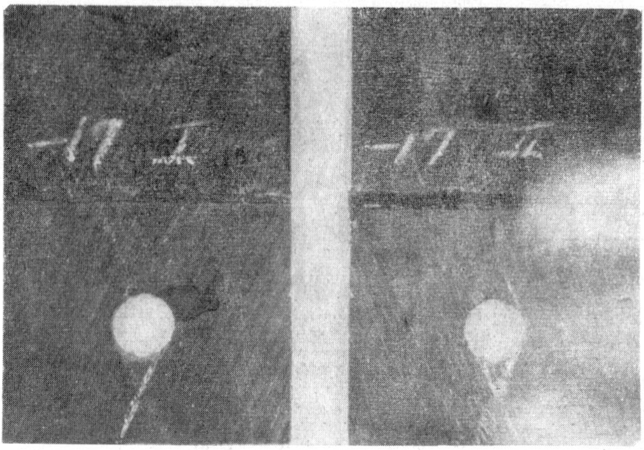

FIG. 1. Failure mode for non unidirectional filament orientations

quasi-isotropic, and 60° respectively. Comparison of typical bearing strength (regardless of failure mode) with specific modulus and comparison of specific strength with specific modulus for boron epoxy is given in Fig. 5.

Based on the test results of refs. 1 and 2 presented in Table 1, it is apparent that composite materials from a bearing or load sustaining standpoint can be used with mechanical attachments, provided the appropriate bearing stresses are considered. However, one must consider, in the design of mechanical joints with composite materials the effects of discontinuous fibers when penetrations are made in a sheet. If the matrix were of shear stiffness and shear strength similar to that of an isotropic material, the shear lag effects would result in a stress distribution shown by the stress trajectories for an isotropic material in Fig. 6. This distribution is a result of the compatibility and equilibrium expressions in a two-dimensional analysis. Stress concentrations exist at the edge of the hole, and if the material is a ductile material, these stress concentrations can be minimized through yielding. However, in the case of a filamentary material with a brittle filament, yielding parallel to the fiber can only occur through the breakage of fibers and subsequent

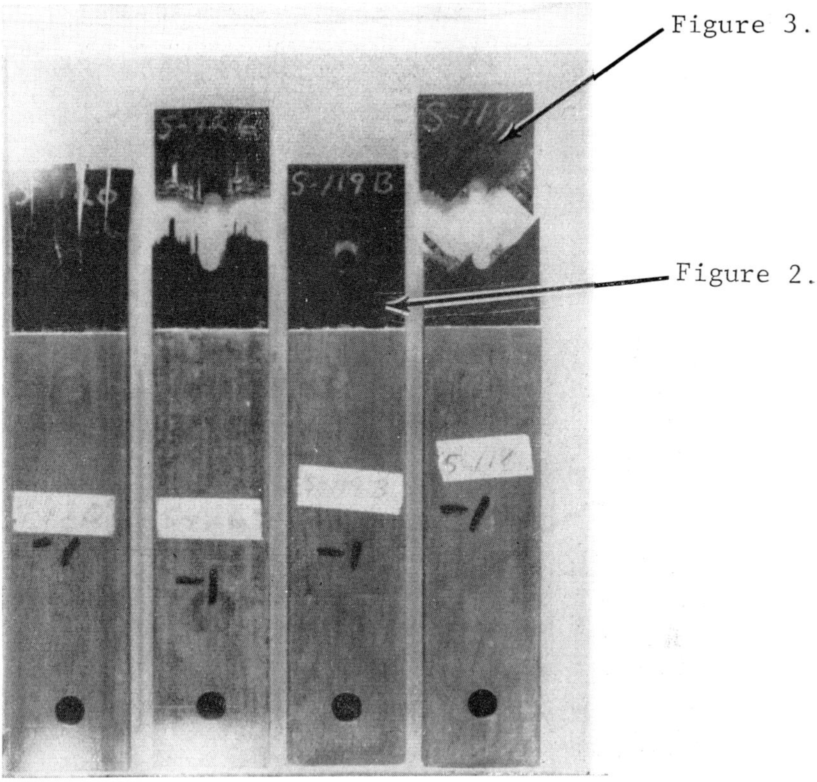

FIGS. 2 and 3. Failure mode for various filament orientations

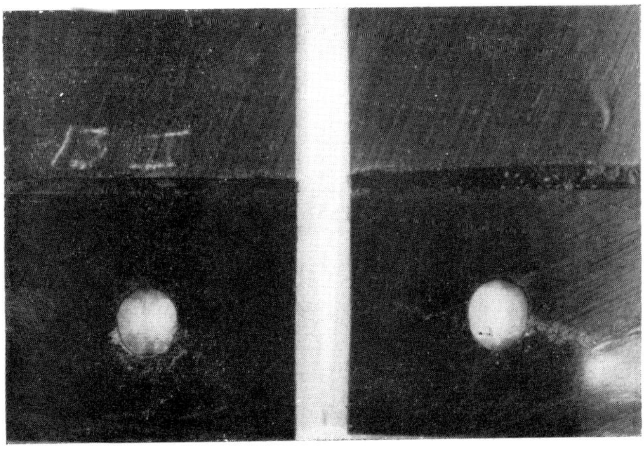

FIG. 4. Failure mode for $\pm 60^{\circ}$ oriented doublers

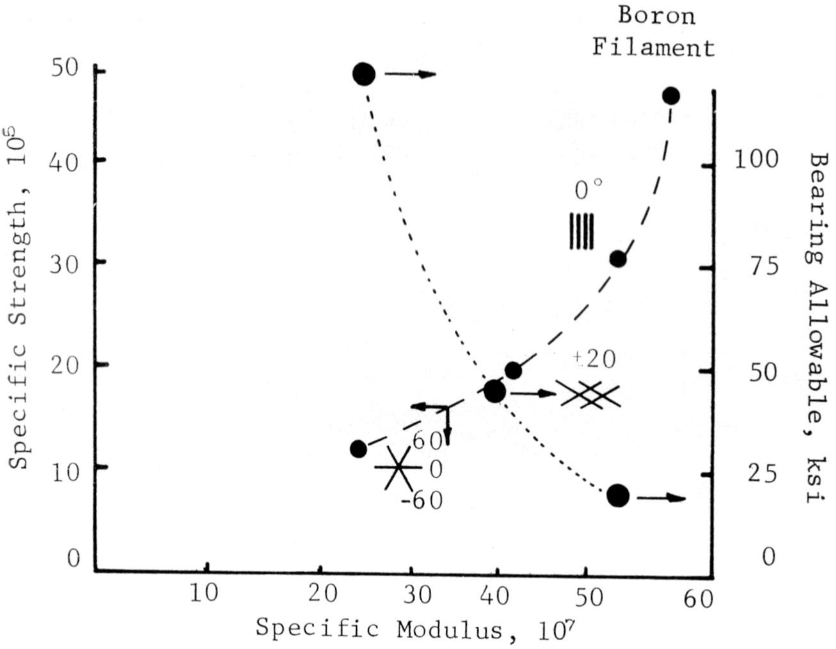

FIG. 5. Bearing strength versus configuration

TABLE 1. BORON LAMINATE—BEARING TEST DATA

No. tests	Nominal thickness in.	D/t ratio	e/D ratio	Layup	F_{bru}, ksi (avg)
3	0.135	1.8	4.8	0°/+60°/−60°	118.0
2	0.084	3.0	1.5	0°/90°	36.0
6	0.068	4.0	1.5	0°/90°	33.5
12	0.052	4.8	1.5	45°/45°	48.0
2	0.072	6.9	2.0	+25°/−25°	35
				+65°/−65°	
2	0.075	6.8	2.0	+25°/−25°	24

shear transfer in the resin. Thus, in both isotropic and filamentary structures a degree of yielding can occur. The magnitude, however, of the resulting peak shear stress is such that permanent weakening can occur in a composite material with penetrations. This is particularly significant when cyclic loading is present. Thus, although an acceptable bearing stress may be present, unacceptable stress concentrations or loss in fatigue life may result in poor efficiency for this type of joint. A considerable amount of work is being done and needs to be done to fully define stress concentrations, fatigue limitations, and optimum orientations for these penetrations.

Another problem in composite materials is the effect of fiber ends at the edge of the hole. This has been experienced in compression testing of unidirectional composites where careful preparation is required to prevent local failure of these fiber ends (brooming). In addition to the effect of the pin load on the fiber ends, for extremely hard filaments (such as silicon carbide, boron, tungsten carbide), the effects of these small-diameter, hard rod ends on the pin itself are unknown, particularly under repetitive loading.

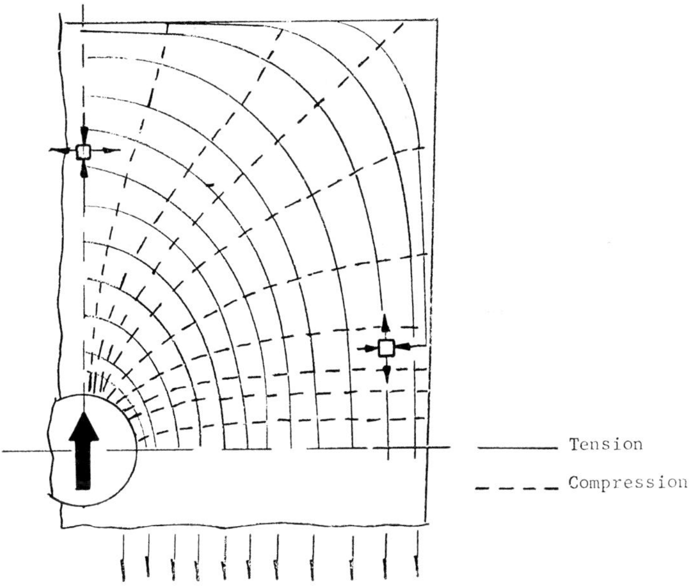

FIG. 6. Typical stress trajectories in a mechanical joint

2. RIVETS. Due to the brittle nature of the filaments, it seems unlikely that the conventional riveting process based upon impact upset could be used, even if the bearing and local stress concentration problems were acceptable.

B. *Lug Attachments*

In the previous section the discussion centered around smaller diameter attachments typically in sheet material. Another type of mechanical attachment is a gross attachment generally considered for larger diameter pins. The typical geometry as shown in Fig. 7 represents the configuration in which $W/D < 2$ and $e/D < 1\frac{1}{2}$. A lug configuration from composite materials must be designed in such a manner that the various high strain points will be simultaneously minimized. The direct bearing, shear tear-out, net tension, and hoop

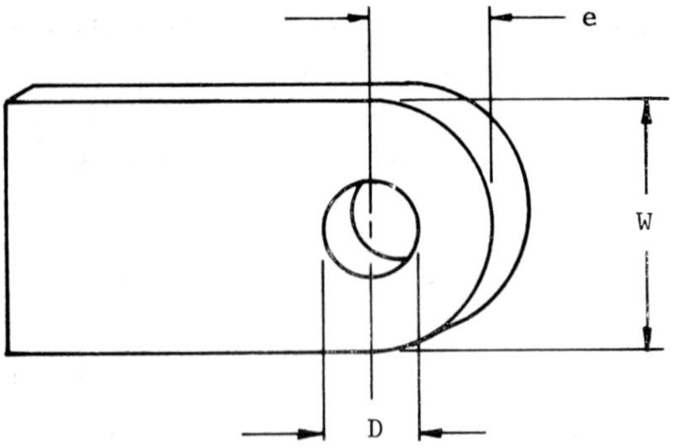

FIG. 7. Lug

tension or transverse tension must all be considered versus the filament orientation. In this case the filamentary nature may be an advantage over isotropic materials in that the filament orientation may be designed with each of these failure modes in mind.

Another approach to the large-diameter lug-type mechanical joint is through the use of metal doublers, either bonded to the surface of the composite or interleafed between the plies of the composite. This type of lug actually is a mix of a mechanical-type joint and an adhesive-bonded joint. The primary mode of load transfer to the composite is through shear as opposed to bearing on the composite plies. An example of this type of configuration is shown in the root end test specimen taken from ref. 3, Fig. 8.

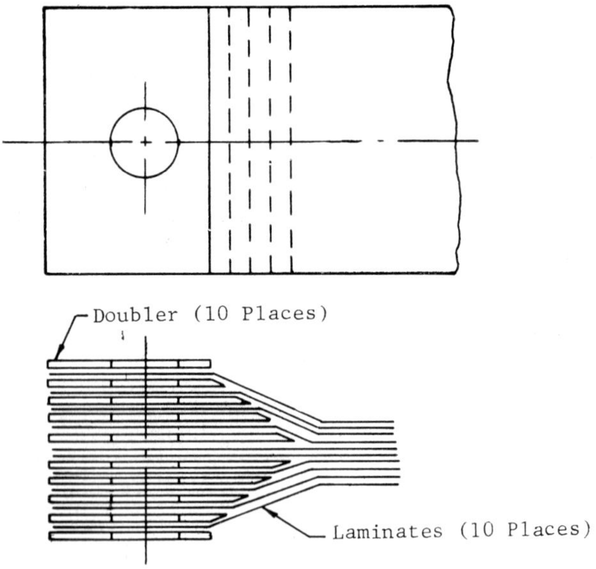

FIG. 8. Joint test specimen

Lugs also present local bearing problems and local stress concentrations which may, through the use of properly oriented filamentary materials, provide a satisfactory lug-type joint. However, in general, direct material substitution from homogeneous material lug design to a composite lug is not acceptable due to the previously mentioned weaknesses in composite materials.

III. ADHESIVE-BONDED JOINTS

A. *Background*

Many of the problems associated with load introduction through mechanical joints are eliminated in an adhesive-bonded joint. However, two problems still exist, which are important to investigate. One is the continuing problem of local stress concentrations, and the other is the interlaminar shear problem. The problem of stress concentrations in adhesive-bonded joints has been studied by many investigators.[4–6] Probably the most often quoted is the analysis of a bonded joint which accounts for joint flexibility, by Goland and Reissner.[6] Although this is an elastic analysis, it does represent the trends and the approximate stress concentrations existing in an adhesive-bonded joint when subjected to various changes in parameters. These adhesive joint analyses typically are limited to the stresses, both shear and normal, occurring in the adhesive layer between two adherends considered as isotropic. Very little work has been accomplished on the analysis of the elastic/plastic stresses in a simple shear joint consisting of an isotropic adherend and a composite adherend, or between two composite adherends.

B. *Shear Transfer—Macro* (Analysis without Regard to Fiber/Matrix Construction)

If one considers the classical shear transfer case and considers the composite material as a homogeneous material of equivalent stiffness (essentially the macro approach with a low G adhesive), the shear transfer characteristics are readily determined. For a joint of the type illustrated in Fig. 9, consisting of two different layups of composite materials with modulus E_1 and E_2 and thickness t_1 and t_2, the elastic shear transfer characteristics are as shown in Fig. 10. In this case, because of the capability to orient the fibers in the composite, various combinations of E_1 and E_2 could be considered. Thus it is possible to shift the peak stresses from one end of the joint to the other, or to balance the peak shear stresses (for the case where E_1t_1 equals E_2t_2). Because of the capability to orient fibers (adjust values of E), it is possible, with additional versatility, to reduce peak shear stresses in the design of joints with composite materials.

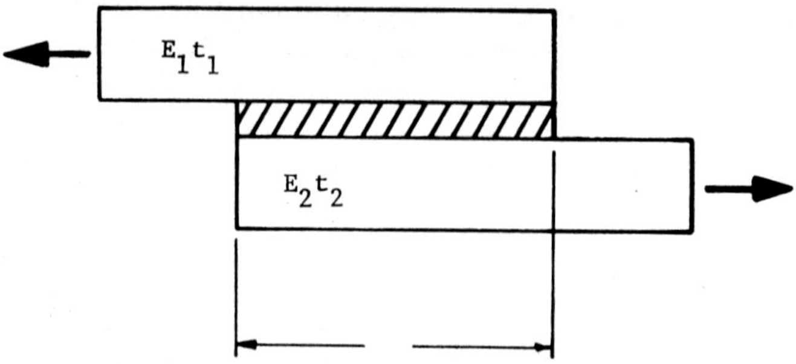

FIG. 9. Simple lap shear model

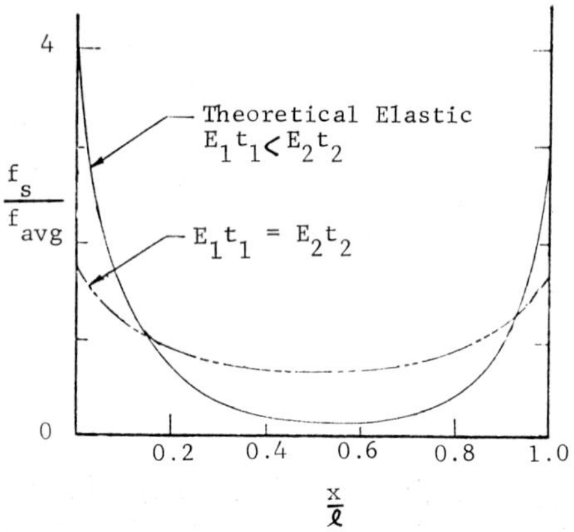

FIG. 10. Shear stress distribution

Since resin materials make good adhesive systems also, adhesive joints made from composite materials often are joined by the use of the matrix material itself. In this case, the assumption that the adhesive is highly flexible with respect to the composite resin is no longer valid. Here, the so-called macro approach of considering a homogeneous adherend must be replaced by an analysis which considers the effect of the various matrix layers between filaments as flexible, a quasi-micro approach.

C. *Shear Transfer—Micro* (Consider Laminate Construction of Material)

In the case of joint systems which consist of adherends adhesive-bonded to composite materials, in which the adhesive stiffness is of the same stiffness as the composite matrix layers, the analysis is considerably more complex. Consider for example the simple double-shear joint shown in Fig. 11. Assume a homogeneous adherent is bonded to a multiple-layered composite material. To simplify the analysis, a uniform shear is considered applied to the

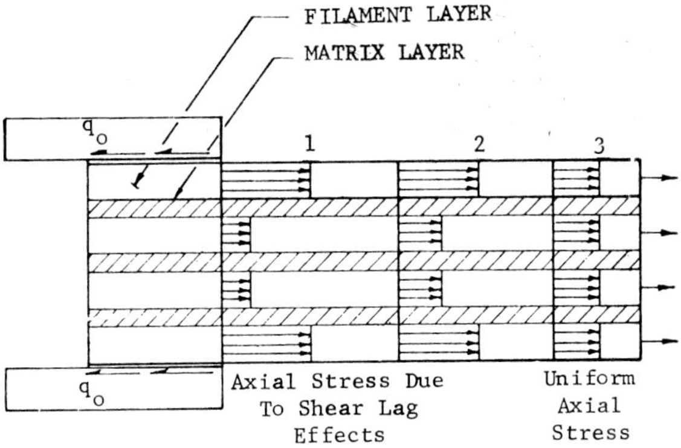

FIG. 11. Multi-ply shear distribution

composite q_0. The question to be answered is: what is the distribution of axial load in each of the plies as a function of the distance along the joint. The loading of composite materials on test specimens is often through end plates adhesive-bonded to the composite specimen. In most cases the joint failed at the end of the homogeneous adherend plates, as shown in Fig. 12. Many investigators reported this failure to be due to weak interlaminar shear characteristics of the composite. However, investigation of the distribution

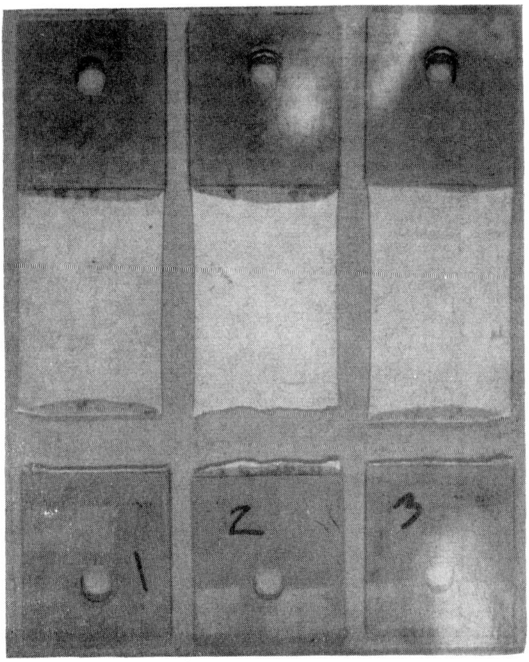

FIG. 12. Tensile failure model

of stresses shows that a higher stress in the top layers of the composite exists, and the failure is due to these stresses in the outer ply in straight tension. In no case does the interlaminar shear between plies of the composite become the critical failure mode for this type of joint. Thus, in the case of this joint the problem is to minimize the amount of axial load concentrated in the top layers of the composite.

1. ANALYSIS OF A MULTI-PLY LAMINATE LOADED IN SHEAR. An analysis was conducted in ref. 7 which resulted in the determination of the stresses in each of the layers of a multi-plied laminate loaded in shear on the outer ply. The analysis considered a model consisting of a series of alternate laminate layers of high- and low-modulus materials simulating essentially the fibrous layers and the resin layers. It was assumed that the modulus of the filamentary material was of higher order than that of the resin. The equations of equilibrium for the joint, without considering the normal stresses, are:

$$\frac{dq_i}{dx}+\frac{dq_{i+1}}{dx} = t_f\frac{d^2f_i}{dx^2}$$

where q_i and q_{i+1} are the shear flows on the ith laminate and f_i is the axial stress in the ith laminate.

The equations of strain compatibility are:

$$\frac{f_i}{E_f} = \frac{t_m}{G_m}\frac{dq_i}{dx}+\frac{f_{i+1}}{E_f}.$$

This results in a set of differential equations of the following form:

$$\frac{d^2f_i}{dx^2} = -b_1^2f_{i-1}+2b_1^2f_i-b_1^2f_{i+1}$$

where

$$b_1^2 = \frac{G_m}{E_ft_mt_f}.$$

A solution, as shown in ref. 7, results in a set of functionals which define the relationship of the stresses in each of the layers as a function of the distance along the joint. A plot of this stress for some typical examples, for a small number of layers, is shown in Fig. 13. It is shown that it is possible by a variation of certain parameters to minimize this high stress which exists in the first layer. However, it is important to point out that the stresses in the first layer are always greater than average, and that the interlaminar shear stresses are always less than the shear stresses which are transferred from the adherend through the adhesive to the composite. Thus, if the resin system used in the composite is used as an adhesive, it would be difficult for the joint to fail between layers by interlaminar shear.

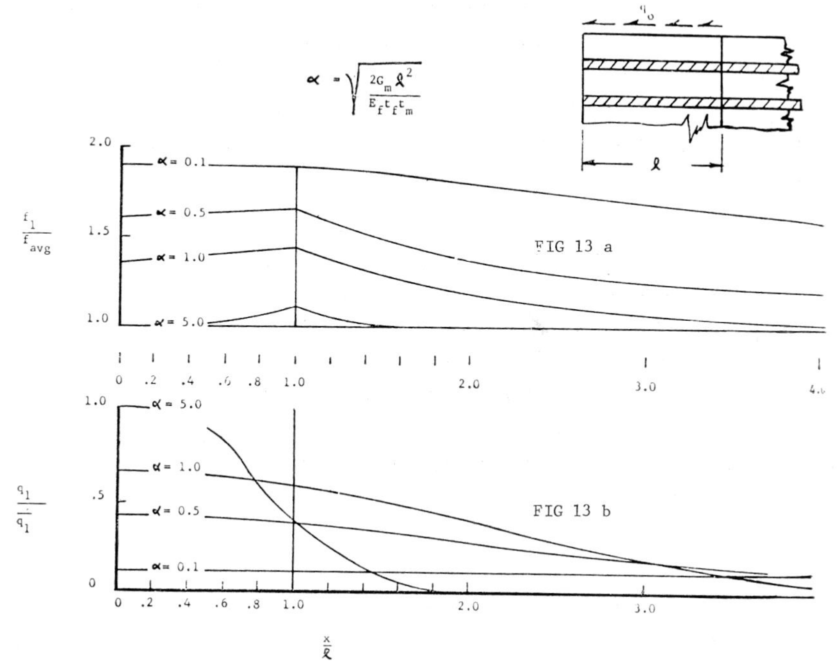

FIG. 13. Interply shear transfer

2. INTERLEAVED DOUBLER JOINTS. Another means of minimizing the amount of peak stress occurring in the first layer of an adhesive joint is to interleave the joint with the adherent material. For a high number of plies, interleaved homogeneous adherends distribute the load to each of the layers more directly. For example, Figs. 14 and 15 illustrate the two ends of a simple lap joint. This configuration was tested in tension and, as expected, resulted in failure at the non-interleaved end at approximately 83% load capability of the net section of the composite. The failure occurred as shown in Fig. 16. This test demonstrated on a single specimen, with the conditions exactly repeated end for end, that the interleaving of homogeneous metal materials

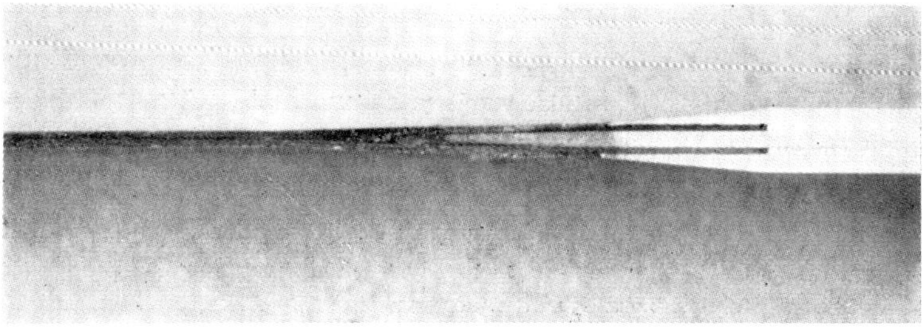

FIG. 14. End joint configuration 1

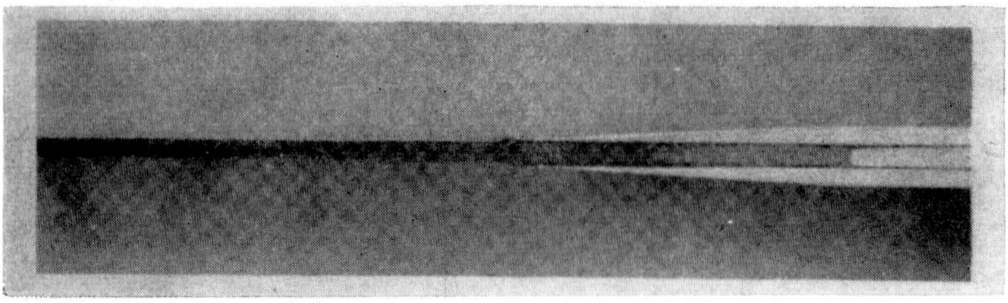

FIG. 15. End joint configuration 2

FIG. 16. Failure mode in tension

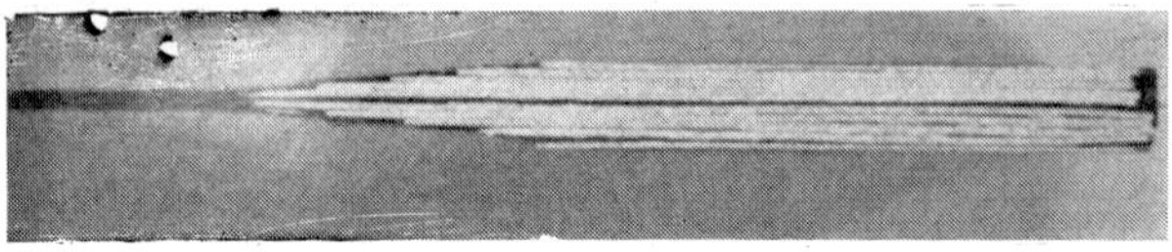

FIG. 17. Multi-ply end joint configuration

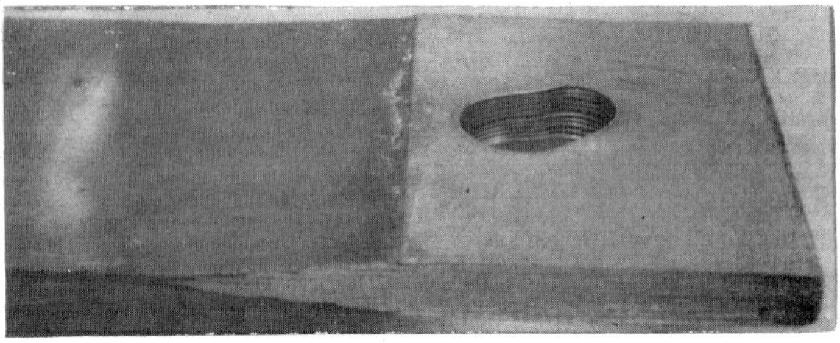

FIG. 18. Bearing failure mode

does decrease the stress concentration, resulting in an improved strength joint. A further increase in interleaving is shown in Fig. 17. Results of this test are given in Fig. 18, which demonstrated bearing failure and a complete lack of a capability to fail through the joint end. Since one of the major uses for lap shear joints is anticipated to be the joining of composite skins in aircraft structures, and since typical skins would be three to four plies thick, the interleaving for lap shear joints of this nature probably would not be required or economically justified.

IV. SUMMARY

In summary, it can be stated that, of the two methods of joining composite materials, the use of pins, rivets or bolts, or the use of adhesive bonding, the adhesive-bonded joint will provide a better quality and more uniform distribution of load introduction to composite materials. There is a place, however, for bolted or pinned joints in composites. More care must be taken in their design.

The primary areas that must be considered in the mechanical joining of polymer matrix composite materials are:

1. The effect of weak interlaminar shear in multi-plied laminates.
2. The effect of weak inplane shear and weak transverse tension on unidirectional composites.
3. The effects of cut fibers causing additional resin stress concentrations.

In the case of adhesive-bonded joints, shear stress concentrations do exist as well as the increased stresses in the adjacent plies, and these must be considered.

We do not have to be afraid of the joining of composite materials either by pins and bolts or by adhesive bonding. However, we must carefully analyze these joints and appropriately predict the allowables of these joints. In addition, further efforts are required in design and analysis for composite joints. Consideration of only homogeneous-type joint designs can lead to additional serious problems in composite material joints.

REFERENCES

1. General Electric Monthly Progress Report No. 2, Contract AF33(615)-3279, December 1965.
2. Whittaker Corporation, Narmco Research & Development Division, Monthly Progress Report No. 6, Contract AF33(615)-3288, March 1966.
3. Filippi, F. J., Keogh, J., Brison, R., Dodd, R., Berg, K. R. *et al.* Application of Advanced Fibrous Reinforced Composite Materials to Helicopter Rotor Blades, Technical Report AFML-TR-66-209, August 1966.
4. Volkersen, O. Shear Stress Distribution in Bonded Joints, *Journal Energie & Technik*, **5** (1953).
5. Matting, A., and Ulmer, K. Stress Distribution in Metal-Bonded Joints, *VDI-Z* (English text), November 1963.
6. Goland, M. and Reissner, E. The Stresses in Cemented Joints, paper presented at the Annual Meeting of ASME, 1943.
7. Berg, K. R. Analysis of Axial Stresses in a Multi-Ply Laminate Loaded in Shear on the Outer Ply, Whittaker Corporation, Narmco Research & Development Division, Technical Note 67-TN-61, 11 April 1967.

EVALUATION OF THE POTENTIAL STRUCTURAL PERFORMANCE OF COMPOSITES

L. A. RIEDINGER, M. H. KURAL and G. W. REED, JR.

Lockheed Missiles and Space Company, Sunnyvale, Calif.

I. INTRODUCTION

Within a material and structure development program hopefully leading to an optimum structural design—whatever it may include—care must be taken on an engineering basis to realize the advantages claimed by efficiency parameters and minimum weight analyses. This paper attempts to show potential performance by means of optimistic mechanical properties expected in the future for various unidirectional composites and their effects on the total composite are emphasized. This analysis indicates that the need remains for better analytical methods, assumptions, approaches, testing, and judgement.

In Fig. 1 an example is shown of a composite containing non-uniform fiber distribution with voids in the matrix material and poor interfacial bonding which is a typical present-day construction. Also shown in Fig. 1 is a eutectic composite which is virtually free of voids and has a hexagonal pattern of fiber arrangement with excellent interfacial bonding. This is an example of the desirable qualities of the future composites; however, an increase in percent volume of fibers is a requirement that must be met.

The primary components treated herein are the cylindrical shell and a tapered box beam since results of these can be qualitatively extended into a broad category of applications. Beginning with properties of various fiber and matrix materials, unidirectional composite properties are calculated. Mechanical properties for various geometric arrangements of multiple layers are obtained by established methods. The effects of geometric arrangements of layers on axial and lateral buckling loads are obtained for comparison. Variations of E_L, E_T, and G_{LT} of the unidirectional composite that are associated with the various methods of micromechanics have a marked effect on the buckling allowable for a cylinder constructed of multiple layers at selected orientations. Accuracy of the macromechanical properties from micromechanics is all important for the final mechanical structural properties.

The buckling investigations of this paper have been based on two different Lockheed methods. In the first method[1] the constitutive relations considered

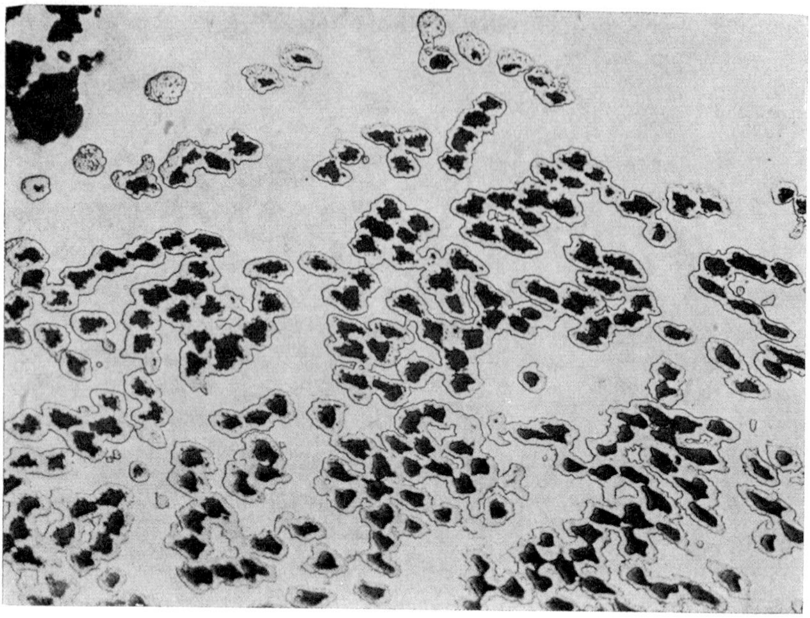

FIG. 1A. Chromized graphite cloth infiltrated with alloyed Au-10 Cr

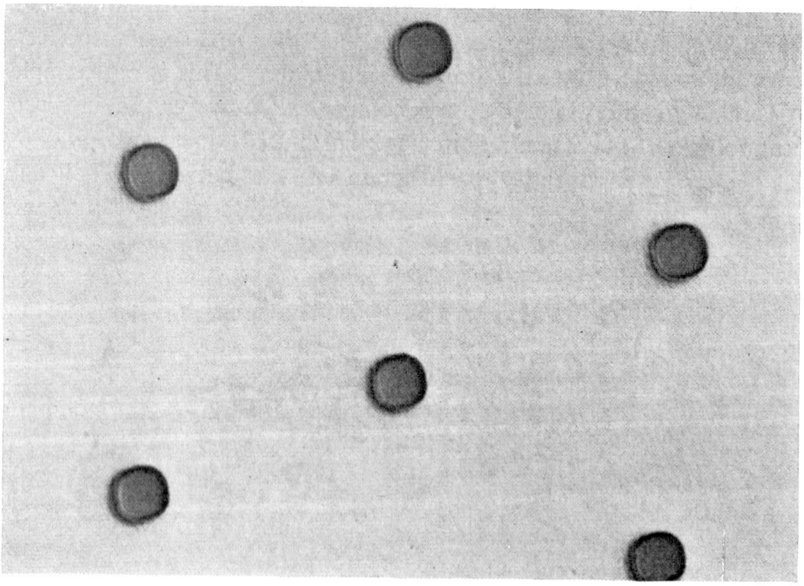

FIG. 1B. Filamentary structure, Mg-Mg_2Si eutectic

are sufficiently general to include shells which are made up of orthotropic layers. The stress–strain relationships are based on the Reissner–Stavsky method for so-called aeolotropic plates. General anisotropy would be another appropriate term. The derivation of the relations as well as the general expressions for elastic constants are given in Appendix A and reported in ref. 2. The second method[(3)] is an orthotropic approach where the shell wall is considered to be homogeneous through the thickness with overall structural properties. The pertinent equations for the buckling analysis are outlined in Appendix B.

A tapered honeycomb sandwich composite wing box beam section is also analyzed using shear and bending loads and dimensions typical of current aircraft. The applicability and the significant weight saving of composite systems is demonstrated. The fastener problem will also be discussed. Spherical pressure hulls will also be treated as will size effect.

The intent of this paper is not to present the development of the analyses methods used in reaching the results and conclusions but rather to indicate some composite potentials including the problem areas. Perhaps some of the results can assist in forming a sense of values for the determination of emphasis ratings for follow-on investigations.

II. DISCUSSION

The composite field is very broad and is comprised of a number of different composite systems including continuous filament composites, whisker composites, eutectic or directionally solidified composites, and composites formed by powder metallurgy. Each of these composite systems has specific advantages and potential applications. Of the many potential composite systems a few appear most promising on the basis of current developments. Some of the more promising composite systems are considered in detail for several applications.

Continuous filament composite systems appear attractive because the continuous filaments provide structural continuity and can be directionally oriented to maximize the structural efficiency. Continuous filaments of high strength, high modulus, and low density are becoming available, providing composites with high specific strengths and moduli relative to metallic engineering materials. Continuous filaments can develop near maximum fiber strength in relatively low-strength matrices because long fiber lengths are available and/or possible. Continuous filaments are easily fabricated into structural components by filament winding or woven fabric lamination processes and are further advanced than the discontinuous whisker composite systems.

Whisker composites appear attractive because of the very high fiber tensile strengths approaching the theoretical maximum of the materials. It would appear that whisker composites have potentially higher specific tensile strength than other composites. However, present fabrication and metallurgical problems have limited the effectiveness of whisker fibers. Currently produced whisker composites have significantly lower specific strengths compared to continuous filament composites. Another difficulty with whisker composites is the problem of orientation. Various processes for alignment of fibers are under study and have had some success. The formation of whisker yarn provides a means of aligning whiskers by winding the yarn. However, whisker alignment at high volume fractions is difficult; also, considerable damage to the fibers may result in compaction. It appears likely that whisker composites will be limited to lower volume fractions than can be obtained with continuous filaments. In the predictions of whisker composite strengths it is assumed that 50% volume fractions can be achieved. This may be somewhat optimistic.

Unidirectionally solidified eutectic composites appear attractive because of the relative ease with which a composite can be prepared. No separate process is needed to manufacture the filaments and special handling problems associated with the filaments are avoided. A good interfacial bond between fiber and matrix is automatically obtained. A primary structural advantage of eutectic composites is their excellent thermal stability which suggests high temperature applications as the area most suitable for exploitation.

A primary disadvantage of the eutectic composites is the volume fraction of the high strength phase which is fixed by the uniqueness of the eutectic composition. Generally these volume fractions are relatively low. However, investigators are studying the possibilities of increasing the volume fractions. Another disadvantage is that the technique is currently not capable of producing specimens with anything more than the most simple geometric shapes, such as rods and plates. However, more recent work gives promise that more complex casting shapes may eventually become available.

In powder metallurgy composites generally a material of high strength or high modulus or both is combined with a lower strength, lower modulus material to produce a composite with a compromise in modulus and strength but with better elongation or impact strength or other properties that may be desired. An example of such a composite is Lockalloy which is about 62% beryllium and 38% aluminum. It produces a composite with a modulus, tensile ultimate and yield strength that closely follows the "Law of Mixtures" and has improved elongation facilitating fabricability, and thereby reducing fabrication costs as compared with beryllium. It has been shown(4) that Lockalloy is far superior to common engineering materials such as aluminum on a weight and cost basis and is nearly equal to beryllium on cost effectiveness without consideration of the advantages of Lockalloy in fabricability.

Composites formed by powder metallurgy may prove to have specific advantages over filamentary composites for specific applications. Generally these materials can be treated in fabrication processes the same way as the common metallic materials. Forming methods and attachment methods would be similar except for some of the restrictions on bend radii, bearing strength, etc., which may be slightly different. In cost effectiveness comparisons of composites the advantages of using existing fabrication techniques with powder metallurgy composites may far outweigh any specific strength advantage of filamentary systems.

III. MECHANICAL PROPERTIES

Several methods of prediction of the elastic properties of unidirectional composites have been proposed,[5–11] but the relative merits of these methods are not readily apparent. Detailed study of the relationships and comparison to actual composite materials is required for a final determination of which methods are most applicable and what additional theoretical development work is required.

It is also not the primary purpose of this paper to evaluate the micromechanical methods available, but rather to evaluate the relative potential performance of total composites using calculated values of moduli and strength by means of refs. 5–12. The resulting basic properties are listed in Tables 1–4.

The composite tensile strengths were computed using the law of mixtures relationship

$$\sigma_c = v_f\sigma_f + v_m\sigma_m^* \tag{1}$$

where σ_m^* is the stress in the matrix at the ultimate strain of the filaments and σ_f is defined by the average tensile strength of the filaments. This has been shown to be in reasonable good agreement with test results.[13–15] However, other investigators have shown σ_f to be better defined by the "bundle strength"[16] and in some experiments, synergistic effects were noted where improved effective fiber strengths resulted in the composite.[17]

The whisker composite tensile strength is also computed by the "law of mixtures"; however, the short lengths of whisker fibers reduce their effective strength as shown by the following modified "law of mixtures" equation.[13]

$$\sigma_c = \sigma_f v_f\left(1-\frac{1}{2\beta}\right)+\sigma^* m(1-v_f) \tag{1a}$$

where $\beta = L/L_C$,

$L_c = \sigma_f D/2\tau$,

L = length of fibers (in.),

D = diameter of fibers (in.),

τ = shear strength of matrix (psi).

TABLE 1. EXAMPLES OF POTENTIAL FIBER MATERIALS

Material	Specific weight lb/in^3	F_{tu} psi$\times 10^3$	E psi$\times 10^6$	F_{tu}/ϱ in.$\times 10^6$	E/ϱ in.$\times 10^6$	Temp. limit °F
Continuous filaments						
Glass	0.090	750–1000	15–20	8.35–11.1	167–222	1000–2500
Boron	0.085	500	60	5.9	706	2000
Silicon carbide	0.119	500	70	4.2	590	2500
Boron carbide	0.090	200	70	2.22	775	3000
Boron nitride	0.065–0.08	200	10	2.46–3.1	123–154	3000
Beryllium	0.066	200	44	3.03	667	1200
Carbon	0.054	200	50	5.5	925	6000
Whiskers						
Al_2O_3	0.144	2000–3800	40–76	26	528	2000
SiC	0.115	1500–3500	71	30	617	2500
Graphite	0.049	2800–4000	82–102	81	1650	6000

TABLE 2. EXAMPLES OF POTENTIAL MATRIX MATERIALS

	Specific weight lb/in.^{3}	F_{tu} ksi	F_{cy} ksi	F_{su} ksi	E psi$\times 10^6$	G psi$\times 10^6$	Temp. limit °F
Organic resins							
Epoxy	0.04–0.05	4–13	15–24	~6.0	0.50	0.185	250
Phenolics	0.050	6–9	12–15	~5.0	0.40	0.148	500
Polyimides	0.051	10–12	24	~8.0	0.45	0.167	800
Metallics							
Aluminum	0.100	60	55	36	10.6	4.0	375
Magnesium	0.064	35	19	19	6.5	2.4	350
Titanium	0.160	160	154	100	16.0	6.1	1200
Molybdenum	0.369	110	104	80	46.0	17.0	3000
Ceramics							
Silica	0.115	~12	~12	~6	10–20	5–10	3000
Alumina	0.125	~20	~20	~10	30–50	15–25	3000

TABLE 3. ESTIMATED POTENTIAL MECHANICAL PROPERTIES OF SELECTED UNIDIRECTIONAL COMPOSITES

Material	Specific weight lb/in³	Longitudinal tensile strength ksi	Transverse tensile strength ksi	Longitudinal compressive strength ksi	Transverse compressive strength ksi	Longitudinal stiffness psi × 10⁻⁶	Transverse stiffness psi × 10⁻⁶	Shear stiffness psi × 10⁻⁶
Filament composite†								
Boron-epoxy	0.076	350	7	325	15	42	2.8	1.7
Boron-Al	0.094	324	60	396	40	40	26.0	10.9
Beryllium-Al	0.079	132	60	302	40	30	23.9	10.8
Silicon carbide-Al	0.111	324	60	460	40	46	27.0	11.1
Silicon carbide-Ti	0.135	364	160	590	170	48.6	36.7	13.7
Boron carbide-Al	0.094	222	60	460	40	46	27	11.1
Carbon-epoxy	0.051	245	7	300	15	30	2.8	1.7
Whisker composites‡								
Al_2O_3-Al	0.122	235–1740	60	380	40	34	23.7	7.34
SiC-Al	0.107	216–1600	60	356	40	32	23.2	7.47
Graphite-Al	0.074	250–1830	60	410	40	37	23.8	7.73
Eutectic composites								
Ti-8.7 Si	0.16	160	80	264	70	24.5	16.0	6.1
Al-33 Cu	0.135	49	49	95	101	13.2	13.1	

† Filament volume fraction 70% for epoxy matrix and 60% for metallic matrix.
‡ Filament volume fraction 50%.

Equation (1a) indicates that discontinuous-length fibers will not contribute 100% of their strength to the composite. However, with lengths many times the critical length, L_c (large aspect ratios), the values calculated will approach the equivalent values obtained for continuous filaments. Weeton and Signorelli[13, 18, 19] believe that it will be relatively easy to produce composites with fibers having L/L_c values such that better than 90% of fiber strengths will be usable. Predictions of whisker composite strength in this paper assumed 90% effectiveness.

The compression behavior of composites has not been as widely investigated as the tensile behavior. The problem has been analyzed theoretically by several investigators[12, 20, 21, 22] and it is concluded that failure of the unidirectional composite occurs when the compressive stress parallel to the filaments reaches the microstability limit of the filaments in one of two modes,

TABLE 4. ESTIMATED MECHANICAL PROPERTIES OF SELECTED ISOTROPIC COMPOSITES

Material	Specific weight lb/in³	Tensile strength ksi	Compressive strength ksi	Extensional stiffness psi×10⁻⁶	Shear stiffness psi×10⁻⁶
Filament composites†					
Boron-epoxy	0.076	110	90	16.4	6.3
Boron-Al	0.094	123	133	31.9	12.5
Beryllium-Al	0.079	76	110	27.3	11.9
Silicon carbide-Al	0.111	122	151	34.3	13.3
Silicon carbide-Ti	0.135	192	251	44.5	17.9
Boron carbide-Al	0.094	98	149	35.6	13.4
Carbon-epoxy	0.051	84	84	14.0	5.4
Whisker composites‡					
Al_2O_3-Al	0.122	100–476	129	29.9	11.4
SiC-Al	0.107	90–441	123	28.8	11.0
Graphite-Al	0.074	110–499	136	31.0	12.1

† Filament volume fraction 70% for epoxy matrix and 60% for metallic matrix.
‡ Filament volume fraction 50%.

the extension mode or shear mode. From ref. 20, the extension mode is given by

$$\sigma_c = 2v_f\left[\frac{v_f E_m E_f}{3(1-v_f)}\right]^{1/2} \tag{2}$$

and the shear mode is given by

$$\sigma_c = \frac{G_m}{1-v_f} \tag{3}$$

based on elastic microstability of the composite. The shear mode is dominant over most volume fractions of technical interest and consequently as pointed out by Rosen[(20)] the matrix shear stiffness G_m is the most important property controlling the compressive strength of the composite.

Predictions of compressive strength based on eq. (3) result in composite stress levels well above the elastic limit of the matrix material which suggests that the compressive strength criteria need to be adjusted for inelastic behavior of the matrix. Reference 9 proposes an elastic-perfectly plastic matrix model using the yield stress as a characteristic stress level and the secant modulus as a measure of effective material stiffness. This results in the following compression micro-instability equation:

$$\sigma_c = \left[\frac{v_f E_f \sigma_y}{3(1-v_f)}\right]^{1/2}. \tag{4}$$

Schuerch[22] proposes that microinstability above the matrix yield stress can be predicted by replacing the transverse Young's modulus and shear modulus with the appropriate tangent moduli of the matrix corresponding to the axial compressive strain in the unidirectional composite. The compressive microstability stress is then defined as

$$\sigma_c = (v_f E_f + v_m E_{m_s})\,\bar{\varepsilon} = \frac{G_m}{1-v_f} = \frac{E_{m_t}}{2(1+v_m)(1-v_f)} \tag{5}$$

where $\bar{\varepsilon}$ and G_m are defined by the effective tangent moduli. Equation (5) generally predicts lower strengths than eq. (4) and also appears to predict strengths which are more consistent with test data. Therefore, this method was used to predict the compressive strengths of the composites in Tables 3 and 4.

The potential mechanical properties of multiple layered composites are computed by conventional methods as described in ref. 11 for elastic properties and ref. 23 for strength properties.

The structural efficiencies of the various continuous filament composites listed in Tables 3 and 4 are compared in Figs. 2, 3 and 4 for the conventional structural efficiency parameters of tensile strength/specific weight, compressive

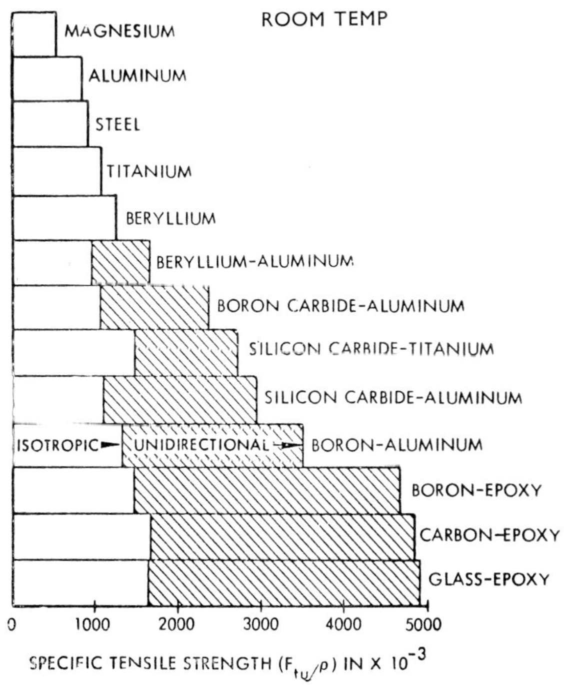

FIG. 2. Comparison of tensile strength efficiency of composite materials

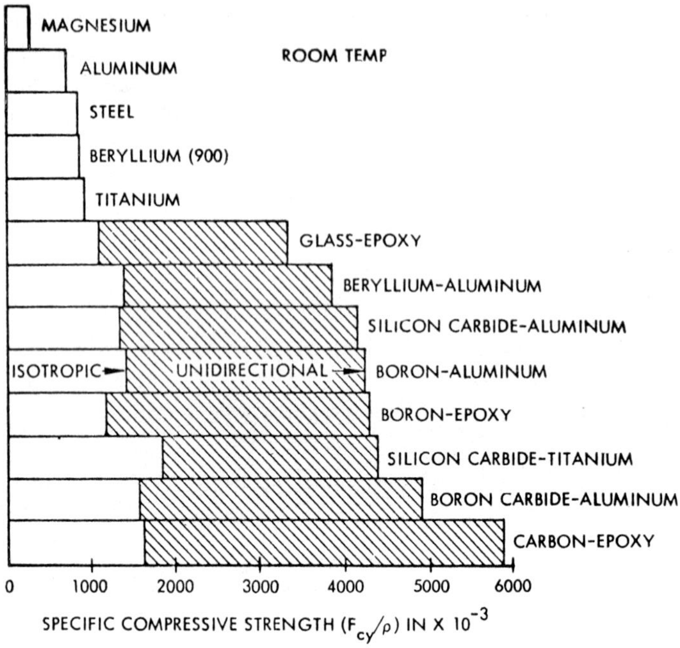

FIG. 3. Comparison of compressive strength efficiency of composite materials

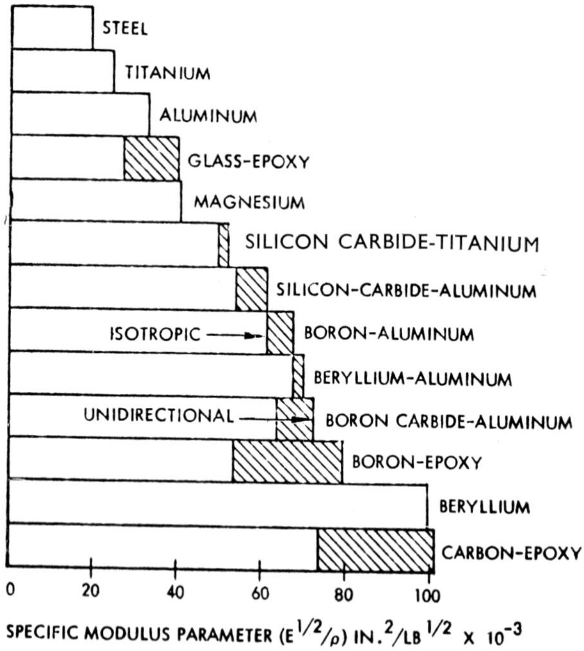

FIG. 4. Comparison of compressive buckling efficiency of composite materials

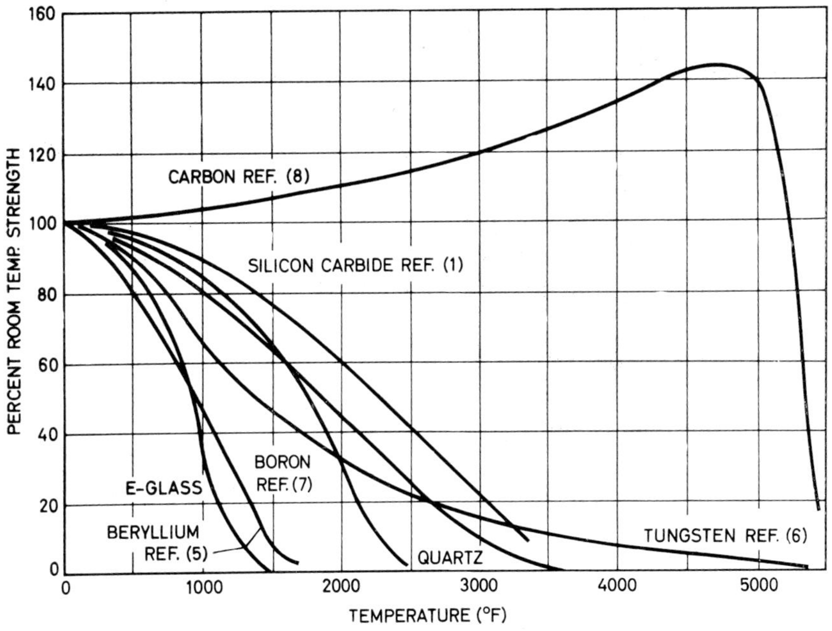

FIG. 5. Effect of temperature on filament strength

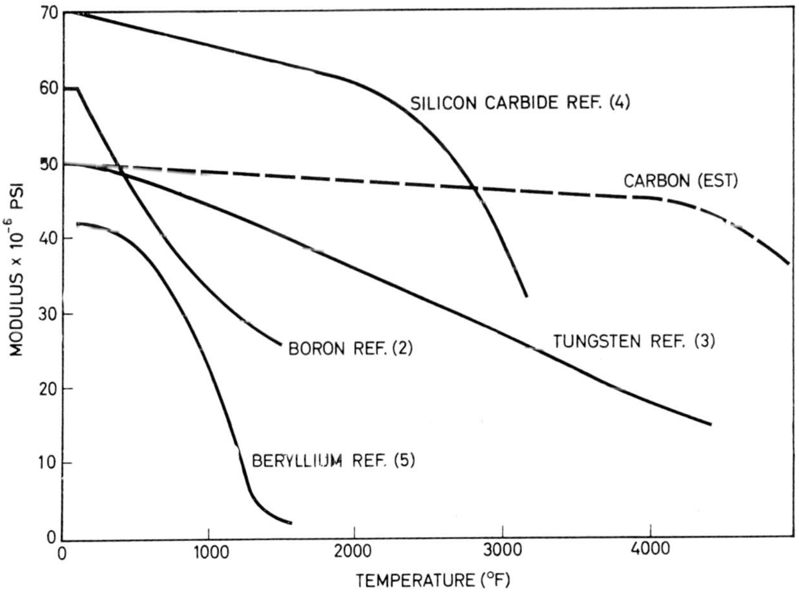

FIG. 6. Effect of temperature on modulus of filaments

yield strength/specific weight and [elastic modulus]$^{1/2}$/specific weight respectively. The shaded area in the composite bar graphs indicate the difference in mechanical properties of the unidirectional composite (Table 3) and the isotropic composite (Table 4). The relative merits of the composites are compared to conventional engineering material and specific advantages in structural efficiency of composites appear to be possible in all applications. However, relative advantages indicated by these simple comparisons may not be realized in actual structural applications because of many detail requirements, secondary support structure requirements, etc. Interactions of mechanical properties such as elastic modulus and compressive yield in buckling criteria may lead to a different relative relationship between the composites than indicated by the simple comparisons shown in Figs. 2, 3, and 4.

Many of the fibers listed in Table 1 have relatively high working temperature limits and as such may be available for high temperature applications. The effect of temperature on filament strength for various filaments is shown in Fig. 5 and effect of temperature on modulus is shown in Fig. 6. It is noted that carbon and silicon carbide appear best at very high temperatures and boron, beryllium and glass may be useful at moderately elevated temperatures.

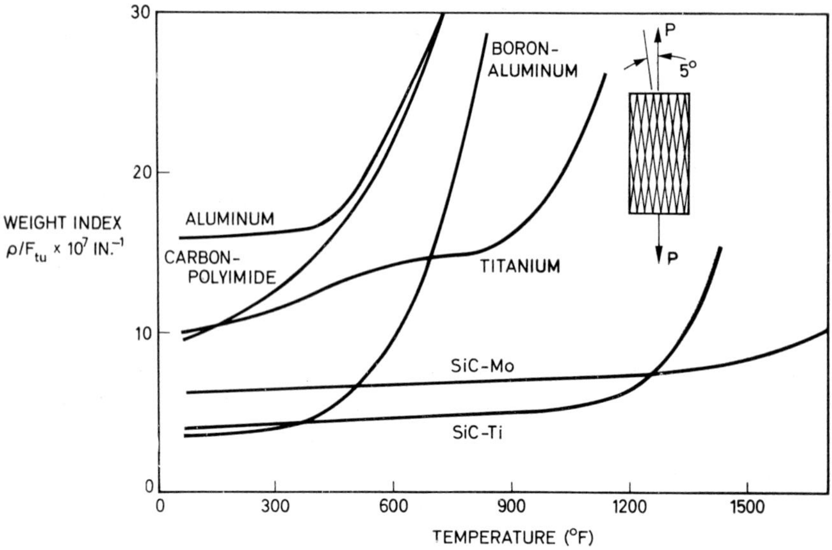

FIG. 7. Effect of temperature on efficiency of composites in tension application

The elevated temperature behavior of the composite depends on the effects of temperature on the matrix as well as the filaments and in Figs. 7 and 8 effects of temperature on the weight efficiency of typical composites in tension and compression respectively are shown. The relative merits of the various composites are evident and it appears that composites may have some definite

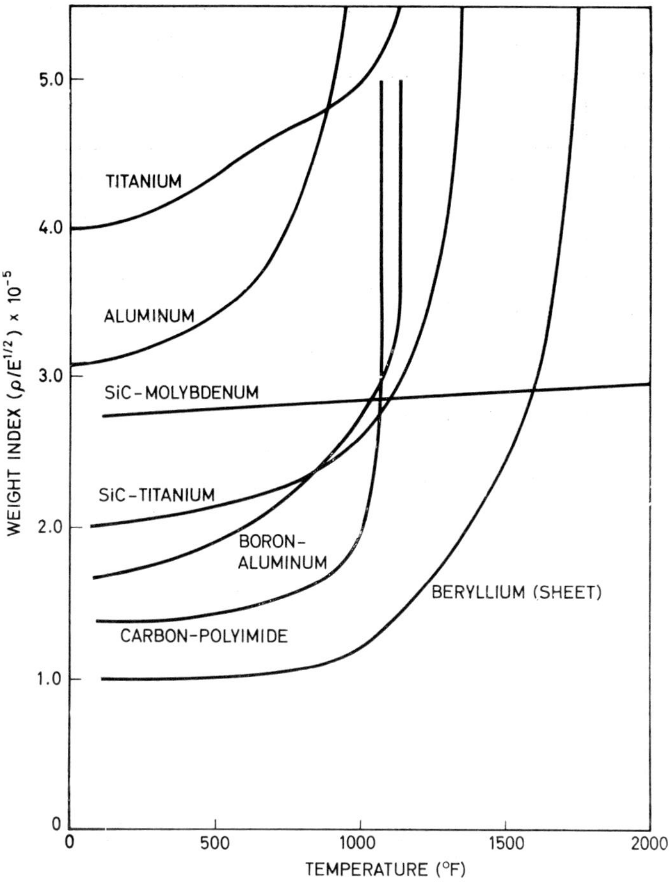

FIG. 8. Effect of temperature on compression buckling efficiency of selected composites

advantage over conventional engineering materials for elevated temperature applications. The elevated temperature applications should be investigated more thoroughly since even greater advantages in using composites are apparent in this application area.

IV. CYLINDERS UNDER LATERAL PRESSURE

The two methods used in the buckling analysis are given in refs. 1 and 3. Method 1 can be referred to as the anisotropic method. The constitutive relations have been derived considering a layered wall construction where each layer is orthotropic but the laminates are oriented in such a way that the axes of the elastic symmetry do not coincide with the adjacent layers. For shells of this type there occurs a coupling phenomenon between in-plane stretching and transverse bending which does not occur in the theory of homogeneous shells such as the one used in method 2.

The derivation of the constitutive relations for method 1 is based on the energy principle. The stress resultant terms have been obtained by varying the strain energy density term with respect to the strains. The presence of more than one layer has been accounted for by summing up all the layers in the strain energy expression. (See Appendix A which is taken from the latest LMSC composite buckling analysis.[2])

The constitutive relations of method 2 are well known in the literature since they do not include the coupling terms of method 1. Here, the orthotropic material is considered homogeneous through the thickness.

As for the comparison of the buckling analyses, both methods are formulated according to the classical shell deflection theory. Neither method includes the effect of transverse shear deflections. The basic approach to the stability analyses has been the same, that is, the analyses have been formulated as an eigenvalue problem, the lowest eigenvalue of the stability equations corresponding to the critical load.

The investigations of the lateral pressure on cylindrical structures is initially treated. The results of both methods are identical for the cylinder used as the example due to the multiplicity of the symmetric layers in the wall construction of the cylinder (16 layers). The buckling allowables derived from elastic constants based on two different sources (Tsai[5] and Ekvall[10]) account for the difference in the results shown in Figs. 9 and 10 as a function of the layup angle. The respective computer programs use the different referenced elastic constants. It can be observed that the determination of the properties is most important as will be discussed later. Also, the maximum allowable pressure corresponds to a pure 90° hoop layup. If the so-called isotropic layup is used (see Fig. 32 in Nomenclature) there would be a weight penalty for a pure lateral loading condition as shown in Figs. 9 and 10. However, the isotropic layup shows the best efficiency for the most practical structures especially with the addition of axial-loading requirements.

No attempt is made here to formulate interaction relationships for combined loading in composites. This is a formidable task in itself and is also reported in ref. 1 (method 1).

The results of the Tsai and Ekvall elastic properties are compared in Figs. 11 and 12 for the boron epoxy and boron aluminum composites for "isotropic" and 90° layups. The results indicate that the higher shear modulus of aluminum has a marked effect on buckling. Figure 13 compares the allowable pressure on a cylinder with a pure hoop layup versus the isotropic case for a boron-epoxy composite.

Various composites for lateral pressure allowables are compared in Figs. 14 and 15 for the 90° and the isotropic layups. On a weight basis, the boron

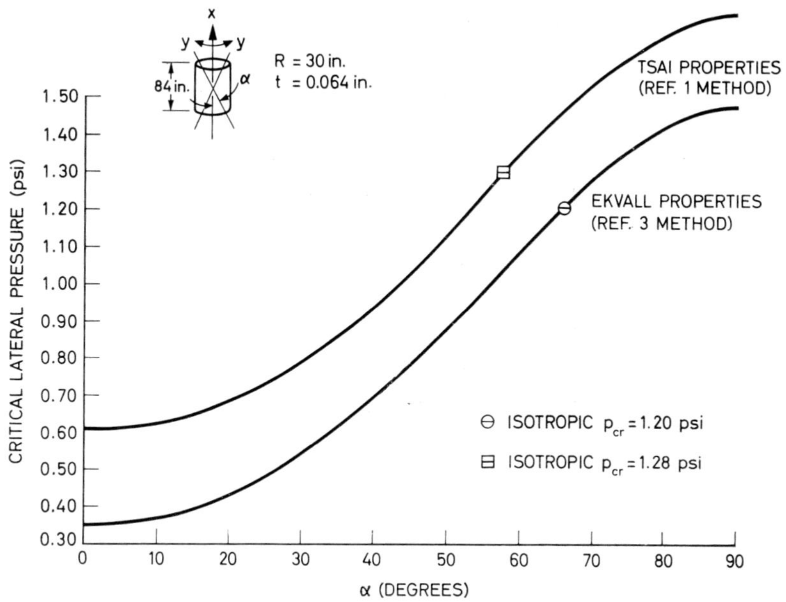

FIG. 9. Boron-epoxy composite cylinders—constant weight

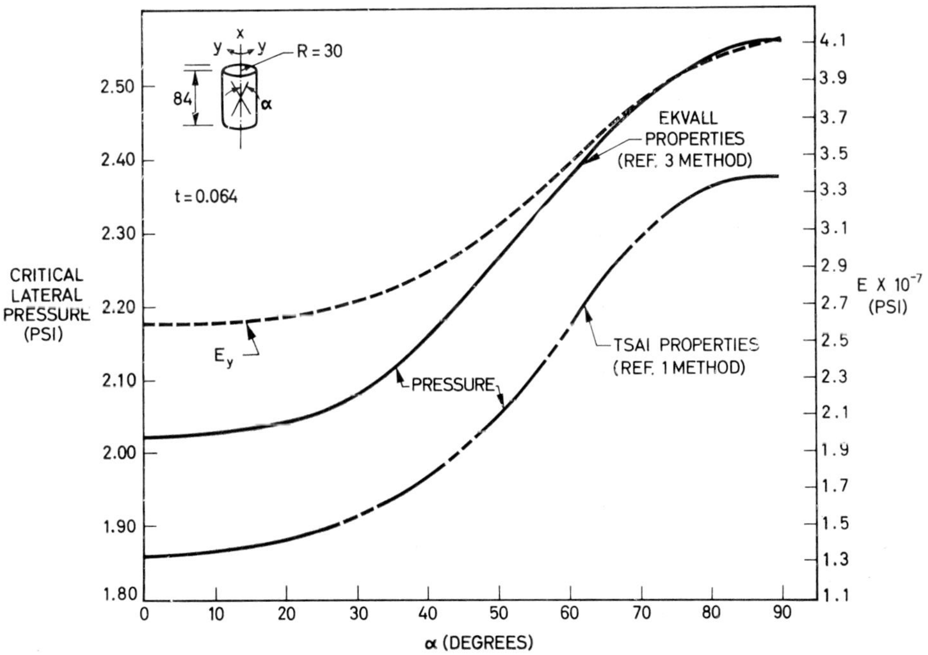

FIG. 10. Boron-aluminum composite cylinders—constant weight

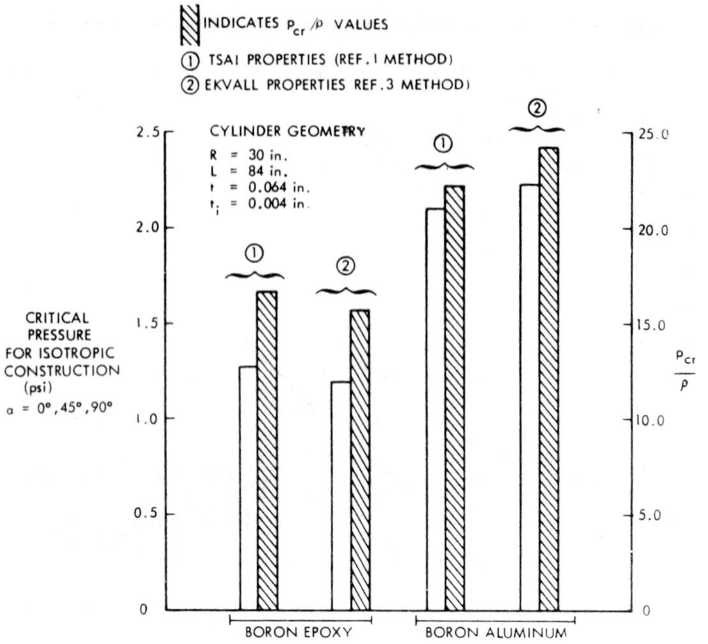

FIG. 11. Buckling of laterally loaded composite cylindrical shells (comparison of two analysis methods—isotropic construction)

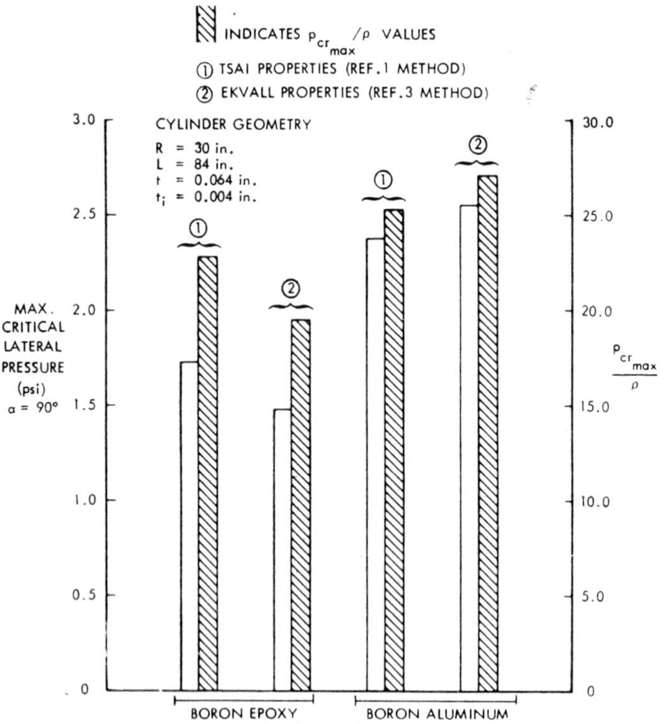

FIG. 12. Buckling of laterally loaded composite cylindrical shells—filaments in hoop direction only

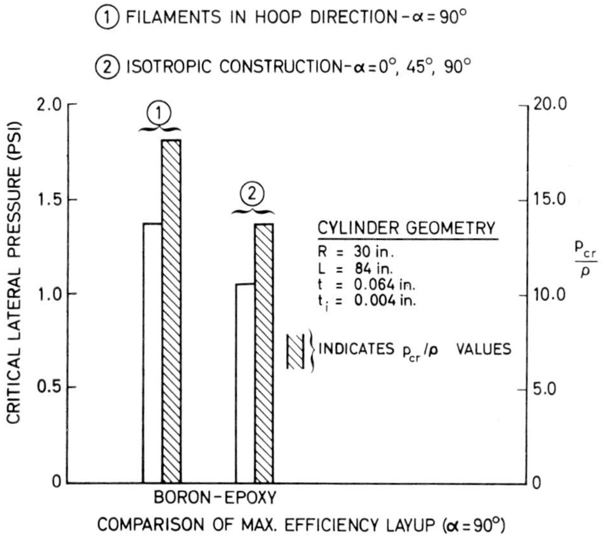

FIG. 13. Comparison of maximum efficiency layup ($\alpha = 90°$) with isotropic construction for a cylindrical shell subjected to lateral pressure

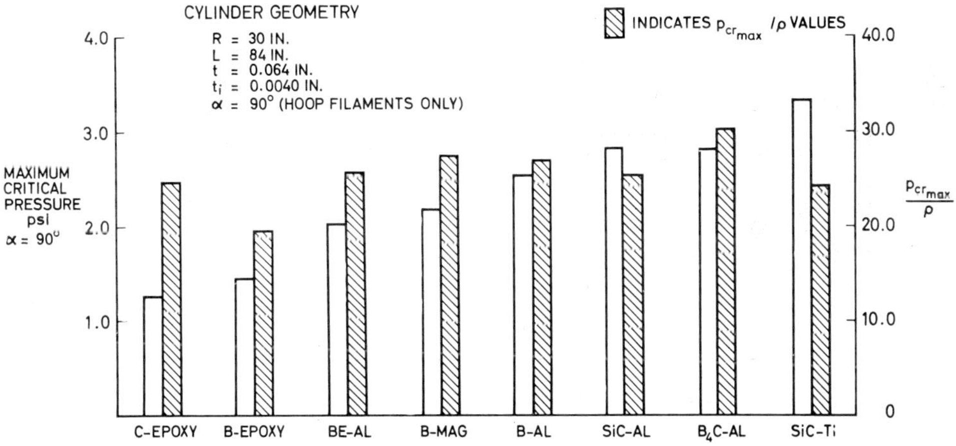

FIG. 14. Comparison of constant geometry composite cylinders using external lateral pressure

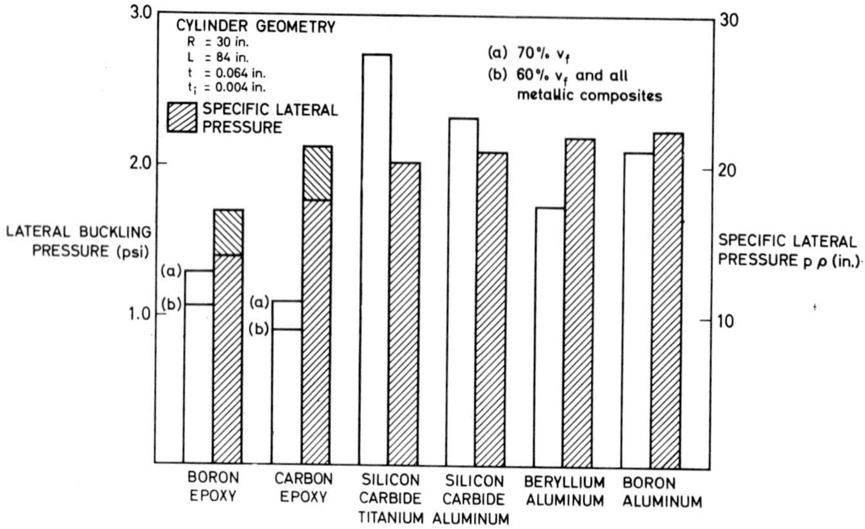

FIG. 15. Specific comparison of materials for cylinder under radial pressure—isotropic construction

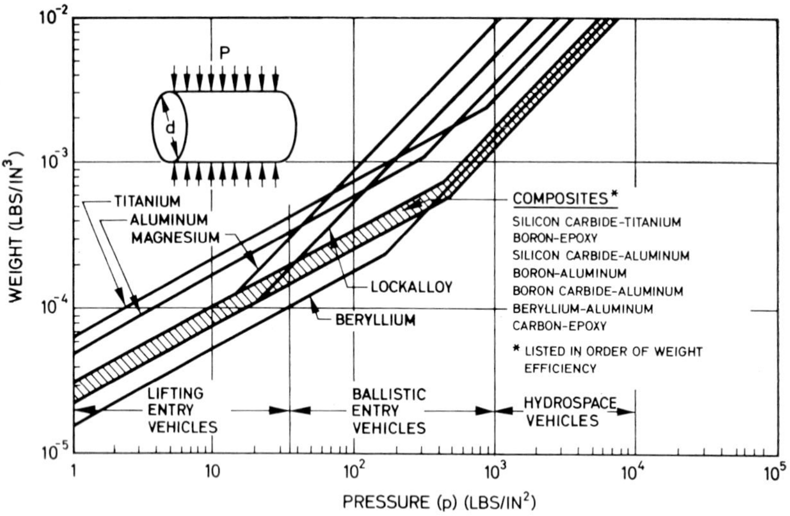

FIG. 16. Material efficiency for cylinders under radial pressure

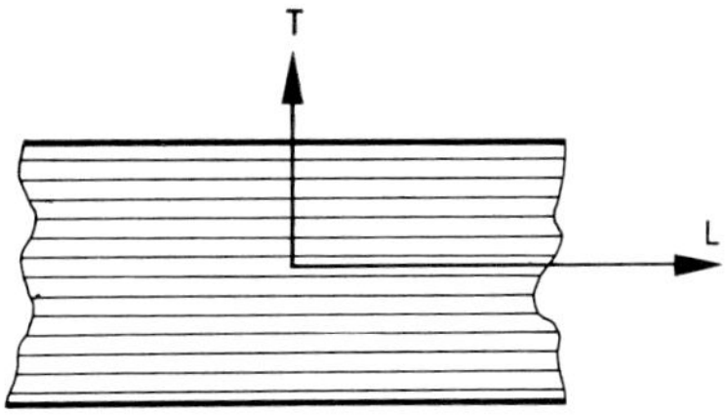

Ekvall	Tsai	Dow, Rosen, Hashin
$G_{LT} = \dfrac{G_m G_f}{R\Phi G_m (1-R) G_f}$ $R = 1.275 V_f$ $\Phi = \displaystyle\int_0^{\pi/2} \dfrac{\sin\theta \, d\theta}{\dfrac{G_m}{G_f} + \sin\theta\left(1 - \dfrac{G_m}{G_f}\right)}$	$G_{LT} = (1-C)\, G_m \dfrac{2G_f - (G_f - G_m)V_m}{2G_m + (G_f - G_m)V_m} + C G_f \dfrac{(G_f + G_m) - (G_f - G_m)V_m}{(G_f + G_m) + (G_f - G_m)V_m}$ $C = 0.2$	$G_{LT} = G_m \dfrac{(G_f + G_m) + (G_f - G_m)V_f}{(G_f + G_m) - (G_f - G_m)V_f}$

FIG. 17. Expressions for G_{LT} under different experimentors

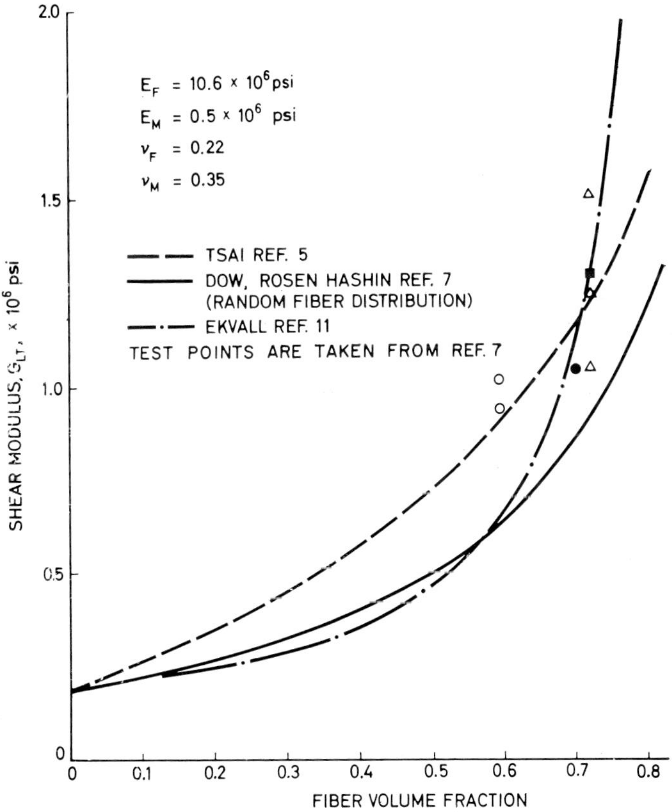

FIG. 18. Comparison of shear modulus predictions under different methods

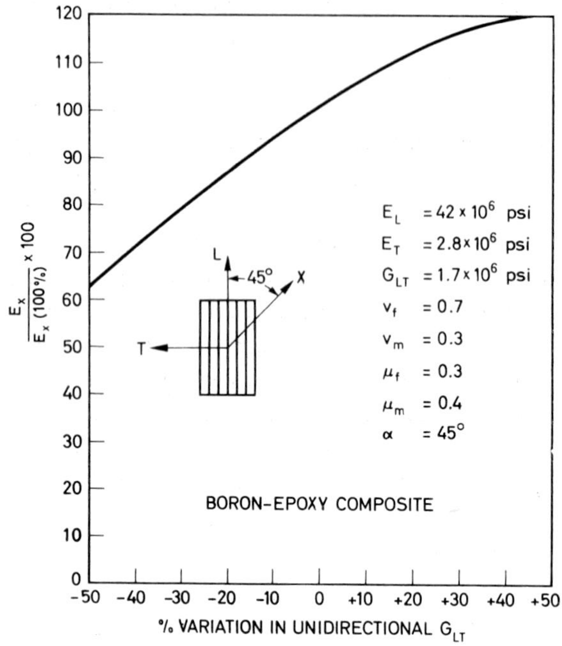

FIG. 19. Unidirectional G-effect on E_x of a total composite

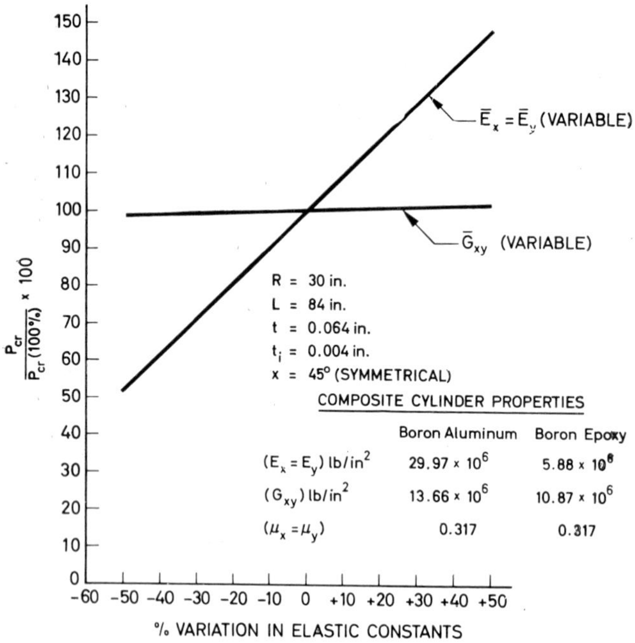

FIG. 20. Variations of critical lateral pressure with elastic constants, $\alpha = 45°$

carbide in aluminum and the boron in magnesium composites are slightly superior. The aluminum matrix composites appear to be superior to the epoxy and titanium types.

In Fig. 16 the composites are compared with metallic construction for optimum designed ring stiffened cylinders under radial pressure using an isotropic buckling criteria. The relative order of composite efficiency is seen to be different from that indicated in Fig. 15 which is based on an anisotropic buckling criteria. This difference can be attributed to the lower shear modulus terms of the orthotropic composite analysis. At high loadings the relative efficiency of the composites is different because a yield criteria dominates. The advantages of the composites are apparent in both the low load range and high load range, but in the high load range associated with deep submersibles the composites have the greater potential. The potential of composites in spherical shell deep submersibles is discussed in a following section.

The shear modulus of an unidirectional composite is a most important property for buckling analyses especially when a multidirectional layup is anticipated. The total composite derives its structural properties from the macro properties of each layer. Three different investigators express the unidirectional composite shear modulus G_{LT} in different forms (Figs. 17 and 18). The method and test results depend on the mathematical model chosen and the subsequent construction of the test specimen. However, a 40% change in G_{LT} of a unidirectional composite results in a 20–30% variation in E_x and E_y as shown in Fig. 19. Thus it is seen that the estimation of G_{LT} of each unidirectional layer is very critical. The allowable lateral pressure for a cylinder with a layup of $\pm 45°$ is in almost direct relation to the variation in $\bar{E}_x$ or $\bar{E}_y$. $\bar{G}_{xy}$ of the total composite (thickness), however, does not affect the allowable pressure significantly. This is shown in Fig. 20.

V. CYLINDERS UNDER AXIAL LOAD

Buckling due to axial load in a cylinder is another important topic with wide application. The axial load analyses in this report have been based on ref. 1 which uses the anisotropic approach. The boron–epoxy and boron–aluminum composites were treated to obtain the optimum angle of layup with respect to the axial direction. This can be seen in Fig. 21.

The boron–aluminum possesses an allowable axial load over twice that of the boron-epoxy. This is again primarily due to the higher shear modulus G_m of the matrix material, aluminum, causing the moduli of elasticity of the total composite to be greater than the boron-epoxy composite. It is also to be noted that for axial buckling the isotropic layup is approximately the same as the 15° layup for boron-aluminum. For the boron–epoxy the isotropic configuration is approximately 30% higher than the 15° layup.

An efficiency comparison for other composites with the isotropic layup for the cylindrical shell is shown in Fig. 22. It can be seen that the aluminum matrix composites, in particular the boron–aluminum cylinder, have the greatest weight advantage.

In Fig. 23 the relative efficiency of composites are compared to metals for optimum stiffened cylinders under axial compression using an isotropic buckling criteria. It is seen that the relative order of composite efficiencies is different from those shown in Fig. 22 where an anisotropic buckling criteria was used. These differences in relative efficiency are due to the lower shear modulus effects included in the anisotropic analysis. Composites are definitely superior to the easily fabricated common metals. However, whether the design and fabrication problems can be universally solved remains to be seen.

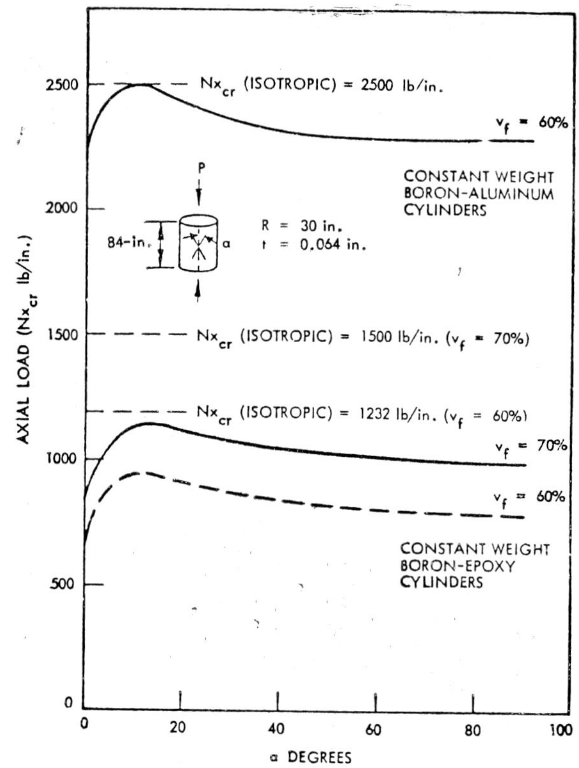

FIG. 21. Effect of layup on the critical load of composite—aeolotropic analysis

The buckling loads given in Figs. 21, 22 and 23 are based on the classical theory and therefore do not represent the true critical values. These loads are expected to be much lower as in the case of isotropic cylinders although there is practically no test data to justify this conclusion. For relative comparison purposes however, the values given in this report are valid.

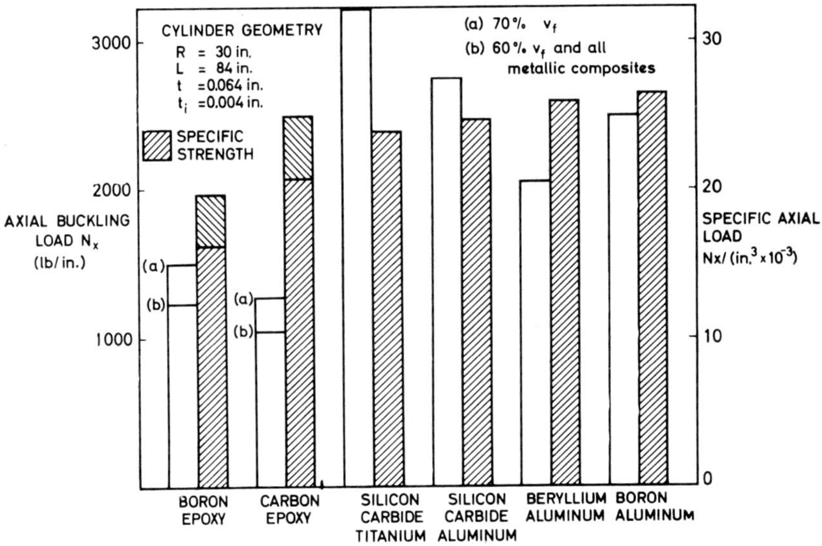

FIG. 22. Specific comparison of materials for cylinder under axial load—isotropic construction

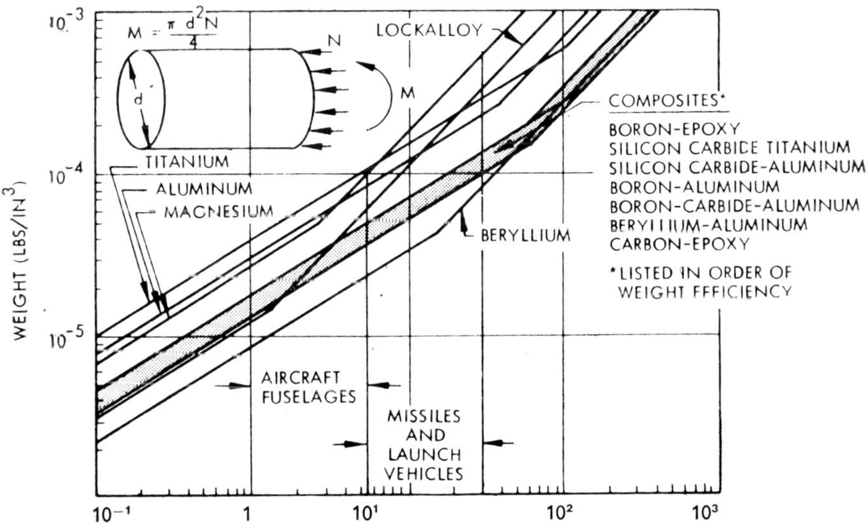

FIG. 23. Material efficiency of stiffened cylinders under axial compression

VI. SPHERICAL SHELLS

Related to the cylinder lateral pressure case is the hydrostatic pressure condition on a sphere. Spherical pressure hulls potentially constructed from various composites have a distinct weight advantage as evidenced by the buoyancy factor in Fig. 24. In other words, for the same weight a deep submersible can reach approximately twice the depth using composites as compared to metals. Silicon carbide in a titanium matrix possesses the greatest efficiency.

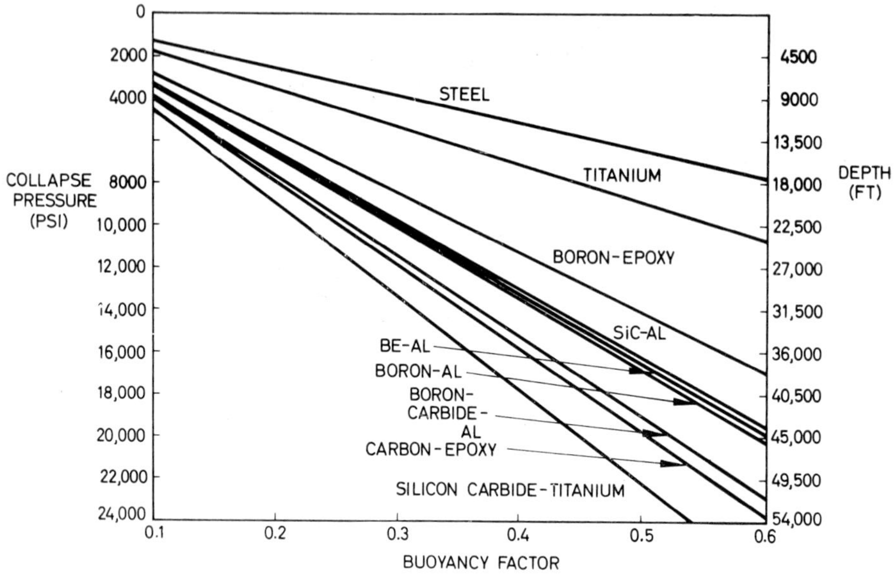

FIG. 24. Collapse depth for special pressure hulls

VII. BOX BEAM ANALYSIS

A typical tapered box beam is also treated to demonstrate the potential in wing and empennage applications of many composites. The various thicknesses of the facing sheets of the sandwich construction and the core height are noted in Fig. 25. Each composite is designed for yield due to high loading, accounting for the various face and total sandwich thicknesses. It can be seen that the optimum web spacing is four for all composites. As in the spherical pressure vessel case the silicon carbide-titanium composite is superior due to the potentially high yield strength.

The fitting problem on a sandwich such as this is a serious one. The theoretical required load is 14,100 lb/in. across the chord of the box beam used as the example. The required fitting is shown in Fig. 26. An aluminum channel possessing the properties as noted, must have a 3 in. long bond length and a thickness of 0.150 in. Tests using a similar thickness and a bond length of 2 in. failed at 5400 lb/in.

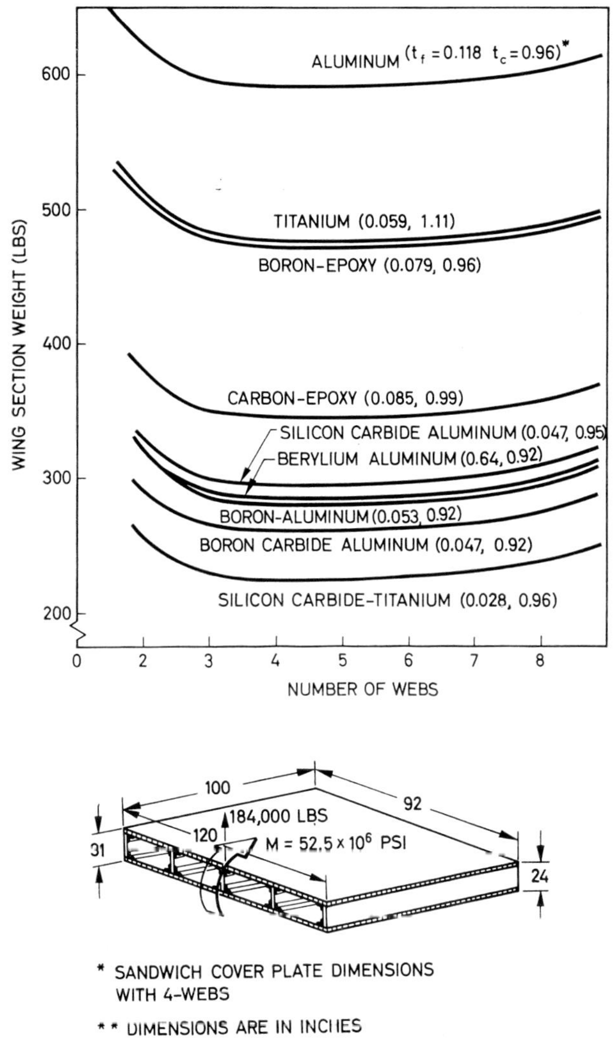

FIG. 25. Optimum weight vs. number of sparwebs for aluminum honeycomb sandwich wing box structure with isotropic composite facing

Figures 27, 28 and 29 show the test specimens. One type of attachment, an epoxy reinforcement in the aluminum core (number four) failed at 2000 lb/in. The other was the bonded channel type, with an expected allowable load of

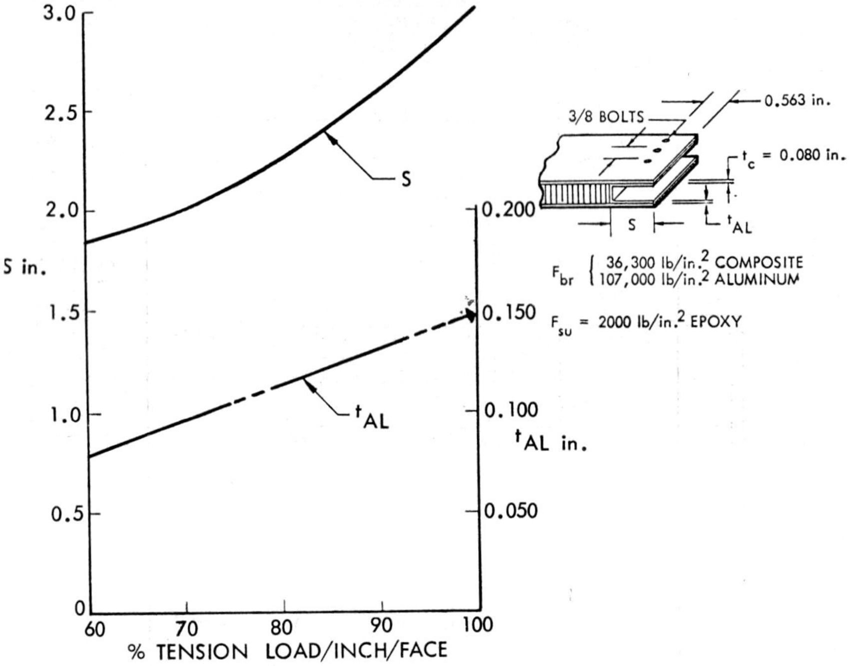

FIG. 26. Design curve for box beam edge attachment

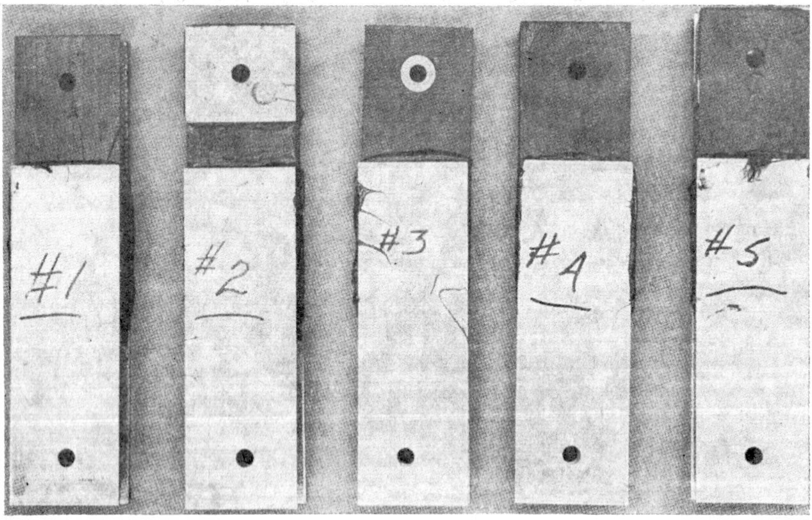

FIG. 27. Boron-epoxy aluminum honeycomb attachment test specimen

FIG. 28. Failure of channel-reinforced specimen showing tear of components at bond area—5400 lb load

FIG. 29. Failure of epoxy-reinforced specimen at 2400 lb

10,000 lb compared to the test failure load of 5400 lb. This is indeed a critical problem.

Providing the fastener problem can be solved, the overall potential for integrally stiffened box beams using composites is shown in Fig. 30, assuming isotropic buckling criterion. Comparing the relative efficiency of composites in Fig. 30 with Fig. 25 differences are noted which are due to the fact that the sandwich construction in Fig. 25 is buckling in the inelastic range while the comparisons in Fig. 30 are for elastic buckling.

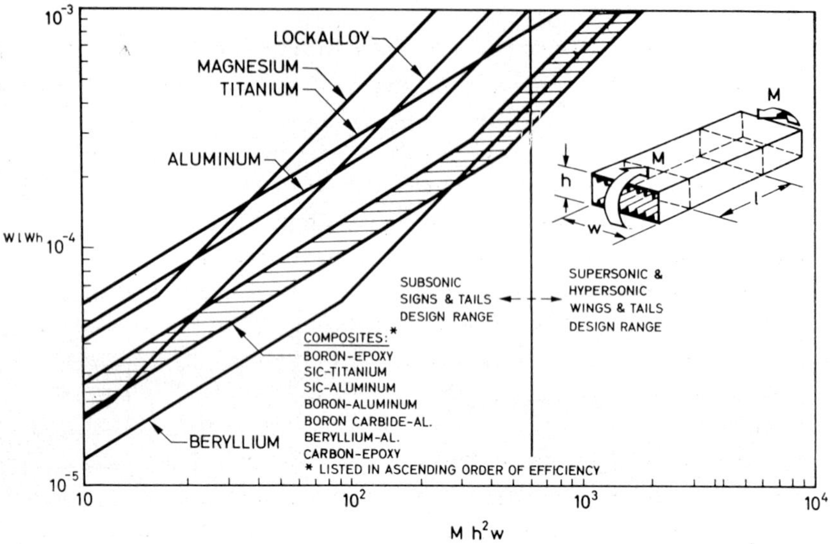

FIG. 30. Material efficiency for stiffened box beam under bending load

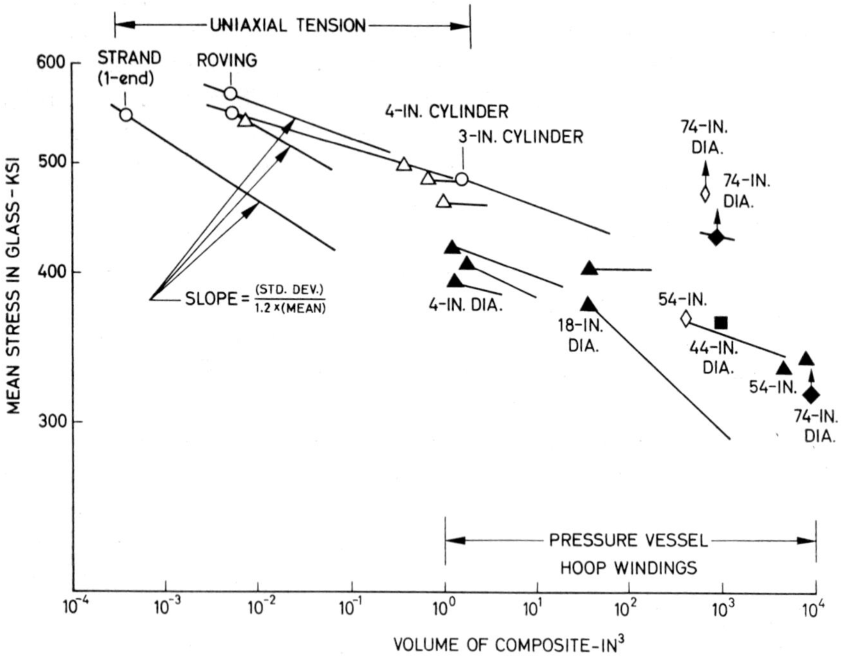

FIG. 31. Mean tensile strength of S-glass composite structures

VIII. SIZE EFFECTS ON COMPOSITE PROPERTIES

A recent investigation at LMSC of tensile strengths of composite structures indicated that the tensile strength of glass filament wound pressure vessels cannot be predicted by the existing rule of mixtures of bundle-theories. The rule of mixtures model does not apply because of the well-known statistical variation and size effects in glass filaments. Bundle theory also requires some further development before it can apply to pressure vessels. The theory does not account for the fact that the load of a broken filament is transferred by shear to only a limited number of neighboring filaments. This effectively determines the number of filaments which act cooperatively as a "bundle" in a pressure vessel. A considerable quantity of strength data now exists for glass-filament structures. The data reveal certain patterns of behavior which improved theoretical models should seek to explain.

Figure 31 shows mean strengths of various small and large test specimens of fiberglass (S994HTS glass in epoxy). The specimens (volume of composite subjected to the "hoop" stress only at which failure occurred) range from a 204-filament 10-in.-long strand up to full-size rocket motors—factor of 10 million in volume. The data show a distinct drop in strength with size, as would be expected if specimens of a given size showed statistical variation. The statistical variation for a particular size should, according to size-effect theory, enable one to predict the mean strength of large or smaller specimens. If a Weibull distribution of composite strength is assumed, this prediction is simply a line through the mean whose slope is determined by the statistical coefficient of variation of the data. Each point which represented more than two tests has a sloping line drawn through it representing the predicted strength of larger specimens. Each line is terminated at a volume corresponding to the total number (and volume) of specimens tested.

Considering the fact that this data was generated over 5 years from several sources representing various designs and techniques, the success of this simple theory is remarkable. The solid symbols represent the hoop windings of pressure vessels and rocket motors for which the axial and hoop strains were about equal. The downward trend corresponds to a coefficient of variation of about 2.5%, which agrees well with the coefficients of much of the data. The open symbols represent specimens with much less strains normal to the hoop fibers. Excluding the strand specimens, these also form a group consistent with a 2.5% coefficient of variation. The drop in strength from the roving to the largest motors is 40%. The drop from simple tension to pressure vessels is 17%; the most plausible reasons for this are: (1) the cracks parallel to filaments which appear in the resin of motors under high pressure and/or (2) the reduction in filament-resin bond strength due to normal tension stress.

The exceptional strength of the few new 74-in. experimental chambers is of course very interesting. This suggests that a considerable improvement in strength of large motors may be quite possible in the near future, and that better theoretical understanding should result in weight saving.

IX. CONCLUSIONS AND RECOMMENDATIONS

1. Standardized nomenclature and definitions are required.
2. Shear modulus G_{LT} in the macro-sense has a significant effect on the Young's modulus, E_x for a unidirectional composite. This in turn has an almost direct percentage effect on the overall buckling strength.
3. Metal matrices are superior for buckling applications.
4. A cylinder with only a hoop layup yields the greatest lateral pressure capability.

   For axial loading the isotropic layup is superior on an equal weight basis to a cylinder with an optimum angle cross-ply layup.
5. Composites used in sandwich type construction for box beams are superior to metals commonly used in the same type of construction.
6. Attachment problems are critical. So far the weight advantages are significantly decreased due to heavy fittings for adequate bearing strength and carry-through structure.
7. Additional research is required for prediction of compressive strength of composites.
8. Buckling studies and tests are required to establish probabilities of buckling failure versus classical values.
9. Failure modes require study including volume effect in tensile applications.

REFERENCES

1. LMSC-806705. Buckling of Orthotropic Cylinders (draft), by G. B. Mah, B. O. Almroth and E. V. Pittner, Lockheed Missiles & Space Company, Sunnyvale, California, 9 January 1967.
2. LMSC 4-17-67-1. Buckling of Shells of Revolution With Various Wall Constructions, by D. Bushnell and B. O. Almroth, Lockheed Missiles & Space Company, 1967.
3. LMSC 3-80-61-45. The Buckling of Orthotropic Cylindrical Shells Under Radial Pressure Varying in the Axial Direction, by M. H. Kural, Lockheed Missiles & Space Company, January 1962.
4. ARMSTRONG, BURNS and CRAWFORD. Large Stiffened Cylinder Weight-Cost Comparisons Between Be-38% Al and Other Materials, *Journal Spacecraft*, **4**, No. 3.
5. TSAI, S. W. Structural Behavior of Composite Materials, NASA Report CR-71, July 1964.
6. TSAI, S. W. Strength Characteristics of Composite Materials, NASA Report CR-224, April 1965.
7. HASHIN, Z. and ROSEN, B. W. The Elastic Moduli of Fiber-Reinforced Materials, *Journal Applied Mechanics*, June 1964.
8. ROSEN, B. W., DOW, N. F. and HASHIN, Z. Mechanical Properties of Fiberous Composites, NASA Report CR-31, April 1964.

9. Dow, N. F., Rosen, B. W. and Hashin, Z. Studies of Mechanics of Filamentary Composites, NASA Report CR-492, June 1966.
10. Ekvall, J. C. Elastic Properties of Orthotropic Monofilament Laminates, Paper No. 61-AV-56 presented at ASME Aviation Conference, Los Angeles, California, 12–16 March 1961.
11. Ekvall, J. C. Structural Behavior of Monofilament Composites, presented at AIAA 6th Structures and Materials Conference, Palm Springs, California, April 5–7, 1965.
12. Schuerch, H. Compressive Strength of B-Metal Composites, NASA-CR-202, April 1965.
13. Weeton and Signorelli. *Fiber Metal Composites*, Chap. 19; ed., Burke, Reed and Weiss, *Strengthening Mechanisms Metal and Ceramics*, Proceedings of the 12th Sagamore Army Materials Research Conference, Aug. 24–27, 1965.
14. Bonnano, F. R. The Fabrication and Mechanical Properties of SiC Filament-Metal Matrix Composites, 10th Nat. SAMPE Symposium on Advanced Fibrous Reinforced Composites, San Diego, California, Nov. 9–11, 1966.
15. Kelly, A. and Tyson, W. R. *Fiber-Strengthened Materials*, Chap. 13, *High Strength Materials*, ed., V. F. Zackay, John Wiley, New York, 1965.
16. Kreider, K. G. and Leverant, G. R. Boron Aluminum Composite Fabricated By Plasma Spraying, 10th National SAMPE Symposium on Advanced Fibrous Reinforced Composites, San Diego, California, Nov. 9–11, 1966.
17. *Proposal for Application of Advanced Fiberous Reinforced Composite Materials*, Vol. I, LMSC-894090, 10 May 1965.
18. McDanels, D. L., Jech, R. W. and Weeton, J. W. Stress-Strain Behavior of Tungsten-Fiber-Reinforced Copper Composites, NASA TN-D-1881, October 1963.
19. Petrasek, Signorelli and Weeton. Metallurgical and Geometrical Factors Affecting Elevated Temperature Tensile Properties of Discontinuous-Fiber Composites, NASA TN-D-3886, March 1967.
20. Rosen, B. W. Mechanics of Composite Strengthening, *Fiber Composite Materials*, S. H. Bush, ed., ASM 1965.
21. Foye, R. L. Compression Strength of Unidirectional Composites, AIAA Paper No. 66-143 AIAA 3rd Aerospace Sciences Meeting, New York, New York, January 24–26, 1966.
22. Schuerch, H. Prediction of Compressive Strength in Uniaxial Boron Fiber-Metal Matrix Composite Materials, *AIAA Journal*, **4**, No. 1, January 1966.
23. MIL-HDBK-17. Plastics for Flight Vehicles, 5 November 1959.
24. Elkins, P. E. *et al.* Modified Silicon Carbide Continuous Filaments, SAMPE Symposium, Vol. 10, 1966.
25. Cook, J. L. and Sakurai, T. T. Stress Rupture and Tensile Test Techniques for Single Boron Filaments at Room and Elevated Temperatures, Presented at 10th National SAMPE Symposium, 1966.
26. Schmidt, F. F. and Ogden, H. R. The Engineering Properties of Tungsten and Tungsten Alloys DMIC Report 191, Battelle Memorial Institute, September 1963.
27. Refractory Ceramics of Interest in Aerospace Structural Applications — A Materials Selection Handbook, RTD-TDR-63-4102, October 1963.
28. Beryllium Design Data, LMSD-484722, Lockheed Missiles & Space Company.
29. Rogers, B. A. *The Properties of Eleven Less Common Metals*, Chap. 9 of *Metals for Supersonic Aircraft and Missiles*, edited by D. W. Grobecker, Published by American Society for Metals, 1958.
30. Herring, H. W. *et al.* Research on Boron Filaments and Boron Reinforced Composites. Presented at 10th National SAMPE Symposium, 1966.
31. France, L. L. *et al.* Mechanical Behavior of a Continuous Filament Carbon Composite, Presented at 10th National SAMPE Symposium, 1966.
32. Yue, A. S. *et al.* Advanced Metallic Composites, LMSC Report 6-77-67-4, January 1967.

NOMENCLATURE

C_{ij}	moduli of elasticity as defined in the constitutive equations (lb/in^2)
D	diameters of fibers (in.)
E_f	modulus of elasticity of fibers (lb/in^2)
E_L	modulus of elasticity of unidirectional composite in the direction of fibers (lb/in^2)
E_m	modulus of elasticity of the matrix (lb/in^2)
E_{m_t}	tangent modulus of elasticity of matrix (lb/in^2)
E_{m_s}	secant modulus of elasticity of matrix (lb/in^2)
E_T	transverse modulus of elasticity of unidirectional composite (lb/in^2)
E_x	modulus of elasticity of unidirectional composite in the x-direction (lb/in^2)
$\bar{E}_x$	modulus of elasticity of total composite in the x-direction (lb/in^2)
E_y	modulus of elasticity of unidirectional composite in the y-direction (lb/in^2)
$\bar{E}_y$	modulus of elasticity of total composite in the y-direction (lb/in^2)
G_{LT}	shear modulus of unidirectional composite in the LT plane (lb/in^2)
G_m	shear modulus of matrix (lb/in^2)
G_{xy}	shear modulus of unidirectional composite in the xy plane (lb/in^2)
$\bar{G}_{xy}$	shear modulus of total composite in the xy plane (lb/in^2)
L	length of fibers (also length of cylinder) (in.)
L_c	$\frac{1}{2}\frac{\sigma_f}{\tau}D$ critical fiber length
M	moment (in.-lb)
m	number of axial waves in buckling pattern
N, N_x	uniform axial load per inch of circumference (lb/in)
n	number of circumferential waves in buckling pattern
S_{ij}	moduli of compliance
p	pressure (psi)
p_{cr}	critical buckling pressure (psi)
R	radius (in.)
t	total thickness of the shell (in.)
t_i	thickness of one layer (in.)
V	shear (lb)
v_f	percent filaments by volume
v_m	percent matrix by volume
x	longitudinal direction
y	circumferential direction
α	the angle between the unidirectional fibers and the x-axis (deg)
β	L/L_C
σ_c	composite strength (psi)
σ_f	average tensile strength of the filaments (psi) (see text)
σ_m	average stress in matrix (psi)
σ_m^*	stress in matrix at the ultimate strain of the filaments (psi)
τ	shear strength of fibers (psi)
μ	Poisson's ratio
ϱ	specific weight (lb/in^3)
ε	middle surface strain
$\varkappa$	change in curvature

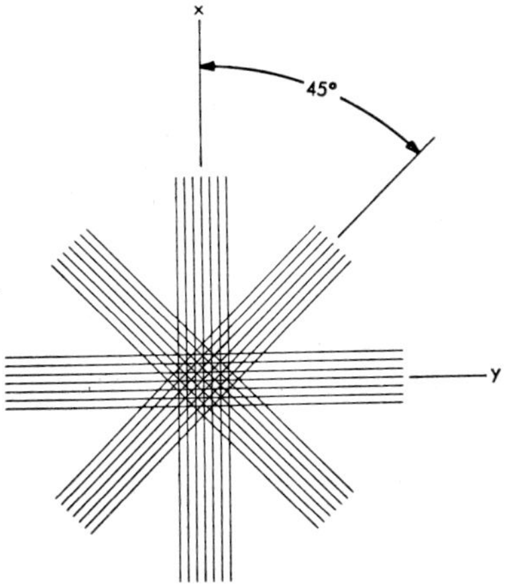

FIG. 32. Isotropic construction

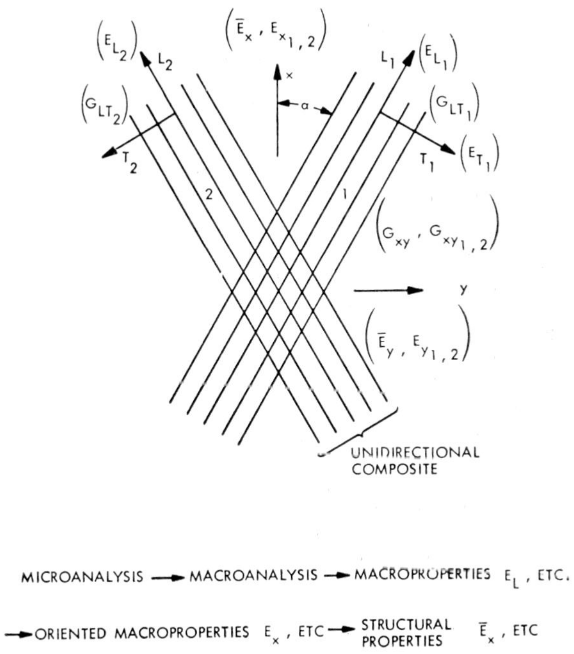

FIG. 33

Appendix A

CONSTITUTIVE RELATIONS FOR FIBER-REINFORCED SHELLS[2]

I. *Fiber-reinforced Cylinders*

In general fiber-reinforced shells are composed of several layers, with the fibers unidirectional in each layer. Because of the fabrication techniques in use, a cylindrical shell will have such layers appear pair wise, one with the winding angle (reference a generator) equal to $+\beta$ and one with the winding angle equal $-\beta$. We will therefore consider such a pair of layers as one layer. The elastic constants for the layer will be expressed in terms of geometrical and physical constants. First, however, we must consider the properties for a layer with unidirectional fibers. This has been done by S. W. Tsai (NASA CR-71, 1964).

Tsai includes two correction factors: the misalignment factor, k, and the contiguity factor, C.

For structures that are wound with prestressed fibers it appears reasonable to assume $k = 1.0$. The contiguity factor is of importance when the stiffness perpendicular to the fiber direction is evaluated ($C = 0$ for isolated filaments and $C = 1$ for contiguous filaments). The factor C is higher, of course, if the matrix content is low, but it is generally closer to 0 than to 1. Tsai uses successfully $C = 0$ in his comparison between theory and test.

The following information is needed for each layer:

E_f = modulus of fibers

ν_f = Poisson's ratio for fibers

E_m = modulus of matrix

ν_m = Poisson's ratio for matrix

t = thickness of layer

β = angle of wind

χ = matrix content (by volume).

By use of the equations by Tsai E_{11}, E_{22}, ν_{12}, and G may be determined. As E_{11} is modulus parallel to the fibers and E_{22} is the modulus perpendicular to the fibers; we have

$$\nu_{21} = \nu_{12} E_{22} / E_{11}$$

and

$$\bar{c}_{11} = E_{11}/(1-\nu_{12}\nu_{21})$$

$$\bar{c}_{22} = E_{22}/(1-\nu_{12}\nu_{21})$$

$$\bar{c}_{12} = \nu_{12}\bar{c}_{22}$$

$$\bar{c}_{33} = G$$

where the $\bar{c}_{ij}$ represent the stiffnesses of the layer in its principal directions.

We let ε_x' and ε_y' represent the strains in the axial and circumferential directions and γ_{xy}' corresponding shear strain.

The quantities, ε_1, ε_2, γ_{12}, are strains with respect to a cordinate system which is rotated an angle β. Then

$$\varepsilon_1 = \varepsilon_x' \cos^2\beta + \varepsilon_y' \sin^2\beta + \gamma_{xy}' \cos\beta \sin\beta$$

$$\varepsilon_2 = \varepsilon_x' \sin^2\beta + \varepsilon_y' \cos^2\beta + \gamma_{xy}' \cos\beta \sin\beta$$

$$\gamma_{12} = \gamma_{xy}'(\cos^2\beta - \sin^2\beta) + 2(\varepsilon_y' - \varepsilon_x') \cos\beta \sin\beta.$$

We will now consider the two sublayers with $\pm\beta$ winding angle; the extensional and bending stiffnesses will be determined.

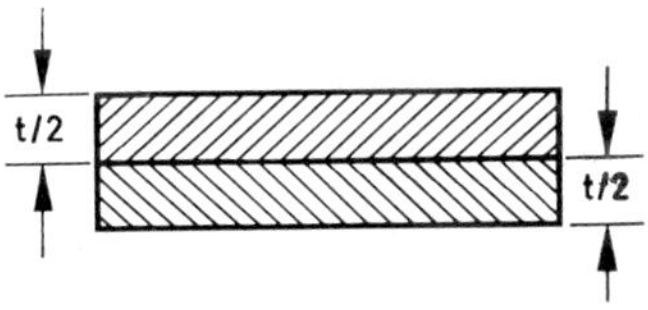

The strains at the middle surface of the double layer are ε_x, ε_y, and γ_{xy} in the directions given by the cylindrical coordinates. The changes of curvatures are $\varkappa_x$, $\varkappa_y$, $\varkappa_{xy}$. Hence in

Layer 1	Layer 2
$\varepsilon_x' = \varepsilon_x + \frac{t}{4}\varkappa$	$\varepsilon_x' = \varepsilon_x - \frac{t}{4}\varkappa_x$
$\varepsilon_y' = \varepsilon_y + \frac{t}{4}\varkappa_y$	$\varepsilon_y' = t_y - \frac{t}{4}\varkappa_y$
$\gamma_{xy}' = \gamma_{xy} + \frac{t}{2}\varkappa_{xy}$	$\gamma_{xy}' = \gamma_{xy} - \frac{t}{2}\varkappa_{xy}$

In Layer 1

$$\varepsilon_1 = \left(\varepsilon_x + \frac{t}{4}\varkappa_x\right)\cos^2\beta + \left(\varepsilon_y + \frac{t}{4}\varkappa_y\right)\sin^2\beta + \left(\gamma_{xy} + \frac{t}{2}\varkappa_{xy}\right)\cos\beta\sin\beta$$

$$\varepsilon_2 = \left(\varepsilon_x + \frac{t}{4}\varkappa_x\right)\sin^2\beta + \left(\varepsilon_y + \frac{t}{4}\varkappa_y\right)\cos^2\beta + \left(\gamma_{xy} + \frac{t}{2}\varkappa_{xy}\right)\cos\beta\sin\beta$$

$$\gamma_{12} = \left(\gamma_{xy} + \frac{t}{2}\varkappa_{xy}\right)(\cos^2\beta - \sin^2\beta) + 2\left(\varepsilon_y + \frac{t}{4}\varkappa_y - \varepsilon_x - \frac{t}{4}\varkappa_x\right)\cos\beta\sin\beta.$$

In Layer 2

$$\varepsilon_1 = \left(\varepsilon_x + \frac{t}{4}\varkappa_x\right)\cos^2\beta + \left(\varepsilon_y - \frac{t}{4}\varkappa_y\right)\sin^2\beta - \left(\gamma_{xy} - \frac{t}{2}\varkappa_{xy}\right)\cos\beta\sin\beta$$

$$\varepsilon_2 = \left(\varepsilon_x - \frac{t}{4}\varkappa_x\right)\sin^2\beta + \left(\varepsilon_y - \frac{t}{4}\varkappa_y\right)\cos^2\beta - \left(\gamma_{xy} - \frac{t}{2}\varkappa_{xy}\right)\cos\beta\sin\beta$$

$$\gamma_{12} = \left(\gamma_{xy} - \frac{t}{2}\varkappa_{xy}\right)(\cos^2\beta - \sin^2\beta) - 2\left(\varepsilon_y - \frac{t}{4}\varkappa_y - \varepsilon_x + \frac{t}{4}\varkappa_x\right)\cos\beta\sin\beta.$$

In addition we have the following changes of curvatures in the reference systems $+\beta$ and $-\beta$ to the generator:

$$\varkappa_1 = \varkappa_x\cos^2\beta + \varkappa_y\sin^2\beta \pm \varkappa_{xy}\cos\beta\sin\beta$$

$$\varkappa_2 = \varkappa_x\sin^2\beta + \varkappa_y\cos^2\beta \pm \varkappa_{xy}\cos\beta\sin\beta$$

$$\varkappa_{12} = \varkappa_{xy}(\cos^2\beta - \sin^2\beta) \pm (\varkappa_x - \varkappa_y)\cos\beta\sin\beta$$

where, when there are double signs, the upper applies to Layer 1 and the lower to Layer 2.

The strain energy density in the two layers is

$$V = \left\{\frac{t}{2}\left[\frac{\varepsilon_1^2 E_{11}}{1-\nu_{12}\nu_{21}} + \frac{\varepsilon_2^2 E_{22}}{1-\nu_{12}\nu_{21}} + \frac{2\nu_{12}E_2}{1-\nu_{12}\nu_{21}}\varepsilon_1\varepsilon_2 + G\gamma_{12}^2\right]\right.$$

$$\left. + \frac{t^3}{8\cdot 12}\left[\frac{E_{11}}{(1-\nu_{12}\nu_{21})}\varkappa_1^2 + \frac{E_{22}}{1-\nu_{12}\nu_{21}}\varkappa_2^2 + \frac{2\nu_{12}E_{21}}{1-\nu_{12}\nu_{21}}\varkappa_1\varkappa_2 + G\varkappa_{12}^2\right]\right\}.$$

A factor of 4 for the last term here was deleted for reasons which will be explained below.

$$
\begin{aligned}
V = \frac{1}{2}\Bigg\{ \frac{t}{2}\Bigg[& \frac{E_{11}}{1-\nu_{12}\nu_{21}} \Bigg\{ \left(\varepsilon_x + \frac{t}{4}\varkappa_x\right)\cos^2\beta + \left(\varepsilon_y + \frac{t}{4}\varkappa_y\right)\sin^2\beta \\
& + \left(\gamma_{xy} + \frac{t}{2}\varkappa_{xy}\right)\cos\beta\sin\beta \Bigg\}^2 + \frac{E_{22}}{1-\nu_{12}\nu_{21}} \Bigg\{ \left(\varepsilon_x + \frac{t}{4}\varkappa_x\right)\sin^2\beta \\
& + \left(\varepsilon_y + \frac{t}{4}\varkappa_y\right)\cos^2\beta + \left(\gamma_{xy} + \frac{t}{2}\varkappa_{xy}\right)\cos\beta\sin\beta \Bigg\}^2 + \frac{2\nu_{12}E_{22}}{1-\nu_{12}\nu_{21}} \\
& \times \Bigg\{ \left(\varepsilon_x + \frac{t}{4}\varkappa_x\right)\cos^2\beta + \left(\varepsilon_y + \frac{t}{4}\varkappa_y\right)\sin^2\beta + \left(\gamma_{xy} + \frac{t}{2}\varkappa_{xy}\right)\cos\beta\sin\beta \Bigg\} \\
& \times \Bigg\{ \left(\varepsilon_x + \frac{t}{4}\varkappa_x\right)\sin^2\beta + \left(\varepsilon_y + \frac{t}{4}\varkappa_y\right)\cos^2\beta + \left(\gamma_{xy} + \frac{t}{2}\varkappa_{xy}\right)\cos\beta\sin\beta \Bigg\} \\
& + G\Bigg\{ \left(\gamma_{xy} + \frac{t}{2}\varkappa_{xy}\right)(\cos^2\beta - \sin^2\beta) + 2\left(\varepsilon_y + \frac{t}{4}\varkappa_y - \varepsilon_x - \frac{t}{4}\varkappa_x\right)\cos\beta\sin\beta \Bigg\}^2 \\
& + \frac{E_{11}}{1-\nu_{12}\nu_{21}} \Bigg\{ \left(\varepsilon_x - \frac{t}{4}\varkappa_x\right)\cos^2\beta + \left(\varepsilon_y - \frac{t}{4}\varkappa_y\right)\sin^{12}\beta \\
& - \left(\gamma_{xy} - \frac{t}{2}\varkappa_{xy}\right)\cos\beta\sin\beta \Bigg\}^2 + \frac{E_{22}}{1-\nu_{12}\nu_{21}} \Bigg\{ \left(\varepsilon_x - \frac{t}{4}\varkappa_x\right)\sin^2\beta \\
& + \left(\varepsilon_y - \frac{t}{4}\varkappa_y\right)\cos^2\beta - \left(\gamma_{xy} - \frac{t}{2}\varkappa_{xy}\right)\cos\beta\sin\beta \Bigg\}^2 + \frac{2\nu_{12}E_{22}}{1-\nu_{12}\nu_{21}} \\
& \times \Bigg\{ \left(\varepsilon_x - \frac{t}{4}\varkappa_x\right)\cos^2\beta + \left(\varepsilon_y - \frac{t}{4}\varkappa_y\right)\sin^2\beta - \left(\gamma_{xy} - \frac{t}{2}\varkappa_{xy}\right)\cos\beta\sin\beta \Bigg\} \\
& \times \Bigg\{ \left(\varepsilon_x - \frac{t}{4}\varkappa_x\right)\sin^2\beta + \left(\varepsilon_y - \frac{t}{4}\varkappa_y\right)\cos^2\beta - \left(\gamma_{xy} - \frac{t}{2}\varkappa_{xy}\right)\cos\beta\sin\beta \Bigg\} \\
& + G\Bigg\{ \left(\gamma_{xy} - \frac{t}{2}\varkappa_{xy}\right)(\cos^2\beta - \sin^2\beta) - 2\left(\varepsilon_y - \frac{t}{4}\varkappa_y - \varepsilon_x + \frac{t}{4}\varkappa_x\right)\cos\beta\sin\beta \Bigg\}^2 \Bigg] \\
& + \frac{t^3}{96}\Bigg[\frac{E_{11}}{(1-\nu_{12}\nu_{21})}(\varkappa_x\cos^2\beta + \varkappa_y\sin^2\beta + \varkappa_{xy}\cos\beta\sin\beta)^2 + \frac{E_{22}}{1-\nu_{12}\nu_{21}} \\
& \times(\varkappa_x\sin^2\beta + \varkappa_y\cos^2\beta + \varkappa_{xy}\cos\beta\sin\beta)^2 + \frac{2\nu_{12}E_{22}}{1-\nu_{12}\nu_{21}} \\
& \times(\varkappa_x\cos^2\beta + \varkappa_y\sin^2\beta + \varkappa_{xy}\cos\beta\sin\beta)(\varkappa_x\sin^2\beta + \varkappa_y\cos^2\beta \\
& - \varkappa_{xy}\cos\beta\sin\beta) + G(\varkappa_{xy}(\cos^2\beta - \sin^2\beta) + (\varkappa_y - \varkappa_x)\cos\beta\sin\beta)^2 \\
& + \frac{E_{11}}{1-\nu_{12}\nu_{21}}(\varkappa_x\cos^2\beta + \varkappa_y\sin^2\beta - \varkappa_{xy}\cos\beta\sin\beta)^2 + \frac{E_{22}}{1-\nu_{12}\nu_{21}} \\
& \times(\varkappa_x\sin^2\beta + \varkappa_y\cos^2\beta - \varkappa_{xy}\cos\beta\sin\beta)^2 + \frac{2\nu_{12}E_{22}}{1-\nu_{12}\nu_{21}}(\varkappa_x\cos^2\beta \\
& + \varkappa_y\sin^2\beta - \varkappa_{xy}\cos\beta\sin\beta)(\varkappa_x\sin^2\beta + \varkappa_y\cos^2\beta - \varkappa_{xy}\cos\beta\sin\beta) \\
& + G(\varkappa_{xy}(\cos^2\beta - \sin^2\beta) - (\varkappa_y - \varkappa_x)\cos\beta\sin\beta)^2 \Bigg]\Bigg\}
\end{aligned}
$$

$$V = \frac{t}{2}\left[\frac{E_{11}}{1-\nu_{12}\nu_{21}}\left\{\varepsilon_x \cos^2\beta+\varepsilon_y \sin^2\beta+\frac{t}{2}\varkappa_{xy}\cos\beta\sin\beta\right\}^2\right.$$

$$+\frac{E_{11}}{1-\nu_{12}\nu_{21}}\left\{\frac{t}{4}\varkappa_x \cos^2\beta+\frac{t}{4}\varkappa_y \sin^2\beta+\gamma_{xy}\cos\beta\sin\beta\right\}^2$$

$$+\frac{E_{22}}{1-\nu_{12}\nu_{21}}\left\{\varepsilon_x \sin^2\beta+\varepsilon_y \cos^2\beta+\frac{t}{2}\varkappa_{xy}\cos\beta\sin\beta\right\}^2$$

$$+\frac{E_{22}}{1-\nu_{12}\nu_{21}}\left\{\frac{t}{4}\varkappa_x \sin^2\beta+\frac{t}{4}\varkappa_y \cos^2\beta+\gamma_{xy}\cos\beta\sin\beta\right\}^2$$

$$+\frac{2\nu_{12}E_{22}}{1-\nu_{12}\nu_{21}}\left\{\varepsilon_x \cos^2\beta+\varepsilon_y \sin^2\beta+\frac{t}{2}\varkappa_{xy}\cos\beta\sin\beta\right\}$$

$$\times\left\{\varepsilon_x \sin^2\beta+\varepsilon_y \cos^2\beta+\frac{t}{2}\varkappa_{xy}\cos\beta\sin\beta\right\}$$

$$+\frac{2\nu_{12}E_{22}}{1-\nu_{12}\nu_{21}}\left\{\frac{t}{4}\varkappa_x \cos^2\beta+\frac{t}{4}\varkappa_y \sin^2\beta+\gamma_{xy}\cos\beta\sin\beta\right\}$$

$$\times\left\{\frac{t}{4}\varkappa_x \sin^2\beta+\frac{t}{4}\varkappa_y \cos^2\beta+\gamma_{xy}\cos\beta\sin\beta\right\}$$

$$+G\left\{\gamma_{xy}(\cos^2\beta-\sin^2\beta)+2\frac{t}{4}\varkappa_y\cos\beta\sin\beta-2\frac{t}{4}\varkappa_x\cos\beta\sin\beta\right\}^2$$

$$\left.+G\left\{\frac{t}{2}\varkappa_{xy}(\cos^2\beta-\sin^2\beta)+2\varepsilon_y\cos\beta\sin\beta-2\varepsilon_x\cos\beta\sin\beta\right\}^2\right]$$

$$+\frac{t^3}{96}\left[\frac{E_{11}}{1-\nu_{12}\nu_{21}}\{(\varkappa_x\cos^2\beta+\varkappa_y\sin^2\beta)^2+\varkappa_{xy}^2\cos^2\beta\sin^2\beta\}\right.$$

$$+\frac{E_{22}}{1-\nu_{12}\nu_{21}}\{(\varkappa_x\sin^2\beta+\varkappa_y\cos^2\beta)^2+\varkappa_{xy}^2\cos^2\beta\sin^2\beta\}$$

$$+\frac{2\nu_{12}E_{22}}{1-\nu_{12}\nu_{21}}\{(\varkappa_x\cos^2\beta+\varkappa_y\sin^2\beta)(\varkappa_x\sin^2\beta+\varkappa_y\cos^2\beta)+\varkappa_{xy}^2\cos^2\beta\sin\beta\}$$

$$\left.+G\left\{\frac{1}{4}\varkappa_{xy}^2(\cos^2\beta-\sin^2\beta)^2+(\varkappa_y-\varkappa_x)^2\cos^2\beta\sin^2\beta\right\}\right]$$

$$N_x = \frac{\partial V}{\partial \varepsilon_x} = t\left[\cos^2\beta \frac{E_{11}}{1-\nu_{12}\nu_{21}}\left(\varepsilon_x \cos^2\beta + \varepsilon_y \sin^2\beta + \frac{t}{2}\varkappa_{xy}\cos\beta\sin\beta\right)\right.$$

$$+\sin^2\beta \frac{E_{22}}{1-\nu_{12}\nu_{21}}\left(\varepsilon_x \sin^2\beta + \varepsilon_y \cos^2\beta + \frac{t}{2}\varkappa_{xy}\cos\beta\sin\beta\right)$$

$$+\frac{\nu_{12}E_{22}}{1-\nu_{12}\nu_{21}}\left\{\cos^2\beta\left(\varepsilon_x \sin^2\beta + \varepsilon_y \cos^2\beta + \frac{t}{2}\varkappa_{xy}\cos\beta\sin\beta\right)\right.$$

$$\left.+\sin^2\beta\left(\varepsilon_x \cos^2\beta + \varepsilon_y \sin^2\beta + \frac{t}{2}\varkappa_{xy}\cos\beta\sin\beta\right)\right\}$$

$$+G(-2\cos\beta\sin\beta)\left\{\frac{t}{2}\varkappa_{xy}(\cos^2\beta - \sin^2\beta)\right.$$

$$\left.\left.+2\varepsilon_y\cos\beta\sin\beta - 2\varepsilon_x\cos\beta\sin\beta\right\}\right]$$

with

$$N_x = c_{11}\varepsilon_x + c_{12}\varepsilon_y + c_{16}\varkappa_{xy}$$

$$c_{11} = t\left[\frac{E_{11}}{1-\nu_{12}\nu_{21}}\cos^4\beta + \frac{E_{22}}{1-\nu_{12}\nu_{21}}\sin^4\beta + \frac{2\nu_{12}\nu_{21}}{1-\nu_{12}\nu_{21}}\cos^2\beta\sin^2\beta\right.$$

$$\left.+4G\cos^2\beta\sin^2\beta\right]$$

$$\left\{\begin{aligned}
c_{11} &= \frac{t}{1-\nu_{12}\nu_{21}}\Big[E_{11}\cos^4\beta + E_{22}\sin^4\beta \\
&\quad + \{2\nu_{12}E_{22} + 4G(1-\nu_{12}\nu_{21})\}\cos^2\beta\sin^2\beta\Big] \\
c_{12} &= \frac{t}{1-\nu_{12}\nu_{21}}\Big[\{E_{11}+E_{22}-4G(1-\nu_{12}\nu_{21})\}\sin^2\beta\cos^2\beta \\
&\quad + \nu_{12}E_{22}(\cos^4\beta + \sin^4\beta)\Big] \\
c_{16} &= \frac{t^2}{2(1-\nu_{12}\nu_{21})}\sin\beta\cos\beta[E_{11}\cos^2\beta + E_{22}\sin^2\beta + \nu_{12}E_{22} \\
&\quad - 2G(1-\nu_{12}\nu_{21})(\cos^2\beta - \sin^2\beta)]
\end{aligned}\right.$$

$$N_y = \frac{\partial V}{\partial \varepsilon_y} = t\left[\sin^2\beta \frac{E_{11}}{1-\nu_{12}\nu_{21}}\left(\varepsilon_x \cos^2\beta + \varepsilon_y \sin^2\beta + \frac{t}{2}\varkappa_{xy}\cos\beta\sin\beta\right)\right.$$

$$+\cos^2\beta \frac{E_{22}}{1-\nu_{12}\nu_{21}}\left(\varepsilon_x \sin^2\beta + \varepsilon_y \cos^2\beta + \frac{t}{2}\varkappa_{xy}\cos\beta\sin\beta\right)$$

$$+\sin^2\beta \frac{\nu_{12}E_{22}}{1-\nu_{12}\nu_{21}}\left(\varepsilon_x \sin^2\beta + \varepsilon_y \cos^2\beta + \frac{t}{2}\varkappa_{xy}\cos\beta\sin\beta\right)$$

$$+\cos^2\beta \frac{\nu_{12}E_{22}}{1-\nu_{12}\nu_{21}}\left(\varepsilon_x \cos^2\beta + \varepsilon_y \sin^2\beta + \frac{t}{2}\varkappa_{xy}\cos\beta\sin\beta\right)$$

$$+(2\cos\beta\sin\beta)G\left\{\frac{t}{2}\varkappa_{xy}(\cos^2\beta-\sin^2\beta)+2\varepsilon_y\cos\beta\sin\beta\right.$$

$$\left.\left.-2\varepsilon_x\cos\beta\sin\beta\right\}\right]$$

with

$$N_y = c_{12}\varepsilon_x + c_{22}\varepsilon_y + c_{26}\varkappa_{xy}$$

$$\left\{\begin{aligned} c_{22} &= \frac{t}{1-\nu_{12}\nu_{21}}[E_{11}\sin^4\beta + E_{22}\cos^4\beta \\ &\quad + \{2\nu_{12}E_{22} + 4G(1-\nu_{12}\nu_{21})\}\sin^2\beta\cos^2\beta \\ c_{26} &= \frac{t^2}{2(1-\nu_{12}\nu_{21})}\sin\beta\cos\beta[E_{11}\sin^2\beta + E_{22}\cos^2\beta + \nu_{12}E_{22} \\ &\quad + 2G(1-\nu_{12}\nu_{21})(\cos^2\beta - \sin^2\beta)] \end{aligned}\right.$$

$$N_{xy} = \frac{\partial V}{\partial \gamma_{xy}} = t\left[\cos\beta\sin\beta\frac{E_{11}}{1-\nu_{12}\nu_{21}}\left\{\frac{t}{4}\varkappa_x\cos^2\beta + \frac{t}{4}\varkappa_4\sin^2\beta\right.\right.$$

$$\left.+\gamma_{xy}\cos\beta\sin\beta\right\} + \cos\beta\sin\beta\frac{E_{22}}{1-\nu_{12}\nu_{21}}\left\{\frac{t}{4}\varkappa_x\sin^2\beta + \frac{t}{4}\varkappa_y\cos^2\beta\right.$$

$$\left.+\gamma_{xy}\cos\beta\sin\beta\right\} + \cos\beta\sin\beta\frac{\nu_{12}E_{22}}{1-\nu_{12}\nu_{21}}\left\{\frac{t}{4}\varkappa_x + \frac{t}{4}\varkappa_y\right.$$

$$\left.+2\cos\beta\sin\beta\,\gamma_{xy}\right\} + G(\cos^2\beta-\sin^2\beta)\left\{\gamma_{xy}(\cos^2\beta-\sin^2\beta)\right.$$

$$\left.\left.+2\frac{t}{4}\varkappa_y\cos\beta\sin\beta - 2\frac{t}{4}\varkappa_x\cos\beta\sin\beta\right\}\right]$$

with

$$N_{xy} = c_{33}\gamma_{xy} + c_{34}\varkappa_x + c_{35}\varkappa_y$$

$$\begin{cases} c_{33} = \dfrac{t}{(1-\nu_{12}\nu_{21})}[(E_{11}+E_{22}+2\nu_{12}E_{22})\cos^2\beta\sin^2\beta \\ \quad + G(1-\nu_{12}\nu_{21})(\cos^2\beta-\sin^2\beta)^2] \\ c_{34} = \dfrac{t^2}{4(1-\nu_{12}\nu_{21})}\cos\beta\sin\beta[E_{11}\cos^2\beta+E_{22}\sin^2\beta+\nu_{12}E_{22} \\ \quad -2G(1-\nu_{12}\nu_{21})(\cos^2\beta-\sin^2\beta)] \\ c_{35} = \dfrac{t^2}{4(1-\nu_{12}\nu_{21})}\cos\beta\sin\beta\,[E_{11}\sin^2\beta+E_{22}\cos^2\beta\nu_{12}E_{22} \\ \quad +2G(1-\nu_{12}\nu_{21})(\cos^2\beta-\sin^2\beta)] \end{cases}$$

$$\begin{aligned} M_x = \frac{\partial V}{\partial \varkappa_x} &= c_{34}\gamma_{xy} + \frac{t^3}{16}\cdot\frac{E_{11}}{1-\nu_{12}\nu_{21}}\cos^2\beta\{\varkappa_x\cos^2\beta+\varkappa_y\sin^2\beta\} \\ &+ \frac{t^3}{16}\cdot\frac{E_{22}}{1-\nu_{12}\nu_{21}}\sin^2\beta\{\varkappa_x\sin^2\beta+\varkappa_y\cos^2\beta\} \\ &+ \frac{t^3}{16}\cdot\frac{\nu_{12}E_{22}}{1-\nu_{12}\nu_{21}}\{\cos^2\beta(\varkappa_x\sin^2\beta+\varkappa_y\cos^2\beta) \\ &+ \sin^2\beta(\varkappa_x\cos^2\beta+\varkappa_y\sin^2\beta)\} + \frac{t^3}{16}\cdot 4G\cos^2\beta\sin^2\beta(\varkappa_x-\varkappa_y) \\ &+ \frac{t^3}{48}\left\{\frac{E_{11}}{1-\nu_{12}\nu_{21}}\cos^2\beta(\varkappa_x\cos^2\beta+\varkappa_y\sin^2\beta)\right. \\ &+ \frac{E_{22}}{1-\nu_{12}\nu_{21}}\sin^2\beta(\varkappa_x\sin^2\beta+\varkappa_y\cos^2\beta) + \frac{\nu_{12}E_{22}}{1-\nu_{12}\nu_{21}}[\cos^2\beta(\varkappa_x\sin^2\beta \\ &\left. +\varkappa_y\cos^2\beta)+\sin^2\beta(\varkappa_x\cos^2\beta+\varkappa_y\sin^2\beta)]+4G\cos^2\beta\sin^2\beta(\varkappa_x-\varkappa_y)\right\} \end{aligned}$$

with

$$M_x = c_{34}\gamma_{xy} + c_{44}\varkappa_x + c_{45}\varkappa_y$$

$$\begin{cases} c_{44} = \dfrac{t^3}{12(1-\nu_{12}\nu_{21})}\{E_{11}\cos^4\beta+E_{22}\sin^4\beta+(2\nu_{12}E_{22}+4G')\sin^2\beta\cos^2\beta\} \\ c_{45} = \dfrac{t^3}{12(1-\nu_{12}\nu_{21})}\{\sin^2\beta\cos^2\beta(E_{11}+E_{22}-4G')+\nu_{12}E_{22}(\cos^4\beta+\sin^4\beta)\} \end{cases}$$

$$\begin{aligned} M_y = \frac{\partial V}{\partial \varkappa_y} &= c_{35}\gamma_{xy} + c_{45}\varkappa_x + \left\{\frac{t^3}{16}\,\frac{E_{22}}{1-\nu_{12}\nu_{21}}\cos^4\beta + \frac{t^3}{16}\,\frac{\nu_{12}E_{22}}{1-\nu_{12}\nu_{22}}\right. \\ &\times\sin^2\beta\cos^2\beta + 4G\,\frac{t^3}{16}\cos^2\beta\sin^2\beta + \frac{t^3}{48}\,\frac{E_{11}}{1-\nu_{12}\nu_{21}}\sin^4\beta \\ &+ \frac{t^3}{48}\,\frac{E_{22}}{1-\nu_{12}\nu_{21}}\cos^4\beta + \frac{t^3}{48}\,\frac{\nu_{12}E_{22}}{1-\nu_{12}\nu_{21}}\sin^2\beta\cos^2\beta \\ &\left. + \frac{t^3}{48}\cdot 4G\cos^2\beta\sin^2\beta\right\}\varkappa_y \end{aligned}$$

with

$$M_y = c_{35}\gamma_{xy}+c_{45}\varkappa_x+c_{55}\varkappa_y$$

$$c_{55} = \frac{t^3}{12(1-\nu_{12}\nu_{21})}[E_{11}\sin^4\beta+E_{22}\cos^4\beta+(\nu_{12}E_{22}+4G')\sin^2\beta\cos^2\beta]$$

$$M_{xy}+M_{yx} = \frac{\partial V}{\partial\varkappa_{xy}} = \left\{\frac{t^3}{16}\frac{1}{(1-\nu_{21}\nu_{12})}[E_{11}\cos^2\beta\sin^2\beta+E_{22}\cos^2\beta\sin^2\beta\right.$$
$$+\nu_{12}E_{22}\cos^2\beta\sin^2\beta]$$
$$+\frac{t^3}{16}\cdot G(\cos^2\beta-\sin^2\beta)^2\cdot 4+\frac{t^3}{48}\frac{1}{(1-\nu_{12}\nu_{21})}[E_{11}\cos^2\beta\sin^2\beta$$
$$+E_{22}\cos^2\beta\sin^2\beta+\nu_{12}E_{22}\cos^2\beta\sin^2\beta]+\frac{t^3}{48}\cdot G(\cos^2\beta$$
$$\left.-\sin^2\beta)^2\cdot 4\right\}\varkappa_{xy}+c_{16}\varepsilon_x+c_{26}\varepsilon_y.$$

A factor of 4 included in the second and fourth terms here is used to compensate for the factor which was previously deleted and also for the loss of torsional stiffness which occurs if two layers are considered separately and thereafter added.

With $M_{xy}+M_{yx} = c_{16}\varepsilon_x+c_{26}\varepsilon_y+c_{66}\varkappa_{xy}$

$$c_{66} = \frac{t^3\cos^2\beta\sin^2\beta}{12(1-\nu_{12}\nu_{21})}[E_{11}+E_{22}+\nu_{12}E_{22}]+\frac{t^3}{3}\cdot G(\cos^2\beta-\sin^2\beta)^2.$$

II. *Layered Shells*

The coefficients derived above define the constitutive relations for one (double) layer as follows:

$$\begin{aligned}
N_x &= c_{11}\varepsilon_x+c_{12}\varepsilon_y+c_{16}\varkappa_{xy}\\
N_y &= c_{12}\varepsilon_x+c_{22}\varepsilon_y+c_{26}\varkappa_{xy}\\
N_{xy} &= c_{33}\gamma_{xy}+c_{34x}+c_{35}\varkappa_y\\
M_x &= c_{34}\gamma_{xy}+c_{44x}+c_{45}\varkappa_y\\
M_y &= c_{35}\gamma_{xy}+c_{45}\varkappa_x+c_{55}\varkappa_y\\
M_{yx}+M_{xy} &= c_{16}\varepsilon_x+c_{26}\varepsilon_4+c_{66}\varkappa_{xy}.
\end{aligned}$$

Here ε_x, ε_y, and γ_{xy} are strains at the midplane of the layer.

The strain energy density in one layer can be written

$$\begin{aligned}
&\tfrac{1}{2}\{\varepsilon_xN_x+\varepsilon_yN_y+\gamma_{xy}N_{xy}+\varkappa_xM_x+\varkappa_yM_y+\varkappa_{xy}(M_{xy}+M_{yx})\}\\
&= \tfrac{1}{2}\{c_{11}\varepsilon_x^2+2c_{12}\varepsilon_x\varepsilon_y+2c_{16}\varepsilon_x\varkappa_{xy}+c_{22}\varepsilon_y^2+2c_{26}\varepsilon_y\varkappa_{xy}+c_{33}\gamma_{xy}^2\\
&\quad+2c_{34}\gamma_{xy}\varkappa_x+2c_{35}\gamma_{xy}\varkappa_y+c_{44}\gamma_x^2+2c_{45}\varkappa_x\varkappa_y+c_{55}\varkappa_y^2+c_{66}\varkappa_{xy}^2\}.
\end{aligned}$$

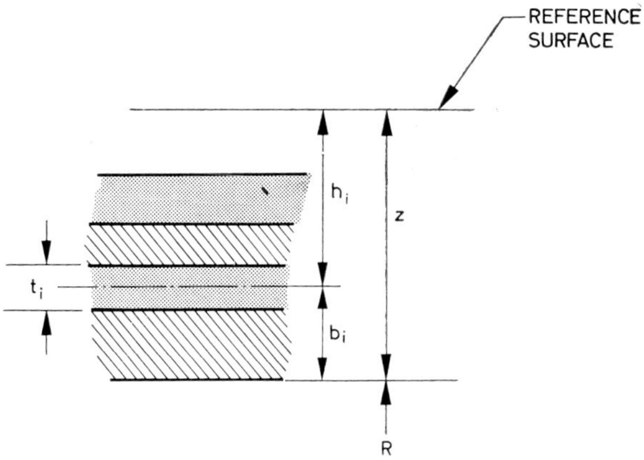

Strains in layer number 1

$$\varepsilon_x^i = \varepsilon_x + h_i \varkappa_x$$
$$\varepsilon_y^i = \varepsilon_y + h_i \varkappa_y$$
$$\gamma_{xy}^i = \gamma_{xy} + 2h_i \varkappa_{xy}$$

where ε_x, ε_y, γ_{xy} now apply at the reference surface.

$$V = \tfrac{1}{2} \sum_{i=1}^{I} [c_{11}^i(\varepsilon_x + h_i\varkappa_x)^2 + 2c_{12}^i(\varepsilon_x + h_i\varkappa_x)(\varepsilon_y + h_i\varkappa_y)$$
$$+ 2c_{16}^i(\varepsilon_x + h_i\varkappa_x)\varkappa_{xy} + c_{22}^i(\varepsilon_y + h_i\varkappa_y)^2 + 2c_{26}^i(\varepsilon_y + h_i\varkappa_y)\,\varkappa_{xy}$$
$$+ c_{33}^i(\gamma_{xy} + 2h_i\varkappa_{xy})^2 + 2c_{34}^i(\gamma_{xy} + 2h_i\varkappa_{xy})\varkappa_x + 2c_{35}^i(\gamma_{xy} + 2h_i\varkappa_{xy})\,\varkappa_y$$
$$+ c_{44}^i\varkappa_x^2 + 2c_{45}^i\varkappa_x\varkappa_y + c_{55}^i\varkappa_y^2 + \tfrac{1}{4}\, c_{66}\varkappa_{xy}^2]$$

$$N_x = \frac{\partial V}{\partial \varepsilon_x} = \sum_{i=1}^{I} [c_{11}^i(\varepsilon_x - h_i\varkappa_x) + c_{12}^i(\varepsilon_y + h_i\varkappa_y) + c_{16}^i\,(\varkappa_{xy})]$$

$$c_{11} = \sum_{i=1}^{I} c_{11}^i,\ c_{12} = \sum_{i=1}^{I} c_{12}^i,\ c_{14} = \sum_{i=1}^{I} -h_i c_{11}^i,\ c_{15} = \sum_{i=1}^{I} +h_i c_{12}^i,\ c_{16} = \sum_{i=1}^{I} c_{16}^i$$

$$N_y = \frac{\partial V}{\partial \varepsilon_y} = \sum_{i=1}^{I} [c_{12}^i(\varepsilon_x + h_i\varkappa_x) + c_{22}^i(\varepsilon_y + h_i\varkappa_y) + c_{16}^i\varkappa_{xy}]$$

$$c_{22} = \sum_{i=1}^{I} c_{22}^i,\ c_{24} = \sum + c_{12}^i h_i,\ c_{25} = \sum + c_{22}^i h_i,\ c_{26} = \sum c_{26}^i$$

$$N_{xy} = \frac{\partial V}{\partial \gamma_{xy}} = \sum_{i=1}^{I} c_{33}^i(\gamma_{xy} + 2h_i\gamma_{xy}) + c_{34}^i\gamma_x + c_{35}^i\gamma_x$$

$$c_{33} = \sum c_{33}^i,\ c_{34} = \sum c_{34}^i,\ c_{35} = \sum c_{35}^i,\ c_{36} = \sum +2h_i c_{36}^i$$

$$M_x = \frac{\partial V}{\partial \varkappa_x} = \sum + h_i c_{11}^i(\varepsilon_x + h_i\varkappa_x) + h_i c_{12}^i(\varepsilon_y + h_i\varkappa_y) + h_i c_{16}^i\varkappa_{xy} + c_{34}^i(\gamma_{xy} + 2h_i\varkappa_{xy})$$
$$+ c_{44}^i\varkappa_x + c_{45}^i\varkappa_y$$

$$c_{44} = \sum (h_i^2 c_1^i + c_{44}^i),\ c_{45} = \sum (h_i^2 c_{12}^i + c_{45}^i),\ c_{46} = \sum + h_i(c_{16}^i + 2c_{34}^i)$$

$$M_y = \frac{\partial V}{\partial \varkappa_y} = \sum +c_{12}^i h_i(\varepsilon_x + h_i\varkappa_x) + c_{22}^i h_i(\varepsilon_y + h_i\varkappa_y) - h_i c_{26}^i \varkappa_{xy} + c_{35}(\gamma_{xy} + 2h_i\varkappa_{xy})$$

$$+c_{45}^i \varkappa_x + c_{55}^i \varkappa_y$$

$$c_{55} = \sum (h_i^2 c_{22}^i + c_{55}^i), \; c_{56} = \sum -h_i(c_{26}^i + 2c_{35}^i)$$

$$M_{yx} + M_{xy} = \frac{\partial V}{\partial \varkappa_{xy}} \sum c_{16}^i(\varepsilon_x + h_i\varkappa_x) + c_{26}^i(\varepsilon_y + h_i\varkappa_7) + h_i c_{33}^i(\gamma_{xy} + 2h_i\varkappa_{xy}) + h_i c_{xy}\varkappa_x$$

$$+h_i c_{25}^i \varkappa_y + c_{66}^i \varkappa_{xy}$$

$$c_{66} = 4\sum (h_i^2 c_{33}^i + \tfrac{1}{4} c_{66}^i)$$

The reference surface is chosen such that:

$$c_{36} = 0 \qquad \sum h_i c_{33}^i = 0$$

$$h_i = z - b_i$$

$$\sum (z_i - b_i) c_{33}^i = 0$$

$$z \sum c_{33}^i = \sum b_i c_{33}^i$$

$$z = \frac{\sum b_i c_{33}^i}{\sum c_{33}^i}.$$

Appendix B

EQUATIONS FOR ORTHOTROPIC ANALYSIS

The equations given in this appendix are taken from ref. 3 where a small deflection buckling analysis is presented for cylindrical orthotropic shells subjected to radial pressures varying in the axial direction. The critical pressure term has been slightly changed to reflect a uniform constant pressure application.

The critical lateral pressure is given by

$$P_{cr} = \frac{D_x t}{R} C_{\min}$$

where $C_{\min}$ is obtained from the following set of homogeneous linear equations (minimized with respect to n):

$$\begin{vmatrix} \frac{U_1}{C} - 2 & 0 & 0 & 0 \ldots \\ 0 & \frac{U_2}{C} - 2 & 0 & 0 \ldots \\ 0 & 0 & \frac{U_3}{C} - 2 & 0 \ldots \\ 0 & 0 & 0 & \frac{U_4}{C} - 2 \ldots \\ \vdots & \vdots & \vdots & \vdots \end{vmatrix} = 0$$

where

$$U_{m=1,\,2} = -\frac{\left(\dfrac{D_x D_\varphi - D_1^2}{D_x^2}\right)\left(\dfrac{m\pi}{\lambda}\right)^4 + \dfrac{h^2}{12R^2}\left[\left(\dfrac{m\pi}{\lambda}\right)^4 + 2n^2\dfrac{(D_1+2G)}{D_x}\left(\dfrac{m\pi}{\lambda}\right)^2 + \dfrac{D_\varphi}{D_x}n^4\right]\left[\left(\dfrac{m\pi}{\lambda}\right)^4 + \left(\dfrac{D_x D_\varphi - D_1^2 - 2D_1 G}{D_x G}\right)\left(\dfrac{m\pi}{\lambda}\right)^2 n^2 + \dfrac{D_\varphi}{D_x}n^4\right]}{\dfrac{1}{2}(1-n^2)\left[\left(\dfrac{m\pi}{\lambda}\right)^4 + \left(\dfrac{D_x D_\varphi - D_1^2 - 2D_1 G}{D_x G}\right)\left(\dfrac{m\pi}{\lambda}\right)^2 n^2 + \dfrac{D_\varphi}{D_x}n^4\right]}$$

$$D_x = \frac{E_x}{1-\mu_{x\phi}\mu_{\phi x}} \qquad D_1 = \frac{E_\phi \mu_{x\varphi}}{1-\mu_{x\varphi}\mu_{\varphi x}} \qquad n = \text{number of circumferential waves}$$

$$D_\varphi = \frac{E_\phi}{1-\mu_{x\varphi}\mu_{\varphi x}} \qquad G = G_{x\varphi} \qquad m = \text{number of axial waves.}$$

OPTIMIZATION OF SHELL STRUCTURES UNDER BENDING AND TORSION LOADS

ULRICH HÜTTER†

THE application of fiber-matrix composites (FMC) for structural members loaded not only, or at least not predominantly, in an uniaxial way (e.g., simple tension or compression) is limited by the fact that matrix materials of all known composites have much lower ultimate stress limits than the fibers themselves.

I. "GEDANKENMODELL"—A FICTIVE MODEL

The resulting consequences can be clarified considering a simple tube model with a thin wall consisting of a two-layer laminate. It is assumed that shear forces between the fibers themselves and between the two layers are so

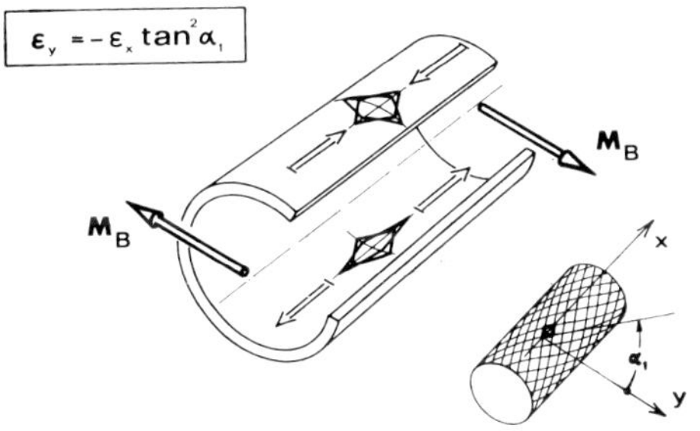

FIG. 1. Deformation of basic rhombic elements

small that they can be neglected. The fibers are screw-lines (helices) which are symmetrical about the tube axis. The fiber thickness is of the magnitude of half the wall thickness (Fig. 1).

This results in an idealized fiber network built up from rhombic elements.

If the crossing fibers of the ply angle cannot make any displacing motion relative to each other, the rhombi will deform under load in such a way that the relation for Poisson's ratio becomes $\nu_{xy} = -\tan^2 \alpha$. Designing α greater than 45 degrees, relatively large bulges at the tube will result when it is subjected to axial compression. It will be heavily contracted when under tensile

† o. Professor Dr. rer. techn., Institut für Flugzeugbau der Technischen Hochschule Stuttgart, Deutsche Forschungsanstalt für Luft- und Raumfahrt.

load. If the tube is bent there occurs bulging in the compression zone of the ply angle.

Twisting the tube, the one layer tries to bulge out while the other layer is forced to contract in diameter.

A laminate not restrained with respect to motion in radial directions shows a similar behavior. In other words when the simplified two-layer model is twisted in one direction, the outer layer bulges out and the inner layer contracts; but when twisted in the opposite direction, both layers will nearly maintain their positions.

Since the free movability of the laminate of such a single-curved shell without shear transfer yields radial movements, it seems logical to expect that a constraint of this radial movability will directly affect the elastic constants and the loading capacity of the shell.

Such a constraint is given by a radial reinforcement of the ply angle accomplished most effectively and simply by circumferential fibers which are wrapped around outside to hinder expansion and inside to support against contraction.

Thus, shells subjected to bending must have at least one layer outside and one inside to support the compression and tension zones.

II. SIMPLIFIED DEMONSTRATION MODEL

Experiments with a drastically idealized two-layer demonstration model show outside the load induction ranges all those phenomena which result from the foregoing considerations. The fibers are represented by thin helically arranged rods which are free to move apart from mutual restraint (Fig. 2).

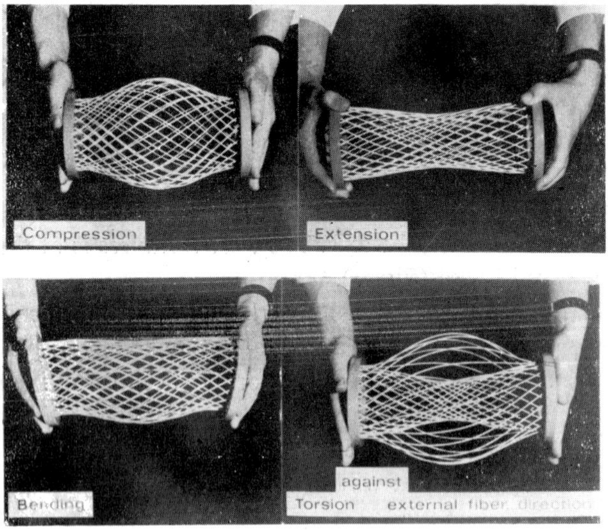

FIG. 2. Deformations of idealized model

The model display indicates that under a pure bending load, displacement of cross-sections results due to lateral extension of the compression zone and lateral contraction of the tension zone. The direction of this displacement is just opposite to that direction of resulting deflection expected and known for structures of homogeneous material.

III. DEFORMATION OF CROSS-SECTION

Admittedly, structures using fiber-matrix composites (FMC) do not behave so drastically as the simple model shown, the fibers of which had no shear-tie at all. However, a deformation like that shown so impressively at the model can be experimentally verified as a shifting of the neutral line against the direction of beam deflection given by ETB (engineers' theory of bending).

This shifting of the neutral line was studied at the Deutsche Forschungsanstalt für Luft- und Raumfahrt (DFL) in Stuttgart, testing crosswise filament wound tubes of fiber-glass-reinforced epoxy-resin[1]. The circle tubes were fitted with lateral reinforcing layers of different thicknesses. Tubes without reinforcement were tested in comparison (Fig. 3).

The specimens of this test series were of such dimensions as to eliminate the possibility of failure due to buckling. The tubes were loaded by a pure bending moment, that is to say there were no transverse forces acting on them. There also were no disturbances inside the measuring range due to load induction.

The shift of the neutral line is defined by the shift angle $\Delta\beta$.

At the unsupported test specimen, the neutral line shifted greatly with

FIG. 3. Test set-up for fiber-glass-reinforced tubes

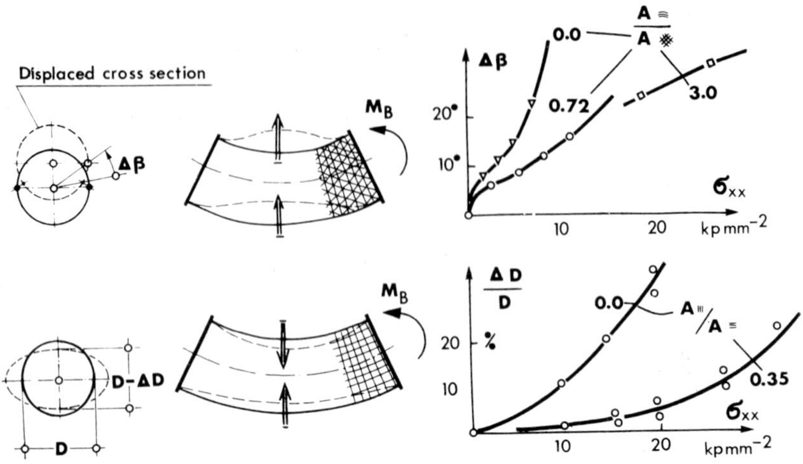

FIG. 4. Cross-section displacement under bending loads

very low values of load intensity, and this increased progressively as load increased. The supported test tubes showed a distinct increase of the shift angle $\Delta\beta$ due to stresses. The greater the ratio of the thickness of the hoop layer relative to the thickness of the angle ply layer, the flatter was the slope of the $\Delta\beta$ variation (Fig. 4).

A similar phenomenon was reported by Grüninger,[(2)] DFL Stuttgart, in 1964. He referred to the influence of hoop layers upon the oval deformation of thin-walled curved circle-tubes as described by Bantlin and von Kármán.

FIG. 5. Shell deformation due to lateral bulging

These latter measurements were made at tubes consisting only of longitudinal FGRP roving layer and the said supporting hoop layer, if any. The results showed that in relation to the stresses in the longitudinal layer, laterally unsupported tubes had a greater tendency of becoming oval under load than those having hoop-layers (Fig. 5).

Recognizable progressive ovalling and increased shifting of the neutral line start at that stress level corresponding to the yield point of the matrix material.

It is remarkable that in the range of the compression zone, the shell deformation due to lateral bulging-out is directed against the shell deformation dependent on ovalling.

IV. "FORMSCHLUSS"—THE FORM-CLOSED CONNECTION OF LAYERS

The loading capacity of a multilayer laminate is dependent on whether the network elements resulting from the fiber layers are free to move or not at the cross points of fibers even with vanishing shear-tie.

A fiber subjected to tension, running along a spatial surface between two points, and having no friction-tie with this surface will remain in its initial position only if the surface curve followed represents the shortest distance between the two points considered.

On a spherical surface, this shortest distance is a great circle. On a tapered shell with straight generatrix, it is a line intersecting all generatrices at an angle changing continuously. These lines can be easily realized by a filament winding technology.

It is required to control angular velocity ω of the mandrel and/or advance velocity $\dot{x}$ of the winding head in such a way that $\dot{x}$ remains a continuous function of the tangential velocity at the first point of contact of the strand,

$$\dot{x}/r \cdot \omega \cdot \cos \overline{Nr} = f(x).$$

At a tapered shell with a circle as cross section, we obtain $f(x) = k_0 + k \cdot x$. A strand pattern meeting such conditions has the typical feature that the centers of curvatures of the cross lines lie upon the surface-normal which intersects the surface at the cross point. Therefore, the resultant of all normal forces

$$dN_{iw} = A_{fi} \cdot \sigma_i \cdot \frac{de_i}{r_i}$$

coincides with the vector of the surface-normal, and cross points of fibres do not shift in the direction of any cross fibers if subjected to loads. We call such a behavior a form-closed connection of fibers (Formschluss). Primary

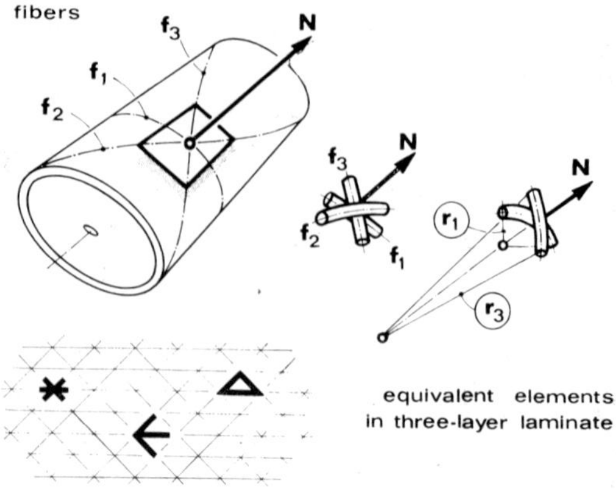

FIG. 6. Equivalent elements in three-layer laminate

condition requires the arrangement of all strand layers in such a way that any radial movement of fibers out of contact is prevented (Fig. 6).

In case these relations are given (i.e., fixed crosspoints assured by formclosed connections) then the laminate quasi like a network of statically determined elements will be able to carry not only longitudinal and transverse loads but also shear loads. The shear strength of the matrix needs only to contribute a small part.

With flat specimens having radii $r_i \rightarrow \infty$ this type of formclose is not feasible. In strength tests on flat specimens it is therefore difficult to realize a lateral

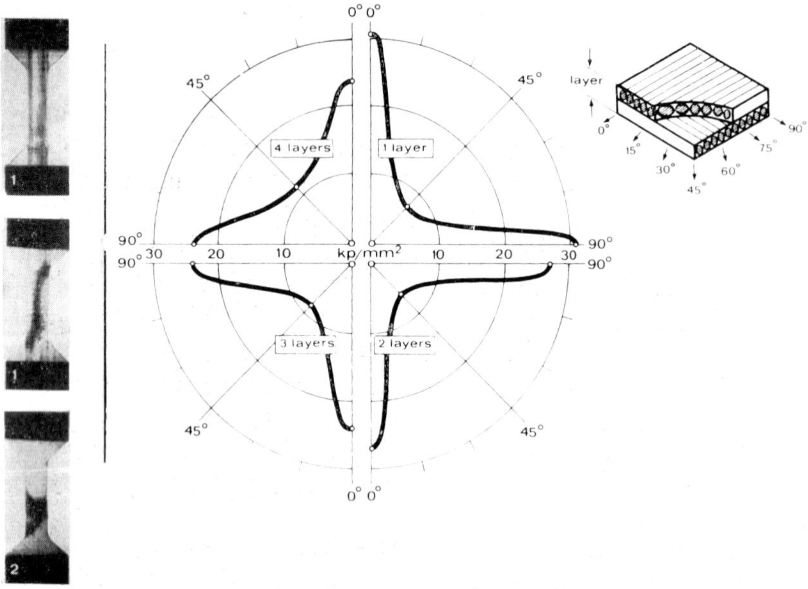

FIG. 7. Ultimate strength of multilayer flat specimen

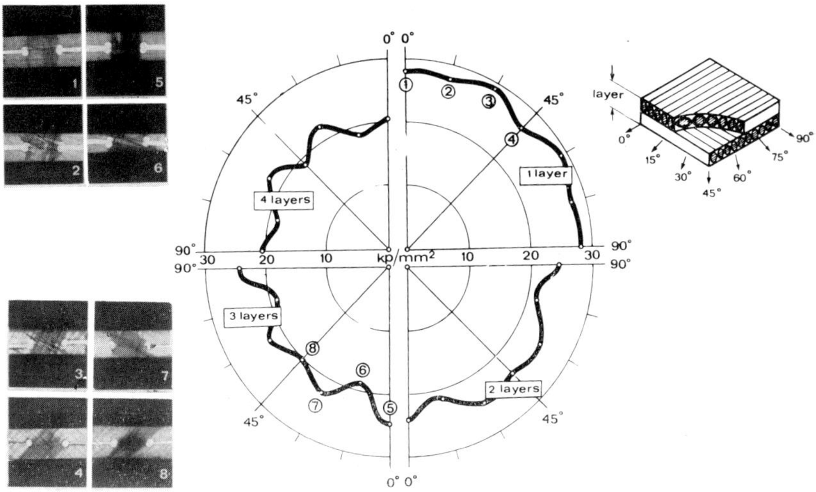

FIG. 8. Ultimate strength of multilayer shells

support with the same stiffness as is given by the hoop layers of a curved shell.

As pointed out earlier,[1, 2] the standard longitudinal test specimen, particularly, does not show the full laminate strength, because a premature failure occurs as a result of secondary effects not present at continuous shell structures (Figs. 7, 8).

Only short specimens at which lateral support is simulated by clamping make it possible to record strength values somewhat representative for the ultimate strength limit of multilayer laminates.

Unfortunately, with such short specimens lateral contraction and shear

FIG. 9. Bending fractures after initial shear failure

deformation are prevented in the clamping range. Therefore, after initial failure due to shear forces between the fibers, bending fractures of the remaining strands can be observed by photomicrography (Fig. 9).

V. THE REPRESENTATIVE ELEMENT

These and similar considerations regarding strength tests with short specimens lead us to maintain a hypothesis which describes the failure modes of laminates by assuming the distortion of certain regular interconnected elements continuously distributed throughout the laminate. This holds especially for form-closed supported laminates.

There is an additional reason for applying such discrete elements in the fundamental analysis of local elastic behavior and failure modes of the shell. It is useful to have already prepared that type of element necessary for matrix methods when computing the static and dynamic behavior of total structures.

With the method to be suggested, all stresses and deformations of such an element are derived from a force vector $F = (F_{xx'}, F_{yy'}, F_{xy})$ which acts on it considering the given ratios of cross-section areas $A_1 : A_2 : A_3 \ldots$

These sections A_i are assumed to be large enough to enclose a certain number of single fibers, but the square roots of the areas shall be very small compared to the wall thickness of the shell. The shortest side length of the elements (e_i) shall be great with respect to the square root of the smallest section area A_i.

Normal forces resulting from the curvature of the element shall be balanced by pressure gradients within the matrix material. These are again transformed into stresses in the plane of the shell (Figs. 10, 11).

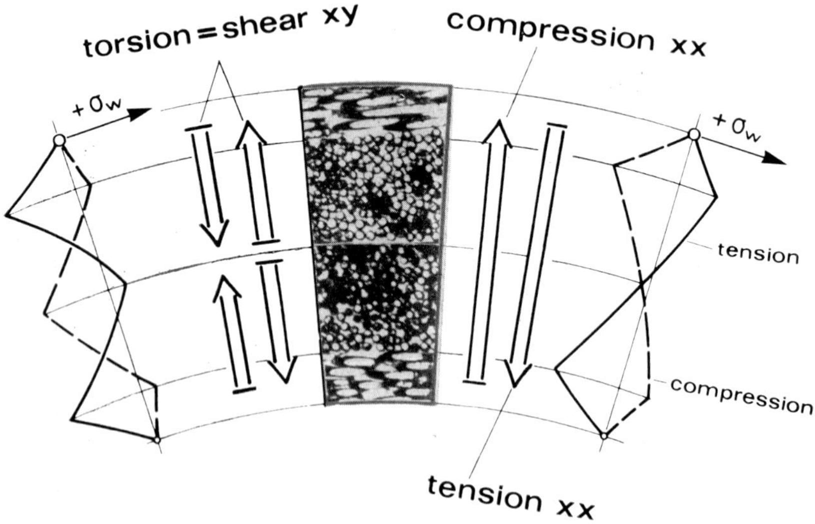

FIG. 10. Stress gradients through the thickness of the shell

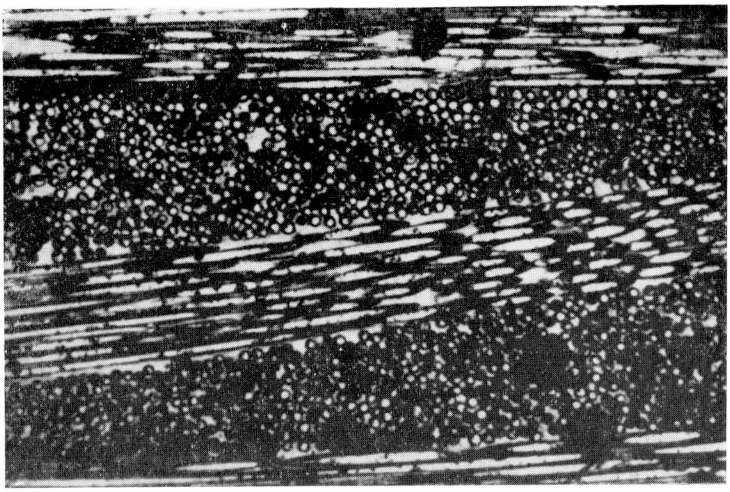

FIG. 11. Micrograph of cross-section

Examining the failure modes caused by these normal forces will enable us to determine maximum stress intensities. In this way we can prevent delamination when the matrix is under tension and the consequent destruction of the supporting effect; as well as yielding of the compressed matrix material and ultimate fiber destruction by hertzian pressure (Fig. 12).

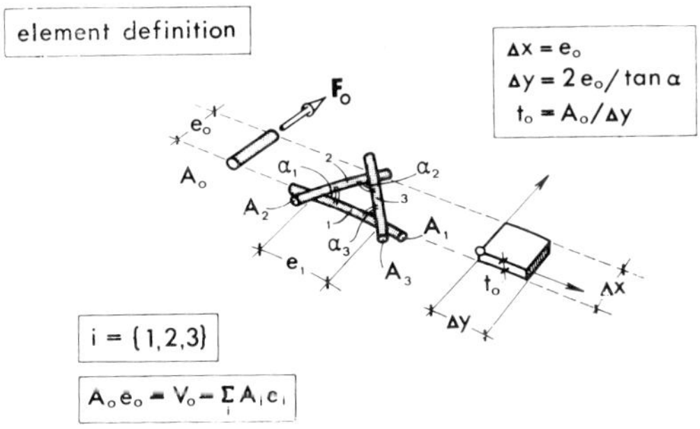

FIG. 12. Element definition

Determining the section area ratios for a triangle element ($A_1 : A_2 : A_3$ or $\sum t_1 \cdot l_1 : \sum t_2 \cdot l_2 : \sum t_3 \cdot l_3$, where $t_{1,2,3}$ = layer thickness, $l_{1,2,3}$ = number of plies per layer), it is necessary to take into account the failure modes of longitudinal strands, and especially which of the strands forming the element considered are loaded by compression and which are loaded by tension resulting from the load-collective. This is important because ultimate stress limit is lower in compression than in tension (Fig. 13).

$$1 > \chi = \frac{\sigma''' \text{ ultimate, compression}}{\sigma''' \text{ ultimate, tension}}.$$

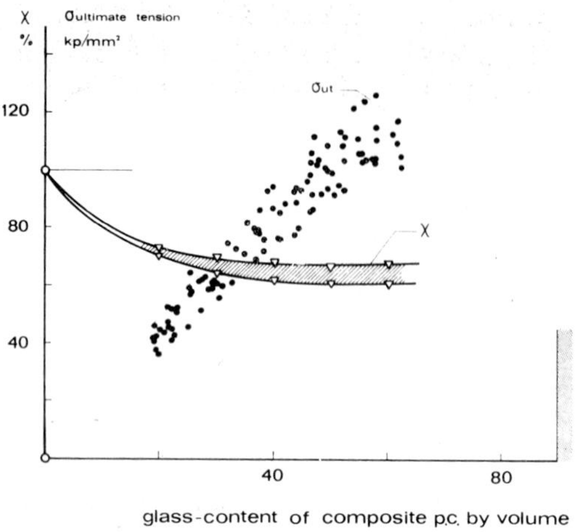

FIG. 13. Ultimate strength as function of fiber volume

The reason for premature failure of straight composite strands under axial compression has been given by Kossira[(3)] (Institut für Flugzeugbau der Technischen Hochschule Stuttgart—IFB). He stated that lateral buckling of fibers no longer elastically supported by the matrix can cause secondary failure and ultimately overall failure if the contact pressure results in local yielding of matrix material (Fig. 14).

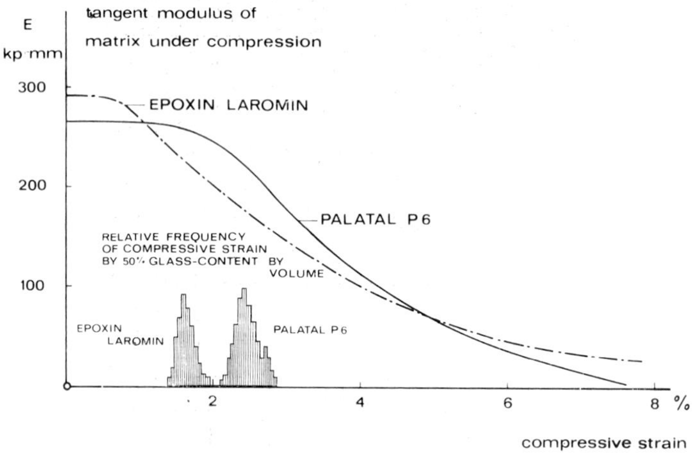

FIG. 14. Compression failure for uniaxial composite

These observations have been proved impressively by comparisons of a macroscopic rubber-plexiglass model with photomicrographs of FGRP specimens (Figs. 15, 16).

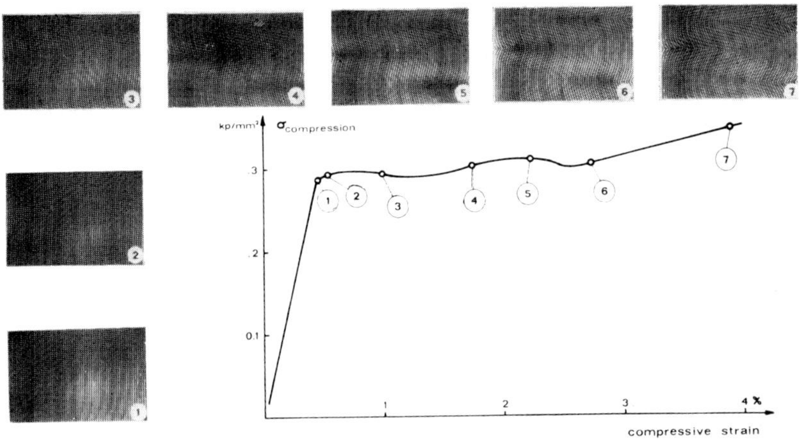

FIG. 15. Fiber instability in rubber-plexiglass model

Distorting the representative epsylon-, star-, or triangle-element, the single strand rods and the longitudinal ones suffer shear and transverse deformations. From this, shear and normal stresses result in the rods acting between the single fibers (Fig. 17).

The total strain energy of the rods computed from all these stresses must be equal to the displacement work of the outer load vector (F). Thus any distortion phenomena contribute to the load capacity of the shell.

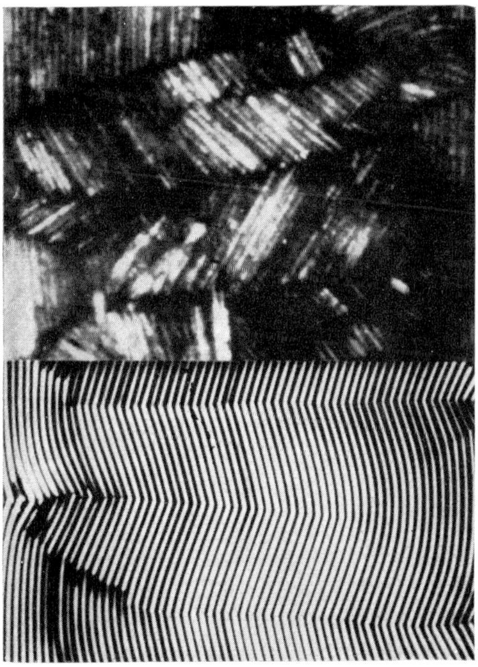

FIG. 16. Photomicrograph of FGRP specimen in compression

In most cases the contribution of matrix forces to the total load capacity is relatively small. Therefore, a fast converging iteration procedure yields sufficiently accurate values of $[\sigma_{ij}]_k$, $[\varepsilon_{ij}]_k$ and $[F_k]$, even if nonlinear relations exist between stress and strain of matrix and/or composite.

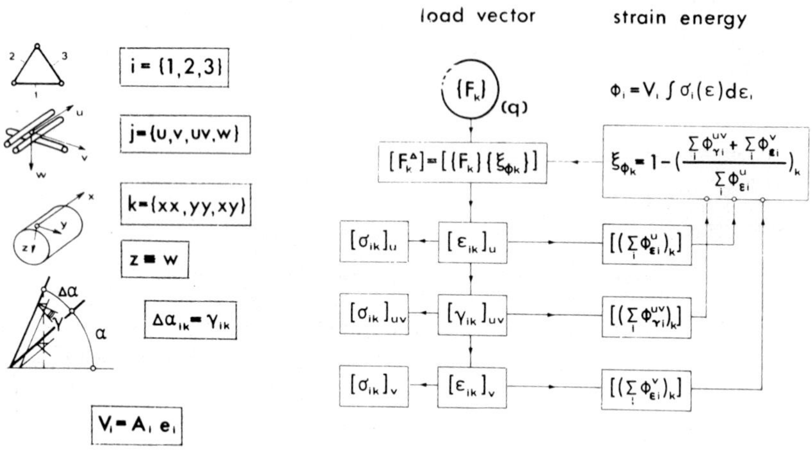

FIG. 17. Internal stress and strain interactions

VI. INFLUENCE OF FAILURE MODES ON THE LOAD CAPACITY OF SHELLS

As these effects add to the loading capability, they all influence the failure limits of the element, too. To determine these limits it will be necessary to investigate all possible failure modes.

The following failure modes can occur in the system:

1. Tension failure of the composite strand rods (direction u).
2. Compression failure of the composite strand rods (direction u).
3. Tension failure of matrix material due to transverse extension by element distortion (direction v).
4. Compression failure of matrix material due to transverse contraction by element distortion (direction v).
5. Shear failure of the matrix material (direction u, v).
6. Local buckling of inner hoop layer by compression of it.
7. Local bending failure of the whole shell due to tangential load components caused by loading of cross-ply layers.
8. Tension failure of the matrix due to radial movement of inner layers apart from outer hoop layers.
9. Compression failure of the matrix by hydrostatic or hertzian contact pressure between laminated layers (direction w).

For each failure mode there is a corresponding discrete ultimate strength. On the other hand each component of the load collective adds to the corresponding stresses and pressures of every rod of the system-element and also of the interlayer range.

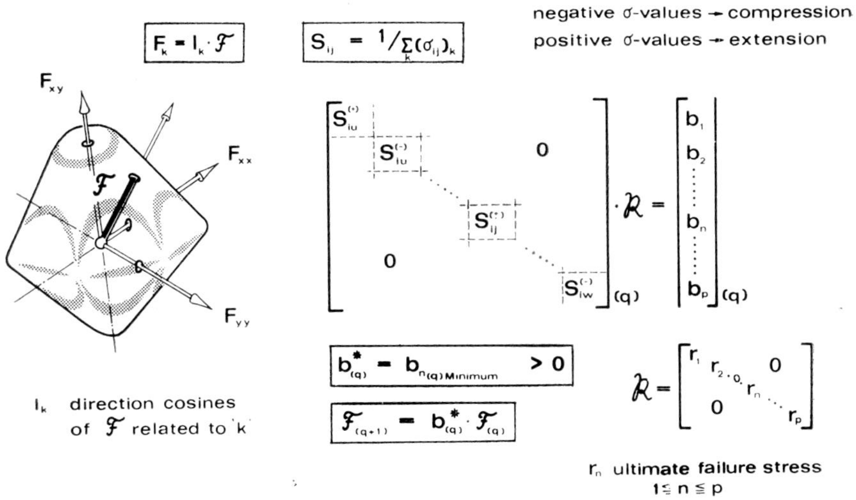

FIG. 18. Relation between load vector and failure limits

The elements of the vector defined in Fig. 18 are reciprocal values of the sum of all stresses computed from the load collective by superposition considering every rod and each failure direction.

$$S_{ij} = \frac{1}{\sum [\sigma_{ij}]_k}.$$

The diagonal matrix R of all failure stresses and pressures multiplied with the diagonal matrix of the (S_{ij})-arrays yields a collective of p coefficients $(b_n) = b_1, b_2, b_3, \ldots, b_p)$, the smallest positive value b^* of which multiplied with the assumed load vector gives that load vector which causes nowhere transgression of any stress critical with respect to failure.

As the interfiber effects are influenced by the absolute magnitude of the vector F, an iteration procedure is necessary to get the exact values of the stresses.

Scanning the failure limits within a three-dimensional vector space of possible load collectives furnishes a three-dimensional spatial surface which is closed but not steady in every range. This spatial surface not only limits the loading capability of the element but describes in addition the failure modes occurring (Fig. 19).

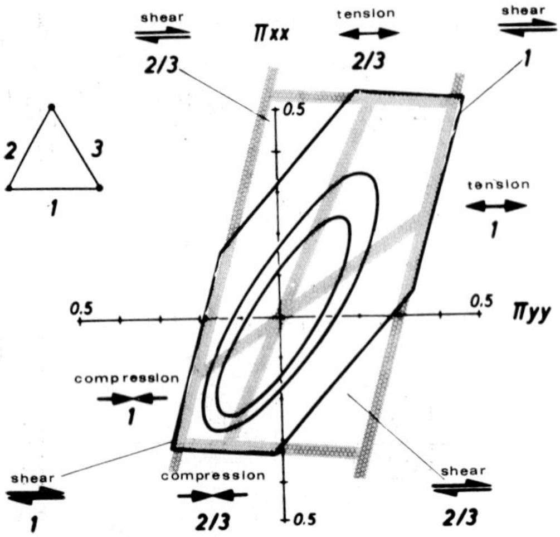

FIG. 19. Failure surface

To estimate the construction qualities of a laminated ply angle shell, we can normalize the failure load vectors using one which is representative of the properties of the unidirectional composite. For that purpose a straight element is used for comparison, the length e_0 and section A_0 of which are selected so as to assure that the volume of the element compared equals that of the representative laminate element:

$$A_0 \cdot e_0 = \sum_i A_i \cdot e_i .$$

Using a suitable reference thickness this corresponds to a rectangular element (see Fig. 16):

$$\Delta x \cdot \Delta y \cdot t_0 .$$

Let $F_0 = A_0 \cdot \sigma'''$ be the element's loading capacity under tension. Then the normalized failure vector can be expressed by its components

$$\pi = [F_k / F_0] = [\pi_k] .$$

With regard to the element, the curvature of the shell in general is relatively insignificant. It influences only the secondary stresses between the laminated layers (failure modes list points 6 to 9).

On the other hand, the ratio of ultimate shear strength of the matrix material to ultimate tension strength of the composite $\vartheta = \tau_{(m)}/\sigma_{(c)}$, is an essential parameter. In this the index (m) stands for matrix material and (c) for composite.

With given components of the composite, i.e., given matrix material and given fiber material, the ratio ϑ is given too (Figs. 20–24).

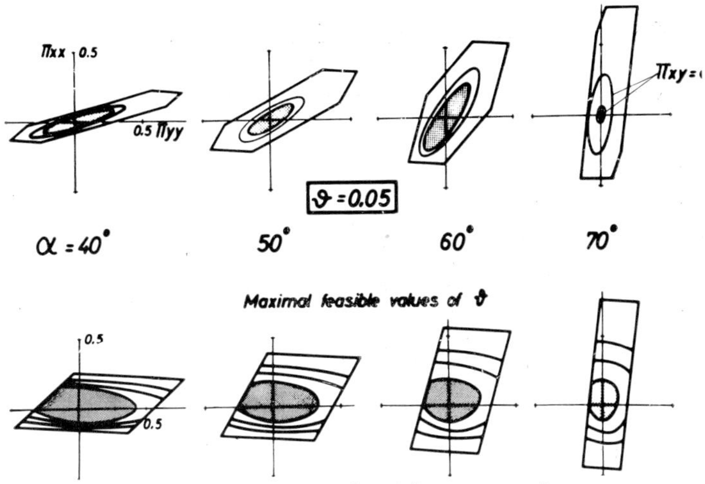

FIG. 20. Failure limits of multilayer composites

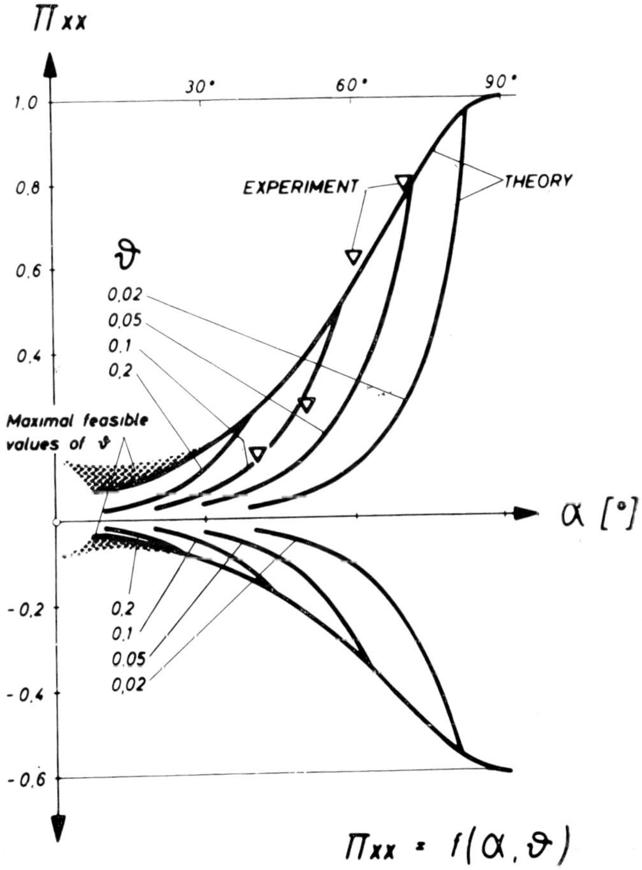

FIG. 21. Π_{xx} as function of ply angle α

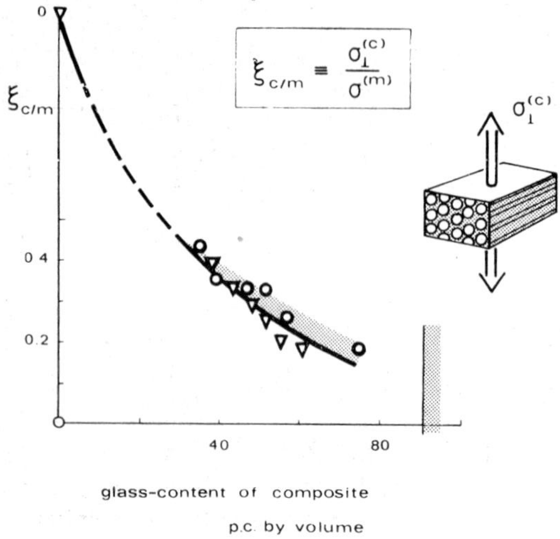

FIG. 22. Influence of glass content on transverse ultimate strength

Adapting the considerations and results explained so far to the optimization of a cylindrical or tapered shell loaded by bending, torsion, and internal or external pressure, we see that mainly the shell-design parameters

(1) area ratio of element rods and

(2) ply angles

are of remarkable influence.

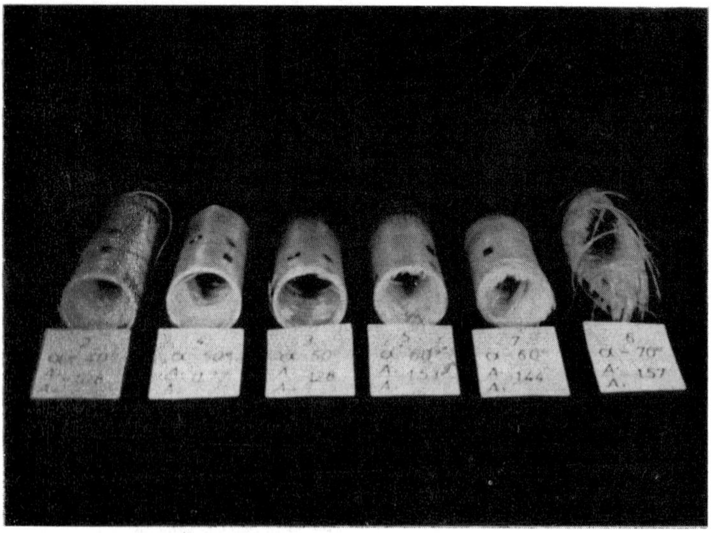

FIG. 23. Test tube specimen after failure under axial load

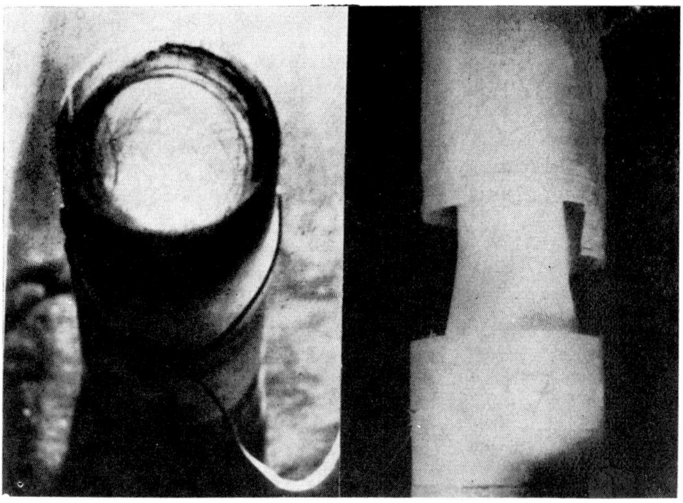

FIG. 24. Failure mode of tube specimen under axial load

For a symmetrically designed structure with three basic layers there is $A_2 = A_3$ and $\alpha_2 = \alpha_3 = \alpha$. Thus we can use α and $\psi = A_2/A_1$ as characteristic geometrical parameters.

If an internal or external pressure does not exist ($F_{yy} = 0$), then this ratio can be determined as $A_2/A_1 = x/\cos\alpha$.

Thereby it is assumed that neither rod 2 nor rod 1 shall fail prematurely when loaded by the load vector F_{xx}.

It seems not rational to transgress this value of ψ, but it may be convenient to stay below this value for a given load-collective. Therefore, a factor $1 \geqslant k \geqslant 0$ is defined. Hence

$$\psi = k \cdot x/\cos\alpha.$$

For a given load collective, it is useful to include this factor k in the optimization considerations.

Besides this appears the additional parameter $\vartheta = \tau_{(m)}/\sigma_{(c)}$ which is influenced by the components.

VII. OPTIMIZATION OF A SHELL

We shall define optimization here as optimizing the structure in such a way that the relation Load capability versus Weight (and/or stiffness versus weight, if maximum load is influenced by deformation) should reach a maximum value. The term load capability shall express and include the envelope of any possible load collective.

In the following it is assumed that a cylindrical or tapered shell of closed circular or elliptical cross-section is subjected to combined bending and torsion loads. Such a shell could be designed from simple angle-ply laminate with hoop reinforcement. However, with certain local ratios of torque M_T versus bending moment M_B, it might be more advantageous to use longitudinal strands being certain that the most feasible distance from the neutral line is utilized.

It goes without saying that the load density on the shell considered needs an overall dimensioning which is appropriate to prevent premature general buckling of the structure under maximum loads. This can be done at low load density by using discrete reinforcing hoops set at small distance instead of a continuous supporting layer.

It is obvious that a short beam of given height H subjected only to pure bending (that is free of transverse loads) will become lightest if the whole load-carrying material is concentrated in flanges of minimum area at the distance H.

Such an idealized beam will carry the maximum bending moment with respect to the cross-section area at given height.

The ratio η_B of maximum bending moment per unit area of the shell versus maximum bending moment per unit area of such an idealized two-

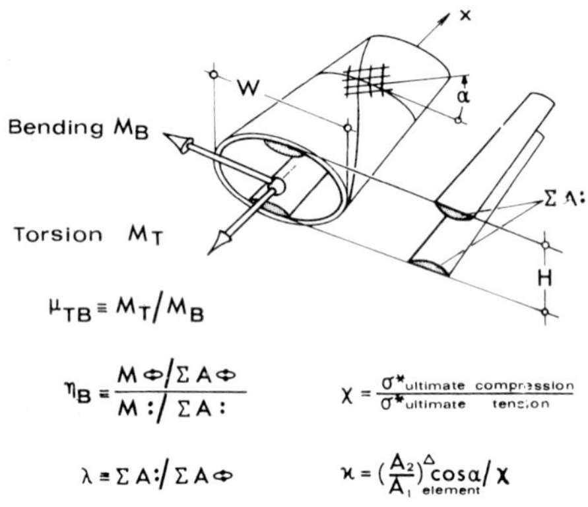

FIG. 25. Parameter definition

flange beam is used in the following optimization to compare different shapes of shells with respect to their loadability versus weight relation. Thus the comparison factor η_B is understood as bending efficiency of the shell beam (Fig. 25).

The bending efficiency of a given shell depends apparently on the intensity of the additional torque. Thus the bending efficiency η_B is computed and plotted as a function of μ_{TB} = torque M_T/bending moment M_B.

The ratio of the summarized flange areas A: (longitudinal strand) versus cross-section area A_0 of the filament-wound laminated shell $\lambda = A:/A_0$ and the ply-angle α are parameters.

As the longitudinal Young's modulus of a cross-ply laminated shell is smaller than the modulus of straight strands, the compatibility conditions do not allow the full utilization of the bending-load capability of the laminated shell. Nevertheless, shells fitted with additional longitudinal composite stringers will be more favorable with respect to minimum weight than pure laminated shells of angle-ply, provided μ_{TB} is < 1 (Fig. 26).

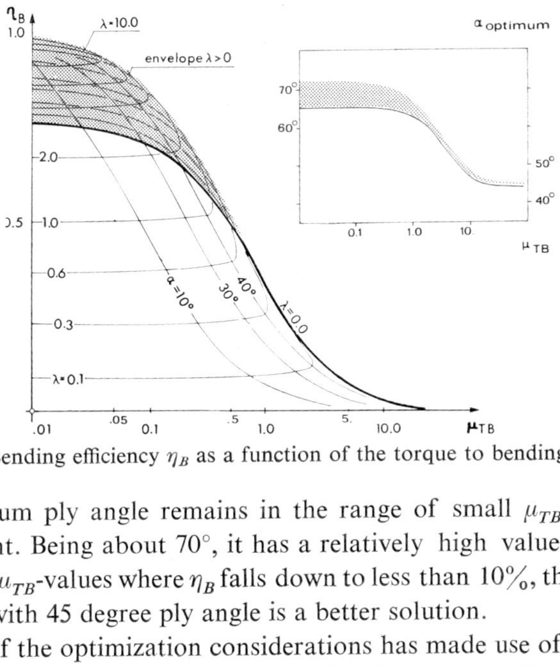

FIG. 26. Bending efficiency η_B as a function of the torque to bending ratio η_{TB}

The optimum ply angle remains in the range of small μ_{TB}-values $\leqslant 0.3$ fairly constant. Being about 70°, it has a relatively high value. To suit extremely high μ_{TB}-values where η_B falls down to less than 10%, the lightly supported shell with 45 degree ply angle is a better solution.

This part of the optimization considerations has made use of the strength-weight relation. The stiffness weight relation has been omitted, but can be included following a similar line of thought. Methods analyzing the stiffness of laminates have been described by many authors, particularly by Wiedemann[(4)] and by Tsai.[(5)]

Our own tests and those of others have confirmed the correspondence between analytical approach and experimental results. It is remarkable that symmetrical laminates with an optimum ply angle of about 70° possess roughly 90% of the stiffness of the orthotropic material.

These results apply particularly to aircraft structures, where the μ_{TB}-value of wings is greater than unity only in the neighborhood of the tips. In the more inward root region of a wing, where weight is greater, this value remains smaller than one-third considering usual velocity ranges as well as lift and moment coefficients.

At fuselage structures loaded mainly by internal pressurization, the lateral reinforcement problem can be readily solved using outer hoop layers.

It is not in the scope of this paper to explain how to design cut-outs for windows, doors and landing-gear bays, etc. so that they can meet the conditions of the foregoing considerations. It can be said only that the supporting layers need not be exactly tangential to the cross-section contour. They may be freely designed with a small slope without losing their supporting properties.

The envelope of optimum values of the η_B versus μ_{TB}-curves of multilayer composites as described in this paper can be compared with corresponding diagrams which have been gained assuming an isotropic material, for example like an advanced light alloy.

The amount of tolerable bending moment per unit weight of a structural member is proportional to the breaking length (Versagenslänge) $L = \sigma/\gamma$ of the material used, where σ is the critical value of ultimate failure stress and γ the materials specific weight.

Therefore the breaking length must be included when making a realistic comparison of the loading capabilities of shells subjected to combined bending and torsion loads when they are constructed from advanced composites or conventional homogeneous materials of comparable high quality (Fig. 27).

The diagram shows the minimum values for the ratio of rupture lengths σ/γ to yield a composite structure of equal weight.

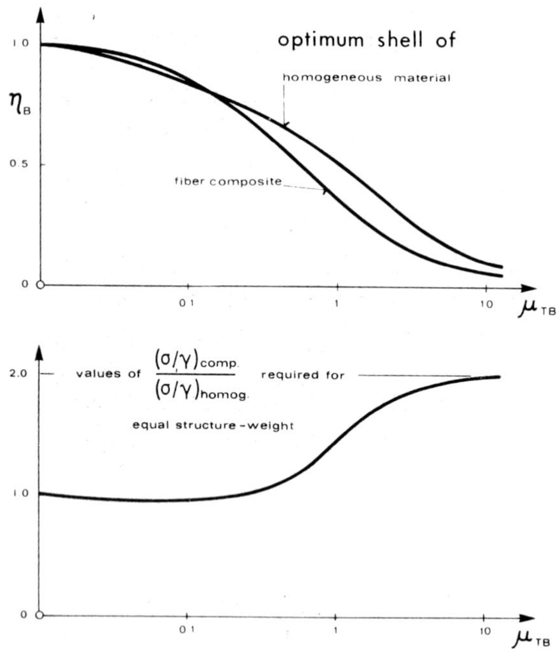

FIG. 27. Optimum values for bending efficiency η_B

VIII. GENERAL REMARKS AND ACKNOWLEDGEMENTS

This has been a somewhat encyclopedical report on a research program still running. There is no doubt that the described method for synthesizing optimum composite lightweight structures can gain more reliability and practicability by introducing additional parameters. More exact studies about failure phenomena and adaption of the methods used in elastomechanics will help toward this goal. Also required are advanced studies to improve the technology of producing very thin-walled strand-laminates.

The thoughts given here have already found their way into practical aircraft construction from FGRP composites. As an example, we can mention the wing shells of the glider H-301 "Libelle". Four "Libelles" flew in the American gliding competition in 1967 and ranked 3, 6, 17, and 19 (Fig. 28).

FIG. 28. Glider H-301 "Libelle"

ACKNOWLEDGEMENTS

The author would like to thank Mr. Klaus D. Schäfer (IFB Stuttgart) as well as Mr. Peter Bamberg and Mr. Helmut Krauss (DFL Stuttgart) who have contributed much material for this paper. They especially made useful suggestions for the theoretical formulation of the problems concerned, helped to mechanize the computations and carried out the experiments. Mr. Martin Weist (IFB Stuttgart) took the photograph of the demonstration model.

Without the kind assistance of these and other members of the IFB and the DFL Stuttgart, it would have been impossible to complete the work in the short time on hand.

REFERENCES

1. HÜTTER, U. Vermessung räumlicher Spannungszustände durch eingebettete Dehnmeßstreifen in Glas/Harz-Konstruktionen, *Kunststoffe*, **53**, 831–838 (1963).
2. GRÜNINGER, P. G. Festigkeits- und Deformationsuntersuchungen an Bauteilen aus Glasfaser-Kunststoff, *Kunststoffe*, **54**, 759–767 (1964).
3. KOSSIRA, H. Der Einfluss von Faserdurchmesser und Glasgehalt auf die Festigkeit harzgebundener Glasfaserstränge, Dissertation TH, Stuttgart 1963.
4. WIEDEMANN, J. Beitrag zum Problem orthotroper Platten ohne allgemeine Neutralebene, Teil I und II, *Luftfahrttechnik/Raumfahrttechnik* **8**, No. 10, 283–289 (1962); **9**, No. 3, 73–82 (1963).
5. TSAI, S. W. and AZZI, V. D. Strength of Laminated Composite Materials, *AIAA Journal*, **4**, No. 2 (1966).

BIBLIOGRAPHY

MATTING, A. and NIEDERSTADT, G.: Zur Berechnung glasfaserverstärkter Kunststoffe, *Kunststoffe*, **57**, 32–38 (1967).

WHITNEY, J. M. and RILEY, J. B.: Elastic Properties of Fiber Reinforced Composite Materials, *AIAA Journal*, 1537–1542 (1966).

CRICHLOW, W. J. and SORENSON, V. S.: Advances in Monofilament Structural Composite Technology, *Journal of Aircraft*, **3**, No. 5, 431–435 (1966).

SHAFFER, BERNHARD W.: The Influence of Filament Orientation on the Material Properties of Reinforced Plastics, *Proceedings 19th Annual Meeting of the SPI, Reinforced Plastics Division*, Sect. 6E, Chicago 1964.

KRAUSS, H.: Untersuchungen von Bauteilprüflingen aus glasfaserverstärktem Kunststoff auf Biegung, Torsion und überlagerter Biegung und Torsion, DFL-Bericht 5/66, unpublished 1966.

HOLISTER, G. S. and THOMAS, C.: Fiber Reinforced Materials, in *Compressive Ultimate Strength*, pp. 108–116, Elsevier Publishing Company Inc., 1966.

LEVENETZ, B.: Compressive Applications of Large Diameter Fiber Reinforced Plastics, *Proceedings 19th Annual Meeting of the SPI, Reinforced Plastics Division*, Sect. 14D, 1964.

ARGYRIS, J. H.: Triangular Elements with Linearly Varying Strain for the Matrix Displacement Method, *Journal of the Royal Aeronautical Society*, **69**, No. 658, 711–713 (1965).

ARGYRIS, J. H.: Matrix Displacement Analysis of Plates and Shells, *Ingenieur-Archiv*, **35**, 102–142 (1966).

MILLER, R. E. JR.: Structural Analysis Flexible Grid Technique for SST Wing Parametric Studies, *Journal of Aircraft*, **2**, 257–261 (1965).

BENSON, N. K.: Die Optimierung von Flugtragwerken: 5. Lehrgang für Raumfahrttechnik, Kapitel 306, Stuttgart 1966.

CONEN, H.: Versteifte Schalen aus Glasfaser-Kunststoff, *Kunststoffe*, **56**, 105–109 (1966).

WIEDEMANN, J.: Optimaldimensionierung von Bauweisen: 3. Lehrgang für Raumfahrttechnik, Kapitel 311, Aachen 1964.

HÜTTER, U.: Glass-Fiber Reinforced Plastics as Structural Material for the Aircraft Industry, The Society of the Plastics Industry, Inc., Technical and Management Conference Reinforced Plastics Division, 16th Annual, Sect. 13-A, Chicago 1961.

OLEESKY, S. S. and MOHR, J. G.: *Handbook of Reinforced Plastics*, Chapter IV, 3, Filament Winding, 1964.

HÜTTER, U.: The Transfer of Force in Highly Stressed Monocoque Bodies of Glass-Reinforced Plastics, *Transactions and Journal of the Plastics Institute*, **30**, 16–27 (1962).

MOFFAT, G. JR.: Reno Retrospective, Soaring, *Journal of the Soaring Society of America*, pp. 20–21 (1966).

COMMENT TO PAPERS BY PROF. U. HÜTTER AND PROF. L. A. SCHMIDT

H. SCHUERCH

BOTH papers (see pp. 527–549 and 553–582) are addressed to the same question: Given a set of loads (or a load envelope), what is the optimal fiber geometry that will support these loads with adequate safety?

While Prof. Hütter's approach appeals to the intuition of the designer, Prof. Schmit relegates the design evolution to digital processing, and therefore retains great generality.

A third and perhaps fruitful approach which can generate optimal fiber layouts is provided by the so-called Mitchell theorem: This states that, given a set of loading conditions, there is a "framework" of linear elements occupying a defined space in which all tension members and all compression members are uniformly stressed. If the space occupied by this framework has no strain absolutely greater than the tensile or compressive strain respectively of these members, then the framework is both of minimum weight and of maximum stiffness for the specified set of loadings.

For planar structures (i.e., for plates), the trajectory field of the "framework" satisfying the Mitchell criterion satisfies the same hyperbolic differential equation which governs the Hencky–von Mises–Prandtl networks of sliplines in ideally plastic flow. A large number of "optimal" fiber structures, therefore, can be generated from an existing body of knowledge in plasticity.

It would be of particular interest to examine, if the digital procedures used by Prof. Schmit indeed tend to converge towards the Hencky–von Mises–Prandtl solutions. A good example for this is the cantilever beam case that has been worked out by Hemp, Cox and their co-workers in England as sketched below:

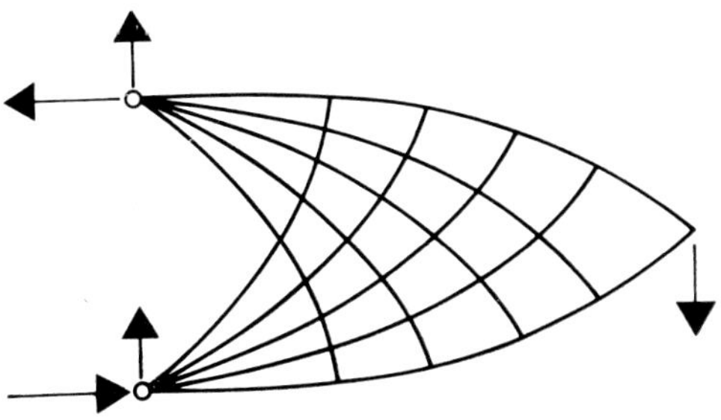

THE STRUCTURAL SYNTHESIS CONCEPT AND ITS POTENTIAL ROLE IN DESIGN WITH COMPOSITES†

LUCIEN A. SCHMIT, JR.‡

ABSTRACT

The use of composite materials in the design of structural systems offers the designer a wide range of new degrees of freedom. With this increase in the number of alternative choices available to the structural engineer comes the responsibility for providing a rational basis for seeking optimum designs. The structural synthesis concept is reviewed as a framework within which the quantifiable portion of the design process can be formulated. A concise statement of a rather general class of structural synthesis problems is recognized as a nonlinear mathematical programming problem. The essential ideas involved in sequential solution methods are illustrated by outlining a feasible directions solution method. It is pointed out that alternative formulations are possible in which the usual statement, as an inequality constrained minimization problem, is transformed into a sequence of unconstrained minimization problems. Extension of the structural synthesis approach to fiber composite structural systems is illustrated in terms of an idealized example. The example problem is a simply supported, rectangular, multi-layer, resin-matrix, fiber composite plate subject to a multiplicity of inplane load conditions. The failure modes considered include gross plate buckling, combined stress failure in a layer and delamination. The objective function is taken as cost of fiber material per unit surface area of plate. Finally, in discussing the potential role of structural synthesis in the design of composite material structural systems some problem areas are identified.

LIST OF SYMBOLS

a_1, a_2	empirical constants
$\mathbf{D}$	vector of design variables
E	modulus
G	shear modulus
$g_j(\mathbf{D})$	jth constraint function
J	total number of constraints
J_c	set of active constraints
K_1, K_2	empirical constants
$K_{\alpha\beta}$	empirical or computable factor (see ref. 21)

† The sponsorship of the Advanced Research Projects Agency, Department of Defense, through Contract No. AF 33(615)-3110 administered by the Air Force Materials Laboratory is acknowledged.

‡ Professor of Engineering and Professor-in-Charge of Solid Mechanics, Structures and Nechanical Design, Engineering Division, Case Institute of Technology.

k	apparent volume ratio
$\bar{k}$	actual volume ratio
L	number of load conditions
$M(\mathbf{D})$	objective function
N	membrane force
N_f	number of fibers per roving end
N_l	number of layers (plies)
r	arbitrary constant in Fiacco–McCormick function
s	scalar in feasible directions method
S	layer strength
$\mathbf{S}$	direction vector
t	time
t_p	layer (ply) thickness
x, y, z	plate coordinate system
α	step length
β	empirical factor
ε	strain
θ	layer orientation angle
λ	set of positive constants in feasible directions method
ν	Poisson's ratio
ϱ	weight density
σ	stress
ϕ	Fiacco–McCormick function
$\Delta\phi$	relative rotation of adjacent layers (plies)
∇	gradient of a function
ε	contained in, or a member of

Superscript

q	trial design index

Subscripts

allow	allowable
C	compression
cr	critical
f	fiber property
i	failure mode index
j	constraint index
k	load condition index
l	layer (ply) property
m	matrix property
p	allowable, (layer) ply property
s	shear

T	tension
tor	torsion
x, y, z	coordinate directions
α	T—tension or C—compression
β	T—tension or C—compression

I. INTRODUCTION

The advent of high performance composite materials offers the structural designer a wide range of new degrees of freedom. These materials have made it possible to think in terms of simultaneous design of structural configuration and structural material. With this dramatic increase in the number of alternative choices available to the structural engineer comes the responsibility for providing a rational basis for seeking an optimum design.

Structural design may be viewed as a complex, multi-level decision-making process involving an interrelated mix of quantifiable and nonquantifiable decisions. The interface between that portion of the structural design process that can be put on a strictly logical basis and that portion that is nonquantifiable is moving forward. Seeking to automate the quantifiable portion of the design process helps to distinguish the strictly logical decisions from the qualitative judgments. The structural synthesis concept provides a framework for formulating the quantifiable portion of the structural design process on which modern optimization techniques can be brought to bear. To the extent that we can clearly define a growing portion of the design process for composite material structural systems it should be possible to quantitatively formulate structural synthesis problems and seek solutions using optimization algorithms.

The fundamental restriction on a structural design is that the probability of survival, when exposed to a statistically described environment over a prescribed life, exceeds a specified minimum. The transition from a statistical to a deterministic design philosophy has traditionally been accomplished by replacing the environment with a discrete set of load conditions and introducing the safety factor concept. In civil engineering practice there is an essential distinction between the limit design philosophy and the elastic design philosophy. In the former the safety factor is applied to the load and collapse is guarded against; in the latter the safety factor is applied to yield stress and unsatisfactory behavior under service load is guarded against. Aircraft structural engineering practice has traditionally included a dual requirement, namely that catastropic failure under ultimate load be prevented and that permanent damage under service load (limit load) be precluded as well. The structural synthesis concept is in principle applicable to structural design based on any one of these philosophies.

Structural synthesis is defined as the rational, directed evolution of a structural system which, in terms of a defined objective, efficiently performs a set of specified functional purposes.[1] The term structural synthesis is used to refer to rational optimum design of structural systems. It is distinct from creative synthesis which is inherently qualitative and involves the bringing together of many concepts in a broad engineering plan.

II. DEFINITIONS AND GENERAL PROBLEM STATEMENT

In order to facilitate a concise statement of a rather general class of structural synthesis problems, several definitions should be agreed upon. A structural system is described by a set of quantities, some of which are viewed as variables during the design process. In formulating a structural synthesis problem, certain quantities are fixed at the outset and these will be called *preassigned parameters*. All those quantities describing the structural system that are not preassigned will be called *design variables*. Based on engineering judgment, the actual environment to which the structural system is exposed is replaced with several distinct sets of mechanical and thermal loads. Each set of loads is referred to as a *load condition* and the several distinct load conditions will be referred to as the *loading system*. A trial design for a structural system is inadequate and the structure is said to fail if the structural behavior does not remain within the confines of stated limits. What constitutes failure or unsatisfactory behavior must be carefully defined and this can be expected to vary from one design task to another. A *failure mode* will be defined as any structural behavior characteristic subject to limitation by the responsible engineer. Finally, the *objective function* will be defined as a function of the design variables, the value of which can be used as a basis for choice between alternative acceptable designs.

A rather general class of structural synthesis problems may now be stated concisely as follows:

Given the *preassigned parameters* and the set of distinct *load conditions* find the vector **D** of *design variables* such that the objective function $M(\mathbf{D})$ is minimized (or maximized) subject to a system of inequality constraints on the design variables

$$g_j(\mathbf{D}) \leq 0 \qquad j = 1, 2, \ldots, J \tag{1}$$

where the functions $g_j(\mathbf{D})$ are such that (1) unsatisfactory behavior in each failure mode in each load condition is prevented and (2) the range of values that the design variables may assume is restricted.†

† These restrictions may be based on fabrication limitation, geometric realizability, or they may be intended to confine the design to the region where the analyses employed are essentially valid.

Stated in this manner, this class of structural synthesis problems is clearly seen to be a nonlinear mathematical programming problem.

The formulation of a structural synthesis problem involves the exercise of engineering judgments in several ways. The adoption of a design philosophy and an objective function are fundamental judgments that must be rendered in stating a structural synthesis problem. The substitution of a loading system including appropriate load conditions for the actual environment is an idealization traditionally employed by structures engineers which requires considerable insight. Selection of analysis techniques to predict the behavior of trial designs so as to guard against each failure mode in each load condition will benefit from previous engineering experience. The fundamental difference between the qualitative judgments leading to the statement of a structural synthesis problem and the quantitative decisions that lead to its solution should be kept in mind.

III. SOLUTION METHODS

From the concise statement of a general class of structural synthesis problems previously given, it is apparent that these are in fact nonlinear programming problems. Therefore, methods for seeking the solution of structural synthesis problems may be drawn from the mathematical programming literature, or alternate methods may be devised by engineers with insight into the special aspects of their design problems.

In principle, any structural synthesis problem can be solved by selecting a large number of values for each design variable over its range of acceptable values and then trying all possible combinations. Of this large number of designs, that which satisfies all of the constraints and has the best value of the objective function can be taken as a good approximation of the optimum design. The trouble, of course, is that this "try them all" parameter survey method is too inefficient to be practical in all but the simplest cases.

Methods of searching for the best design can be classified into simultaneous and sequential methods. The simultaneous methods are characterized by the fact that all trial designs are selected before the analysis of any of the designs is started. Thus, in theory it would be possible to execute the analysis of all trial designs simultaneously using many computers. On the other hand, sequential methods are characterized by the fact that future trial designs may be generated using the results of previous trial designs. The sequential methods are of the following form:

$$\mathbf{D}^{(q+1)} = \mathbf{D}^{q} + \alpha_q \mathbf{S}^{q} \tag{2}$$

where the vector $\mathbf{D}^{(q+1)}$ represents the $(q+1)$th trial design, the vector $\mathbf{D}^q$ represents the qth design, and the qth design modification is represented by

$\alpha_q \mathbf{S}^q$. Structural synthesis studies have been based almost exclusively upon sequential methods. A review of nonlinear programming procedures for design optimization is given in ref. 2. Also, a recent numerical survey of nonlinear programming methods will be found in ref. 3. A wide variety of sequential design modification techniques exist; and new methods, as well as variations on established techniques, continue to be offered in the literature. By and large, the question of which is the most efficient method of optimization is an open one, and it is likely that the answer to this question depends on the form of the particular problem at hand.

To illustrate, a class of sequential methods known as methods of feasible directions[4] will be briefly discussed. If at the design represented by $\mathbf{D}^q$ the functions $g_j(\mathbf{D}^q) < 0$ for all $j = 1 \rightarrow J$, then $\mathbf{S}^q$ can be taken as $-\nabla M(\mathbf{D}^q)$ (for minimization). The objective function may now be viewed as a function of a single variable α that is

$$M(\mathbf{D}^q + \alpha \mathbf{S}^q) = f(\alpha). \tag{3}$$

Let the α which minimizes $M(\mathbf{D}^q + \alpha \mathbf{S}^q)$ be denoted α^*. Then α is taken as α^* if none of the constraints are violated, i.e. $g_j(\mathbf{D}^q + \alpha^* \mathbf{S}^q) \leq 0$ for all $j = 1 \rightarrow J$. Otherwise, α is taken as the value nearest to α^* that does not violate the inequality constraints. This situation is depicted schematically in Fig. 1.

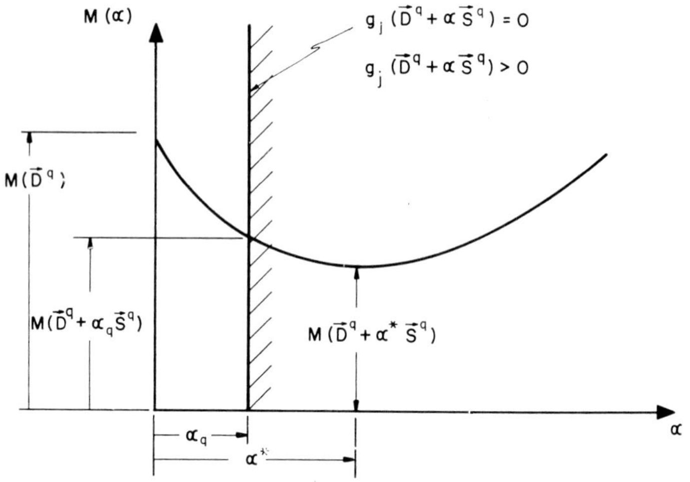

FIG. 1. Constrained one-dimensional minimum

The foregoing procedure usually causes $g_j(\mathbf{D}^{q+1}) = 0$ for some set of values j denoted as the set J_c. The constraints with indices in the set J_c are referred to as active or critical constraints. When a design in the sequence at which some of the constraints are active (i.e. $g_j = 0$ for $j \varepsilon J_c$) is reached, then the direction of steepest descent ($\mathbf{S} = -\nabla M$) will not generally be a direction in which further design modification can be made without constraint violation. It is possible, however, to state in terms of inequality constraints the conditions that a direction **S** must satisfy so that the function M will be reduced without violating the active constraints.

Let $\mathbf{D}^{q+1}$ represent a design such that $g_j(\mathbf{D}^{q+1}) = 0$ for $j \varepsilon J_c$. The next design in the sequence is to be determined from the expression

$$\mathbf{D}^{q+2} = \mathbf{D}^{q+1} + \alpha_{q+1}\mathbf{S}^{q+1}. \tag{4}$$

The direction $\mathbf{S}^{q+1}$ must not violate any of the active constraints at the current design point $\mathbf{D}^{q+1}$; that is, it must be feasible:

$$\mathbf{S}^{(q+1)} \cdot \nabla g_j(\mathbf{D}^{q+1}) \leqslant 0 \qquad \text{for } j \varepsilon J_c \tag{5}$$

where $\nabla g_j(\mathbf{D}^{q+1})$ denotes the gradient to the jth active constraint at $\mathbf{D}^{q+1}$. Also the direction $\mathbf{S}^{q+1}$ must improve the value of the objective function (decrease it in the case of minimization); that is, it must be usable:

$$\mathbf{S}^{q+1} \cdot \nabla M(\mathbf{D}^{q+1}) \leqslant 0 \tag{6}$$

where $\nabla M(\mathbf{D}^{q+1})$ denotes the gradient to the objective function at $\mathbf{D}^{q+1}$. A direction **S** satisfying the inequality constraints stated in eqs. (5) and (6) is called a usable-feasible direction. A schematic diagram of the usable-feasible sector is shown in Fig. 2.

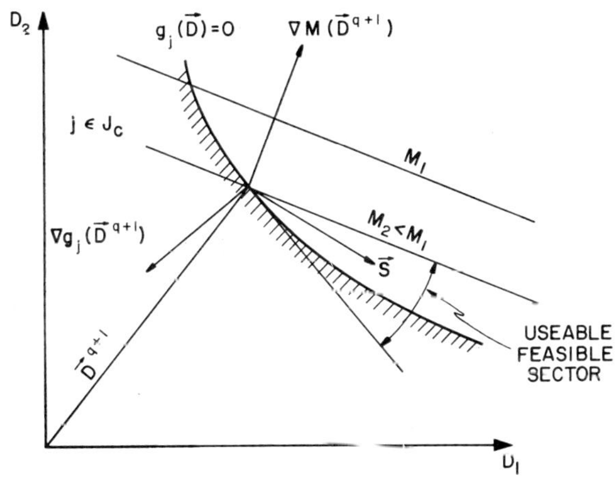

FIG. 2. Schematic of usable-feasible sector

A method of feasible directions is actually any method for determining a direction $\mathbf{S}^{q+1}$ such that the inequality constraints stated in eqs. (5) and (6) are satisfied. The determination of a "good" usable-feasible direction can be accomplished by formulating and solving the following programming problem:

Given $\nabla g_j(\mathbf{D}^{q+1})$ for $j \varepsilon J_c$ and $\nabla M(\mathbf{D}^{q+1})$ as well as a set of positive constants λ_j.

Find $\mathbf{S}^{q+1}$ and a scalar s such that

$$\begin{aligned} &(a)\ \mathbf{S}^{q+1} \cdot \nabla g_j(\mathbf{D}^{q+1}) + \lambda_j s \leqslant 0 \\ &(b)\ \mathbf{S}^{q+1} \cdot \nabla M(\mathbf{D}^{q+1}) + s \leqslant 0 \\ &(c)\ \mathbf{S}^{q+1} \cdot \mathbf{S}^{(q+1)} \leqslant 1 \quad \text{or} \quad |\mathbf{S}| < 0 \end{aligned} \tag{7}$$

while $s \rightarrow$ max.

This direction finding problem is substantially simpler than the original problem in that the objective function and all of the inequality constraints (with the exception of (c)) are linear. Once a direction $\mathbf{S}^{q+1}$ has been established, the problem is again reduced to minimizing a function of a single variable α subject to the inequality constraints $g_j(\mathbf{D}^{q+1} + \alpha \mathbf{S}^{q+1}) \leqslant 0$ for all $j = 1 \rightarrow J$.

It is possible to transform the basic inequality constrained minimization problem, representing a broad class of structural synthesis problems, into a sequence of unconstrained minimization problems. This approach was put forward by Carroll[5] and was developed further by Fiacco and McCormick.[6] In this method a new function $\phi(\mathbf{D})$ is defined as follows,

$$\phi_k(\mathbf{D}) = M(\mathbf{D}) - r_k \sum_{j=1}^{J} 1/g_j(\mathbf{D}), \tag{8}$$

where r_k is an arbitrary positive constant. Assume that an initial design $\mathbf{D}^0$ is available such that all of the constraints are satisfied $[g_j(\mathbf{D}^0) < 0; j = 1 \rightarrow J]$. Then for a particular choice of r (initially large) starting from $\mathbf{D}^0$, and minimizing the function $\phi_1(\mathbf{D})$ leads to a design $\mathbf{D}^1$ which satisfies the inequality constraints; that is, $g_j(\mathbf{D}^1) < 0$ for $j = 1 \rightarrow J$. The second unconstrained minimization is started from $\mathbf{D}^1$ with $r_2 < r_1$, and the minimum of the function $\phi_2(\mathbf{D})$ yields a design $\mathbf{D}^2$ such that min $\phi_2(\mathbf{D}) = \phi_q(\mathbf{D}^2)$ and $g_j(\mathbf{D}^2) < 0$ for $j = 1 \rightarrow J$. The smaller r_k is, the closer $\mathbf{D}^k$ can approach the solution; namely the $\mathbf{D}$ such that $M(\mathbf{D}) \rightarrow$ min and $g_j(\mathbf{D}) \leqslant 0$ for $j = 1 \rightarrow J$. The "penalty function" $-r_k \sum_{j=1}^{J} 1/g_j(\mathbf{D})$ serves to insure that the inequality constraints will continue to be satisfied as the minimum of $M(\mathbf{D})$ is sought. The smaller r_k is taken, the less the "penalty function" interfers with the true minimum of $M(\mathbf{D})$. However, small values of r_k lead to functions $\phi_k(\mathbf{D})$ that are difficult to minimize; therefore, the value of r_k is reduced sequentially.

The penalty function approach has appeal for several reasons. It can solve structural synthesis problems by employing any of the algorithms for seeking the unconstrained minimum of a function of many variables. A particularly powerful algorithm for unconstrained minimization is that presented by Davidon[7] and further developed by Fletcher and Powell.[8] The literature of unconstrained minimization will not be reviewed here; however, it will be noted that an extensive list of references on this subject can be found in refs. 9, 10, or 11. A second apparent advantage of the penalty function approach is that for the sequence of designs generated as r_k is decreased, none of the constraints are active until the optimum design is found. Qualitatively speaking, the technique tends to keep the sequence of designs away from all the constraints; that is to say, the method tends to generate a sequence of designs that "funnel" down the middle of the acceptable region. Finally, it should be mentioned that transforming the inequality constrained minimization problem into a sequence of unconstrained minimization problems appears to be an effective method of dealing with problems where the constraint functions are time dependent, that is $g_j(t, \mathbf{D}) \leq 0$ for $j = 1 \rightarrow J$. When the inequality constraints depend upon time as well as the design variables, the penalty function can be constructed in the form of an integral over the time period of interest as follows:

$$\phi_k(\mathbf{D}) = M(\mathbf{D}) - r_k \sum_{j=1}^{J} \left[\frac{1}{t_{\max}} \int_0^{t_{\max}} \frac{dt}{g_j(t, \mathbf{D})} \right]. \tag{9}$$

The effect of this extension of the sequential unconstrained minimization technique, suggested by Zoutendijk,[3] is to introduce into the penalty function the influence of each constraint over the time period of interest, rather than just the influence of each constraint at the time when it is most critical.

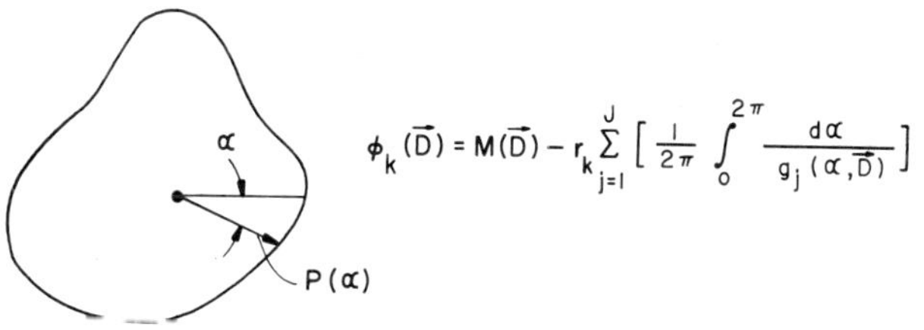

FIG. 3. Envelope loading

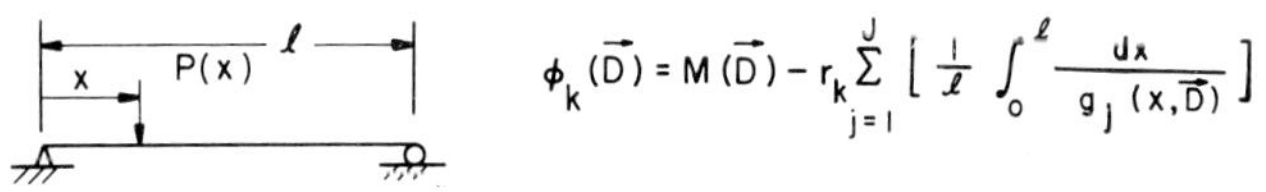

FIG. 4. Moving load

This integral penalty function formulation can also be employed to deal with envelope type loadings (see Fig. 3) or moving load situations (see Fig. 4).

Having in mind the general framework within which structural synthesis problems are formulated as well as an understanding of a few of the solution algorithms that may be employed, we will now consider an example involving composite materials.

IV. EXAMPLE PROBLEM

Consider a simply supported rectangular multilayer resin-matrix fiber composite plate as shown in Fig. 5. Let the planform dimensions (a, b) of the plate be preassigned parameters. The matrix material is assumed to be

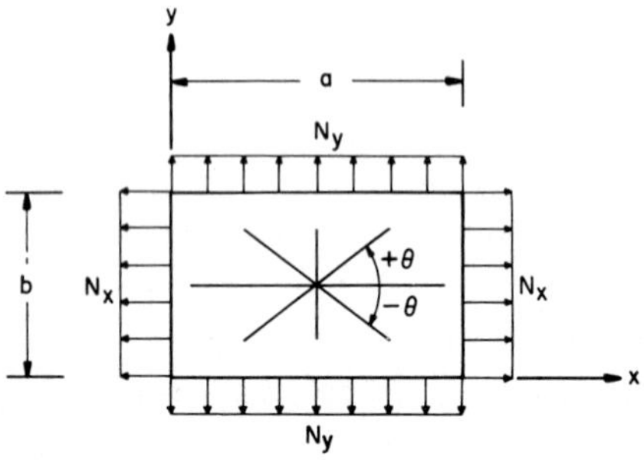

FIG. 5. Simply supported fiber composite plate

isotropic, and the following matrix properties are considered preassigned parameters:

ϱ_m	weight density
E_m	modulus of elasticity
ν_m	Poisson's ratio
S_{mc}	compressive strength
ε_{mpt}	allowable tensile strain
ε_{mpc}	allowable compressive strain
ε_{mps}	allowable shear strain
ε_{mptor}	allowable shear strain (torsion test).

The fiber material is assumed to be transversely isotropic (see Fig. 6 for orientation of fiber coordinate system), and the following properties of the fiber material are treated as preassigned parameters:

ϱ_f	weight density of fiber
N_f	number of fibers per roving end
d_f	equivalent fiber diameter
G_{f12}	shear modulus in the 1–2 plane
$E_{f22} = E_{f33}$	transverse modulus of elasticity
$\nu_{f12} = \nu_{f13}$	absolute ratio of the fiber transverse strain to fiber longitudinal strain due to longitudinal stress†
$\nu_{f23} = \nu_{f32}$	Poisson's ratio in the plane of transverse isotropy
S_{ft}	bundle tensile strength of fiber.

† The remaining Poisson's ratios are determined from the reciprocal theorem.

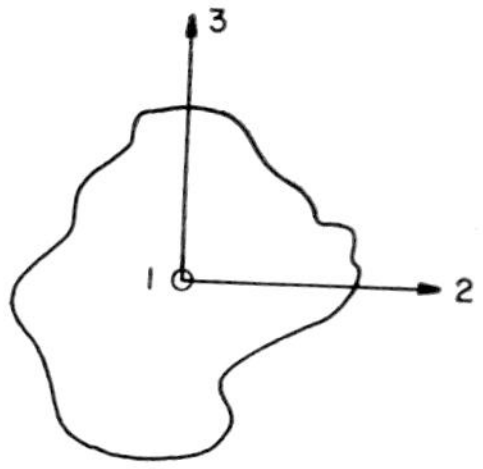

FIG. 6. Fiber coordinate system

The apparent volume fraction of fiber k_f, the estimated volume fraction of voids in the matrix k_v and several empirical factors depending on the fabrication process are also considered to be preassigned parameters.

This example problem considers three design variables. The number of layers (N_l) is taken as a design variable. The minimum number of layers is set at ten and the plate thickness is built up by adding two layers at a time symmetrically with respect to the reference plane, as depicted in Fig. 7. The

FIG. 7. Balanced layup pattern

layer pattern is such that linear bending-membrane coupling is precluded; however, the gross membrane behavior of the plate is symmetric only when the number of layers is an integer multiple of four (i.e., 12, 16, 20, etc.). The angle orienting the fiber direction in each layer (θ) is taken as the second design variable. This design variable is permitted to assume any value over the range from $-90°$ to $90°$. The longitudinal modulus of elasticity of the fiber material (E_{f11}) is treated as the third design variable.† This design variable is permitted to assume any value over the range from 10×10^6 psi to 100×10^6 psi.

† It should be noted that it is not realistic to assume that the fiber material properties treated as preassigned parameters remain fixed as E_{f11} varies. This deficiency could be removed by expressing the pertinent fiber material properties (other than E_{f11}) as preassigned functions of the design variable E_{f11}.

The plate is subject to a multiplicity of load conditions each of which is given by specifying the uniform membrane loads N_x and N_y.

Three failure modes are considered and each of these must be guarded against in each load condition. Let L denote the number of distinct load conditions in the load system. Then the inequality constraints $g_j(\mathbf{D}) \leq 0$ for $j = 1 \rightarrow J$ can be written in the alternate form

$$g_{ik}(\mathbf{D}) \leq 0 \qquad k = 1 \rightarrow L; \quad i = 1, 2, 3 \tag{10}$$

where it is understood that

$$j = k+(i-1)L \qquad j = 1 \rightarrow J \quad \text{and} \quad J = 3L. \tag{11}$$

The simple constraints on the range of values that the design variables may assume are numbered $j = 3L+1$ through $j = 3L+6$.

The first failure mode ($i = 1$) considered is gross buckling of the composite plate. The gross buckling failure mode is represented by

$$g_{1k}(\mathbf{D}) = \frac{N_{xk}}{(N_x)_{crk}} + \frac{N_{yk}}{(N_y)_{crk}} - 1 \tag{12}$$

where N_{xk} and N_{yk} represent the given applied loads in the x and y directions respectively in the kth load condition. The $(N_x)_{crk}$ and the $(N_y)_{crk}$ can be determined for a particular trial design (N_l, θ, E_{f11}) and load condition by a procedure outlined in Appendix A. The essential point to be recognized here is that $g_{1k}(\mathbf{D})$ is computable according to a specified procedure given the preassigned parameters, the load condition k, and a trial design [$\mathbf{D}$, i.e. a set of values for N_l, θ, and E_{f11}]. The values of $(N_x)_{crk}$ and $(N_y)_{crk}$ depend upon the critical buckling mode shape and the flexural rigidities of the composite plate. The flexural rigidities in turn depend upon the design variables (N_l, θ, E_{f11}) and the preassigned parameters. Furthermore, if $g_{1k}(\mathbf{D})$ has a positive value, gross buckling failure in the kth load condition is predicted and the trial design is unacceptable. It should also be noted that when the gradient $\nabla g_{ik}(\mathbf{D})$ is required, it is approximated by using central finite difference representations for each of the components.

The second failure mode ($i = 2$) considered is combined stress failure in a layer. Layers at $\pm\theta$ are checked. When N_l is an integer multiple of four, the results for $+\theta$ are identical with those for $-\theta$. When N_l is not an integer multiple of four, the $-\theta$ layers are found to be critical. This second failure mode ($i = 2$) provides the constraints $j = L+1$ through $j = 2L$ and it is represented by

$$g_{2k}(\mathbf{D}) = \left(\frac{\sigma_{l11k}}{S_{l11\alpha k}}\right)^2 + \left(\frac{\sigma_{l22k}}{S_{l22\beta k}}\right)^2 + \left(\frac{\sigma_{l12k}}{S_{l12k}}\right)^2 - K_{\alpha\beta}\frac{\sigma_{l11k}}{|S_{l11\alpha k}|}\frac{\sigma_{l22k}}{|S_{l22\beta k}|} - 1 \tag{13}$$

where σ_l denotes layer stress, S_l denotes layer strength, the one (1) direction coincides with the fiber direction in the layer, the two (2) direction is transverse to the fiber direction in the plane of the layer, and the subscripts α and β are read as T (Tension) or C (Compression) depending upon whether the corresponding numerator (layer stress) is tensile (+) or compressive (−). Given the preassigned parameters, the load conditions k, and a trial design $D(N_l, \theta, E_{l11})$, the layer stresses and the layer strengths are determined according to the procedure outlined in Appendix A. If for $+\theta$ or $-\theta$ the function $g_{2k}(\mathbf{D})$ is greater than zero, then the analysis predicts that failure under combined stress occurs in the kth load condition. When any of these combined stress failure modes $j = L+1 \rightarrow 2L$ become active, the gradient $\nabla g_j(\mathbf{D})$ for $j \varepsilon J_c$ is required, and it is approximated using central finite difference representations for each of the components.

The third failure mode ($i = 3$) considered is a delamination shear failure mode. Basically this requirement limits the relative rotation between two adjacent layers. Relative rotations between layers imply high shear strains in the matrix material of the interlayer layers. The constraints $j=2L+1$ through $j=3L$ are provided by this third failure mode and it is represented by

$$g_{3k}(\mathbf{D}) = \frac{|\Delta\phi_{jk}|}{(\Delta\phi_j)_{\text{allow}}} - 1 \tag{14}$$

where $\Delta\phi_{jk}$ is the relative rotation between adjacent layers $j+1$ and j in the kth load condition and $(\Delta\phi_j)_{\text{allow}}$ is the allowable relative rotation. When the preassigned parameters, the load condition k and a trial design $\mathbf{D}(N_l, \theta, E_{f11})$, are specified, the relative rotations and the allowable relative rotation can be determined according to the procedure outlined in Appendix A. Whenever $g_{3k}(\mathbf{D})$ is greater than zero the analysis predicts that delamination failure occurs. When a delamination failure mode is active and the gradient is needed, it is determined using a central finite difference approximation for each of the components.

The objective function employed for the example problem is intended to represent cost of fiber material per unit surface area of plate. The cost of fiber is assumed to increase linearly with longitudinal fiber modulus (E_{f11}); hence the objective function may be expressed as follows,

$$M(\mathbf{D}) = \varrho_f \bar{k}_f t_p (K_1 + K_2 E_{f11}) N_l, \tag{15}$$

where ϱ_f is the weight density of the fiber material, $\bar{k}_f$ is the actual fiber volume fraction, t_p is the thickness of a layer, $(K_1+K_2E_{f11})$ expresses the cost per unit weight of fiber material as a linear function of longitudinal modulus, and N_l is the number of layers. Note that this objective function is independent of the design variable θ. It should also be noted that the layer thickness (t_p) is assumed to depend upon the equivalent fiber diameter (d_f), the number of fibers per roving end (N_f), and the apparent volume fraction of fiber (k_f) as follows:

$$t_p = \frac{d_f}{2}\left[\frac{\pi N_f}{k_f}\right]^{1/2}. \tag{16}$$

Numerical results have been obtained using a feasible directions method algorithm. The specific numerical values used for the preassigned parameters in this numerical example are $a=20$ in.; $b=16$ in.

$$\varrho_m = 0.043 \text{ lb/in}^3; \quad E_m = 0.54\times 10^6 \text{ lb/in}^2 \quad \nu_m = 0.36$$

$$S_{mc} = 21{,}000 \text{ lb/in}^2$$

$$\varepsilon_{mpt} = 0.020; \quad \varepsilon_{mpc} = 0.050; \quad \varepsilon_{mps} = \varepsilon_{mp\,\text{tor}} = 0.045$$

- - - - - - - - -

$$\varrho_f = 0.052 \text{ lb/in}^3; \quad N_f = 1440; \quad d_f = 0.000291 \text{ in}$$

$$G_{f12} = 2.0\times 10^6 \text{ lb/in}^2; \quad E_{f22} = E_{f33} = 2.0\times 10^6 \text{ lb/in}^2$$

$$\nu_{f12} = \nu_{f13} = 0.20; \quad \nu_{f23} = \nu_{f32} = 0.25$$

$$S_{ft} = 180{,}000 \text{ lb/in}^2$$

- - - - - - - - -

$$k_f = 0.40; \quad k_v = 0.0$$

- - - - - - - - -

$$\bar{\beta}_m = 1.55 \text{ (eq. A20)}; \quad \tilde{\bar{\beta}}_f = 1.0 \text{ (eq. A21)}$$

$$\beta_m = 1.26 \text{ (eq. A29a)}; \quad \beta_f = 1.00 \text{ (eq. A29b)}$$

$$\beta_{fT} = 1.0; \quad \beta_{mT} = 1.0 \text{ (eq. A33)}$$

$$\beta_{mC} = 1.0; \quad \beta_{fC} = 0.1 \text{ (eq. A34)}$$

$$a_1 = 13.3; \quad a_2 = 31{,}900 \text{ lb/in}^2 \text{ (eq. A35)}$$

$$\beta_{12} = 0.4 \text{ (eq. A36)}$$

$$\beta_{22T} = 0.1 \text{ (eq. A40)}$$

$$\beta_{22C} = 0.9 \text{ (eq. A46)}$$

$$\beta_{\text{del}} = 33.0 \text{ (eq. A51)}$$

- - - - - - - - -

$K_1 = 450$ cost units/lb of fiber material

$K_2 = 5.0$ cost units/lb of fiber material $- 10^6$ lb/in^2 of longitudinal fiber modulus

The load system for this numerical example consists of three distinct load conditions specified as follows:

Loads \ Condition	$k = 1$	$k = 2$	$k = 3$
N_{xk} lb/in	-1000	-1200	$+700$
N_{yk} lb/in	-500	$+600$	$+700$

Three distinct initial trial designs were selected. For each of these initial designs a sequence of designs was generated using the structural synthesis program. The terminal design associated with each of these three initial trial designs are presented in Table 1. Furthermore, the constraint function values plus unity (i.e., $g_{ik}(\mathbf{D})+1$) for the three terminal designs are tabulated for each failure mode for each load condition in Table 1. Note that

$$g_{ik}(\mathbf{D}) \leqslant 0 \tag{17}$$

is equivalent to

$$g_{ik}(\mathbf{D})+1 \leqslant 1; \tag{18}$$

therefore tabulated values of $(g_{ik}(\mathbf{D})+1)$ near unity represent active or critical constraints. The three terminal designs are distinct but they are not widely separated and the objective function values for these three terminal designs differ by a maximum 4.3%. In each case the plate buckling failure mode in load condition one is critical; that is the constraint $g_{11}(\mathbf{D})\approx 0$ is active. It is also observed, from the summary of results in Table 1, that for each of the three terminal designs the layer combined stress failure mode in load condition two [i.e., $g_{22}(\mathbf{D})$] is near to being a critical or active constraint.

While this example is rather idealized, it nevertheless exhibits many of the features characteristic of more realistic composite material structural system design problems. The example involves only three design variables; however, the number of layers (N_l) is a discrete sizing variable, the layer orientation angle (θ) is a configuration variable, and the longitudinal fiber modulus E_{f11} may be classified as a material variable. The example also considers the loading system to be made up from several distinct loading conditions and a balanced optimum design is sought. The failure modes considered are intended to guard against buckling failure, combined stress failure and delamination. Finally, the objective function employed (minimization of cost of fiber material per unit surface area of plate) is intended to suggest that minimization of total structural weight need not necessarily be the goal of every structural optimization capability.

TABLE 1. SUMMARY OF NUMERICAL RESULTS

Design	N_l	$\theta°$	$E_{f11}\times 10^{-6}$	$M(\mathbf{D})$	$(g_{ik}+1)$								
					11	21	31	12	22	32	13	23	33
Start 1	40	65	70	8.99									
Terminal 1	18	50	28.0	2.98	1.00	0.14	0.12	0.58	0.90	0.32	—	0.46	0.03
Start 2	50	45	50.0	9.38									
Terminal 2	18	45	29.5	3.02	1.00	0.08	0.17	0.49	0.87	0.56	—	0.30	0.00
Start 3	34	30	30.0	4.50									
Terminal 3	20	40	20.7	3.11	1.00	0.03	0.05	0.48	0.84	0.27	—	0.51	0.03

Note: Machine running time from start to terminal in each case was less than 15 min. 1107 ALGOL. Based on past experience run times would be less the 3 min. IBM 7094 FORTRAN.

V. PROBLEMS AND POTENTIAL

During the last four years substantial progress has been reported in the structural synthesis area. Large scale capabilities for dealing with the minimum weight optimum design problem based on linear, static, finite-element structural analyses have been reported [for example, see refs. 12 and 13]. Effort has also been directed toward exploratory structural synthesis studies of small systems based on analyses involving dynamics,[14] aeroelasticity,[15] thermoelasticity,[16] and kinematics.[17] A more comprehensive review of the literature is given in ref. 18. The extensive activity and growing capability in the structural synthesis area suggest that the extension of these ideas and the associate techniques into the realm of structural design with composite materials can be pursued with a reasonable expectation of success.

There are, however, some pitfalls and problem areas that should be mentioned. First of all, there is the relative minima problem. The sequential redesign methods are, at best, only able to find a local optimum design. The relative minima problem is characterized by the existence of two or more designs (**D**) each of which has no acceptable designs of better or equal merit within some finite neighborhood about them. In general, it is not possible to test whether a proposed optimum design is in fact a global or absolute optimum. One approach to building some confidence that a proposed optimum is a global optimum is to run multiple sequential synthesis paths from a sampling of distinct initial trial design points. The extent to which relative minima will be present in structural design with composite materials remains to be assessed. However, prior experience suggests that configuration variables (such as θ) and discrete variables (such as N_l) may lead to relative minima. The existence of relative minima may be viewed as a manifestation of the continuing need for creative control on the part of the engineer. Thus, the selection of alternative initial trial designs may, in many instances, be advantageously made by the engineer, rather than by having recourse to the techniques of experiment design.

The existence of structural synthesis capablities should not be misconstrued to represent a substitute for creative design. They are, in fact, only a tool for trying to get the most out of any particular design concept. They may also permit an evaluation of alternative concepts based on a comparison of optimum designs for each concept. The need to devise practical design concepts that capitalize upon the new degrees of freedom offered by composite materials is widely recognized, and optimization efforts should not be permitted to inhibit the generation of imaginative and creative design concepts.

In extending the structural synthesis concept to design with composite materials, attention should be given to selecting and generating analysis methods that are adequate to guard against failure, yet well suited to embedment within a redesign process. The goal of employing analyses that adequately predict behavior without being unnecessarily complex is, of course, not new, but rather it is a traditional responsibility of the structural engineer. What is, to some degree, unique is the opportunity that exists in the composite materials area to concurrently develop analysis and synthesis methods. In the past optimum design methodology has generally evolved only after a firmly established analysis capability has been in existence for some time. Assuming that both macro- and micro-stress analysis methods for composite materials are still in a relatively fluid status, the opportunity to attempt an evolution in parallel of structural analysis and synthesis procedures exists.

Prior experience suggests that as the range of design variables and the scope of the behavior constraints broaden, new formulations of the combined analysis-synthesis problem are created. For example, consider the use of a nonlinear, static, structural analysis as a basis for minimum weight design. Then, since each analysis will probably be conducted using an iterative procedure, the thought comes to mind to formulate the problem so that redesign can take place simultaneously with the iterative steps of the analysis. Thus, the integrated approach to structural analysis and synthesis presented in refs. 19 and 20 is an example of innovations in formulation that can be stimulated by the needs of particular problem classes. The generation and evaluation of various formulation and solution algorithm combinations may well lead to a substantial advance in computational efficiency, thus broadening the applicability of the structural synthesis approach.

Much remains to be done and clearly there are difficulties to be overcome. However, the prospect of contributing to the effective simultaneous design of structural configuration and material by extending the structural synthesis concept is bright and intriguing.

The use of composite materials in the fabrication of structural systems significantly widens the range of possible designs open to the engineer. These new degrees of freedom offer the prospect of achieving marked advances in the development of high performance structural systems. However, to realize this promise will require the rational and responsible management of these new degrees of freedom.

Such venerable guides to materials evaluation and selection as specific modulus, specific strength, and structural indices will continue to be valuable tools under certain circumstances. However, in many applications the pertinent structural response will depend upon numerous engineering material properties, and acceptable behavior will be required in several distinct load conditions. Modern structural concepts will often be statically indeterminate leading to significant behavioral interaction between the individual elements of the structural system. The need to consider the influence of numerous material properties, multiple load conditions, and structural component interaction becomes particularly apparent when thinking in terms of tailoring a composite material by varying its composition and constituents. It is thought that extending the structural synthesis approach to include composition and material design variables will provide a comprehensive and rational tool to assist the engineer in the intelligent exploitation of the potential offered by composite materials.

The structural synthesis approach can be employed to assess the sensitivity of optimum designs to perturbations in the load conditions or preassigned parameters. These automated optimum design capabilities can provide a tool for evaluating an existing set of constituent materials (or a class of existing constituent materials) in terms of a specified structural task. In principle, at least, it is conceivable that this approach could also be employed to assess the merits of hypothetical sets of constituent materials for a specified structural task. Thus, in the future structural engineers may be able to quantitatively specify goals, in terms of attainable combinations of engineering properties, toward which materials research and development should be directed.

In conclusion, it is suggested that the formulation of the quantifiable portion of the structural design process within the framework of the structural synthesis concept and the application of mathematical programming algorithms to obtain solutions can be expected to play an increasing role in structural design with composite materials as our understanding of the behavior of these new materials continues to improve.

APPENDIX A

List of Symbols

A	constant defined by eq. (A28a)
$\tilde{A}$	constant defined by eq. (A43)
a	plate dimension along the x-axis
a_1 and a_2	empirical factors in eq. (A35)
B	constant defined by eq. (A28b)
$\tilde{B}$	constant defined by eq. (A44)
b	plate dimension along the y-axis
C	constants defined by eqs. (A29a) and (A29b)
$\tilde{C}$	constants defined by eqs. (A20) and (A21)
d_f	equivalent fiber diameter
E	modulus of elasticity
G	shear modulus
H	interlayer-layer empirical coefficient
K	empirical coefficient in combined strength constraint
k	apparent volume ratio (fiber, matrix)

$\bar{k}$	actual volume ratio (fiber, matrix)
k_v	actual volume ratio (voids)
m	mode shape x-direction
m^*	critical mode shape x-direction
N	membrane force
N^x	defined by eq. (A2)
N^y	defined by eq. (A4)
$(N_x)_r$	defined by eq. (A3)
$(N_y)_r$	defined by eq. (A5)
del	delamination
f	fiber properties
i	behavior constraint index; layer index
j	combined constraint index; interlayer–layer index
k	load condition index
m	matrix properties
p	allowable limit in matrix strains
s	shear
T	tension
tor	torsion
v	voids
x, y	associated with plate axes
1, 2	associated with layer directions
α	tension or compression in combined strength constraint
β	tension or compression in combined strength constraint

Arrays

A_c	composite axial stiffness defined by eq. (A32)
C_{fl}	fiber-layer relations defined by eq. (A24)
C_{ml}	matrix-layer relations defined by eq. (A25)
D_c	composite flexural rigidities defined by eq. (A7)
E_{li}	strain-stress law for ith layer defined by eq. (A9)
E_f	strain-stress law for fiber defined by eq. (A26)
E_m	strain-stress law for matrix defined by eq. (A27)
n	mode shape y-direction
n^*	critical mode shape y-direction
p	constant defined by eq. (A39)
S	allowable strength for layer
x, y, z	plate coordinate axes and corresponding distances
1, 2, 3	layer coordinate axes
β, $\tilde{\beta}$	empirical factors associated with layer elastic constants and layer strengths
δ	interlayer-layer thickness defined by eq. (A53)

ε	strain due to applied load
θ	layer orientation angle
ν	Poisson's ratio
σ	stress due to applied load
ϕ_μ	matrix-strain magnification factor
$\Delta\phi$	relative rotation of adjacent layers defined by eq. (A48)

Superscripts

x	associated with N^x in eq. (A2)
y	associated with N^y in eq. (A4)
T	transpose of matrix
-1	inverse of matrix

Subscripts

C	compression
c	composite properties
cr	critical
R_{li}	transformation matrix for the ith layer defined by eq. (A8)
S_j	transformation matrix for the jth interlayer layer defined by eq. (A11)

Vectors

D	design variables
{N}	membrane applied loads
$\{\varepsilon\}$	strains due to applied loads
$\{\sigma\}$	stresses due to applied loads

Failure Mode Analysis

The constraint functions $g_j(\mathbf{D})$ are in general complicated functions of the design variables. In fact, in many cases, the $g_j(\mathbf{D})$ are functions only in the sense that a set of rules exists for computing their value given the argument **D**. This appendix contains a summary of the rules used to evaluate the constraint functions representing the three failure modes to be guarded against in the example problem described in section IV. The derivation and discussion of assumptions underlying these expressions will be found in ref. 21.

Failure mode $i = 1$—Gross buckling of the composite plate under biaxial loading

$$g_{1k}(\mathbf{D}) = \frac{N_{xk}}{(N_x)_{crk}} + \frac{N_{yk}}{(N_y)_{crk}} - 1 \tag{A1}$$

where N_{xk} and N_{yk} are the applied loads in the kth load condition (see Fig. 5), and $(N_x)_{crk}$ and $(N_y)_{crk}$ are determined according to the following procedure:

$$N^x = \frac{(N_x)_r}{1+\left(\frac{N_{yk}}{N_{xk}}\right)\left(\frac{a}{b}\right)^2\left(\frac{n}{m}\right)^2} \tag{A2}$$

where (a, b) are the planform dimensions of the plate and (m, n) are the integers describing the mode shape $\sin m\pi \times \sin n\pi y/ab$ and

$$(N_x)_r = +\frac{\pi^2(D_{c11}D_{c22})^{1/2}}{b^2}\left[\left(\frac{D_{c11}}{D_{c22}}\right)^{1/2}\left(\frac{mb}{a}\right)^2 + (D_{c12}+D_{c21}+4D_{c33})\frac{n^2}{(D_{c11}D_{c22})^{1/2}} + \left(\frac{D_{c22}}{D_{c11}}\right)^{1/2}\left(\frac{an^2}{bm}\right)^2\right] \tag{A3}$$

where the D_{c11}, D_{c22}, D_{c12}, D_{c21}, and D_{c33} are the flexural rigidities of the composite plate. Also let N^y be defined by

$$N^y = \frac{(N_y)_r}{1+\left(\frac{N_{xk}}{N_{yk}}\right)\left(\frac{b}{a}\right)^2\left(\frac{m}{n}\right)^2} \tag{A4}$$

where

$$(N_y)_r = +\frac{\pi^2(D_{c11}D_{c22})^{1/2}}{a^2}\left[\left(\frac{D_{c22}}{D_{c11}}\right)^{1/2}\left(\frac{na}{b}\right)^2 + (D_{c12}+D_{c21}+4D_{c33})\frac{m^2}{(D_{c11}D_{c22})^{1/2}} + \left(\frac{D_{c11}}{D_{c22}}\right)^{1/2}\left(\frac{m^2b}{an}\right)^2\right]. \tag{A5}$$

If $N_{xk}<N_{yk}<0$ (i.e., both compressive) or $N_{xk}<0$ and $N_{yk}\geqslant 0$, then find the critical buckling mode shape (m^*, n^*) by seeking values of (m, n) that make N^x take on its smallest positive value [see eq. (A2)]. The values of $(N_x)_{crk}$ and $(N_y)_{crk}$ are then given by the negative of $(N_x)_r$ and $(N_y)_r$ respectively using (m^*, n^*) for (m, n) in eqs. (A3) and (A5).

If $N_{yk} < N_{xk} < 0$ (i.e. both compressive) or $N_{yk}< 0$ and $N_{xk}\geqslant 0$ then find the critical buckling mode shape (m^*, n^*) by seeking values of (m, n) that make N^y take on its smallest positive value [see eq. (A4)]. The values of $(N_x)_{crk}$ and $(N_y)_{crk}$ are then given by the negative of $(N_x)_r$ and $(N_y)_r$ respectively using (m^*, n^*) for (m, n) in eqs. (A3) and (A5).

If N_{xk} and $N_{yk}\geqslant 0$, then buckling cannot occur and $g_{1k}(\mathbf{D})$ is set equal to -1. This buckling analysis is adapted from that given in ref. 22.

The dependence of the flexural rigidities, D_{c11}, D_{c22}, D_{c12}, D_{c21}, and D_{c33} upon the preassigned parameters, empirical factors and the design variables is expressed as follows. Let the matrix of flexural rigidities for the composite plate $[D_c]$ be defined by

$$[D_c] = \begin{bmatrix} D_{c11} & D_{c12} & 0 \\ D_{c21} & D_{c22} & 0 \\ 0 & 0 & D_{c33} \end{bmatrix}. \tag{A6}$$

Then $[D_c]$ is determined from the expression

$$[D_c] = \sum_{i=1}^{n} \left\{ \int_{Z_i}^{Z_{i+1}} z^2 [R_{li}]^T [E_{li}] [R_{li}] \, dz \right\} + \sum_{j=1}^{n-1} z_j^2 H_j [S_j] \tag{A7}$$

where in this expression the index i refers to the ith layer of the composite and the index j refers to the j interlayer-layer of the composite (see Fig. A1). The transformation matrix $[R_{li}]$ between the layer axes of orthotropy and the plate planform axes x–y is given by

$$[R_{li}] = \begin{bmatrix} \cos^2\theta_i & \sin^2\theta_i & \frac{1}{2}\sin 2\theta_i \\ \sin^2\theta_i & \cos^2\theta_i & -\frac{1}{2}\sin 2\theta_i \\ -\sin 2\theta_i & \sin 2\theta_i & \cos 2\theta_i \end{bmatrix} \tag{A8}$$

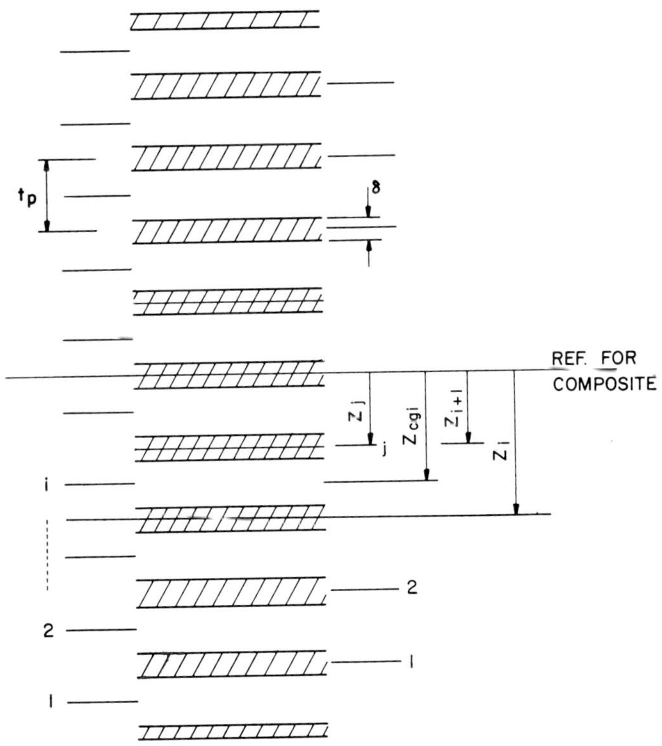

FIG. A1. Idealized geometry of multi-layer composite

and the elastic properties of the ith layer are contained in $[E_{li}]$ as follows:

$$[E_{li}] = \begin{bmatrix} \frac{1}{E_{l11}} & -\frac{\nu_{l21}}{E_{l22}} & 0 \\ -\frac{\nu_{l12}}{E_{l11}} & \frac{1}{E_{l22}} & 0 \\ 0 & 0 & \frac{1}{E_{l12}} \end{bmatrix} \quad \text{(A9)}$$

The second term in eq. (A7) represents the contribution to the flexural rigidities of the resistance to relative rotation between layers (ref. 23) where

$$H_j = 0.0093G_m \quad \text{(A10)}$$

and

$$[S_j] = \tfrac{1}{4}\begin{bmatrix} S_{11j} & S_{12j} & S_{13j} \\ S_{21j} & S_{22j} & S_{23j} \\ S_{31j} & S_{32j} & S_{33j} \end{bmatrix} \quad \text{(A11)}$$

and

$$S_{11j} = S_{22j} = (\sin 2\theta_{j+1} - \sin 2\theta_j)^2 \quad \text{(A12)}$$

$$S_{33} = (\cos 2\theta_{j+1} - \cos 2\theta_j)^2 \quad \text{(A13)}$$

$$S_{12j} = S_{21j} = -(\sin 2\theta_{j+1} - \sin 2\theta_j)^2 \quad \text{(A14)}$$

$$S_{23} = S_{32} = -S_{31} = -S_{13} = (\sin 2\theta_{j+1} - \sin 2\theta_j)\times (\cos 2\theta_{j+1} - \cos 2\theta_j). \quad \text{(A15)}$$

The layer elastic constants are based upon a semi-empirical micromechanics theory (see ref. 21), and their dependence upon the preassigned parameters, empirical factors, and the design variables follows. The layer shear modulus E_{l12} is given by

$$E_{l12} = \frac{G_m}{\left(\frac{G_m}{C_f^2 G_{f12}}\right)\bar{k}_f + \frac{\bar{k}_m}{C_m^2}} \quad \text{(A16)}$$

where G_m is the matrix-material shear-modulus, $\bar{k}_f$ is the actual volume fraction of fibers:

$$\bar{k}_f = (1-k_v)k_f, \quad \text{(A17)}$$

$\bar{k}_m$ is the actual volume fraction of matrix

$$\bar{k}_m = (1-k_v)k_m, \quad \text{(A18)}$$

and it is noted that

$$k_m = 1-k_f-k_v. \quad \text{(A19)}$$

The symbols $\tilde{C}_m$ and $\tilde{C}_f$ are defined by

$$\tilde{C}_m = \left(\frac{\bar{k}_m}{k_m}\right)\tilde{\beta}_m \tag{A20}$$

and

$$\tilde{C}_f = \left(\frac{\bar{k}_f}{k_f}\right)\tilde{\beta}_f \tag{A21}$$

where $\tilde{\tilde{\beta}}_f$ and $\tilde{\tilde{\beta}}_m$ are empirical factors. The remaining layer-elastic-constants are determined from

$$[E_{li}] = \bar{k}_f[C_{fl}]^T\,[E_f]\,[C_{fl}]+\bar{k}_m[C_{ml}]^T\,[E_m]\,[C_{ml}] \tag{A22}$$

where

$$[E_{li}] = \begin{bmatrix} \frac{1}{E_{l11}} & -\frac{\nu_{l21}}{E_{l22}} & -\frac{\nu_{l31}}{E_{l33}} \\ -\frac{\nu_{l12}}{E_{l11}} & \frac{1}{E_{l22}} & -\frac{\nu_{l32}}{E_{l33}} \\ -\frac{\nu_{l13}}{E_{l11}} & -\frac{\nu_{l23}}{E_{l22}} & \frac{1}{E_{l33}} \end{bmatrix} \tag{A23}$$

$$[C_{fl}] = \begin{bmatrix} \frac{1}{A\bar{k}_m E_m} & \frac{1}{A}\left(\frac{\nu_{f21}}{C_f E_{f22}}-\frac{\nu_m}{C_m E_m}\right) & \frac{1}{A}\left(\frac{\nu_{f31}}{C_f E_{f33}}-\frac{\nu_m}{C_m E_m}\right) \\ 0 & \frac{1}{C_f} & 0 \\ 0 & 0 & \frac{1}{C_f} \end{bmatrix} \tag{A24}$$

$$[C_{ml}] = \begin{bmatrix} \frac{1}{B\bar{k}_f E_{f11}} & \frac{1}{B}\left(\frac{\nu_m}{C_m E_m}-\frac{\nu_{f21}}{C_f E_{f22}}\right) & \frac{1}{B}\left(\frac{\nu_m}{C_m E_m}-\frac{\nu_{f31}}{C_f E_{f33}}\right) \\ 0 & \frac{1}{C_m} & 0 \\ 0 & 0 & \frac{1}{C_m} \end{bmatrix} \tag{A25}$$

$$[E_f] = \begin{bmatrix} \frac{1}{E_{f11}} & -\frac{\nu_{f21}}{E_{f22}} & -\frac{\nu_{f31}}{E_{f33}} \\ -\frac{\nu_{f12}}{E_{f11}} & \frac{1}{E_{f22}} & -\frac{\nu_{f32}}{E_{f33}} \\ -\frac{\nu_{f13}}{E_{f11}} & -\frac{\nu_{f23}}{E_{f22}} & \frac{1}{E_{f33}} \end{bmatrix} \tag{A26}$$

$$[E_m] = \frac{1}{E_m}\begin{bmatrix} 1 & -\nu_m & -\nu_m \\ -\nu_m & 1 & -\nu_m \\ -\nu_m & -\nu_m & 1 \end{bmatrix} \tag{A27}$$

and it is noted that

$$A = \left(\frac{1}{E_{f11}} + \frac{1}{E_m}\frac{\bar{k}_f}{\bar{k}_m}\right) \tag{A28a}$$

$$B = \left(\frac{1}{E_m} + \frac{1}{E_{f11}}\frac{\bar{k}_m}{\bar{k}_f}\right) \tag{A28b}$$

$$C_m = \left(\frac{\bar{k}_m}{k_m}\right)\beta_m \tag{A29a}$$

$$C_f = \left(\frac{\bar{k}_f}{k_f}\right)\beta_f. \tag{A29b}$$

Failure mode i = 2—Combined stress in a layer

$$g_{2k}(\mathbf{D}) = \left(\frac{\sigma_{l11k}}{S_{l11\alpha k}}\right)^2 + \left(\frac{\sigma_{l22k}}{S_{l22\beta k}}\right)^2 + \left(\frac{\sigma_{l12k}}{S_{l12k}}\right)^2 - K_{\alpha\beta}\frac{\sigma_{l11k}}{|S_{l11\alpha k}|}\frac{\sigma_{l22k}}{|S_{l22\beta k}|} \tag{A30}$$

where the layer stresses for the kth load condition may be expressed as follows,

$$\{\sigma_{li}\}_k = [E_{li}]\,[R_{li}]\,[A_c]^{-1}\,\{N_{ck}\}, \tag{A31}$$

where $[E_{li}]$ is given by eq. (A22), $[R_{li}]$ is given by eq. (A8).

$$\{N_{ck}\} = \begin{Bmatrix} N_{xk} \\ N_{yk} \\ 0 \end{Bmatrix} \tag{A31a}$$

$$\{\sigma_{li}\}_k = \begin{Bmatrix} \sigma_{l11k} \\ \sigma_{l22k} \\ \sigma_{l12k} \end{Bmatrix} \tag{A31b}$$

and the composite membrane-stiffness is expressed as follows,

$$[A_c] = \sum_{i=1}^{n}\int_{z_i}^{z_{i+1}} [R_{li}]^T\,[E_{li}]\,[R_{li}]\,dz + \sum_{j=1}^{n-1} H_j[S_j], \tag{A32}$$

where $[R_{li}]$ is defined by eq. (A8), $[E_{li}]$ is defined by eq. (A22), $[S_j]$ is defined by eq. (A11) and $H_j = 0.0093\ G_m$.

The tensile longitudinal layer strength S_{l11T}, which is independent of the load conditions k, is given by the rule of mixtures as follows:

$$S_{l11T} = S_{ft}\left(\beta_{ft}\bar{k}_f + \beta_{mT}\bar{k}_m\frac{E_m}{E_{f11}}\right). \tag{A33}$$

The β_{fT} and the β_{mT} are empirical factors, the actual fiber volume ratio $\bar{k}_f$ is given by eq. (A17) and the actual matrix volume fraction $\bar{k}_m$ is given by eq. (A18). The compressive longitudinal layer strength S_{l11C} is given by either the rule of mixtures as follows:

$$S_{l11C} = S_{mC}\left(\beta_{mC}\bar{k}_m+\beta_{fC}\bar{k}_f\frac{E_{f11}}{E_m}\right) \tag{A34}$$

where β_{mC} and β_{fC} are empirical factors, or

$$S_{l11C} = a_1 S_{l12}+a_2 \tag{A35}$$

which is based upon debonding of the matrix material around the fibers. The a_1 and a_2 are empirical constants given in ref. 24. The inplane layer-shear failure strength S_{l12} is given by

$$S_{l12} = \beta_{12}\left[\frac{\varepsilon_{mps}}{\phi_{\mu_{12}}}\right]G_m \tag{A36}$$

where β_{12} is an empirical factor determined from shear tests. The matrix shear-strain magnification-factor $\phi_{\mu_{12}}$, based on an extension of the results reported in refs. 25 through 29, is given by

$$\phi_{\mu_{12}}= \beta_v\left[\frac{1}{1-p\left(1-\dfrac{\bar{k}_m k_f}{k_m\bar{k}_f}\dfrac{G_m}{G_{f12}}\right)}\right] \tag{A37}$$

where

$$\beta_v = \frac{1}{1-(4k_v/\pi k_m)^{1/2}} \tag{A38}$$

$$p = (4\bar{k}_f/\pi)^{1/2} \tag{A39}$$

and $\bar{k}_f$, $\bar{k}_m$ and k_m are defined in eqs. (A17), (A18), (A19) respectively. The transverse-tensile layer-strength S_{l22Tk} is given by

$$S_{l22T_k} = \beta_{22T}\left[\frac{\varepsilon_{mpT}}{\phi_{\mu_{22k}}}\right]E_{l22} \tag{A40}$$

where β_{22T} is an empirical factor. The transverse-tensile strain magnification factor in the matrix $\phi_{\mu 22k}$ is given by

$$\phi_{\mu 22k} = \begin{cases}\beta_v\left[\dfrac{1}{1+p(\tilde{A}-1)}\right]\left[1+p(\nu_{f12}-\tilde{B})\dfrac{\varepsilon_{l11k}}{\varepsilon_{l22k}}\right] & \text{if} \quad \varepsilon_{l22k}\neq 0\\ 1 & \text{if} \quad \varepsilon_{l22k}=0\end{cases} \tag{A41}$$

where the layer strains for the kth load condition are given by

$$\{\varepsilon_{li}\}_k = [R_{li}][A_c]^{-1}\{N_{ck}\} \tag{A42}$$

in which $[R_{li}]$, $[A_c]$, p, β_v and $\{N_{ck}\}$ are defined in eqs. (A8), (A32), (A39), (A38) and (A31a) respectively. Furthermore,

$$\tilde{A} = \left[\frac{\bar{k}_m k_f}{k_m \bar{k}_f} \frac{E_m}{E_{f22}} \times \frac{(1-\nu_{f12}\nu_{f21})}{(1-\nu_m^2)}\right] \quad \text{(A43)}$$

$$\tilde{B} = \left[\nu_m \times \frac{\bar{k}_m k_f}{k_m \bar{k}_f} \frac{E_m}{E_{f22}} \times \frac{(1-\nu_{f12}\nu_{f21})}{(1-\nu_m^2)}\right] \quad \text{(A44)}$$

$$\nu_{f21} = E_{f22}\frac{\nu_{f12}}{E_{f11}} \quad \text{(A45)}$$

and the dependence of ν_{f21} upon the design variables E_{f11} should be noted. The transverse-compressive layer-strength S_{l22Ck} is given by

$$S_{l22Ck} = \beta_{22C}\left[\frac{\varepsilon_{mpC}}{\phi_{\mu 22k}}\right] E_{l22} \quad \text{(A46)}$$

where β_{22C} is an empirical factor and $\phi_{\mu 22k}$ is given by eq. (A41).

Failure mode i = 3—Delamination

$$g_{3k}(\mathbf{D}) = \frac{|\Delta\phi_{jk}|}{(\Delta\phi_j)_{\text{allow}}} - 1 \quad \text{(A47)}$$

where $\Delta\phi_{jk}$ is the relative rotation between adjacent layers $j+1$ and j (see Fig. A1) in the kth load condition and $(\Delta\phi_j)_{\text{allow}}$ is the allowable relative rotation. The relative rotation $\Delta\phi_{jk}$ is determined from the expression

$$\Delta\phi_{jk} = \tfrac{1}{2}(\varepsilon_{yyk}-\varepsilon_{xxk})(\sin 2\theta_{j+1}-\sin 2\theta_j)+\tfrac{1}{2}\varepsilon_{xyk}(\cos 2\theta_{j+1}-\cos 2\theta_j) \quad \text{(A48)}$$

where the composite strains referred to the planform coordinate system (for the kth load condition) ε_{xxk}, ε_{yyk}, and ε_{xyk} are given by

$$\{\varepsilon_{ck}\} = [A_c]^{-1}\ \{N_{ck}\} \quad \text{(A49)}$$

where

$$\{\varepsilon_{ck}\} = \begin{Bmatrix}\varepsilon_{xxk}\\ \varepsilon_{yyk}\\ \varepsilon_{xyk}\end{Bmatrix}. \quad \text{(A50)}$$

$\{N_{ck}\}$ is given by eq. (A31a) and $[A_c]$ is given by eq. (A32).

The allowable relative rotation may be expressed as

$$(\Delta\phi_j)_{\text{allow}} = \beta_{\text{del}}\left(\frac{\varepsilon_{mp\,\text{tor}}}{\phi_{\mu\,\text{del}}}\right) \quad \text{(A51)}$$

where the delamination strain magnification factor $\phi_{\mu\,\text{del}}$ is given by

$$\phi_{\mu\,\text{del}} = \frac{0.0093}{\delta} \quad \text{(A52)}$$

in which δ is the thickness of the interlayer-layer and depends upon the apparent fiber volume fraction k_f and the equivalent fiber diameter d_f as follows:

$$\delta = \left[\left(\frac{\pi}{4k_f}\right)^{1/2} - 1\right] d_f. \tag{A53}$$

The factor β_{del} is to be determined from delamination tests and $\varepsilon_{mp\,tor}$ is the allowable matrix-shear-strain from a torsion test.

REFERENCES

1. SCHMIT, L. A. Structural Design by Systematic Synthesis, *Proc. 2nd Nat. Conf. on Electr. Comp., Str. Div. ASCE*, pp. 105–132 (1960).
2. KOWALIK, J. Nonlinear Programming Procedures and Design Optimization, *Acta Polytechnica Scandinavica, Math. and Comp. Mach.* Series No. 13, Trondheim, 1966.
3. ZOUTENDIJK, G. Nonlinear Programming: A Numerical Survey, *J. SIAM Control,* **4,** No. 1, pp. 194–210 (1966).
4. ZOUTENDIJK, G. *Methods of Feasible Directions*, Elsevier Press, Amsterdam, 1960.
5. CARROLL, C. W. The Created Response Surface Technique for Optimizing Nonlinear, Restrained Systems, *Oper. Res.* **9**, No. 2, pp. 169–184 (1961).
6. FIACCO, A., and MCCORMICK, G. P. The Sequential Unconstrained Minimization Technique for Nonlinear Programming, A Primal-Dual Method, *Manag. Sci.* **10**, No. 2, pp. 360–365 (1964).
7. DAVIDON, W. C. Variable Metric Method for Minimization, A.E.C. Res. and Dev. Rept. ANL-5990 (1959).
8. FLETCHER, R. and POWELL, M. J. D. A Rapidly Convergent Descent Method for Minimization, *Computer J.* **6**, 163–168 (1963).
9. MCNUNN, J. and JORGENSEN, G. A Review of the Literature on Optimization Techniques and Minimax Structural Response Problems, TR 66-5, Dept. of Aero and Engr. Mech., University of Minnesota 1966.
10. ZIMMERMAN, J. R. *Mathematical Optimization Techniques*, Vol. I, *Computers in Engineering Design Education*, Univ. of Michigan, pp. 142–188 (1966).
11. BOGNER, F. K., MALLETT, R. H., MINICH, M. D. and SCHMIT, L. A. Development and Evaluation of Energy Search Methods of Non-linear Structural Analysis, AFFDL TR-65-113 (1965).
12. GELLATELY, R. A., GALLAGHER, R. H. and LUBERACKI, W. A. Development of a Procedure for Automated Synthesis of Minimum Weight Structures, FDL-TDR-64-141 (1964).
13. GELLATELY, R. A. Development of Procedures for Large Scale Automated Minimum Weight Structural Design, AFFDL-TR-66-180 (1966).
14. SCHMIT, L. A. and FOX, R. L. Synthesis of a Simple Shock Isolator, NASA CR 55 (1964).
15. SCHMIT, L. A. and THORNTON, W. A. Synthesis of an Airfoil at Supersonic Mach Number, NASA CR 144 (1965).
16. SCHMIT, L. A. and KINSER, D. E. The Synthesis of a Laminated Plate for High Temperature Application, EDC Report No. 2-64-7, Case Instit. of Tech. (1964).
17. FOX, R. L. and WILLMERT, K. D. Optimum Design of Curve-Generating Linkages with Inequality Constraints, *Trans. ASME, J. of Engr. for Industry*, **89,** Series B, No. 1, pp. 144–152 (1967).
18. SCHMIT, L. A. A Review of Some Structural Synthesis Studies, International Symposium on the Use of Computers in Structural Engineering, Univ. of Newcastle upon Tyne 1966.

19. SCHMIT, L. A. and FOX, R. L. An Integrated Approach to Structural Synthesis and Analysis, *J. AIAA*, **3**, No. 6, pp. 1104–1112 (1965).
20. FOX, R. L. and SCHMIT, L. A. Advances in the Integrated Approach to Structural Synthesis, *J. Spacecraft and Rockets*, **3**, No. 6, pp. 858–866 (1966).
21. CHAMIS, C. C. Structural Synthesis of Multilayer Fiber Composite Panels, Ph.D. Thesis, Case Institute of Technology, in preparation.
22. SCHMIT, L. A., KICHER, T. P. and MORROW, W. M. Structural Synthesis Capability for Integrally Stiffened Waffle Plates, *J. AIAA*, **1**, No. 12, pp. 2820–2836 (1963).
23. HARRINGTON, R. A., CHAMIS, C. C. and TATARZYCKI, E. Design Criteria on Filament Wound Structures Subjected to Combined Loading, B. F. Goodrich Research Report to ABL, AD 451216 (1964).
24. FRIED, N. The Response of Orthogonal Filament Wound Material to Compressive Stress, *Proc. 20th Annual SPI Conf.*, Section 1-C (1965).
25. KIES, J. A. Maximum Strains in Resin of Fiberglass Composites, NRL 5752 (1962).
26. SCHULZ, J. C. Maximum Stresses and Strains in the Resin of a Filament Wound Structure, *Proc. 18th Annual SPI Conf.*, Section 7-D (1963).
27. GRESZCZUK, L. B. Theoretical and Experimental Studies on Properties and Behavior of Filamentary Composites, Douglas Paper No. 3550 (1965).
28. GRESZCZUK, L. B. Effects of Voids on Strength Properties of Filamentary Composites, *Proc. 22nd Annual SPI Conf.*, Section 20-A (1967).
29. ROGERS, C. W., THORNTON, H. R., CORNISH, R. H. and LASDAY, A. H. Application of Advanced Fibrous Reinforced Composite Materials to Airframe Structures, AFML-TR-66-313, Vol. II (1966).

ADVANCED CONCEPTS FOR COMPOSITES[†]

H. SCHUERCH[‡]

I. LIST OF SYMBOLS

a	radius of circular inclusion (in.)
A, B, C, D	integration constants
E	Young's modulus (psi)
F^2	Hencky–von Mises distortion energy failure criterion $(\text{psi})^2$
K_1, K_2	non-dimensional elastic composite constants
r	radial coordinate (in.)
s	applied stress (psi)
z	axial coordinate (in.)
δ	angle between principal stress and radial direction
μ	Poisson's ratio
Φ	Airy stress function
σ	normal stress (psi)
θ	azimuth coordinate angle
τ	shear stress (psi)

Subscripts

f	refers to inclusion
m	refers to matrix
r	radial stress component
1, 2, 3	refers to principal stresses
θ	tangential stress component

II. SUMMARY

Stress function methods are used to determine the field of stress concentrations produced by cylindrical inclusions in an otherwise homogeneous matrix material subject to an applied stress transverse to the inclusion axis. It is found that the stress field generated by single cylindrical inclusions can be completely characterized by two non-dimensional constants that depend upon

† This work has been conducted with the financial support of the National Aeronautics and Space Administration.

‡ President, Astro Research Corporation, Santa Barbara, California.

the relative elastic properties of the composite. Additional results are given for the case of an inclusion in the form of a spherical void.

The stress field is presented graphically in the form of stress surfaces for a number of selected cases. The calculated stress fields are interpreted in terms of fracture mechanics for non-ideally loaded fiber composites, and for several types of polycrystalline ceramics.

Based upon the analytical studies, guidelines are developed for composites using advanced types of microgeometry. Several promising approaches are reduced to the fabrication of laboratory test specimens.

III. INTRODUCTION

The useful "design strength" of high performance filamentary composite materials is frequently disappointingly low because, in practical applications, stresses in directions other than along the fiber axis are incurred. Such non-ideal loadings may involve normal stresses applied in directions transverse to the fiber axis, such as in the case of biaxially stressed, cross-laminated shells, pressure vessels, etc. Similar conditions occur due to geodesic fiber curvature, non-columnated fibers, or due to conditions prevailing at points of concentrated loadings, attachments or individually discontinuous fibers.

In some instances, these non-ideal applications are avoidable by appropriate design techniques. In many others, however, ideal design configurations need to be compromised for practical reasons, and the subject of the materials sensitivity to non-ideal loading becomes an appropriate matter of concern. Hence, the question of stress concentrations generated by the presence of inclusions in an otherwise homogeneous "matrix" body is of central interest to the mechanical functioning and failure mechanisms of composite materials, made with either ductile or non-ductile matrix materials. In addition, the nature of the stress pattern caused by inclusions of either high or low rigidity can conceivably shed light upon the strength and failure phenomena of polyphase, non-ductile materials, such as dense or porous polycrystalline ceramics, or glassy materials with included flaws in the form of voids, or hard crystallites. Finally, a thorough understanding of geometrical microdesign effects upon fracture origin in synthetically textured materials provides guides for the development of improved composites, particularly with regard to their sensitivity to non-ideal loading conditions.

IV. ANALYSIS OF STRESSES DUE TO CIRCULAR INCLUSIONS

To obtain a detailed and quantitative insight into the stresses prevailing in the vicinity of an inclusion, the case of cylindrical inclusion with a circular section embedded in an infinite slab of otherwise homogeneous matrix material, subject to uniaxial stress, has been analyzed (Fig. 1). Assuming Hookean materials, this problem lends itself to a rigorous closed form solution for the two limiting cases of plane stress and plane strain. The results of this analysis may also be considered as a first approximation for multiple inclusions, spaced several diameters from each other, rendering the mutual influence of adjacent inclusions relatively unimportant or allowing reasonable approximations by superposition of interfering stress patterns. The closed form solutions, furthermore, can aid in the development of digital computer

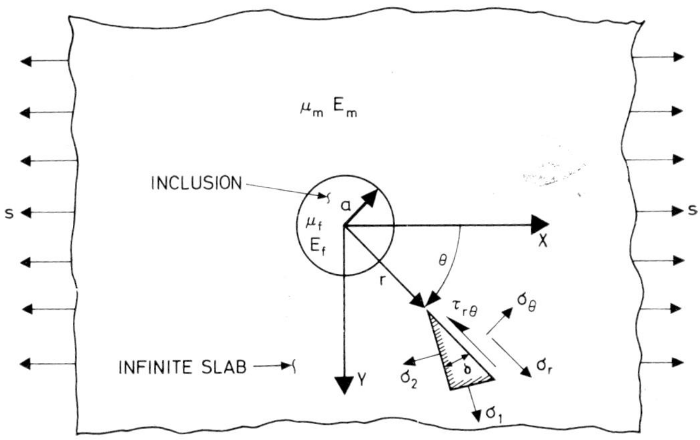

FIG. 1. Infinite body with circular inclusion

programs which are capable of handling arbitrary geometrical arrays of inclusions, by providing checks for accuracy and formulations for initial conditions.

Considering the symmetry of the problem shown in Fig. 1, the Airy stress function satisfying the biharmonic differential equation in polar coordinates is:[1]

$$\Phi = A_0 \log_e r + B_0 r^2 \log_e r + C_0 r^2 + \sum_{n=1}^{\infty} (A_{2n} r^{2n} + B_{2n} r^{2n+2} + C_{2n} r^{-2n} + D_{2n} r^{-2n+2}) \cos 2n\theta. \tag{1}$$

Since the stress remains finite for $r \to \infty$, the coefficient B_0 has to be zero. Only the first term ($n=1$) of the Fourier series needs to be used. Thus the stress function reduces to

$$\Phi = A_0 \log_e r + C_0 r^2 + \left[A_2 r^2 + B_2 r^4 + C_2 \frac{1}{r^2} + D_2 \right] \cos 2\theta \tag{1a}$$

and the stresses are obtained from the derivatives as follows:

$$\left.\begin{aligned}\sigma_r &= \frac{1}{r}\frac{\partial\Phi}{\partial r}+\frac{1}{r^2}\frac{\partial^2\Phi}{\partial\theta^2} = \frac{A_0}{r^2}+2C_0-\left[2A_2+6C_2\frac{1}{r^4}+4D_2\frac{1}{r^2}\right]\cos 2\theta\\ \sigma_\theta &= \frac{\partial^2\Phi}{\partial r^2} = -\frac{A_0}{r^2}+2C_0+\left[2A_2+12B_2r^2+6C_2\frac{1}{r^4}\right]\cos 2\theta\\ \tau_{r\theta} &= -\frac{\partial}{\partial r}\left(\frac{1}{r}\frac{\partial\Phi}{\partial\theta}\right) = \left[2A_2+6B_2r^2-6C_2\frac{1}{r^4}-2D_2\frac{1}{r^2}\right]\sin 2\theta.\end{aligned}\right\} \quad (2)$$

Both sets of six coefficients A_0, C_0, A_2, B_2, C_2, D_2, for the inclusion and the matrix can be determined by boundary conditions. After some algebraical transformations, it can be shown that the solutions can be expressed in terms of two groupings, K_1 and K_2, of the relative elastic properties of matrix and inclusion, where:

for plane stress: / *for plane strain:*

$$K_1 = \frac{E_f(1-\mu_m)-E_m(1-\mu_f)}{E_f(1+\mu_m)+E_m(1-\mu_f)}\,;\quad K_1 = \frac{E_f(1-2\mu_m)(1+\mu_m)-E_m(1-2\mu_f)(1+\mu_f)}{E_f(1+\mu_m)+E_m(1-2\mu_f)(1+\mu_f)}$$

$$K_2 = \frac{E_f(1+\mu_m)-E_m(1+\mu_f)}{E_f(3-\mu_m)+E_m(1+\mu_f)}\,;\quad K_2 = \frac{E_f(1+\mu_m)-E_m(1+\mu_f)}{E_f(3-4\mu_m)(1+\mu_m)+E_m(1+\mu_f)}$$

and where E is the Young's modulus, μ is the Poisson's ratio; the subscript f refers to the inclusion, and the subscript m refers to the matrix. With these factors, the stress distribution (2) reduces to two sets of expressions:

inside the inclusion ($r < a$)

$$\begin{aligned}\frac{\sigma_r}{s} &= \frac{1}{2}(1+K_1)+\frac{1}{2}(1+K_2)\cos 2\theta\\ \frac{\sigma_\theta}{s} &= \frac{1}{2}(1+K_1)-\frac{1}{2}(1+K_2)\cos 2\theta\\ \frac{\tau_{r\theta}}{s} &= -\frac{1}{2}(1+K_2)\sin 2\theta\\ \frac{\sigma_z{}^*}{s} &= \mu_f(1+K_1)\end{aligned} \quad (2a)$$

and in the matrix ($r > a$)

$$\begin{aligned}\frac{\sigma_r}{s} &= \frac{1}{2}\left(1+K_1\frac{a^2}{r^2}\right)+\frac{1}{2}\left[1-K_2\left(3\frac{a^4}{r^4}-4\frac{a^2}{r^2}\right)\right]\cos 2\theta\\ \frac{\sigma_\theta}{s} &= \frac{1}{2}\left(1-K_1\frac{a^2}{r^2}\right)-\frac{1}{2}\left[\left(1-3K_2\frac{a^4}{r^4}\right)\right]\cos 2\theta\\ \frac{\tau_{r\theta}}{s} &= -\frac{1}{2}\left[1+K_2\left(3\frac{a^4}{r^4}-2\frac{a^2}{r^2}\right)\right]\sin 2\theta\\ \frac{\sigma_z{}^\dagger}{s} &= \mu_m\left(1+2K_2\frac{a^2}{r^2}\cos 2\theta\right).\end{aligned} \quad (2b)$$

† Only for plane strain.

Hence, the stress field is completely characterized by the two non-dimensional composite elastic constants K_1 and K_2, which in turn, are unique functions of the modulus ratio, the Poisson's ratios of both constituents, and the assumptions regarding the third principal stress (or strain). This reduction of parameters aids considerably in the detailed evaluation of stress fields.

The corresponding principal stresses to the stress distribution given in eqs. (2a) and (2b) are obtained by the well-known transformations:

$$\sigma_{1,2} = \tfrac{1}{2}(\sigma_r+\sigma_\theta) \pm \tfrac{1}{2}[(\sigma_r-\sigma_\theta)^2+4\tau_{r\theta}^2]^{1/2}$$

and the angle δ between the direction of the principal stress (stress trajectory) and the radial direction is obtained from

$$\tan 2\delta = \frac{2\tau_{r\theta}}{\sigma_r-\sigma_\theta}.$$

With these transformations, the principal stress magnitudes and trajectory angles inside the inclusion ($r < a$) become:

$$\left.\begin{aligned} \frac{\sigma_1}{s} &= 1+\frac{1}{2}(K_1+K_2) \\ \frac{\sigma_2}{s} &= \frac{1}{2}(K_1-K_2) \\ \delta &= -\theta. \end{aligned}\right\} \tag{3a}$$

Hence the stress distribution inside the inclusion is uniform but biaxial, the principal axes being aligned parallel and normal to the axis of applied stress. In view of the very complicated stress field in the matrix surrounding the inclusion, this result is, at first sight, quite surprising. It should also be observed that this result holds for multiple inclusions only if these are widely spaced.

The principal stresses and their trajectory angle in the matrix are obtained from eq. (2b):

$$\left.\begin{aligned} \frac{\sigma_1}{s} &= \frac{1}{2}\Bigg\{1+\frac{a^2}{r^2}(2K_2\cos 2\theta+K_1\cos 2\delta)+\left[1-K_2\left(3\frac{a^4}{r^4}-2\frac{a^2}{r^2}\right)\right]\times \\ &\qquad \cos 2\theta\cos 2\delta-\left[1+K_2\left(3\frac{a^4}{r^4}-2\frac{a^2}{r^2}\right)\right]\sin 2\theta\sin 2\delta\Bigg\} \\ \frac{\sigma^2}{s} &= \frac{1}{2}\Bigg\{1+\frac{a^2}{r^2}(2K_2\cos 2\theta-K_1\cos 2\delta)-\left[1-K_2\left(3\frac{a^4}{r^4}-2\frac{a^2}{r^2}\right)\right]\times \\ &\qquad \cos 2\theta\cos 2\delta+\left[1+K_2\left(3\frac{a^4}{r^4}-2\frac{a^2}{r^2}\right)\right]\sin 2\theta\sin 2\delta\Bigg\} \\ \tan 2\delta &= -\frac{\left(1+K_2\left[3\frac{a^4}{r^4}-2\frac{a^2}{r^2}\right]\right)\sin 2\theta}{\left(1-K_2\left[3\frac{a^4}{r^4}-2\frac{a^2}{r^2}\right]\right)\cos 2\theta+K_1\frac{a^2}{r^2}}. \end{aligned}\right\} \tag{3b}$$

Finally, for the investigation of failure in ductile components of the composite, the distortion energy, F^2 (Hencky–von Mises[(2)]), is of interest:

$$F^2 = \tfrac{1}{2}(\sigma_1-\sigma_2)^2+(\sigma_2-\sigma_3)^2+(\sigma_3-\sigma_1)^2. \quad (4)$$

This expression becomes for plane stress

$$F^2 = \sigma_1^2+\sigma_2^2-\sigma_1\sigma_2 \quad (4a)$$

and for plane strain

$$F^2 = (\sigma_1^2+\sigma_2^2)(1-\mu+\mu^2)-\sigma_1\sigma_2(1+2\mu-2\mu^2)$$

or

$$F^2 = 4\tau_{\max}^2+\sigma_1\sigma_2.$$

V. EVALUATION OF SOLUTIONS

A. *Stresses Inside Inclusion*

As seen from eq. (3a), the stress condition inside the inclusion is uniform, biaxial for plane stress and triaxial for plane strain. The principal stress trajectories, therefore, form a rectangular pattern as shown in Fig. 2. Also shown in Fig. 2 are the values for the stress concentration factors σ/s in

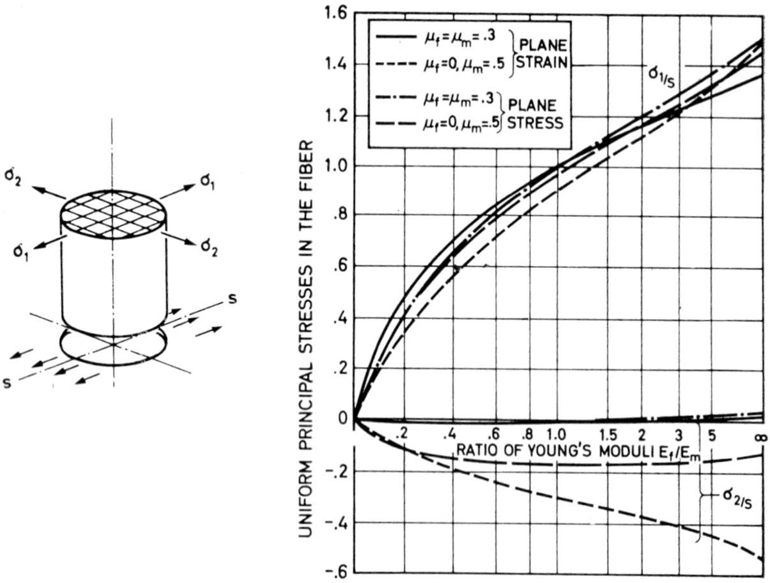

FIG. 2. Stress field inside a cylindrical inclusion

function of the modulus ratio, $\frac{E_f}{E_m}$, for two typical combinations of Poisson's ratio, and for conditions of either plane stress or plane strain.

It will be observed that rigid inclusions act as stress concentrators, experiencing a direct principal stress of approximately 1.5 times the stress applied far away from the inclusion. Stress reversal is observed, amounting to values as high as -0.5. The existence of stress reversal is significant for cases where tensile fracture can originate within the inclusion due to an applied compressive stress. This may occur, particularly in the case of fibers with present or incipient longitudinal cracks, such as observed, for instance, in composite boron fibers.[3, 4]

B. *Stresses in Matrix*

While the condition of stress in the inclusion is uniform, the stress in the matrix in the vicinity of the inclusion is quite complex.

Three limiting cases have been evaluated representing the effect of rigid inclusions, of voids and of differential Poisson's ratios.

Table 1 summarizes the three cases and lists the corresponding values for K_1 and K_2 for both plane stress and plane strain conditions.

TABLE 1. COMPOSITE ELASTIC CONSTANTS USED FOR EVALUATION

	Plane stress		Plane strain	
	K_1	K_2	K_1	K_2
Hole: $E_f/E_m = 0$	-1	-1	-1	-1
Rigid inclusion: $E_f/E_m = \infty$, $\mu_m = 0.3$	0.539	0.4815	0.4	0.5556
Different Poisson's ratios $E_f/E_m = 1$, $\mu_f = 0$; $\mu_m = 0.5$	-0.2	0.1492	-0.40	0.20

The stress field for these three cases has been completely evaluated. A difficulty arises in the visual presentation of the stress tensor which is required to develop an intuitive insight into the micromechanical aspects of the stress field and its implications upon choice of improved geometries. For this reason, a somewhat unconventional presentation in the form of two principal

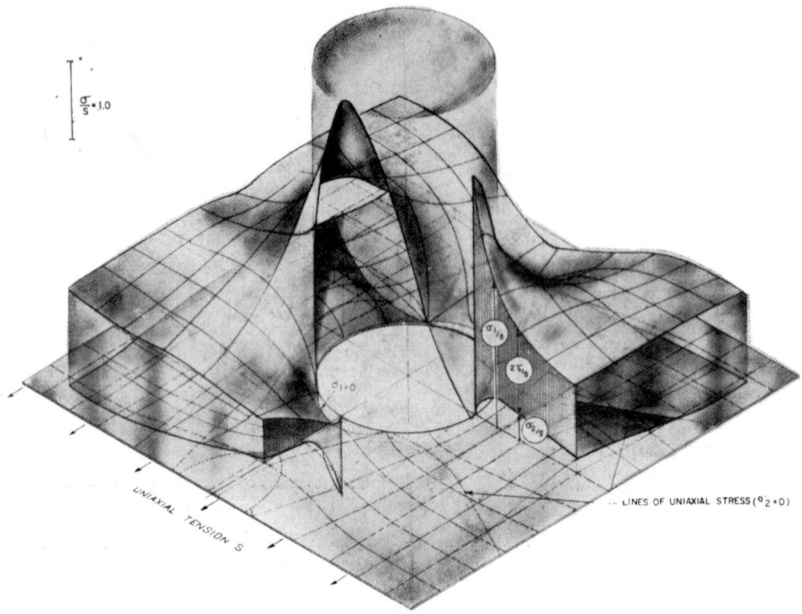

FIG. 3. Stress distribution around a cylindrical void, $\mu_m = 0.3$

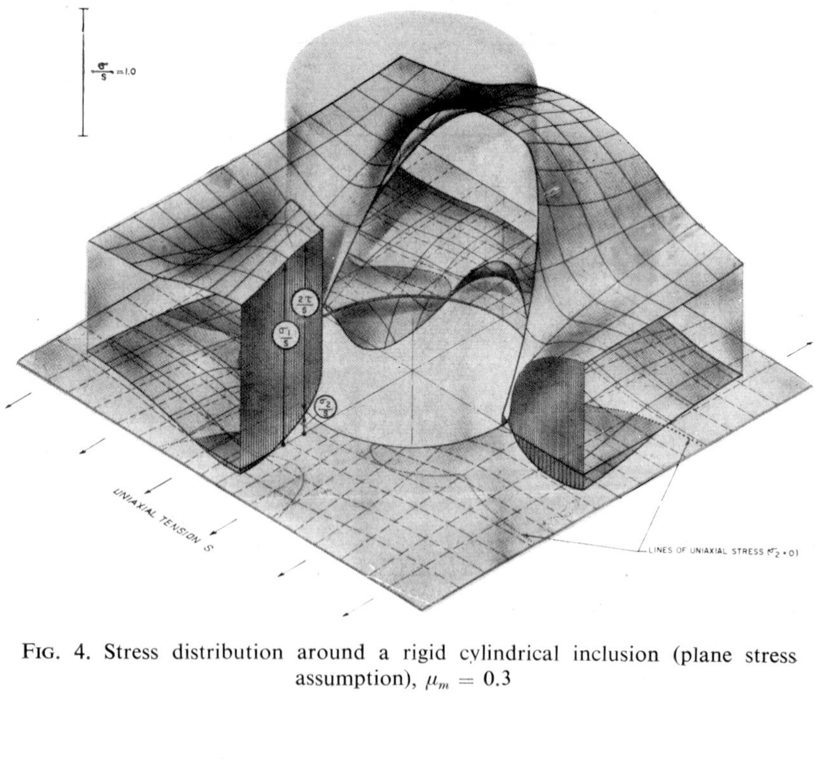

FIG. 4. Stress distribution around a rigid cylindrical inclusion (plane stress assumption), $\mu_m = 0.3$

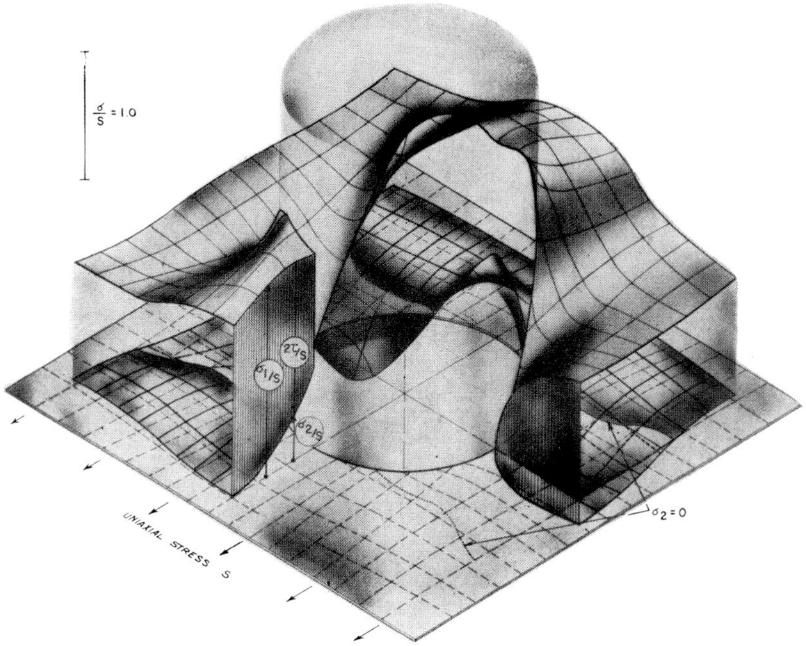

FIG. 5. Stress distribution around a rigid cylindrical inclusion (plane strain assumption), $\mu_m = 0.3$

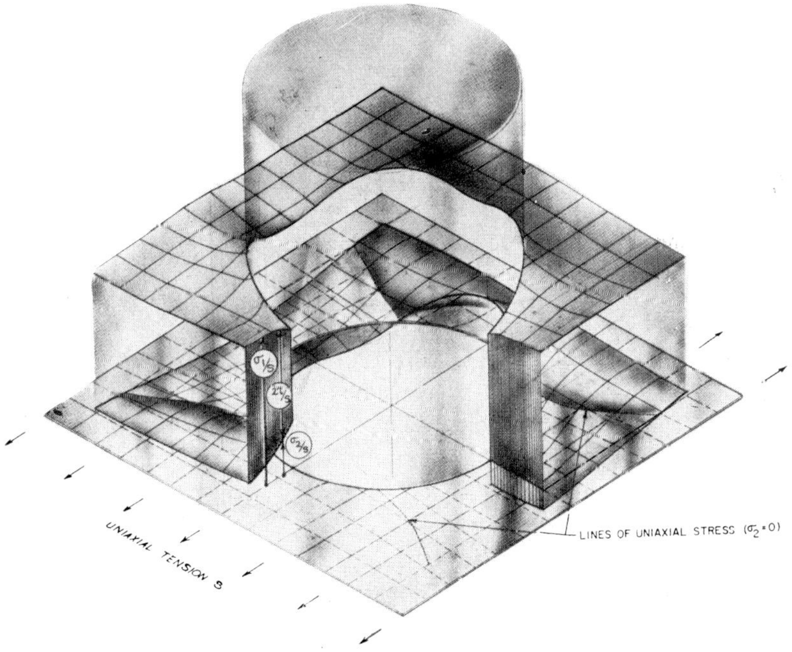

FIG. 6. Stress distribution around a cylindrical inclusion with different Poisson's ratio (plane stress assumption), $E_f/E_m = 1$, $\mu_f = 0$, $\mu_m = 0.5$

stress surfaces with parametric lines representing the principal stress trajectories has been chosen. Hence, the results are mapped in the form of two "stress surfaces" for σ_1 and σ_2, and their associated stress trajectories in Fig. 3 for a cylindrical void, Figs. 4 and 5 for the rigid inclusion, and Figs. 6 and 7 for a cylindrical inclusion of equal modulus but different Poisson's ratio.

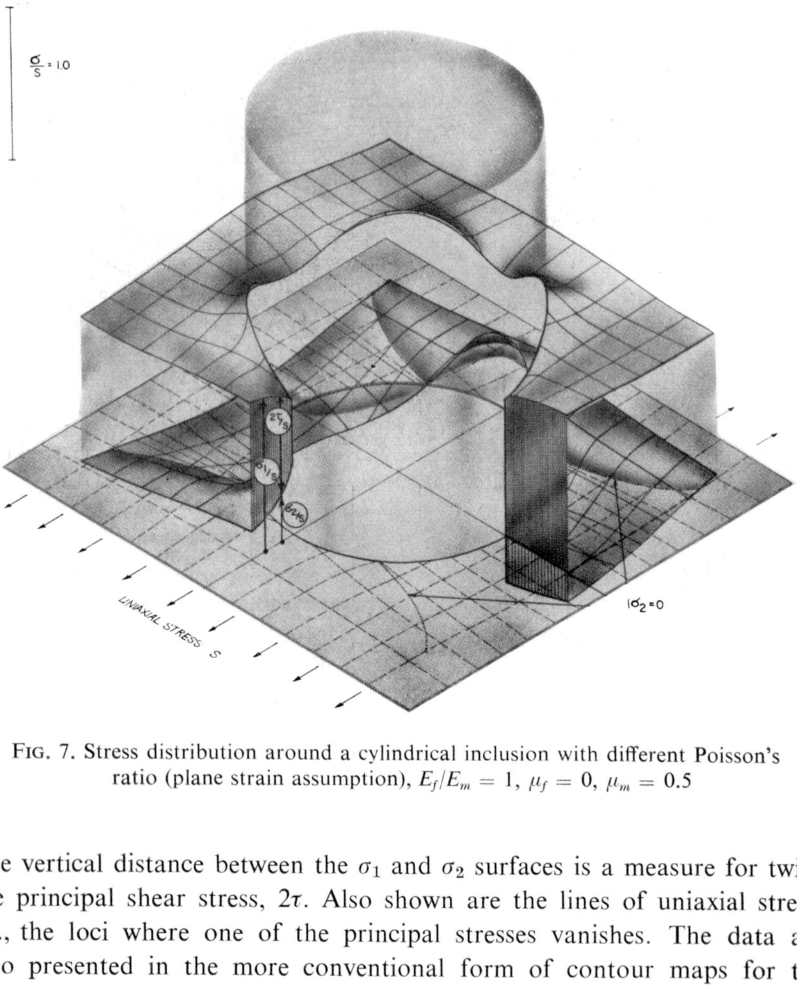

FIG. 7. Stress distribution around a cylindrical inclusion with different Poisson's ratio (plane strain assumption), $E_f/E_m = 1$, $\mu_f = 0$, $\mu_m = 0.5$

The vertical distance between the σ_1 and σ_2 surfaces is a measure for twice the principal shear stress, 2τ. Also shown are the lines of uniaxial stress, i.e., the loci where one of the principal stresses vanishes. The data are also presented in the more conventional form of contour maps for the principal stresses, and for the Hencky–von Mises distortion energy in Figs. 8 through 13.

For the purpose of comparison, and for the purpose of assessing the cascading effect discussed in section VI, the case of a spherical void has also been evaluated.

It is assumed that a homogeneous matrix is subject to a uniform uniaxial applied stress at a great distance from the spherical void. The evaluation is based upon the solutions given by Southwell and Gough[5]. Results are presented for the meridional plane in the form of contour maps for the three principal stresses in Fig. 14, and in Fig. 16 in the form of the radial stress

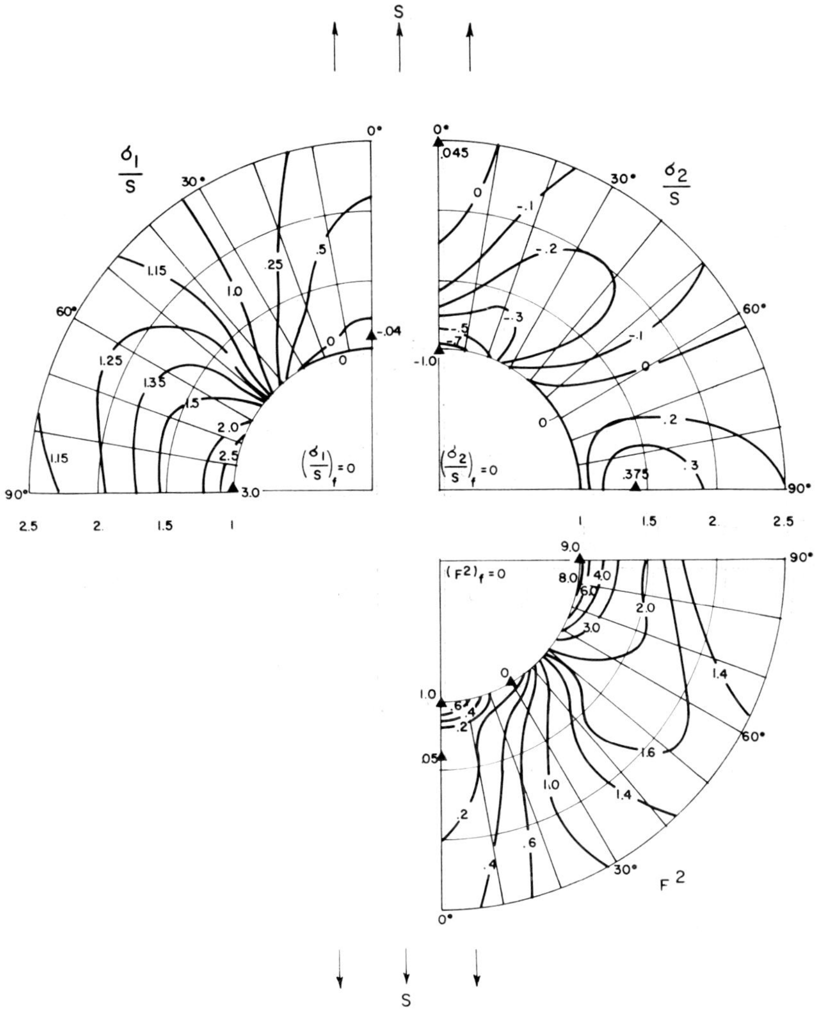

FIG. 8. Stress distribution around a cylindrical void (plane stress assumption), $\mu_m = 0.3$

distribution for the rotationally symmetrical pattern in the equatorial plane of the spherical void.

The case for the rigid inclusion with plane stress assumption (Fig. 4) shows an interesting singularity in the form of a point of biaxial isotropic stress, located a very short distance (i.e., about 0.4% of the fiber radius) outside the inclusion interface. At this point, a discontinuity of stress trajectories occurs. A detail of this singularity is shown in Fig. 15. A similar point is observed at the surface of the void (Fig. 3), where both principal stresses vanish.

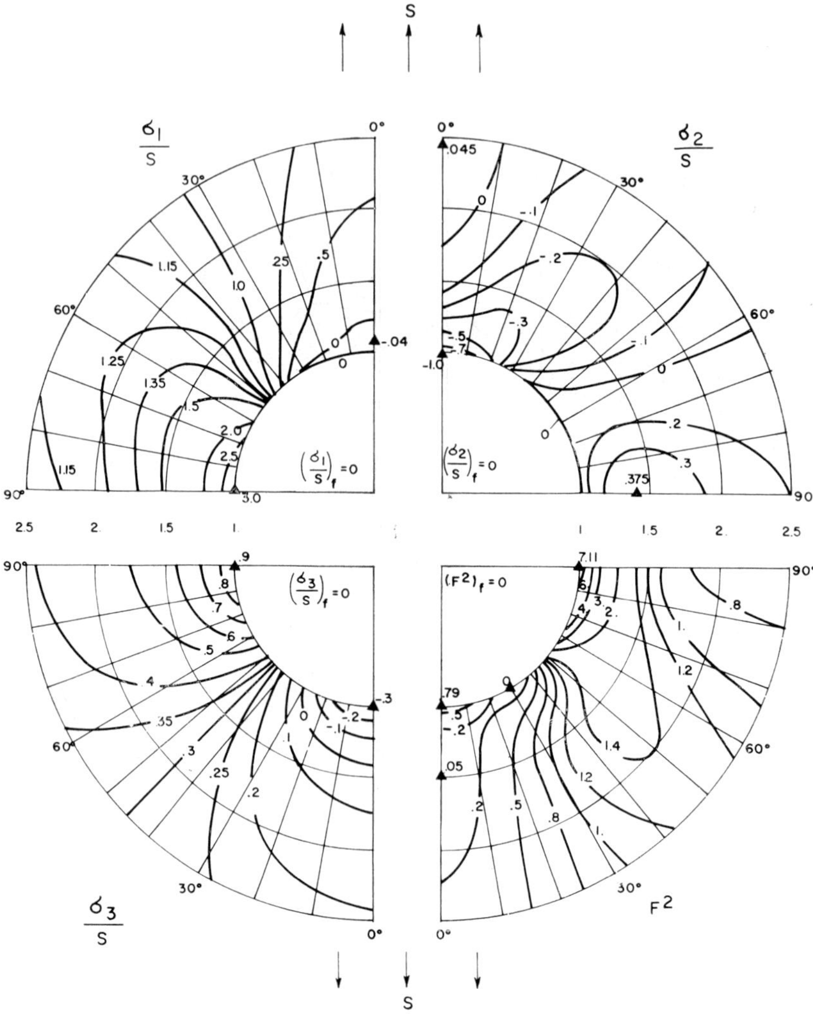

FIG. 9. Stress distribution around a cylindrical void (plane strain assumption), $\mu_m = 0.3$

The stress singularity for the rigid inclusion is of interest because it is located within a distance from the fiber which may be sufficiently small to be affected by the discrete nature of the molecular structure of the matrix material. For a conventional glass fiber with a radius of 2μ, for instance, the distance between surface and stress singularity is only 80Å, which is well within the size range of polymeric molecules.

The presence of both direct and reverse stress concentration factors is

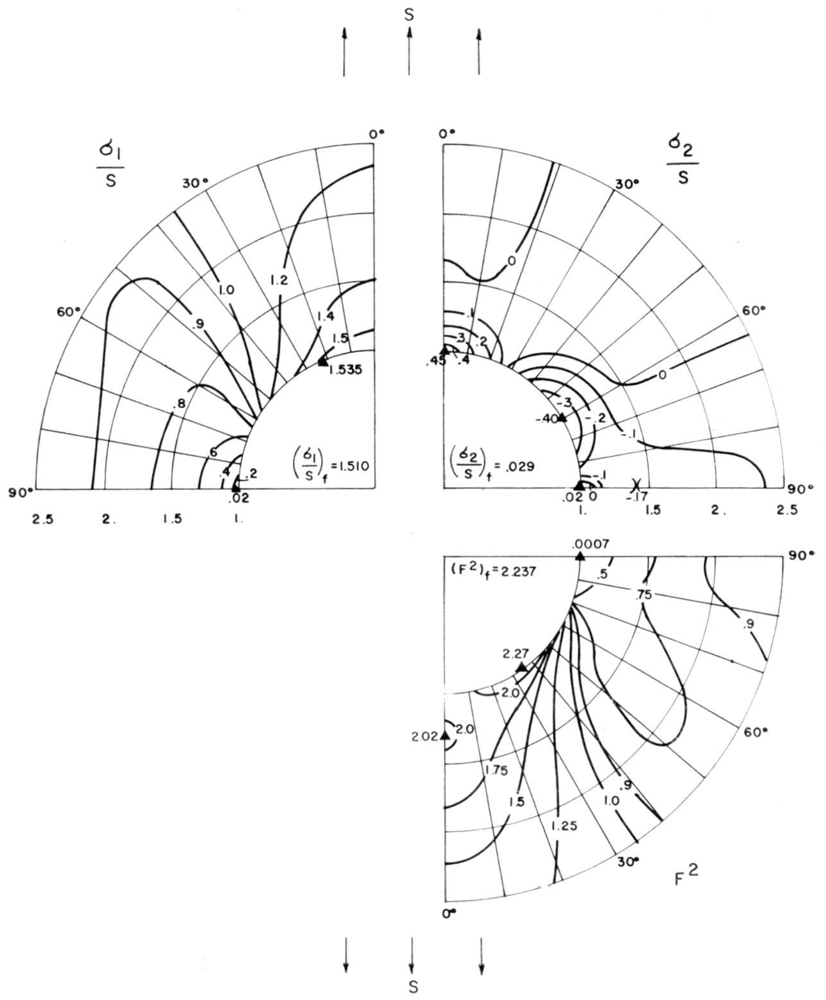

FIG. 10. Stress distribution around a rigid cylindrical inclusion (plane stress assumption), $\mu_m = 0.3$

observed—i.e., the presence of the inclusion causes the stress to be magnified and also to be locally reversed. Thus, for instance, application of compression to the material will cause localized areas to be subject to tension. Stress reversal peaks are also observed in the case of the cylindrical void and of the

TABLE 2. MAGNITUDE AND LOCATION OF EXTREME STRESS CONCENTRATION

Inclusion	Condition	σ_1/s	θ	r/a	σ_2/s	θ	r/a	σ_3/s	θ	r/a	F^2/s^2	θ	r/a
Void, $E_f/E_m = 0$ $\mu_m = 0.3$	Plane stress	3.0 −0.04	90° 0°	1.0 1.1	0.375 −1.0	90° 0°	1.3 1.0				9.0	90°	1.0
	Plane strain	3.0 −0.04	90° 0°	1.0 1.1	0.375 −1.0	90° 0°	1.3 1.0	0.90 −0.30	90° 0°	1.0 1.0	7.11	90°	1.0
	Spherical void	2.046 −0.04	90° 0°	1.0 1.07	0.205 −0.682	90° 0°	1.3 1.0	0.136 −0.682	90° 0°	1.0 1.0	7.85	90°	1.0
Rigid, $E_f/E_m = \infty$ $\mu_f = \mu_m = 0.3$	Plane stress	1.535 0.02	21° 90°	1.0 1.0	0.45 −0.40	0° 60°	1.0 1.0				2.27 2.02	35° 0°	1.0 1.3
	Plane strain	1.55 1.52	20° 0°	1.0 1.1	0.63 −0.46	0° 62°	1.0 1.0	0.633 −0.033	0° 90°	1.0 1.0	2.0 1.44	40° 0°	1.0 1.4
Varying Poisson's ratio $E_f/E_m = 1$	Plane stress	1.09 1.04	35° 0°	1.0 1.6	0.315 −0.175	0° 72°	1.0 1.0				1.24 1.15	45° 90°	1.0 1.5
$\mu_f = 0$ $\mu_m = 0.5$	Plane strain	1.14 1.035	35° 0°	1.0 1.7	0.50 −0.30	0° 90°	1.0 1.0	0.70 0.30	0° 90°	1.0 1.0	1.29 1.20	55° 90°	1.0 1.2

rigid inclusion in areas removed from the interface. Table 2 summarizes the magnitude and locations of the extremes for the principal stress and the Hencky–von Mises distortion energies.

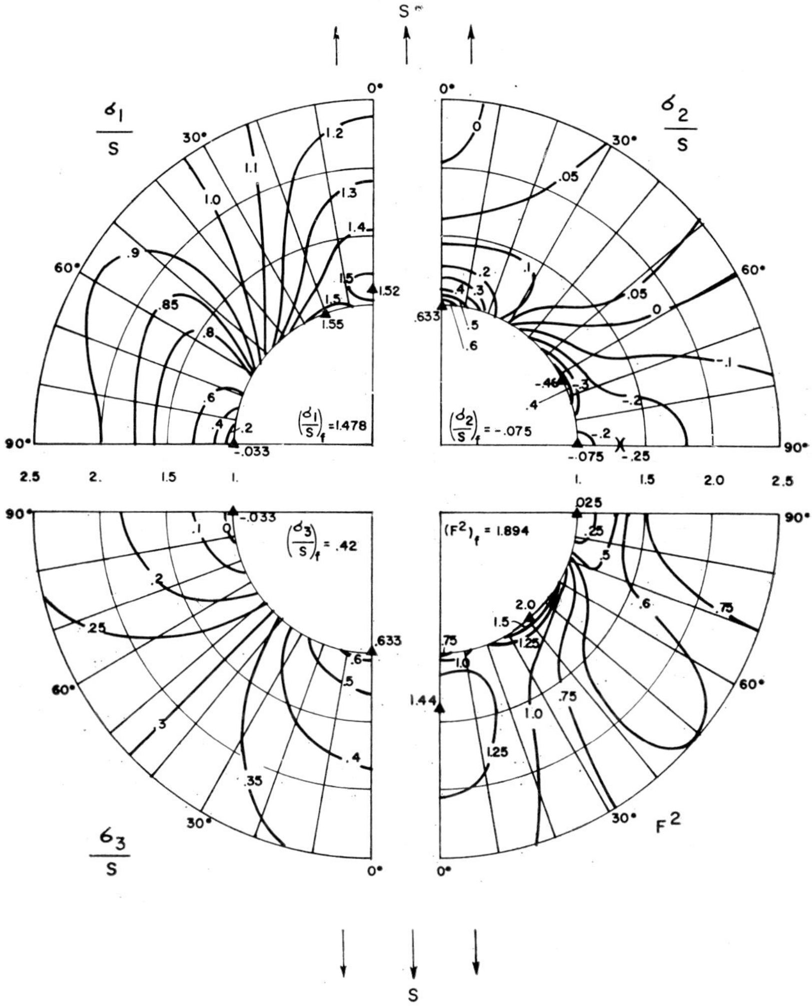

FIG. 11. Stress distribution around a rigid cylindrical inclusion (plane strain assumption), $\mu_m = 0.3$

The effect of relative modulus upon magnitude and location of the extremes of principal stress at the inclusion interface is shown in Figs. 17 and 18 for the assumption of plane strain and Poisson's ratios of 0.3 for both constituents. The largest principal stress is found at $\theta = 90°$, and the largest stress reversal at $\theta = 0°$, for the cases where $\frac{E_f}{E_m} < 1$, i.e., for the cases of "soft" inclusions. In the case where $\frac{E_f}{E_m} > 1$ (hard inclusions), the stress extremes occur at azimuth angles ranging from 23° to 21° for largest principal stress

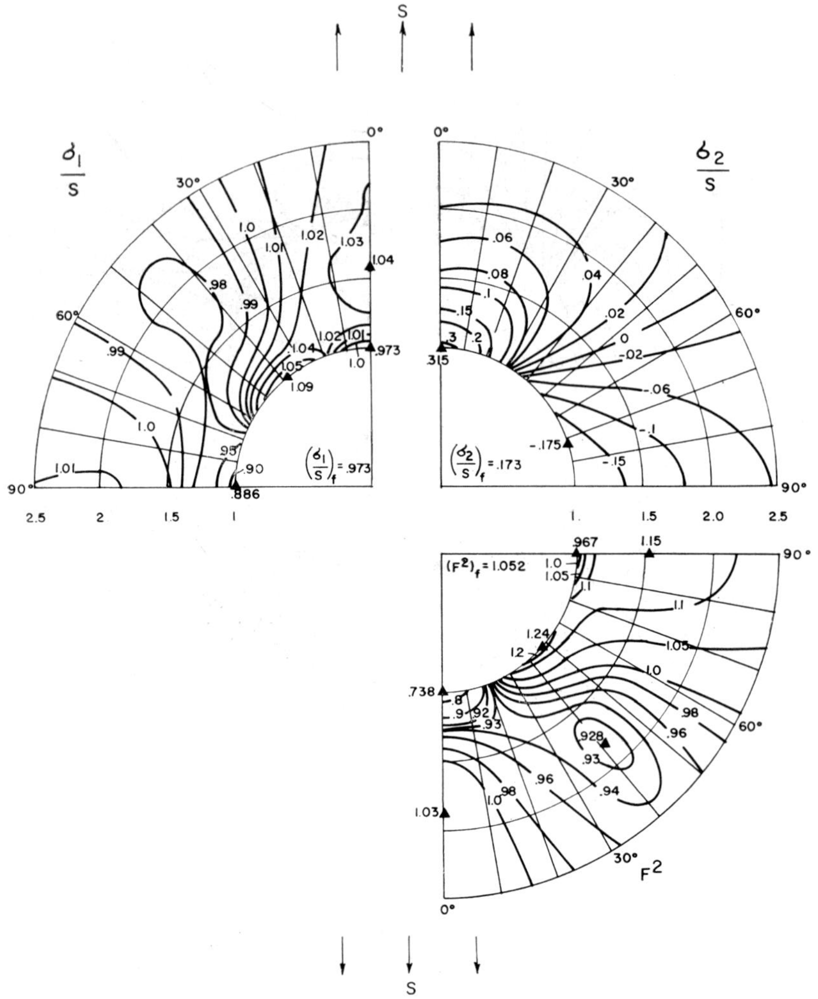

FIG. 12. Stress distribution around a cylindrical inclusion with different Poisson's ratio (plane stress assumption), $\mu_f = 0$, $\mu_m = 0.5$

and at azimuth angles ranging from 54° to 62° for the largest stress reversal.

The distribution of the principal stress magnitudes, and of the stress components tangential and normal to the interface for the case of a rigid inclusion ($E_f/E_m = \infty$) are shown in Fig. 19. The tangential and normal stress components may be used for the prediction of interface failure by means of a shear-tension interaction failure hypothesis commonly employed in the

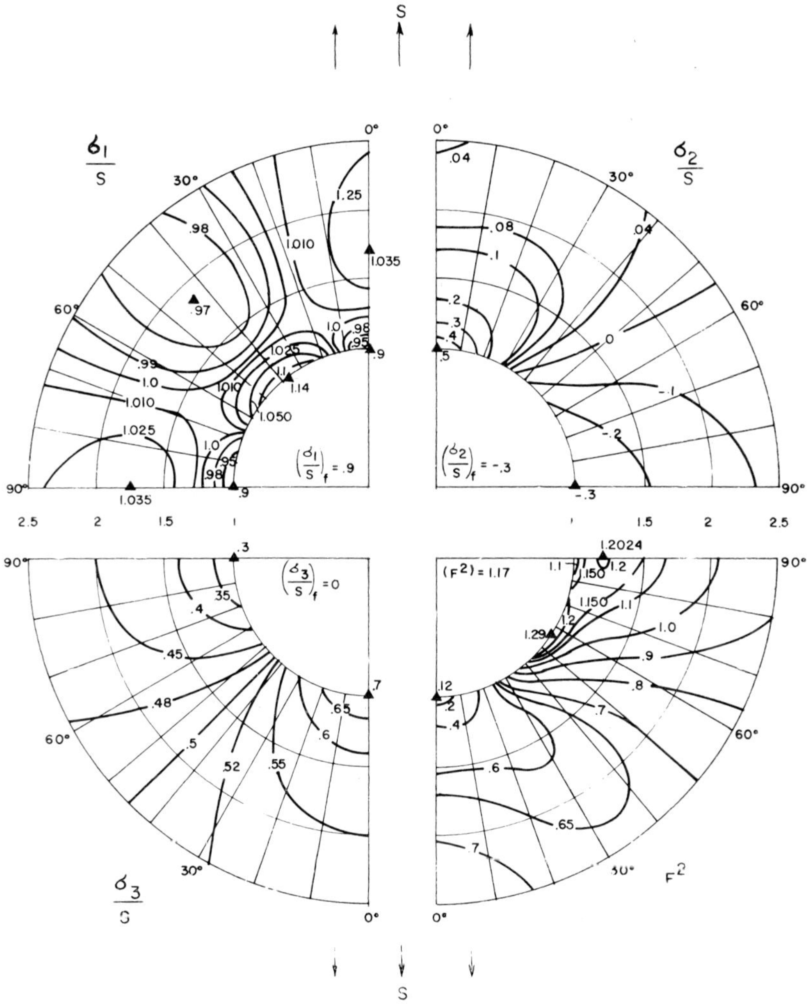

FIG. 13. Stress distribution around a cylindrical inclusion with different Poisson's ratio (plane strain assumption), $\mu_f = 0$, $\mu_m = 0.5$

strength analysis of thin bonded or brazed joints. The principal interface stresses may be used in the case where an "interface complex" exists of sufficient thickness to allow applying classical failure criteria (such as the Hencky–von Mises or the maximum principal stress criterion).

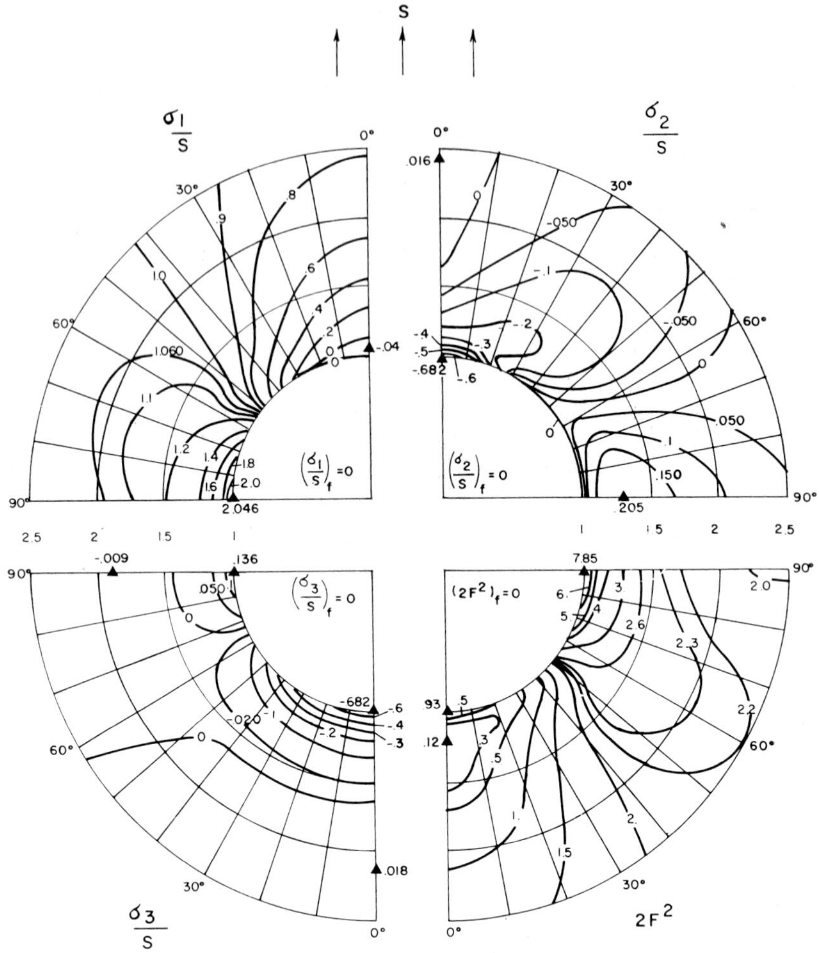

FIG. 14. Stress distribution around a spherical void, meridional plane, $\mu_m = 0.3$

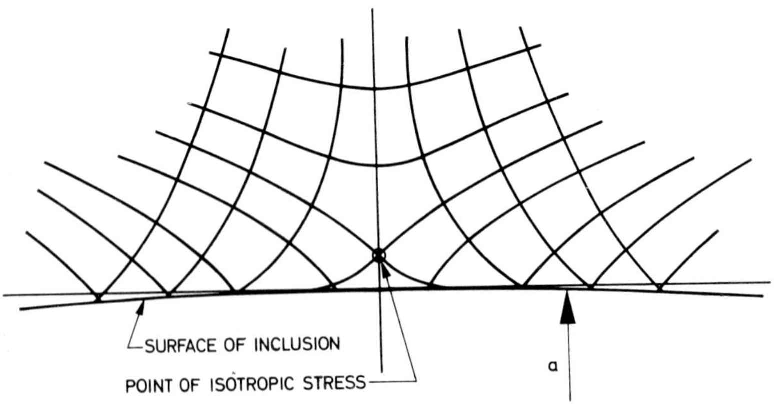

FIG. 15. Singularity of principal stress trajectories in vicinity of inclusion surface

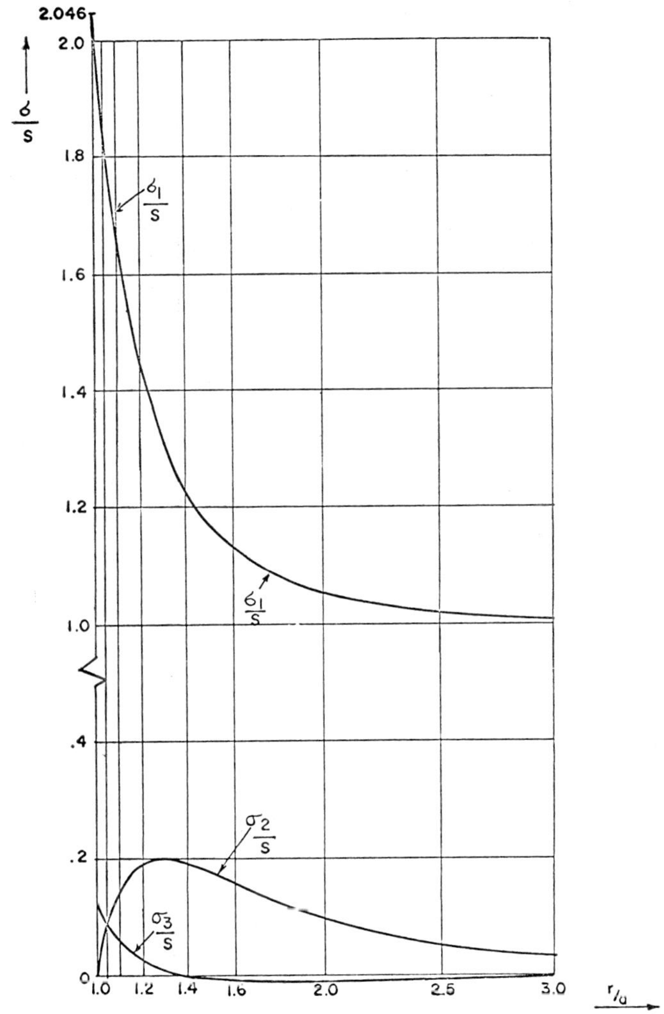

FIG. 16. Stress distribution around a spherical void, equatorial plane, $\mu_m = 0.3$

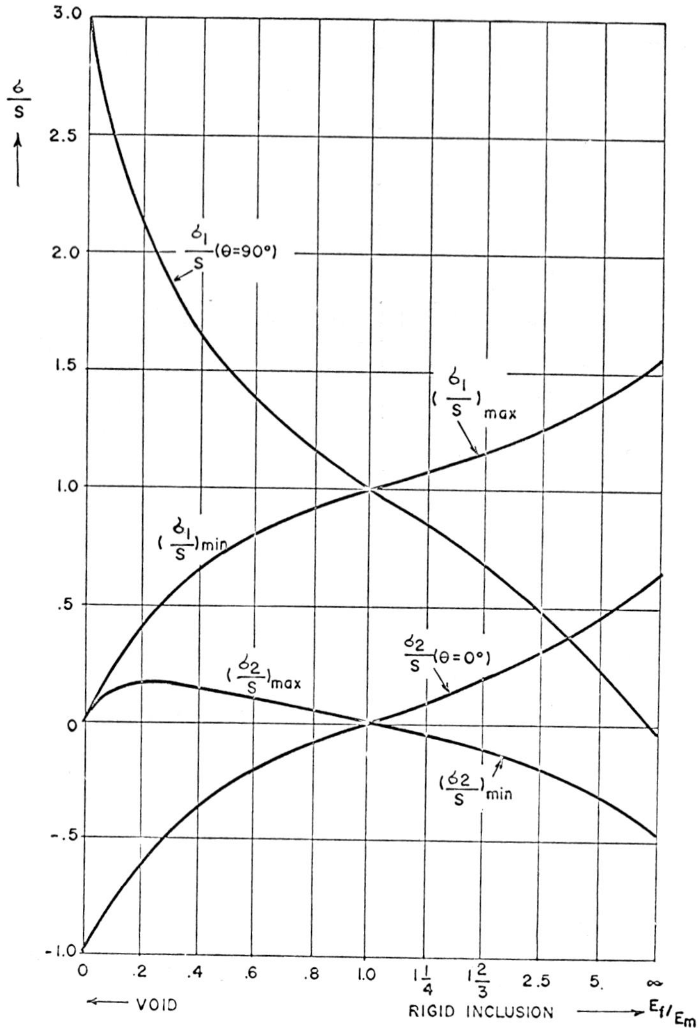

FIG. 17. Magnitude of extreme principal stress at interface (plane strain assumption, $\mu_f - \mu_m = 0.3$)

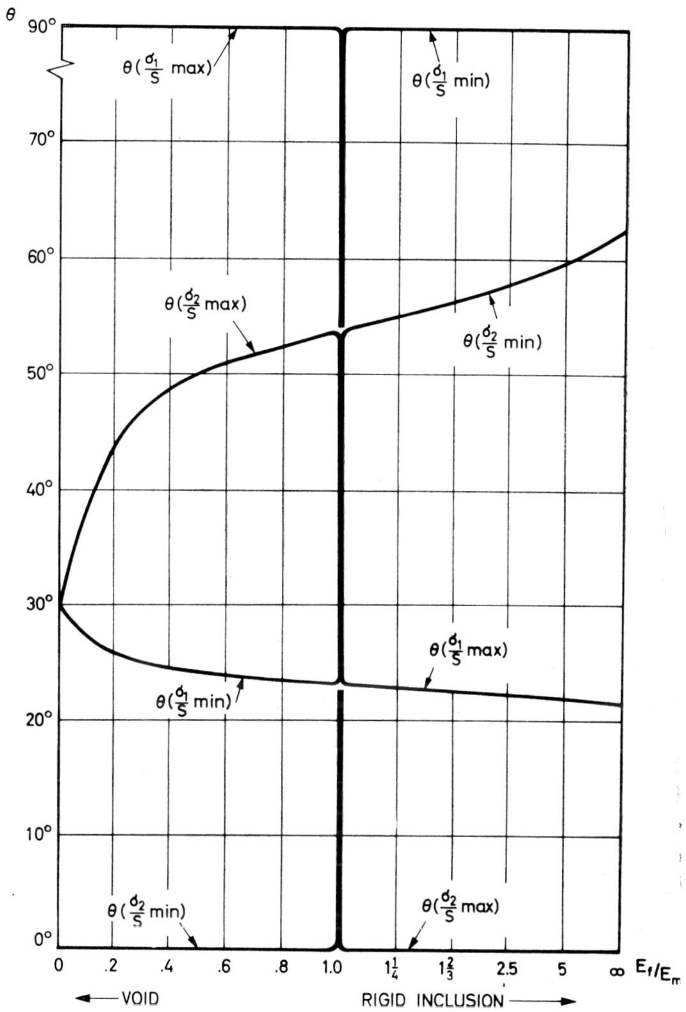

FIG. 18. Location of extreme principal stress at interface (plane strain assumption, $\mu_f - \mu_m = 0.3$)

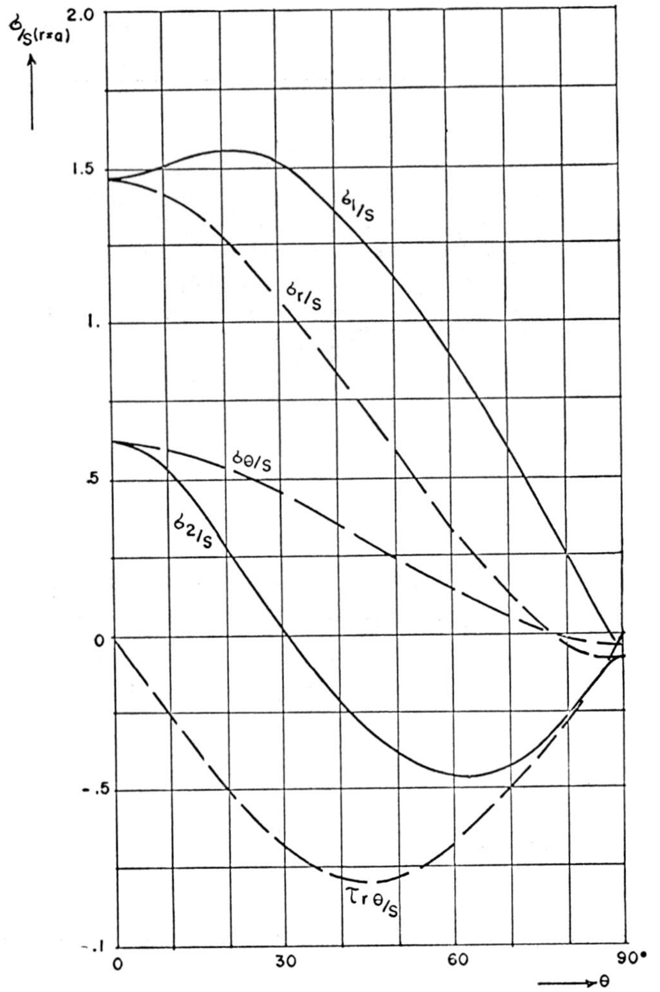

FIG. 19. Stresses at interface of rigid cylindrical inclusion (plane strain assumption)

VI. DISCUSSION

A. *Failure Mechanics of Composites*

The presence of "hard" inclusions generates stress concentrations, localized primarily at the interface between inclusion and matrix. This effect is compounded by Poisson's ratio differences between matrix and inclusion. A

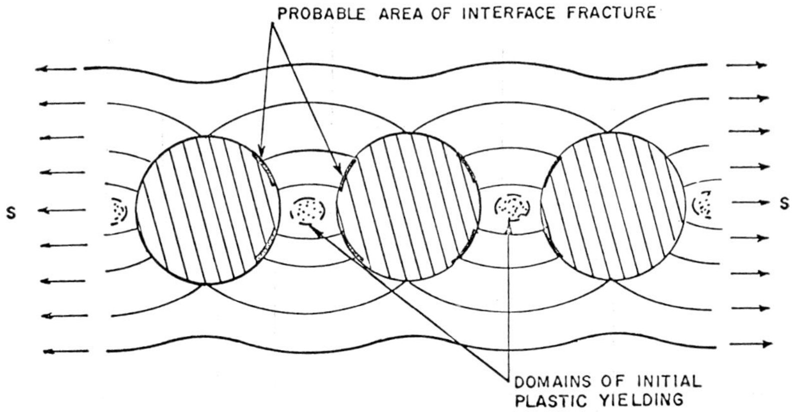

FIG. 20a. Array of hard inclusion exhibiting "series failure mode"—plastic yield aggravates stress in critical failure area

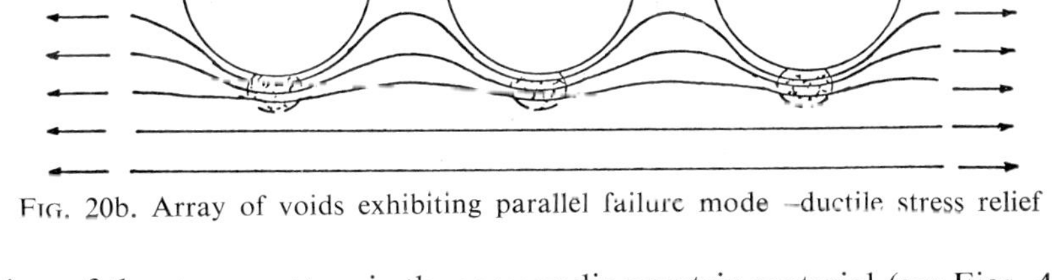

FIG. 20b. Array of voids exhibiting parallel failure mode—ductile stress relief

review of the stress pattern in the surrounding matrix material (see Figs. 4, 5, 10 and 11) suggests that sub-linear elastic behavior, ductile yielding, etc., of the matrix will not relieve but will rather aggravate such stress concentrations, contrary to the ductile stress relief found in the case of "soft" inclusions, notches, and voids in a strong parent material. This is due to the unfavorable "series" characteristics of the "hard" inclusion case (Fig. 20a), which attracts higher stresses into the critical interface areas as the surrounding matrix begins to weaken due to yielding. This behavior is the reverse of the "parallel" character of multiple connected domains represented by a matrix with "soft" inclusions or with included voids (Fig. 20b).

The domains of significant stress concentrations at the interface are generally small compared with the diameter of the inclusion. Thus, ductile relief of those areas will generally be limited; significant plastic flow (in metallic matrices) may be expected to predominate in areas somewhat removed from the interface. Areas of maximum stresses and of highest distortion energy are also associated with unusually large local stress gradients. The effect of large stress gradients upon yield and fracture boundaries will require the formulation and experimental verification of appropriate criteria. This should constitute a fruitful field for future research.

In addition to the stress concentrations at the interface, hard inclusions will also be subject to internal stresses higher than the average applied stress (see Fig. 2). The stress condition internal to the inclusion, moreover, is biaxial and subject to transverse stress reversal. Thus, even in the case of lateral compression of the composite, tensile transverse stresses may occur. These can be the original cause of premature fracture, particularly in the case of brittle, hard fibers that may be subject to built-in stresses and longitudinal cracks. The same statement, of course, applies to applied transverse tension where, particularly in the case of closely packed fibers, significant tensile stresses may be generated by an unfavorable cross-sectional fiber geometry.

An interesting and somewhat unexpected side product of this investigation refers to the theory of brittle fracture in uniaxial compression: The stress reversals occurring either on voids or on hard inclusions appear to provide a rational, and possibly, quantitative explanation for the observed ratios of tensile and compressive strengths in brittle materials. The reasoning is as follows: Assume that the brittle material will fail if a limiting principal tensile stress is exceeded. If this material contains inclusions, then the ratio of tensile to compressive applied gross stress will equal the ratio of the direct to reverse stress concentration extremes. This ratio has been plotted in Fig. 21 for the case of plane strain and cylindrical inclusions in function of modulus ratio. The ranges shown in Fig. 21, varying from 1/3 to 1/4 for voids and soft inclusions (porous ceramics), and from 1/16 to 1/8 for moderately hard inclusions (dense ceramics), compare well with experienced strength ratios of a wide range of technical ceramic materials. Experimental data taken from ref. 6 are listed in Table 3 for the purpose of comparison.

The possibility of a "cascading" effect of progressively smaller inclusions may be suggested. Consider, for instance, the case of a hard inclusion: Chemical or metallurgical reactions between fiber and matrix (for instance, in the case of metallic matrices with refractory fibers), or solute gases in polymeric matrices will often cause the formation of microscopic voids at the interface.

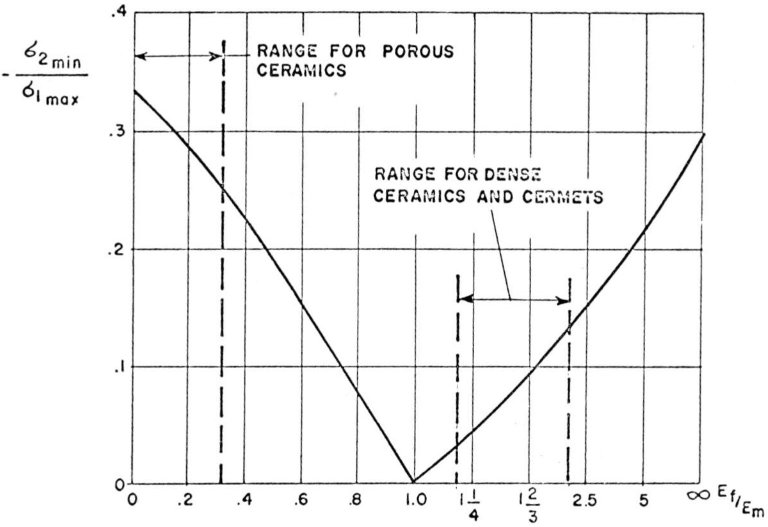

FIG. 21. Ratio of stress concentration extremes

TABLE 3. TECHNICAL STRENGTH OF CERAMIC MATERIALS

Technical strength	Compressive ksi	Tensile ksi	Ratio tens/compr
Dense Ceramics:			
Steatite ($2MgO \cdot SiO_2$)	80	8.5	0.11
Fosterite ($2MgO \cdot SiO_2$)	85	10	0.12
Zircon ($ZrO_2 \cdot SiO_2$)	100	12	0.12
Titania (TiO_2)	100	7.5	0.075
Alumina 85%	140	20	0.14
Alumina 96%	400	25	0.062
	Compressive ksi	**Bending ksi**	**Ratio bend/compr**
Porous Ceramics:			
Alumina — Highly porous	30	10	0.33
Alumina — Moderately porous	56	21	0.37
Alumina — Crushable	7	2.5	0.36
Magnesia	8	2	0.25
Magnesium silicate	10	5	0.5

Thus, a situation now arises, where in addition to the stress field generated by the hard inclusion, a "subfield" exists in the domains of the still smaller spherical voids. If these voids are located in areas of either positive or reverse stress concentrations, a multiplicative effect upon either will be observed. The magnitude of this multiplicative or "cascading" effect can be estimated from the data given in Table 2. Conversely, the cascading effect may conceivably be employed to provide favorable stress field interactions by judicious choice of both relative sizes and packing geometries. Examples for this approach are discussed in the following section.

B. *Guidelines for the Microdesign Development of Improved Composites*

Several avenues of improving the transverse failure characteristics of fiber composites are indicated by the present analysis.

The resistance to fracture due to transverse strains can be improved by appropriate choice of cross-sectional geometry. The use of elliptical fiber sections has been suggested by Rosen *et al.*[7] and experimental work on ma-

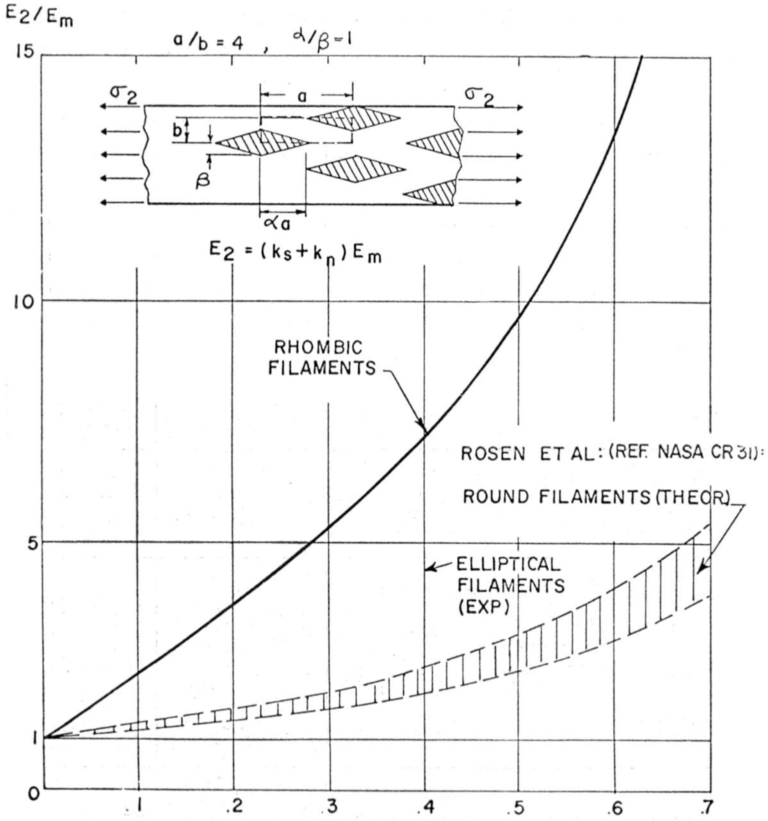

FIG. 22. Transverse modulus, E_2, of composites with round, elliptical and rhombic fiber sections

croscopic model composites has confirmed predicted improvements in transverse modulus. In addition, an approximate analysis of the stress field in the matrix of a composite made from moderately closely packed fibers of rhombic section showed considerable promise in producing uniform stress fields and good lateral compliance by simple shearing deformation of the matrix even for completely rigid fibers.

An analysis was conducted in predicting the transverse modulus of laminates made from fibers with rhombic sections for a range of packing densities. The results are summarized in Fig. 22. Experimental data for elliptical fiber composites are also shown in Fig. 22 and, in comparison, the theoretically available range for round fibers, based upon work published in ref. 7.

VII. EXPERIMENTAL STUDIES

A. *Composite Materials with Shaped Reinforcements*

Glass fibers with quasi-elliptical sections have been obtained in experimental quantities from DeBell and Richardson Inc.[8] The cross-section of a laminate made from these fibers in an epoxy resin is shown in Fig. 23. The desired nesting geometry was not fully achieved, primarily because the fiber section was oblong-rectangular with rounded edges rather than the desired elliptical section.

Due to difficulties in obtaining satisfactory solid fibers with shaped sections, this approach was abandoned in favor of a two-stage laminating process, in which shaped rods or tape is prepared in special dies from a glass-resin laminate. These elements then are subsequently bonded by an elastomeric matrix to form the final composite.

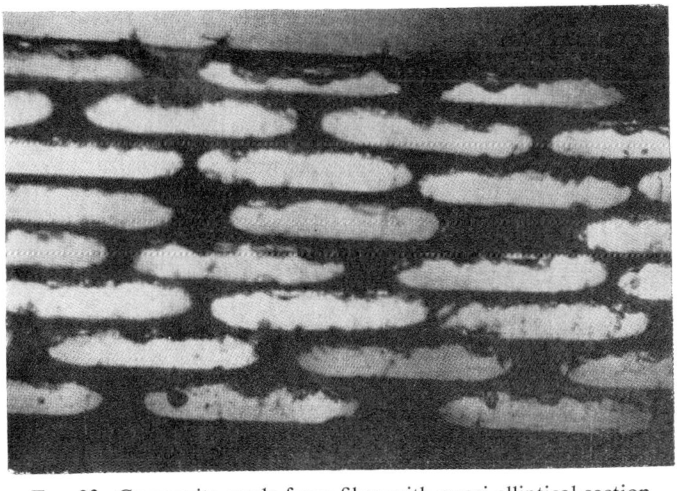

FIG. 23. Composite made from fiber with quasi-elliptical section

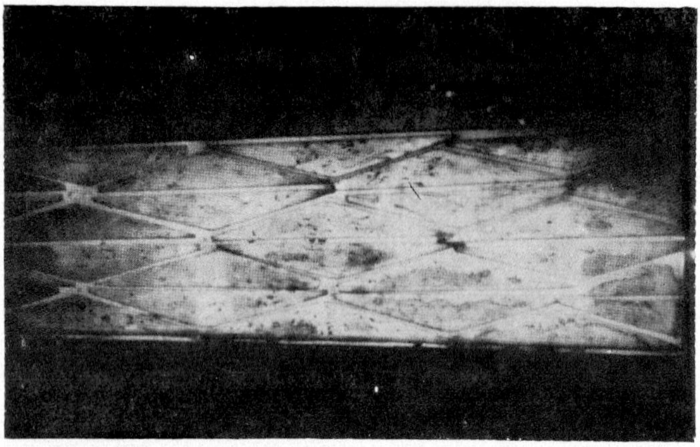

FIG. 24. Two-stage composite with rhombic tape reinforcement

A two-stage composite material of rhombic texture[9] was prepared to demonstrate the rhombic geometry. For this purpose, tape of flat triangular section was prepared in a shaped mandrel from standard S-glass, epoxy-impregnated roving. This tape was subjected to cure and then assembled in a rhombic pattern, using a compliant binder in the form of a high strength polyurenthane elastomer.† No attempts were made to obtain optimum adhesion between the pre-cured rods and the elastomer other than normal precautions insuring reasonable cleanliness. A section of the resulting laminate is shown in Fig. 24.

Another approach of utilizing microgeometrical design parameters to improve the transverse stress performance of composites is exemplified in

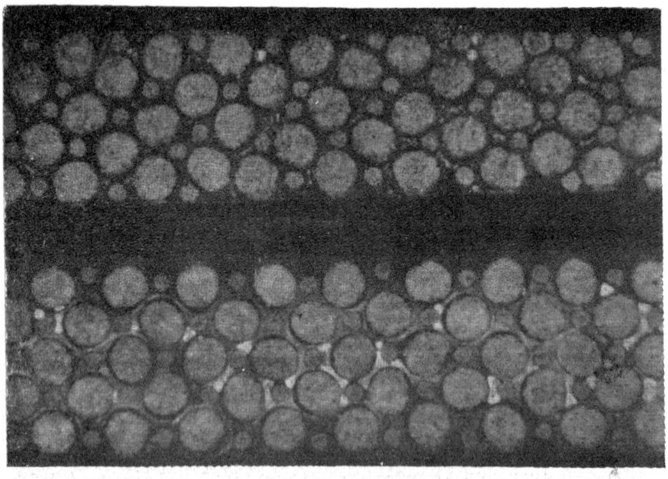

FIG. 25. Two-stage composites with aggregate fiber diameters

† Adiprene L—Trademark: E. I. du Pont de Nemours & Co.

FIG. 26. Cross-woven mat of graded fiberglass rods prior to lamination

Fig. 25. Here, commercially available cylindrical rods of two diameters, made from S-glass fibers with epoxy binders, were aggregated in two typical patterns and laminated in an elastomer matrix. The pattern geometry was selected to improve packing density and, hopefully, to reduce the danger of premature transverse fracture.

An additional feature of these composites is a transverse reinforcing system, by which the individual rods were woven into a mat-like fabric shown in

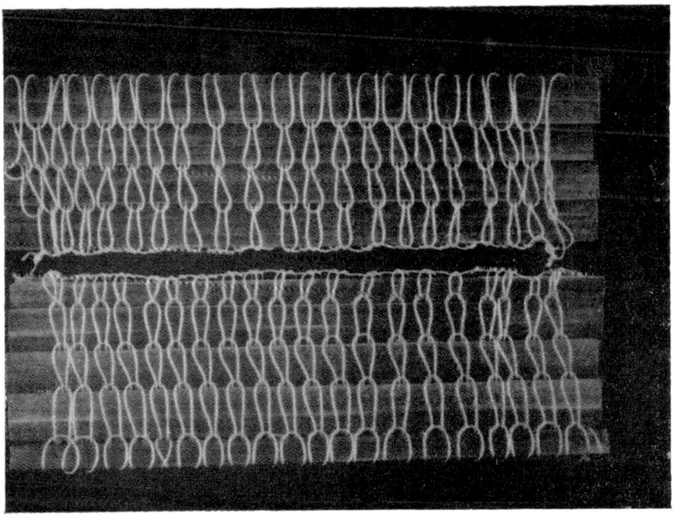

FIG. 27. Cross-woven mat of rhombic fiberglass tape prior to lamination

Figs. 26 and 27. The process by which this mat is constructed is described in detail in ref. 10. Because they are very flexible, the transverse filaments forming the cross-weave do not reduce the rod columnation; however, they contribute materially to the transverse strength of the laminate. By this means, the transverse properties can be controlled to any desired degree without necessarily resorting to conventional cross-lamination. As a further advantage, the pre-fabricated mat greatly improves the ease by which practical aggregate composites of the type shown can be fabricated.

B. *Test Results*

It was desired to examine the relative performance of the two-stage composites described above under conditions of highly non-ideal loading. Such a test is intended to provide a measure for the materials insensitivity to "poor design", yielding a conceptual equivalent to "ductility" in conventional metallic materials.

For this purpose, test specimens representing a crude joint design have been prepared. The test specimens were mounted into a double clevis-type test fixture, as shown in Fig. 28, and subject to tensile forces applied at the

FIG. 28. Shear-bearing test fixture

bolts by the test fixture. Spreading of the bolt centers was recorded in function of applied loads. The loading provided in this manner generates three types of critical stress conditions in the laminate sample:

A transverse tensile stress in the vicinity of the bolt bearing surface that can result in catastrophic axial splitting (cleavage) of the test specimen.

Shearing stresses along the bolt hole edge surfaces reacted by the tensile force in the specimen center that can result in progressive shear-out.

Axial compressive stresses adjacent to the bolt surface that can result in bearing failure.

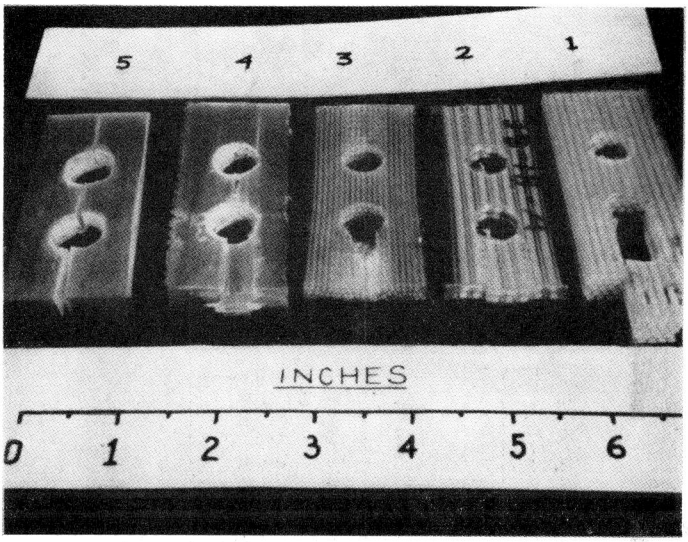

FIG. 29. Fracture-type of five shear-bearing joint test specimens

The fracture appearance of five typical test specimens is shown in Fig. 29. Compositions and observed mechanical properties at failure are tabulated in Table 4. Stresses at maximum bearing loads are given in two columns, one based upon total bearing area (i.e., specimen thickness times bolt diameter), and one based upon the glass fiber section in the bearing area only. Stress–displacement diagrams for the five samples are shown in Fig. 30.

The maximum load at failure based upon total bearing area was generally lower in the two-stage composites than in the standard laminate. However, since the two-stage composites fail always in progressive rather than in catastrophic modes, it can be expected that careful surface preparation of the rods with primers compatible with the epoxy-glass-adiprene system would materially improve their failure load levels. Also, it will be noted that the maximum bearing stress capability based upon glass content of the two-stage composites was generally higher than that provided by the standard laminate.

TABLE 4. PROPERTIES OF SHAPED TWO-STAGE COMPOSITES

No.	Type	Composition			Mechanical properties		
		Glass vol. %	Epoxy vol. %	Adiprene and void content vol. %	Elongation at max. load in.	Bearing stress at failure based on Total section psi	Bearing stress at failure based on Glass section psi
1	Round rod Aggregate 1 : 1 0.090/0.035 in.	26.7	20.0	53.3	0.281	4730	17,500
2	Round rod Aggregate 1 : 1 0.090/0.035 in.	12.8	9.6	79.6	0.350	3520	27,500
3	Round rod 0.035 in.	6.8	5.1	88.1	1.180	3700	54,400
4	Rhombic rod 1/2×1/4 in.	61.0	20.0	19.0	0.005	6860	11,200
5	Standard laminate (ref. sample)	76	24		0.002	8550	11,200

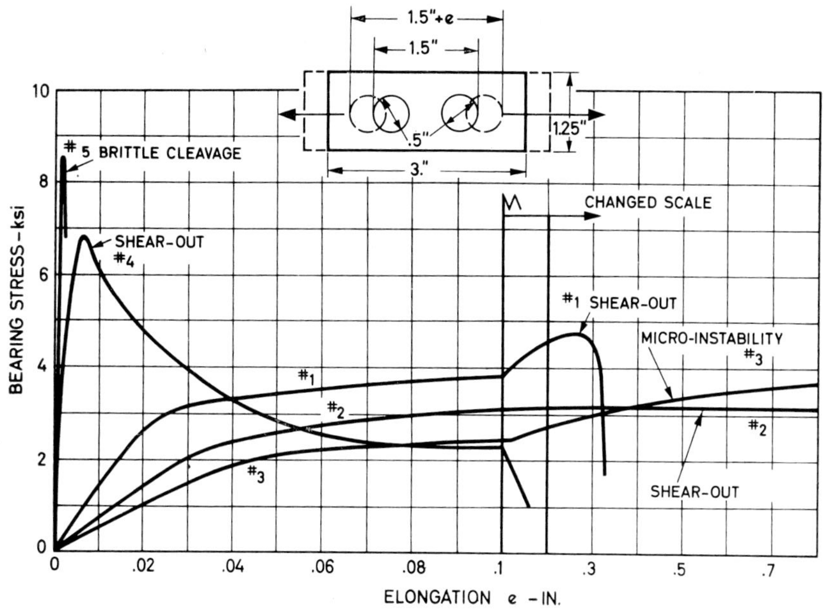

FIG. 30. Bearing stress vs. elongation

The data shown in Table 4 and Fig. 30 indicate a major improvement in elongation at maximum load by use of the shaped rod composites as compared with the reference sample made from identical constituent materials but in the standard laminate form. The difference in failure mechanisms are also reflected in the striking difference of failure appearance, ranging from the totally brittle, longitudinal cleavage observed in the standard laminate reference sample to progressive shear-out (samples 1, 3, and 4), and compressive microinstability in the bearing area (sample 3).

VIII. CONCLUSIONS

The micromechanical model subjected to analysis here represents a very much simplified version for conditions that may exist in actual cases. Evidently, for a complete characterization of the stresses prevailing in the vicinity of inclusions several other factors may have to be considered in addition to the effect of transverse stress. These include more complex applied loading patterns, non-linear materials response, stress interaction effects from neighboring inclusions, axial stress effects, and particularly the presence of "built-in" stresses originating, for instance, from differential thermal expansion or from interface interaction causing localized volumetric changes in the matrix-inclusion complex. The primary purpose of the work presented, therefore, is to demonstrate the utility of a rigorous analytical approach to the understanding of failure and strength in textured composite materials rather than to yield directly applicable data for any one particular case. Nevertheless, several qualitative conclusions may be stated, based upon the foregoing results:

The presence of hard inclusions formed by fibers in a relatively soft matrix can generate serious stress concentrations that may form the sites of origin for premature failure in actual, non-idealized structural useage of these materials.

The recognition of appropriate failure criteria provides guides towards a more judicious use of microdesign concepts, particularly with regard to the geometry of individual fibers and the packing arrangement.

Significant benefits appear to be available by departing from simple round fibers of uniform diameter in more or less accidental packing arrangements. In particular, fibers with graded diameters and rhombic fiber sections in either single or two-stage composites show promise of reducing the susceptibility of potentially high strength composites to premature failure.

A practical process of introducing a two or three-dimensional secondary reinforcement for control of packing geometry and improved transverse strength without degradation of the primary, unidirectional composite strength has been demonstrated in principle.

REFERENCES

1. GOODIER, J. N. and TIMOSHENKO, S. *Theory of Elasticity*, McGraw-Hill. 1951.
2. VON MISES, R. Mechanik der festen Koerper im plastisch-deformablen Zustand, *Nachr. K. Ges. Wiss., Goettingen*, pp. 582–592 (1913).
3. WITUCKI, R. Boron Filaments, NASA CR-96, September (1964).
4. WITUCKI, R. Research on High Modulus, High Strength Filaments and Composites, Astro Research Corporation Report ARC-R-203, November (1965).
5. SOUTHWELL, R. V. and GOUGH, H. J. On Concentration of Stress in the Neighborhood of a Small Spherical Flaw; and the Propagation of Fatique Fractures in Elastically Isotropic Materials, *Phil. Mag.* (7), pp. 71–97 (1926).
6. American Lava Corporation. ALSIMAG, Chart 611.
7. ROSEN, B. W. *et al.* Mechanical Properties of Fibrous Composites, NASA CR-31, April (1964).
8. DEBELL & RICHARDSON, Inc. Studies of Hollow Multipartitioned Ceramic Structures, Fourth Quarterly Progress Report, June 16, 1965–September 16, 1965.
9. U.S. Patent No. 3,121,451. Isotensoid Structures, February 18, 1964.
10. SCHUERCH, H. Joint Design for Flexible Space Structure, Paper No. VI-1, *Proceedings of the Society of Aerospace Materials and Process Engineers Ninth National Symposium, Dayton, Ohio, November 1965.*

DISCUSSION

J. Dundurs
(*Northwestern University*)

When a composite body consisting of two isotropic and elastic phases is loaded by prescribed surface tractions, generally stress depends on *three* dimensionless parameters formed from the elastic constants. Dr. Schuerch has observed that the stress induced by a circular inclusion in a field of uniform tension is completely characterized by *two* composite parameters. The purpose of my comments is to call attention to the fact that the important reduction in the dependence of stress on elastic constants is a consequence of a general result in plane elasticity.

Since these results hold regardless of the topographic features of the composite, be it a matrix reinforced with fibers, or be it a laminate, it is convenient to adopt a notation different from that used by Dr. Schuerch. Thus, I shall use suffixes 1 and 2 to distinguish between the elastic constants of the two phases. Although it matters not in principle how the two materials are labeled, it is suggested, nevertheless, that a matrix be taken as material 1 and the fibers as material 2. Furthermore, it is expedient to deal with both plane strain and plane stress simultaneously by employing $\varkappa = 3-4\nu$ for plane strain and $\varkappa = \dfrac{(3-\nu)}{(1+\nu)}$ for plane stress, where ν stands for Poisson's ratio. Denoting the shear modulus by G and setting $\Gamma = G_2/G_1$, the composite constants I propose are

$$\delta = \frac{\Gamma\varkappa_1+1}{\varkappa_2+1}, \quad \eta = \frac{\Gamma+\varkappa_2}{\varkappa_2+1}.$$

If we let G_1, $\varkappa_1$ be the elastic constants for the matrix, and G_2, $\varkappa_2$ those for the fiber, the parameters used by Dr. Schuerch become

$$K_1 = \frac{\delta-\eta}{2\eta-1}, \quad K_2 = \frac{\eta-1}{\delta}.$$

The reduction in the dependence of stress on the elastic constants takes place when the geometry and loading of the composite body is such that it is in a state of either plane strain or plane stress. If the composite body has holes, an additional requirement is that the vector sums of tractions on holes vanish for every individual hole. It can be shown† that then, under the condition of the body being subjected to prescribed tractions, *the stress depends only on the two composite constants δ and η*. In terms of Cartesian components of stress, this result reads

$$\sigma_{xx} = \sigma_{xx}(x, y;\ \delta, \eta),$$
$$\sigma_{xy} = \sigma_{xy}(x, y;\ \delta, \eta),$$
$$\sigma_{yy} = \sigma_{yy}(x, y;\ \delta, \eta).$$

It may also be added that the reduction in the dependence of stress on elastic constants takes place regardless of the mechanical conditions at the interfaces, be it perfect adhesion, slip or local separation.

The results also carry over to more than two materials. If, for instance, the composite consists of three materials, dimensional analysis predicts that stress depends on *five* parameters involving the elastic constants. As such we could take Γ_{21}, Γ_{31}, $\varkappa_1$, $\varkappa_2$ and $\varkappa_3$, where $\Gamma_{21} = G_2/G_1$ and $\Gamma_{31} = G_3/G_1$. Under the same restrictions as for the composite consisting of two phases, however, it can be shown that the stress is a function of only the following *four* combinations of elastic constants:

$$\delta_{21} = \frac{\Gamma_{21}\varkappa_1+1}{\varkappa_2+1}, \quad \eta_{21} = \frac{\Gamma_{21}+\varkappa_2}{\varkappa_2+1},$$

$$\delta_{31} = \frac{\Gamma_{31}\varkappa_1+1}{\varkappa_3+1}, \quad \eta_{31} = \frac{\Gamma_{31}+\varkappa_3}{\varkappa_3+1}.$$

H. SCHUERCH

(*Astro Research Corporation*)

I wish to thank Professor Dundurs for his most enlightening comments. The fact that the reduction of materials variables in the case of single inclusion is only a special case of a more general theorem was suspected by myself and my co-workers, but, in a brief search, we have been unable to establish a definitive proof. The general theorem given by Professor Dundurs will be most valuable for further studies in this field, particularly, where more complex three-dimensional geometries will be involved.

The notation K_1 and K_2, used in the present paper, was chosen as the most convenient grouping appropriate to the particular problem of single cylindrical inclusions that has been treated here.

† Cf. J. Dundurs, Effect of Elastic Constants on Stress in a Composite under Plane Deformation, *Journal of Composite Materials*. in press.

SESSION V

FAILURE MECHANICS

STRENGTH OF UNIAXIAL FIBROUS COMPOSITES

B. WALTER ROSEN

Space Sciences Laboratory,
General Electric Company,
King of Prussia, Pa.

ABSTRACT

Alternate existing theoretical models for the analysis of the tensile strength of parallel filament composites are examined. Included are analyses of different failure mechanisms. A single failure mechanism for all fibrous composites does not appear to exist. Rather, the failure mode, as well as the failure level, is dependent upon both fiber and matrix characteristics. Failures due to longitudinal and transverse crack propagation, statistical accumulation of internal fractures, and matrix flow are all possible. Models for these failure modes are considered and interactions between them are described. Quantitative comparisons between the different analyses are presented. Although only limited experimental data are available for composites whose constituents are well characterized, these are also compared to the appropriate analytical results.

The relationship between the compressive strength of uniaxial composites loaded parallel to the filaments and the constituent properties is briefly reviewed.

I. INTRODUCTION

The high tensile strength of contemporary uniaxial fibrous composites loaded in the fiber direction is well known and widely utilized. In contrast, the present understanding of the tensile failure mechanism and the influence thereon of constituent properties is limited. One aim of the present paper is to relate several existing analyses of uniaxial tensile strength which together provide the elements of a rational theory for the tensile failure of fibrous composites.

Recent analyses and tests of uniaxial composites have indicated that their strength in compression is also attractively high. Accordingly the mechanics of composite compressive failure are also reviewed.

Other studies of composites have demonstrated that weaknesses in the direction transverse to the fibers limit the efficiency of composites used in thin plate and shell structures. A relatively unexplored area is that of structural design to utilize uniaxially stressed composites. The studies described herein may contribute to the practicality of such composite structures by defining methods to improve uniaxial composite strength.

II. TENSILE STRENGTH

Considerable attention has been devoted to the definition of composite tensile strength as a function of fiber strength. A recent comprehensive review of much of this literature is presented in ref. 1. Principal attention in that review was directed toward composites in which both constituents are ductile metals. Many contemporary high strength filaments are brittle and the present review is directed toward an understanding of composites utilizing such fibers.

A logical starting point for a discussion of composite tensile strength is the prominent "rule of mixtures" which states that

$$P^* = \sum_{i=1}^{n} v_i P_i \tag{1}$$

where P^* is the effective composite property,
v_i is the volume fraction of the ith phase of an n phase composite,
P_i is the value for the ith phase, of the property in question.

For example, the composite density is given by the weighted average (on a volume basis) of the constituent densities, so that for a fibrous composite

$$\varrho^* = v_f \varrho_f + v_b \varrho_b \tag{2}$$

where ϱ^* is the composite density,
v_f, v_c are fiber and binder volume fractions,
ϱ_f, ϱ_b are fiber and binder densities.

Similarly the strength of a uniaxial fibrous composite in the fiber direction is frequently written as:

$$\sigma^* = v_f \sigma_f + v_b \sigma_b \tag{3}$$

where σ^* is composite strength,
σ_f, σ_b are fiber and binder strengths.

When σ_f and σ_b are interpreted as the average stress in the loading direction in each phase, at the point of incipient failure, the above equation is unassailable. The problem associated with the use of eq. (3) is produced by the assumption that the fiber strength as measured by testing individual fibers can be used directly in eq. (3). This widely accepted assumption seems somehow to have gained acceptance as a basic postulate in the theory of the mechanics of fibrous composites. In the following discussion, some effort will be devoted to an attempt to discredit this approach. These studies will show that only for a limited class of composites are the average fiber stresses at composite failure, and the average strength of a set of single fiber tests, one and the same quantity. In addition to this somewhat negative approach, which has been motivated by the widespread acceptance of a fallacious argument, this review will also present a positive approach to the definition of the relationship between the fiber strength to be used in eq. (3) and the properties of fibers tested separately.

Figure 1 shows several simple variations of composite strength with fiber volume fraction. Three straight lines radiate from the point (1.0, σ_f). These represent eq. (3) for σ_b equal in turn to: (1) the matrix strength in uniaxial tension; (2) the tension stress in the matrix under uniaxial tension when the strain is equal to the fiber strength divided by the fiber Young's modulus;

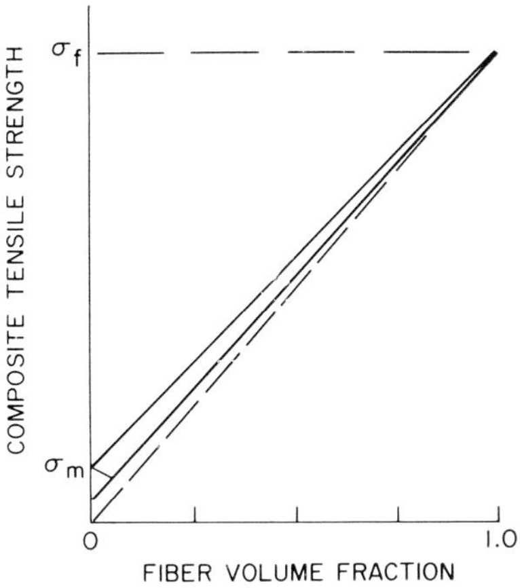

FIG. 1. Characteristic variation of tensile strength of uni-directionally reinforced composite with volume fraction of reinforcement

and (3) zero. To indicate the significance of these plots, consider two constituents having the same Poisson's ratios. Relative axial displacement of the two ends of such a composite would result in a uniform state of strain throughout the composite and a uniform uniaxial state of stress in each phase. The upper curve implies that the failure load is the sum of the failure loads of the individual phases tested separately. The central curve implies that the failure load is the sum of that carried by the fibers plus the matrix load at the fiber failure strain (assumed elastic). The short curve connecting the matrix strength point to the center curve shows the case in which it is assumed that the fibers fail first and that the failure load is the load carried by the matrix alone. Much has been made of this region and the volume fraction at the intersection of this latter line with the central line is denoted as the critical volume fraction by some authors. That is, fibers add to or detract from the matrix strength as their volume fraction is above or below this volume fraction respectively. In actuality, the strength of the fibers themselves is so much higher than the strength of the matrix that the rule of mixtures is well represented by the lowest straight line of Fig. 1 which runs between zero strength at zero volume fraction and some fiber strength at 100% volume fraction. In general, the direct contribution of the matrix strength to the composite strength is a small one. The effect of the matrix properties on composite strength, however, is quite pronounced, as will be shown shortly.

The above discussion of the validity of the rule of mixtures is invalidated by the facts that, for many practical composites, the matrix stress distribution is non-uniform and the fiber strength is not a unique quantity. For composites in which both phases do not have the same Poisson's ratio the state of stress in each phase is neither uniaxial nor uniform. For brittle fibers the strength can only be characterized by a statistical distribution function such

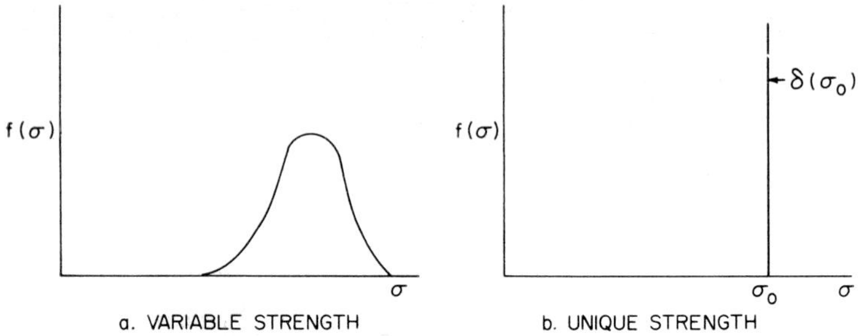

FIG. 2. Illustrative statistical distribution functions for single fiber strengths

as that shown in Fig. 2a; whereas the rule of mixtures would only apply to a fiber having a strength characterized by a Dirac delta function distribution as shown in Fig. 2b. For glass and boron filaments and for many of the other high strength filaments, the distribution of Fig. 2a is the one that is appropriate. For these fibers, depending upon the magnitude of the variation of these fiber strengths, there are different possible failure mechanisms. It is the treatment of these failure mechanisms for composites containing brittle fibers that the following discussion emphasizes. However, it is of interest to examine first the results obtained for a distribution function like the one of Fig. 2b. This is illustrated in Fig. 3 which presents some of the data obtained by McDanels *et al.*[2, 3] for tungsten fibers in a copper composite. These fibers are of uniform strength, the matrix is ductile, and the composites fail along the straight line variation represented by the rule of mixtures between zero strength and the single fiber strength.

For those filaments which have a broad statistical distribution function, one must expect that the behavior of composites can only be understood in a statistical nature. For illustrative purposes, it is convenient to treat fibers characterized by a Weibull distribution function. This is chosen because certain fibers are experimentally observed to be characterized by Weibull distributions—some over a portion of the range of gauge length of the test specimen and others over the entire range. Also the length dependence of the strength of a brittle fiber is conveniently demonstrated with a Weibull strength distribution function $f(\sigma)$, chosen here in the form:

$$f(\sigma) = L\alpha\beta\sigma^{\beta-1} \exp(-L\alpha\sigma^{\beta}) \tag{4}$$

where L is the fiber length,
σ is the fiber strength,
α, β are statistical parameters.

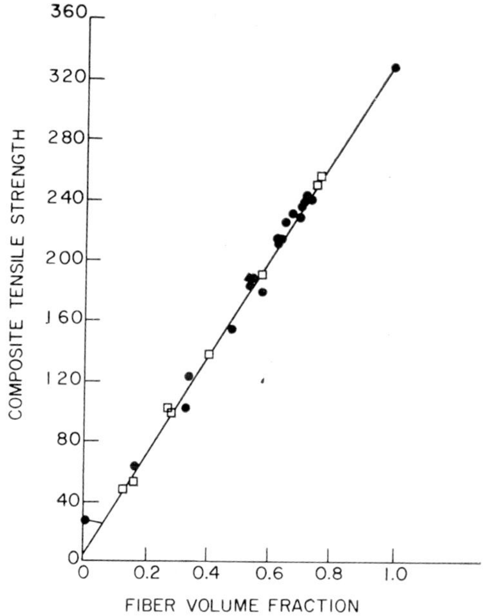

FIG. 3. Experimental data for strengths of composites fabricated from fibers of uniform strengths (from McDanels *et al.*[2,3])

The kth moment, M_k, of this statistical distribution function, defined by:

$$M_k = \int_0^\infty \sigma^k f(\sigma)\, d\sigma, \tag{5}$$

can be used to define the mean fiber strength:

$$\bar{\sigma} = M_1 \tag{6}$$

and the standard deviation:

$$s = [M_2 - M_1^2]^{1/2}. \tag{7}$$

For the Weibull distribution of eq. (4), substitution into eqs. (6) and (7) yields:

$$\bar{\sigma} = (\alpha L)^{-1/\beta}\Gamma\left(1+\frac{1}{\beta}\right)^{1/2} \tag{8}$$

and

$$s = (\alpha L)^{-1/\beta}\left[\Gamma\left(1+\frac{2}{\beta}\right)-\Gamma^2\left(1+\frac{1}{\beta}\right)\right]^{1/2}. \tag{9}$$

The coefficient of variation, μ, for this distribution is given by:

$$\mu \equiv \frac{s}{\bar{\sigma}} = \frac{\left[\Gamma\left(1+\frac{2}{\beta}\right)-\Gamma^2\left(1+\frac{1}{\beta}\right)\right]^{1/2}}{\Gamma\left(1+\frac{1}{\beta}\right)}. \tag{10}$$

It is now possible to assign some physical significance to the parameters of the distribution function of eq. (4). First of all, it is seen that the coefficient of variation is a function only of the parameter, β. Indeed it can be shown that eq. (10) is approximated to within 3% (for $0.05 \leqslant \mu$ 0.50) by:

$$\mu = \beta^{-0.92}. \tag{11}$$

Thus β is essentially an inverse measure of the coefficient of variation, and for practical fibers characterized by eq. (4):

$$\beta > 1. \tag{12}$$

From eq. (8) it can now be seen that for a unit length

$$\bar{\sigma}_1 = k\alpha^{-\frac{1}{\beta}} \tag{13}$$

where

$$k = \Gamma\left(1+\frac{1}{\beta}\right) \tag{14}$$

and from (12) and (14):

$$0.88 \leqslant k \leqslant 1.0.$$

Thus the quantity $\alpha^{-1/\beta}$ may be viewed as a reference stress level.

With these definitions, eq. (8) may be plotted in the form shown in Fig. 4. Here, the tensile strength of single fibers of length L is shown as a function of some normalized fiber length for different values of β. The upper curve, where β is infinity, is the spike distribution function, that is, the delta function. For this case, all the fibers are the same strength and there is no length dependence. For β of 10, which is approximately a 12% coefficient of variation, an order of magnitude change in the fiber length produces about a 20% drop in its average strength and for β of 4, and order of magnitude change in length produces almost a 50% drop in strength. This is the first dilemma encountered in the study of brittle fibers; namely, the question of what is the reference strength to be used in eq. (3) when strength is length dependent.

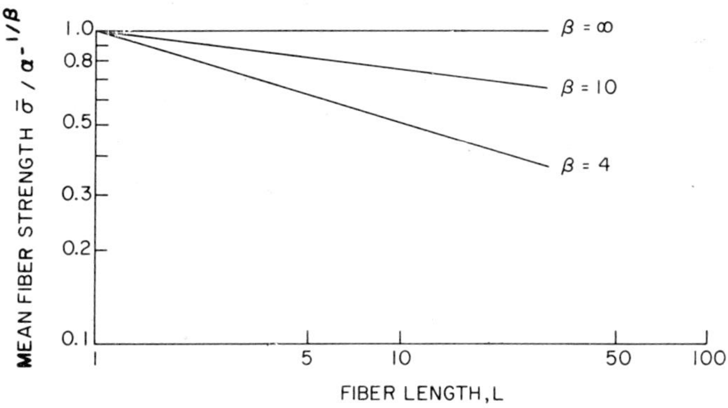

FIG. 4. Strength length variation of fiber strengths following Weibull statistical distribution functions

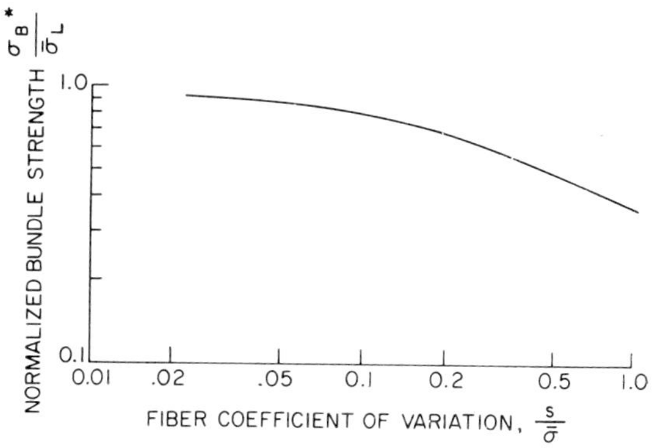

FIG. 5. Variation of fiber bundle strength with fiber coefficient of variation

A second aspect of this statistical approach arises from the fact that a composite contains a bundle of filaments. These filaments are acting in parallel and it is well known that the strength of a bundle of filaments, whose members are not of uniform strength, is not equal to the mean fiber strength. Following the work of Coleman,[4] one obtains the results shown in Fig. 5. This shows the variation of bundle strength (again a statistical characterization, so that the bundle strength is in this case measured by the mode of the strength) normalized with respect to the mean strength of individual fibers of the same length as the bundles. This is plotted as a function of the fiber population's coefficient of variation and the general result is that the average strength of the bundle is somewhat less than the average strength of the fibers, and is a decreasing function of the coefficient of variation of the fibers.

The approach to the tensile strength problem proposed in ref. 5 is to idealize the fibrous composite into a statistical model that will permit the utilization of the statistical information obtained from the individual fiber tests. Parratt[6] originally proposed that distributed internal fractures occurred prior to composite failure. His suggested mechanism of failure in a fibrous composite is represented in Fig. 6. There is an accumulation of fiber breaks, occurring as the load increases, resulting in the fibers breaking into shorter lengths having increasing strength until the fibers are so short that the shear stresses

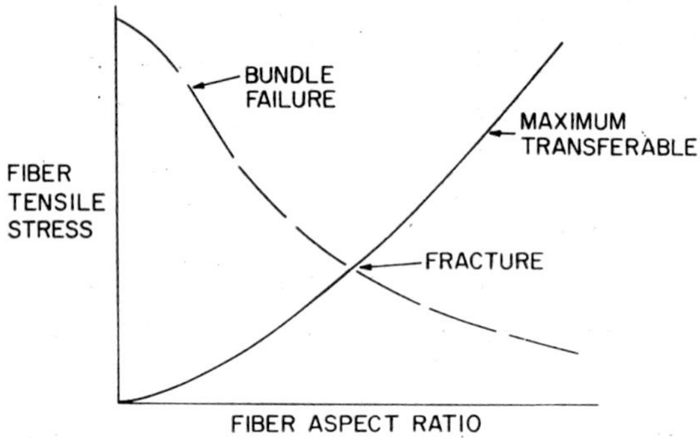

FIG. 6. Influence of fiber length on maximum average stress of fibers in a uniaxial composite

at the interface between fiber and matrix are not able to transfer sufficient load into the fibers to fracture them. Composite failure would then occur by fiber pullout. In ref. 5, it was shown that a failure mechanism that will occur before this is simply a weakening of one cross-section which results from a statistical accumulation of fiber fractures. This will be referred to as the cumulative weakening failure model. To understand this mechanism we treat composites having fibers of non-uniform strength. In the composite it is expected that some fiber breaks will occur at low stress levels. It then becomes important to evaluate the effect upon the stress fields of one of these breaks. This subject will not be reviewed because it has been discussed by various investigators[7-11] using different approximations to the problem. Basically the result is as shown schematically in Fig. 7. At the point of the fiber break, the extensional stress in the fiber drops to zero, as the distance along the fiber away from the break increases, the stress builds back up again, by virtue of shear transmitted across the interface between the two constituents. These localized shear stresses can be quite high and indeed for many binders it is logical to expect that there will be either a yielding of the matrix material or perhaps an interface failure between the two constituents. It is also important to recognize that this local drop in the stress in one filament will produce a load concentration in the adjacent filaments. Note that there are two different kinds of stress concentrations of importance: there is the load concentration in the filament, expressed in terms of the average stress carried in the filament,

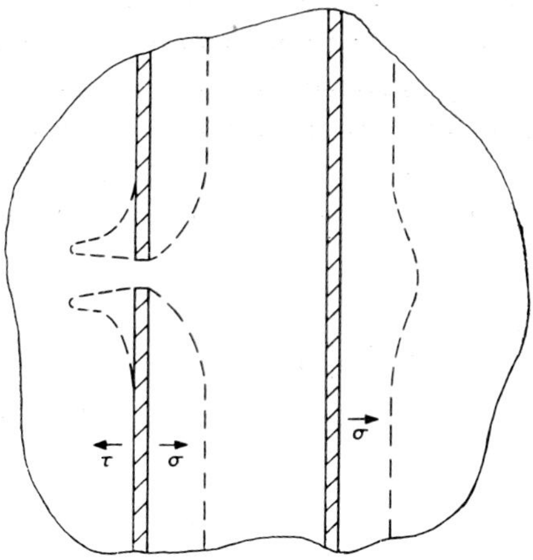

FIG. 7. Perturbation of fiber extensional stresses and interface shear stresses in the vicinity of a fiber break

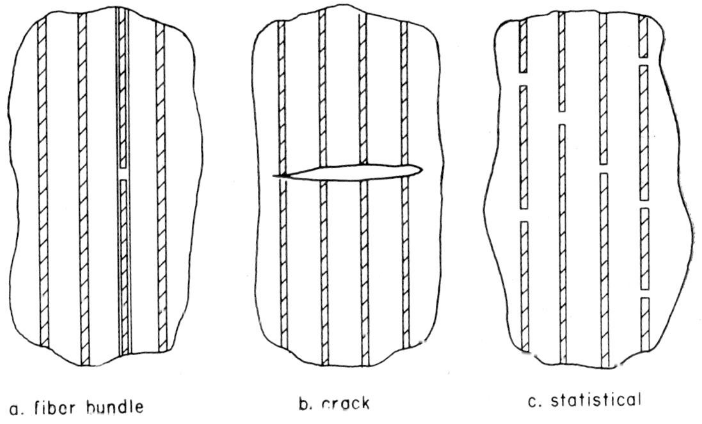

FIG. 8. Possible composite tensile failure modes

and there is the true or micro-stress concentration at the end of the crack which propagates through the matrix from one broken fiber over to the next fiber. Either the average stress concentration factor or the local stress concentration factor can result in a failure of the adjacent filament, and perhaps of the composite.

As a result of this first fiber break, there are several alternate subsequent effects that may occur (Fig. 8). After the first fiber break, the interface shear stresses may separate the broken fiber from the remaining composite by propagating along the length of the fiber (Fig. 8a). When this happens the fiber is totally ineffective and the composite is acting as if it were a bundle of filaments. For this case the strength of the composite is not enhanced by the presence of a matrix material. Alternately, the crack from the first break may propagate out in a direction normal to the fibers across the other filaments, as a result of the local stress concentrations, leading to immediate composite fracture. When both of these crack propagation effects are suppressed, cumulative damage results, whereby cracks do not propagate through adjacent fibers and as the load increases additional fiber fractures occur at other sites. An indication of the relative possibility of the occurrence of crack propagation as compared with the cumulative damage failure mode can be obtained by examining the question of average stress concentrations.

Hedgepeth[11] utilized a shear lag model wherein the rate of change of the load in the fiber is equated to the shear forces being transmitted uniformly through the matrix from the adjacent filaments. The equilibrium equation may be written as:

$$E_f A \frac{\partial^2 u_n}{\partial x^2} + \frac{G_b h}{d}(u_{n+1} - 2u_n + u_{n-1}) = m \frac{\partial^2 u_n}{\partial t^2} \tag{15}$$

where $E_f A$ is the fiber extensional stiffness,
u_n axial displacement of nth fiber,
x an axial coordinate,
$G_b h$ the binder shear stiffness,
d the fiber spacing,
m the mass per unit length associated with a filament,
t the time.

Following ref. 11, eq. (15) may be rewritten, for the static case, in non-dimensional form as:

$$\frac{\partial^2 U_n}{\partial \xi^2} + U_{n+1} - 2U_n + U_{n-1} = 0. \tag{16}$$

Hedgepeth obtained solutions for the case of r adjacent broken fibers. Thus he had boundary conditions of the form:

$$\left.\begin{aligned} U_n(0) &= 0 \qquad && n < 0;\ n \geqslant r \\ P_n(0) &= 0 && 0 \leqslant n \leqslant r-1 \\ P_n(\infty) &= 1 && \end{aligned}\right\} \tag{17}$$

where P_n is the dimensionless load in the nth fiber. The solution is obtained by solving the influence function problem for one broken fiber. That is, for boundary conditions:

$$\left.\begin{aligned} U_0(0) &= 1 \\ U_n(0) &= 0 \qquad n \neq 0. \\ P_n(\infty) &= 0 \end{aligned}\right\} \tag{18}$$

If we denote the non-dimensional forces which result from the unit problem defined by eqs. (18), $L_n(\xi)$, the solution to the original problem defined by boundary conditions (17), for r adjacent broken filaments is given by

$$P_n(\xi) = 1 + \sum_{m=0}^{r-1} L_{n-m}(\xi)\, U_m(0) \tag{19}$$

and the maximum stress concentration is given by

$$P_r(0) = 1 + \sum_{m=0}^{r-1} L_{r-m}(0)\, U_m(0) \tag{20}$$

where
$$L_n(0) = \frac{4}{(4n^2-1)\pi}. \tag{21}$$

The numerical values obtained from these equations in ref. 11 are shown in Table 1. A single broken fiber in this two dimensional model produces a stress concentration factor of 1.33. As the number of adjacent broken fibers increases, the stress concentration factor goes up so that a mechanism exists for generating a crack propagation failure through the composite, transverse to the filaments. Also shown in Table 1 are the dynamic overshoot factors obtained in ref. 11. These in the limit approach a 27% overshoot of the static stress concentration. In the cumulative weakening model, it is of importance to assess the influence of distributed fiber fractures and it is possible to extend

TABLE 1. STRESS CONCENTRATION FACTORS
(from ref. 11)

Number of broken fibers	Average stress concentration factor	Dynamic overshoot
1	1.33	1.15
2	1.60	1.19
3	1.83	1.20
4	2.03	
5	2.22	
6	2.39	
∞	—	1.27

$$k_r = \frac{4\cdot 6\cdot 8\ \dots\ (2r+2)}{3\cdot 5\cdot 7\ \dots\ (2r+1)}$$

the previous work to show what happens when the composite contains periodically broken fibers. Thus, if every nth fiber is broken at the same cross-sectional location in a composite, the boundary conditions (17) are replaced by:

$$\left.\begin{array}{ll} P_n(0) = 0 & n = fr \\ & f = \text{non-negative integer} \\ U_n(0) = 0 & n \neq fr. \end{array}\right\} \tag{22}$$

Thus, the resulting force distribution is obtained from:

$$P_n(\xi) = 1 + \sum_{m=-\infty}^{\infty} L_{n-m}(\xi)\, U_m(0). \tag{23}$$

From eqs. (22) and (23) it is seen that

$$0 = 1 + \sum_{m=-\infty}^{\infty} L_{f(r-m)}(0) U_{fm}(0). \tag{24}$$

By symmetry the value of U_{fm} is independent of m. Therefore using eq. (21) and (24) and setting $r = 0$:

$$U_{fm} = -\frac{\pi}{4 \sum\limits_{m=-\infty}^{\infty} \dfrac{1}{(4f^2m^2-1)}}. \tag{25}$$

The maximum load concentration factor is given by $P_{fr\pm 1}$ (0), or for simplicity, from eqs. (21), (23) and (25):

$$P_1(0) = 1 - \frac{\sum\limits_{m=-\infty}^{\infty} \dfrac{1}{[4(1-fm)^2-1]}}{\sum\limits_{m=-\infty}^{\infty} \dfrac{1}{[4f^2m^2-1]}} \tag{26}$$

it can be shown that

$$P_1(0) = 1 + \frac{\sin \dfrac{\pi}{f} \tan \dfrac{\pi}{2f}}{\cos \dfrac{\pi}{f} - \cos \dfrac{2\pi}{f}}. \tag{27}$$

These load concentration factors are presented in Table 2. Obviously if every other fiber is broken, the load concentration factor is 2 as shown and if every third fiber in this two-dimensional model is broken it is 1.5. If every fth fiber is broken as f goes to infinity, eq. (27) yields the previous result for a single broken fiber, namely, 1.33. There is only a small increase in the stress concentration factor as the crack density builds up. That is, the interaction of the stress concentration factors does not become a serious problem until

TABLE 2. PERIODICALLY BROKEN FIBERS

Fraction of fibers broken, $1/f$	Average stress concentration factor
1/2	2.00
1/3	1.50
1/4	1.41
1/5	1.38
1/6	1.37
1/10	1.34

the crack density becomes very high. The value for a single fiber, 1.33, increases only to 1.50 when the fraction of broken fibers is as high as 1/3.

These results are for the two-dimensional case, while for the practical material, it is the three-dimensional load concentration that is important. A simple comparison between the two cases can be provided by considering the three-dimensional problem for a hexagonal array of fibers. Such a composite can be considered to be composed of a series of identical elements containing a central broken fiber surrounded by six unbroken fibers. Evidently, by symmetry, the load concentration factor is 1.17, while eq. (27), for $f = 7$, yields a factor of 1.36. Thus the stress concentration effect is substantially reduced for the three-dimensional case. The influence of these load concentration effects upon composite failure will be treated further after the cumulative damage model of refs. 5 and 12 are reviewed.

The cumulative damage model considers that, in the vicinity of an individual break, a portion of each fiber may be considered ineffective, as discussed previously. The composite may then be considered to be composed of layers of dimension equal to the ineffective length. Any fiber which fractures within this layer will be unable to transmit a load across the layer. The applied load at that cross-section is then assumed to be uniformly distributed among the unbroken fibers in each layer. The effective load concentrations which would introduce a non-uniform redistribution of these loads is not considered initially. A segment of a fiber within one of these layers may be considered as a link in the chain which constitutes an individual fiber. Each layer of the composite is then a bundle of such links and the composite itself a series of such bundles as shown in Fig. 9. Treatment of a fiber as a chain of links is appropriate to the hypothesis that fracture is due to local imperfections. The links may be considered to have a statistical strength distribution such as that of Fig. 2 which is equivalent to the statistical flaw distribution along the fibers. The realism of such a model is demonstrated by the length dependence of fiber strength. That is, longer chains have a high probability of having a

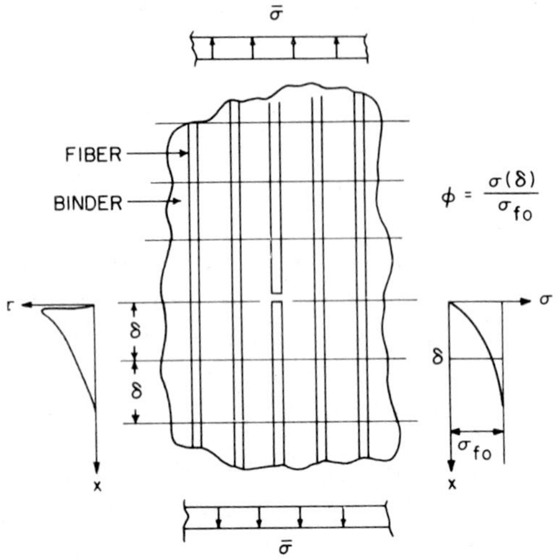

FIG. 9. Fiber-reinforced composite-tensile failure model

weaker link than shorter chains, and this is supported by experimental data for brittle fibers (for example, Fig. 10 with the data of ref. 13) which demonstrate that mean fiber strength is a monotonically decreasing function of fiber length. For this model it is first necessary to define a link dimension by consideration of the perturbed stress field in the vicinity of a broken fiber. This length, δ, was defined in ref. 5 as the value at which the stress (elastic or inelastic) reached some fraction, φ, of the undisturbed fiber stress. Friedman[14] defines this length as that point at which a step change from zero fiber stress to the undisturbed fiber stress yields an average fiber stress equal to the true average fiber stress. The former result for the elastic case (corrected for an algebraic error) is:

$$\frac{\delta}{d_f} = 1.15\left[\frac{1}{2}\left(\frac{E_f}{G_b}\right)\frac{1-v_f^{1/2}}{v_f^{1/2}}\right]^{1/2}. \tag{28}$$

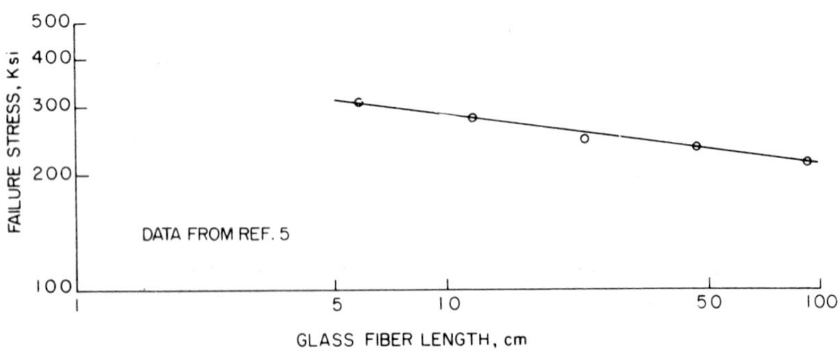

FIG. 10. Effect of length on mean fiber strength of E-glass fibers (from ref. 13)

The ref. 14 elastic result differs only in the coefficient, and is:

$$\frac{\delta}{d_f} = \left[\frac{1}{2}\left(\frac{E_f}{G_b}\right)\frac{1-v_f^{1/2}}{v_f^{1/2}}\right]^{1/2}. \quad (29)$$

It is then necessary to define the statistical strength distribution of the individual links which can be obtained indirectly from the experimental data for the fiber-strength-length relationship. A fiber is considered to consist of a chain of links. Any consistent combination of link length and link strength distribution that reproduces the fiber strength data is acceptable. With link length defined, the statistical distribution of link strength is obtained from experimental fiber strength distributions. Consider links characterized by the distribution function $f(\sigma)$ and the associated cumulative distribution $F(\sigma)$ where

$$F(\sigma) = \int_0^\sigma f(\sigma)\, d\sigma. \quad (30)$$

For n such links forming a chain that fails when the weakest link fails, the distribution function $g(\sigma)$ for the chain is defined by

$$g(\sigma) = nf(\sigma)[1-F(\sigma)]^{n-1}. \quad (31)$$

From this, the cumulative distribution function, $G(\sigma)$ for the fibers is obtained

$$G(\sigma) = \int_0^\sigma g(\sigma)\, d\sigma. \quad (32)$$

Therefore

$$G(\sigma) = 1-[1-F(\sigma)]^n. \quad (33)$$

The solution of the inverse problem is desired. That is, given the fiber data, $g(\sigma)$ and $G(\sigma)$, define the link data for a link length, δ. From eq. (33)

$$F(\sigma) = 1-[1-G(\sigma)]^{1/n} \quad (34)$$

and thus from eqs. (30) and (34)

$$f(\sigma) = [g(\sigma)/n][1-G(\sigma)]^{(1/n)-1}. \quad (35)$$

As an example, consider fibers characterized by a strength distribution of the Weibull type

$$g(\sigma) = L\alpha\beta\sigma^{\beta-1} \exp(-L\alpha\sigma^\beta). \quad (36)$$

This form has been shown to characterize the experimental length, L, to strength, σ, relationship of fibers. Use of eq. (36) in eqs. (32) and (35) yields

$$f(\sigma) = \alpha\delta\beta\sigma^{\beta-1} \exp(-\alpha\delta\sigma^\beta) \quad (37)$$

where $L = n\delta$.

The constants α and β can be determined from the experimental data for mean strength and coefficient of variation by using eqs. (11) and (13).

These results can then be used in the statistical study of a series of bundles (see ref. 15) and utilized to define the distribution for the strength of the fibrous composite. It has been shown[11] that for a large number of filaments, N, the strength of a bundle approaches a normal distribution with mean strength:

$$\bar{\sigma}_B = \sigma_m[1-F(\sigma_m)] \tag{38}$$

and standard deviation

$$\Psi_B = \sigma_m \left\{ \frac{F(\sigma_m)}{N} [1-F(\sigma_m)] \right\}^{1/2} \tag{39}$$

where the maximum stress, σ_m, is defined by:

$$\frac{d}{d\sigma} \{\sigma[1-F(\sigma)]\}_{\sigma=\sigma_m} = 0. \tag{40}$$

The composite is now a chain, the strength of whose elements are given by the normal distribution function $\omega(\sigma_B)$. The strength of a chain containing m links of this population is characterized by the distribution function $\lambda(\sigma_c)$:

$$\lambda(\sigma_c) = m\omega(\sigma_c)[1-\Omega(\sigma_c)]^{m-1} \tag{41}$$

where
$$\Omega(\sigma_c) = \int_0^{\sigma_c} \omega(\sigma)\, d\sigma.$$

For fibers which can be characterized by a Weibull distribution of the strength of a series of individual fibers of a specified gage length tested to failure, it is possible to represent the fiber reinforced composite strength with a closed form solution. In particular, the statistical mode of the composite tensile strength, σ_t^*, is found from eq. (41) by setting

$$\frac{d\lambda}{d\sigma_c} = 0. \tag{42}$$

The result is:

$$\sigma_t^* = v_f(\alpha\delta\beta e)^{-1/\beta}. \tag{43}$$

Note that this result is equal to the mean strength of the bundle of fibers of length, δ, for very large N.

Certain results of this analysis are presented in Fig. 11. Here the composite strength, normalized with respect to the mean strength, $\bar{\sigma}_L$, of individual fibers of length, L, is plotted as a function of the coefficient of variation, μ, of individual fiber strength values. The variation, μ, is the standard deviation divided by the mean fiber strength. Curves are shown for various values of the ratio of fiber test length to "ineffective length". It is seen that for reference fibers of length equal to the "ineffective length", that is for the basic bundle of fiber links of the model described previously, the statistical mode of the composite strength is lower than the mean fiber strength by an amount which

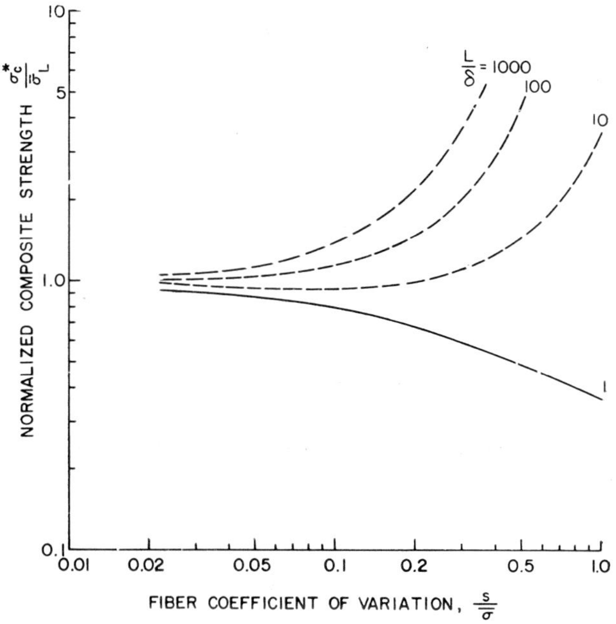

FIG. 11. Effect of fiber coefficient of variation and reference fiber length to mean fiber strength

increases as the coefficient of variation of the strength of the individual fibers increases. For more practical length ratios, L/δ (that is, for single fibers tested at gage lengths large compared to the "ineffective length"), the analysis indicates that composite strength is larger than the mean fiber strength. In order to explain these results, consider the size effect on composite and on fiber individually. The practical composite specimen is large as compared with the fiber link. That is, a composite cross-section contains many fibers and the composite length is many times the ineffective length. For this composite, the strength is insensitive to changes in composite dimension. Thus the composite strength defined by eq. (43) is valid for practical composites regardless of size. However, an individual fiber has a strength which is a strong function of length. There is an increasing likelihood of encountering serious flaws as fiber length increases and hence mean fiber strength decays monotonically with increasing fiber test length. A very long fiber can have a low strength, but when this fiber is incorporated into a composite the fiber break which occurs at a low stress level will have a relatively unimportant effect on composite strength. Thus the ratio of composite strength to mean fiber strength is strongly dependent upon the length of the individual fiber specimens. Note that the normalized strength ratio is close to unity for coefficients of variation as large as 15%. However, the fact that the magnitudes of composite strength and mean fiber strength are close to one another should not be taken to justify the hypothesis that composite performance is some fraction of fiber performance. The frequently quoted "strength conversion" results becloud the importance of fiber coefficient of variation and fiber test length.

Another way of viewing the significance of the result of the statistical tensile failure model arises from the fact that the statistical mode of the composite strength is equal to the mode of a bundle of filaments whose length is equal to the characteristic of "ineffective" length. Thus, the difference between the strengths of a composite and a bundle of filaments is equal to the difference between the strength of a bundle of fibers whose length is equal to the composite length and that of a bundle of fibers whose length is equal to the "ineffective" length. Since the latter is on the order of 10 fiber diameters and the former many multiples of the latter, this is a large effect. A plot of bundle strength, σ_b^*, normalized with respect to composite strength, σ_t^* (the statistical modes of both) is presented in Fig. 12 as a function of the ratio of composite length, L, to "ineffective" length, δ, for a fixed value of the coefficient of variation, μ. The merits of the compositing process are apparent herein.

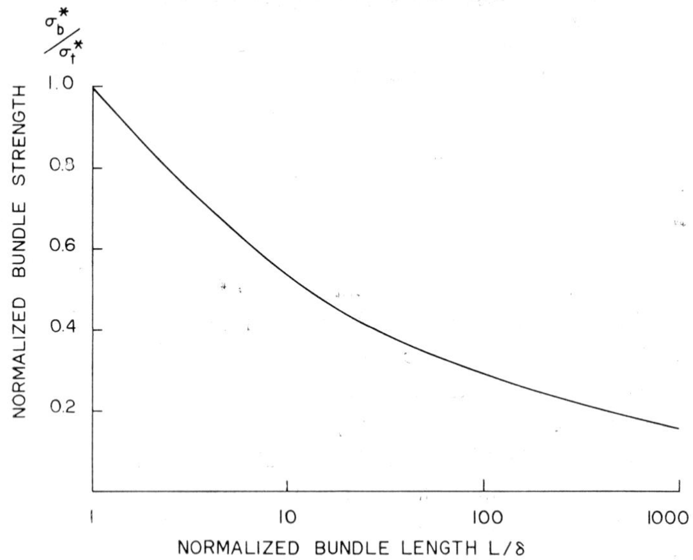

FIG. 12. Effect of length on fiber bundle strength

Next, one would like to investigate other failure mechanisms and determine whether there would be an interaction with the previous failure mechanisms. An aid in this study is provided by some early data on glass-reinforced plastic models.[17] A new interpretation of these data has recently been presented by Zweben.[18] The data were obtained on a specimen having a single layer of glass filaments embedded in an epoxy, loaded in tension, and observed photo-elastically during test so that the internal failure mechanisms could be identified. The specimens are shown schematically in Fig. 13 and an actual specimen

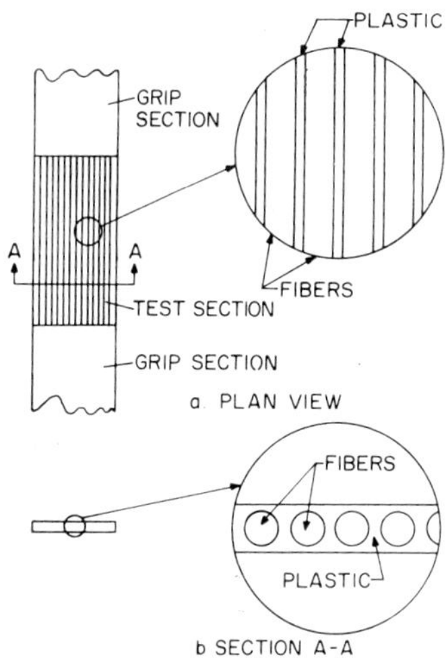

FIG. 13. Experimental tensile failure specimen

is shown in Fig. 14. Dark areas represent zero stress or low stress, whereas bright areas represent high stress. This photograph was taken at 99% of the ultimate load of this particular composite. The distributed fiber fractures are clearly visible and it is clear that the ineffective lengths are short. The characteristic accumulation of breaks as a function of load is shown in Fig. 15 for a typical specimen. Gerard and Becker [19] have suggested that the existence of this cumulative damage is perhaps a means whereby one can assess, by proof-testing, the stress level to which composites may go. Indeed if these are consistent representations, the technique may be a useful one. Figure 16 shows the number of breaks in a series of six fiber specimens that were nominally identical to the one of Fig. 15. The dotted curve is the curve obtained by Zweben for the expected number of breaks in these specimens. This is based on the use of the experimentally observed ineffective length and the distribution function obtained by testing the fibers individually. The expected number of breaks appears to agree reasonably well with the experimental data at higher load levels.

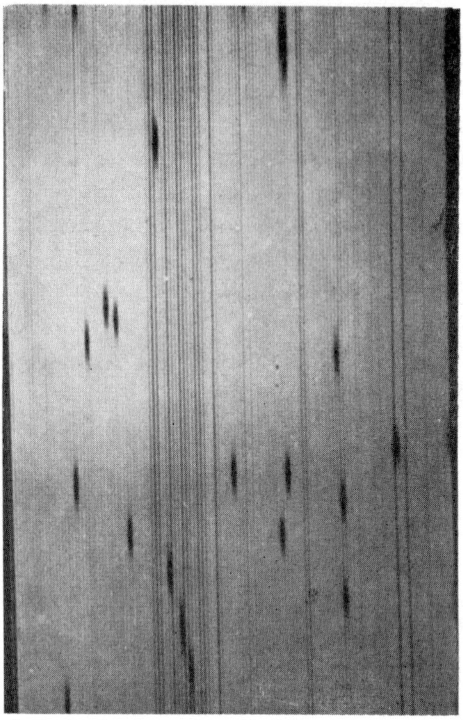

FIG. 14. Typical photoelastic view of tensile failure specimen at 99% of ultimate strength. Fibers are 0.0035 in. diameter E-glass

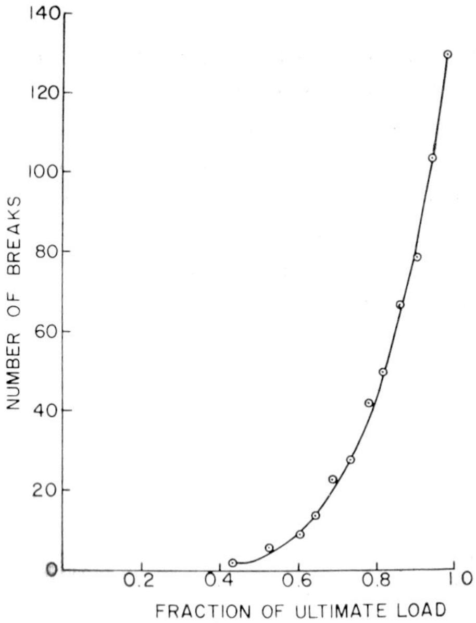

FIG. 15. Number of fiber breaks in specimen similar to that of Fig. 14 as a function of applied load

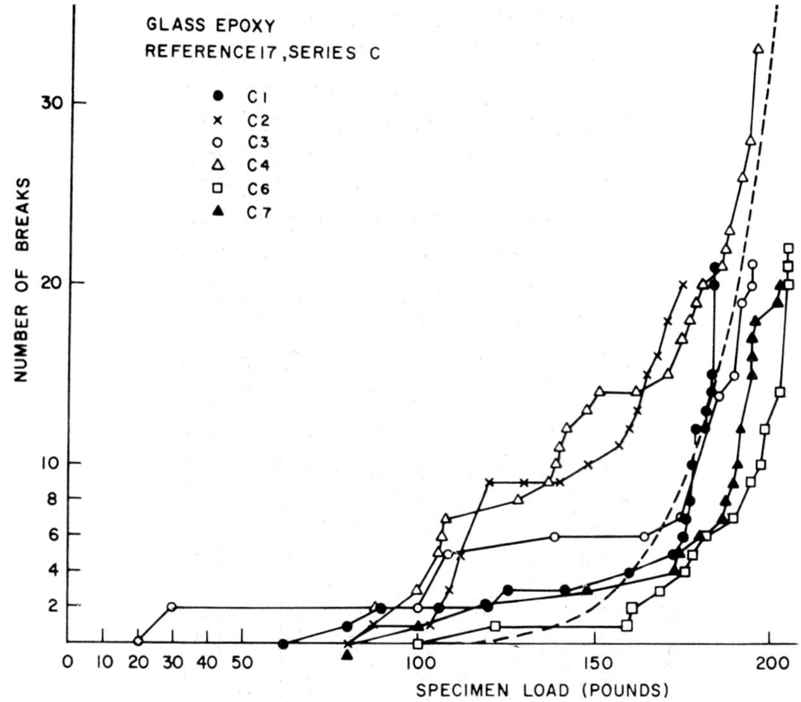

FIG. 16. Number of breaks as a function of applied load (ref. 17, Series C)

The results for these test specimens as compared to the statistical theory are shown in Fig. 17. Composite strength is plotted as a function of the ineffective length ratio. Two different matrix materials were used with identical fibers. The matrix materials differ substantially in their stiffness and strength so that there are different ineffective lengths. These are experimentally observed ineffective lengths and experimentally observed strength values. An alternate evaluation of these tests is presented by Zweben.[18] He examines the possibility of a crack propagation starting to occur at the higher stress levels in the

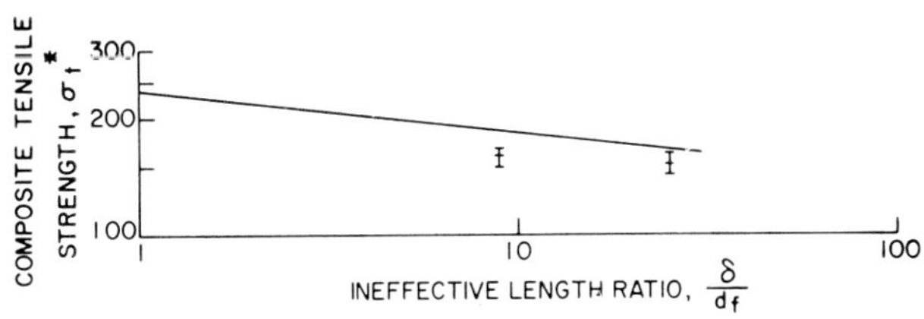

FIG. 17. Comparison of theoretical and experimental composite tensile strengths from data of ref. 17

composite. This is based upon the stress concentrations that are shown in the two-dimensional results of Table 1. Figure 18 is a plot of the expected number, E_n, of groups of broken fibers as a function of the average fiber stress. The band is the range of the experimental data for the specimens shown in Fig. 16. The cumulative damage model predicts a stress higher than the range of experimental failures. The expected number of single breaks in the particular experimental specimen increases rapidly prior to failure as shown previously in Fig. 16. The probability of getting a pair of breaks adjacent to one another, produced by getting the first break resulting from the fiber strength characteristics and the second break resulting from the increased probability of failure in the adjacent fibers because of the stress concentration effect was treated in ref. 18. This result is also shown in Fig. 18 and it is seen that the

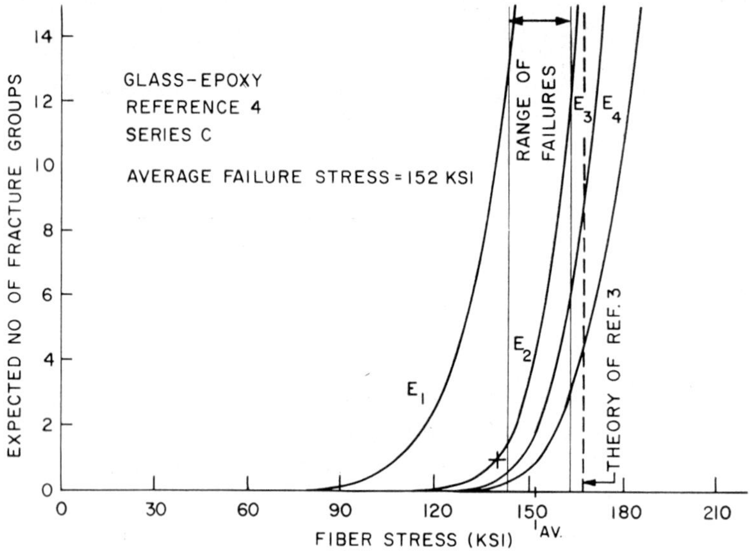

FIG. 18. Expected number of groups of fractures as a function of applied load (ref. 4, Series C)

expected number of multiple fiber fractures rises sharply in the observed range of composite failure. These preliminary results give evidence that the statistical model can be extended to cover the crack propagation phenomenon at incipient failure.†

It is evident that a continuous fiber composite becomes a discontinuous fiber composite prior to failure. Thus, the methods described above can be applied to the important question of discontinuous fibrous composites. Friedman[14, 20] applied the statistical model described previously to a set of fibers which had an initial finite length distribution. Results were normaliz-

† For further discussion of the load concentration effect, see ref. 26.

ed so that the mean fiber length was unity and the fiber lengths were uniformly distributed between $1 - \bar{y}$ and $1 + \bar{y}$ in the sample problem reviewed here. The resulting normalized composite strength, as a ratio to a composite containing continuous fibers of the same strength distribution, is plotted in Fig. 19. This is plotted as a function of this mean fiber length normalized with respect to the ineffective length that is used in the statistical model. The upper curve, for $\bar{y} = 0$, that is, uniform short fibers shows that the strength of the composite builds up quite rapidly as the average fiber length increases and for fiber lengths

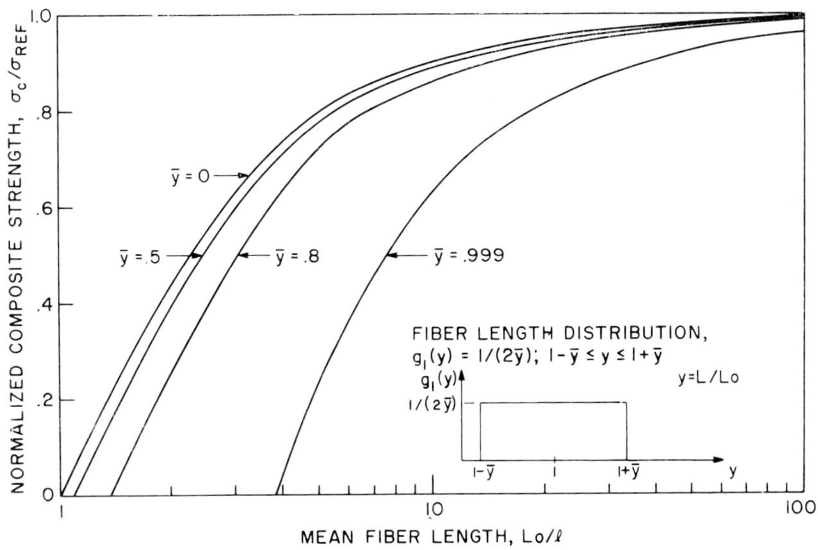

FIG. 19. Effect of mean fiber length on normalized composite strength for discontinuous fiber-reinforced composites (from ref. 14)

equal to ten ineffective lengths, the composite is about 90% of the strength of a continuous composite. Ineffective lengths are on the order of 10 diameters, thus even a fiber 100 diameters in length is a long fiber approaching the efficiency of the continuous fiber composite.

These results can be compared with the experimental data for discontinuous fibrous composites obtained by Kelly and Tyson.[21] Figure 20 shows those data for composites having different initial length to diameter ratios with best fit lines drawn through the data. The ratios of the strength of these discontinuous fiber composites to the strength of the continuous fiber composite made from the same constituents is plotted in Fig. 21 on the curve $\bar{y} = 0$ of Fig. 19. In order to do this, it is necessary to adjust for the difference in definition of ineffective length in refs. 20 and 21 (as shown in Fig. 21). These are experimental points, both in terms of ineffective length and strength level.

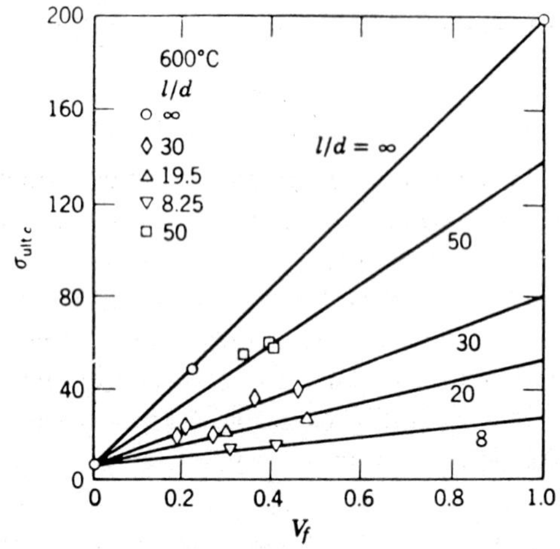

FIG. 20. Experimentally measured values of strengths of discontinuous fiber-reinforced composites (from Kelly and Davies[21])

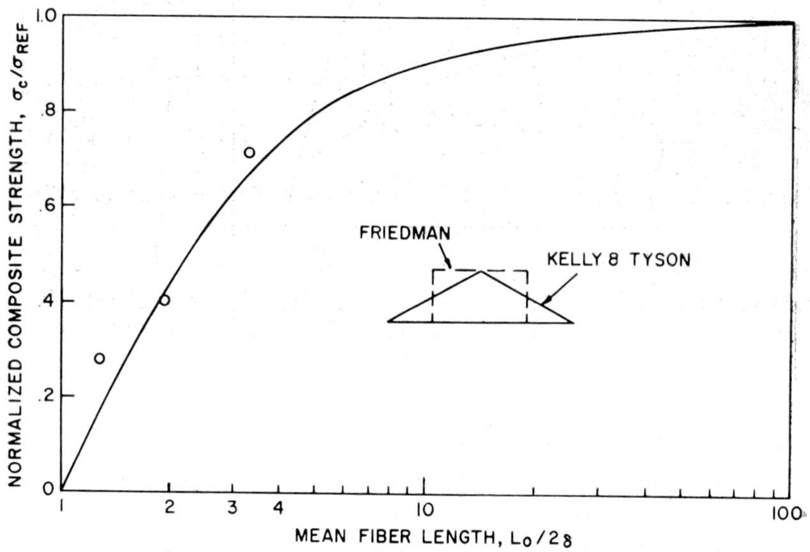

FIG. 21. Variation of strength with fiber length of discontinuous fiber-reinforced composites based on data of Kelly and Davies,[1] and comparison and calculations of Friedman[20]

Again it appears, that the statistical model offers a means to assess the behavior of the composite. In the case of the discontinuous fibrous composite, the introduction of a number of ends in the fabrication process is equivalent to an initial distribution of breaks, as far as the statistics of failure are concerned.

III. COMPRESSIVE STRENGTH

Recent studies[22, 23] have demonstrated the high compressive strength that can be achieved by uniaxial fibrous composites under a load parallel to the filaments. Several analyses of the failure mechanics for this case have been presented and they will be reviewed briefly in this section. It was suggested[24] that the mechanism of failure was a micro-instability of the fibers in a fashion analogous to the buckling of a column on an elastic foundation. It has been demonstrated that this is practical even for a brittle material, such as glass. For example, Fig. 22 from ref. 4 shows single *E*-glass filaments embedded in

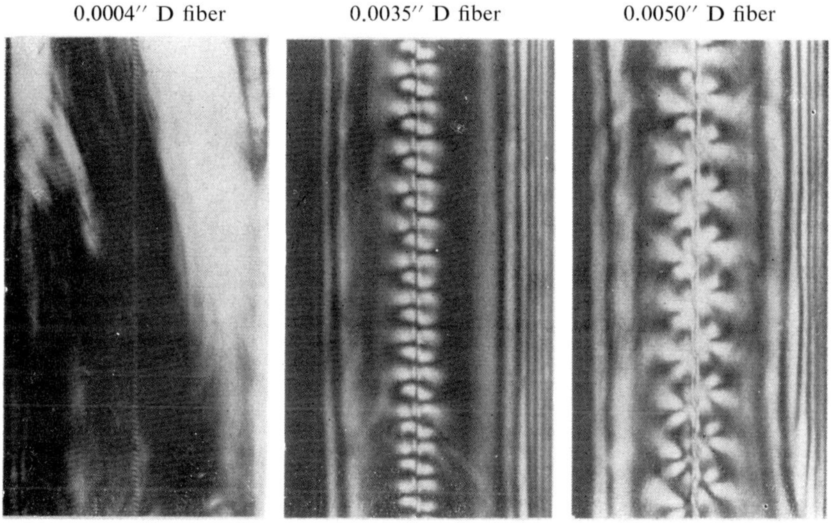

FIG. 22. Photoelastic stress pattern for three individual E-glass fibers imbedded in an epoxy matrix

blocks of Epoxy which have been cooled to produce a compressive strain. The three separate specimens contain fibers of 0.0050 in., 0.0035 in. and 0.0004 in. The photoelastic stress pattern emphasizes the repetitive nature of the deformation pattern for each fiber and supports the contention that the deformations result from an elastic instability.

Analyses of this instability were performed independently in refs. 4 and 22. The analyses treat a layered two-dimensional medium wherein the fibers are assumed to buckle sinusoidally. When adjacent fibers buckle 180° out of phase with each other, the major matrix strains are extensional; when they buckle in phase, the major strains are shear strains. These two modes are denoted the extensional and shear modes respectively.

The resulting compressive strength for the shear mode is given by

$$\sigma_c = \frac{G_b}{1-v_f}. \tag{44}$$

The associated strain is

$$G_c = \frac{1}{v_f(1-v_f)}\left(\frac{G_b}{E_f}\right). \tag{45}$$

The equivalent results for the extension mode are given by

$$\sigma_c = v_f\left[\frac{2E_fE_bv_f}{3(1-v_f)}\right]^{1/2} \tag{46}$$

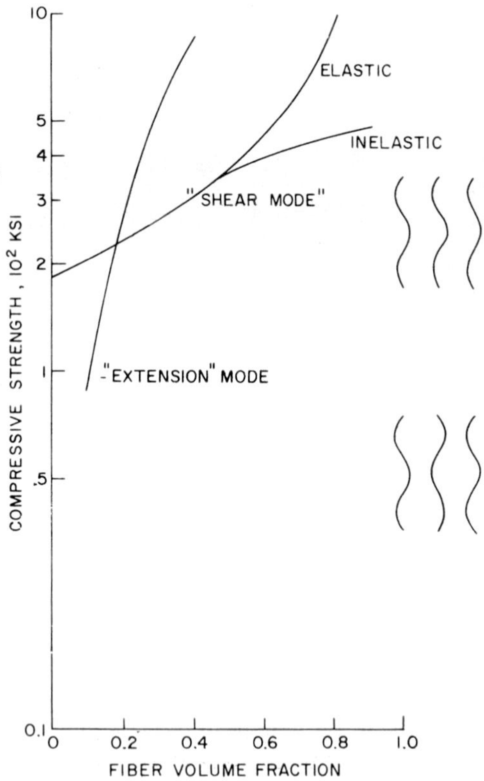

FIG. 23. Compressive strength of glass-reinforced epoxy composites

and

$$G_c = \left[\frac{2v_f}{3(1-v_f)}\right]^{1/2} \left(\frac{E_b}{E_f}\right)^{1/2}. \tag{47}$$

The stresses of eqs. (44) and (46) are plotted in Fig. 23 for glass reinforced plastics as a function of fiber volume fraction. The figure shows that, except for small fiber volume fractions, the shear mode governs composite strength. The failure strain defined by eqs. (45) and (47) are plotted in Fig. 24 which shows that for many composites, the computed failure strain levels exceed the proportional limit strain for the matrix. In an effort to assess the significance of this result the matrix was treated as an elastic perfectly plastic material and the secant modulus was used to define matrix stiffness. For the generally dominant shear mode, this led to the following result, in place of eq. (44):

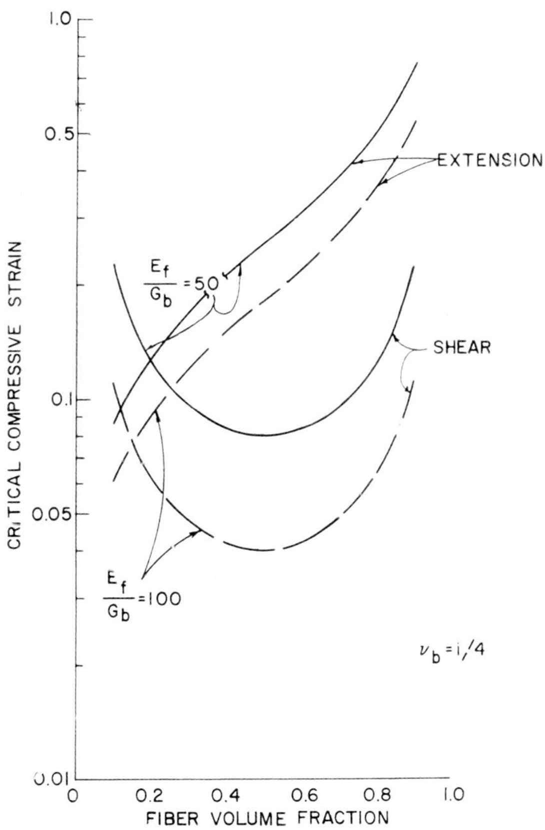

FIG. 24. Critical compressive strain for fibrous composites

$$\sigma_c = \left[\frac{v_f E_f \sigma_y}{3(1-v_f)}\right]^{1/2} \tag{48}$$

where σ_y is the matrix yield stress level.

The stresses of eq. (48) are labelled "inelastic" on Fig. 23.

It should be noted that the internal instability of a layered medium was initially treated in ref. 25. Since the fiber composite instability was studied in refs. 12 and 22 by a two-dimensional approximation of the material geometry, the result of ref. 25 is directly comparable. This result can be written in the form:

$$\sigma_c = \frac{G_b}{V_b}\left[\frac{1+n\dfrac{v_b}{v_f}}{1+n^2+\dfrac{n(v_b-v_f)^2}{2v_b v_f}}\right]^{1/2} \tag{49}$$

where $$n = \frac{G_b}{G_f}.$$

When the fiber to binder modulus ratio is large n approaches zero and eq. (49) reduces to eq. (44).

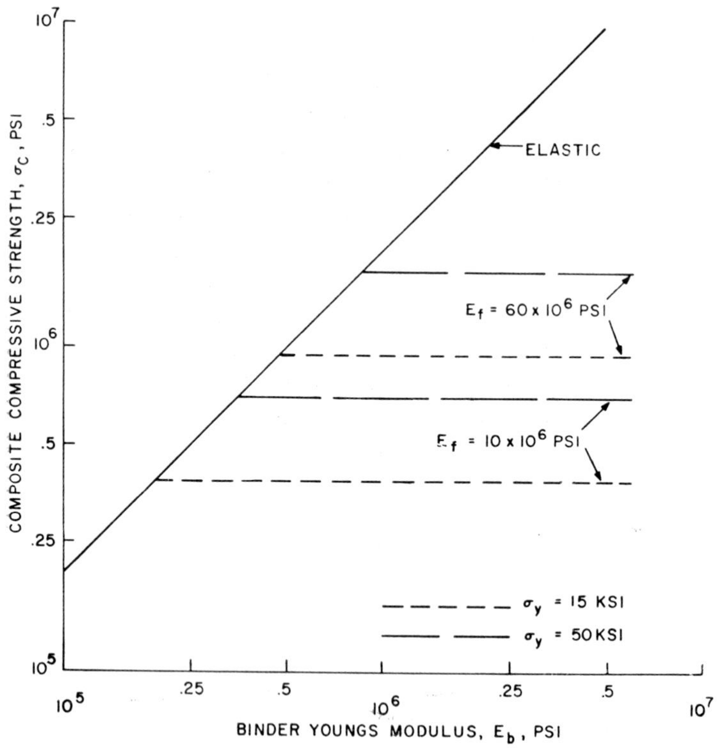

FIG. 25. Influence of matrix properties upon composite compressive strength

The results of these studies, including the elastic-plastic binder can be summarized as shown (Fig. 25). This is a plot of compressive strength of the composite as a function of the binder Young's modulus. For the elastic case, the binder Young's modulus, is the only governing parameter. However, for the inelastic case, there are strength cutoffs which depend both upon the fiber modulus and upon the binder strength. For example, the compressive strength of composites of glass filaments in an epoxy cannot be improved by increasing the binder Young's modulus unless the binder strength is increased to exceed 50 ksi. For higher modulus fibers such as boron or carbon fibers in an epoxy matrix, improvements in composite strength cannot be achieved by increasing binder strength until the binder stiffness is also increased. The message presented in this figure is simply that the nature of changes made in matrix properties to improve the compressive strength of composites of a given fiber depends upon the base of reference. In some cases performance is limited by a binder yield stress at a given fiber modulus, whereas for other cases a gain in compressive strength could be achieved by improving the binder modulus. This approach is intended to present guidance for reasonable changes in matrix properties to yield improved composite strength.

IV. CONCLUSIONS

This paper has attempted to demonstrate the existence of rational methods for the evaluation of failure of uniaxial fiber composites. In tension, these composites were shown to have the potential for strengths normally associated only with very short lengths of fine filaments. Methods for extending this analysis to include discontinuous fibers and to treat other failure mechanisms were indicated. Although only simple fiber strength distributions and ineffective length computations were utilized, the failure model can treat any more realistic fiber-strength distribution function and can utilize any more precise ineffective length assessment. Similarly the simple compressive strength evaluations also show very high potential strengths and these are not in serious disagreement with available experimental data for glass-epoxy composites.

The fact that these uniaxial tensile and compressive strengths are so high, and the fact that in many plate and shell applications the low transverse strengths are very detrimental, seems to point toward the utilization of uniaxial composites. It is valid to ask the question of how one can build composite structures that are only stressed in the fiber direction and only stressed in tension or compression. The indicated potential strength and stiffness of uniaxial composites seems to call for an assessment of the efficiency of truss structures or of thin plate or shell structures with sturdy composite stiffeners or of geodesic dome structures. The concept of load transmission via members stressed in uniform, uniaxial tension and compression may function hand-in-hand with the use of composite materials to permit the construction of large structures of substantially greater efficiency than present techniques and materials alone.

REFERENCES

1. KELLY, A. and DAVIES, G. J. The Principles of the Fiber Reinforcement of Metals, *Metallurgical Reviews*, **10**, 1965.
2. MCDANELS, D. L., JECH, R. W. and WEETON, J. W. *Metal Progress*, p. 118, 1960.
3. MCDANELS, D. L., JECH, R. W. and WEETON, J. W. NASA TN D-1881, 1963.
4. COLEMAN, B. D. On the Strength of Classical Fibres and Fibre Bundles, *Journal of the Mechanics and Physics of Solids*, **7**, 1958.
5. ROSEN, B. W. Tensile Failure of Fibrous Composites, *AIAA Journal*, **2**, No. 11, November 1964.
6. PARRATT, N. J. Defects in Glass Fibers and Their Effect on the Strength of Plastic Mouldings, *Rubber and Plastics Age*, March 1960, pp. 263–266.
7. DOW, N. F. Study of Stresses Near aDiscontinuity in a Filament-Reinforced Composite Material, General Electric Company, Space Sciences Laboratory, Space Mechanics Memo No. 102, January 1961 (also TIS R63SD61, August 1963).
8. COX, H. L. The Elasticity and Strength of Paper and Other Fibrous Materials, *British Journal of Applied Physics*, Aug. 1951.
9. SADOWSKY, M. A. Transfer of Force by High Strength Flakes in a Composite Material, Watervliet Arsenal, WVT-RR-6105-R, June 1961.
10. ROSEN, B. W., DOW, N. F. and HASHIN, Z. Mechanical Properties of Fibrous Composites, NASA CR-31, August 1963.
11. HEDGEPETH, J. M. Stress Concentrations in Filamentary Structures, NASA TN D-882, May 1961.
12. ROSEN, B. W. Mechanics of Composite Strengthening, in *Fiber Composite Materials*, American Society for Metals, Metals Park, Ohio, 1965.
13. METCALFE, A. G. and SCHMITZ, G. K. Effect of Length on the Strength of Glass Fibers, ASTM 1964, Preprint No. 87, June 1964.
14. FLOM, D. G., MAZZIO, V. F. and FRIEDMAN, E. Whisker Reinforced Resin Composites, Air Force Report, AFML-TR-66-362, 1967.
15. GUCER, D. E. and GURLAND, J. Comparison of the Statistics of Two Fracture Modes, *Journal of the Mechanics and Physics of Solids*, pp. 365–373, 1962.
16. DANIELS, N. W. The Statistical Theory of the Strength of Bundles of Threads, *Proceedings of the Royal Society*, p. 405, 1945.
17. DOW, N. F. and ROSEN, B. W. Evaluations of Filament-Reinforced Composites for Aerospace Structural Applications, NASA CR-207, April 1965.
18. DOW, N. F., ROSEN, B. W., SHU, L. S. and ZWEBEN, C. H. Design Criteria and Concepts for Fibrous Composite Structures, NASA Final Report, NASw-1377, July 1967.
19. BECKER, H. and GERARD, G. Failure Law for Filamentary Composites, *AIAA Journal*, **4**, No. 12, Dec. 1966.
20. FRIEDMAN, E. A Tensile Failure Mechanism for Whisker Reinforced Composites, SPI-RPD 22nd Annual Conference, Paper 4A, February 1967.
21. KELLY, A. and TYSON, W. R. Fiber-Strengthened Materials, in *High Strength Materials*, V. F. Zackay, ed., J. Wiley, New York, 1965b.
22. SCHUERCH, H. Prediction of Compressive Strength in Uniaxial Boron Fiber-Metal Matrix Composite Materials, *AIAA Journal*, **4**, Jan. 1966.
23. FRIED, N. and WINIANS, R. C. Compressive Strength of Parallel Filament-Reinforced Plastics Development of a New Test Method, in ASTM STP 327, Baltimore, Md., March 1963.

24. Dow, N. F. and Gruntfest, I. J. Determination of Most Needed Potentially Possible Improvements in Materials for Ballistic and Space Vehicles, General Electric Company, Space Sciences Laboratory, TIS R60SD389, June 1960.
25. Biot, M. *Mechanics of Incremental Deformations*, John Wiley and Sons, Inc., New York, 1965.
26. Zweben, C. and Rosen, B. W. A Statistical Theory of Material Stregth with Application to Composite Materials, *AIAA* Paper 69–123, Jan. 1969.

THE CONTRIBUTION TO THE WORK OF FRACTURE OF A COMPOSITE MATERIAL OF "PULL-OUT" OF FIBERS

G. A. COOPER and A. KELLY

Department of Metallurgy, University of Cambridge

ABSTRACT

Experiments to investigate the contribution to the work of fracture of the pulling out of fibers from the matrix are described for the system of copper reinforced with tungsten wires.

WHEN a fiber-reinforced material fails in tension, a contribution to the work of fracture may arise from the pulling out of the fibers from holes in the matrix. According to simple theories of this process,[1, 2] a metallic matrix, reinforced by parallel short fibers which fails by pull-out of the fibers, will show a failure stress which increases linearly with the length of overlap of adjacent fibers (l, see Fig. 1) and a work of fracture which increases as the square of this quantity. In the simple theories it is assumed that the plastic properties of the matrix and of the fiber–matrix interface remain constant during pull-out, so that both work hardening and failure in shear are neglected. We have attempted to investigate pull-out by a series of very simple experiments.

Composites of tungsten wires in vacuum-cast copper were made of cross-sectional dimensions 13×0.5 mm, reinforced with eighteen 0.5 mm diameter tungsten wires (a volume-fraction of 43%). Each of the reinforcing wires was broken near the center of the composite, but the breaks were staggered so as to give an effect of interlocking combs (Fig. 1). The amount of overlap l was varied from specimen to specimen. Tests were also carried out on composites containing the same volume-fraction of continuous reinforcement.

At room temperature, the "critical transfer length" l_c of the tungsten–copper system is about three fiber diameters[3] and so tests were carried out at higher temperatures, in order to give a longer transfer length and thus make the effects of pull-out more noticeable. Tests were made at 250°C under silicone oil and at 600°C in argon, at which temperatures the tungsten reinforcement is ductile. Testing was carried out in an Instron machine, at a

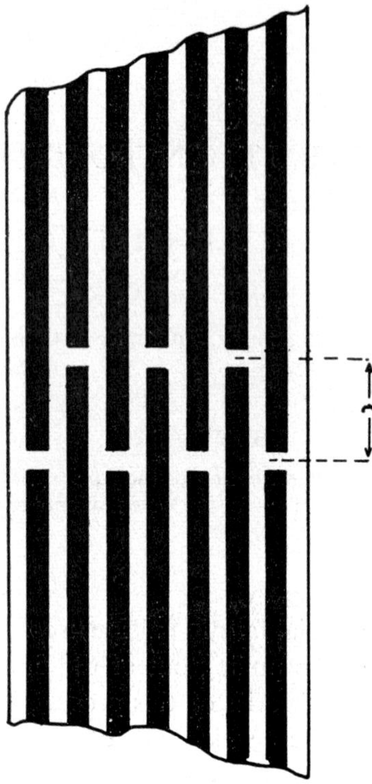

FIG. 1. Schematic view of pull-out test specimen, showing l, the overlap distance

strain rate of 0.05 cm/min. The ultimate load and total area under the load-extension curve were found for each specimen, from which the nominal breaking stress and work of fracture were calculated. The results for the tests at 250°C are shown in Figs. 2 and 3, and for 600°C in Figs. 4 and 5. The breaking stress and work of fracture are plotted against the dimensionless parameter l/d, where d is the fiber diameter. At both temperatures, the breaking stress increased approximately linearly as the overlap distance increased, until the fibers were too deeply interleaved to slide apart: fracture then occurred without pull-out, along one of the two lines of fiber breaks. The work of fracture increased until this same critical overlap distance, and thereaefte made a discontinuous change to a constant value.

Figures 2 and 3 show the variation of breaking stress and work of fracture for varying degrees of overlap at 250°C. Of the two specimens with zror overlap, the circular point represents a specimen made by the usual method, whereas the triangular point refers to a specimen originally cast with continuous wires. After removal from the mold, the wires were broken on one cross-section by bending over a knife-edge. This procedure produced a specimen in which there was no bond at the ends of the tungsten wires, but in

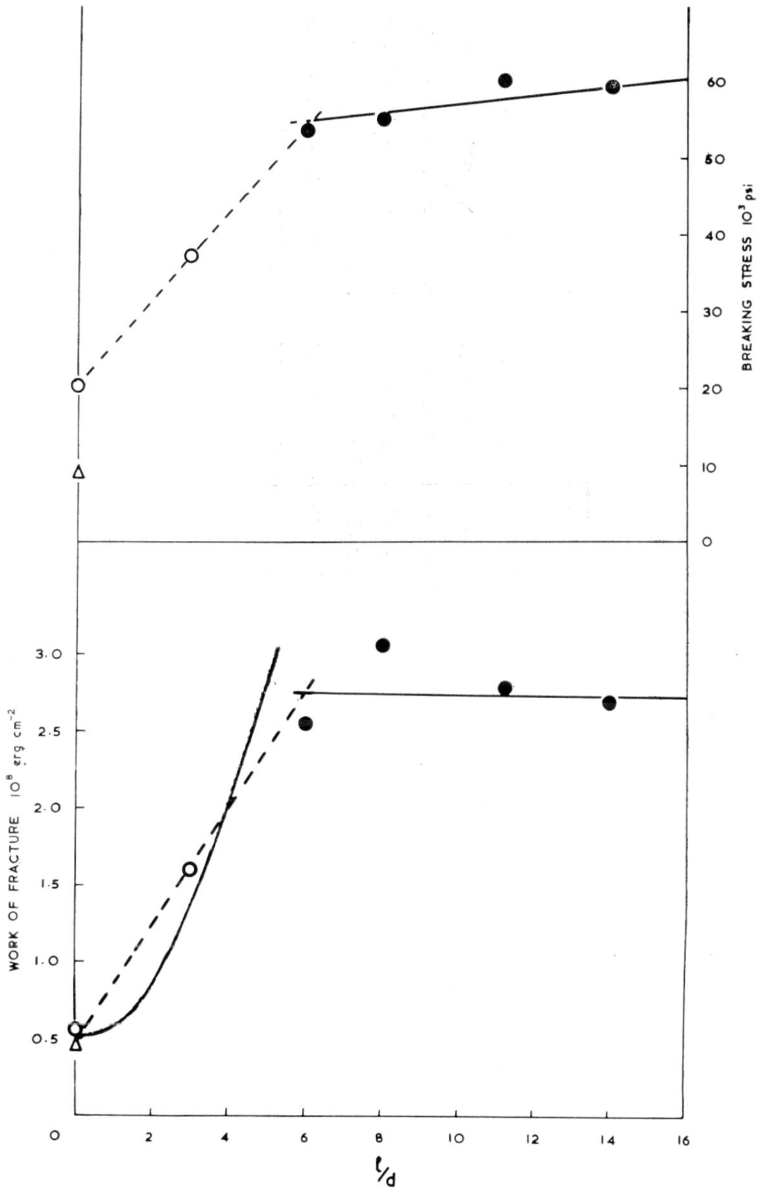

FIG. 2 (top). Variation of breaking stress with overlap at 250°C

FIG. 3 (bottom). Variation of work of fracture with overlap at 250°C

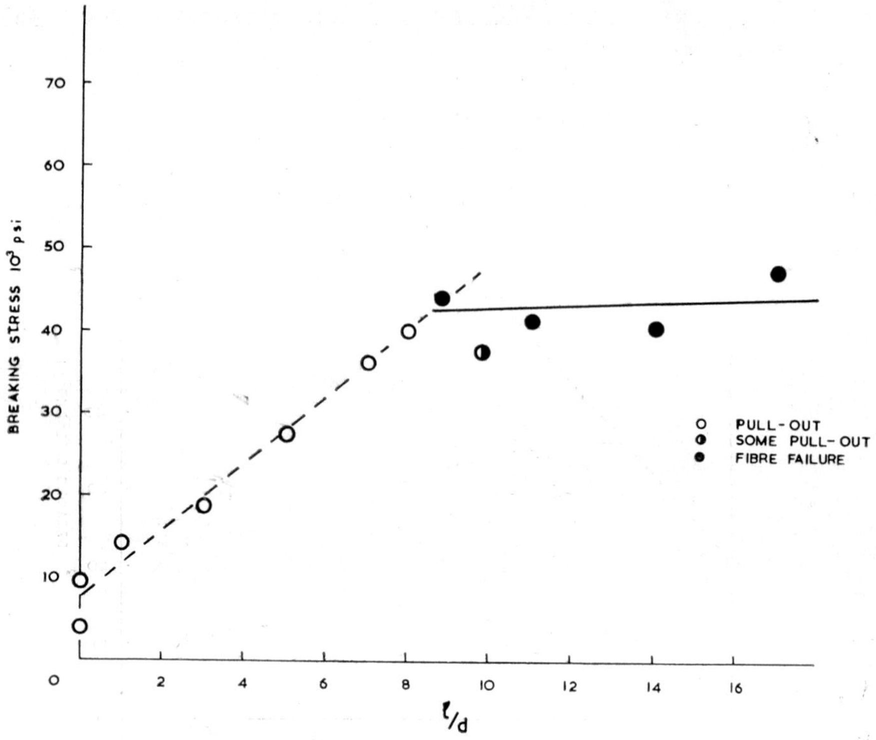

FIG. 4. Variation of breaking stress with overlap at 600°C

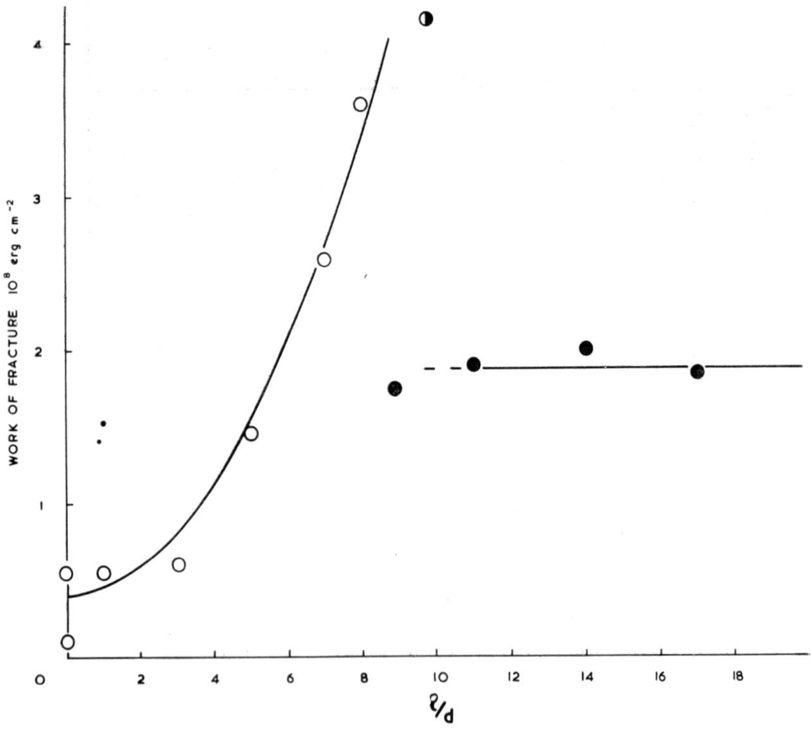

FIG. 5. Variation of work of fracture with overlap at 600°C

which the copper matrix was little affected. In Figs. 2 and 3, the work of fracture is not greatly altered by the lack of end-bond, but the breaking stress is reduced to 0.40 of the value obtained with an end bond; the simplest theory predicts that the ratio should be $(1-V_f)$, i.e. 0.57.

Figure 6 shows the appearance after fracture of specimens with l/d 3 and 6, respectively. The former failed by pull-out. The latter shows more complex behavior. Pull-out had begun, as there is a row of holes right through the

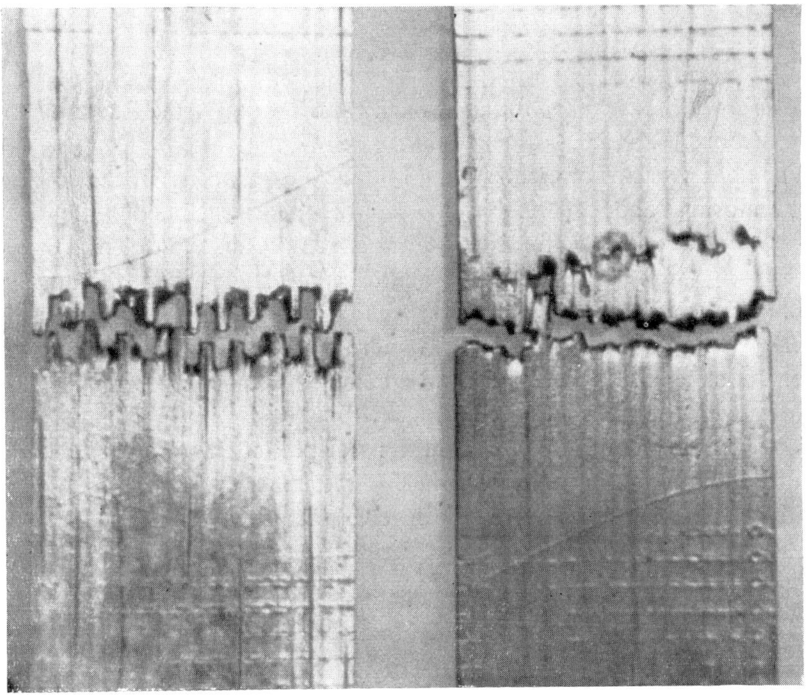

FIG. 6. Pull-out at 250°C. At left $l/d = 3$; at right, $l/d = 6$.

specimen, along the line of fiber ends far from the final fracture surface, but subsequently the specimen broke by failure of the fibers at the other row of fiber breaks. This behavior arises because of work hardening in the matrix and consequent increase in the shear strength of the matrix parallel to the fibers over the length l. This work hardening increases the load necessary to continue pull-out (even though there is a decrease in the length of overlap) so that the breaking stress of the composite is attained along one of the rows of fiber breaks (which are weak sections).

The critical overlap necessary for the fibers to be broken, rather than to pull out, is approximately six fiber diameters.

During pull-out, provided the shear stress τ_y is constant, the nominal breaking stress of the composite is given by:

$$\sigma = \frac{Nd\tau_y l}{A}$$

where l is the amount of overlap, d equals the thickness of the specimen (in this case equal to the fiber diameter), A is the cross-sectional area and N is the number of regions of shearing contact. In these specimens, there were 18 fibers, so $N = 17$, and we obtain from the slope of Fig. 2,

$$\tau_y = 8700 \text{ psi.}$$

This is within the range found by Kelly and Tyson[4] from fracture surface observations (7.4–13.6×10^3 psi).

For fiber overlaps of greater than 3 mm (l/d 6), the UTS of the composite was uniform at about 60,000 psi. To account for this we note that the strength of the continuously reinforced material is 92,000 psi. We expect a strength equal to the sum of half this latter figure, representing the continuous fibers, plus half the strength of a specimen containing end-bonded fibers with zero overlap, to account for the cohesion across the end faces of the separated fibers. Such a sum gives a calculated breaking stress of 56,000 psi, in reasonable agreement with observation.

The work of fracture under conditions of pull-out was found to increase uniformly with the degree of overlap. If there is no work-hardening in the matrix, we may write at any stage in the pulling-out process,

$$F(x) = Nd\,\tau_y x$$

where x is the instantaneous overlap distance. Thus

$$\int_0^l Nd\,\tau_y x\,dx$$

is the work done during pull-out. Integrating, we obtain for the work of fracture per unit area of the composite,

$$W = \frac{Nd^3\tau_y}{2A}\left(\frac{l}{d}\right)^2 + \text{const.}$$

The work done will thus be a quadratic function of the overlap distance, and the constant represents the work of fracture not attributable to pull-out, i.e. the work done in separating the matrix when there is zero overlap. There is no way of directly estimating the constant, but a value for $\frac{Nd^3\tau_y}{2A}$ may be obtained from the measured variation of fracture stress with overlap distance if we assume no work-hardening in the matrix. In the case of the tests at 250°C it is not possible to draw a "best parabola" through the two points corresponding to overlaps of 0 and 1.5 mm ($l/d = 0$ and 3) but taking $\tau_y = 8700$ psi from Fig. 2, a predicted line has been constructed, with equation

$$W = 5\times 10^7 + 9.9\times 10^6 \left(\frac{l}{d}\right)^2 \text{ ergs/cm}^2.$$

The constant is the measured work done at zero overlap. The constructed curve is not inconsistent with the observations, but due to the small number of points only qualitative agreement with theory is found.

The value for the work of fracture of the composite for overlaps greater than the critical shows a constant high value (2.75×10^8 ergs/cm^2). This is because the reinforcement is ductile at 250°C. In the case of a composite reinforced with a *brittle* fiber without a pronounced length-strength dependence, we should expect an approximately plane fracture surface, and so a work of fracture falling to approach that for a "zero overlap" specimen.

At 600°C, the matrix is weaker in shear than at 250°C, and pull-out occurs over greater distances (Fig. 7). As with the tests at 250°C, the breaking stress of the composite increased uniformly until the critical pull-out distance was reached. The fracture stress thereafter remained constant, at a slightly lower value than that measured in the 250°C tests ($\sim$44,000 psi, Fig. 4). The continuously reinforced composite was found to have a UTS of 72,000 psi and so, noting that the breaking stress of a "zero overlap" specimen was 9500 psi, and following the same arguments as above, we obtain a value of $\sim$41,000 psi as the expected breaking stress for overlapped specimens when l/d is greater than the critical value. The agreement between theory and experiment is satisfactory.

The gradient of the line in the pre-critical region of Fig. 4 ($l/d < 9$) is 4000 psi. This gives a value for the shear flow stress of the matrix of 6100 psi, which is somewhat larger than the value found by Kelly and Tyson[4] (3.2–4.6×10^3 psi).

The work-of-fracture data are more informative than those obtained at 250°C. A "best parabola" fit to the data obtained under "pull-out" conditions in Fig. 5 has equation

$$W = 4\times 10^7 + 4.8\times 10^6 \left(\frac{l}{d}\right)^2 \text{ erg/cm}^2.$$

The constant 4.8×10^6 gives a value of τ_y of 4200 psi. This is smaller than the value found from the data shown in Fig. 4, and agrees better with the values obtained by Kelly and Tyson.[4]

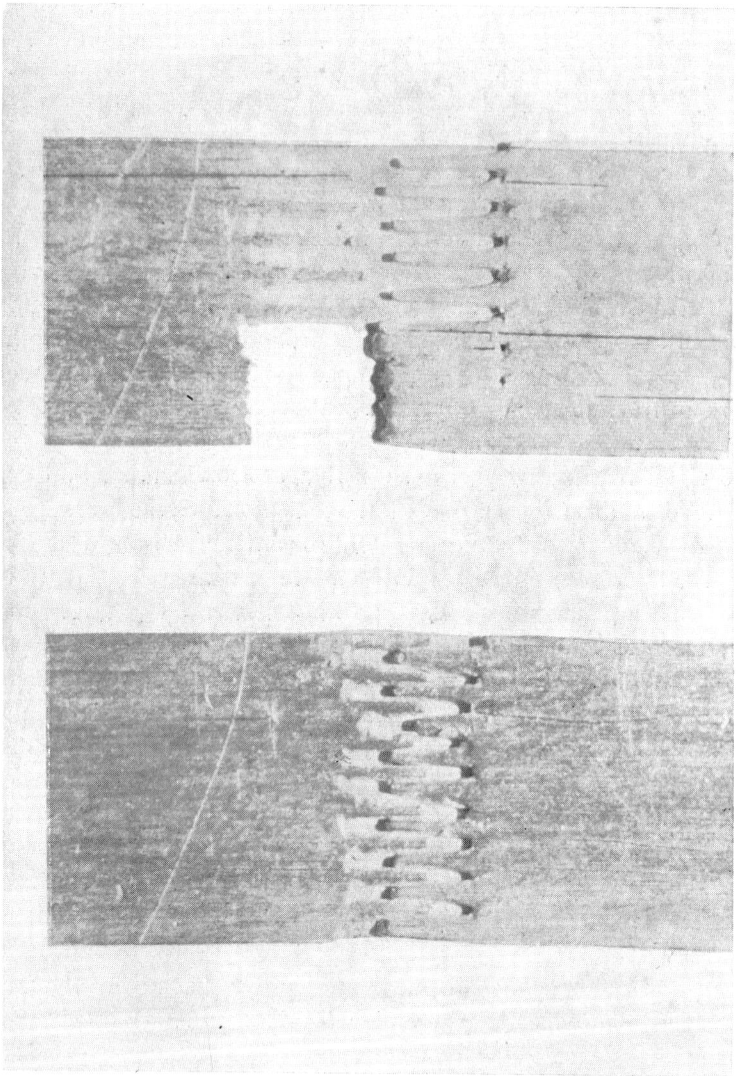

FIG. 7. Pull-out at 600°C. At left, $l/d = 7$; at right, $l/d = 10$

It is to be expected that τ_y derived from measurement of the work of fracture could be lower than the value obtained from the data in Fig. 4. The latter value is derived from the breaking stress of the composite and so is a *maximum* value, whereas the former, derived from total fracture work, must represent some *average* value of the shear stress during pull-out.

As in the case of the tests at 250°C, the work of fracture becomes approximately constant, for l/d greater than 10 (when fiber fracture occurs). The value of the work of fracture in this case is higher than would be expected in the case of a brittle reinforcement, but, as discussed above, is due to the ductility of the reinforcement.

Despite the ductility of the reinforcing fibers, the data shown in Fig. 5 and particularly the points with l/d between 6 and 10 establish experimentally that if care is taken to maximize the amount of pull-out, then the work of fracture *exceeds* the sum of that of the fibers and that of the matrix. For values of (l/d) less than 10 the stresses on the composite are not such as to produce any significant plastic flow of the fibers. The contribution of the fibers to the total work of fracture is therefore negligible in this region and we see therefore that the work of fracture is increased by a factor of about ten by arranging to increase the distance over which pull-out occurs. The experiments described here have, for convenience in using the copper tungsten system, been carried out with a reinforcing fiber which shows some ductility, but the importance of pull-out in increasing the work of fracture of composites containing brittle fibers has been demonstrated.

Qualitatively simple models (1), (2) of the pull-out process agree with the experiment but Fig. 6 demonstrates that work hardening of the matrix must be considered if a reliable estimate of the distances over which pull-out will occur is to be made from the plastic properties of the matrix.

REFERENCES

1. COTTRELL, A. H. *Proc. Roy. Soc.*, **A282**, 2 (1964).
2. KELLY, A. *Strong Solids*, pp. 160 ff. Clarendon, Oxford, 1966.
3. TYSON, W. R. Ph.D. Thesis, Cambridge, 1964.
4. KELLY, A. and TYSON, W. R. *J. Mech. Phys. Solids*, **13**, 329 (1965).

EXPERIMENTAL INVESTIGATIONS ON THE DYNAMIC STRENGTH OF COMPOSITES†

S. R. Bodner and J. M. Lifshitz

Technion—Israel Institute of Technology, Haifa, Israel

ABSTRACT

A number of experimental techniques for investigating various dynamic properties of composite materials are reviewed. In order to study filler-binder interaction during straining, especially the onset of unbonding, an apparatus was developed for the continuous measurement of the dynamic modulus and damping during constant rate of strain tests. The relation of changes in the dynamic modulus and damping to the development of failure mechanisms is discussed. Various methods of dynamic visco-elastic characterization and strength determination are reviewed. These include stress wave techniques for the investigation of the fracture strength at high loading rates. The various experiments were performed on samples of an unfilled and quartz particle filled epoxy resin. The results of these studies are also reported.

I. INTRODUCTION

Experimental procedures for material characterization and strength determination at high rates of loading have been extensively developed in recent years. These techniques are largely concerned with the time dependent response of metals and plastics under both uniaxial and multiaxial loading. A number of the procedures for the characterization and strength determination of viscoelastic materials have been presented in the preceding Symposium on Naval Structural Mechanics.[1] Surveys of experimental methods for investigating the constitutive equations of metals at high rates of loading have been presented recently at an ASME Conference.[2] Most of these techniques are equally applicable to composite materials and can serve to provide much of the essential information.

It appears, however, that certain information relating to the binder-filler interaction would require experimental methods particularly suited to this problem. For example, it would be useful to measure the changes in the dynamic modulus and damping of the composite as it is strained. Such measurements should be sensitive to microstructural changes such as un-

† The research reported in this paper has been sponsored by the Air Force Office of Scientific Research under Contract AF 61(052)-951 through the European Office of Aerospace Research (OAR), United States Air Force.

bonding of filler particles from the matrix or fracturing of the matrix. They would complement the results of dilatational tests in such investigations.

The present paper discusses an experimental program for determining the dynamic mechanical characteristics of a quartz particle filled epoxy resin. For this composite, the filler action is primarily hydrodynamic and does not supply the strength enhancement associated with fiber reinforcement. Nevertheless, the specimens can serve to illustrate the experimental techniques which are equally applicable to fiber reinforced materials.

A dynamic viscoelastic characterization of filled and unfilled specimens was performed using some of the methods described by Kolsky in his survey.[3] The dynamic strength of the composite was examined by the stress wave technique, that is, the stress and strain necessary to cause fracture by spalling were determined. Stress strain relations at relatively low strain rates were also obtained.

An apparatus was developed to continually measure the changes in the dynamic modulus and damping during constant rate of straining tests. The apparatus is capable of measuring either the torsional, extensional or bending modulus and the associated damping value. It could also be used in creep or stress relaxation tests. The results of these measurements were found to be very informative on the failure processes in the composite.

II. PREPARATION OF SPECIMENS

An epoxy mix of 1 : 1 Shell Epicote 815 and Versamide 140 (by weight) was used as the matrix material. Natural silica particles (Ottawa sand, ASTM specification C-109) was used as the filler. The sand particles had a mean diameter of 0.4 mm and the density was approximately 2.63 g/cm^3. The elastic constants of this sand appear in ref. 8 as well as other references. The filled specimens used in the tests described in this paper had a 1 : 1.5 matrix to filler weight ratio which corresponds to a volumetric filler content of 38%.

To prepare the filled specimens, the three ingredients were mixed under vacuum for about one and a half hours after which the liquid mix was poured into cylindrical Teflon molds. Two axial springs in series with two seals were compressed into the molds to provide pressure and to avoid leakage. To prevent settling of the heavier sand particles during the initial curing phase, the molds were slowly rotated about their horizontal axis for 24 hr at 23°C. The molds were then cured at 100°C for 4 hr and slowly oven cooled for 24 hr, after which the specimens were removed from the molds. The pure epoxy specimens were prepared in a similar manner but were not rotated during the initial curing phase.

Aluminum end pieces were bonded to the tensile specimens by means of a cast epoxy connecting section. The heat treatment was then repeated and the specimens were tested eight days afterwards. Tensile failure of these specimens always occurred somewhere in the middle of the length and not at the ends. The tensile test specimens and those used for the superimposed vibration tests were 9.8 mm in diameter and 11 cm long between the end fillets. The specimens used for viscoelastic characterization by the resonant bar technique were 1.27 cm in diameter and 50 cm long. Those used for the dynamic strength tests (by explosive spalling) were 1.27 cm in diameter and generally about 17.5 cm in length.

The glass transition temperature of the epoxy was estimated to be about 40°C from associated tests. The presence of sand filler is not expected to significantly alter the glass transition temperature, e.g., Schwarzl.[4] All the tests were conducted at 23°C so that the material was in the glassy state.

III. EXPERIMENTAL PROCEDURES

A. *Dynamic Viscoelastic Characterization*

The resonant beam and rod technique of Bodner and Kolsky,[5] Fig. 1, was used to obtain the dynamic longitudinal modulus and damping over the frequency range from 50 cps to 2000 cps. The procedure is based upon determining the natural frequencies of the specimens from forced vibration tests. Since the damping of the material is fairly small, the classical elastic relations (with suitable correction factors[5]) could be used to compute the dynamic longitudinal modulus from the natural frequencies. Coils on the specimen ends in the vicinity of magnetic fields were used to drive the specimen and to detect the motion (position A, Fig. 1, for lateral modes and position B for longitudinal vibrations). The internal damping was measured from the decay of the free vibrations at a natural frequency. Lateral vibration modes were determined to 1300 cps (first to fifth harmonics). The first harmonic in the longitudinal mode was a little over 2000 cps.

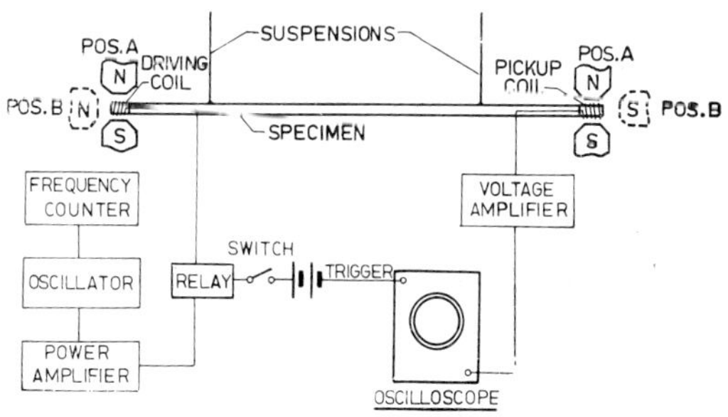

FIG. 1. Experimental arrangement for forced and free vibrations to obtain dynamic viscoelastic properties

Stress pulse transit times were used to determine the dynamic modulus at frequencies in the range of $5-7\times10^4$ cps. Small lead azide charges were detonated at one end of the specimen and the passage of the stress pulses was observed by strain gauges and by an air gap condenser microphone (Fig. 2). The results for the dynamic longitudinal modulus E and the associated damping (logarithmic decrement Δ) for the filled and unfilled specimens are shown in Figs. 3 and 4.

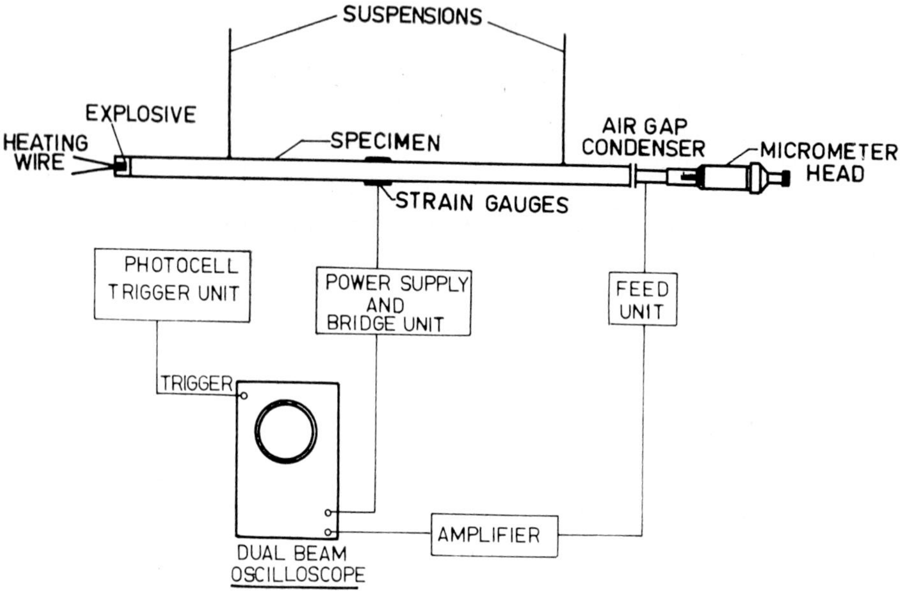

FIG. 2. Experimental arrangement for pulse propagation and determination of dynamic fracture strength by explosive spalling

B. *Strength Properties*

Tensile tests at various rates of straining were performed on filled and unfilled specimens using an Instron testing machine. The displacement of the ends could be used as the strain measure with fairly good accuracy. Typical stress–strain curves are shown in Fig. 5.

The strength at a very high rate of loading was examined by detonating an explosive charge (lead azide) at one end of a rod of the specimen material and measuring the transient strain values prior and subsequent to spalling fracture at the opposite end by means of strain gauges (Fig. 2). By varying the length of the specimen rods and the location of the gauges, it was possible to measure the strain value of the stress pulse retained in the rod immediately after spalling fracture occurred near the free end of the specimen. This would

FIG. 3. Dependence of dynamic extensional modulus (E) on frequency for filled and unfilled specimens

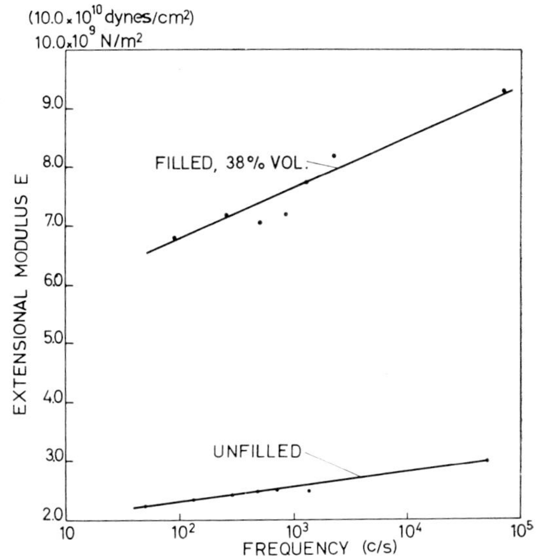

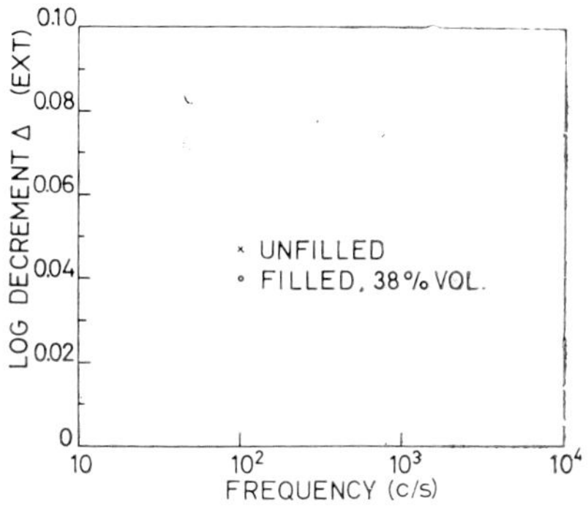

FIG. 4. Frequency dependence of internal damping of filled and unfilled specimens

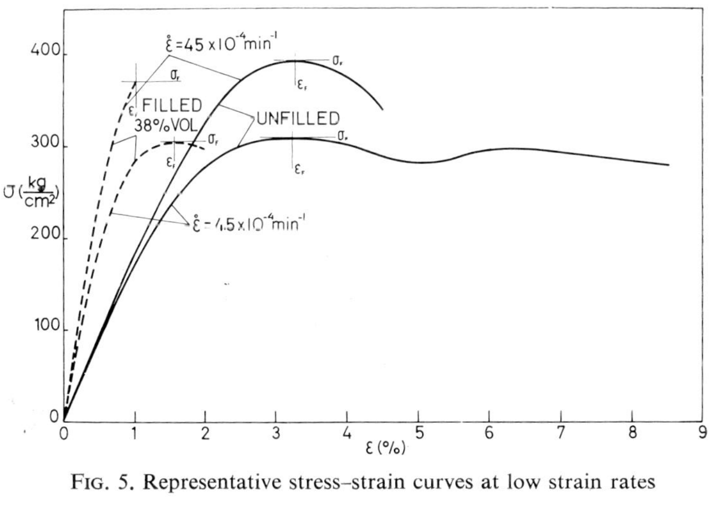

FIG. 5. Representative stress–strain curves at low strain rates

be very close to the tensile fracture strain. Since the material is almost completely elastic at these high rates, this would also give the fracture stress value. It is known that wire resistance strain gauges generally experience a response time lag, but the amplitude rather than the arrival time is important in these tests. For the time of transit measurements, the observation of the elapsed time of arrival of the stress pulse front between successive reflections would effectively eliminate the inherent time error.

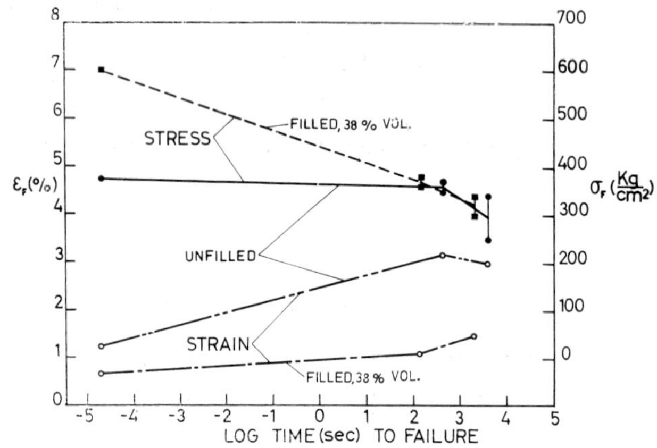

FIG. 6. Dependence of the failure stress and strain on the loading time

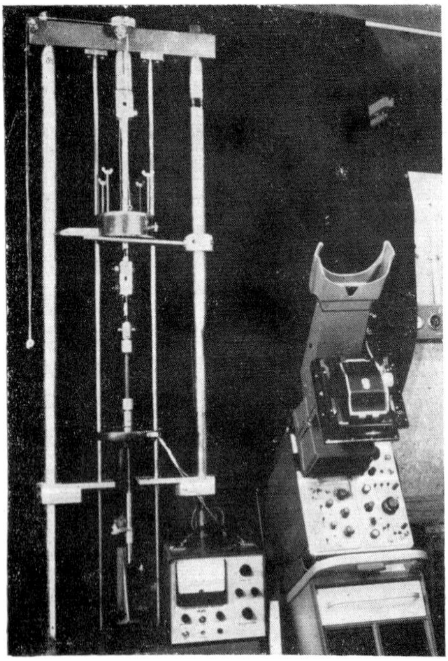

FIG. 7. Drop tester for dynamic stress–strain measurements

The results of the strength tests are plotted in Fig. 6 in terms of the time to reach failure. The slow rate tests on the Instron machine and the stress pulse tests enable over seven decades of time to be represented. To obtain measurements at times of about 10^{-2} sec, various high rate of loading machines can be used. A number of examples of such machines are described in ref. 6. A simple version of such a machine is a drop tester in which a weight drops from a height onto a platform attached to the lower end of the specimen. A load cell connects the upper end of the specimen to the supporting frame and an LVDT could be used to measure the displacement of the platform. The load cell output and the specimen deformation (platform displacement) are simultaneously measured throughout the rapid loading history. An apparatus of this type with a specimen of the composite material in place is shown in Fig. 7.

C. *Changes in Dynamic Moduli and Damping During Straining*

The method used to measure the changes in the dynamic moduli and associated internal damping during straining was based upon utilizing the specimen as the spring component of a simple one degree of freedom spring mass system. The apparatus is based upon one used by Bodner[7] to examine stress induced resonances in metals. Two specimen rods were rigidly connected to a central inertial mass as shown in Fig. 8. The specimens and central mass therefore form the spring mass system. The system could be forced to vibrate in either the torsional, extensional, or bending mode by appropriately locating the magnets with respect to the driving coils on the extension arms of the central mass. The central mass and the extension arms weighed 300 g and had a polar moment of inertia of 8.1 g cm sec^2 for the torsional vibration tests, while a 1600 g mass was used for the extensional vibration tests. The response of the system is indicated by the voltage generated by the output coils. The natural frequency and damping of the system in each of the modes could thereby be readily determined. The dynamic modulus was computed from the natural frequency value using the elastic relation since the damping was sufficiently small. The changes in the specimen dimensions due to straining were considered in computing the moduli values.

Measurements of the natural frequency and the damping could be made while the specimens were strained in an Instron machine (Fig. 9). The frequency of maximum response to imposed vibrations was continually followed during the tensile test. An electronic counter was necessary to accurately determine the changes in the resonant frequency. The driving voltage was cut at frequent intervals and the vibration decay was recorded on an oscillographic recorder or on an oscilloscope.

Some care had to be taken to avoid extraneous resonances and damping sources so that it was necessary to effectively isolate the vibrating system from the testing machine. All the internal connections of the vibrating system were either completely integral or were very rigid. Large, rigid grips were used to hold the system in the testing machine. The natural frequencies of the extension arms on the central mass were designed to be well outside the range of the natural frequencies of the specimen-mass system. The magnets were held on a rigid platform which moved with the Instron crosshead to maintain the proper distance to the extension arms.

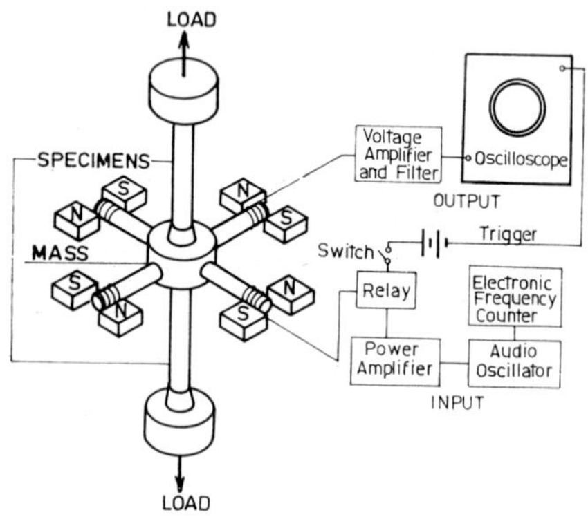

FIG. 8. Schematic of the experimental arrangement for measuring the resonance frequency and damping of strained specimens

FIG. 9. Apparatus for the measurement of the natural frequency and damping during constant rate of strain tests

The measured values of the torsional and extensional moduli and damping as a function of axial strain are shown in Figs. 10, 11, 12 and 13 for two straining rates. Measurements were taken on both filled and unfilled specimens. The initial natural frequency in the torsional mode was 18 cps for the unfilled and 30 cps for the filled specimens. The extensional natural frequency was 226 cps for the unfilled and 368 cps for the filled specimens.

A related experimental technique was developed by Payne[9] who investigated carbon black filled rubbers. His procedure was to subject the specimens to sinusoidal stresses of increasing amplitude and to measure the in and out of phase stress to strain ratios. A decrease of the dynamic modulus and an increase of damping with increasing vibration amplitude was observed. Some difficulties inherent in this procedure are that the strain rate imposed on the specimen varies appreciably during a cycle and that the material behavior ceases to be linear beyond moderate vibration amplitudes. The representation of the large amplitude results in terms of linear viscoelastic parameters is therefore inappropriate. In the present technique, the imposed vibration amplitudes are very small and well within the linear viscoelastic range. The vibrations serve solely to measure the material properties and are not intended to influence those properties. The rate of straining of the specimens is therefore controlled by the tensile loading device.

IV. DISCUSSION

A. *Dynamic Viscoelastic Characterization*

From Fig. 3, the dynamic extensional moduli of the filled and unfilled specimens are seen to vary linearly with the logarithm of the frequency over three decades of frequency. The ratio of the modulus of the filled material to that of the pure epoxy is almost exactly three over the complete range. The damping values (Fig. 4) show some scatter but are approximately independent of frequency and only slighty different for the two materials. The linear variation of the moduli with the logarithm of the frequency was shown by Kolsky[3, 10] to hold for a linear viscoelastic material for which the damping is independent of frequency and not too large.

The most important result from Figs. 3 and 4 is that the time dependent properties of the composite are almost exactly that of the matrix material. The filler appears to have little influence on the viscoelastic properties as indicated by the value and frequency dependence of the damping and the frequency dependence of the moduli. The only significant effect of the filler is the threefold increase in the modulus value. These results complement those of Schwarzl[4] who observed that, for a filled rubber tested at a fixed frequency, the damping, glass transition temperature, and temperature variation of the dynamic modulus were all independent of the value of the filler content.

These results are also in agreement with the theoretical predictions of Hashin[11] for the case of a composite of completely bonded, rigid particles in an incompressible viscoelastic matrix. The theory indicates that for this case the ratio of the time dependent dynamic shear modulus of the composite $G_2(t)$, or, alternatively, the relaxation modulus, to the corresponding parameter of the pure matrix $G_0(t)$ is independent of the time. That is,

$$G_2(t)/G_0(t) = \psi, \tag{1}$$

so that the time dependence of the composite is identical to that of the pure matrix. The nondimensional parameter ψ is a function of the phase geometry only and can be determined from an elastic analysis. For the present specimens, the ratio of the elastic modulus of the filler to that of the matrix is about 30 and Poisson's ratio of the matrix is 0.42. These parameter values reasonably satisfy the requirements of rigid particles in an incompressible matrix on which eq. (1) is based. The analysis of ref. 11 also indicates that for these conditions the damping of the composite is the same as that of the matrix over the complete frequency range. The damping results shown in Fig. 4 show some scatter but are in essential agreement with this prediction.

The threefold increase in the modulus value was compared to a number of analytical predictions for the modulus of elastic composites. For the silica sand, $E_1 = 7.4 \times 10^5$ kg/cm², $\nu_1 = 0.25$,[8] while the measured G_0 of the unfilled epoxy at 18 cps was 0.8×10^4 kg/cm² and ν_0 for this material was measured to be 0.42. These values lead to $G_1/G_0 = 30$. The analysis of van der Poel,[12] for a filler volume concentration of 38%, gives $G_2/G_0 = 3.0$ where G_2 is the shear modulus of the composite. The analysis of ref. 12 also predicts that $K_2/K_0 = 1.61$ and $E_2/E_0 = 2.9$. The results in Fig. 3 give an almost constant value of 3.0 for E_2/E_0 over the frequency range. Although E_0 changes with frequency while E_1 remains constant, the computed value of E_2/E_0 from an elastic analysis depends strongly on the volume filler content and less on the value of E_1/E_0 if that ratio is sufficiently high.

The predictions of other theories, summarized by Schwarzl,[4] for this example are as follows:

Einstein:	$E_2/E_0 = 1.95$
Guth and Smallwood:	$E_2/E_0 = 3.98$
Eilers and Van Dijk:	$E_2/E_0 = 3.70$
Brinkman:	$E_2/E_0 = 3.31$
Hashin (lower bound):	$G_2/G_0 = 1.72$
Hashin (upper bound):	$G_2/G_0 = 17.6$

Of the more generally used theoretical predictions, the van der Poel analysis[12] shows the best agreement with the present experimental results. It appears, however, that ref. 12 contains some theoretical inconsistencies (indicated to the authors by Dr. Z. Hashin) so that the results should be regarded with caution until the questions are resolved. Nevertheless, the analysis of ref. 12 seems to lead to good empirical answers.

The preceding results indicate that an elastic analysis of the strengthening effect of the filler may be adequate for an elastic filler in a viscoelastic matrix if the ratio of the modulus of the filler to that of the matrix is high and the matrix is almost incompressible. That is, for such materials the viscoelastic properties of the unfilled matrix by itself may be a sufficiently accurate representation of the time dependent behavior of the composite. The evaluation of the limitations of this conclusion would require further experimentation.

B. *Strength Properties*

Compared to an equal volume of unfilled material, the particle filled composite contains a smaller percentage of ductile matrix which can elongate to accomodate the same overall strain. In addition, the presence of well bonded, hard filler particles in a matrix leads to strain concentrations in the vicinity of the inclusions. The composite would therefore be expected to fail at lower overall strain values than the unfilled epoxy. This result can be seen from Figs. 5 and 6. At the lower strain rates (longer times to failure), the failure strain of the composite is about 40% that of the unfilled material.

The failure strain for the unfilled specimens was taken to be at the first maximum in the stress strain curve, while that of the composite specimens was either at the maximum or at a preceding fracture (Fig. 5). The different characteristics of the failure strain-loading time curves for the filled and unfilled materials seem to be due to the composite failing before the maximum is reached for the shorter loading times. The failure strain of the composite is also less sensitive to the time of loading (Fig. 6).

A composite of relatively large, hard particles in a soft matrix, while having a higher elastic modulus than the matrix, would not necessarily have greater strength. This is shown by the stress-strain curves of Fig. 5 (obtained at low strain rates) where the strengths of the filled and unfilled specimens are about the same. The non-strengthening characteristics of relatively large and moderately spaced rigid inclusions in a ductile matrix under static loading has been noted and discussed by Drucker.[13] The fracture strength of such materials is essentially that of the matrix. From Fig. 6 it is seen that the strengths of the filled and unfilled materials are almost the same at the longer loading times. The results for the failure strength of the composite specimens showed a fair amount of scatter.

As the loading time to failure decreases, the failure stress increases and the failure strain decreases (Fig. 6). An interesting observation in these results is that the strength of the composite is appreciably greater than that of the unfilled epoxy at the very high loading rates. The strength of the unfilled material is almost independent of the loading time, while that of the composite increases with increasing strain rate. Failure at the high strain rates occurs elastically, and the higher elastic modulus of the composite may be a more significant factor under these conditions. The assumed quasi-static fracture mechanism of vacuole formation and growth in the matrix around the hard particles and subsequent tearing,[14] is likely impeded at the higher loading rates. The failure strain of the composite still remains lower than that of the pure matrix at the high rates (by about 1/2).

C. *Changes in Dynamic Modulus and Damping During Straining*

Since the inclusion of hard filler particles appreciably increases the dynamic modulus of the matrix (by threefold for the 38% filler volume in the present specimens), any separation of the particles from the matrix should be reflected by relatively large changes in the dynamic modulus. In addition, any relative motion between the filler and matrix would be expected to appreciably increase the internal damping. The changes in the dynamic moduli and damping during straining should, therefore, be indicative of binder-filler separation or of fracturing of the matrix for both particle and fiber reinforced composites.

The results shown in Figs. 10, 11, 12 and 13 are for the unfilled and the quartz particle filled specimens strained at two different rates. The parameters G_i and E_i refer to the initial values of the shear and extension moduli (differ-

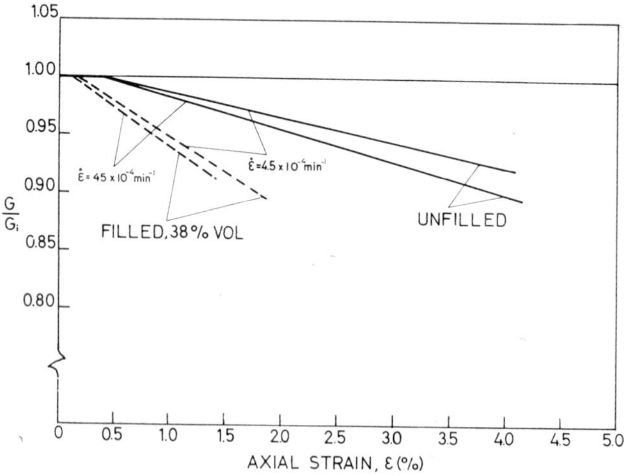

FIG. 10. Change of dynamic shear modulus with axial strain

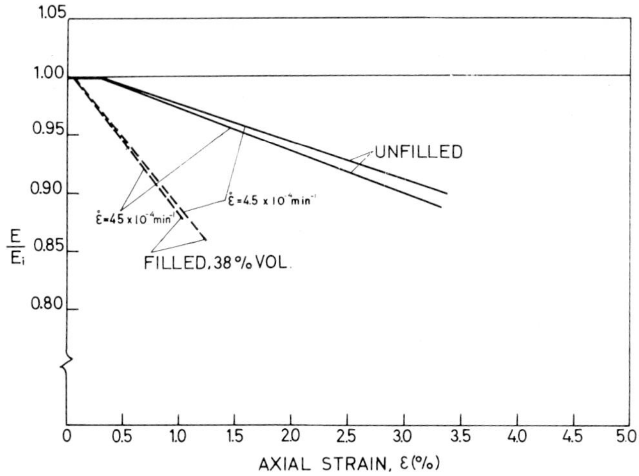

FIG. 11. Change of dynamic extensional modulus with axial strain

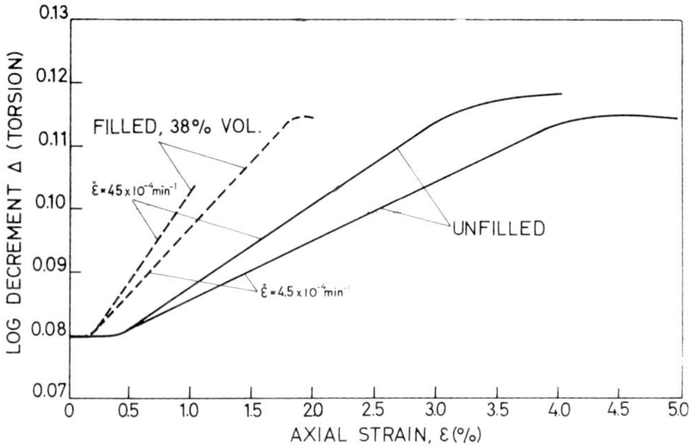

FIG. 12. Change of internal damping of torsional vibrations with axial strain

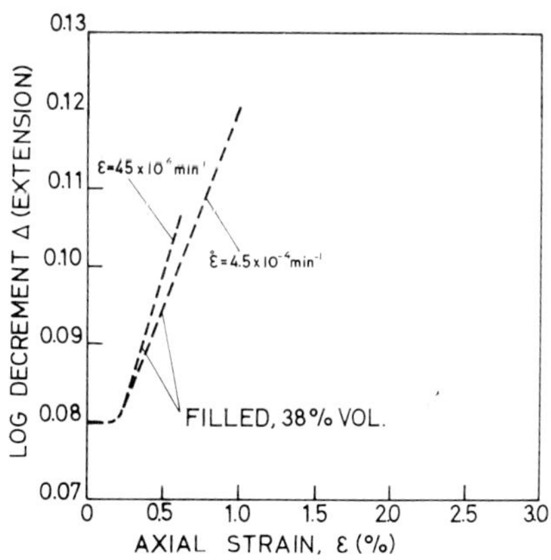

FIG. 13. Change of internal damping of extensional vibrations with axial strain

ing, of course, for the filled and unfilled specimens). The moduli values at a strain ε are designated by G and E. No drastic changes in the ratios G/G_i, E/E_i, and in Δ occurred. Up to the failure strain, the moduli decreased by about 10% and the damping Δ increased by about 40%. This indicates that it is unlikely that particle separation took place prior to overall failure. The total changes up to failure of E/E_i, G/G_i, and Δ are about the same for both the filled and unfilled specimens. The parameters of the filled specimens change more rapidly with axial strain in about the same proportion as the ultimate strains, i.e., about 2.5 times more rapidly for the filled specimens.

These results tend to support the explanation of the failure mechanism as the generation and growth of vacuoles in the matrix. The presence of vacuoles would lower the dynamic modulus while the damping would increase due to the friction generated on the exposed surfaces. Whereas complete separation of the filler particles would lower the modulus about threefold, vacuole formation in the matrix would bring about only the observed 10% decrease. The lower matrix content and strain concentrations in the composite would enhance vacuole formation and therefore increase the rate of change of the parameters with the axial strain. The similarity of the total changes in the moduli and damping for the filled and unfilled specimens indicates that the ultimate failure mechanism is probably the same for both materials.

An interesting aspect of these results was that the dynamic modulus and damping remained essentially constant over an initial range of axial strain. The changes did not commence until after about 0.14% axial strain was imposed on the composite specimens and about 0.35% on the unfilled specimens. These values appeared to be independent of the applied slow strain rates, but differed somewhat for the extensional and torsional measurements. A possible interpretation of these strain values are that they indicate the strains at which microcracking (formation of vacuoles) commences in the matrix. These strain limits for the filled and unfilled specimens are in about the same proportion as the ultimate strains (2.5 to 1). They are appreciably lower than the ultimate strains (by about 1/7), and lower than the limits of linear viscoelastic behavior for superposition of axial strains. Associated tests showed that superposition of axial stress-strain curves at different strain rates held to strains of 0.25% for the composite and 0.80% for the unfilled specimens. These tests are, of course, not as sensitive as the vibration measurements.

The investigation of fiber reinforced materials by this technique should provide useful information on the failure processes. This is the planned direction of future investigations.

V. CONCLUSIONS

The frequency dependence of the dynamic modulus and the value of the internal damping of a quartz particle filled composite were found to be essentially the same as that of the pure epoxy matrix. This result agrees with that predicted by Hashin[11] for an incompressible matrix with completely bonded rigid particle inserts. The relative increase in the dynamic modulus of the composite over that of the pure matrix was adequately predicted by the elastic analysis of van der Poel.[12]

The failure strengths of the filled and unfilled materials were found to be about the same at low rates of loading, and the strain to fracture in the composite was appreciably lower. The strength of the composite increased at high rates of loading while that of the pure epoxy remained about the same.

The total percentage changes in the dynamic moduli and the damping from the onset of loading to failure were found to be the same for the pure epoxy and the composite. The dynamic moduli and damping remained constant up to certain axial strain values which can be interpreted as the conditions for the onset of vacuole formation in the matrix. These characteristics and the relatively low value of the changes tend to indicate that filler-binder separation did not occur prior to overall failure. Failure at the low strain rates appeared to be due to the generation and growth of vacuoles in the matrix.

ACKNOWLEDGEMENTS

The authors would like to thank Mr. Assa Rotem for his contributions to the experimental program. The technical assistance of Messrs. I. Wachtel and D. Vardi is also appreciated.

REFERENCES

1. *Mechanics and Chemistry of Solid Propellants*, Proceedings of the Fourth Symposium on Naval Structural Mechanics, ed. A. C. Eringen, H. Liebowitz, S. L. Koh, J. M. Crowley, Pergamon Press, Oxford 1967.
2. *Behavior of Materials Under Dynamic Loading*, ed. N. J. Huffington Jr., ASME, New York 1965.
3. KOLSKY, H. Experimental Studies of the Mechanical Behavior of Linear Viscoelastic Solids, *Mechanics and Chemistry of Solid Propellants*, p. 357 (1967).
4. SCHWARZL, F. R. On Mechanical Properties of Unfilled and Filled Elastomers, *Mechanics and Chemistry of Solid Propellants*, p. 503 (1967).
5. BODNER, S. R. and KOLSKY, H. Stress Wave Propagation in Lead, *Proc. Third U.S. National Congress of Applied Mechanics*. ASME, New York, p. 495 (1958).
6. *High Speed Testing*, vol. 5, ed. A. G. H. Dietz and F. R. Eirich, Interscience Pub., New York 1965. Especially relevant are papers by P. I. Donnelly and R. H. Ralston, and by E. McAbee and M. Chmura.
7. BODNER, S. R. On Anomalous Resonances in Stressed Crystalline Solids, *Proc. Fourth International Congress on Acoustics*, Copenhagen 1962.

8. HIRSCH, T. J. *Proc. ACI*, **58**, 611 (1961).
9. PAYNE, A. R. The Dynamic Properties of Carbon Black Loaded Natural Rubber Vulcanizates, Part I, *J. Applied Polymer Science*, **6**, 57 (1962).
10. KOLSKY, H. The Propagation of Stress Pulses in Viscoelastic Solids, *Phil. Mag*. **1**, 693 (1956).
11. HASHIN, Z. Viscoelastic Behavior of Heterogeneous Media, *J. Appl. Mech*. **32**, 630 (1965).
12. VAN DER POEL, C. On the Rheology of Concentrated Dispersions, *Rheologica Acta* **1**, 198 (1958).
13. DRUCKER, D. C. Some Remarks on Flow and Fracture, *Developments in Mechanics*, Proc. Eighth Midwestern Mechanics Conference, Pergamon Press, Oxford, p. 201 (1965).
14. WIEGAND, J. H. The Failure Mechanisms of Solid Propellant Grains, *Mechanics and Chemistry of Solid Propellants*, p. 539 (1967).

STABILITY OF PRESTRAINED LAMINATED MEDIA†

W. J. DOUGLAS
Keystone Computer Associates

and

W. JAUNZEMIS
Pennsylvania State University

I. INTRODUCTION

Much of the current research in theoretical elasticity has been concerned with the theory of finite strain, which explores various noteworthy effects that are not present in the classical linearized theory of elasticity. In the limit of infinitesimal deformations many of the important distinctions and interesting effects are obscured, hence the linear theory of elasticity gives little insight into the behavior of elastic materials capable of large deformations.

The solutions of problems involving finite strain are difficult to obtain, therefore, it is natural to extend the available solutions by considering the superposition of additional, small deformations upon finitely deformed bodies. This yields a generalization of the classical theory inasmuch as the reference configuration is no longer a *natural* (i.e., stress-free) state; rather, it is itself the result of a finite deformation from a natural state. In this way the influence of finite deformation on the incremental stresses and deformations can be studied. Furthermore, the additional deformations may serve as a tool for studying the propagation of weak waves in prestrained bodies, as well as for investigating the stability of equilibrium of finitely deformed bodies.

According to the intuitive notion of stability, a small disturbance should produce only a small departure from a stable equilibrium state. In particular, a disturbance may induce small oscillations about the equilibrium state, or even become damped out, depending on the nature of forces that act on the body.‡

Let us mark by superscripts 0 and 1, respectively, variables referred to an equilibrium state and another state selected as follows. We imagine a body jolted out of its equilibrium state by a disturbance D which imparts the kinetic energy K^0 to the body. Initially, then, the total energy is E^0+K^0, where E^0

† This research was supported by the United States Air Force Office of Scientific Research.

‡ Because internal dissipation is absent in elastic bodies, the damping can only be produced by certain "nonconservative" surface loads.

is the strain energy in the equilibrium state. Denoting by W_{01} the work done by external forces in the passage from state 0 to state 1, we obtain from the energy balance that

$$\Delta E = E^1 - E^0 = W_{01} - \Delta K,$$

where $\Delta K = K^1 - K^0$. The condition

$$\Delta E \geqq W_{01} \tag{1}$$

(i.e., $\Delta K < 0$) then characterizes stability (with respect to D), in the sense that a body can depart from a stable equilibrium only at the expense of the kinetic energy obtained from the disturbance D. Neutral equilibrium exists if $\Delta E - W_{01} = 0$ for some displacements (other than translations) and is non-negative, in general.

When dealing with finite deformations, it is expedient to refer all fields associated with the current (in general, unknown) state to a known reference state (which is usually some "initial" state). The Cartesian coordinates of particles in the reference state are denoted by X_1, X_2, X_3. Further, we denote by V the open region occupied by the body in the reference state, by ∂V the boundary of V and by $\mathbf{N}$ the unit normal vectors of ∂V. To a stress vector $\mathbf{t}$ associated with an area da in the current state there corresponds the Piola–Kirchhoff stress (or, "nominal stress") $\tilde{\mathbf{t}}$ assigned to the reference state dA of da, thus $\mathbf{t}\, da = \tilde{\mathbf{t}}\, dA$. The Piola–Kirchhoff stress tensor $\tilde{\mathbf{T}}$ is then defined by†

$$\tilde{t}_i(\mathbf{N}) = \tilde{T}_{ji} N_j. \tag{2}$$

In terms of $\tilde{\mathbf{T}}$, the equations of motion then read

$$\frac{\partial \tilde{T}_{ji}}{\partial X_j} + \varrho^0 b_i = \varrho^0 \ddot{x}_i, \tag{3}$$

where ϱ^0 is the mass density in the reference state, and b_i, $\ddot{x}_i$ are the current body force density and acceleration, respectively.[4]

With the notations introduced above, the stability criterion (1) can be expressed as follows[10]

$$\int_V \Delta E\, dm > \int_0^t dt \left[\int_{\partial V} \tilde{t}_i \dot{x}_i\, dA + \int_V b_i \dot{x}_i\, dm \right],$$

where $dm = \varrho^0\, dV$, E is the strain-energy density per unit mass and $\dot{\mathbf{x}}$ is any motion satisfying the kinematical constraints. We shall deal with *dead loads* $\tilde{t}_i(\mathbf{X}_i, t) = \tilde{t}_i(\mathbf{X})$, $b_i(\mathbf{X}_i, t) = b_i(\mathbf{X})$ only, in which case

$$\int_V \Delta E\, dm \geqq \int_{\partial V} \tilde{t}_i u_i\, dA + \int_V b_i u_i\, dm,$$

$$u_i = x_i(\mathbf{X}_i, t) - x_i(\mathbf{X}_i, 0).$$

With the aid of (2), and assuming that the state 0 is one of equilibrium, thus

$$\frac{\partial \tilde{T}_{ji}}{\partial X_j} + \varrho^0 b_i = 0,$$

† We use the conventional index notation; in particular, summation over any pair of repeated indices is understood.

we finally obtain

$$\int_V (\Delta W - \tilde{T}_{ji}G_{ij})\, dV \geqq 0, \quad G_{ij} = \frac{\partial u_i}{\partial X_j}, \tag{4}$$

$W = \varrho^0 E$ being strain-energy density per unit volume of the reference state. The *linearized* stability criterion is obtained by writing

$$\Delta W = \tilde{T}_{ji}G_{ij} + \tfrac{1}{2}\, C_{ijkm}G_{ij}G_{km},$$
$$\tilde{T}_{ji} = \left.\frac{\partial W}{\partial G_{ij}}\right|_{\mathbf{G}=\mathbf{0}}, \quad C_{ijkm} = \left.\frac{\partial^2 W}{\partial G_{ij}\,\partial G_{km}}\right|_{\mathbf{G}=\mathbf{0}}, \tag{5}$$

so that (4) becomes

$$\tfrac{1}{2}\int_V C_{ijkm}G_{ij}G_{km}\, dV \geqq 0. \tag{6}$$

Stability is a *global* condition: it is affected not only by internal stresses and deformations but also by the boundary conditions. Therefore, stability criteria are, as a rule, global too. This does not mean, however, that *local* criteria of the type

$$C_{ijkm}A_iA_kB_jB_m \geqq 0, \tag{7}$$

where **A**, **B** are any vectors, are without interest. One of the classical results is Hadamard's theorem,(5) according to which (6) implies (7). This leads us to conjecture that, at least for homogeneous initial strain (i.e., when C_{ijkm} are constants), a violation of (7) may correspond to instability.

The instability may be an elastic "internal buckling", or else an "incremental flow". To elaborate these points, let us take an Euler column of length L and bending stiffness EI. If the column is subjected to a compressive load $P > 0$, then the equation of equilibrium,

$$y'' + k^2 y = 0, \quad k^2 = \frac{P}{EI},$$

admits both a trivial solution $y = 0$ and a "standing wave" solution $y = A \sin ky + B \cos ky$. The boundary conditions determine the characteristic *value* of $P_c > 0$ for which a nontrivial solution exists. In particular, for a column with pinned ends, thus $y(0) = y(L) = 0$, we obtain

$$y = A \sin kx, \quad P_c = \frac{\pi^2 EI}{L^2}.$$

For $P < 0$ (tensile loading), on the other hand, the nontrivial solution is $y = A \sinh kx + B \cosh kx$, and the boundary conditions now require that $A = B = 0$. That is, for $P > 0$ the column may become unstable, and the nature of the differential equation changes at $P = 0$.

To obtain a problem in which the amplitude A is determined, we apply moments M at the pinned ends. The equation of equilibrium is then

$$y''+k^2y-\frac{M}{EI} = 0.$$

The solution, for *any P*, is given by

$$y = \frac{M}{P}\left[1-\cos kx-\frac{1-\cos kl}{\sin kl}\sin kx\right],$$

and the ratio of M to the slope θ at $x = 0$ is found to be

$$\mu \equiv \frac{M}{\theta} = kEI\frac{\sin kL}{\cos kL-1}.$$

The modulus μ has the limit $2EI/L$ at $k = 0$, and tends to zero as P approaches P_c. Of course, the latter case cannot be interpreted literally, because it lies outside the scope of the linear theory. Still, it gives an indication as to how the vanishing of the "effective modulus" μ could be interpreted, namely, as an elastic instability. To determine the stage at which the elastic deformation is terminated by the onset of plastic flow (before or after the critical load is reached), the yield behavior of the material must be known.

In analogy to buckling of columns, we may explore conditions for the existence of standing wave solutions in prestrained, laminated bodies. It is natural to expect that instabilities will again correspond, in a linearized theory of small deformations superposed on large, to the vanishing of certain "effective moduli".

II. ELASTIC RESPONSE OF PRESTRAINED BODIES

Our first task will be to develop the kinematics of incremental deformations, and to derive the incremental stress–strain relations for general elastic media. Subsequently, we shall consider special cases in which more specific results can be obtained (e.g., transversely isotropic media).

Let us consider a body in a natural state and let X_i denote the Cartesian coordinates of a typical particle P of the body. Next, we suppose that the body is deformed from its natural state C_N into a new state C_0 in which P has the Cartesian coordinates ξ_i. This deformation is described by

$$\xi_i = \xi_i(\mathbf{X}). \tag{8}$$

The local deformation is then characterized by the deformation gradients

$$\bar{F}_{ij} = \frac{\partial \xi_i}{\partial X_j}, \quad \bar{\mathbf{F}} = \frac{\partial \boldsymbol{\xi}}{\partial \mathbf{X}}. \tag{9}$$

Finally, we permit additional, infinitesimal displacements $\mathbf{u}$ of the particles from positions $\boldsymbol{\xi}$ to final positions $\mathbf{x}$ (Fig. 1):

$$x_i = \xi_i + u_i, \ \mathbf{x} = \boldsymbol{\xi} + \mathbf{u}. \tag{10}$$

Differentiating (10) with respect to initial coordinates X_i, we obtain the chain rule

$$\frac{\partial x_i}{\partial X_j} = \frac{\partial \xi_i}{\partial X_j} + \frac{\partial u_i}{\partial \xi_k}\frac{\partial \xi_k}{\partial X_j} \tag{11}$$

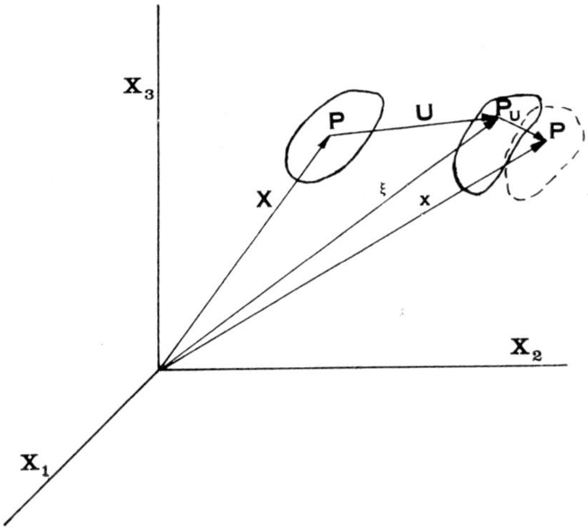

FIG. 1. Small deformation superposed on large deformation

or

$$\delta\mathbf{F} = \mathbf{F} - \overline{\mathbf{F}} = \mathbf{H}\overline{\mathbf{F}} \tag{12}$$

where

$$H_{im} = \frac{\partial u_i}{\partial \xi_m} \tag{13}$$

are the incremental displacement gradients. Introducing the Cauchy–Green deformations tensors

$$\mathbf{B} = \mathbf{F}\mathbf{F}^T, \quad \mathbf{C} = \mathbf{F}^T\mathbf{F} \tag{14}$$

we obtain from

$$\delta\mathbf{B} = \mathbf{B} - \overline{\mathbf{B}} = \mathbf{H}\overline{\mathbf{B}} + \overline{\mathbf{B}}\mathbf{H}^T \tag{15}$$

$$\delta\mathbf{C} = \mathbf{C} - \overline{\mathbf{C}} = \overline{\mathbf{F}}^T(\mathbf{H} + \mathbf{H}^T)\overline{\mathbf{F}} \tag{16}$$

where terms of $0(\mathbf{H}^2)$ have been neglected.

The stress relation for elastic media with a strain–energy function ("hyperelastic media") is given by ref. 1.

$$T_{ij} = \frac{2}{J} F_{ik} \frac{\partial W}{\partial C_{km}} F_{jm} = \frac{1}{J} F_{ik} F_{jm} \frac{\partial W}{\partial E_{km}}, \quad \mathbf{T} = \frac{2}{J} \mathbf{F} \frac{\partial W}{\partial \mathbf{C}} \mathbf{F}^T \tag{17}$$

where $\mathbf{T}$ is the Cauchy stress ("true stress"), W is the strain energy per unit volume in the reference state, and

$$J = \det \mathbf{F} = (\det \mathbf{C})^{1/2} = (\det \mathbf{B})^{1/2}, \ 2\mathbf{E} = \mathbf{C} - \mathbf{1}. \tag{18}$$

The incremental response is given by

$$\delta T_{ij} = \left[\frac{\partial T_{ij}}{\partial F_{km}} \right]_{\overline{F}} \delta F_{km}, \tag{19}$$

so that from (17) we obtain

$$\delta T_{ij} = 2 \left[\frac{\partial J^{-1}}{\partial F_{sn}} \frac{\partial W}{\partial C_{km}} F_{ik} F_{jm} + J^{-1} \frac{\partial^2 W}{\partial C_{km} \partial C_{pq}} \frac{\partial C_{pq}}{\partial F_{sn}} F_{ik} F_{jm} + \right.$$

$$\left. + J^{-1} \frac{\partial W}{\partial C_{km}} \frac{\partial (F_{ik} F_{jm})}{\partial F_{sn}} \right]_{\overline{\mathbf{F}}} H_{sr} \overline{F}_{rn}. \tag{20}$$

Making use of the relation

$$\frac{\partial F_{ij}}{\partial F_{km}} = \delta_{ik} \delta_{jm} \tag{21}$$

we obtain

$$\frac{\partial J^{-1}}{\partial F_{mn}} = -J^{-1} F_{mp} C^{-1}_{pn} = -J^{-1} F^{-1}_{mn} \tag{22}$$

and introducing the infinitesimal strain $\tilde{\mathbf{E}}$ and rotation $\tilde{\mathbf{R}}$

$$\tilde{\mathbf{E}} = \tfrac{1}{2} (\mathbf{H} + \mathbf{H}^T), \ \tilde{\mathbf{R}} = \tfrac{1}{2} (\mathbf{H} - \mathbf{H}^T) \tag{23}$$

(20) becomes

$$\delta T_{ij} = -\tilde{E}_{kk} \overline{T}_{ij} + \tilde{E}_{ik} \overline{T}_{kj} + \overline{T}_{ik} \tilde{E}_{kj} + \tilde{R}_{ik} \overline{T}_{kj} - \overline{T}_{ik} \tilde{R}_{kj} + A_{ijkm} \tilde{E}_{km}$$

$$\delta T_n = -(tr\tilde{\mathbf{E}})\overline{\mathbf{T}} + \tilde{\mathbf{E}}\overline{\mathbf{T}} + \overline{\mathbf{T}}\tilde{\mathbf{E}} + \tilde{\mathbf{R}}\overline{\mathbf{T}} - \overline{\mathbf{T}}\tilde{\mathbf{R}} + \mathbf{A}[\tilde{\mathbf{E}}], \tag{24}$$

where

$$A_{ijkm} = \frac{4}{J} F_{in} F_{jr} \frac{\partial^2 W}{\partial C_{nr} \partial C_{pq}} F_{kp} F_{mq} \bigg|_{\overline{\mathbf{F}}} = \frac{1}{J} F_{in} F_{jr} \frac{\partial^2 W}{\partial E_{nr} \partial E_{pq}} F_{kp} F_{mq}. \tag{25}$$

Note that in the absence of initial stress (i.e., when $\overline{\mathbf{T}} = \mathbf{0}$, $\overline{\mathbf{F}} = \mathbf{1}$, $J = 1$) (24) and (25) reduce to the classical relations

$$\delta T_{ij} = A_{ijkm} \tilde{E}_{km} \tag{26}$$

$$A_{ijkm} = 4 \frac{\partial^2 W}{\partial C_{ij} \partial C_{km}} = \frac{\partial^2 W}{\partial E_{ij} \partial E_{km}}. \tag{27}$$

The strain energy must satisfy two invariance conditions. The first of these is a condition of "isotropy of space," which requires that W be independent of the orientation of the body in space. Thus, if the current coordinates are subjected to an orthogonal transformation $\mathbf{x} \to \mathbf{x}^+ = \mathbf{Qx}$, $\mathbf{F}^+ = \mathbf{QF}$, then $\overline{W}(\mathbf{F}) = \overline{W}(\mathbf{QF})$. Expressing $\mathbf{F}$ as a product $\mathbf{RU}$ of an orthogonal tensor $\mathbf{R}$ with the deformation tensor $\mathbf{U} = \sqrt{\mathbf{C}}$, and letting $\mathbf{Q} = \mathbf{R}^T$, we arrive at $\overline{W}(\mathbf{F}) = \overline{W}(\mathbf{U}) \equiv W(\mathbf{C})$. Because $\mathbf{C}^+ = (\mathbf{F}^+)^T\mathbf{F}^+ = \mathbf{F}^T\mathbf{Q}^T\mathbf{Q}F = \mathbf{C}$, the condition of isotropy of space is now automatically satisfied by virtue of $\mathbf{C}$ being insensitive to a rotation of the current configuration.

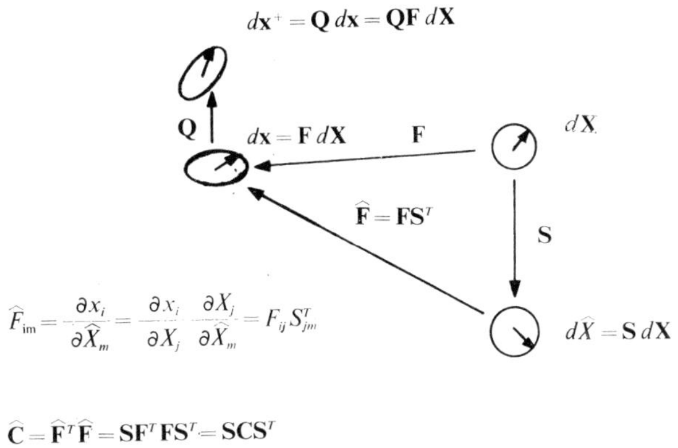

FIG. 2. Mapping scheme

The strain-energy function must be also form-invariant to symmetry operations $\mathbf{S}$ performed in any undistorted state of the body.

That is, to a current state $d\mathbf{x}$ there corresponds an "equivalence class" of initial states $d\mathbf{X}$ (related by the symmetry operations $\mathbf{S}$) such that the mapping from $d\mathbf{X}$ to $d\mathbf{x}$ will yield the same strain energy—*cf.* Fig. 2.

The symmetry operation may be imagined as rotation of the undeformed specimen, which leaves the response of the specimen unchanged:

$$W(\mathbf{C}) = W(\mathbf{SCS}^T).$$

The last invariance requirement is satisfied by replacing the dependence of $\overline{W}$ upon $\mathbf{C}$ by a dependence of $\overline{W}$ upon invariants I_α of $\mathbf{C}$ with respect to the group $\mathbf{s}$: $\{\mathbf{S}\}$ of the symmetry operations. The form-invariance of W to symmetry transformations is then automatically fulfilled by virtue of I_α being insensitive to the symmetry transformations. For isotropic media, these are the principal invariants

$$\mathrm{I} = I_1 = C_{ii}, \quad \mathrm{II} = I_2 = \tfrac{1}{2}(\mathrm{I}^2 - C_{ij}C_{ij}), \quad \mathrm{III} = I_3 = \varepsilon_{ijk}C_{i1}C_{j2}C_{k3}$$
$$\mathrm{I} = tr\mathbf{C}, \quad \mathrm{II} = \tfrac{1}{2}(\mathrm{I}^2 - tr\mathbf{C}^2), \quad \mathrm{III} = \det \mathbf{C}, \tag{28}$$

because the strain-energy function must be form-invariant under all orthogonal transformations. For anisotropic media the group of transformations is more restricted, and correspondingly, there is a greater number of invariants. If there are N such invariants, I_β, $\beta=1, 2, \ldots, N$, then using the notation

$$\frac{\partial W}{\partial I_\beta} = W_\beta, \tag{29}$$

we have

$$\frac{\partial W}{\partial C_{mn}} = \sum_{\beta=1}^{N} W_\beta \frac{\partial I_\beta}{\partial C_{mn}} \tag{30}$$

and

$$\frac{\partial^2 W}{\partial C_{mn}\,\partial C_{pq}} = \sum_{\beta=1}^{N}\sum_{\gamma=1}^{N} W_{\beta\gamma}\frac{\partial I_\beta}{\partial C_{mn}}\frac{\partial I_\gamma}{\partial C_{pq}} + \sum_{\beta=1}^{N} W_\beta \frac{\partial^2 I_\beta}{\partial C_{mn}\,\partial C_{pq}}. \tag{31}$$

For transversely isotropic media the strain-energy function depends on the three principal invariants as well as on two others defined by [(2)]

$$\mathrm{IV} = I_4 = C_{33} \quad \mathrm{V} = I_5 = C_{3i}C_{3i} - \mathrm{IV}^2 = C_{31}^2 + C_{32}^2, \tag{32}$$

thus

$$W = W(I_1, I_2, I_3, I_4, I_5). \tag{33}$$

Differentiation of (28) and (32) yields

$$\frac{\partial \mathrm{I}}{\partial C_{km}} = \delta_{km}, \quad \frac{\partial \mathrm{II}}{\partial C_{km}} = I\delta_{km} - C_{km}, \quad \frac{\partial \mathrm{III}}{\partial C_{km}} = \mathrm{III}\, C_{km}^{-1} \tag{34}$$

and

$$\frac{\partial\, \mathrm{IV}}{\partial C_{km}} = \delta_{3k}\delta_{3m}, \quad \frac{\partial \mathrm{V}}{\partial C_{km}} = \delta_{3k}C_{3m} + C_{3k}\delta_{3m} - 2\,\mathrm{IV}\,\delta_{3k}\delta_{3m}. \tag{35}$$

Substituting (30), (34), and (35) into (17) we obtain

$$T_{ij} = T_{ij}^I + \frac{2}{J}\left[(W_4 - 2\,\mathrm{IV}\,W_5)F_{i3}F_{j3} + W_5(F_{i3}F_{jk}C_{3k} + F_{ik}F_{j3}C_{3k})\right] \tag{36}$$

where

$$T_{ij}^I = \frac{2}{J}\left[(\mathrm{II}\,W_2 + \mathrm{III}\,W_3)\delta_{ij} + W_1 B_{ij} - \mathrm{III}\,W_2 B_{ij}^{-1}\right] \tag{37}$$

is the isotropic part of the stress relation.

Further differentiation of (34) and (35) yields

$$\frac{\partial^2 \mathrm{I}}{\partial C_{mn}\,\partial C_{pq}} = 0, \quad \frac{\partial^2 \mathrm{II}}{\partial C_{mn}\,\partial C_{pq}} = \delta_{mn}\delta_{pq} - \frac{1}{2}(\delta_{mp}\delta_{nq} + \delta_{mq}\delta_{np})$$

$$\frac{\partial^2 \mathrm{III}}{\partial C_{mn}\,\partial C_{pq}} = \mathrm{III}\left[C_{pq}^{-1}C_{mn}^{-1} - \frac{1}{2}(C_{pm}^{-1}C_{qn}^{-1} + C_{pn}^{-1}C_{qm}^{-1})\right] \tag{38}$$

and

$$\frac{\partial^2 \mathrm{IV}}{\partial C_{mn}\partial C_{pq}} = 0, \quad \frac{\partial^2 \mathrm{V}}{\partial C_{mn}\partial C_{pq}} = \frac{1}{2}\,\delta_{3m}(\delta_{3p}\delta_{nq}\delta_{3q}\delta_{np}) + \\ + \frac{1}{2}\,\delta_{3n}(\delta_{3p}\delta_{mq}+\delta_{mp}\delta_{3q}) - 2\delta_{3m}\delta_{3n}\delta_{3p}\delta_{3q}. \tag{39}$$

Substitution of (34), (35), (38), and (39) into (31) and (18) yields

$$\begin{aligned} A_{ijkm} = \frac{4}{J}\,F_{ir}F_{jn}\Big\{ & W_{11}\delta_{rn}\delta_{pq} + W_{12}(2\mathrm{I}\delta_{rn}\delta_{pq} - \\ & -\delta_{rn}C_{pq} - C_{rn}\delta_{pq}) + W_{13}\,\mathrm{III}\,(\delta_{rn}C^{-1}_{pq} + C^{-1}_{rn}\delta_{pq}) + \\ & + W_{14}(\delta_{rn}\delta_{3p}\delta_{3q} + \delta_{3r}\delta_{3n}\delta_{pq}) + W_{15}[\delta_{rn}(\delta_{3p}\delta_{3q} + \\ & + C_{3p}\delta_{3q} - 2\,\mathrm{IV}\,\delta_{3p}\delta_{3q}) + (\delta_{3r}C_{3n} + C_{3r}\delta_{3n} - 2\,\mathrm{IV}\,\delta_{3r}\delta_{3n})\delta_{pq}] + \\ & + W_{22}(\mathrm{I}\delta_{rn} - C_{rn})\,(\mathrm{I}\delta_{pq} - C_{pq}) + W_{23}\,\mathrm{III}\,[(\mathrm{I}\delta_{rn} - C_{rn})\,C^{-1}_{pq} + \\ & + C^{-1}_{rn}(\mathrm{I}\delta_{pq} - C_{pq})] + W_{24}[(\mathrm{I}\delta_{rn} - C_{rn})\delta_{3p}\delta_{3q} + \delta_{3r}\delta_{3n}(\mathrm{I}\delta_{pq} - C_{pq})] + \\ & + W_{25}[(\mathrm{I}\delta_{rn} - C_{rn})\,(\delta_{3p}C_{3q} + C_{3p}\delta_{3q} - 2\,\mathrm{IV}\,\delta_{3p}\delta_{3q}) + \\ & + (\delta_{3r}C_{3n} + C_{3r}\delta_{3n} - 2\,\mathrm{IV}\,\delta_{3r}\delta_{3n})\,(\mathrm{I}\delta_{pq} - C_{pq})] + \\ & + W_{33}\,\mathrm{III}^2\,C^{-1}_{rn}C^{-1}_{pq} + W_{34}\,\mathrm{III}\,(C^{-1}_{rn}\delta_{3p}\delta_{3q} + \delta_{3r}\delta_{3n}C^{-1}_{pq}) + \\ & + W_{35}\,\mathrm{III}\,[C^{-1}_{rn}(\delta_{3p}C_{3q} + C_{3p}\delta_{3q} - 2\,\mathrm{IV}\,\delta_{3p}\delta_{3q}) + \\ & + (\delta_{3r}C_{3n} + C_{3r}\delta_{3n} - 2\,\mathrm{IV}\,\delta_{3r}\delta_{3n})\,C^{-1}_{pq}] + \\ & + W_{44}\delta_{3r}\delta_{3n}\delta_{3p}\delta_{3q} + W_{45}[\delta_{3r}\delta_{3n}(\delta_{3p}C_{3q} + C_{3p}\delta_{3p} - 2\,\mathrm{IV}\,\delta_{3p}\delta_{3q}) + \\ & + (\delta_{3r}C_{3n} + C_{3r}\delta_{3n} - 2\,\mathrm{IV}\,\delta_{3r}\delta_{3n})\delta_{3p}\delta_{3q}] + \\ & + W_{55}(\delta_{3r}C_{3n} + C_{3r}\delta_{3n} - 2\,\mathrm{IV}\,\delta_{3r}\delta_{3n})\,(\delta_{3p}C_{3q} - 2\,\mathrm{IV}\,\delta_{3p}\delta_{3q}) \\ & + W_2\left[\delta_{rn}\delta_{pq} - \frac{1}{2}\,(\delta_{rp}\delta_{nq} + \delta_{rq}\delta_{np})\right] + \\ & + W_3\,\mathrm{III}\left[C^{-1}_{rn}C^{-1}_{pq} - \frac{1}{2}\,(C^{-1}_{rp}C^{-1}_{nq} + C^{-1}_{rq}C^{-1}_{np})\right] + \\ & + W_5\left[\frac{1}{2}\,\delta_{3p}(\delta_{3p}\delta_{nq} + \delta_{3q}\delta_{np}) + \frac{1}{2}\,\delta_{3n}(\delta_{3p}\delta_{rq} + \delta_{3q}\delta_{rp}) - \right. \\ & \left. -2\delta_{3r}\delta_{3n}\delta_{3p}\delta_{3q}\right]\Big\}\,F_{kp}F_{mq}. \end{aligned} \tag{40}$$

In the natural state, i.e., when $\mathbf{F} = \mathbf{C} = \mathbf{1}$, we obtain

$$\begin{aligned} A_{ijkm} = {} & \lambda\delta_{ij}\delta_{km} + \mu(\delta_{ik}\delta_{jm} + \delta_{jk}\delta_{im}) + \alpha(\delta_{ij}\delta_{3k}\delta_{3m} + \\ & + \delta_{3i}\delta_{3j}\delta_{km}) + \beta(\delta_{3i}\delta_{3k}\delta_{jm} + \delta_{3i}\delta_{3m}\delta_{jk} + \delta_{3j}\delta_{3k}\delta_{im} + \delta_{3j}\delta_{3m}\delta_{ik}) \\ & + \gamma\delta_{3i}\delta_{3j}\delta_{3k}\delta_{3m}, \end{aligned} \tag{41}$$

where

$$\begin{aligned}\lambda &= 4(W_{11}+4W_{12}+2W_{13}+4W_{22}+4W_{23}+W_{33}+W_2+W_3)\\ \mu &= -2(W_2+W_3)\\ \beta &= 2W_5\\ \alpha &= 4(W_{14}+4W_{24}+2W_{34})\\ \gamma &= 4W_{44}.\end{aligned} \tag{42}$$

Here, A_{ijkm} represents the fourth rank tensor of linear transversely isotropic response from a natural state.†

It is evident from examination of (24) and (40) that the response strongly depends on the initial state. In the natural state the response is described by (26), where **A** has certain initial symmetries. However, the incremental stress depends on the initial stress and deformation; consequently, when subsequent response is referred to the prestrained configuration, the material may lose the symmetries found in the natural state. For certain stresses and deformations, the initial symmetry may be preserved and the same type of response will occur from the prestrained configuration.‡

III. INCREMENTAL EQUATIONS OF MOTION

It is necessary now to obtain the equations of motion for small deformations superposed on large ones. Denoting differentiation with respect to intermediate coordinates ξ_i by a colon (:), the equation of motion for the intermediate state is

$$\bar{T}_{ij:j}+\bar{\varrho}\bar{b}_i = \bar{\varrho}\ddot{\xi}_i \tag{43}$$

where $\bar{\varrho}$ and $\bar{\mathbf{b}}$ represent the density and body forces, respectively, in the intermediate state. Differentiation with respect to final coordinates is denoted by a comma (,) so that for the total motion we have

$$T_{ij,\,j}+\varrho b_i = \varrho\ddot{x}_i \tag{44}$$

where ϱ and **b** are the values of density and body forces in the final state.

With the chain rules

$$\frac{\partial}{\partial x_j} = \frac{\partial}{\partial \xi_k}\xi_{k,\,j}, \quad \frac{\partial}{\partial \xi_j} = \frac{\partial}{\partial x_k}x_{k:j} \tag{45}$$

we obtain from (10) that

$$\frac{\partial}{\partial x_j} = \frac{\partial}{\partial \xi_j}-u_{k,\,j}\frac{\partial}{\partial \xi_k}, \quad \frac{\partial}{\partial \xi_j} = \frac{\partial}{\partial x_j}+u_{k:j}\frac{\partial}{\partial x_k}. \tag{46}$$

† For a discussion of the representation of transversely isotropic tensors see Jaunzemis.[4]

‡ For a discussion of stresses and deformations which preserve aelotropy see Truesdell and Noll[5] or Douglas.[3]

The total stress is denoted by

$$T_{ij} = \bar{T}_{ij} + \delta T_{ij} \tag{47}$$

so that, by (46), we arrive at

$$T_{ij,j} = \bar{T}_{ij:j} - \bar{T}_{ij:k} u_{k,j} + \delta T_{ij,j}. \tag{48}$$

The intermediate and final densities can be related by[4]

$$\bar{\varrho} \doteq \varrho(1+u_{k:k}) \doteq \varrho(1+u_{k,k}). \tag{49}$$

To obtain the incremental equations of motion, we substitute (48) into (44):

$$\bar{T}_{ij:j} - \bar{T}_{ij:k} u_{k,j} + \delta T_{ij,j} + \varrho(\bar{b}_i + \delta b_i) = \varrho(\ddot{\xi}_i + \ddot{u}_i). \tag{50}$$

Next, we note that, by (43) and (49),

$$\bar{T}_{ij:j} + \varrho \bar{b}_i + \varrho H_{kk} \bar{b}_i = \varrho \ddot{\xi}_i + \varrho H_{kk} \ddot{\xi}_i. \tag{51}$$

Subtracting (51) from (50) and making further use of (46), we obtain the equations of incremental motion (referred to the final coordinates),

$$\delta T_{ij,j} - \bar{T}_{ij,k} u_{k,j} + \varrho(\delta b_i - u_{k,k} \bar{b}_i) = \varrho(\ddot{u}_i - u_{k,k} \ddot{\xi}_i). \tag{52}$$

Because differentiation with respect to ξ_i and x_i is interchangeable to within higher order terms, and noting that

$$\varrho = \bar{\varrho}(1-u_{k:k}) = \bar{\varrho}(1-u_{k,k}), \tag{53}$$

the equations of motion referred to the intermediate coordinates ξ_i are given by

$$\delta T_{ij:j} - \bar{T}_{ij:k} u_{k:j} + \bar{\varrho}(\delta b_i - u_{k:k} \bar{b}_i) = \bar{\varrho}(\ddot{u}_i - u_{k:k} \ddot{\xi}_i) \tag{54}$$

where all terms of 0 ($\mathbf{H}^2$) have been neglected.

If the intermediate configuration is an equilibrium state, and if body forces are absent, then (52) and (54) reduce to

$$\delta T_{ij,j} - \bar{T}_{ij,k} u_{k,j} = \varrho \ddot{u}_i \tag{55}$$

$$\delta T_{ij:j} - \bar{T}_{ij:k} u_{k:j} = \bar{\varrho} \ddot{u}_i. \tag{56}$$

It is evident that the familiar form of the equations of motion is recovered only when the underlying stress field is homogeneous, in which case we obtain

$$\delta T_{ij,j} = \varrho \ddot{u}_i, \quad \delta T_{ij:j} = \bar{\varrho} \ddot{u}_i. \tag{57}$$

To obtain the displacement equations of motion, the incremental stress–strain relations (23), (24) are substituted into (55), thus

$$\begin{aligned} &-u_{k,kj} \bar{T}_{ij} - u_{k,k} \bar{T}_{ij,j} + u_{i,kj} \bar{T}_{kj} + u_{i,k} \bar{T}_{kj,j} + \bar{T}_{ik} u_{j,kj} + \\ &+ A_{ijkm,j} u_{k,m} + A_{ijkm} u_{k,mj} = \varrho \ddot{u}_i \end{aligned} \tag{58}$$

which may be written

$$L_{ijkm} u_{k,mj} + M_{ijk} u_{k,j} = \varrho \ddot{u}_i, \tag{59}$$

where

$$L_{ijkm} = -\bar{T}_{ij}\delta_{km}+\bar{T}_{mj}\delta_{ik}+\bar{T}_{im}\delta_{jk}+A_{ijkm}, \tag{60}$$

and

$$M_{ijk} = \bar{T}_{im,\,m}\delta_{kj}+\bar{T}_{jm,\,m}\delta_{ki}+A_{imkj,\,m}. \tag{61}$$

It is evident from (59) that the incremental equations of motion are quite different from the classical equations because they involve both first and second derivatives of displacements. The coefficients L_{ijkm} depend on the underlying stress and deformation whereas the coefficients M_{ijk} depend on the gradients of stress and deformation. Thus, an inhomogeneous initial stress leads to inhomogeneous incremental equations of motion.

To complete this section we take up the problem of boundary conditions. Displacement boundaries B_u offer no problems; recalling

$$x_i = \xi_i+u_i, \tag{10)R}$$

it follows that here it is necessary to prescribe the values of **u**. In dealing with traction boundaries, B_T, however, the usual results of the classical theory must be generalized to include the effect of initial stresses. For the intermediate and final state we have that

$$\bar{t}_i = \bar{T}_{ji}\bar{n}_j, \qquad t_i = T_{ji}n_j. \tag{62}$$

Setting

$$t_i = \bar{t}_i+\delta t_i, \quad n_j = \bar{n}_j+\delta n_j \tag{63}$$

and using (12) we obtain

$$\delta t_i = \delta T_{ji}\bar{n}_j+\bar{T}_{ji}\delta n_i+\delta T_{ji}\delta n_j. \tag{64}$$

To evaluate $\delta\mathbf{n}$, we begin with [1]

$$dA_j = n_j\,dA = \bar{J}\xi_{i,\,j}\,d\bar{A}_i = \bar{J}\xi_{i,\,j}\,d\bar{A}\bar{n}_i, \tag{65}$$

which relates the intermediate configuration of a material surface element $d\bar{\mathbf{A}}$ with its final configuration $d\mathbf{A}$ where, in the limit of small displacements,

$$\bar{J} = \det|\xi_{i,\,j}| = 1+H_{kk}. \tag{66}$$

With (10) and (66), we obtain

$$n_j\,dA = d\bar{A}(\bar{n}_j+H_{kk}\bar{n}_j-H_{ij}\bar{n}_i-H_{kk}H_{ij}\bar{n}_i), \tag{67}$$

and, neglecting terms of order $\mathbf{H}^2$ and higher,

$$\delta n_j = n_j-\bar{n}_j = H_{kk}\bar{n}_j-H_{ij}\bar{n}_i. \tag{68}$$

It remains to substitute this into (64) (again retaining only terms linear in **H**), thus

$$\delta t_i = (\delta T_{ji}+H_{kk}\bar{T}_{ji}-\bar{T}_{ki}H_{jk})\bar{n}_j, \tag{69}$$

which reduces to the form (62) only when $\bar{\mathbf{T}} = \mathbf{0}$. We record result (69) in direct notation by

$$\delta \mathbf{t} = [\delta \mathbf{T} + (\text{tr } \mathbf{H})\bar{\mathbf{T}} - \mathbf{T}\mathbf{H}^T] \cdot \bar{\mathbf{n}}. \tag{70}$$

IV. STABILITY

A discussion of stability in finite strain is somewhat different from that which is encountered in classical elasticity theory. Euler's work and the subsequent researches of others have led to the interpretation of elastic instability as primarily a geometric phenomenon occurring in rods, plates, and shells due to the large ratios of length to thickness of such bodies. More recently, however, other types of instabilities have been proposed and investigated, particularly an "internal buckling" and other associated instabilities defined by Biot,[(6)] and "material instabilities" that may occur in finite shear, for instance (Truesdell[(7)]). The study of such instabilities complements, and is related to, other methods of exploring the nature of elastic response. Of these, we mention non-uniqueness (interpreted as incipient instability), and the study of "plausibility requirements" which have been put forth by Coleman and Noll,[(8)] Baker and Ericksen,[(9)] and by Truesdell and Noll[(5)] to restrict elastic behavior to being intuitively "reasonable." (These conditions are analogous to requiring that bulk modulus and shear modulus of the linear theory be positive.)

We may define local stability either in terms of the *strong ellipticity* (S–E) condition ("superstability"), according to which a fourth rank tensor $\mathbf{L}$ must satisfy

$$L_{ijkm}A_iA_kB_jB_m > 0 \tag{71}$$

for arbitrary non-zero vectors $\mathbf{A}$ and $\mathbf{B}$, or in terms of the weaker Hadamard's inequality[(7)]

$$L_{ijkm}A_iA_kB_jB_m \geqq 0. \tag{71}^*$$

If $\mathbf{L}$ is the elasticity tensor, then, among other things, (71)* ensures the existence of real wave speeds for any direction of propagation. As shown by Truesdell and Noll,[(5)] in a homogeneously reformed body (71)* also implies stability for the displacement boundary value problem (but not necessarily for the traction boundary value problem). In any case, because (71)* is a necessary condition it provides a simple check of stability.

The existence of standing wave solutions in homogeneously strained bodies means that the equations of equilibrium are a hyperbolic system, rather than an elliptic system. This change from one type to another can be investigated, for *any* state of initial strain, by considering the existence of surfaces of discontinuity ("characteristic surfaces") in the second derivatives of displacements. The basic compatibility relations for surfaces of discontinuity all stem from Hadamard's lemma.

Let $x_i = x_i(s)$ be a space curve, and let $\phi(\mathbf{x})$ be a generic tensor field. Along the space curve, ϕ admits the total derivative $d\phi/ds$, which may be evaluated by the chain rule

$$\frac{d\phi}{ds} = \phi_{,i} \frac{dx_i}{ds}. \tag{72}$$

According to Hadamard's lemma, this chain rule holds also if the space curve lies in a surface of discontinuity, S, of ϕ (Fig. 3), i.e. when the one-sided limits ϕ^+ and ϕ^- on S do not coincide,

$$[\![\phi]\!] = \phi^+ - \phi^- \neq 0. \tag{73}$$

Thus

$$\frac{d\phi^\pm}{ds} = \phi^\pm{}_{,i} \frac{dx_i}{ds}, \tag{74}$$

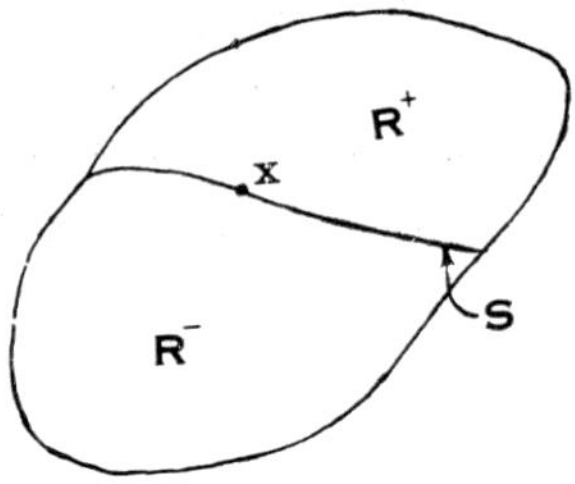

FIG. 3. Definitions for characteristic surfaces

from which it follows that the tangential derivative of a jump equals the jump of the tangential derivative(1)

$$\frac{d}{ds}[\![\phi]\!] = \left[\!\!\left[\frac{d\phi}{ds}\right]\!\!\right]. \tag{75}$$

Introducing the Cartesian components D_k of the surface gradient by

$$D_k = \partial_k - n_k D, \tag{76}$$

where

$$D = \mathbf{n} \cdot \nabla \tag{77}$$

denotes differentiation in a direction normal to the surface, we can express Hadamard's lemma as follows,

$$D_k[\![\phi]\!] = [\![D_k\phi]\!]. \tag{78}$$

Now let

$$[\![\phi]\!] = A, \quad [\![D\phi]\!] = B, \quad [\![D^2\phi]\!] = C. \tag{79}$$

Then $[\![\phi_{,i}]\!] = D_i A + n_i B$; in particular, if $A = 0$, then $[\![\phi_{,i}]\!] = n_i B$. Similarly, assuming that $A = B = 0$, we find that $[\![\phi_{,ij}]\!] = C n_i n_j$. For instance, assuming that displacements and displacement gradients are continuous, but that second gradients need not be continuous, we obtain

$$[\![u_{i,jk}]\!] = \bar{A}_i n_j n_k, \quad \bar{A}_i = [\![D^2 u_i]\!] = [\![u_{i,mn}]\!] n_m n_n. \tag{80}$$

The displacement equations of equilibrium are, from (59),

$$L_{ijkm} u_{k,mj} + M_{ijk} u_{k,j} = 0 \tag{81}$$

where L_{ijkm} and M_{ijk} are coefficients depending on the initial deformation. Assuming that the initial deformation is smooth and that there are no jumps in the first displacement gradients, we obtain from (80) and (81)

$$L_{ijkm} n_m n_j A_k = 0 \tag{82}$$

which is a homogeneous system of equations for the amplitudes A_k. For non-trivial solutions of (82) to exist, it is necessary that

$$\det (L_{ijkm} n_j n_m) = 0. \tag{83}$$

A solution of (83) is not a simple matter, because of the large number of independent parameters. By choosing certain directions **n**, however, one may attempt to find deformations and stresses for which (83) is satisfied; conversely, one may choose deformations, **F**, and then attempt to find the characteristic directions **n**.

As an example, let us consider a laminated medium which is characterized by the transversely isotropic response given by (33) and (36). Let us assume that the initial stress is hydrostatic,

$$\bar{\mathbf{T}} = -p\mathbf{1} \tag{84}$$

so that, by (60),

$$L_{ijkm} n_j n_m = -p\delta_{ik} + A_{ijkm} n_j n_m. \tag{85}$$

The application of hydrostatic stress does not change the nature of the incremental response but does change the magnitudes of the elastic moduli.[3] Specifically, the tensor A_{ijkm} is now given by

$$\begin{aligned} A_{ijkm} = {} & \bar{\lambda}\delta_{ij}\delta_{km} + \bar{\mu}(\delta_{ik}\delta_{jm} + \delta_{jk}\delta_{im}) + \\ & + \bar{\alpha}(\delta_{ij}\delta_{3k}\delta_{3m} + \delta_{3i}\delta_{3j}\delta_{km}) + \bar{\beta}(\delta_{3i}\delta_{3k}\delta_{jm} \\ & + \delta_{3i}\delta_{3m}\delta_{jk} + \delta_{3j}\delta_{3k}\delta_{im} + \delta_{3j}\delta_{3m}\delta_{ik}) + \bar{\gamma}\delta_{3i}\delta_{3j}\delta_{3k}\delta_{3m}, \end{aligned} \tag{86}$$

where the moduli are functions of the deformation

$$\mathbf{F} = \begin{bmatrix} A & 0 & 0 \\ 0 & A & 0 \\ 0 & 0 & B \end{bmatrix}. \tag{87}$$

Substitution of $F_{ij} = A\delta_{ij} + (B-A)\delta_{i3}\delta_{j3}$, (85), and (86) into (83) then yields

$$\begin{aligned} \det [& (\bar{\mu} + n_3^2\bar{\beta} - p)\delta_{ik} + (\bar{\lambda} + \bar{\mu}) n_i n_k + (\bar{\alpha} + \bar{\beta})\, n_3(\delta_{3i} n_k + \delta_{3k} n_i) + \\ & + (\bar{\beta} + n_3^2\bar{\gamma})\delta_{3i}\delta_{3k}] = 0 \end{aligned} \tag{88}$$

or

$$\begin{vmatrix} [\bar{\mu}+n_3^2\bar{\beta}+(\bar{\lambda}+\bar{\mu})n_1^2-p] & (\bar{\lambda}+\bar{\mu})n_1n_2 & (\bar{\lambda}+\bar{\mu}+\bar{\alpha}+\bar{\beta})n_1n_3 \\ (\bar{\lambda}+\bar{\mu})n_1n_2 & [\bar{\mu}+n_3^2\bar{\beta}+(\bar{\lambda}+\bar{\mu})n_2^2-p] & (\bar{\lambda}+\bar{\mu}+\bar{\alpha}+\bar{\beta})n_2n_3 \\ (\bar{\lambda}+\bar{\mu}+\bar{\alpha}+\bar{\beta})n_1n_3 & (\bar{\lambda}+\bar{\mu}+\bar{\alpha}+\bar{\beta})n_2n_3 & \bar{\mu}+\bar{\beta}+(\bar{\lambda}+\bar{\mu}+2\bar{\alpha} \\ & & +3\bar{\beta}+\bar{\gamma})n_3^2-p \end{vmatrix} = 0. \tag{89}$$

In order to obtain more explicit results it is necessary to consider various directions $\mathbf{n}$. For instance, if $\mathbf{n} = (0, 0, 1)$, we obtain

$$p = \bar{\mu}+\bar{\beta}, \quad p = \bar{\lambda}+2\bar{\mu}+2\bar{\alpha}+4\bar{\beta}+\bar{\gamma} \tag{90}$$

as solutions. For $\mathbf{n} = (1, 0, 0)$, we obtain

$$p = \bar{\mu}, \quad p = \bar{\lambda}+2\bar{\mu}, \quad p = \bar{\mu}+\bar{\beta}. \tag{91}$$

To proceed further with the discussion it is necessary to make some simplifying assumptions concerning the strain energy function. Let us assume that the strain energy is a *linear* function of the five invariants, which is but an extension of the Mooney-Rivlin equation for isotropic materials. If this is the case, then all second derivatives of W with respect to the invariants vanish. Substituting (87) into (40) and using the linearity assumption we obtain

$$\begin{aligned} \bar{\lambda} &= 4\left(\frac{A^2}{B}W_2+A^2BW_3\right), \qquad \bar{\mu} = -2\left(\frac{A^2}{B}W_2+A^2BW_3\right), \\ \bar{\alpha} &= \frac{4(B^2-A^2)}{B}W_2, \\ \bar{\beta} &= -2\frac{(B^2-A^2)}{B}W_2+2BW_5, \\ \bar{\gamma} &= 2B\frac{(-3B^2-A^2)}{A^2}W_5 \end{aligned} \tag{92}$$

The condition, $(90)_1$, becomes

$$p = 2B\ (W_5-A^2W_3-W_2). \tag{93}$$

Also, by (36), we have that

$$p = \frac{-2}{A^2B}[(A^4+A^2B^2)W_2+A^4B^2W_3+A^2W_1] \tag{94}$$

and

$$p = \frac{-2}{A^2B}[2A^2B^2W_2+A^4B^2W_3+B^2W_1+B^2W_4]. \tag{95}$$

Equating (94) and (95), we get

$$B^2 = \frac{A^2(W_1+A^2W_2)}{A^2W_2+W_1-W_4} \tag{96}$$

so that A may be taken as the independent parameter. Equating (93) and (94), on the other hand,

$$B^2 W_5 + A^2 W_2 + W_1 = 0 \tag{97}$$

and eliminating B^2, we get

$$(A^2 W_2 + W_1 + A^2 W_5 - W_4)(A^2 W_2 + W_1) = 0. \tag{98}$$

Hence, the two solutions are

$$A^2 W_2 + W_1 + A^2 W_5 - W_4 = 0, \quad A^2 W_2 + W_1 = 0 \tag{99}$$

which means that when

$$A^2 = \frac{W_4 - W_1}{W_2 + W_5}, \qquad A^2 = -\frac{W_1}{W_2} \tag{100}$$

characteristics can exist.

An interpretation of this instability may be obtained from the incremental stress relations. Using (49) and (86), we obtain for the case of hydrostatic loading,

$$\begin{aligned} \delta T_{11} &= (\bar{\lambda}+2\bar{\mu}-p)\tilde{E}_{11}+(\bar{\lambda}-p)\tilde{E}_{22}+(\bar{\lambda}+\bar{\alpha}-p)\tilde{E}_{33} \\ \delta T_{33} &= (\bar{\lambda}+\bar{\alpha}-p)\tilde{E}_{11}+(\bar{\lambda}+\bar{\alpha}-p)\tilde{E}_{22}+ \\ &\quad +(\bar{\lambda}+2\bar{\mu}+2\bar{\alpha}+4\bar{\beta}+\bar{\gamma}-p)\tilde{E}_{33} \\ \delta T_{12} &= 2(\bar{\mu}-p)\tilde{E}_{12}, \quad \delta T_{13} = 2(\bar{\mu}+\bar{\beta}-p)\tilde{E}_{13}. \end{aligned} \tag{101}$$

It is evident that the condition $(90)_2$ coincides with the vanishing of the incremental elastic coefficient of $\tilde{E}_{33}$ in $(101)_2$. Similarly, $(91)_2$ implies the vanishing of the coefficient of $\tilde{E}_{11}$ in $(101)_1$. The remaining two conditions $(90)_1$ and $(91)_3$ imply that the incremental shear moduli vanish. The latter conditions mean that the material has no resistance to superposed incremental shearing stresses once the large initial pressures have been applied. Similarly, the former conditions imply that the material may undergo incremental normal strains without the application of normal stresses.

The relation of the instability to wave propagation may be seen by considering a moving surface of discontinuity in the prestrained material. If ξ is an equilibrium position then ξ_i can be interpreted as material coordinates. For an observer moving together with the surface, $\xi_i = R_i(t)$, $V_i = dR_i/dt$. A jump $[\![\phi]\!]$ is a function of time only; however, to carry a differentiation with respect to time inside the brackets, this differentiation must be interpreted as executed by an observer following the surface of discontinuity, thus

$$\frac{d}{dt}[\![u_i]\!] = [\![\dot{u}_i + V_j \dot{u}_{i:j}]\!] \tag{102}$$

$$\frac{d}{dt}[\![u_{k:m}]\!] = [\![\dot{u}_{k:m} + V_n u_{k:mn}]\!]. \tag{103}$$

In particular, if

$$[\![u_i]\!] = 0, \quad [\![u_{k:m}]\!] = 0 \tag{104}$$

then

$$[\![\ddot{u}_i]\!] = -V_j[\![u_{i:j}]\!] = V_j V_m[\![u_{i:jm}]\!] = 0. \tag{105}$$

Now let $V_j = Vn_j$, where V is the normal speed of propagation. Then from (80) and

$$L_{ijkm}[\![u_{k:mj}]\!] = \varrho[\![\ddot{u}_i]\!] = \varrho V^2 n_j n_m[\![u_{i:jm}]\!]$$

we obtain

$$L_{ijkm}n_m n_j A_k = QA_i, \quad Q = \varrho V^2. \tag{106}$$

The acoustic tensor

$$Q_{ik} = L_{ijkm}n_m n_j \tag{107}$$

describes the squared wave speeds of propagation. These squared speeds will be positive only if **Q** is positive (i.e., only if the principal invariants of **Q** are positive). The resulting characteristic equation is written

$$Q^3 - \mathrm{I}_{\mathbf{Q}}Q^2 + \mathrm{II}_{\mathbf{Q}}Q - \mathrm{III}_{\mathbf{Q}} = 0. \tag{108}$$

If $\mathrm{III}_{\mathbf{Q}} = \det L_{ijkm}n_m n_j = 0$, then one root of (108) is $Q = 0$, meaning that at least one principal wave speed is zero. Therefore, instability coincides with the existence of at least one direction along which the speed of propagation of a plane wave is zero.

These results can be given further meaning by considering natural vibrations of homogeneously prestrained materials. Assuming a solution

$$u_i = A_i \sin n_k x_k \cos \omega t \tag{109}$$

we obtain from (59) (since $M_{ik} = 0$)

$$L_{ijkm}n_m n_j A_k = \varrho\omega^2 A_i \tag{110}$$

which is of the same form as (106), and thus the natural frequencies are obtained from eq. (108). The condition, $\det L_{ijkm}n_j n_m = 0$ implies a solution $\omega = 0$, or $u_i = A_i \sin n_k x_k$. Therefore, such "standing wave" solutions can exist when the underlying state of stress and deformation reach the critical value. If, for subsequent further loading $\det L_{ijkm}n_j n_m < 0$, a solution $\omega^2 < 0$ exists, which means that displacements become unbounded with increasing time. This definition of instability is, of course, well known.

As was pointed out before, an instability may amount to incipient incremental flow, in which case we may speak of a *material instability*. Truesdell[(7)] has recently discussed this for the problem of simple shear. We shall discuss equivalent results for transversely isotropic media.

Let us assume a finite simple shear defined by (Fig. 4)

$$x_1 = X_1, \quad x_2 = X_2, \quad x_3 = KX_1+X_3. \tag{111}$$

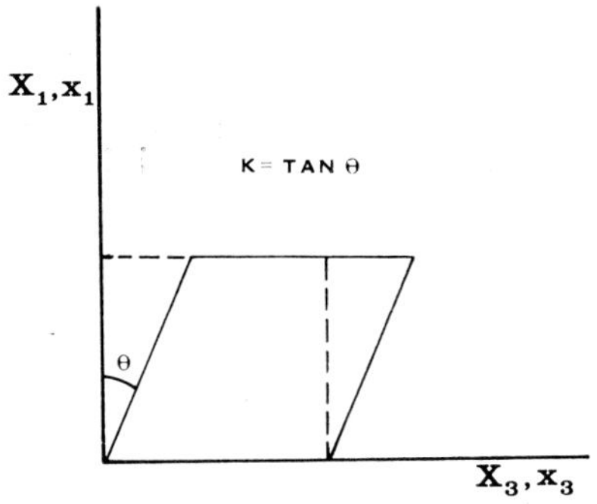

FIG. 4. Definitions for simple shear problem

Then the deformation measures are given by

$$\mathbf{F} = \begin{bmatrix} 1 & 0 & 0 \\ 0 & 1 & 0 \\ K & 0 & 1 \end{bmatrix}, \quad \mathbf{B} = \begin{bmatrix} 1 & 0 & K \\ 0 & 1 & 0 \\ K & 0 & 1+K^2 \end{bmatrix}, \quad \mathbf{B}^{-1} = \begin{bmatrix} 1+K^2 & 0 & -K \\ 0 & 1 & 0 \\ -K & 0 & 1 \end{bmatrix},$$

$$\mathbf{C} = \begin{bmatrix} 1+K^2 & 0 & K \\ 0 & 1 & 0 \\ K & 0 & 1 \end{bmatrix}. \tag{112}$$

According to (36), the stresses are:

$$T_{11} = 2\,(W_1+2W_2+W_3), \quad T_{22} = 2\,[W_1+(2+K^2)W_2+W_3],$$
$$T_{33} = 2\,[(1+K^2)W_1+(2+K^2)W_2+W_3+W_4+K^2W_5]$$
$$T_{13} = 2\,(W_1+W_2+W_5)K, \quad T_{12} = T_{23} = 0. \tag{113}$$

The shear modulus, $\hat{\mu}$, is defined by

$$T_{13} = \hat{\mu}K \tag{114}$$

so that

$$\hat{\mu} = 2(W_1+W_2+W_5). \tag{115}$$

The vanishing of the shear modulus is termed *flow in shear*. Hence, *flow in shear* occurs when

$$W_1+W_2+W_5 = 0. \tag{116}$$

Collapse in shear occurs when the shear stress attains a maximum value, i.e., $\partial T_{13}/\partial K = 0$. Since K is the only deformation parameter we can write

$$\frac{d\hat{\mu}}{dK}K+\hat{\mu} = 0 \tag{117}$$

for *collapse in shear*, or equivalently

$$K\frac{d}{dK}(W_1+W_2+W_5)+W_1+W_2+W_5 = 0. \tag{118}$$

To make the example more explicit, let us assume a form of the strain energy which is analogous to that which was used by Truesdell.[7] Since the deformation is isochoric (i.e., volume preserving) W is independent of III (III $\equiv$ 1), and we can use an extension of the Mooney–Rivlin incompressible material. Therefore, we assume

$$W = \tfrac{1}{2}\mu[(\tfrac{1}{2}+\beta)(\mathrm{I}-3)-(1/2\gamma)(\mathrm{I}-3)^2+(\tfrac{1}{2}-\beta)(\mathrm{II}-3)+ \\ +\alpha\,(\mathrm{IV}-1)+\delta\,(\mathrm{IV}-1)^2+\zeta(\mathrm{V}-1)] \tag{119}$$

where

$$\mu > 0, \quad -\tfrac{1}{2} \leq \beta \leq \tfrac{1}{2}, \quad \gamma \neq 0, \quad \alpha,\ \delta,\ \zeta \text{ are const.}$$

With (28) and (32), we get

$$\mathrm{I} = \mathrm{II} = 3+K^2, \quad \mathrm{IV} = 1, \quad \mathrm{V} = K^2 \tag{120}$$

$$W = \tfrac{1}{2}\mu\left(K^2-\frac{K^4}{2\gamma}+(K^2-1)\zeta\right). \tag{121}$$

By (115),

$$\hat{\mu} = \mu\left(1-\frac{K^2}{\gamma}+\zeta\right) \tag{122}$$

and *flow in shear* occurs when

$$K^2 = \gamma(1+\zeta). \tag{123}$$

Now using (114), we obtain

$$T_{13} = \mu\left(1-\frac{K^2}{\gamma}+\zeta\right)K \tag{124}$$

and *collapse in shear* occurs when

$$\mu\left(1+\zeta-\frac{3K^2}{\gamma}\right) = 0$$

or

$$K^2 = \frac{\gamma}{3}(1+\zeta). \tag{125}$$

Furthermore, from (121) it follows that

$$\frac{dW}{dK^2} = \tfrac{1}{2}\mu\left[1-\frac{K^2}{\gamma}+\zeta\right] = 0$$

when

$$K^2 = \gamma(1+\zeta),$$

which means that the strain energy is maximum at *flow in shear*.

The relation between this material instability in shear and local instability may be examined in a straightforward manner if the materials are assumed to be isotropic. For isotropic media the incremental shear stress-deformation relation may be written in the principal coordinates of the initial deformation[(3)]

$$\delta T_{13} = 2(V_3^2 W_1 + V_1^{-2})H_{13} + 2(V_1^2 W_1 + V_3^{-2} W_2)\, H_{31}. \tag{126}$$

In the principal coordinate system, (112) becomes

$$\mathbf{B} = \begin{bmatrix} V^2 & 0 & 0 \\ 0 & 1 & 0 \\ 0 & 0 & V^{-2} \end{bmatrix}. \tag{127}$$

hence (126) may be written

$$\delta T_{13} = 2(W_1 + W_2)(V^{-2} H_{13} + V^2 H_{31}). \tag{128}$$

For isotropic media $W_5 = 0$, so that from (116) and (128) it is evident that $\hat{\mu} = 0$ implies $\delta T_{13} \equiv 0$ for arbitrary incremental deformations $\mathbf{H}$. Consequently, *flow in shear* implies local instability in the sense that the incremental shear moduli vanish. However, for other kinds of stress–strain relations an instability may be an elastic, "internal" buckling. It seems that, as in the theory of Euler columns, the precise nature of internal buckling could be found only from a complete solution of the nonlinear equations.

V. SUMMARY

The object of this paper was to investigate, and interpret, the instabilities that may occur in finitely deformed laminated media. Starting with the example of column buckling, analogous conditions were obtained for the existence of "standing wave" solutions in laminated media. The occurrence of instability was also related to the vanishing of natural frequencies and wave speeds of propagation. A further interpretation, namely, that of material instability, was illustrated for the problem of simple shear. The method used was that of small deformations superposed on large ones. It was noted that the incremental stress–strain relations and equations of motion are considerably different from the equations obtained when the ground state is a natural one. In particular, the effective elastic moduli may vanish, which would correspond to a loss of stability. Although specific results were obtained only for initial hydrostatic compression, the method can be applied to arbitrary initial stresses and deformations. As is the case with any other linear stability theory, it cannot, in general, predict the behavior after instability has occurred.

REFERENCES

1. TRUESDELL, C. and TOUPIN, R. A. The Classical Field Theories, *Flugge's Handbuch der Physik*, vol. III, part 1, Springer-Verlag, 1960.
2. SMITH, G. F. and RIVLIN, R. S. The Strain-Energy Function for Anisotropic Materials, *Trans. Amer. Math. Soc.* **88,** 175–193 (1958); Stress-Deformation Relations for Anisotropic Solids, *Arch. Rat. Mech. Anal.* **1,** 107–112 (1957).
3. DOUGLAS, W. J. *Elastic Response of Pre-Strained Bodies*, Ph.D. Thesis, Pennsylvania State University, Dept. of Eng. Mech., 1966.
4. JAUNZEMIS, W. *Continuum Mechanics*, The Macmillan Company, 1967.
5. TRUESDELL, C. and W. NOLL. The Non-Linear Field Theories of Mechanics, *Encyclopedia of Physics*, vol. III, part 3, Springer-Verlag.
6. BIOT, M. A. Internal Buckling Under Initial Stress in Finite Elasticity, *Proc. Roy. Soc. Lond.*, **A273,** 306–328 (1963).
7. TRUESDELL, C. Instabilities of Isotropic Perfectly Elastic Materials in Simple Shear, *Proc. 11th International Congress of Applied Mechanics, Munich, 1964*, Springer-Verlag, 1966.
8. COLEMAN, B. D. and NOLL, W. On the Thermostatics of Continuous Media, *Arch. Rat. Mech. Anal.* **4**/2, 97–128 (1959).
9. BAKER, M. and ERICKSEN, J. L. Inequalities Restricting the Form of the Stress Deformation Relations for Isotropic Elastic Solids and Reiner–Rivlin Fluids, *J. Wash. Acad. Sci.*, **44,** 33–35 (0000).
10. BEATTY, M. F. Some Static and Dynamic Implications of the General Theory of Elastic Stability, *Arch. Rat. Mech. Anal.* **19,** 167–188 (1965).

MECHANICAL PROPERTIES OF WIRE-REINFORCED GRAINS

DR. R. A. HELLER†

Virginia Polytechnic Institute

ABSTRACT

A large variety of experiments have been performed to evaluate the mechanical properties of simulated, wire-reinforced solid propellants and to examine typical failure modes of such materials.

It has been found that while axial compressive strength, hoop strength, fatigue, and torsion strength increase with aluminum content, axial tensile strength decreases for the tested configurations. Elastic moduli increase with wire content and relations between the various moduli are adequately described by orthotropic relationships. The experimental results are afflicted by an unusual amount of scatter.

Failure modes are of four types: (1) In axial tension, failure takes place by binder separation unless the reinforcing helix is very steep. (2) Hoop tension failure of the wire as a result of stress concentration on the inner surface produces premature failure under internal pressure. (3) In axial compression, longitudinal and diametral bending failure may be attributed to a lack of radial reinforcement. (4) In vibration tests heat generation due to interlaminar shear and consequent thermal softening of the binder causes failure by delamination.

I. INTRODUCTION

To increase the performance of two-phase solid rocket motors beyond that of currently operational systems in terms of higher combustion pressures, higher accelerations and faster burning rates, experimental programs are in progress for the development of wire-reinforced grains.

In order to develop appropriate mathematical analyses(1) for such helically-reinforced structures and to determine the significance of various physical and mechanical variables and parameters, an extensive test program has been carried out on inert grain samples. Information on real grains may be found in refs. 2 and 3.

These samples were supplied by Rocketdyne, Division of North American Aviation Inc. in 1962; they were produced in a pilot project with flat winding angles (62°–83°), an aluminum wire content ranging from about 5 to 25% by weight, a potassium chloride filler (65–85% by weight), and polybutadiene binder.

† Formerly Deputy Director, Institute for the Study of Fatigue and Reliability, Columbia University.

In these experiments the strength, moduli of elasticity, Poisson's ratios and static and dynamic creep rates were measured under a variety of loading conditions. In addition to the determination of mechanical constants a series of experiments concerned with mechanisms of failure were conducted. It was, however, not possible to obtain direct measurements of some constants because of the limitations imposed by the cylindrical-tubular configuration of the available samples. The findings indicate that a considerable improvement to withstand higher pressures is realized by this method of reinforcement; however, new problems arise from the lack of radial reinforcement. Of these interlaminar shear failures, radial shrinkage failures, and tangential buckling are the most significant.

The large scatter of test results which may be only partly due to the lack of quality control in a pilot project or to the prolonged storage of the samples, is another area of concern.

The importance of stress- and/or strain-rate dependance of the significant mechanical properties is suggested by several experiments.[4] However, a more thorough study of these effects using a larger number of samples is indicated.

The behavior of the orthotropic cylinders in a vibrating environment was also examined. While the cylinders exhibit a high oscillating-load-carrying capacity in comparison to unreinforced cylinders, heat production and failures occur due to softening as a result of laminar shear cycling.

II. DESCRIPTION OF EXPERIMENTS

Inert solid propellant grain models, with various concentrations of potassium chloride filler and polybutadiene binder helically reinforced with aluminum wire (see Fig. 1), were tested under the following conditions:

Type of test	Code
Static axial tension	T
Dynamic axial tension	DT
Static axial compression	C
Dynamic axial compression	DC
Longitudinal bending	B
Hoop tension	HT
Diametral bending	HB
Diametral fatigue	F
Internal pressure	P
Torsion	TO
Axial compressive creep	CC
Axial compressive relaxation	R

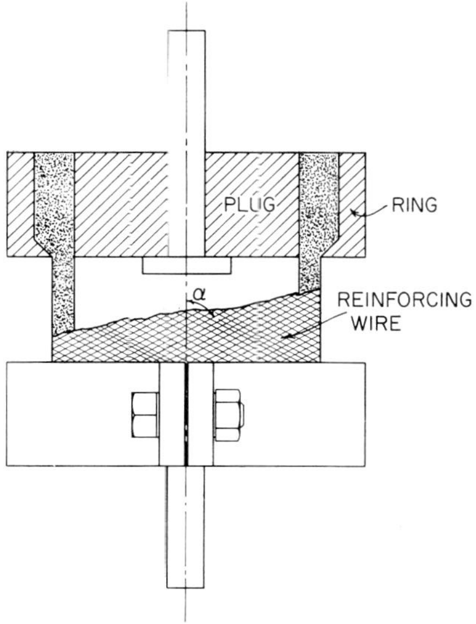

FIG. 1. Wire reinforced grain in axial tension fixture

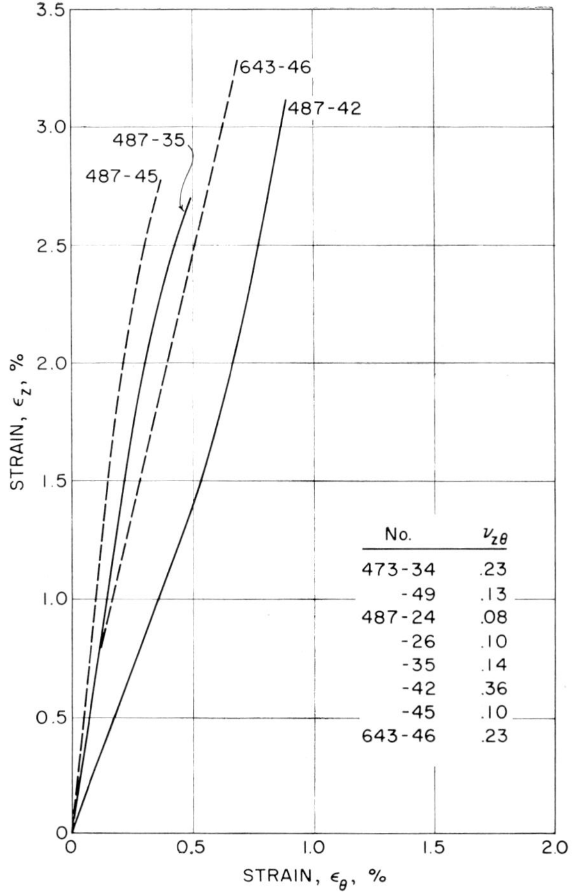

FIG. 2. Axial compressive strain vs. tangential strain on outer surface ($6\frac{1}{2}$ in. O.D. cylinders in axial compression tests)

The tested specimens and their ingredients are listed in Table 1. All tubes were reinforced with 0.005 in. diameter aluminum wire having an average modulus of elasticity $E_w = 12.6 \times 10^6$ psi, yield strength $\sigma_{yw} = 49{,}000$ psi, ultimate strength $\sigma_{uw} = 74{,}200$ psi, and 0.014 in. elongation in a 2 in. gage length.

All $6\frac{1}{2}$ in. diameter grains were helically reinforced, but some of the $3\frac{1}{2}$ in. diameter specimens had in addition a grid type reinforcement running in the longitudinal and transverse directions. For these no winding angle is listed in Table 1.

III. AXIAL TENSION AND COMPRESSION

Six and one-half inch O.D. specimens were cut to 7 in. length with a 2 in. long centrally reduced (6 in. O.D.) section; after being fitted with internal end-plugs and external end rings (see Fig. 1), they were tested at a low strain rate in axial tension. Three and one-half inch O.D. specimens, having a centrally reduced section $6\frac{1}{8}$ in. long, $2\frac{3}{4}$ in. O.D. and 2 in. I.D., were tested in a similar manner at both low and high strain rates.

Elongation within the gage length was assumed to be equal to the cross-head motion of the testing machine.

Typical stress–strain curves based on the total area of the composite are shown in ref. 4 for large diameter specimens at low strain rates (static tests) and at a variety of strain rates for small diameter tubes.

While the two types of experiments are not actually comparable because of the difference in load rates, there is some indication of a size effect. Smaller tubes show both higher strengths and higher elastic moduli than large ones.

A marked strain-rate dependence is indicated in the longitudinal direction. A three-orders-of-magnitude increase in strain-rate approximately doubles the elastic modulus and ultimate strength.

Static compression tests were also performed on $6\frac{1}{2}$ in. diameter, $3\frac{1}{2}$ in. long cylinders. The longitudinal strain as well as the change in outside diameter were recorded in these experiments.

The diameter change was used to compute the tangential surface-strain $\varepsilon_\theta = \Delta(\text{O.D.})/\text{O.D.}_0$, which is plotted versus the longitudinal strain in Fig. 2. The slope of these diagrams is equal to $\gamma_{z\theta}$, Poisson's ratio of tangential to longitudinal strain produced by a longitudinal force.

In order to determine the variation of the modulus of elasticity with increasing amount of reinforcement the measured values have been plotted on semilogarithmic paper versus the aluminum percentage by weight. To account for different winding angles, α, the modulus E_z is also presented as a function of aluminum content multiplied by the cosine of the angle in Fig. 3.

TABLE 1. DESCRIPTION OF SPECIMENS

No.	O.D. in.	I.D. in.	Wire cont. % wt.	Wire angle degrees	Binder cont. wt.	Filler cont. wt.	Tests performed
473–23	6.5	5.0	19.1	72	10.5	70.4	P
24	6.5	5.0	19.7	72	10.4	69.9	F, B
26	6.5	4.9	20.7	72	12.7	66.6	F, T
27	6.5	4.9	7.5	72	12.	80.5	F, P, HT, HB
28	6.5	4.9	10.8	72	11.6	77.6	P
31	6.5	5.0	9.3	72			P
32	6 5	4.9	10.5	72	14.	75.5	F, T
34	6.5	4.9	15.	83	14.	71.	F, C
35	6.5	4.9	12.6	83			P
36			17.6	83			
37	6.5	4.9	14.3	83	14.	71.7	P, HT, HB
42	6.5	5.0	9.2	83	14.5	75.9	P
43	6.5	4.9	14.2	83	13.6	72.2	C
44	6.5	4.9	12.3	83	14.	73.7	F, C
47			9.8	83	14.4	75.8	
48	6.5	5.0	14.5	83	13.6	71.9	R, CC
49	6.5	4.9	14.	72	14.	72.	T, C
50	6.5	4.9	10.	72	14.	76.	T
487–24	6.5	4.9	14.	72	11.	75.	F, T, C
25			8.	72	15.	77.	R
26	6.5	4.9	18.	65	13.	69.	T, C
35	6.5	4.9	11.	72	14.	75.	C
42	6.5	4.9	9.	72	15.	77.	F, C
45	6.5	4.9	10.	62	13.	77.	T, C
491–10	6.5	4.9	24.6	72	9.6	65.6	P, HT, HB
11	6.5	4.9	5.8	83	10.4	83.8	F, P, HT, HB
12	2.73	2.02	8.		11.9	80.1	B
13	2.76	2.03	6.1		12.2	81.7	TO
15	2.80	2.03	14.5		11.1	74.4	B
20	6.5	4.9	8.8		11.9	79.3	F
643–07	2.75	2.00	21.6		11.7	66.7	P, T
14			5.6		12.3	82.1	
34	2.74	2.02	5.7		12.3	82.0	P, DT
38	2.76	2.03	21.7		11.0	67.3	TO
39	2.75	2.00	14.6		11.1	74.3	DT
41	2.75	2.00	9.1		11.8	79.1	T
44			10.3		11.7	78.0	
46	6.5	4.9	8.8	83	11.8	79.4	T, C
47	6.5	4.9	15.9	83	10.9	73.2	P, HT, HB
DP16	6.5	4.9	0		40.	60.	T, C, F, P

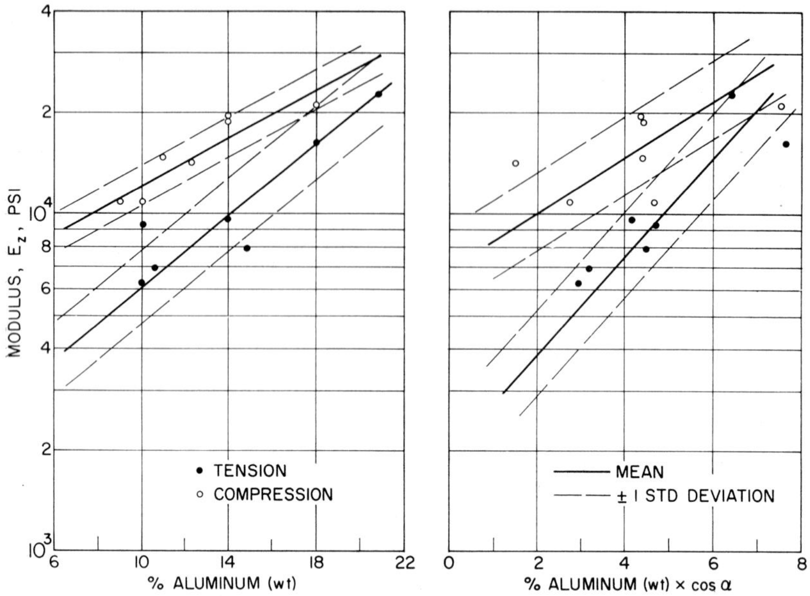

FIG. 3. Axial tensile and compressive elastic moduli as functions of reinforcement

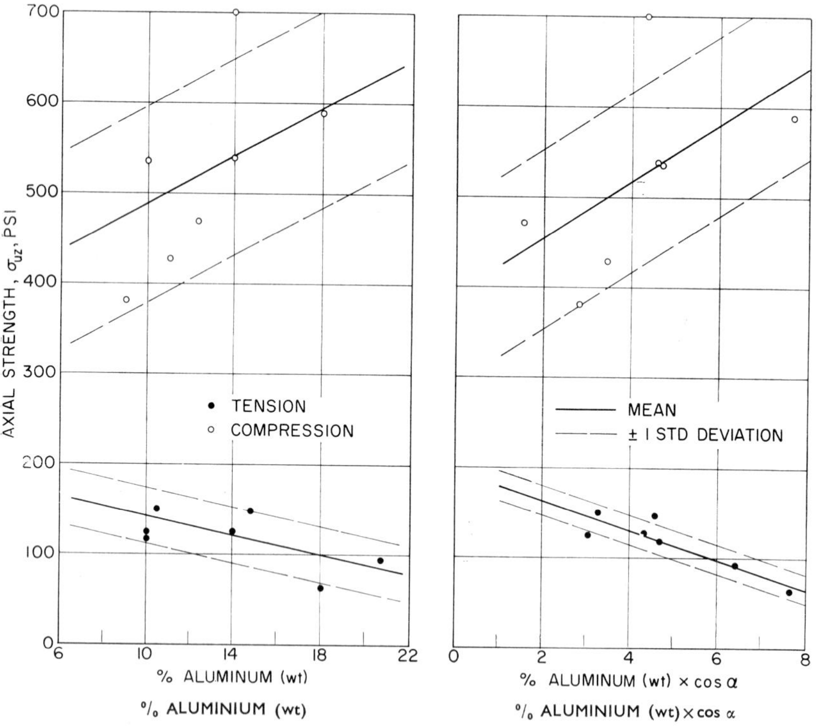

FIG. 4. Axial, ultimate, tensile and compressive strengths as functions of reinforcement

The direction of winding, within the tested range of 62° to 83° does not have as great an influence on the longitudinal modulus as does the amount of reinforcement.

The test results show a wide scatter around a mean least square, regression line; this is the result of lack of quality control in the manufacturing process rather than errors in measurement. The compressive moduli are greater than the tensile ones, because in compression reinforcing wires are in contact with each other. These results seem to contradict the trend reported by Rocketdyne,[3, 5] who found the tensile modulus to be higher than the compressive one for a winding angle of 55°. In view of the steeper angle and the fact that only single numbers were reported without an indication of statistical scatter, the results may not be directly compared.

It is interesting to note that unreinforced, KCl filled, polyurethane (Aerojet DP-16) has an elastic modulus of about 1500 psi, a value approached by the tensile modulus lines of Fig. 3 for zero percent aluminum.

Failure strength as a function of aluminum content, and of effective aluminum content, i.e., $\% \, \mathrm{Al} \times \cos \alpha$ is presented in Fig 4. Failure in axial tension and compression is governed by two different phenomena.

For flat winding angles the tensile load is carried almost entirely by the binder. Since an increase in aluminum content reduces the binder surface between adjacent wires, a reduction in strength with increasing amount of reinforcement is not surprising. The unreinforced material (DP-16) has a tensile strength of about 230 psi. Failure is caused by separation normal to the wires.

In compression the binder between the reinforcement is squeezed out and wire to wire contact is established. As a consequence the compressive strength increases with wire content. Axial compression produces an increase in wall thickness (i.e., the outer diameter increases and the inner decreases) with a compressive tangential stress developing at the inner surface.

The material fails when this compressive stress becomes high enough to buckle the inner surface circumferentially, in a manner similar to the effects of external radial pressure, causing annular separation.

The test results are again scattered. Plotting effective aluminum, $\% \mathrm{Al} \times \cos \alpha$, produces a better fit for tensile strength; for compression, on the other hand, no improvement is visible.

IV. HOOP STRENGTH AND TANGENTIAL MODULUS

Various methods are available for the determination of hoop strength and tangential elastic modulus of thin- and thick-walled rings and tubes. While both the experimental technique and the analysis of data is simpler for thin-walled, narrow rings[6] the information obtained from such tests must be used with caution.

In most cases thin rings are machined from thick-walled tubes. Because the distribution of reinforcement through the wall thickness is not uniform, the thin ring may have a different wire content than the nominal percentage in the original grain. The variation in aluminum content is illustrated in Fig. 5; the number of wires in a segment of 1/2 in. width of a sample were counted with the aid of a microscope.

The choice of ring width introduces a second uncertainty. In a narrow, helically-reinforced ring, the load may be carried entirely by bond strength

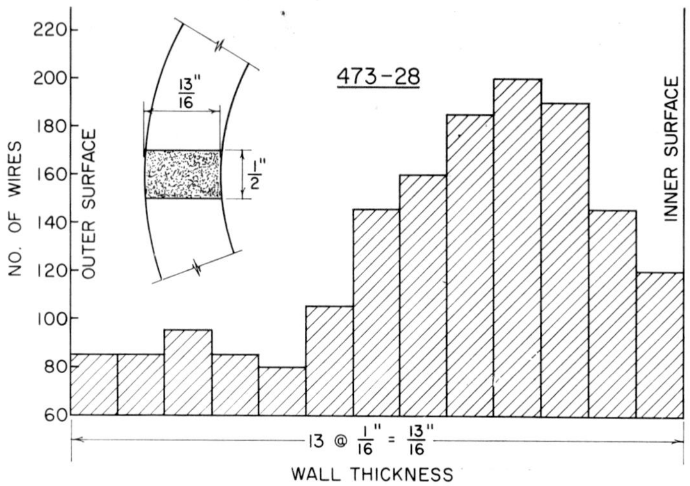

FIG. 5. Hystogram of reinforcement distribution through wall thickness of cylinder

between matrix and reinforcement, while in a wide ring a large proportion of the load is supported by hoop tension in the wire.

The experiments have shown that hoop strength increases with ring width. To develop the full strength of the composite a minimum width of 2 in. was found to be necessary for the flat winding angles (72–83°) tested. For steeper angles even wider rings may be required.

Ring tension tests were performed utilizing two semicircular inserts wrapped in Teflon, in $4\frac{7}{8}$ in. I.D., $t = 1/8$ in. thick rings of width, b, varying between 2 to 3 in. Above a 2 in. width the strength and modulus were found to have statistical variations independent of width. The change in distance between the inserts, Δe, and total force, P, were measured; from these the hoop stress and hoop strain were computed as $\sigma_\theta = P/2bt$ and $\varepsilon_\theta = 2\Delta e/\pi D_{av}$.

A second method, ring bending, was also used to advantage to determine the modulus based on elastic ring theory according to which $\sigma_\theta = 0.955\ D_{av}P/bt^2$

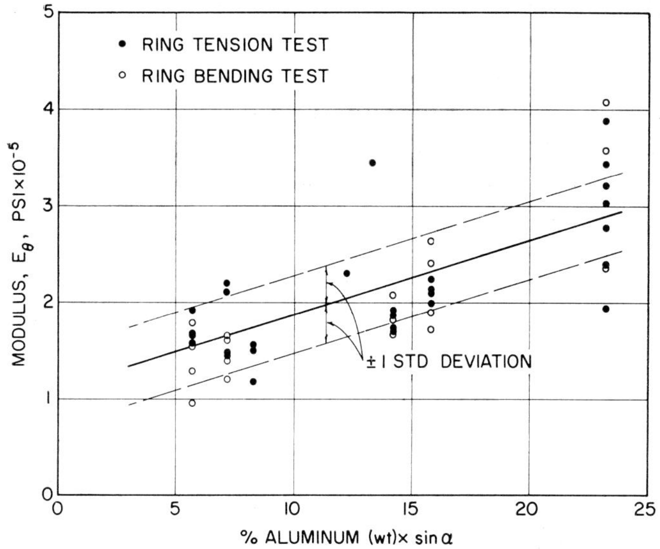

FIG. 6. Tangential elastic modulus as a function of reinforcement

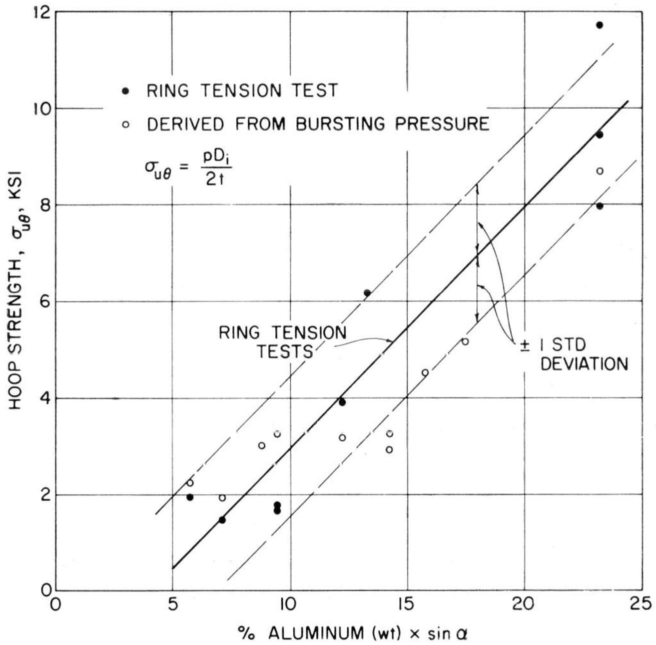

FIG. 7. Ultimate hoop strength as a function of reinforcement. Mean line and standard deviation refers to ring tension tests only

and $\varepsilon_\theta = 4.26 + \Delta e/D^2_{av}$.[7] Values obtained utilizing this method were not different from results of ring tension tests.

The increase in modulus with percent of effective aluminum is shown in Fig. 6, where a least square regression line has been drawn through all the available data together with lines indicating one standard deviation. Since the modulus of the unreinforced (0% Al) propellant is about 1500 psi and that of an all aluminum ring (100% Al) about 10^7 psi, extrapolation of the straight line beyond the test range would lead to an over-estimation for low aluminum content and under-estimation for heavily reinforced grains.

The change of ultimate hoop strength, obtained in ring tension tests, with increasing aluminum content is shown in Fig. 7. A considerable statistical dispersion is evident. Test results based on bursting pressure of thick-walled cylinders, to be described later, are also shown in Fig. 7 but were not used in calculating the regression line and standard deviation.

V. INTERNAL PRESSURE

To correlate thin-ring test results with thick-walled tube behavior internal pressurization tests were performed on $4\frac{7}{8}$ in. I.D., $6\frac{1}{2}$ in. O.D., tubes with lengths ranging from 11 to 14 in.

A rubber bag was inserted in the cylinders whose ends were closed by loose fitting retainer plugs. Leather seals prevented the rubber bag from squeezing out between retainer plug and specimens. A teflon liner protected the bag from being punctured by the aluminum reinforcement and the whole assembly was pressurized with water.

Radial displacement transducers were attached at three points along the length of the specimen, and strain gages were glued to the inner and outer surfaces in both the transverse and longitudinal directions.

Since the longitudinal force component was absorbed by the loose fitting retainer plugs, the specimen was under radial pressure through most of its length and was free to contract longitudinally.

Because the strengthening effects of strain-gages on the specimens have been observed earlier, tangential strains were determined from transducer measurements of the change in outer diameter, while strain gage test results were utilized only for correlation with other strain gage measurements.

Bursting strength as a function of effective aluminum content is presented in Fig. 8. The increase in strength is essentially linear and when the line is extrapolated to 0% aluminum it approaches the bursting pressure of an unreinforced (DP-16), geometrically similar cylinder. Based on thick-walled cylinder theory[7] and a 230 psi tensile strength this value is about 60 psi, while $p = 69$ psi is obtained, on the basis of thin-wall theory. The bursting pressures were also used to compute the hoop strength of the composite using thin-wall isotropic cylinder equations. The computed values fit the experimental results of thin rings reasonably well and are shown for comparison in Fig. 7. Thick cylinder strength calculations would indicate a 16% higher strength for the cylinders used.

At failure the stress distribution through the wall is reasonably uniform.

The relative effectiveness of the reinforcement is evaluated by assuming that the same volume of aluminum is used as a solid shell at the average diameter of the cylinder. The bursting pressure of such isotropic thin cylinders is shown in Fig. 8 as a dashed line.

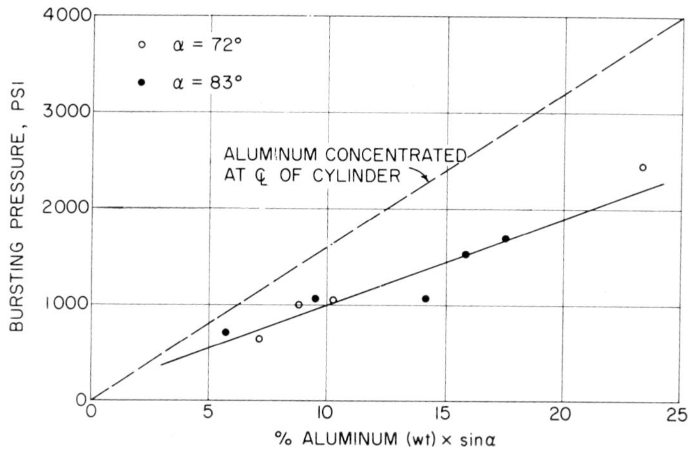

FIG. 8. Bursting pressure as a function of reinforcement

The ratio of orthotropic to isotropic bursting pressures is a measure of the efficiency of this method of reinforcement. Computed values have been plotted in Fig. 9 in increasing order on normal probability paper. It is seen that the reinforcement has a mean efficiency of about 65% with a coefficient of variation of 17%. The relatively low strength of the composite can be partially attributed to the distribution of aluminum wire through the wall thickness illustrated in Fig. 5.

Because a high percentage of reinforcement near the inner boundary increases the theoretical stress concentration,[8] which is not sufficiently relieved on a strain-hardening reinforcement, failure takes place prematurely.

A wire distribution increasing towards the outer surface or reinforcement with variable yield stress[1] (low inside, high outside) would produce a stronger composite cylinder.

The over-all compliance $1/C$, which is the slope of the pressure-displacement curve is presented in Fig. 10 as a function of effective aluminum content. Since strain has been measured utilizing transducers and attached strain

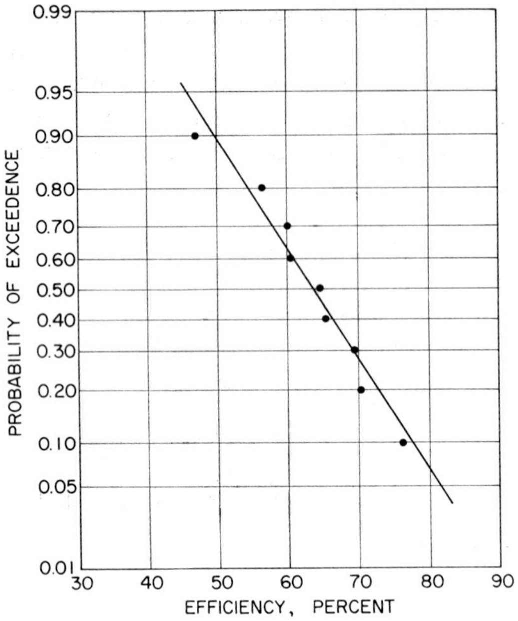

FIG. 9. Statistical distribution of the burst–strength–efficiency in cylinders

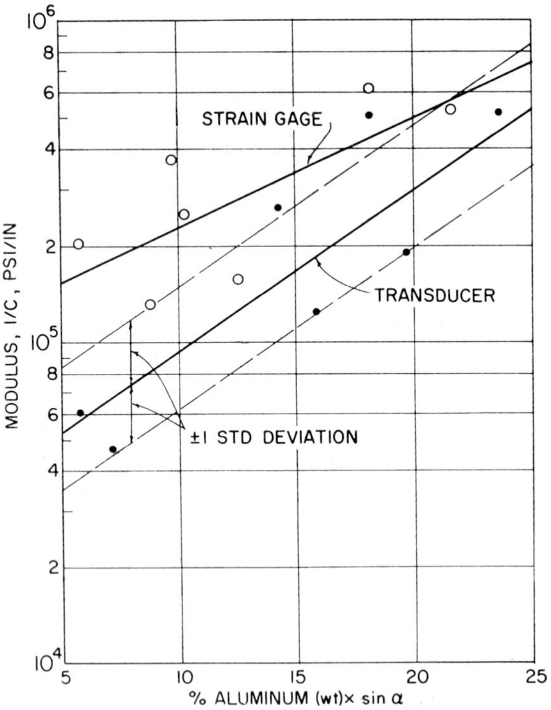

FIG. 10. Pressure–displacement compliance as a function of reinforcement

gages, both types of measurement are shown on the figure, with separate least square regression lines. The reinforcing effect of strain gages is again evident. The difference between the two lines diminishes, as expected, for high concentration of reinforcement. The relationship between C and the other orthotropic constants is nonlinear and may be written in terms of the notation used in ref. 1 as

$$u_b = \frac{2\alpha_{11}b^n}{nK_{0n}a^{n-1}}p = Cp$$

where u_b is the radial displacement of the outer surface whose radius is b, a is the inner radius, p the internal pressure, n the constant of orthotropy:

$$n = \sqrt{\frac{\alpha_{11}}{\alpha_{22}}} = \sqrt{\frac{1/E_r(1-\nu_{rz}\nu_{zr})}{1/E_\theta(1-\nu_{\theta z}\nu_{z\theta})}}$$

with the E's and ν's being orthotropic elastic moduli and Poisson's ratios, and

$$K_{0n} = \left(\frac{b}{a}\right)^{2n} - 1.$$

In addition the symmetry conditions

$$\nu_{r\theta}E_\theta = \nu_{\theta r}E_r; \; \nu_{zr}E_r = \nu_{rz}E_z; \; \nu_{z\theta}E_\theta = \nu_{\theta z}E_z$$

have to be satisfied.

Utilizing some experimentally obtained values from Figs. 2, 3, 6, 10 and 11 for a grain with about 10% aluminum content ($E_\theta \doteq 2\times10^5$ psi, $\nu_{\theta z} \doteq 0.5$, $\nu_{z\theta} < 0.1$, $1/C \doteq 1.5\times10^5$ psi/in.), the above equations yield $n = 14$ and $\alpha_{11} = 10^{-3}$. Since α_{11} is approximately the reciprocal of the radial modulus, $E_r \doteq 10^3$ psi results. ν_{rz} and ν_{zr} cannot be computed without additional data.

Considering the fact that there is no radial reinforcement the modulus in that direction is essentially that of the matrix. For unreinforced (DP-16) KCl-filled polyurethane, this value is about 1000 to 1500 psi.

Electrical strain gages attached to the specimen in the longitudinal and tangential directions on both inner and outer surfaces have strengthening effects on the material and consequently these measurements may not be used for modulus determination without correction. Because it may be assumed that the stiffening effect is the same in both directions the readings may be used to determine Poisson's ratios.

Longitudinal versus tangential strain curves whose slopes are Poisson's ratios $\nu_{\theta z}$ (change in length in the z direction due to a force in the direction of internal pressure) are presented in Fig. 11. The measured values are all around 1/2 and in no instance has a value much larger than that been obtained. The ratios are in direct contradiction with the results of ref. 3 where

$\nu_{\theta z} = 2.17$ has been reported together with $E_z = 26{,}200$ psi, $E_\theta = 60{,}000$ psi and $\nu_{z\theta} = 0.283$.

Strains at the inner and outer surfaces of pressurized cylinders were compared, because strain gages were used, only where their ratios are meaningful. The measured values, $\varepsilon_{\theta a}/\varepsilon_{\theta b} \doteq 3$, are of the right order of magnitude based on the orthotropic cylinder equations of ref. 1.

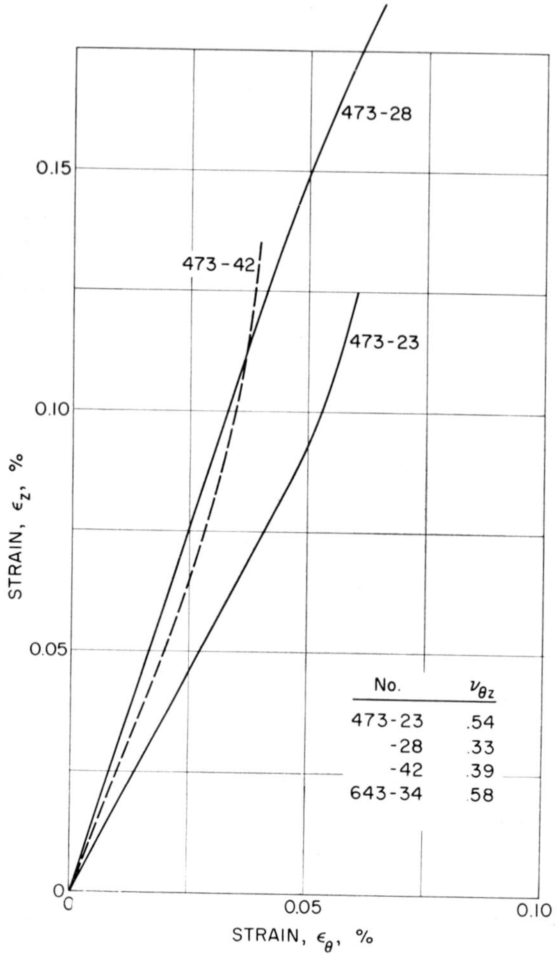

FIG. 11. Axial strain (compressive) vs. tangential-strain (outer surface) under internal pressurization

VI. RESPONSE TO OSCILLATING LOADS

The response of the composite to oscillating loads was investigated by performing vibration tests on 1 in. wide, $6\frac{1}{2}$ in. O.D., $4\frac{7}{8}$ in. I.D. rings. Specimens were inserted in a Baldwin-Sontag fatigue machine applying a 30 cps sinusoidally varying load to opposite ends of the diameter. The steady state deflection amplitude was measured a few seconds after the experiment was started to eliminate the effects due to heating.

Several wider rings were also tested for comparison; however, since even large changes of diameter produce only small local strains, no appreciable differences were found.

Dynamic load–deflection curves were drawn for each specimen and are presented in Fig. 12. For purposes of comparison a curve is shown for the unreinforced polyurethane (DP-16).

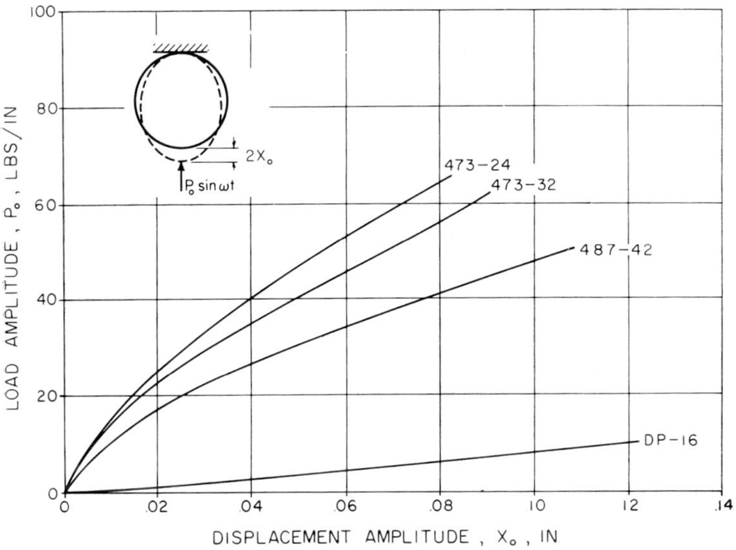

FIG. 12. Dynamic load–deflection curves on diametrally loaded rings

All curves show a nonlinear behavior. Deflections are greater than those predicted on the bases of either thin- or thick-wall isotropic ring equations. Both the non-linearity and the large deflections are the results of interlaminar shear deformations.

Oscillating shear displacement produces a considerable amount of heat.[9] Temperature measurements along the ring center-line are, therefore, good indicators of the position of maximum shearing. Thermocouples were inserted into a ring which was subsequently cycled with a constant, diametral, load-amplitude of 50 lb/in. as shown in Fig. 13.

The temperature was measured at nine points along the center line of the cross-section after 15,000, 30,000 and 60,000 load cycles. Temperature distribution curves together with curves showing the stress distribution based on isotropic theory are presented in Fig. 13. It is evident that maximum temperatures and consequently maximum shearing occur at points where neither the direct stress nor the elementary theory shear stress are maximum.

Since orthotropic equations also predict that maximum shear stresses should occur at the point of load application, a different approach, based on sandwich beam theory and Timoshenko beam equations, is being examined. Preliminary calculations for a three-layer, fixed-end, sandwich beam indicate that interlaminar shear stresses are greatest not at the supports, as assumed in elementary beam theory, but at points about 1/6 of the length

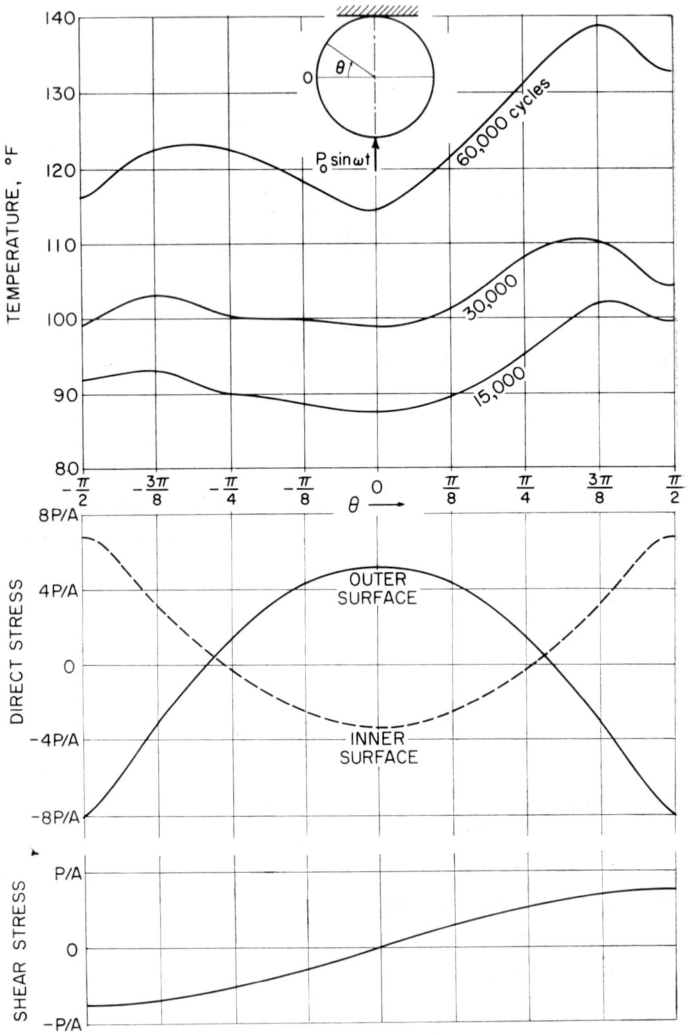

FIG. 13. Measured temperature distribution around circumference during cyclic test; direct stress and shear stress distribution based on thick-walled elastic ring theory

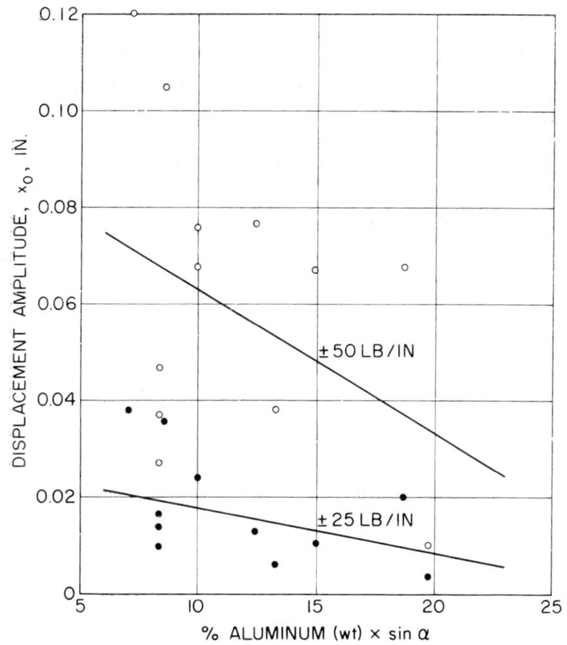

FIG. 14. Dynamic displacement as a function of reinforcement

away from them. The exact position of the largest shear stresses is a function of the sandwich layer thicknesses and the ratio of elastic moduli.[10]

To show the dynamic stiffening effect of the reinforcement, least square regression lines are presented in Fig. 14, for the diametral displacement amplitude as a function of the applied sinusoidal load and effective aluminum content. The data exhibit very large scatter.

VII. DYNAMIC CREEP

The amplitude of motion increases when a constant-amplitude load cycle, as described in the previous section, is applied to specimens over an extended period of time. The rate of increase in deflection depends on the amount of reinforcement and the amplitude of the applied load cycles. Heavily reinforced rings exhibit a small amount of this "dynamic creep" which, for moderate load amplitudes eventually reaches a limit as shown in Fig. 15. Rings with less reinforcement exhibit increasing deformation until failure takes place by delamination at the ring center-line at points about 70° above the horizontal diameter.

It is interesting to note that rings made of the unreinforced matrix material do not exhibit such cycle-dependent increase in deformation, since no interlaminar shearing takes place in an isotropic material.

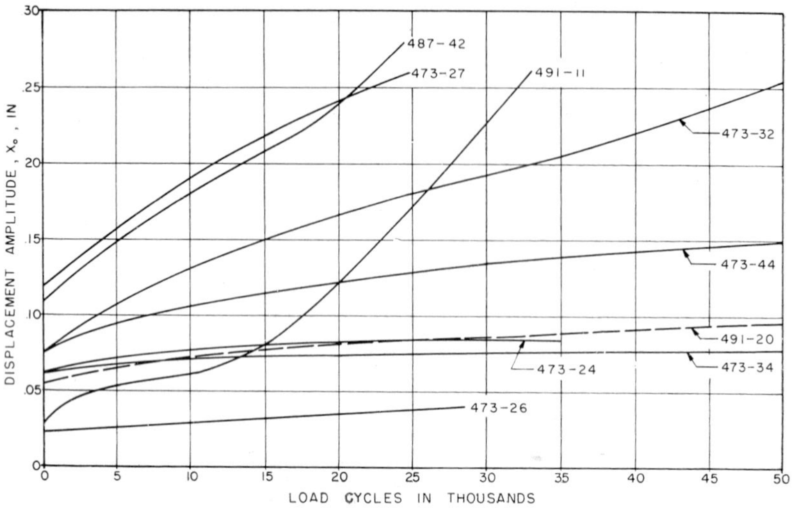

FIG. 15. Dynamic creep curves under a constant amplitude load of $P_0 = \pm 50$ lb/in. (30 cps)

If the combination of load amplitude and aluminum content is such that the deformation does not increase beyond a limiting value complete recovery takes place shortly after unloading. If, however, considerable creep has occurred, recovery is not complete.

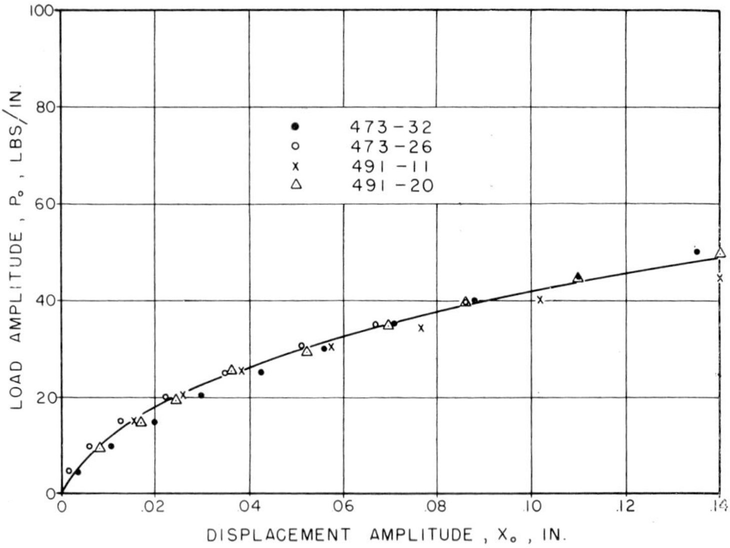

FIG. 16. Dynamic load-displacement curves after failure (delamination) in dynamic creep

The failed ring, after delamination has taken place, acts as a series of concentric thin rings and still has some load carrying capacity. Because the shear stresses are no longer present the amount of reinforcement has little effect on strength and all rings have the same load–deflection curves as shown in Fig. 16.

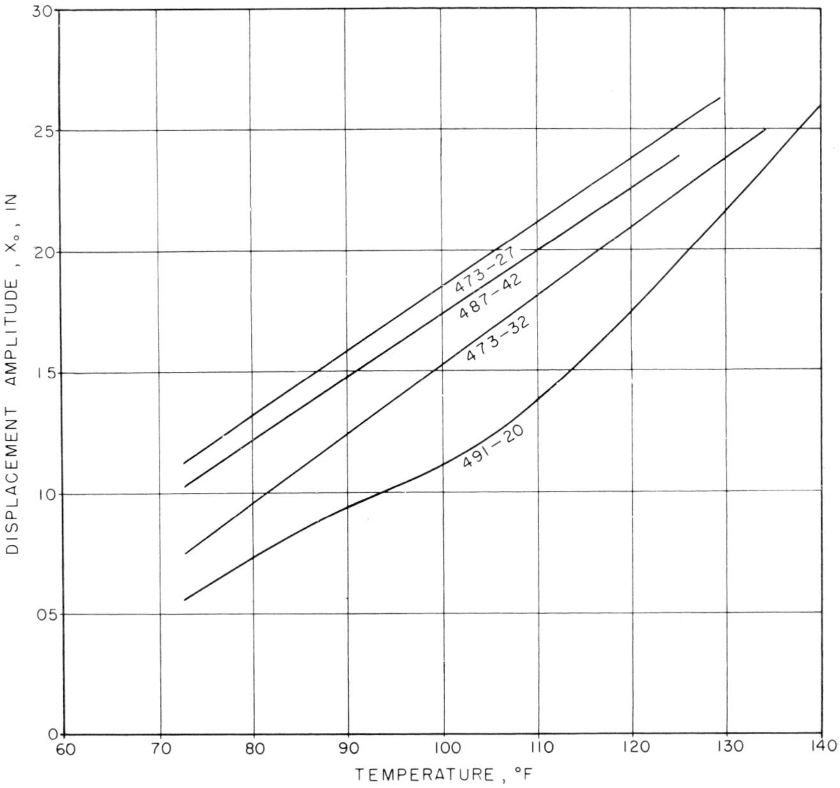

FIG. 17. Dynamic displacement as a function of temperature rise at point where eventual delamination occurred ($P_0 = \pm 50$ lb/in.)

As mentioned earlier, the observed weakening of the material under cyclic loading is attributed to thermal softening of the matrix material as a result of large interlaminar shear. The dynamic deflection is a linear function of the temperature rise in most cases from the beginning of the experiment to failure (Fig. 17).

In highly reinforced rings the thermal conductivity of the aluminum helps to remove excess heat from high shear regions; and under moderate loads a heat balance is established so that no further temperature increase occurs and consequently, no change in deflection takes place.

VIII. CONCLUSIONS

A large variety of experiments have been performed to evaluate the mechanical properties of simulated, wire-reinforced solid propellants and to examine typical failure modes of such materials.

It has been found that while axial compressive strength, hoop strength, fatigue, and torsion strength increase with aluminum content, axial tensile strength decreases for the tested configurations. Elastic moduli increase with wire content and relations between the various moduli are adequately described by orthotropic relationships.[1] The experimental results are afflicted by an unusual amount of scatter which may be attributed to several factors: (1) tests were performed over a 3-year period during which material properties have changed; (2) specimens were produced during the early stages of development and hence lack adequate quality control; (3) wire distribution, curing techniques, binder and filler content were not considered in the usual evaluation of test results.

Because of the scatter some curves would fit equally well on linear and logarithmic plots and consequently no reliable empirical equations can be established. For the same reason extrapolation beyond the test region is not recommended.

Failure modes are of four types:

1. Axial tension strength is governed by the strength of the binder and failure takes place by binder separation unless the reinforcing helix is steep.
2. Hoop strength is influenced by wire content and distribution. Failure of the wire in hoop tension as a result of stress concentration on the inner surface produces failure of the grain.
3. In axial compression and diametral bending, failure may be attributed to a lack of radial reinforcement, which results in buckling of the inner surface in the case of axial compression and delamination by interlaminar shear in the case of diametral bending.
4. In vibration tests heat generation due to interlaminar shear and consequent thermal softening of the binder causes failure by delamination. The position of maximum shear in a laminated tube cannot be calculated using either isotropic or orthotropic equations without paying special attention to the interlaminar shear stress distribution and the weakness in shear of the material. Work is still in progress towards establishing such relations.[10]

While a considerable amount of information has been obtained a complete characterization of the composite would require further data in the following areas:

1. Viscoelastic response of the material including the influence of temperature, thermal shock and cycling, load rate and storage.

2. Damping and complex modulus.
3. Shrinkage stresses.
4. Thermal expansion and conductivity.
5. Optimal wire distribution, material, and winding angle.
6. Biaxial behavior.
7. Interlaminar shear stress distribution.
8. Effects of stress concentration created by grain shapes, cut wires and thermal stress.
9. Statistical aspects and reliability.

ACKNOWLEDGMENTS

This work has been supported by the Office of Naval Research under Contract No. Nonr 266(78) and has been carried out at the Civil Engineering Research Laboratories of Columbia University.

REFERENCES

1. FREUDENTHAL, A. M. Deformation and Failure Analysis of Reinforced Grains, in *Mechanics and Chemistry of Solid Propellants*, Fourth Symp. on Naval Struct. Mech. (1965), pp. 463–483, Pergamon, New York (1967); Report No. 28 Contract No. Nonr 266(78), Columbia University, June 1965.
2. *Inter Agency Chemical Rocket Propulsion Group, Working Group on Mechanical Behavior*, CPIA Publication No. 119, V. II, (Confidential) DOD. No. AD-376-133, 1966.
3. CAPUTO, A. A., MINNICH, B. H. and ALEXANDER, E. L. *Technical Summary Report*, Research on the Response of Reinforced Solid Propellants, Rept. No. R-6651, Contract No. Nonr 4988 (00) FBM. Rocketdyne Div. of North Amer. Aviation Inc. 1966. (Confidential).
4. HELLER, R. A. Mechanical Properties of Wire Reinforced Grains, Report No. 38, Contract No. Nonr 266(78), Columbia University, 1967.
5. CROWLEY, J. M. Progress in the Navy Contract Research Program on Mechanical Behavior of Solid Propellant Grains, CPIA Publication No. 119, V.I. p. 702, DOD: No. AD-489-917 (1966).
6. FRIED, N. *Survey of Methods of Test for Parallel Filament Reinforced Plastics*, Am. Soc. Testing and Mats., Special Technical Publication No. 327, p. 19 (1962).
7. TIMOSHENKO, S. *Strength of Materials*, Vol. II, p. 81, Van Nostrand, New York 1948.
8. BIENICK, M., SPILLERS, W. R. and FREUDENTHAL, A. M. Non-homogeneous Thick-walled Cylinders under Internal Pressure, *J. Am. Rocket Soc.* **32**, 1269 (1962).
9. PROSEN, S. P. *et al.* Compression, Fatigue and Stress Studies on NOL Ring Specimens, Ref. 6, p. 118.
10. HELLER, R. A. Inter-laminar Shear Stress in Sandwich Beams, *Proc. Soc. Experimental Stress Anal.* **26**, No. 2 (1969); Report No. 33, Contract No. Nonr 266(78), Columbia University, 1967.

THE EVALUATION OF THE STRENGTH OF FIBER-REINFORCED COMPOSITES

L. S. SHU

Research Scientist,
Space Sciences Laboratory,
General Electric Company,
King of Prussia, Pa.

ABSTRACT

The strength of unidirectional fiber-reinforced composites is evaluated quantitatively in terms of the strength of their constituents. Methods of limit analysis of plasticity are applied to obtain bounds for the limit loads which characterize the strength. The same technique can be used to evaluate the strength of fibrous composite laminates.

I. INTRODUCTION

One of the most important aspects of the theory of mechanical behavior of fibrous composites is the analysis of the strength of the material. A comprehensive review of the current work in this area was given by Tsai.[1] Among the theories proposed to predict the strength of the composites is the theory of Tsai and Azzi.[2, 3] According to the theory, the strength of the laminated fibrous composites as well as the constituting unidirectional fibrous composite layers can be predicted once the basic strength characteristics of the constituting layers are known. However, they were unable to obtain these basic strength characteristics of the composite from the knowledge of the strength of its constituents. Therefore, experiments were performed on the composite to supply the necessary information. In the present paper, an attempt is made to determine the basic strength characteristics of the unidirectional fibrous composite as functions of strength and geometry of its constituents. A relationship of this type is necessary in the improvement of the strength of the composites by proper choice of potential constituents. Theorems of limit analysis of plasticity are used to obtain bounds that characterize the basic strength properties. The strength of laminated composites can also be evaluated by the same technique.

II. ANISOTROPIC STRENGTH OF COMPOSITES

In the Tsai–Azzi strength theory of composites[2, 3] a unidirectional fibrous composite, considered as quasi-homogeneous and anisotropic, possesses a set of principal axes of anisotropy. Referred to this set of axes, Hill's generalized von Mises' yield condition[4] has the form:

$$2f(\tau_{ij}) \equiv F(\tau_{22}-\tau_{33})^2+G(\tau_{33}-\tau_{11})^2+H(\tau_{11}-\tau_{22})^2+2L\tau_{23}^2 \\ +2M\tau_{31}^2+2N\tau_{12}^2 = 1 \tag{1}$$

where the coefficients F, G, H, L, M, N are parameters characterizing the state of anisotropy and τ_{ij} are components of the stress tensor referred to the principal axes of anisotropy $x_1x_2x_3$ where the x_1 axis is parallel to the fibers.† In general, the yield condition (1) can be represented by a surface in a stress space. A stress point within the yield surface represents a "safe" stress state. Yielding can occur if the stress point is on the yield surface. Since unidirectional fibrous composite layers in laminates are thin compared with their lateral dimensions, only τ_{11}, τ_{22} and τ_{12} are considered non-zero if the x_3 axis is along the thickness of the layer. Furthermore, since unidirectional fibers are randomly located in a composite layer, it can be assumed to be transversely isotropic. Then the yield condition (1) reduces to

$$\left(\frac{\tau_{11}}{X_1}\right)^2+\left(\frac{\tau_{22}}{X_2}\right)^2+\left(\frac{\tau_{12}}{X_{12}}\right)^2-\frac{\tau_{11}\tau_{22}}{X_1^2} = 1 \tag{2}$$

where X_1, X_2 and X_{12} are the normal yield stress in the direction of the fibers, the normal yield stress in the direction transverse to the fibers and the yield stress in axial shear of the composite, respectively. These are the three basic strength characteristics of the unidirectional fibrous composites. Once these are known, the yield condition (2) can be employed to determine whether a combined state of (plane) stress can cause failure of the composite. For a thin laminate under combined axial and shear forces, bending and twisting moments, the three basic strength characteristics of the unidirectional layers can determine by the yield condition (2), as the applied loads increase monotonically, the successive failure of the constituting layers as well as the intensity of the applied loads at which the laminate fails as a result of failure of all layers.

Therefore, according to Tsai–Azzi's theory, X_1, X_2 and X_{12} are the key quantities for the failure analysis of fiber-reinforced composites. In the next section, effort is made to evaluate these quantities analytically in terms of the strength and geometry of the constituents.

† Henceforth, unless otherwise specified, $i, j = 1, 2, 3$; summation on repeated indices is implied.

III. LIMIT ANALYSIS OF UNIDIRECTIONAL FIBROUS COMPOSITES

The composite material under consideration consists of a relatively soft matrix material in which stiffer fibers are embedded. Both materials are assumed to be elastic-perfectly plastic and satisfy the von Mises yield criterion. As shown in Fig. 1, referred to an orthogonal Cartesian coordinate system, a typical unidirectional fibrous composite is taken to be a cylinder with rectangular cross-section. Circular fibers running from base to base of the specimen are in x_1 direction. Limit analysis of such a specimen with various

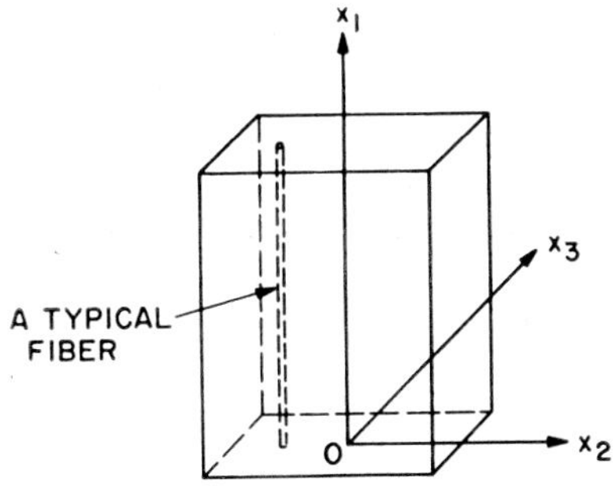

FIG. 1. Composite specimen (fibers in X_1 direction)

types of arrangements of fibers in matrix under various types of surface loading was carried out by Shu and Rosen.(5) In this paper, only the "random array" geometry is considered: circular fibers of various diameters are randomly located in the specimen. Each of them can be surrounded entirely by a concentric cylinder of matrix material. A cylinder consisting of a fiber of radius r_f and the outer matrix-shell of radius r_b is called a composite cylinder. It is assumed that a constant $\beta \equiv r_f/r_b$ can be chosen so that the composite cylinders are non-overlapping. The entire specimen can then be considered as an assemblage of composite cylinders and the remaining matrix volume. The lateral boundary of the specimen may touch or cut through some fibers. In both cases the associated composite cylinders are "incomplete". Since in practice fiber diameters are very small compared with the transverse dimensions of the specimen, the total number of fibers in the interior of the specimen is much larger than the total number of those possibly on the lateral boundary. Hence, the total volume of "incomplete" composite cylinders is much smaller than that of the "complete" cylinders.

If V, V_1 and V_2 denote, respectively, the total volumes of the specimen, the composite cylinders and the remaining matrix in the specimen, the following obvious relation holds:

$$V = V_1 + V_2. \tag{3}$$

In the case where the entire specimen is occupied by composite cylinders, $V_2 = 0$. Following Hashin and Rosen[6] this distribution is called the "random array".

The von Mises yield criterion which the fiber and matrix materials are assumed to obey has the following form:[7]

$$\frac{s_{ij}s_{ij}}{2} \leqslant k^2 \tag{4}$$

where s_{ij} are components of the stress deviator and k is the yield stress in simple shear for the fiber material (denoted by k_f) or for the matrix material (denoted by k_b).

Under the conditions of plane strain perpendicular to the x_1-axis, von Mises yield criterion (4) reduces to

$$(\tau_{22}-\tau_{33})^2+4\tau_{23}^2 \leqslant 4k^2 \tag{5}$$

where τ_{22}, τ_{33} and τ_{23} are components of the stress tensor in the transverse plane.

In order to evaluate the three basic strength characteristics of the composite specimen shown in Fig. 1, the specimen is subjected to axial shear stress τ_{12}, transverse tensile stress τ_{22} and longitudinal tensile stress τ_{11}, respectively. The upper and lower bound theorems of limit analysis of plasticity[8, 9] will be used to obtain bounds for the limit loads τ_{12}^L, τ_{22}^L and τ_{11}^L which represent X_{12}, X_2 and X_1, respectively.

Case 1. Axial shear stresses τ_{12} applied on the boundary of the composite specimen.

According to the lower bound theorem, a uniform shear stress field τ_{12} can be chosen as the statically admissible stress field. Since τ_{12} can nowhere violate the yield condition (4) for both matrix and fibers, it follows that a lower bound for the limit load τ_{12}^L is

$$(\tau_{12}^L)_L = k_b.$$

For upper bound construction, a kinematically admissible velocity field is chosen as follows:

(a) In V_2 (and thus also on the boundary of the composite cylinders) and on the entire lateral boundary of the specimen,

$$\begin{Bmatrix} u_1 \\ u_2 \\ u_3 \end{Bmatrix} = \begin{Bmatrix} 0 \\ \gamma_1 x_1 \\ 0 \end{Bmatrix} \tag{6}$$

where γ_1 is a positive number.

(b) In any composite cylinder, referred to a local cylindrical polar coordinate system,

$$\begin{Bmatrix} u_1 \\ u_r \\ u_\theta \end{Bmatrix} = \begin{Bmatrix} -\gamma_1 r \cos\theta \\ \gamma_1 x_1 \cos\theta \\ -\gamma_1 x_1 \sin\theta \end{Bmatrix} \qquad \text{for } 0 \leqslant r \leqslant r_f$$

$$= \begin{Bmatrix} \dfrac{\gamma_1 \beta^2}{1-\beta^2}\left(1-\dfrac{r_b^2}{r^2}\right) r \cos\theta \\ \gamma_1 x_1 \cos\theta \\ -\gamma_1 x_1 \sin\theta \end{Bmatrix} \qquad \text{for } r_f \leqslant r \leqslant r_b. \tag{7}$$

The velocity field (7) in a composite cylinder is the elastic displacement solution to the boundary value problem with the boundary condition (6) prescribed.† The problem is the same as that which was formulated by Hashin and Rosen[6] with the modification that for $0 \leqslant r \leqslant r_f$, the velocity field is associated with rigid body motion.

With this velocity field constructed for the entire specimen, the dissipation density function and the rate of external work done can be obtained[5] to yield an upper bound for τ_{12}^L:

$$\frac{(\tau_{12}^L)_U}{k_b} = \frac{1}{\pi(1-\beta^2)} \int_\beta^1 \int_0^{2\pi} R \sqrt{\left(1+\frac{\beta^4}{R^4}\right) + 2\frac{\beta^2}{R^2} \cos\theta}\, d\theta\, dR. \tag{8}$$

The above expression is for "random array" in which $V_2 = 0$ and the fiber-volume fraction $v_f = \beta^2$.

The integral in (7) is integrated numerically for different fiber-volume fractions. The result is shown in Fig. 2 where $(\tau_{12}^L)_U/k_b$ in (8) is plotted as a function of v_f ($0 < v_f < 1$, $v_f = \beta^2$). Note that particularly in Fig. 2

$$\lim_{v_f \to 0} \frac{(\tau_{12}^L)_U}{k_b} = 1,$$

$$\lim_{v_f \to 1} \frac{(\tau_{12}^L)_U}{k_b} = \frac{4}{\pi}.$$

† The velocity field in any "incomplete" composite cylinder is defined by solving similar elastic displacement boundary value problems for the "incomplete" composite cylinder. However, since the volume of the "incomplete" composite cylinder is small, the difference between their actual contribution to the dissipation function and that obtained by treating all cylinders as "complete" is negligible. This approximation is implied in the subsequent analysis wherever a similar situation arises.

From the above result, it is concluded that X_{12} is at most about 27% above and at least the same as the yield stress in shear for the matrix.

Case 2. Transverse tensile stresses τ_{22} applied on the boundary of the composite specimen.

For lower bound construction, a uniform tensile stress τ_{22} throughout the specimen is chosen as a statically admissible stress field. Since the von Mises yield condition (4) can nowhere be violated in the specimen, the lower bound associated with the constructed statically admissible stress field is

$$(\tau_{22}^L)_L = \sqrt{3}k_b. \tag{9}$$

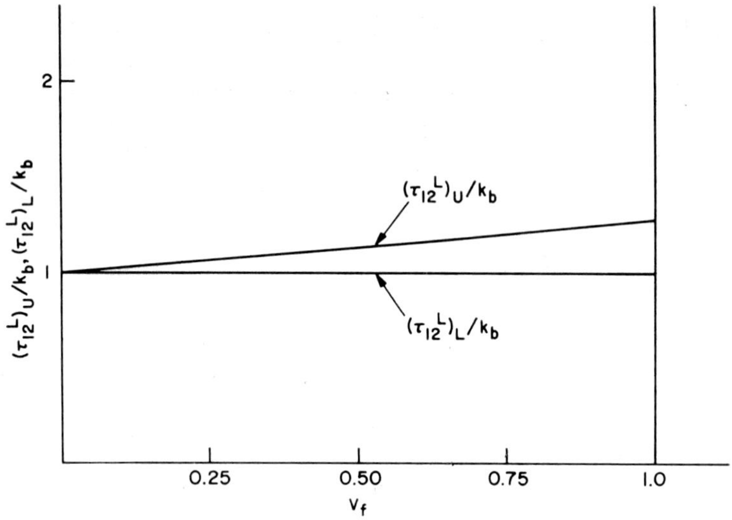

FIG. 2. Upper and lower bounds for τ_{12}^L

For upper bound construction, a kinematically admissible velocity field is chosen as follows:

(a) In V_2 (and thus also on the boundary of the composite cylinders) and on the boundary of the specimen,

$$\begin{Bmatrix} u_1 \\ u_2 \\ u_3 \end{Bmatrix} = \begin{Bmatrix} 0 \\ \dfrac{\gamma_2}{2}x_2 \\ -\dfrac{\gamma_2}{2}x_3 \end{Bmatrix} \tag{10}$$

where γ_2 is any positive number.

(b) In any composite cylinder, referred to the local coordinate system,

$$\begin{Bmatrix} u_1 \\ u_3 \\ u_3 \end{Bmatrix} = \begin{Bmatrix} 0 \\ \dfrac{\gamma_2}{2} x_2 \left[A_1 + A_2 \dfrac{x_2^2 + 3x_3^2}{2r_f^2} + A_3 \dfrac{r_f^4(-x_2^2 + 3x_3^2)}{r^6} + A_4 \dfrac{r_f^2(x_2^2 - x_3^2)}{2r^4} \right] \\ -\dfrac{\gamma_2}{2} x_3 \left[A_1 + A_2 \dfrac{3x_2^2 + x_3^2}{2r_f^2} + A_3 \dfrac{r_f^4(3x_2^2 - x_3^2)}{r^6} + A_4 \dfrac{r_f^2(-x_2^2 + x_3^2)}{2r^4} \right] \end{Bmatrix}$$

$$\text{for } r_f \leq r \leq r_b \qquad (11)$$

$$= \begin{Bmatrix} 0 \\ \dfrac{\gamma_2}{2} x_2 \left(B_1 + B_2 \dfrac{x_2^2 + 3x_3^2}{2r_f^2} \right) \\ -\dfrac{\gamma_2}{2} x_3 \left(B_1 + B_2 \dfrac{3x_2^2 + x_3^2}{2r_f^2} \right) \end{Bmatrix} \quad \text{for } 0 \leq r \leq r_f$$

where

$$A_1 = \frac{1}{C} \left\{ 4(1-\eta)(1+\eta)\beta^4 - 3(1-\eta)(1+\eta)\beta^2 + \frac{(1+\eta)^2}{\beta^2} \right\}$$

$$A_2 = \frac{1}{C} \{ 4(1-\eta)(1+\eta)(1-\beta^2)\beta^2 \}$$

$$A_3 = \frac{1}{C} \left\{ (1-\eta) \left[[(1-\eta)\beta^2 + \frac{(1+\eta)}{\beta^2} \right] \right\}$$

$$A_4 = \frac{1}{C} \left\{ 4(1-\eta) \left[-2\eta + (1+\eta) \left(1 + \frac{1}{\beta^6} \right) \right] \beta^4 \right\}$$

$$B_1 = \frac{1}{C} \left\{ 8(1-\eta)\beta^4 - 6(1-\eta)\beta^2 + \frac{2(1+\eta)}{\beta^2} \right\}$$

$$B_2 = \frac{1}{C} \{ 8(1-\eta)\beta^2(1-\beta^2) \}$$

and $C = (1-\eta)^2\beta^6 + 4(1-\eta)(1+\eta)\beta^4 - 6(1-\eta)(1+\eta)\beta^2 + 4(1-\eta)(1+\eta) + (1+\eta)^2/\beta^2.$

The velocity field (11) in a composite cylinder is the incompressible elastic displacement solution to the boundary value problem with the boundary condition (10) prescribed. The problem was formulated by Hashin and Rosen[6] in evaluating the effective plane strain shear modulus of elastic fiber-reinforced composites. The constant η in (11) considered as the displacement solution to the elastic problem is the ratio of the elastic shear moduli of the fiber and matrix materials. However, it can be considered as merely a parameter when (11) is used as a kinematically admissible velocity.

For the case of the "random array", an application of the upper bound theorem gives an upper bound

$$\frac{(\tau_{22}^L)_U}{2k_b} = \frac{k_f}{k_b} I_1 + I_2 \tag{12}$$

where

$$I_1 = 2\int_0^{\beta} R \left| \frac{8(1-\eta)\beta^4 - 6(1-\eta)\beta^2 + \dfrac{2(1+\eta)}{\beta^2} + 12(1-\eta)(1-\beta^2)R^2}{C} \right| dR$$

and

$$I_2 = \frac{1}{\pi}\int_{\beta}^{1}\int_{0}^{2\pi} \frac{R}{|C|} \sqrt{\Phi_1(\eta,\beta,R) + \Phi_2(\eta,\beta,R)\cos\theta}\; d\theta\, dR$$

in which

$$\Phi_1(\eta,\beta,R) = \left\{4(1-\eta)(1+\eta)\beta^4 - 3(1-\eta)(1+\eta)\beta^2 + \frac{(1+\eta)^2}{\beta^2} + 6(1-\eta)(1+\eta)(1-\beta^2)R^2\right\}^2 + \frac{(1-\eta)^2\beta^8}{R^4}\left\{-\frac{3}{R^2}\left[\frac{(1+\eta)}{\beta^2} + (1-\eta)\beta^2\right] + 2\beta^2\left[-2\eta + (1+\eta)\left(1+\frac{1}{\beta^6}\right)\right]\right\}^2$$

$$\Phi_2(\eta,\beta,R) = 2\left\{4(1-\eta)(1+\eta)\beta^4 - 3(1-\eta)(1+\eta)\beta^2 + \frac{(1+\eta)^2}{\beta^2} + 6(1-\eta)(1+\eta)(1-\beta^2)R^2\right\}\frac{(1-\eta)\beta^4}{R^2}\left\{-\frac{3}{R^2}\left[\frac{(1-\eta)}{\beta^2} + (1-\eta)\beta^2\right] + 2\beta^2\left[-2\eta + (1+\eta)\left(1+\frac{1}{\beta^6}\right)\right]\right\}$$

and $\beta^2 = v_f$.

For any finite k_f/k_b and for $\eta \to \infty$, (12) becomes

$$\frac{(\tau_{22}^L)_U}{2k_b} = \frac{2}{\pi(1-\beta^2)^3}\int_{\beta}^{1}\int_{0}^{2\pi} R\sqrt{\Psi_1(\beta,R) + \Psi_2(\beta,R)\cos\theta}\; d\theta\, dR \tag{13}$$

where

$$\Psi_1(\beta,R) = \frac{\beta^4}{R^4}\left[(\beta^4+\beta^2+1) - \frac{3}{2}(\beta^2+1)\frac{\beta^2}{R^2}\right]^2 + \left[\frac{-(4\beta^4+\beta^2+1)}{2} + 3\beta^2R^2\right]^2$$

and

$$\Psi_2(\beta,R) = 2\frac{\beta^2}{R^2}\left[(\beta^4+\beta^2+1) - \frac{3}{2}(\beta^2+1)\frac{\beta^2}{R^2}\right]\left[-\frac{(4\beta^4+\beta^2+1)}{2} + 3\beta^2R^2\right].$$

The right-hand side of expression (13) was obtained by Shu and Rosen[5] for an upper bound of the limit load (normalized with respect to k_b) for transverse shear stresses applied on composites reinforced with rigid fibers. In Fig. 3, $\frac{(\tau_{22}^L)_U}{2k_b}$ is plotted as a function of v_f according to (13). It is a monotonically increasing function of v_f with $\lim_{v_f \to 0} \frac{(\tau_{22}^L)_U}{2k_b} = 1$ and $\lim_{v_f \to 1} \frac{(\tau_{22}^L)_U}{2k_b} = \infty$.

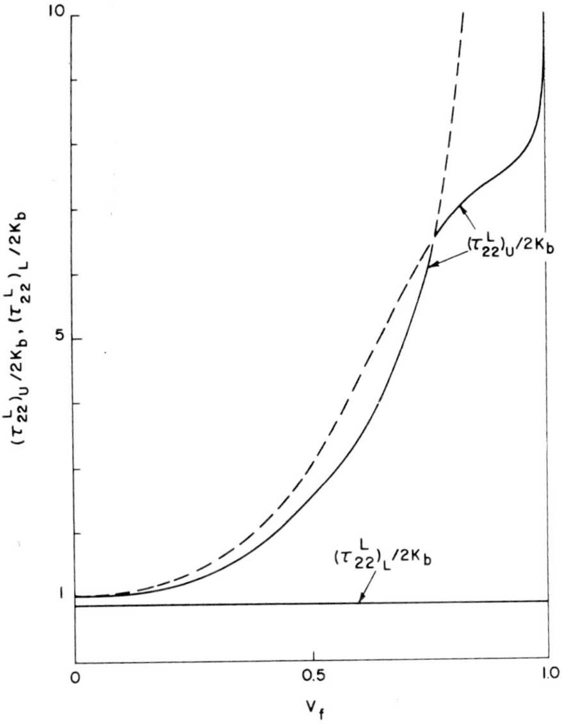

FIG. 3. Upper and lower bounds for τ_{22}^L

On the other hand, for any finite η, $\frac{(\tau_{22}^L)_U}{2k_b}$ can be plotted as a function of v_f by evaluating numerically I_1 and I_2 in (12). For example, for $\eta = 100$, $\frac{(\tau_{22}^L)_U}{2k_b}$ is also a monotonically increasing function of $v_f \left(\lim_{v_f \to 0} \frac{(\tau_{22}^L)_U}{2k_b} = 1 \right.$ and $\left. \lim_{v_f \to 1} \frac{(\tau_{22}^L)_U}{2k_b} = \frac{k_f}{k_b} \right)$, which is shown in Fig. 3, with $k_f/k_b = 10$. It is observed that for fiber volume fractions smaller than about 75%, the upper bound for the limit load τ_{22}^L is higher for $\eta = 100$ than for $\eta \to \infty$, but for higher volume fraction, the opposite is true. Therefore, for any fiber-volume fraction between 0 and 1, the lower value of the bounds obtained from $\eta = 100$ and $\eta \to \infty$ will give a better estimate of τ_{22}^L and it is shown in Fig. 3 in solid line as a function of v_f.

It is interesting to note that for $\eta = 1$, (12) reduces to the following simple form

$$\frac{(\tau_{22}^L)_U}{2k_b} = 1+v_f\left(\frac{k_f}{k_b}-1\right) \tag{14}$$

which is commonly known as the "rule of mixtures". From Fig. 3, it is seen that the straight line represented by (14) will be higher than the chosen curve for $\frac{(\tau_{22}^L)_U}{2k_b}$ for $0<v_f<1$. Therefore, it is concluded that the "rule of mixtures" used in this case would overestimate the composite strength.

In Fig. 3, $\frac{(\tau_{22}^L)_L}{2k_b}$ obtained from (9) is also shown. The difference between the upper and lower bounds for $v_f \to 0$ is due to the fact that in the upper bound construction, a plane strain velocity field is used as a kinematically admissible velocity field.

Case 3. Longitudinal tensile stresses τ_{11} applied on the boundary of the composite specimen.

The study of longitudinal strength of unidirectional fibrous composites is extensive. Various models and failure mechanisms have been proposed in the literature. In the present work, the yield strength in axial tension is obtained by the construction of very simple velocity and stress fields.

For upper bound construction, in the entire region of the composite specimen, a kinematically admissible velocity field is chosen as

$$\begin{Bmatrix} u_1 \\ u_2 \\ u_3 \end{Bmatrix} = \begin{Bmatrix} \varepsilon_0 x_1 \\ -\frac{\varepsilon_0}{2} x_2 \\ -\frac{\varepsilon_0}{2} x_3 \end{Bmatrix}.$$

An application of the upper bound theorem gives

$$(\tau_{11}^L)_U = \tau_f v_f + \tau_b v_b \tag{15}$$

or

$$\frac{(\tau_{11}^L)_U}{\tau_b} = 1+v_f\left(\frac{\tau_f}{\tau_b}-1\right)$$

where τ_f and τ_b are the yield stresses in tension for the fiber and matrix, respectively and $v_b = 1-v_f$.

Relation (15) is known as the "rule of mixtures" for the prediction of the strength of a composite. If τ_f and τ_b are given, $\frac{(\tau_{11}^L)_U}{\tau_b}$ is a linear function of fiber-volume fraction v_f.

For lower bound construction, if the applied stress τ_{11} is assumed to be uniformly distributed on the boundary of the composite specimen, then a lower bound $(\tau_{11}^L)_L$ is equal to τ_b. However, if we assume that the tensile stress τ_{11} is not uniformly distributed on the boundary surface and we are only interested in the average stress intensity on the boundary surface that causes failure, then a higher lower bound can be obtained. In fact, a statically admissible stress field can be chosen as follows:

In the region occupied by fibers,

$$\tau_{ij} = \begin{bmatrix} \tau_f & 0 & 0 \\ 0 & 0 & 0 \\ 0 & 0 & 0 \end{bmatrix}. \tag{16}$$

In the region occupied by the matrix,

$$\tau_{ij} = \begin{bmatrix} \tau_b & 0 & 0 \\ 0 & 0 & 0 \\ 0 & 0 & 0 \end{bmatrix}. \tag{17}$$

It is obvious that the stress field expressed in (16) and (17) satisfies the requirements to be statically admissible.

The average traction corresponding to this stress field is therefore

$$\tau_f v_f + \tau_b v_b \tag{18}$$

which can be taken as $(\tau_{11}^L)_L$.

According to (15) and (18), the upper and lower bounds for τ_{11}^L coincide. Therefore, X_1 can be determined from the "rule of mixtures."

From the above results, it is observed that bounds for τ_{22}^L are far apart, especially for high fiber-volume fractions. Further effort should be made to improve the bounds in order to have a better estimate of the limit load. For lower bound construction, uniform stress distribution used as statically admissible stress fields can only give lower bounds which are the corresponding matrix yield stresses. In order to obtain higher lower bounds, one has to assume applied tractions to be non-uniformly distributed on the boundary surface according to the properties of the fibers and matrix. Equilibrium stress fields can then be constructed in equilibrium with the applied tractions. Then the lower bound theorem can be applied to obtain higher lower bounds. To decide the distribution of the applied tractions on the boundary and to construct an equilibrium stress field in the body is not at all easy in general. The success in the construction of such a statically admissible stress field for Case 3 is due to the simplicity of geometry and loading conditions.

IV. FIBROUS COMPOSITE LAMINATES

A common type of fibrous composite laminate used in engineering structures are thin plates consisting of unidirectional fiber-reinforced composite layers of identical mechanical properties but with different orientation and in general with different thickness. The strength of this type of laminates has been recently studied by a number of investigators.[(1)] In this section, it is shown that the kinematically admissible velocity fields constructed in the preceding section can be used to obtain upper bounds of limit loads for in-plane loading applied on laminates.

To demonstrate the method, consider a laminate composed of unidirectional fibrous laminae subjected to in-plane shear stress τ_{12}. The problem is

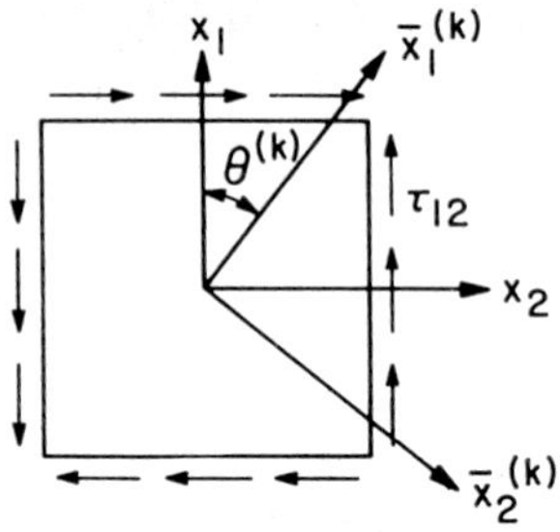

FIG. 4. The kth layer in a laminate

to find an upper bound for the limit load τ_{12}^L. A typical constituting layer, the kth layer, in which fibers are all running in $\bar{x}_1^{(k)}$ direction, is shown in Fig. 4. Depending on the orientation of fibers in the layer, the principal axes $\bar{x}_1^{(k)}\bar{x}_2^{(k)}\bar{x}_3^{(k)}$ can be defined by an angle $\theta^{(k)}$ measured from the laminate axes $x_1x_2x_3$.

For upper bound construction, a kinematically admissible velocity field is constructed in the following manner:

The velocity field $(u_1, u_2, u_3) = (0, \gamma_1 x_1, 0)$ is assigned in V_2 of each layer, referred to $x_1x_2x_3$ axes. In the kth layer referred to its principle axes $\bar{x}_1^{(k)}\bar{x}_2^{(k)}\bar{x}_3^{(k)}$, the above velocity field is transformed into the following form:

$$\begin{Bmatrix} \bar{u}_1^{(k)} \\ \bar{u}_2^{(k)} \\ \bar{u}_3^{(k)} \end{Bmatrix} = \begin{Bmatrix} \gamma_1 \sin\theta^{(k)} \cos\theta^{(k)} \bar{x}_1^{(k)} - \gamma_1 \sin^2\theta^{(k)} \bar{x}_2^{(k)} \\ \gamma_1 \cos^2\theta^{(k)} \bar{x}_1^{(k)} - \gamma_1 \sin\theta^{(k)} \cos\theta^{(k)} \bar{x}_2^{(k)} \\ 0 \end{Bmatrix}. \tag{19}$$

The right-hand side of equation (19) can be decomposed into four parts so that

$$\bar{\mathbf{u}}^{(k)} = \bar{\mathbf{u}}_1^{(k)} + \bar{\mathbf{u}}_2^{(k)} + \bar{\mathbf{u}}_3^{(k)} + \bar{\mathbf{u}}_4^{(k)} \tag{20}$$

where

$$\bar{\mathbf{u}}_1^{(k)} \equiv \begin{Bmatrix} 0 \\ \gamma_1 \cos^2 \theta^{(k)} \bar{x}_1^{(k)} \\ 0 \end{Bmatrix}$$

$$\bar{\mathbf{u}}_2^{(k)} \equiv \begin{Bmatrix} 0 \\ -\dfrac{\gamma_1}{2} \sin \theta^{(k)} \cos \theta^{(k)} \bar{x}_2^{(k)} \\ \dfrac{\gamma_1}{2} \sin \theta^{(k)} \cos \theta^{(k)} \bar{x}_3^{(k)} \end{Bmatrix}$$

$$\bar{\mathbf{u}}_3^{(k)} \equiv \begin{Bmatrix} \gamma_1 \sin \theta^{(k)} \cos \theta^{(k)} \bar{x}_1^{(k)} \\ -\dfrac{\gamma_1}{2} \sin \theta^{(k)} \cos \theta^{(k)} \bar{x}_2^{(k)} \\ -\dfrac{\gamma_1}{2} \sin \theta^{(k)} \cos \theta^{(k)} \bar{x}_3^{(k)} \end{Bmatrix}$$

and

$$\bar{\mathbf{u}}_4^{(k)} \equiv \begin{Bmatrix} -\gamma_1 \sin^2 \theta^{(k)} \bar{x}_2^{(k)} \\ 0 \\ 0 \end{Bmatrix}.$$

Except for multiplicate factors and the reference coordinate system, $\bar{\mathbf{u}}_1^{(k)}$, $\bar{\mathbf{u}}_2^{(k)}$, and $\bar{\mathbf{u}}_3^{(k)}$ are the same as those velocity fields constructed in V_2 for the three different cases of loading in the preceding section. Furthermore,

$$\bar{\mathbf{u}}_4^{(k)} \equiv \begin{Bmatrix} -\gamma_1 \sin^2 \theta^{(k)} \bar{x}_2^{(k)} \\ 0 \\ 0 \end{Bmatrix} = \begin{Bmatrix} 0 \\ -\gamma_1 \sin^2 \theta^{(k)} \bar{x}_1^{(k)} \\ 0 \end{Bmatrix} + \begin{Bmatrix} -\gamma_1 \sin^2 \theta^{(k)} \bar{x}_2^{(k)} \\ \gamma_1 \sin^2 \theta^{(k)} \bar{x}_1^{(k)} \\ 0 \end{Bmatrix} \quad (21)$$

where the first part of the decomposition has the same form as $\bar{\mathbf{u}}_1^{(k)}$ while the second part represents a rigid body motion of V_2 as a whole.

The velocity field chosen in V_1 is also composed of four parts: The first part is due to $\bar{\mathbf{u}}_1^{(k)}$, which is the same as (7) with the modification that γ_1 in (7) is replaced by $\gamma_1 \cos^2 \theta^{(k)}$. The second part is due to $\bar{\mathbf{u}}_2^{(k)}$ which is the same as (11) with the modification that γ_2 is now replaced by $-\gamma_1 \sin\theta^{(k)} \cos \theta^{(k)}$. The third part is due to $\bar{\mathbf{u}}_3^{(k)}$ which is exactly the same as $\bar{\mathbf{u}}_3^{(k)}$ in V_2. The fourth part is due to $\bar{\mathbf{u}}_4^{(k)}$ which can be obtained in the same way as in Case 1 in the preceding section. However, a moment's reflection reveals that the strain rate associated with $\bar{\mathbf{u}}_4^{(k)}$ is exactly the same as that associated with $\bar{\mathbf{u}}_1^{(k)}$ if $\gamma_1 \cos^2 \theta^{(k)}$ is replaced by $-\gamma_1 \sin^2 \theta^{(k)}$ since the additional rigid body motion contributes nothing to the value of the strain rate. Now, after the kinematically admissible velocity field is constructed as described above, the associated strain rate and dissipation density function can be calculated without difficulty. Moreover, the rate of external work done can also be calculated so that an upper bound for τ_{12}^L can be obtained.

As an example, consider a laminate composed of two unidirectional fibrous composite layers of equal thickness with $\theta^{(1)} = 0$ and $\theta^{(2)} = -\pi/2$.

In this simple geometry of lamination, the kinematically admissible velocity constructed above reduces to a very simple form.

For layer 1, the velocity field (19) reduces to

$$\begin{Bmatrix} \bar{u}_1^{(1)} \\ \bar{u}_2^{(1)} \\ \bar{u}_3^{(1)} \end{Bmatrix} = \begin{Bmatrix} 0 \\ \gamma_1 \bar{x}_1^{(1)} \\ 0 \end{Bmatrix} \text{ in } V_2. \tag{22}$$

For layer 2, the velocity field (19) reduces to

$$\begin{Bmatrix} \bar{u}_1^{(2)} \\ \bar{u}_2^{(2)} \\ \bar{u}_2^{(2)} \end{Bmatrix} = \begin{Bmatrix} -\gamma_1 \bar{x}_2^{(2)} \\ 0 \\ 0 \end{Bmatrix} \text{ in } V_2. \tag{23}$$

Then, following the principles described above, it is easy to obtain the velocity fields and the associated strain rate fields for both layers. After some manipulation, it turns out that

$$\frac{(\tau_{12}^L)_U}{k_b} = \frac{1}{\pi(1-\beta^2)} \int_\beta^1 \int_0^{2\pi} R \sqrt{\left(1+\frac{\beta^4}{R^4}\right)+2\frac{\beta^2}{R^2}\cos\theta}\; d\theta\, dR$$

or the "random array" geometry.

The above equation is exactly (8), which means that the upper bound $(\tau_{12}^L)_U$ is not higher for this type of cross-ply laminates than for the unidirectional fibrous composite.

Similarly, it is easy to show that the three types of velocity fields constructed in the preceding section can be used to construct upper bounds for limit loads for laminates subjected to any in-plane stresses.

V. CONCLUSION

The strength of unidirectional fibrous composites has been evaluated in terms of the properties of their constituents. The evaluation is in the form of bounds. Further effort is necessary to obtain closer bounds in the case of transverse tension. Improvement of bounds also means better estimation of the strength of laminates. Laminates under bending and twisting moments have not been considered here.

Finally, it is observed from the analysis that the strength of fibrous composites is strongly influenced by the properties of the matrix material.

ACKNOWLEDGEMENT

The author wishes to express his gratitude to Dr. B. Walter Rosen, Consulting Engineer of the Space Sciences Laboratory, General Electric Company, for many valuable discussions. This paper is based upon work performed under NASA Contract NASw-1377.

REFERENCES

1. TSAI, S. W. A Review of Strength Theories of Composite Materials, presented at the Major Status Report—Advanced Composites, Washington, D.C., September 1966.
2. AZZI, V. D. and TSAI, S. W. Anisotropic Strength of Composites, *Experimental Mechanics*, **5**, 283–288 (1965).
3. TSAI, S. W. and AZZI, V. D. Strength of Laminated Composite Materials, *AIAA Journal*, **4**, No. 2, 296–301 (1966).
4. HILL, R. A Theory of the Yielding and Plastic Flow of Anisotropic Metals, *Proceedings of the Royal Society of London*, **A281**, 297 (1948).
5. SHU, L. S. and ROSEN, B. W. Strength of Fiber-Reinforced Composites by Limit Analysis Methods, *Journal of Composite Materials*, **1**, No. 4, 366–381 (1967).
6. HASHIN, Z. and ROSEN, B. W. The Elastic Moduli of Fiber Reinforced Materialss, *Journal of Applied Mechanics*, **31E**, 223–232 (1964).
7. See for example, PRAGER, W. and HODGE, JR., P. G. *Theory of Perfectly Plastic Solids*, John Wiley and Sons, Inc. 1965.
8. DRUCKER, D. C., GREENBERG, H. J. and PRAGER, W. The Safety Factor of an Elastic-Plastic Body in Plane Strain, *Journal of Applied Mechanics*, **18**, 371–378 (1951b.)
9. KOITER, W. T. General Theorems for Elastic Plastic Solids, Chapter IV in *Progress in Solid Mechanics*, Sneddon and Hill, eds., North-Holland 1960.

ELASTIC INTERFACE

F. W. SCHMIEDESHOFF

Benet Research and Engineering Laboratories Watervliet Arsenal

ABSTRACT

The non-homogeneous cylinder under internal pressure is evaluated with the modulus expressed as a series of Fermi–Dirac distribution functions. This leads to a model describing not only the increased rigidity for fibers but also the continuous change of the modulus between the matrix and the fibers. The incompressible solid is studied to make the non-homogeneous solution more tractable.

LIST OF SYMBOLS

σ_r	radial stress
σ_θ	hoop stress
β	constant
c_1	constant
c_2	constant
E_m	matrix modulus
E_f	fiber modulus
ν	Poisson's ratio
E	Young's modulus
ε_r	radial strain
ε_θ	hoop strain
U	radial displacement
a	inside radius
b	outside radius

I. INTRODUCTION

The interface was requested to be analyzed by the Chairman as a subject for this conference. In general the interface is normally idealized by assuming perfect bonding which neglects the influence of properties changing with position in the bonded region. The cylinder is evaluated as loading is applied at a surface resulting in the transfer of stresses to adjacent fibers as well as a geometry which can be evaluated experimentally in contrast to the tensile test for composites which has proven difficult for fibrous composite materials.

II. ANALYSIS

We have the following stress–strain relations in cylindrical coordinates:

$$\varepsilon_r = \frac{1}{E}[\sigma_r - \nu(\sigma_\theta + \sigma_z)]$$
$$\varepsilon_\theta = \frac{1}{E}[\sigma_\theta - \nu(\sigma_r + \sigma_z)] \tag{1}$$
$$\varepsilon_z = \frac{1}{E}[\sigma_z - \nu(\sigma_r + \sigma_\theta)]\,.$$

For simplicity the displacement in the longitudinal direction is assumed to vanish. On this basis, $\varepsilon_z = 0$, $\sigma_z = \nu(\sigma_r + \sigma_\theta)$.

Equations (1) then reduce to

$$\varepsilon_r = \frac{(1-\nu^2)}{E}\left[\sigma_r - \frac{\nu\sigma_\theta}{1-\nu}\right]$$
$$\varepsilon_\theta = \frac{(1-\nu^2)}{E}\left[\sigma_\theta - \frac{\nu\sigma_r}{1-\nu}\right]. \tag{2}$$

Assuming the solid is incompressible ($\nu = 0.5$) results in the following expressions,

$$\varepsilon_r = \frac{3}{4E}[\sigma_r - \sigma_\theta]$$
$$\varepsilon_\theta = \frac{3}{4E}[\sigma_\theta - \sigma_r] \tag{3}$$

and therefore

$$\varepsilon_r + \varepsilon_\theta = 0. \tag{4}$$

The strains in the radial and hoop directions are expressed respectively in terms of the radial displacement as follows,

$$\varepsilon_r = \frac{dU}{dr}, \quad \varepsilon_\theta = \frac{U}{r}. \tag{5}$$

From eqs. (3) the stress difference can be expressed as a function of the radial displacement through equation (5).

$$\sigma_r - \sigma_\theta = \frac{4E}{3}\frac{dU}{dr} = -\frac{4E}{3}\frac{U}{r}. \tag{6}$$

From eqs. (4) and (5) we obtain the following differential equation,

$$\frac{dU}{dr} + \frac{U}{r} = 0 \tag{7}$$

which has the solution

$$U = \frac{C_1}{r}. \tag{8}$$

Equilibrium equations yield the following for the conditions imposed:

$$\frac{d\sigma_r}{dr} + \frac{\sigma_r - \sigma_\theta}{r} = 0. \tag{9}$$

Using eqs. (6) and (8) eq. (9) becomes

$$\frac{d\sigma_r}{dr} = \frac{4E}{3}\frac{C_1}{r^3}. \tag{10}$$

Assuming E is a function of the radius only, then we obtain from eq. (10)

$$\sigma_r = \frac{4C_1}{3}\int_a^r \frac{E\,dr}{r^3} + C_2. \tag{11}$$

From eq. (6) we have

$$\sigma_\theta = \sigma_r + \frac{4E}{3}\frac{U}{r}.$$

Substituting from eqs. (8) and (11) we obtain the hoop stress

$$\sigma_\theta = \frac{4C_1}{3}\int_a^r \frac{E\,dr}{r^3} + C_2 + \frac{4E}{3}\frac{C_1}{r^2}. \tag{12}$$

We shall assume that the cylinder is subjected to internal and external pressurization. Accordingly at $r = a$, $\sigma_r = -P_0$ which yields $C_2 = -P_0$ and at $r = b$, $\sigma_r = -P_1$ which yields

$$C_1 = \frac{P_0 - P_1}{\frac{4}{3}\int_a^b \frac{E\,dr}{r^3}}.$$

With these values for C_1 and C_2 we finally obtain

$$\sigma_r = (P_0 - P_1)\frac{\int_a^r \frac{E\,dr}{r^3}}{\int_a^b \frac{E\,dr}{r^3}} - P_0 \tag{13}$$

$$\sigma_\theta = \frac{(P_0 - P_1)}{\int_a^b \frac{E\,dr}{r^3}}\left[\int_a^r \frac{E\,dr}{r^3} + \frac{E}{r^2}\right] - P_0. \tag{14}$$

Only the internal pressurization problem will be analyzed. Therefore for $P_1=0$ we obtain,

$$\bar{\sigma}_r = \frac{\sigma_r}{P_0} = \frac{\int_a^r \frac{E\,dr}{r^3}}{\int_a^b \frac{E\,dr}{r^3}} - 1$$

$$\bar{\sigma}_\theta = \frac{\sigma_\theta}{P_0} = \frac{\int_a^r \frac{E\,dr}{r^3} + \frac{E}{r^2}}{\int_a^b \frac{E\,dr}{r^3}} - 1. \qquad (15)$$

For additional simplicity let $\bar{r}=r/a$ and assume $b/a=2$. The wall ratio chosen is considered fairly large which is a value required to maintain elastic conditions in high strength cylinders for pressures of the order of 50,000 psi. These conditions yield the following for the stresses,

$$\bar{\sigma}_r = \frac{\int_1^{\bar{r}} \frac{E(\bar{r})\,d\bar{r}}{\bar{r}^3}}{\int_1^2 \frac{E(\bar{r})\,d\bar{r}}{\bar{r}^3}} - 1 \qquad (16)$$

$$\bar{\sigma}_\theta = \frac{1}{\int_1^2 \frac{E(\bar{r})\,d\bar{r}}{\bar{r}^3}} \left\{ \int_1^{\bar{r}} \frac{E(\bar{r})\,d\bar{r}}{\bar{r}^3} + \frac{E(\bar{r})}{\bar{r}^2} \right\} - 1. \qquad (17)$$

At this point in the analysis the position dependency of the modulus must be specified so as to describe the continuous variation from matrix to fiber as well as the repetitive variations required to define fiber reinforcement.

In Fig. 1 the function $g_1\,(r)$ is the Fermi–Dirac function. In the limit as $\beta \to 0$ we obtain a step function. The function $g_2(r)$ describes the function for $r_2^* > r_1^*$. The difference of these functions yields the function $f_{1,2}$ which describes a symmetrical continuously varying function about the mid point between r_2^* and r_1^*. By suitable adjustment of the ordinate we can define the function $f_{n+1,n+2}$ shown in Fig. 2 where E_f is the modulus of the fiber and E_m represents the modulus of the matrix.

For a material which describes the repetitive nature of reinforcement as well as the continuous variation of the modulus through an interface we arrive at the following expression,

$$E = E_m + (E_f - E_m) \sum_{n=1,3,\ldots}^{m} \left\{ \frac{1}{e^{\frac{(\bar{r}-\bar{r}^*_{n+1})}{\bar{\beta}}}+1} - \frac{1}{e^{\frac{(\bar{r}-\bar{r}^*_{n})}{\bar{\beta}}}+1} \right\} \tag{18}$$

where $\bar{r} = r/a$, $\bar{\beta} = \beta/a$.

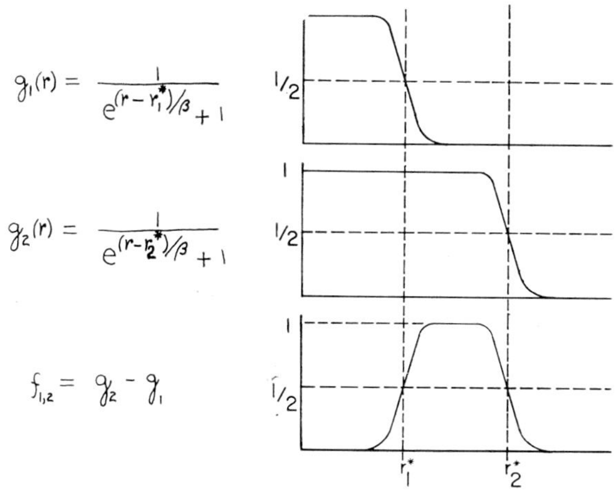

FIG. 1. Distribution function

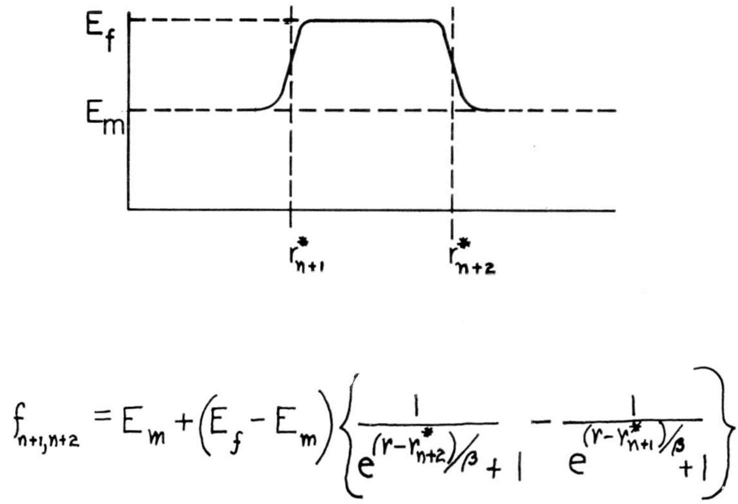

FIG. 2. Modulus as a function of position

Substituting (18) into (16) and (17) yields for the stresses,

$$\bar{\sigma}_\theta = \frac{1}{3/8+\left(\frac{E_f}{E_m}-1\right) I_{12}} \left\{ \frac{1}{2}\left(1+\frac{1}{\bar{r}^2}\right) + \left(\frac{E_f}{E_m}-1\right)\left[I_{1\bar{r}} + \frac{1}{\bar{r}^2} \sum_{n=1,3,\ldots}^{m} \frac{1}{e^{\frac{(\bar{r}-\bar{r}^*_{n+1})}{\bar{\beta}}}+1} - \frac{1}{e^{\frac{(\bar{r}-\bar{r}^*_n)}{\bar{\beta}}}+1}\right]\right\} - 1 \tag{19}$$

$$\sigma_r = \frac{\frac{1}{2}\left(1-\frac{1}{\bar{r}^2}\right) + \left(\frac{E_f}{E_m}-1\right) I_{1\bar{r}}}{\frac{3}{8}+\left(\frac{E_f}{E_m}-1\right) I_{12}} - 1 \tag{20}$$

$$\text{where} \quad I_{1\bar{r}} = \int_1^{\bar{r}} \left\{ \sum_{n=1,3,\ldots}^{m} \frac{1}{e^{\frac{(\bar{r}-\bar{r}^*_{n+1})}{\bar{\beta}}}+1} - \frac{1}{e^{\frac{(\bar{r}-\bar{r}^*)}{\bar{\beta}}}+1} \right\} \frac{d\bar{r}}{\bar{r}^3} \tag{21}$$

and $I_{12} = I_{1\bar{r}}$ for $\bar{r} = 2$.

III. DISCUSSION

The modulus is characterized by $\beta=0.001$ which defines a sharp interface and $r^*_{n+1}-r^*_n = 0.01$ which defines the effective fiber thickness. The distance between fibers is then approximately 0.082 for two cases investigated, notably five equally spaced fibers in the first half thickness and then ten fibers with the same spacing through the entire thickness. The ratio of fiber modulus to matrix modulus was taken as ten and $\bar{\beta} = \beta$.

The stresses for these two conditions are shown respectively in Figs. 3 and 4. For the case when fiber and matrix moduli are equal, the maximum hoop stress occurs at $\bar{r}=1$ with a value of 1.66. With fiber reinforcement this value is reduced to 0.7 at the internal surface. The stiff fibers carry most of the load requiring an order of magnitude increase in strength if this is to be realized. In addition the interface is also required to carry large stresses when the modulus increases from the matrix to the fiber.

Figure 4 shows the stresses at the fibers. The introduction of five additional fibers has little effect upon the stresses. A reduction of about 8% in peak hoop stress can be realized at the fiber.

The analysis conducted on the incompressible solid can be considered only as a trend when dealing with more realistic materials. However, by choosing various combinations of Fermi–Dirac distribution functions the effects of softening at the interface can also be studied for a number of fiber widths and volume fractions. The softening of the interface has been suggested by others as a means for reducing stress concentrations at fiber tips and increasing

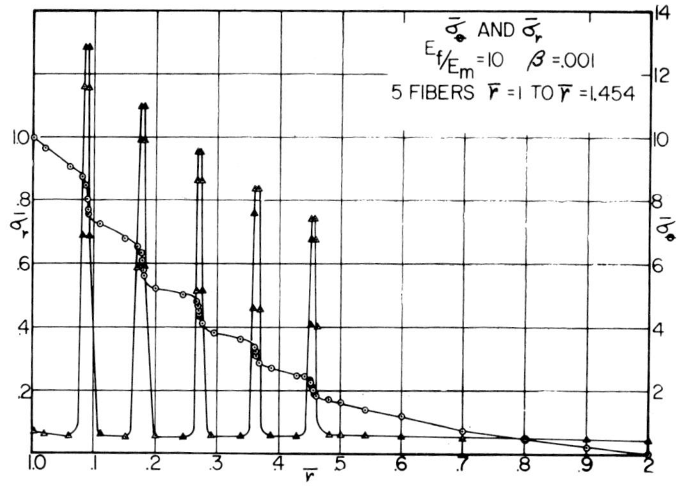

FIG. 3. Hoop stress and radial stress for five fibers with radius of 1 to 1.454

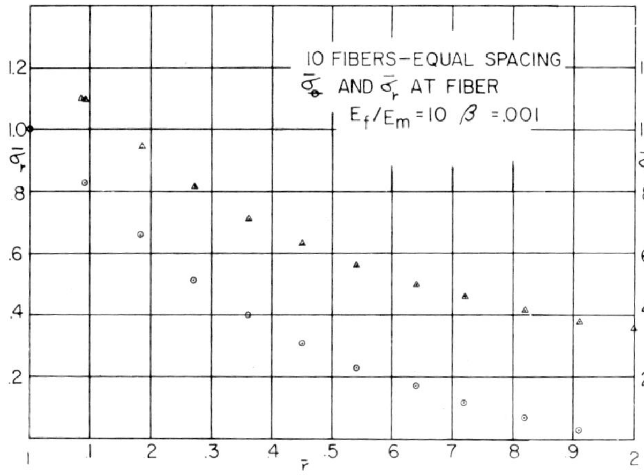

FIG. 4. Hoop stress and radial stress for ten fibers equally spaced through the entire thickness

fatigue life. The analysis can also be extended to include the elastic-plastic behavior which would predict the formation of plastic zones and residual stresses after the release of the pressure.

IV. ACKNOWLEDGEMENTS

Mr. David Priest programmed the numerical integration for the stresses as well as determining the intervals to give accurate results.

Discussions with Mr. S. Beck were extremely helpful in arriving at the problem and solution.

SESSION VI

ENVIRONMENTAL INTERACTION

ENVIRONMENTAL FACTORS IN THE DESIGN OF COMPOSITE MATERIALS

STEPHEN W. TSAI†

Professor of Engineering,
Washington University,
St. Louis, Missouri

ABSTRACT

Some of the environmental factors which affect the structural performance of composite materials are discussed. Various possible mechanisms of environmental factors are examined, together with relevant experimental data. Finally, an analytical approach for predicting environmental effects, based on rate theory, is considered.

I. INTRODUCTION

The superior strength and stiffness of composite materials are seldom used to full advantage in structural applications. This is due to uncertainty of the material behavior under cyclic and impact loading and various environmental factors, e.g. exposure to water, water vapor or other corrosive environments, change in temperature, and long-term physical and chemical stability. Since reliable theory and experimental data are practically nonexistent, superior properties of composite materials are severely penalized by the use of unusually large margins of safety in actual design. Thus, the immediate future of composite materials as a class of engineering materials may depend more on improved reliability of present-day composites than production of new composites. Universal acceptance of composites as high performance materials will depend very much on the confidence of the designer and user of the materials. Until the degradation of composite materials by various environmental factors is better understood so that corrective measures can be taken, the true potential of such materials cannot be realized.

The apparent causes of degradation may involve several factors:

A. Loss of strength of the reinforcing fibers by a stress-corrosion mechanism.
B. Degradation of the fiber–matrix interface resulting in loss of adhesion and interfacial bond strength.
C. Permeability of the matrix material to corrosive agents, such as water vapor which affects both A and B above.

† Now Chief Scientist, Air Force Materials Laboratory, Wright-Patterson Air Force Base, Ohio.

D. Normal viscoelastic dependence of matrix modulus and strength on time and temperature.
E. Combined action of temperature and moisture accelerated degradation.

As a result of these environmental factors, the utility of composite materials is terminated when the stiffness is reduced sufficiently to cause structural instability and/or failure or rupture of the material is induced.

II. MATHEMATICAL FRAMEWORK

Before each of the environmental factors is examined in detail, it may be useful to define a general framework through which the relevance of the environmental factors to the structural performance of composite materials can be properly established. One important characteristic of composite materials is their macroscopic anisotropy. The properties measured in the longitudinal direction of a unidirectional composite are in general quite different from those in the transverse direction. Any attempt in understanding the effect of environmental factors must take account of the inherent anisotropy. One theory which has been found experimentally to yield reasonably accurate strength prediction is Hills' criterion for orthotropic materials.[1,2] For two-dimensional orthotropic bodies, this theory states that the strength depends on three principal strengths, viz., the longitudinal, transverse and shear strengths. Degradation of the principal strengths by a given environmental factor will not be to the same degree, e.g., stress corrosion of fibers will probably degrade the longitudinal strength more than the transverse and shear strengths on a percentage basis, and "plasticization" of the matrix will probably affect the transverse and shear strength more than the longitudinal strength. Thus a crucial question from a designer's viewpoint is how to assess the environmental factors on the total performance of the composite materials, which in the most common configuration are in the form of laminated anisotropic bodies. A large reduction of the longitudinal strength alone may or may not have significant effects on the total performance of the composite. This points toward the need of a mathematical framework into which theoretical and experimental efforts can be properly integrated. A great deal of existing work on the environmental effects has been concerned with segmented problem areas, and it is difficult to assess their relevance to the total performance of composites.

Although current mechanics theory of composites has not provided a reasonably realistic yet tractable mathematical model of the general process of degradation, the need of a consistent framework should be fully appreciated. The elastic properties of composite materials can be conveniently divided into micro and macromechanics. Micromechanics study takes into account the local heterogeneity, which reduces the study to the problem of multiple isotropic inclusions, whereas macromechanics study is concerned with quasi-homogeneous anisotropic layered media. The bridge between micro- and macromechanics is the unidirectional layer.

A similar approach of micro- and macromechanics for the study of environmental factors should be adopted. Basically, the behavior of a unidirectional composite is orthotropic and can thus be described by four principal elastic constants and three principal strengths. The number of the elastic constants is exact in the mathematical sense but that of the principal strengths is only approximate and depends on the theory used. It should be fruitful to describe the environmental interaction in terms of the degradation of the orthotropic elastic constants and strengths.

III. FIBER STRENGTH

In principle the strength of materials is limited by the magnitude of the forces that bind atoms together. In practice the "strength" of most solids is determined or limited by imperfections or flaws. In glasses and other amorphous solids, the important imperfections are surface cracks or other flaws that grow under the influence of stress and chemical attack. It is common experimental knowledge that, when subjected to appropriate environments, many hard amorphous or crystalline solids exhibit a type of failure in which the strength is markedly influenced by the time during which the load is applied. For example, it is a known fact that humid atmospheres reduce the breaking strength of silicate glasses, and the strength of many plastics are impaired when immersed in detergent solutions and solvents.

Since breaking stresses are generally small in comparison with the theoretical limits, the failure phenomena must include reference to defects or flaws, and since failure is time-dependent, it is natural to conclude that the defects are subject to change during the test. For example the strength of silicate glasses are notably sensitive to abrasion and it is generally accepted that surface flaws are controlling influences. Most workers have found it logical to assume that delayed failure under constant load is caused by the growth of these flaws under the influence of a reactive environment to a critical size at which the stress concentration at the most critical flaw is sufficient for spontaneous failure. The occurrence of fracture depends, therefore, on the state of the material. If there is a flaw of sufficient dimensions which depends on the state of stress and the bulk properties of the material, prompt fracture is certain; and if there is no such flaw, the sample is sure to survive until a flaw qualifies through flaw growth, stress increase, or time dependent changes in bulk properties.

In general, details of stress corrosion mechanisms for different materials and environments are not well understood. However, all systems have similar attributes or consequences:

1. In inert environments or at low temperatures, where the reaction rate of corrosion processes should be negligibly slow, the breaking strengths of materials become independent of the duration of load and always reach a relatively high value.
2. Equivalently high strengths are observed if the loading rate is rapid with respect to the reaction rate.
3. Exposing these materials to reactive environments before, but not during a test, generally has little effect on test results, suggesting that the corrosion rate is accelerated by the stress.
4. Exposure of these materials to reactive environments during a test leads to delayed rupture at strengths substantially reduced with respect to (1) and (3) above.
5. There is a continuous influence of temperature on the relationship between the time of loading and the failure stress. In general a continuous loss of strength occurs with increasing temperatures.

Thus the usage of brittle high strength fibrous materials is limited by a time dependent distribution of flaw sizes. If only strength degradation of fibers occurs in a fibrous composite, the longitudinal strength of a unidirectional composite will be seriously affected while producing only a slight alteration of the transverse and shear strengths.

It is not surprising to discover that the coating of fibers or their incorporation into a matrix provides a means of exploiting the properties of these high strength reinforcement materials. The coating or "coupling agent" acts to protect the fibers from abrasion or other sources of surface flaws during fabrication as well as providing, along with the matrix, a barrier between the aggressive environment and the reinforcement. High strength reinforcements are also developed with specific efforts devoted to achieving a chemical composition which is inert with respect to the anticipated service environment.

Extensive data on the effect of water or water vapor on glass-reinforced composites are available (see papers on this subject elsewhere in this volume) and will not be covered here. Typical results for a glass-epoxy composite have been reported by Boller(3) and are shown in Fig. 1. In these experiments the unidirectional composites were under constant uniaxial stress at constant temperatures in a water environment as indicated in Fig. 1. We find the 50% relative humidity and water immersion experiments lead to a continuous loss of strength with increasing time, the rate of strength loss being greatly accelerated by the immersion experiments. It is tempting to suggest that the total immersion curve represents the extremely long time asymptote for the humid aging environment characterizing the interaction between water-silica glass, time, and stress. However, such a simple correlation must be tempered with the knowledge that excessive absorption of water in the matrix induces a mechanical damage to the matrix which may or may not be realized in long-term exposures to other atmospheres containing reduced concentrations of water vapor. This point is discussed in more detail in a later section.

Results of water-boil tests of boron-epoxy (Narmco 5505) composites were reported by General Dynamics/Fort Worth Division.[4] After boiling in tap water for 2 hr, the longitudinal strength of the unidirectional composite

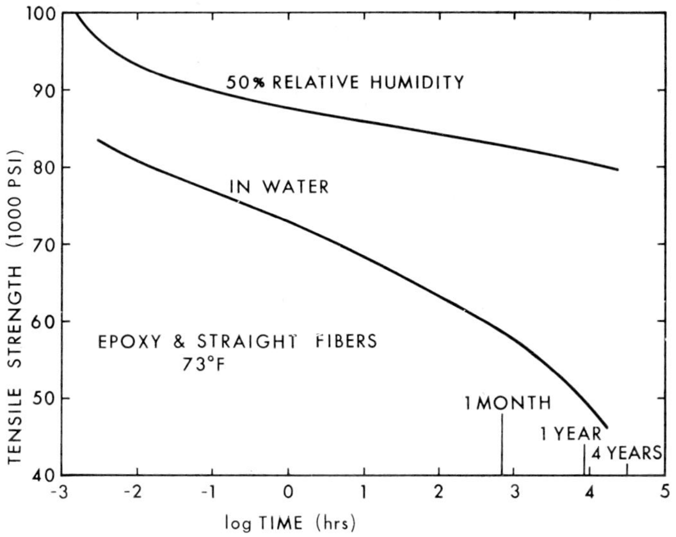

FIG. 1. Stress rupture data of Boller for a laminated composite of unidirectional epoxy glass plates in different water environments

was degraded by a maximum of 6%. Presumably, degradation of longitudinal strength of boron-epoxy composites in humid atmospheres will be substantially less than glass-epoxy systems. However, further data are needed.

Qualitative information regarding the importance of the stress-corrosion process can be obtained through measurements of the time dependent longitudinal strength and stiffness; however, other factors also influence the test data so that it is rather difficult to obtain quantitative information. Detailed investigations will examine the relationships of relative humidity or water concentration in the composite with time to rupture at a particular value of stress and will record how the experimental variables change with changes in temperature. It is an obvious experimental fact that the failure of these systems is strongly influenced by a time-dependent stress-corrosion phenomenon.

IV. INTERFACE

Degradation of composite properties due to environmental effects may be attributed to the loss of adhesion and bond strength at the fiber–matrix interface. It has been shown that the passage of water along the interface is at a rate much higher than permeation through the matrix in a glass-epoxy laminate. If passage of water destroys the interfacial bond, this would result in some form of degradation in the composite properties. Studies of interfacial bonds are rather crude at the present time but we do have some qualitative information: (a) the addition of coupling agents promoting adhesion between phases results in a slight increase in longitudinal stiffness and strength when tested in the dry state; and (b) when tested under wet conditions the untreated fibrous composites possess low strength which increases strongly with coating agent concentration until the "dry" strength is approached.

Failure of the interface bond is termed dewetting. Although this phenomenon has been observed for some time in the study of reinforced elastomers and solid rocket motors, it cannot be predicted analytically with any reliability. The "interfacial strength" may be easier to determine experimentally from the transverse and shear properties of a unidirectional composite than the longitudinal properties. Again, the intrusion of "environmental" factors will greatly affect the analysis as the variability of the interface properties will depend upon the time scale of observation.

V. MATRIX EFFECTS

A. *Irreversible Temperature Effects*

In many instances of military engineering interest, composite systems will be exposed to excessive temperatures for varying periods of time. In general, organic matrix materials are unstable with respect to increased temperature and undergo a chemical breakdown which we shall denote as thermal degradation. If the degradation reactions persist for a sufficient length of time or if they are sufficiently rapid, substantial chemical degradation will occur such that the matrix material is volatilized off in the form of a vapor. Of course, such drastic events compromise the mechanical integrity of the composite system and provide an upper bound on temperature in materials technology. One semiquantitative procedure to follow and characterize the degradation process is the measure of the amount of volatiles given off by the material as a function of time and temperature or simply to record the weight loss as a function of time and temperature. Such data[3] are shown in Fig. 2 for a phenolic-glass system. At all temperatures of 300°F and above, decomposition

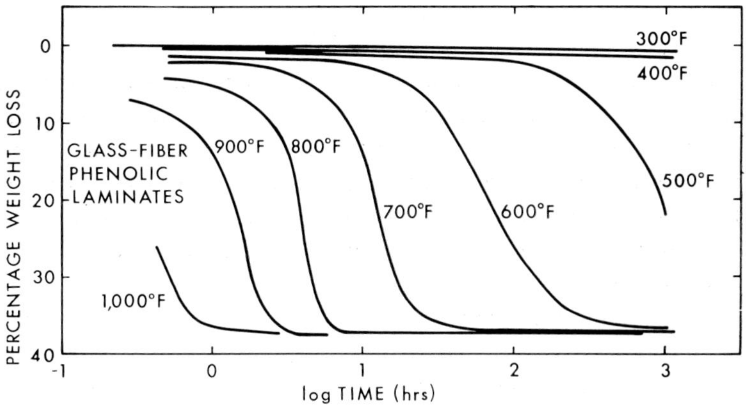

FIG. 2. Deterioration as weight loss at elevated temperatures for a phenolic glass laminate

of the matrix is occurring at a "rate" which is uniformly accelerated with increasing temperature. If the weight loss curves are shifted relative to each other along the log t axis, the curves overlap so as to form a single smooth curve (Fig. 3). The shift distances along the log time scale are denoted by the symbolism log a_D. In Fig. 3, the data was shifted[(5)] to the 300°F curve and one can readily see that the data obtained at different temperatures defines a composite curve representative of the decomposition process. Hence, it appears that a time–temperature superposition applies to this type of data.

It is expected that a temperature–time superposition will be valid in any rate process for which the temperature influences only the rate constant k. In such a case the process under consideration will be a function of the produce kt where t is the time for which the process has been going on. Even

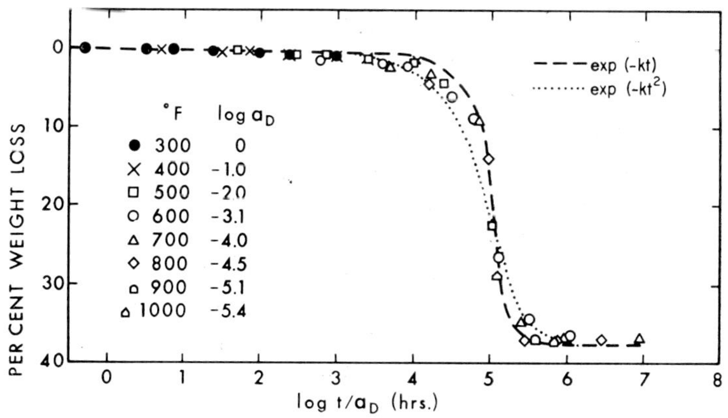

FIG. 3. Master curve of weight loss vs. reduced time constructed from Fig. 2 employing the log a_D terms on the graph

when the temperature enters the mathematical function describing the behavior in a more complicated way than this, it is still usually true that the major temperature dependence is through the rate constant. By calling k_0 the rate constant at the reference temperature (i.e. the temperature for the data to which the other data are shifted; in this case 300°F) and k the rate constant at any other temperature, then $\log (k/k_0)$ is given by the amount by which the data must be shifted along the $\log t$ axis; our $\log a_D$. The values of $\log a_D$ or $\log (k/k_0)$ are often plotted against reciprocal temperature with the following interpretation.

If the rate constant takes the form

$$k = A \exp(-\Delta H/RT) \tag{1}$$

then the ratio of the change in rate constants for temperatures T and T_0 is given by

$$\log (k/k_0) = \log a_D = -\frac{\Delta H}{2.303R}\left(\frac{1}{T} - \frac{1}{T_0}\right) \tag{2}$$

where ΔH is the activation energy and R is the gas constant. Thus a plot of $\log a_D$ against $1/T$ should give rise to a straight line of slope $\Delta H/2.303R$. In Fig. 4 a plot of this nature has been constructed from the data of Fig. 3 and an apparent activation energy of 7.85 kcal/mole was computed. This activation energy is considered to be of the proper order of magnitude for the process involved.

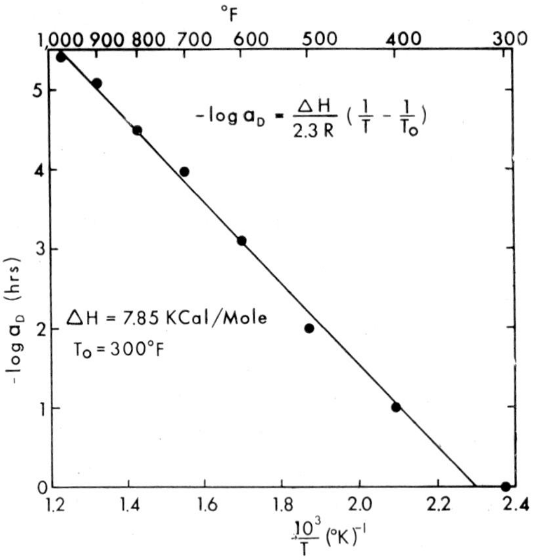

FIG. 4. Determination of the activation energy from an Arrhenius plot of the shift factors determined in Fig. 3

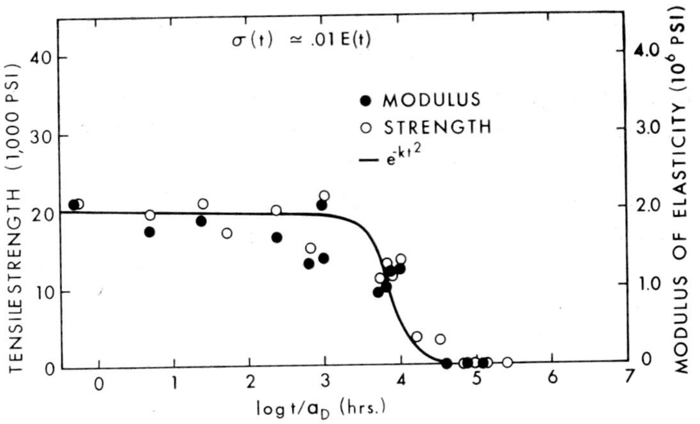

FIG. 5. Composite curves of the modulus and strength for the phenolic glass composite against reduced time during the degradation reaction

Returning to Fig. 3 we see that the experimental data can be approximated by either

$$\frac{w(t)-w(\infty)}{w(0)-w(\infty)} = 1-\exp(-kt)$$

or (3)

$$\frac{w(t)-w(\infty)}{w(0)-w(\infty)} = 1-\exp(-kt^2)$$

where $w(t)$ is percent weight loss at time t, $w(\infty)$ at infinite time, in this case 37%, and $w(0)$ is condition at $t = 0$, or zero. Figure 3 clearly shows that exposure of this phenolic-glass composite to 300°F for times in excess between 5.0×10^3 and 10^4 hr leads to a significant change in matrix properties through thermal decomposition. If one now wishes to estimate the extent of degradation at, for example, 500°F, one simply shifts the entire master curve of Fig. 3 to the left by two decades and so on.

If matrix decomposition occurs, this must ultimately lead to a loss in stiffness and strength which would parallel the loss of the matrix integrity. In fact, Boller has obtained data in tension, compression and flexure for the phenolic-glass composite samples which were soaked at the same temperatures presented in Fig. 2 for different periods of time. The data generally shows the same trends as presented in Fig. 2. If there is a deterministic relationship between matrix decomposition and mechanical response, then the data can be shifted by the same log a_D values evaluated in Fig. 4 to obtain master curves of stiffness and strength loss. The results are shown in Fig. 5 where one can also see that the mechanical property loss is following the same rate law, exp $(-kt^2)$, as the decomposition reaction. Also illustrated in Fig. 5 is the proportionality between the loss of modulus and the parallel loss in strength or

$$\sigma(t) \simeq 0.01\ E(t). \tag{4}$$

Thus it appears that we may ultimately be able to develop rather simple deterministic relations between thermal stability and matrix modulus to predict composite stiffness. In turn we can then develop a cause–effect relationship between the time–temperature dependent stiffness and the "strength" of the composite. Hopefully this would then give us a continuous theory going from the physical chemistry of composite materials to the mechanical response of a structural unit.

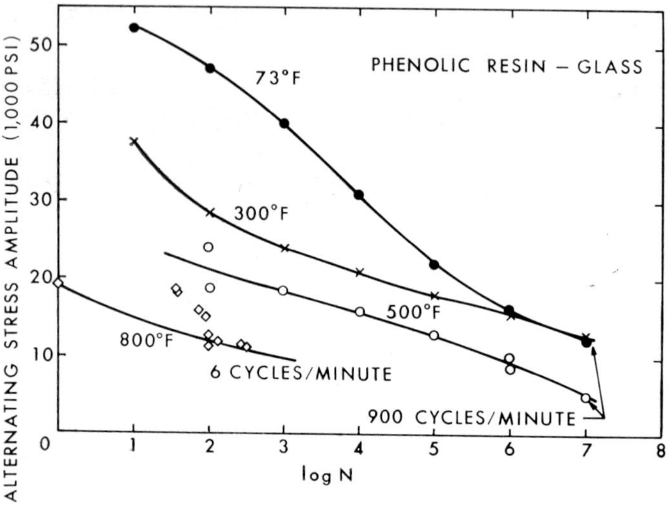

FIG. 6. Fatigue curves for phenolic-glass composite determined at various temperatures

We may further reinforce the ideas presented here by considering some data obtained by Stevens[(6)] in fatigue experiments conducted at different temperatures. Figure 6 represents experimental results for a phenolic-glass composite. These results are typical for this type of experiment. An increase in temperature clearly accelerates the rate of fatigue damage imparted at a given stress level. For example, a maximum stress of 20,000 psi gives instantaneous rupture at 800°F, rupture after 250 cycles at 500°F, after 1.78×10^4 cycles at 300°F and 2.0×10^5 cycles at 73°F. The data presented in Fig. 6 strongly suggests a simple time–temperature relationship. In Fig. 7, the fatigue data have been superimposed employing the values of $\log a_D$ determined experimentally by the construction of Fig. 3. As can be seen, the superposition is remarkably good considering the normal statistical scatter of lifetime data and defines an almost complete master fatigue curve known as an S–N curve in conventional engineering practice. Clearly, an understanding of the implications of these types of procedures in material and engineering design will constitute a necessary step in the further progression of the science of composite materials.

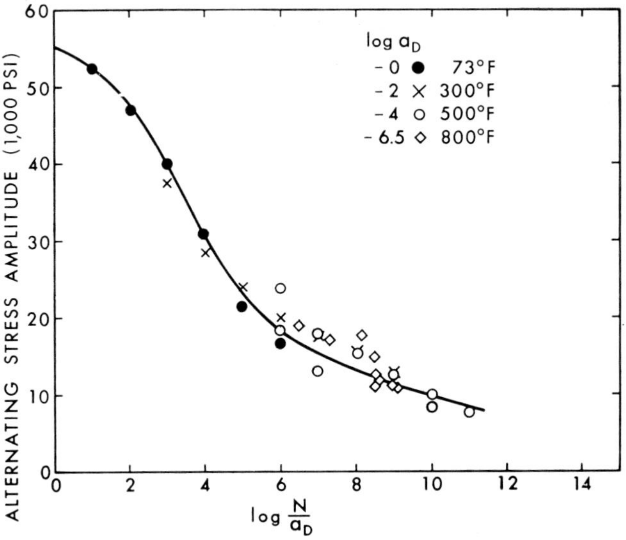

FIG. 7. Fatigue master curve constructed with the same shift factors, log a_D, as employed in Fig. 3

B. *Reversible Temperature Effects*

In the last section we illustrated temperature induced processes which lead to an irreversible change in composite properties through rate processes. In this section we shall be concerned with rate processes which are reversible with changes in temperature. This area of investigation is called viscoelasticity. If we restrict ourselves to levels of stress and strain which allow the material to respond in a linear manner, we are dealing with linear viscoelasticity. It is absolutely essential that one is aware of the differentiation between reversible and irreversible phenomena. For example, if we measure the modulus of a typical cold setting unfilled epoxy at room temperature, it exhibits a modulus of roughly 300,000 psi. If, however, the temperature is increased to say 200°F the modulus will be only 140,000 psi. On returning to room temperature we may regain the 300,000 psi modulus. In comparison with the earlier decomposition problem, an exposure of the material to say 500 or 600°F would result in a permanent change in the structure of the material with the natural consequence that on returning to the standard temperature, the material will exhibit a modulus value which is different from the original modulus. The same remarks are equally true regarding strength properties where there can be large changes in strength due to both reversible or irreversible processes.

The normal viscoelastic response of typical organic matrix material exhibits properties paralleling the data shown in Figs. 2 and 3. For example, consider the simple experiment in which we apply a step increase in length of a slender rod of our material and measure the surface traction of stress required to maintain the specimen in the extended configuration. The ratio of this stress to the computed constant strain is the relaxation modulus of time dependent Young's modulus for an isotropic material. If data are obtained as a function of time and temperature and the time dependent modulus is plotted as ordinate with log t as abscissa, then we obtain a plot very similar in appearance to Fig. 2. The temperature region of rapid strength loss is denoted as the glass transition temperature and is approximately 230°F for a cold setting epoxy. Just as with the decomposition data, a time–temperature superposition can be employed to obtain a master curve of modulus against reduced time (log t/a_T) as in Fig. 3. See for example the recent data of Kaelble[7] or Theocaris.[8] For reversible viscoelastic data, the time–temperature shift distance is denoted by log a_T and is well approximated by eq. (2). For epoxy resins below the transition temperature, ΔH is in the range of 20 kcal/mole. Thus, viscoelastic response has a much stronger temperature dependence than the decomposition reactions. More importantly, the viscoelastic data are spread out over a much larger time scale than Fig. 3. While the steep descent of Fig. 3 occurs in two decades, the transition in modulus due to viscoelastic processes requires twenty decades in time or more. Consequently, modulus data is represented as an exponential series

$$E(t) = \sum E_n e^{-t} \tau_n \tag{5}$$

instead of a simple exponential function. The quantity τ_n is the "relaxation time" of the "nth" term of the series and is the inverse of a rate constant. Analytically, time–temperature reduction means $t/\tau_n \rightarrow t/\tau_n a_T$. Thus knowing the reduced master curve of $\log E(t)$ against log t/a_T and the analytical form of a_T, the modulus is known for any other temperature.

The next problem is to relate the time–temperature dependent material properties to the constitutive equations describing time-independent response. For example, the linear elastic constitutive equation for an isotropic solid is given as

$$\sigma_{ij} = [K-(2/3)G]\Delta\delta_{ij}+2Ge_{ij} \tag{6}$$

where $\delta_{ij} = 0$ if $i \neq j$,
$= 1$ if $i = j$,
σ_{ij} = stress tensor,
e_{ij} = strain tensor,
K = bulk modulus,
G = shear modulus,
Δ = volumetric strain.

Now we shall assume that the components of the stress and strain tensors are time-dependent, and replace the material parameters K and G by time-dependent material functions $K(t)$ and $G(t)$. If $\varrho(t)$ is a component of the stress or strain tensor, we define its Carson transform $\overset{*}{\varrho}$ by the relation

$$\overset{*}{\varrho} = p \int_0^\infty e^{-pt}\varrho(t)\, dt. \tag{7}$$

Similarly, the Carson transform $\overset{*}{m}(p)$ of a time-dependent material function $m(t)$ is defined as

$$\overset{*}{m}(p) = p \int_0^\infty e^{-pt}m(t)\, dt. \tag{8}$$

The relationship involving time between the components of the stress tensor and the components of the strain tensor will now be in the form

$$\overset{*}{\sigma}_{ij} = [\overset{*}{K}(p)-(2/3)\overset{*}{G}(p)]\, \overset{*}{\Delta}\delta_{ij}+2G(p)\overset{*}{e}_{ij}. \tag{9}$$

This can be considered as the constitutive equation in classical linear isotropic viscoelastic theory, and is the time dependent generalization of eq. (6). In order to apply eq. (9) to the solution of boundary value problems, it is necessary to invert this equation. The right-hand side involves the products of Carson transforms, the inversion of which leads to a convolution integral. Thus, inversion of eq. (9) gives

$$\sigma_{ij}(t) = \int_0^t \left\{\left[K(t-\theta)-\frac{2}{3}\, G(t-\theta)\right] \frac{d\Delta\,(\theta)}{d\theta}\,\delta_{ij} \right.$$
$$\left. +2G(t-\theta)\,\frac{de_{ij}(\theta)}{d\theta}\right\} d\theta. \tag{10}$$

This equation is called the superposition equation of Boltzmann. The superposition principle is of the utmost importance for the description of viscoelastic behavior under small stress and strains. It tells how many and what kind of measurements are necessary to characterize the viscoelastic behavior completely. It tells further how the results of different forms of excitation history can be related to each other; for instance, how we can predict the result of a dynamic stress–strain response from a creep experiment and conversely. The great power of the superposition principle lies in its generality as it is not restricted to any special material: all its consequences remain valid independent of the micro-structure of the material under consideration. The existence of the elastic-viscoelastic correspondence principle is a direct and natural consequence of the superposition equation (10).

Engineering stress analysis of viscoelastic materials employs the "correspondence rule" wherein transforms can be taken of the time-dependent variables and forcing functions to obtain an associated elasticity problem in the transformed variables. The solution of this problem, when transformed back into real time variables gives the desired result; in other words, the same techniques employed in the exercise of transforming eq. (6) into eq. (10). The use of the correspondence rule is of course dependent upon our ability to solve the associated elasticity boundary value problem; if the elasticity solution is intractable, the viscoelastic solution will be more so. The actual mathematical details of the treatment can also be quite difficult at times. While the above technique provides a reasonable treatment for isothermal analysis, the problem of transient thermal viscoelastic stress analysis is still without a general solution. The difficulty here is the sensitivity of viscoelastic material response to temperature which results in a simultaneous change in material properties with the transient temperature distribution through the body.

These general procedures are also applicable to anisotropic viscoelastic bodies. The details of the viscoelastic theory and stress analysis are reviewed by Halpin(9), Schapery(10), and Hilton.(11) There is, unfortunately, a serious lack of reference material available at this time dealing with viscoelastic response and analysis of composite systems with the associated critical comparison against the appropriate experimental data. Further complicating the problem is the fact that composite systems all exhibit a strong nonlinear response for which we lack the suitable analytical tools. These nonlinearities are particularly perplexing when one attempts to develop a rigorous analysis of the ultimate properties of the system.†

Just the presence of linear viscoelastic process greatly complicates the evaluation of the expected structural integrity of a specific design. In all instances, the rupture or failure of viscoelastic solids is strongly influenced by time dependent phenomena.(12) Now if we assume that it is feasible to execute a stress analysis for a given loading, specified material, type of anisotropy and geometrical configuration, it is still necessary to combine the stress distribution with some statement of a criterion for failure. With very few exceptions, viscoelastic materials are experimentally analyzed in a uniaxial stress state. While these efforts give invaluable information relating the rupture stress or strain to the "reduced time scale" t/a_T or reduced strain or loading rate $1/\bar{R}_{a_T}$,‡ they do not provide the necessary information for the prediction of viscoelastic failure in a multiaxial stress–strain state. This is a serious problem as most geometries of engineering interest are in a state such that the stress and strain distribution is multiaxial. While experience with isotropic solids exhibiting very small or no time dependence does indicate that the octahedral stress or von Mises criterion can correlate uniaxial and multiaxial states,

† See Ashton, J. E., *J. Composite Materials*, **2**, 116 (1968).

‡ $\bar{R}$ = time rate factor related to de/dt or $d\sigma/dt$, where $\bar{R}$ is approximately equal to time.

no such relationship has been verified for viscoelastic materials. Instead of a simple unique correlation between uniaxial and multiaxial states for a time–temperature independent body, the association of multiaxial states for viscoelastic materials must also include strain rate $1/\bar{R}_{a_T}$ or time scale t/a_T. As a consequence there cannot be a single failure surface characterizing fracture but a multitude of closed surfaces, each surface characterizing the multiaxial state for a specific value of $1/\bar{R}_{a_T}$ or t/a_T. The strength problem is particularly complicated for the anisotropic body as workers have not, at this time, been able to experimentally verify the general multiaxial criterion of Hill[1] or others for a time–temperature independent anisotropic body.

In addition, time dependence in the fracture process complicates the analysis of the statistics of fracture in that we must now discuss fracture in terms of a time-dependent probability distribution of flaw strength as indicated in our discussion of stress corrosion. Furthermore, time dependence in material or fracture response leads to the phenomenon of fatigue or cumulative damage wherein the lifetime of the material is a strong function of the stress–strain–temperature history of the sample or structure. In other words, how do we predict a fatigue failure from constant strain rate or constant load failure data? Finally, viscoelastic response necessarily requires energy dissipation in response to vibrational excitation. This energy dissipation can lead to large cumulative heating which can result in accelerated fatigue damage or strength loss and in some cases to chemical decomposition of the matrix materials. This phenomenon can be directly related to an interaction between the imaginary part of the complex shear modulus, or loss tangent, and the natural frequency of the design configuration. Design attention should be given to this problem when composite materials are employed in structures subjected to dynamic loads.†

C. *Permeation*

Permeation of water or other agents through the matrix will have two effects on the properties of the matrix. The first effect is that, for instance, water will swell the hard glassy matrix material and accelerate rate process. In other words, the absorption of water vapor is equivalent to increasing the ambient temperature of the material. This process is termed "plasticization" in chemical industry technology. Thus, if we repeat our stress relaxation experiment at constant temperature but at different concentrations of water in the matrix, we shall obtain data[9] similar to Fig. 2 (where temperatures on the curve are now concentrations of plasticizing agent). This experiment may be performed, for example, at constant temperature and differing relative humidities. Superposition of the data similar to the operation in Fig. 3 can be

† See Schultz, A. B. and Tsai. S. W., *J. Composite Materials* **3**, 434 (1969).

performed with the time-plasticizer concentration parameter a_c being given as[9]

$$\log a_c = \frac{-C_3(1-C/\varrho_2)}{C_4+(1-C/\varrho_2)} \tag{11}$$

where C is the concentration of swelling agent, ϱ_2 is the density of the polymeric matrix at temperature T, and C_3 and C_4 are empirical constants to be determined by the experiment. A simultaneous change in temperature and swelling agent concentration could then be approximated by the reduced time $t/a_T a_c$. Again, inducing time dependence by a swelling agent leads to a viscoelastic response which can be handled within the framework of linear viscoelasticity, and composite response can, in principle, be predicted by employing the correspondence principle.[10] It is also obvious that the plasticizing effects cause a time or rate dependent decrease in ultimate properties.

Now, the above remarks suggest that swelling of an epoxy or phenolic matrix by water can be treated in a straightforward manner. However, this is not quite true as there are additional effects which are irreversible and separate from the plasticizing action described above. The plasticizing action leads to reversible property changes, while the absorption of swelling agent and subsequent redrying of a specimen of matrix material quite often results in permanent irreversible property loss. To understand how this irreversible loss occurs, let us examine the recent results of Alfrey, Gurnee and Lloyd.[13] In this paper they point out that hard glassy crosslinked polymers (matrix materials) when placed in contact with a solvent or swelling agent give an "anomalous" diffusion effect not explicable in terms of Fickian process. Experimentally one observes that as the absorbed agent penetrates into the material, a sharp advancing boundary separates the inner hard rigid core from the outer swollen softer shell. Behind the front, which advances at constant velocity, the material seems to be in an equilibrium state with respect to swelling agent concentration. Because the swollen material forms a shell around the unswollen materials and is directly connected to the unswollen core, the core will be placed under a state of triaxial stress. If one performs a stress analysis on a cylindrical specimen, it can be shown[13] that the core is under a large tensile stress which increases sharply as the swelling front advances into the material. When the swelling stresses reach a critical value in the core, a fracture occurs. In some cases this effect can be predicted analytically.[13]

In addition to matrix rupture, this mechanism could also lead to the rupture of the matrix–fiber interface and if very high swelling stresses are produced, fiber rupture might occur. Thus we have a means whereby irreversible damage is inflicted by the swelling process which is clearly different than the plasticizing effect described above. Moreover, the swelling rupture of the resin and fiber–resin interface exposes the fibers to a high solvent atmosphere which when coupled with the swelling induced tension in the fibers can lead to severe stress corrosion of the reinforcement.

These above remarks would appear to explain the "water boil" tests and why these tests can inflict such severe damage upon samples which are stored in boiling water for only a few hours. But our discussion also raises the question as to whether "water boil" is an accurate diagnostic accelerated aging test. It is fair question to ask if the storage of a composite panel in a normal relative humidity atmosphere and at temperatures of conventional engineering interest would ever be subjected to the same apparent magnitude of internal stress as occur in the "water boil" test. In any event, the loss of composite properties due to water or swelling agent permeation of the composite matrix is a highly coupled mechanism and much work must be done before useful analytical treatment of the various processes can become available for engineering design.

Since it appears that both large stress and strain[9] and/or permeation of water through a resin matrix produces voids, it is of interest to see whether we can predict the gross behavior of the undirectional composite as a function of void content. It is assumed that voids are randomly distributed spherical cavities in the matrix. The effect of voids on the composite properties can be predicted using the properties of a degraded matrix, i.e., a porous matrix. There are a number of equations for calculating the effective stiffness and strength of a porous continuum, e.g., Mackenzie,[14] Karner,[15] and Hashin.[16] All theories predict various degrees of degradation as void content increases. Coble and Kingman[17] showed that the Mackenzie equation[14] agreed well with the data of sintered alumina. If we use this equation, i.e.

$$1-\frac{\bar{G}}{G_m} = \frac{5(3K_m+4G_m)}{9K_m+8G_m}P+AP^2, \tag{11}$$

where $\bar{G}$ = shear modulus of the porous matrix; G_m, K_m = shear and bulk moduli of the pure matrix; P = volume fraction of pores in the matrix; A = constant; and assume the matrix materials to be elastic and isotropic, then

$$\begin{aligned} K_m &= E_m/3(1-2\nu_m) \\ G_m &= E_m/2(1+\nu_m). \end{aligned} \tag{12}$$

Combining eqs. (11) and (12) and after rearranging,

$$\bar{G}/G_m = 1-\frac{15(1-\nu_m)}{7-5\nu_m}P-AP^2. \tag{13}$$

The constant A can be determined by setting

$$\bar{G} = 0 \quad \text{when } P = 1.00. \tag{14}$$

Furthermore, pores in the matrix can be related to the voids in the composite as follows:

$$P = v_v/v_m \tag{15}$$

where v_v = void content in the composite, v_m = matrix content. By combining eqs. (13), (14) and (15),

$$\bar{G}/G_m = 1 - \frac{15(1-\nu_m)}{7-5\nu_m}\left(\frac{v_v}{v_m}\right) + \frac{8-10\nu_m}{7-5\nu_m}\left(\frac{v_v}{v_m}\right)^2. \tag{16}$$

Comparing this equation with experimental data reported by General Dynamics[4], the material properties that correspond to boron-epoxy composites are:

$$\begin{aligned} \nu_m &= 0.35 \\ v_m &= 0.38 \end{aligned} \tag{17}$$

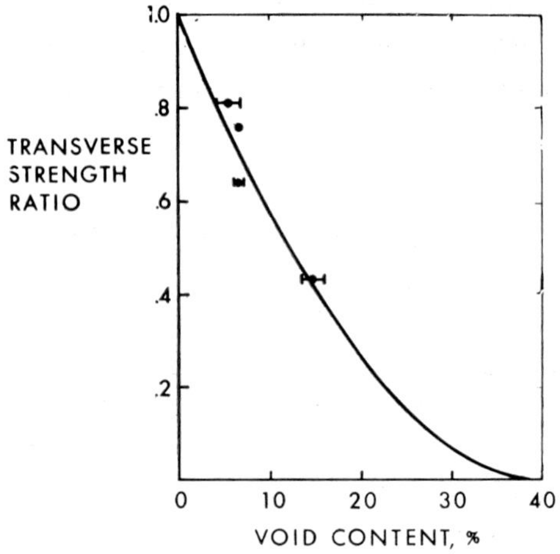

FIG. 8. Dependence of transverse strength ratio on void content of boron epoxy composite

where it is assumed that the fiber content is 55%, and glass 7%. It is also assumed for the purpose of comparison that the matrix material (Narmco 5505) is such that the ratio of the $\bar{G}/G_m$ is the same as the ratio of the composite transverse strength to the matrix strength. With these assumptions, the theoretical prediction based on eq. (16) is shown as a solid line in Fig. 8 along with the corresponding experimental data and their variations. Thus, the effect of voids on the transverse strength of unidirectional composites is quite drastic. The effect of voids on the longitudinal strength should be negligible which agrees with the measured data also reported in ref. 4. The formation of voids due to the combined action of stress and permeation will lead to a strong nonlinear viscoelastic response[9] which will be largest in the shear and transverse properties and least in the longitudinal direction of a unidirectional composite.

VI. CONCLUSIONS

Preliminary evidence as indicated here establishes the feasibility of using conventional rate process theory, as employed in other areas of material science, for the correlation and prediction of environmental processes on the engineering response of composite structures. The mechanisms discussed should lay the groundwork for establishing a unified experimental program to define a more comprehensive and realistic evaluation of the performance of structural composites than the empirical approach.

VII. ACKNOWLEDGEMENTS

This work is supported in part by the Air Force Materials Laboratory contract AF 33(615)-3430, and the Office of Naval Research and the Advanced Research Project Agency contract ONR No. N00014-67-C-0218. The author wishes to thank Dr. J. C. Halpin for his helpful suggestions and assistance.

REFERENCES

1. HILL, R. *Mathematical Theory of Plasticity*, Oxford University Press, 1950.
2. AZZI, V. D. and TSAI, S. W. Anisotropic Strength of Composites, *Experimental Mechanics*, Sept. 1965.
3. BOLLER, K. H. Strength Properties of Reinforced Plastic Laminates at Elevated Temperatures, Wright Air Development Center Tech. Report 59–569 (1960).
4. Fort Worth Division of General Dynamics, Structural Airframe Application of Advanced Composite Materials, Air Force Materials Laboratory Contract No. AF 33(615)-5257, Third Quarterly Progress Report, March 1967.
5. HALPIN, J. C. Unpublished results.
6. STEVENS, G. H. Fatigue Tests of Phenolic Laminate at High Stress Levels and Elevated Temperatures, Forest Products Lab. Report No. 1884 (1961).
7. KAELBLE, D. H. Dynamic and Tensile Properties of Epoxy Resins, *J. Appl. Poly. Sci.* **9**, 1213 (1965).
8. THEOCARIS, P. S. *Rheology Acta*, **2**, 92 (1962).
9. HALPIN, J. C. Introduction to Viscoelasticity, in *Composite Materials Workshop*, edited by S. W. Tsai, J. C. Halpin, and N. J. Pagano, Technomic Publishing Co., Inc., Stamford, Conn. 1968.
10. SCHAPERY, R. A. Stress Analysis of Viscoelastic Composite Materials, in *Composite Materials Workshop*, edited by S. W. Tsai, J. C. Halpin and N. J. Pagano, Technomic Publishing Co., Inc., Stamford, Conn. 1968.
11. HILTON, HARRY H. Viscoelastic Analysis, in *Engineering Design for Plastics*, edited by Eric Baer, Reinhold Publishing, New York 1964.
12. HALPIN, J. C. *J. Composite Materials*, **1**, 64 (1967); J. C. Halpin and Gerhard Jacoby, Fracture of Amorphous Polymeric Solids, *Macro Molecular Reviews*, edited by Peterlin, Zimm and Goodman, Interscience, New York 1968.
13. ALFREY, TURNER JR., GURNEE, E. F. and LLOYD, W. G. Diffusion in Glass Polymers, *J. Polymer Science*, Part C, **12**, 249 (1966).
14. MACKENZIE, J. K. *J. Proc. Phy. Soc. (London)*, **63B**, 2 (1956).
15. KARNER, E. H. *J. Proc. Phy. Soc. (London)*, **69B**, 802 (1956); **69B**, 808 (1956).
16. HASHIN, Z. *Bull. Res. Council, Israel*, **56**, 46 (1955).
17. COBLE, R. L. and KINGMAN, W. D. *J. Am. Ceram. Soc.* **39**, 377 (1956).

HIGH-TEMPERATURE ABLATION INTERACTION

F. L. SCHUYLER

Space Sciences Laboratory,
General Electric Company

and

E. FRIEDMAN†

Re-centry and Environmental Systems Division,
General Electric Company

ABSTRACT

The objective of this work is to initiate a self-consistent, integrated heat shield analysis. Considered is a conical anisotropic structure which is immersed in a quasi-steady hypersonic environment. The internal properties of the structure are permitted to vary through its thickness. A compressible, chemically reacting, non-similar boundary layer solution is obtained and intimately coupled with the structure's heat transfer and mass loss. Using a transient heat conduction theory and postulating mechanical behavior to be characterized by an elastic-perfectly plastic solid, the structure is analyzed with respect to its overall thickness change (due to vaporization ablation), temperature distribution, thermally-induced stress and deformation fields, surface temperature, heat transfer rate and regions of plastic flow.

LIST OF SYMBOLS

C	specific heat of solid, ft²/sec² °K
c_p	specific heat at constant pressure, ft²/sec² °K
$\bar{c}_p$	mean specific heat at constant pressure, ft²/sec² °K
c_s	specific heat of solid, ft²/sec² °K
D	mass diffusion coefficient (Fick's law), ft²/sec
f, f_1	yield functions
F_1, F_2	deformation functions
g	description of state of stress (−1, elastic; =0, plastic)
h_i	specific enthalpy of ith species, $\int_{540}^{T} c_{p_i} dT + h_i^0$, ft²/sec²
$h_{f_i}^0$	specific enthalpy of formation, ft²/sec²
k_{p_i}	chemical equilibrium function
k	thermal conductivity, lbf/sec °F
L_0	thickness of structure, ft

† Formely with Space Sciences Laboratory, General Electric Company.

$(L_{is})_v$	energy required for gasification of ith species of shield, ft^2/sec^2
M	molecular weight, lb/mole
$\dot{m}$	mass loss of shield, lbf sec/ft^2
p	pressure, lbf/ft^2
Q_{cond}	$\left(\frac{k\varrho}{\varrho_r}\frac{\partial T}{\partial Y}\right)_{w,g} = \left(k\frac{\partial T}{\partial \bar{y}}\right)_{w,g}$ (see eq. 26), lbf/ft sec
Q_s	entire right side of eq. (26), lbf/ft sec
$\mathcal{R}$	universal gas constant, (ft^2 lb)/(sec^2 mole °K)
$\dot{r}$	recession rate of shield, ft/sec
r	ablated thickness of shield $= \int \dot{r}\, d\tau$, ft
T	temperature, °K
u	x component of gas velocity, ft/sec
v	y component of gas velocity, ft/sec
v_i	y component of diffusion velocity, ft/sec
U	free stream velocity, ft/sec
V	Dorodnitsyn transformed v, ft/sec
w_i	rate of chemical production, lbf sec/ft^4
x	coordinate parallel to shield surface, ft
y	coordinate perpendicular to shield surface fixed in space, directed into solid, ft
$\bar{y}$	coordinate perpendicular to shield surface fixed in space, directed out into gas, ft
Y	Dorodnitsyn variable, ft
Y_0	reference temperature yield stress, lbf/in^2
Y_1	coordinate perpendicular to solid surface fixed on solid surface and directed into solid
Y_2	yield stress normal to shield surface, $= 1/\sqrt{2F}$, lbf/in^2
z	coordinate parallel to shield surface, perpendicular to flow, ft
α	coefficient of thermal expansion, 1/°K
α_i	mass fraction of ith species
δ	boundary layer thickness, ft
ε	strain component parallel to shield surface
ε_1^T, ε_2^T	thermal expansion functions
$\Delta x, \Delta Y, \Delta Y_1$	increments in x, Y and Y_1 directions, ft
λ	proportionality factor in flow rule, lbf/in^2 sec
μ	viscosity coefficient, lbf sec/ft^2
ϱ	density, lbf sec^2/ft^4
σ	stress component parallel to shield surface, lbf/in^2
σ_{ij}	component of stress, lbf/in^2
τ	time, sec
τ_w	viscous shear stress at shield's surface, lbf/ft^2

$\Delta\tau$	increment in time, sec
$(\dot{})$	time derivative $= \partial/\partial\tau$, 1/sec

Yield parameters F, G, H, L, M, N, $(\text{in}^2/\text{lbf})^2$
Elastic constants E_1, E_2, ν_{11}, ν_{12}, ν_{21}, G_1

Subscripts

a	position at cold surface of shield
fs	free stream
g	gas
i	ith species
is	ith component of solid's species
j	index referring to any time
$j, 1$	index referring to solid surface at any time
m, n	index referring to any x and y position
$m, 1$	index referring to any x position along the surface, $\bar{y} = 0$
r	reference value (free stream when constant)
s	solid
w	gas–solid interface
w_o	reference temperature (taken here as 4550 °K)

I. INTRODUCTION

The complexity of re-entry vehicle systems continues to increase. This is due to the demand for improved performance under increasingly severe flight environments. Advanced material systems offer the greatest potential and challenge for attaining these required improvements. Advanced filament reinforced composite materials, for example, have been shown to offer significant payoff potential in many re-entry vehicle applications, and it is expected that entire re-entry systems can benefit from these new materials.

Many re-entry vehicles serving military as well as non-military purposes encounter environments during flight, the severity of which cannot yet be simulated in presently available experimental facilities. Reliable experimental data for such flight regimes can, therefore, be obtained only from actual flight measurements; it is not necessary to emphasize the expense of such undertakings.

For these reasons, flight and laboratory experiments must be preceded by carefully planned analyses delineating the parameters which are important and, therefore, should be measured in order to obtain maximum pertinent information with minimum effort. This then requires a capability to determine and predict analytically the behavior of re-entry structures under the severe re-entry conditions encountered in actual flights.

The outer shells of manned and unmanned re-entry vehicles generally consist of layers of different materials designed with an exterior heat shield and an interior load carrying structure with various protection and bond layers suitably located between the two primary layers. This poses a serious reliability problem. In addition, it also represents an inefficient use of the material layers, requiring at times a 20 to 30% increase in re-entry weight[1]. Therefore, the required analysis must not only be able to analyze a given heat shield but it should also be capable of yielding, in its final formulation, a completely integrated, monolithic re-entry heat shield structure in which the material properties, through the thickness of the structure, are continuously distributed in an optimum manner. This optimization would obviously enhance the shield's properties in areas dictated by environmental influences.

Before a synthesis and optimization capability can be developed, however, it is first necessary to develop a capability to analyze, in a consistent manner, a given heat shield structure. In addition, this must be accomplished taking into account the effects of the various environmental and material phenomena which are encountered during typical re-entry conditions.

The work described in this paper is a first step toward this capability, and consists of two parts: (a) an aerothermochemical part—concerned with the (unified ablation) analysis of the boundary layer which is formed on the ablating shield, together with the shield's thermal field and dimensional changes; and (b) a thermostructural part—concerned with the evaluation of thermally induced stresses and deformations in the structure.

II. BACKGROUND

Although there are countless numbers of papers dealing with thermal strain phenomena, solid heat conduction phenomena, and heat and mass transfer coupled boundary layer phenomena, there are only a few investigations which have alluded to the possibility of integrating the boundary layer and solid heat conduction phenomena into a unified, self-consistent analysis, and even fewer studies of the possibility of integrating *all* the phenomena.

The boundary layer on an ablating wall has been well treated by many investigators, e.g., Gilbert and Scala,[2,3] Lees,[4] and Baron[5] to mention only a few. Heat conduction in a solid with surface recession has also been well treated. Carslaw and Jaeger[6] is a standard reference. More recent, however, is the excellent review of solid-phase, transient, thermal conduction presented by Price, *et al.*[7] In ref. 7, solid surface radiation and combustion is treated, as well as in depth heat generation. When possible, closed form, analytical solutions are presented.

Coupling of the fluid mechanics and solid heat conduction phenomena is treated under certain approximations in, for example, refs. 8–11. Sharma *et al.*[8] couple a transient, one-dimensional solid heat conduction analysis with an analysis of the one-dimensional flow of ablation gases away from the wall.

Allen and Suess[9] treat solid heat conduction phenomena using a transient two-dimensional approach. The boundary condition at the solid–gas interface is specified in the sense that correlation formulae between heat and mass transfer are used. Blockage of thermal energy due to mass transfer away from the surface is included in a convective heat transfer coefficient. The solid is assumed to be composed of only one material at any axial location, although the material may vary from one location to the other and each may have its thermal properties vary with temperature.

In a 1964 lecture series, Ostrach[10] reviewed in detail the studies and analyses performed by many investigators including himself. Besides reviewing these analyses, he also performed detailed comparisons between them. His primary attention was devoted to melting ablation where a liquid layer is formed on the surface of the solid. The solution of the temperature distribution through a finite thickness solid with a receding surface is discussed. Similarly, the liquid phase and gaseous phase flow fields are discussed. Although not explicitly performed, some discussion is carried out concerning coupling of the flow field and solid.

To conclude this discussion of the solid–fluid coupling, the work of Schuyler and Torda[11] should be noted. Their work is the basis for the present investigation. They obtain, numerically, the solution to the laminar, compressible, chemically reacting (nonequilibrium) boundary layer which exists on a subliming infinitely thick wall. The steady state heat transfer and temperature distribution within the wall are determined simultaneously with the boundary layer and inviscid flow fields. That is, the flow field and solid heat conduction are intimately coupled. Of special interest to the present problem is that the surface temperature, species mass fraction, heat transfer rate, sublimation rate and surface shear stress are determined *a posteriori*. In addition, chemical reactions are permitted to occur throughout the entire boundary layer thickness and no assumptions concerning similarity laws are made. All thermal properties are permitted to vary as functions of temperature, pressure and chemical composition. Numerical stability criteria have been carefully examined and rules established for obtaining rapid numerical convergence.

Although the physical model in ref. 11 was flow through a cylindrical chamber with combustible walls, the boundary layer solution procedure was subsequently extended by Schuyler to flow past a cone. Furthermore, the method of coupling the solid to the fluid is directly usable in the present problem.

With regard to analyses of stresses and deformations in structures and structural elements subjected to thermal environments, there has been a large volume of work carried out using the assumption of perfectly elastic material behavior. For problems in which the structure is subjected to severe temperatures resulting from aerodynamic heating, this idealization has been shown to be unrealistic and inelastic effects should be accounted for.[12,13] Therefore, for the present paper, the material is assumed to behave as an elastic-perfectly plastic solid, with solid phase viscous effects neglected. In future problems, however, viscosity should be accounted for, especially at high temperatures and for processes of longer time duration.

To date, a limited number of papers have been written on the elastoplastic response of structures under transient thermal conditions. Those studies that have been carried out, (e.g., refs. 14–19) for structures in which the effects of curvature are not considered, have made use of very idealized thermal conditions, and have considered the material to be homogeneous and isotropic.

The papers by Weiner,[14] Landau and Weiner,[15] and Yüksel[16] consider constant–thickness structures with temperature independent properties. Landau, Weiner, and Zwicky[17] treated a viscoelastic-plastic material with a temperature dependent yield condition. It was found that, for the particular problem considered, the effect of viscosity is negligible. Structures undergoing continuous phase transformation were investigated by Weiner and Boley,[18] who analyzed a solidifying slab using the temperature solution given by Neumann (see Carslaw and Jaeger[6]), and by Friedman and Boley[19], who estimated the thermomechanical response of a melting plate.

With this brief review of past work concerning the individual aspects necessary for an integrated analysis (fluid mechanics, heat transfer, and solid mechanics), let us now couple them to form a self-consistent structural analysis procedure. To be kept in mind is that the primary objective of the present work is to perform the first, and by far the most important, step in the development of a theory which will be capable of analyzing and synthesizing integrated heat protection and structural systems for modern, high performance re-entry vehicles.

III. SOLUTION PROCEDURE

A. *Physical Model*

A sharp pointed cone traveling at re-entry speeds will be studied. The flow field will be assumed to be quasi-steady, laminar, and axisymmetric. Because it is much more difficult to numerically program, equilibrium chemistry is assumed. (The difficulty with nonequilibrium chemistry is not with numerical programming but with the validity of the reaction rate expressions.) The fluid will be assumed to be representative of air (actually, nitrogen will be used) and species diffusion will be included. The gas phase and ablation characteristics of the cone material will be based on the properties of graphite (a commonly used heat shield material). In the present study, the effects of curvature of the conical shield structure are neglected with respect to the structure's temperature distribution, stresses and deformations. The farther the distance from the cone tip, the more realistic is this assumption. In addition, the material is taken to behave as an anisotropic elastic-perfectly plastic solid on a macroscopic level, with its yield surface being temperature dependent and its thermomechanical properties, including the yield surface, varying continually through the thickness. With respect to the temperature, stress, and strain distribution through the cone, a transient, one-dimensional analysis will be performed. The heat shield structure is taken to be traction-free at both faces and at the edges.† As a result, the calculated stresses and deformations are only dependent on the temperature transients.

The problem of the structure's mechanical response is formulated in terms of a strain increment, or strain rate, theory, rather than a total strain theory. This treatment permits the computation of stress rates at a given time in terms of known or previously evaluated quantities at that time. In this manner, the solution for stress rates is dependent only on conditions at the time in question and is essentially independent of past history. Integration of these stress rates over time makes possible the calculation of the complete profile of mechanical response (i.e., stresses, deformations, regions of plastic flow, etc.).

All the essential features of an integrated analysis are considered. All gas phase thermal properties are treated as functions of temperature and the properties of the solid vary throughout the cone thickness as well as with temperature.

B. *Mathematical Model*

1. FLOW FIELD. The equations which govern the steady flow of a compressible, chemically reacting viscous fluid over an axisymmetric heat shield surface are given, with respect to the Dorodnitsyn plane, as follows:

$$x = x,\ Y = \int \varrho/\varrho_r\, d\bar{y},\ V = v\varrho/\varrho_r + u\frac{\partial Y}{\partial x}$$

Global continuity equation

$$\frac{\partial u}{\partial x} + \frac{\partial V}{\partial Y} + \frac{u}{x} = 0 \tag{1}$$

Species continuity (diffusion) equation

$$\varrho u \frac{\partial \alpha_i}{\partial x} + \varrho V \frac{\partial \alpha_i}{\partial Y} = w_i + \frac{\varrho}{\varrho_r}\frac{\partial}{\partial Y}\left(\frac{D\varrho^2}{\varrho_r^2}\frac{\partial \alpha_i}{\partial Y}\right) \tag{2}$$

Momentum equation

$$u\frac{\partial u}{\partial x} + V\frac{\partial u}{\partial Y} = -\frac{1}{\varrho}\frac{\partial p}{\partial x} + \left[\varrho\mu\frac{\partial^2 u}{\partial Y^2} + \frac{\partial \varrho\mu}{\partial Y}\frac{\partial u}{\partial Y}\right]\frac{1}{\varrho_r^2} \tag{3}$$

† It will subsequently be shown that the surface tractions on the heated surface due to the flow field are negligible in comparison to the thermally induced internal stresses.

Energy equation

$$\varrho \bar{c}_p \left[u \frac{\partial T}{\partial x} + V \frac{\partial T}{\partial Y} \right] = \frac{\varrho}{\varrho_r^2} \frac{\partial \varrho k}{\partial Y} \frac{\partial T}{\partial Y} + \frac{k \varrho^2}{\varrho_r^2} \frac{\partial^2 T}{\partial Y^2}$$

$$+ \frac{\varrho}{\varrho_r} \frac{\partial T}{\partial Y} \Sigma c_{p_i} \varrho^2 \frac{D}{\varrho_r} \frac{\partial \alpha_i}{\partial Y} - \Sigma h_i w_i + u \frac{\partial p}{\partial x} + \left(\frac{\varrho}{\varrho_r} \right)^2 \mu \left(\frac{\partial u}{\partial Y} \right)^2 \quad (4)$$

$$p = \varrho \mathcal{R} \Sigma \frac{\alpha_i}{M_i} T \quad (5)$$

$$\bar{c}_p = \Sigma c_{p_i} \alpha_i . \quad (6)$$

The Dorodnitsyn plane is used solely for the easing of the numerical stability criteria.

2. Gas phase chemical model. The gas model is an equilibrium mixture of air (represented by nitrogen, N_2) and sublimated heat shield material (represented by the characteristics of graphite, C). The chemical equations of importance in the gas phase are:

$$\begin{aligned} 3\,C &\rightleftharpoons C_3 \\ N_2 &\rightleftharpoons 2\,N \\ N + C &\rightleftharpoons CN. \end{aligned} \quad (7)$$

It is to be noted that if air was actually used (i.e., a mixture of oxygen and nitrogen) the solution procedure would be exactly the same. Only the physical size of the computer program would be increased because of the increased number of species.

At the solid surface, the heterogeneous chemical reaction is also represented by an equilibrium reaction,

$$C(s) \rightleftharpoons C(g). \quad (8)$$

3. Solid phase heat conduction. The temperature distribution through the cone is determined from the one-dimensional heat conduction equation (see sketch below):

$$\varrho_s c \frac{\partial T}{\partial \tau} = \frac{\partial}{\partial y} \left(k_s \frac{\partial T}{\partial y} \right). \quad (9)$$

The thickness coordinate is changed from a stationary system (y) to a normalized moving system (Y_1),

$$Y_1 = \frac{y - \int \dot{r}\, d\tau}{L_0 - \int \dot{r}\, d\tau} \quad (10)$$

where $\dot{r}$ is the regression rate of the cone's surface and L_0 is the shield's initial thickness.

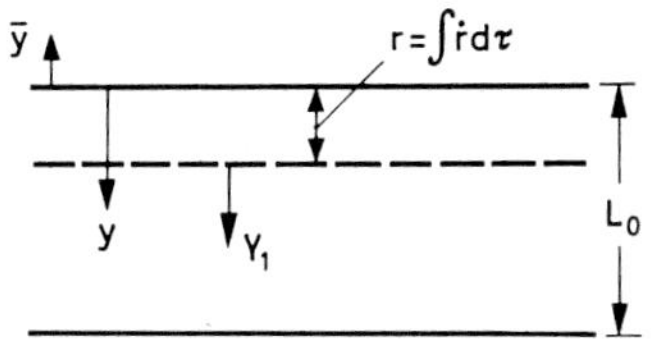

The resulting equation is:

$$\left[\frac{k_s}{(L_0-\int \dot{r}\,d\tau)^2}\right]\frac{\partial^2 T}{\partial Y_1^2}+\left[\frac{1}{(L_0-\int \dot{r}\,d\tau)^2}\frac{\partial k_s}{\partial Y_1}+\frac{\varrho_s c\dot{r}(1-Y_1)}{(L_0-\int \dot{r}\,d\tau)}\right]\frac{\partial T}{\partial Y_1}=\varrho_s c\frac{\partial T}{\partial \tau}. \tag{11}$$

This transformation has the effect of maintaining the validity range of the Y_1 equation between 0 and 1. Whereas for the y-coordinate, the range of concern, except for time zero, would be between some time varying value $\int \dot{r}d\tau$ and L_0. The usefulness of the transformation will be seen when the equations are transformed to finite differences and their stability criteria are investigated.

4. Solid constitutive equations. In the following, stress–strain relations in both the elastic and plastic regions of the structure are presented. Associated with these expressions are criteria for determining whether a given surface within the structure is in an elastic or a plastic region.

In an elastic-perfectly plastic solid, the strain rate is taken to be the sum of an elastic part and a plastic part, the latter being non-zero, of course, only when a plastic state of stress exists. Therefore, we may write

$$\dot{\varepsilon}_{ij} = \dot{\varepsilon}^E_{ij}+(1-g)\,\dot{\varepsilon}^P_{ij} \tag{12}$$

where $\dot{\varepsilon}^E_{ij}$ and $\dot{\varepsilon}^P_{ij}$ are the elastic and plastic strain rates, respectively, and g is unity or zero, depending on whether an elastic or a plastic state of stress exists.

The yield condition and associated flow rules used are those suggested by Hill,[20] for a material with three mutually orthogonal planes of symmetry at each point, and are analogous to those of von Mises for isotropic materials. The yield condition, in the form $f(\sigma_{ij}, T) = 0$, is as follows:

$$\begin{aligned}2f(\sigma_{ij}, T) = {} & F(T)(\sigma_{yy}-\sigma_{zz})^2+G(T)(\sigma_{zz}-\sigma_{xx})^2+H(T)(\sigma_{xx}-\sigma_{yy})^2\\ & +2L(T)\sigma_{yz}^2+2M(T)\sigma_{zx}^2+2N(T)\sigma_{xy}^2-1 = 0\end{aligned} \tag{13}$$

where the x, y, and z axes are the principal axes of anisotropy, which remain fixed in direction throughout the material. Note that the yield function is independent of hydrostatic stress, and can be expressed solely as a function of the deviatoric stress components. The six temperature dependent parameters are related to the yield stresses in the three principal directions of anisotropy and the three yield stresses in shear with respect to these directions.[20] The components of the plastic strain rate tensor are related to the stress components through the expression

$$\dot{\varepsilon}_{ij}^{P} = \lambda \frac{\partial f}{\partial \sigma_{ij}}. \tag{14}$$

For the problem presently under consideration, rotational symmetry of the anisotropy is taken in the plane of the heat shield, which is parallel to the axes x and z. Thus, $F = H$, $L = N$, and $M = F+2G$. The stress–strain relations are then as follows, the subscripts 1 and 2 referring to directions in the plane of the shield (x-z plane), and perpendicular to the shield (y direction), respectively:†

$$\begin{aligned}
\dot{\varepsilon}_{xx} &= \frac{1}{E_1}\dot{\sigma}_{xx} - \frac{\nu_{12}}{E_2}\dot{\sigma}_{yy} - \frac{\nu_{11}}{E_1}\dot{\sigma}_{zz} + \dot{\varepsilon}_1^T + (1-g)\lambda[-G(\sigma_{zz}-\sigma_{xx}) + F(\sigma_{xx}-\sigma_{yy})] \\
\dot{\varepsilon}_{yy} &= -\frac{\nu_{21}}{E_1}\dot{\sigma}_{xx} + \frac{1}{E_2}\dot{\sigma}_{yy} - \frac{\nu_{21}}{E_1}\dot{\sigma}_{zz} + \dot{\varepsilon}_2^T + (1-g)\lambda[-F(\sigma_{xx}-\sigma_{yy}) + F(\sigma_{yy}-\sigma_{zz})] \\
\dot{\varepsilon}_{zz} &= -\frac{\nu_{11}}{E_1}\dot{\sigma}_{xx} - \frac{\nu_{12}}{E_2}\dot{\sigma}_{yy} + \frac{1}{E_1}\dot{\sigma}_{zz} + \dot{\varepsilon}_1^T + (1-g)\lambda[-F(\sigma_{yy}-\sigma_{zz}) + G(\sigma_{zz}-\sigma_{xx})] \\
\dot{\varepsilon}_{yz} &= \frac{1}{2G_1}\dot{\sigma}_{yz} + (1-g)\,\lambda L\sigma_{yz} \qquad\qquad (15) \\
\dot{\varepsilon}_{zx} &= \frac{1+\nu_{11}}{E_1}\dot{\sigma}_{zx} + (1-g)\lambda(F+2G)\,\sigma_{zx} \\
\dot{\varepsilon}_{xy} &= \frac{1}{2G_1}\dot{\sigma}_{xy} + (1-g)\,\lambda L\sigma_{xy}
\end{aligned}$$

where, $\dot{\varepsilon}_i^T = \frac{\partial}{\partial T}[\varepsilon_i^T(y, T)]\,\dot{T}$ $(i = 1, 2)$ are the rates of thermal expansion.

Note also that $\frac{\nu_{12}}{E_2} = \frac{\nu_{21}}{E_1}$.

The criteria for determining whether an elastic ($g = 1$) or a plastic ($g = 0$) state of stress exists are as follows:

$$\begin{aligned}
g(y, \tau) &= 1, \quad \text{if} \quad f < 0, \quad \text{or if} \quad f = 0 \text{ and } \dot{f} < 0 \\
g(y, \tau) &= 0, \quad \text{if} \quad f = 0 \quad \text{and} \quad \dot{f} = 0.
\end{aligned} \tag{16}$$

These three possible combinations of f and $\dot{f}$ represent, respectively, an elastic state of stress, elastic unloading from a previously plastic state of stress, and continuation of plastic flow.

† It is an inherent assumption of this formulation that the various elastic coefficients (E_1, ν_{12}, etc.) are independent of temperature. This simplification is a reasonable one for problems involving plastic flow.[21]

5. BOUNDARY CONDITIONS. There are three boundaries of concern: first, the inviscid flow field ($\bar{y} \geqslant \delta$); second, the gas–solid interface ($\bar{y} = 0$); third, the cold side of the heat shield ($\bar{y} = -L_0$). At these boundaries, the following conditions must be satisfied:

$$\text{@ } \bar{y} \geqslant \delta, \left.\begin{aligned} u &= U_{fs} \\ p &= p_{fs} \\ T &= T_{fs} \end{aligned}\right\} \text{ which, for a cone, are constant} \tag{17}$$

$$\Sigma\alpha_i = \alpha_{N_2} + \alpha_N = 1.0$$

$$\text{@ } \bar{y} = 0, \quad u = 0. \tag{18}$$

Also, we have conservation of mass and energy across this ($\bar{y} = 0$) interface. Conservation of mass will help us determine the amount of mass ablating from the wall. Because the wall is impermeable, the total mass transfer at the wall of the "element" air (nitrogen) is zero. That is, the total mass transfer of N_2 and N must be zero at the solid–gas interface. We may, therefore, write:

$$v_w \equiv \frac{\dot{m}_w}{\varrho_w} = \frac{1}{\varrho_w}\left[\dot{m}_C + \dot{m}_{C_3} + \frac{M_C}{M_{CN}}\,\dot{m}_{CN}\right]_w \tag{19}$$

where

$$(\dot{m}_i)_w = \left[\alpha_i(\varrho v) - \varrho D\frac{\partial\alpha_i}{\partial y}\right]_w. \tag{20}$$

Therefore,

$$v_w = \frac{\dfrac{\varrho D}{\varrho_r}\left[\dfrac{\partial\alpha_C}{\partial Y} + \dfrac{\partial\alpha_{C_3}}{\partial Y} + \dfrac{M_C}{M_{CN}}\dfrac{\partial\alpha_{CN}}{\partial Y}\right]}{\left[1 - \left(\alpha_C + \alpha_{C_3} + \dfrac{M_C}{M_{CN}}\,\alpha_{CN}\right)\right]} \tag{21}$$

and

$$V_w = \frac{(\varrho v)_w}{\varrho_r}. \tag{22}$$

The recession rate, $\dot{r}$, of the wall is

$$\dot{r} = \frac{\varrho_r V_w}{\varrho_s}. \tag{23}$$

The conservation of energy across the interface is a balance of the heat conducted to the interface through the gas, the heat conducted away from the interface through the solid, the heat convected and diffused away from the interface through the gas and finally the heat absorbed by the solid to affect the phase change (from solid to gas). This may be written as follows:

$$\left[\left(k\frac{\partial T}{\partial \bar{y}}\right)_s = \left(k\frac{\partial T}{\partial \bar{y}}\right)_g - \sum_{i=i}\alpha_i\varrho h_i(v+v_i) + \varrho_s\dot{r}\sum_{i=is}\alpha_i(h_i - L_{iv})_g\right]_w \tag{24}$$

where

$$v_i \equiv -\frac{D}{\alpha_i}\frac{\partial \alpha_i}{\partial \bar{y}}$$
$$(h_{is}-L_{is_v})_g \equiv (h_{is})_s . \tag{25}$$

This last definition states that the energy (enthalpy) of the solid species in the gas phase minus the energy required for the phase change is equal to the energy of the solid species in the solid phase. Transforming to the coordinate systems of the solid and gas yields,

$$\left\{\left[-\frac{k}{(L_0-\int \dot{r}\,d\tau)}\frac{\partial T}{\partial Y_1}\right]_s = \left[\frac{k\varrho}{\varrho_r}\frac{\partial T}{\partial Y}\right]_g + \frac{\varrho^2 D}{\varrho_r}\Sigma h_i \frac{\partial \alpha_i}{\partial Y} - \varrho v \Sigma\left[\alpha_i h_i - \alpha_{is}(h_{is}-L_{is_v})_g\right]\right\}_w . \tag{26}$$

This relationship will help us determine the temperature at the interface.

The combined use of the conservation of mass and energy (eqs. (21) and 26)) help determine the interface species distribution.

The boundary stresses at the solid's ablating surface, though considered to be negligible for the problem of determining internal stresses and deformations, are given by:

Normal stress $= p = p_{fs}$

Tangential stress $= \tau_w = \left(\mu\frac{\partial u}{\partial \bar{y}}\right)_w = \left(\frac{\varrho\mu}{\varrho_r}\frac{\partial u}{\partial Y}\right)_w$ (27)

@ $y = L_0$

$T = T_a$

Surface traction $= 0.$ (28)

C. *Method of Solution*

To determine the stress distribution through the shield structure, it is necessary to know both the temperature history throughout the structure and the rate of recession of the heated surface. This requires, therefore, that the fluid mechanics and solid heat conduction (unified ablation) problem be solved first. Results from this analysis are then used as input to the solid mechanics problem. The end result will yield the structure's stress and deformation history, the mass loss rate (ablation) of the structure, and the species distribution throughout the flow field.

1. UNIFIED ABLATION. (a) *Boundary layer.* The problem here is to solve the quasi-steady boundary layer flow field simultaneously with the solid's transient temperature distribution. The problem is highly non-linear and intimately coupled. Numerical methods will be employed.

Before presenting the numerical scheme, let us briefly discuss the coupling of the chemistry with the flow field. Equilibrium chemistry has been assumed. Therefore, the local species concentration is determined from the diffusion equation (eq. 2) as a function of local temperature, temperature gradient, and pressure. The effect of chemistry on the flow field is through the term $\Sigma w_i h_i$ in the energy equation (eq. 4). Although the species production rate w_i is not known *a priori*, we can substitute its dependence on species gradients into the energy equation from the diffusion equation. Knowing the following temperature and pressure dependent relationships between mass fraction and equilibrium constant,

$$\alpha_{C_3} = \alpha_C^3(\overline{M}p)^2(M_{C_3}/M_C^3)K_{p_{C_3}} \tag{29}$$

$$\alpha_N = \left[\frac{\alpha_{N_2}M_N^2}{(\overline{M}p)M_{N_2}}K_{p_N}\right]^{1/2} \tag{30}$$

$$\alpha_{CN} = \alpha_C\alpha_N(\overline{M}p)\,(M_{CN}/M_CM_N)K_{p_{CN}} \tag{31}$$

$$\overline{M} = \frac{1}{\dfrac{\alpha_C}{M_C}+\dfrac{\alpha_{C_3}}{M_{C_3}}+\dfrac{\alpha_N}{M_N}+\dfrac{\alpha_{CN}}{M_{CN}}+\dfrac{\alpha_{N_2}}{M_{N_2}}} \tag{32}$$

$$\alpha_C+\alpha_{C_3}+\alpha_N+\alpha_{CN}+\alpha_{N_2} = 1 \tag{33}$$

we may finally write the quantity w_i in terms of $\dfrac{\partial\alpha_i}{\partial Y}$, α_i, p, $\dfrac{\partial p}{\partial x}$, T and $\dfrac{\partial T}{\partial x}$. That is,

$$w_i = \varrho u\left(R_i+S_i\frac{\partial T}{\partial x}\right)+\varrho V\frac{\partial\alpha_i}{\partial Y}-\frac{\varrho}{\varrho_r}\frac{\partial}{\partial Y}\left(\frac{D\varrho^2}{\varrho_r}\frac{\partial\alpha_i}{\partial Y}\right) \tag{34}$$

where R_i and S_i are functions of α_i, p, $\partial p/\partial x$ and T. The primary result of these algebraic manipulations is that the only x derivative in the energy equation is that associated with the temperature (the quantity $\partial p/\partial x$ is always given and for a cone is zero). As we will see, this permits a straightforward procedure to be used for calculating the temperature at successive x-stations in the flow field.

Equations (1) to (4) may now be solved numerically. Following the procedure of Schuyler and Torda,[11] the equations are put in finite difference form. The momentum, energy, and species diffusion equations are replaced by forward differences for the x-coordinate and central differences for the Y-coordinate. The continuity equation is handled slightly differently. Although central differences for Y are used, backward differences must be employed for the x derivative. The equations in finite difference form are:

Continuity

$$V_{m+1,\,n+1} = V_{m+1,\,n} - \frac{\Delta Y}{2\Delta x}\left[u_{m+1,\,n+1} - u_{m,\,n+1} + u_{m+1,\,n} - u_{m,\,n}\right] - \frac{\Delta Y}{2x}\left[u_{m+1,\,n+1} + u_{m+1,\,n}\right] \tag{35}$$

Momentum

$$u_{m+1,\,n} = u_{m,\,n} + \frac{\Delta x}{u_{m,\,n}\,\Delta Y^2}\left\{(u_{m,\,n+1} - u_{m,\,n-1})\left[\left(\frac{(\varrho\mu)_{m,\,n+1} - (\varrho\mu)_{m,\,n-1}}{4\varrho_r^2}\right) - \frac{V_{m,\,n}\,\Delta Y}{2}\right] - \frac{\Delta Y^2}{\varrho_{m,\,n}}\frac{\Delta p}{\Delta x} + \frac{(\varrho\mu)_{m,\,n}}{\varrho_r^2}\left[u_{m,\,n+1} - 2u_{m,\,n} + u_{m,\,n-1}\right]\right\} \tag{36}$$

Species continuity

$$\alpha_{i_{m+1,\,n}} = \alpha_{i_{m,\,n}} + \frac{\Delta x}{u_{m,\,n}\,\Delta Y^2}\left\{(\alpha_{i_{m,\,n+1}} - \alpha_{i_{m,\,n-1}})\left[\left(\frac{(\varrho^2 D)_{m,\,n+1} - (\varrho^2 D)_{m,\,n-1}}{4\varrho_r^2}\right) - \frac{V_{m,\,n}\,\Delta Y}{2}\right] + \frac{w_{i_{m,\,n}}\,\Delta Y}{\varrho_{m,\,n}} + \frac{(D\varrho^2)_{m,\,n}}{\varrho_r^2}\left[\alpha_{i_{m,\,n+1}} - 2\alpha_{i_{m,\,n}} + \alpha_{i_{m,\,n-1}}\right]\right\} \tag{37}$$

Energy

$$\begin{aligned} T_{m+1,\,n} = T_{m,\,n} + \frac{\Delta x}{u_{m,\,n}\left(\bar{c}_{p_{m,\,n}} + \Sigma \dfrac{h_{f_i}^0 S_i}{T_{m,\,n}^2}\right)} \Bigg\{ \bar{c}_{p_{m,\,n}} \Bigg[\left(\frac{\varrho_{m,\,n}}{\varrho_r}\right)^2 \frac{D_{m,\,n}}{\bar{c}_{p_{m,\,n}}} \Sigma \left(\frac{\partial \alpha_i}{\partial Y} c_{p_i}\right)_{m,\,n} \\ - V_{m,\,n} + \frac{1}{\varrho_r^2 \bar{c}_{p_{m,\,n}}}\left(\frac{(\varrho k)_{m,\,n+1} - (\varrho k)_{m,\,n-1}}{2\Delta Y}\right)\Bigg] \left[\frac{(T_{m,\,n+1} - T_{m,\,n-1})}{2\Delta Y}\right] \\ + \frac{(k\varrho)_{m,\,n}}{\varrho_r^2}\left[\frac{T_{m,\,n+1} - 2T_{m,\,n} + T_{m,\,n-1}}{\Delta Y^2}\right] + \left(\frac{u}{\varrho}\right)_{m,\,n}\frac{\Delta p}{\Delta x} \\ + \frac{(\varrho\mu)_{m,\,n}}{\varrho_r^2}\left[\frac{u_{m,\,n+1} - u_{m,\,n-1}}{2\Delta Y}\right]^2 - u_{m,\,n}\,\Sigma h_{f_i}^0 R_i \\ + \left[\frac{1}{\varrho_r^2}\left(\frac{(\varrho^2 D)_{m,\,n+1} - (\varrho^2 D)_{m,\,n-1}}{2\Delta Y}\right) - V_{m,\,n}\right]\left[\Sigma h_{f_i}^0 \left(\frac{\partial \alpha_i}{\partial Y}\right)_{m,\,n}\right] \\ + \left(\frac{\varrho_{m,\,n}}{\varrho_r}\right)^2 D_{m,\,n}\left[\Sigma h_{f_i}^0 \left(\frac{\partial^2 \alpha_i}{\partial Y^2}\right)_{m,\,n}\right]\Bigg\}. \end{aligned} \tag{38}$$

It is seen that on the left side of each of these equations, there is only one term and only it is associated with the downstream location $m+1$. Everything on the right side is known. That is, at station m the values of all quantities are known. The order of solution of these equations is $T_{m+1,\,n}$; $U_{m+1,\,u}$; $\alpha_{i_{m+1,\,n}}$; $V_{m+1,\,n}$.

The question of numerical stability criteria will be discussed later. Let us first discuss the solid heat conduction analysis.

(b) *Heat conduction in the solid.* Here, as for the boundary layer solution, numerical methods are used. For the Y_1-derivatives central differences are used and for the time (τ) derivatives forward differences are used. The resulting formula is:

$$T_{n,j+1} = T_{n,j} + \frac{\Delta\tau}{(\varrho_s c)_{n,j}} \left\{ \frac{k_{n,j}}{(L_0 - \int \dot{r}\, d\tau)_j^2} \left[\frac{T_{n+1,j} - 2T_{n,j} + T_{n-1,j}}{\Delta Y_1^2} \right] \right.$$

$$\left. + \left[\frac{(k_{n+1,j} - k_{n-1,j})}{(L_0 - \int \dot{r}\, d\tau)_j^2 2\Delta Y_1} + \frac{(\varrho c \dot{r})_{n,j}(1 - Y_1)}{(L_0 - \int \dot{r}\, d\tau)_j} \right] \left[\frac{T_{n+1,j} - T_{n-1,j}}{2\Delta Y_1} \right] \right\}_s . \tag{39}$$

It is seen that all terms on the right side are evaluated at the current time (j). Only the left side is associated with the future time ($j+1$). Before coupling the boundary layer solution with the solid heat conduction solution, numerical stability criteria must be established for the above equations (eqs. 36–39).

(c) *Numerical stability criteria.* The procedure used is exactly the same as presented in ref. 11.

For the gas phase, the stability criteria associated with the momentum equation (eq. 36) were chosen as the primary criteria. (Criteria obtained by considering the energy and diffusion equations (37 and 38) are very similar and are presented in ref. 11.) The result is:

$$\left[\Delta Y < \frac{2\mu}{V\varrho} \left(\frac{\varrho}{\varrho_r} \right)^2 \right]_{m,n} \tag{40}$$

$$\left[\Delta x < \frac{u \Delta Y^2 \varrho}{2\mu} \left(\frac{\varrho_r}{\varrho} \right)^2 \right]_{m,n} . \tag{41}$$

For each x-location (m) the values of Δx and ΔY are held constant. Δx, however, is permitted to vary from one x-location to the next.

For the solid phase, we obtain:

$$\Delta\tau < \left\{ \frac{\varrho c}{2k} \left[L_0 - \int \dot{r}\, d\tau \right]^2 \Delta Y_1^2 \right\}_s . \tag{42}$$

ΔY_1 is chosen to be constant (this is possible only because, $0 \leqslant Y_1 \leqslant 1$). It is seen, however, that $\Delta\tau$ varies with time because of the term $\int \dot{r}\, d\tau$. For slow recession rates, the variation of $\Delta\tau$ will be small.

(d) *Solid–gas boundary conditions.* Of prime concern here are the expressions for conservation of mass and energy (eqs. 21 and 26). In finite difference form these are, respectively:

$$(v_w)_{m+1,1} = -[(\varrho D)_{m+1,1}/\Delta Y \varrho]\Big\{(\alpha_{C_{m+1,2}} - \alpha_{C_{m+1,1}}) + (\alpha_{C3_{m+1,2}} - \alpha_{C3_{m+1,1}}) + \frac{M_C}{M_{CN}}(\alpha_{CN_{m+1,2}} - \alpha_{CN_{m+1,1}})\Big\}\Big/\left[1 - \left(\alpha_C + \alpha_{C3} + \frac{M_C}{M_{CN}}\alpha_{CN}\right)_{m+1,1}\right] \tag{43}$$

$$\left[\frac{-k_{1,j+1}}{(L_0 - \int \dot{r}\, d\tau)_{j+1}\Delta Y_1}(T_{2,j+1} - T_{1,j+1})\right]_s = \Big\{\Big[\frac{(k\varrho)_{m+1,1}}{\varrho_r \Delta Y}(T_{m+1,2} - T_{m+1,1})\Big] + \frac{(\varrho^2 D)_{m+1,1}}{\varrho_r \Delta Y}\Sigma[h_{i_{m+1,1}}(\alpha_{i_{m+1,2}} - \alpha_{i_{m+1,1}})] - (\varrho v)_{m+1,1}\Sigma[(\alpha_i h_i)_{m+1,1} - (\alpha_{i_{s_{m+1,1}}}(h_{i_s} - L_{i_{s_v}})_{m+1,1})]\Big\}_g \tag{44}$$

(e) *Coupling the boundary layer and the solid heat conduction solutions.* The boundary conditions of paragraph (d) are used for this purpose.

From the boundary layer solution, the value of the temperature at $(m+1, 2)$, (i.e., a ΔY up from the surface), has already been calculated. Similarly, from the solid heat conduction solution, the value of the temperature at $(2, j+1)$ has also been calculated. The pressure at the surface is also known (equal to p_{fs}). In addition, the values of the mass fractions at the surface are known as functions of temperature and pressure from the equilibrium constants. It is observed, therefore, that only the surface temperature is unknown. Because of the very intimate coupling between the relationships for conservation of mass and energy (eqs. 21 and 26) at the surface, the surface temperature cannot be determined explicitly. An iterative scheme must be used. The procedure, however, is quite simple. Guess a surface temperature. Calculate all thermodynamic properties corresponding to this temperature and also all mass fractions. The mass fractions are obtained from eqs. (29)–(33) together with the solid reaction (eq. 8) given by:

$$\alpha_C = \frac{M_C K p_C}{(\bar{M} p)}.$$

From eq. (43), the value of v_w (i.e., $v_{m+1,1}$) may now be substituted into the conservation of energy expression, eq. (44). Both the left side and right side of the resultant equation have been evaluated and it is now only necessary to check if they are equal. If not, guess another value for surface temperature. Perform a linear iteration until both sides are equal. Convergence to the proper solution has proven to be quite rapid. The procedure is performed at each surface location for each $\Delta\tau$. The final result is a time varying surface temperature, species concentration, and mass loss. The time history of the structure's temperature field is also obtained. It is the surface mass loss and structure temperature history which are used as inputs to the solid mechanics problem.

2. Mechanical response of solid. In determining internal stresses resulting from a transient one-dimensional temperature field in a structure free of boundary tractions, it is reasonable to assume the following:

$$\begin{aligned} \sigma_{xx} &= \sigma_{zz} = \sigma(y, t) \\ \sigma_{yy} &= \sigma_{yz} = \sigma_{zx} = \sigma_{xy} = 0. \end{aligned} \tag{45}$$

For an elastic structure, the solution is known to be of this form.[21] For an elastic-perfectly plastic structure it can be shown that uniqueness of solution is attainable for stress components of this form, provided the boundary conditions are properly satisfied. (An extension of the uniqueness theorem for stress rates given in ref. 21, to account for anisotropy and receding boundaries (ablation), can easily be carried out.) Equations (15) now reduce to:

$$\begin{aligned} \dot{\varepsilon}_{xx} &= \dot{\varepsilon}_{zz} = \dot{\varepsilon}(y, t) = \frac{1-\nu_{11}}{E_1}\dot{\sigma}+\dot{\varepsilon}_1^T+(1-g)\,\lambda F\sigma \\ \dot{\varepsilon}_{yy}(y, t) &= -\frac{2\nu_{21}}{E_1}\dot{\sigma}+\dot{\varepsilon}_2^T-2(1-g)\,\lambda F\sigma \\ \dot{\varepsilon}_{yz} &= \dot{\varepsilon}_{zx} = \dot{\varepsilon}_{xy} = 0. \end{aligned} \tag{46}$$

It is easily shown[20] that F, the only remaining yield parameter of eq. (13), is related to the yield stress Y_2 in the direction normal to the plane of the shield by the following:†

$$F = \frac{1}{2Y_2^2}. \tag{47}$$

The yield condition (eq. 13) then becomes equivalent to:

$$f_1(\sigma, T) = \sigma^2 - Y_2^2 = 0 \tag{48}$$

where $f_1 = 2Y_2^2 f$.

In elastic regions ($g = 1$), it is seen from the first of eqs. (46) that

$$\dot{\sigma} = \frac{E_1}{1-\nu_{11}}(\dot{\varepsilon}-\dot{\varepsilon}_1^T) \tag{49}$$

while in plastic regions ($g = 0$), the requirement that $f_1 = \dot{f}_1 = 0$ (or $f = \dot{f} = 0$ from eqs. (16)) results in:

$$\dot{\sigma} = (sgn\ \sigma)\,\dot{Y}_2 \tag{50}$$

where $(sgn\ \sigma) = \dfrac{\sigma}{|\sigma|} = \dfrac{\sigma}{Y_2}$ and $\dot{Y}_2 = \dfrac{\partial}{\partial T}[Y_2(y, T)]\,\dot{T}$.

† It is of interest to note that even though the non-zero stress components act in the plane of the shield, the yield function f, depends on the yield stress in the normal direction. This is due to an isotropic state of stress existing in the plane of the shield while the yield condition depends essentially on the maximum shear stress developed.

In summary, then, the expression for stress rate and the criteria for determining g are:

$$\dot{\sigma} = g \frac{E_1}{1-\nu_{11}} (\dot{\varepsilon} - \dot{\varepsilon}_1^T) + (1-g)(sgn\, \sigma)\dot{Y}_2. \tag{51}$$

and

$$\begin{aligned} g(y, \tau) &= 1, \quad \text{if} \quad f_1 < 0, \text{ or } \quad \text{if} \quad f_1 = 0 \quad \text{and} \quad \dot{f}_1 < 0 \\ g(y, \tau) &= 0, \quad \text{if} \quad f_1 = 0 \quad \text{and} \quad \dot{f}_1 = 0 \end{aligned} \tag{52}$$

respectively. The criteria of eqs. (52) are completely equivalent to those of eqs. (16).

These criteria, as they presently stand, may not, however, be readily amenable to numerical computation since $\dot{f}_1$ is a function of stress rate $\dot{\sigma}$. This potential difficulty arises from the stress rate being discontinuous in time during transition from an elastic to a plastic state of stress (loading), or vice versa (unloading). It is possible, however, to develop these criteria in terms of quantities that are at all times continuous functions of time. This is done by making use of the requirement of positive plastic power in which, for $g = 0$,

$$\sigma_{ij}\dot{\varepsilon}_{ij}^P > 0. \tag{53}$$

For the problem at hand, this inequality simply reduces to

$$\sigma\dot{\varepsilon}^P > 0. \tag{54}$$

By making use of the first of eqs. (46), it is easily shown that this results in

$$\sigma\dot{\sigma} < \frac{E_1}{1-\nu_{11}} \sigma(\dot{\varepsilon} - \dot{\varepsilon}_1^T) \tag{55}$$

in regions where $g = 0$. If $g = 1$, $\dot{\sigma} = \dfrac{E_1}{1-\nu_{11}} (\dot{\varepsilon} - \dot{\varepsilon}_1^T)$. The revised criteria for g are then as follows:

$$\begin{aligned} g = 1, \quad & \text{if} \quad \sigma^2 - Y_2^2 < 0, \\ & \text{or if} \quad \sigma^2 - Y_2^2 = 0 \quad \text{and} \quad \frac{E_1}{1-\nu_{11}} \sigma(\dot{\varepsilon} - \dot{\varepsilon}_1^T) - Y_2\dot{Y}_2 < 0 \\ g = 0, \quad & \text{if} \quad \sigma^2 - Y_2^2 = 0 \quad \text{and} \quad \frac{E_1}{1-\nu_{11}} \sigma(\dot{\varepsilon} - \dot{\varepsilon}_1^T) - Y_2\dot{Y}_2 > 0. \end{aligned} \tag{56}$$

Thus, at a given time, the procedure for calculating stress rates throughout the thickness of the structure is determined in terms of material property data (all of which can be dependent on the thickness dimension y, with ε_1^T and Y_2 being dependent also on T), temperature, stress, and strain rate. Since stress is determined by integration of stress rates computed at previous times, and is thus a known quantity, it is only the last quantity—strain rate—which is unknown *a priori*. To determine strain rate, use is made of the two compatibility conditions that are not satisfied trivially, and the boundary conditions on the edges of the structure. The compatibility conditions to be satisfied are:

$$\frac{\partial^2 \varepsilon_{xx}}{\partial y^2} = \frac{\partial^2 \varepsilon_{zz}}{\partial y^2} = \frac{\partial^2 \varepsilon(y, \tau)}{\partial y^2} = 0. \tag{57}$$

Thus,

$$\dot{\varepsilon}(y, \tau) = \dot{F}_1(\tau) + \dot{F}_2(\tau) y \tag{58}$$

is the form that the strain rate must take. For the present problem, the boundaries of the structure are considered to be free; its edges are therefore tractionless. This condition can be approximated in the St. Venant sence by setting the resultant force and moment equal to zero. Defining the location of the ablating surface by $y = r \equiv \int \dot{r}\, d\tau$, these conditions are:

$$\begin{aligned} &\int_{r(\tau)}^{L_0} \sigma(y, \tau)\, dy = 0 \\ &\int_{r(\tau)}^{L_0} \sigma(y, \tau)\, y\, dy = 0. \end{aligned} \tag{59}$$

Since the above equations are satisfied at all times

$$\begin{aligned} \frac{d}{d\tau} \int_{r(\tau)}^{L_0} \sigma(y, \tau)\, dy &= \int_{r(\tau)}^{L_0} \dot{\sigma}(y, \tau)\, dy - \dot{r}(\tau)\, \sigma(r, \tau) = 0 \\ \frac{d}{d\tau} \int_{r(\tau)}^{L_0} \sigma(y, \tau)\, y\, dy &= \int_{r(\tau)}^{L_0} \dot{\sigma}(y, \tau)\, y\, dy - r(\tau)\, \dot{r}(\tau) \sigma(r, \tau) = 0. \end{aligned} \tag{60}$$

Substituting the expression for $\dot{\varepsilon}$ (eq. 58) into the expression for $\dot{\sigma}$ (eq. 51), and then substituting the result into eqs. (60), the solution of the latter two simultaneous equations for $\dot{F}_1$ and $\dot{F}_2$ is

$$\begin{aligned} \dot{F}_1 &= -\frac{R_2[S_2 - r\dot{r}\sigma(r)] - R_3[S_1 - \dot{r}\sigma(r)]}{R_1 R_3 - R_2^2} \\ \dot{F}_2 &= \frac{R_1[S_2 - r\dot{r}\sigma(r)] - R_2[S_1 - \dot{r}\sigma(r)]}{R_1 R_3 - R_2^2} \end{aligned} \tag{61}$$

where

$$R_i(\tau) = \int_{r(\tau)}^{L_0} \frac{E_1}{1 - \nu_{11}}\, g y^{i-1}\, dy, \qquad i = 1, 2, 3$$

$$S_j(\tau) = \int_{r(\tau)}^{L_0} \left[\frac{E_1}{1 - \nu_{11}}\, g \dot{\varepsilon}_1^T - (1 - g)(sgn\, \sigma)\, \dot{Y}_2 \right] y^{j-1}\, dy, \; j = 1, 2.$$

Equations (51), (56), (58) and (61) are the expressions needed to determine stress rates throughout the shield structure at a given time. Though these expressions are explicit in nature and no differential or integral equations are involved, it is, practically speaking, necessary to evaluate the unknown stress rates by making use of a digital computer. A brief outline of the calculation procedure employed follows.

At a given time station τ_j, the material properties are known quantities, while the distributions of temperature and temperature rate, amount of material ablated, and ablation (recession) rate are obtained from the results of the unified ablation analysis. The stress distribution through the thickness of the structure is also known from integration of stress rates computed at previous times. The only quantities unknown *a priori* at τ_j are the strain rate (determined from eq. (58) by $\dot{F}_1(\tau_j)$ and $\dot{F}_2(\tau_j)$), the $g(y, \tau_j)$, and the $\dot{\sigma}(y, \tau_j)$. The one difficulty arising in the numerical computation of these quantities is that the expressions for $\dot{F}_1$ and $\dot{F}_2$ are dependent on the $g(y)$ (eqs. 61), while the criteria for determining the $g(y)$ are in turn dependent on strain rate (eqs. 56). This is resolved by employing a simple iterative scheme. As a first approximation, the values of $\dot{F}_1$ and $\dot{F}_2$ at the previous time station τ_{j-1}, are used to evaluate the $g(y)$ at τ_j (eqs. 56). New values of $\dot{F}_1(\tau_j)$ and $\dot{F}_2(\tau_j)$ are then computed from eqs. (61). The $g(y)$ are recalculated using these new values, and the process is repeated until the $\dot{F}_1$ and $\dot{F}_2$ converge. Convergence is rather rapid due to g changing its value at relatively few y stations from τ_{j-1} to τ_j. When the final convergent values of $\dot{F}_1$, $\dot{F}_2$, and the $g(y)$ are calculated, the distribution of stress rates in the shield structure is then computed directly from eq. (51).

Stresses are calculated by integration of the stress rates, so that the stress distribution at some time $\tau = \tau^*$ may be determined from the following:

$$\sigma(y, \tau^*) = \sigma(y, 0) + \int_0^{\tau^*} \dot{\sigma}(y, \tau)\, d\tau \tag{62}$$

where $\tau = 0$ is the time at which the initial state (temperature and stress) of the body is specified.

It is interesting to note that the formulation of expressions for stress rates, which are dependent only on quantities at the time in question, is independent of the time scale. It is only when numerically integrating stress rates to obtain stresses that a smaller time interval will result in a somewhat more accurate solution.

IV. SAMPLE CALCULATION

As an example problem, the integrated solution for a 20° half angle cone traveling at a Mach number of 20 at an altitude of 140,000 ft is obtained. The shield structure of concern is 0.1 ft thick. The "air" is taken to be nitrogen, the gaseous phase of the cone material is based on graphite, and the solid phase is based on a glass fiber reinforced plastic.

A. *Thermodynamic, Chemical Kinetic, and Thermomechanical Variables*

For purposes of illustration, thermodynamic variables in the gas phase have been assumed to depend on temperature but not species. In the solid phase, variation with position as well as temperature is assumed. We therefore have for the gas phase,

$$k = a_k T^{1/2} \qquad \text{lbf/sec °R}$$

$$D = a_D \frac{T^{3/2}}{p} \qquad \text{ft}^2/\text{sec}$$

$$\mu = a_\mu T^{1/2} \qquad \text{lbf sec/ft}^2$$

$$c_p = a_{cp} + b_{cp} T \qquad \text{ft}^2/\text{sec}^2\ \text{°R}$$

$$L_{is_v} = \text{constant} \qquad \text{ft}^2/\text{sec}^2$$

$$h = \Sigma \alpha_i h_i = \Sigma \alpha_i c_p (T - 540) + h_{f_i}^0 \qquad \text{ft}^2/\text{sec}^2.$$

In the solid,

$$k_s = \left(1 - 0.6 \frac{y}{L_0}\right) 5.0583 \times 10^{-2}\, T^{0.157}$$

$$\varrho_s = \left(1 - 0.4 \frac{y}{L_0}\right) 4.968944$$

$$c = \left(1 + 0.2 \frac{y}{L_0}\right) 626.451\, T^{0.38}.$$

For the homogeneous and heterogeneous chemical reactions, the equilibrium constants are given by

$$K_{p_i} = \exp\left[(a_i - b_i/T)/0.43429\right].$$

The thermomechanical properties of the solid structure are taken to be representative of those of a glass-fiber-reinforced plastic. Variations in the volume concentration of the reinforcement make possible variations in these properties through the thickness of the shield. The scarcity of material property data at temperatures considerably higher than room temperature makes it extremely difficult to obtain reliable estimates of the temperature variation of the thermal expansion function ε_1^T, and the yield stress Y_2—the two temperature-dependent mechanical properties of interest. This unfortunate fact serves to accentuate the obvious need for obtaining considerably more high temperature experimental data of this sort than is presently available. Since the example problem outlined here is meant primarily to illustrate the use of the integrated approach and to present some typical results, these two properties, though varying with the thickness dimension y, are taken for the sake of simplicity to vary linearly with temperature, as follows:

$$\varepsilon_1^T(y, T) = \alpha_1(y)(T-T_a)$$

$$Y_2(y, T) = Y_0(y)\frac{T_{w0}-T}{T_{w0}-T_a}$$

where $\alpha_1(y)$ is the reference temperature coefficient of thermal expansion, $Y_0(y)$ the reference temperature yield stress, T_a the temperature maintained at the "cold" surface of the shield structure (used as the reference temperature), and T_{w0} the wall temperature approached during the latter stages of the ablation process, when a steady state temperature distribution is approached. Note that the yield stress goes to zero at this temperature.

In addition, the thickness dependent properties are taken to vary linearly with y in such a manner as to minimize the severity of the initial stress distribution, $\sigma(y, 0)$. These properties are of the following form:

$$\frac{E_1}{1-\nu_{11}} = a_{E_1}\left(1+0.5\frac{y}{L_0}\right) \qquad \text{lbf/in}^2$$

$$\alpha_1 = a_{\alpha_1}\left(1-0.5\frac{y}{L_0}\right) \qquad /°\text{K}$$

$$Y_0 = a_{Y_0}\left(1+0.5\frac{y}{L_0}\right) \qquad \text{lbf/in}^2.$$

It is expected that variations of this degree through the thickness are attainable for reinforced plastics.

B. *Input Data*

The constants used in the thermodynamic, chemical and mechanical expressions are as follows:

$p_{fs} = 630.2$ lbf/ft$^2 \approx 0.3$ atm	$b_{c_p} = 0.0416$
$U_{fs} = 20{,}600$ ft/sec	$h^0_{fN} = 3.64\times10^8$ ft^2/sec^2
$T_{fs} = 3000$ °K	$h^0_{f_{N_2}} = 0.0$ ft^2/sec
$(\alpha_{N_{2fs}}) = 0.978$ calculated	$h^0_{fCN} = 1.92\times10^8$ ft^2/sec^2
$a_k = 2.1\times10^{-4}$	$h^0_{f_c} = 6.43\times10^8$ ft^2/sec^2
$a_D = 6.0\times10^{-5}$	$h^0_{f_c} = 2.375\times10^8$ ft^2/sec^2
$a_\mu = 2.6\times10^{-8}$	$L_{is_v} = 7.5\times10^6$ ft^2/sec^2
$a_{cp} = 7500$	$\varrho_1 = 4.969$
$\varepsilon_k = 0.4$	$a_{sc} = 0.025$
$\varepsilon_c = 0.6$	$a_{E_1} = 5.0\times10^6$ lbf/in^2
$\varepsilon_\varrho = 1.2$	$a_{\alpha_1} = 20.0\times10^{-6}$/°K
$T_a = 600$ °R	$a_{Y_0} = 10.0\times10^3$ lbf/in^2
$T_{w0} = 4550$ °K	$L_0 = 0.1$ ft.

Species	a_i	b_i°R
N	7.0	9.0×10^4
N_2	—	—
CN	−6.63	-7.59×10^4
C	8.12	6.69×10^4
C_3	−14.77	-12.83×10^4

C. *Results of Sample Problem*

As already discussed, the results required from the unified ablation analysis for the thermostructural analysis are the shield's recession rate and internal temperature history.

Typical of these results are the curves in Figs. 1, 2, 3 and 4. Figure 1 shows the variation, along the cone's surface of the shield's recession rate, mass loss

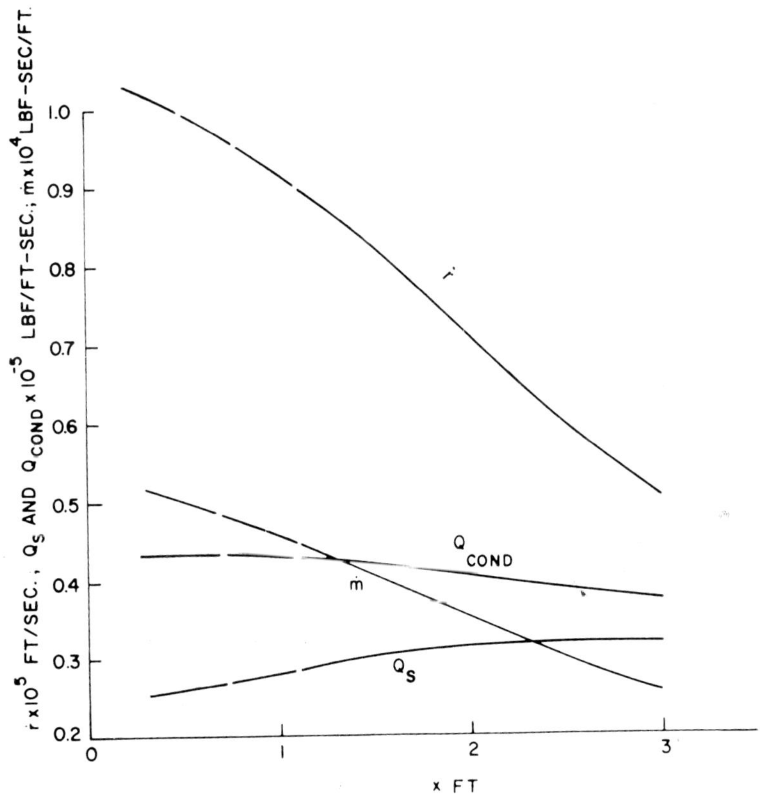

FIG. 1. Typical shield recession rate, heating rate and mass loss rate along cone surface at $\tau = 40$ sec

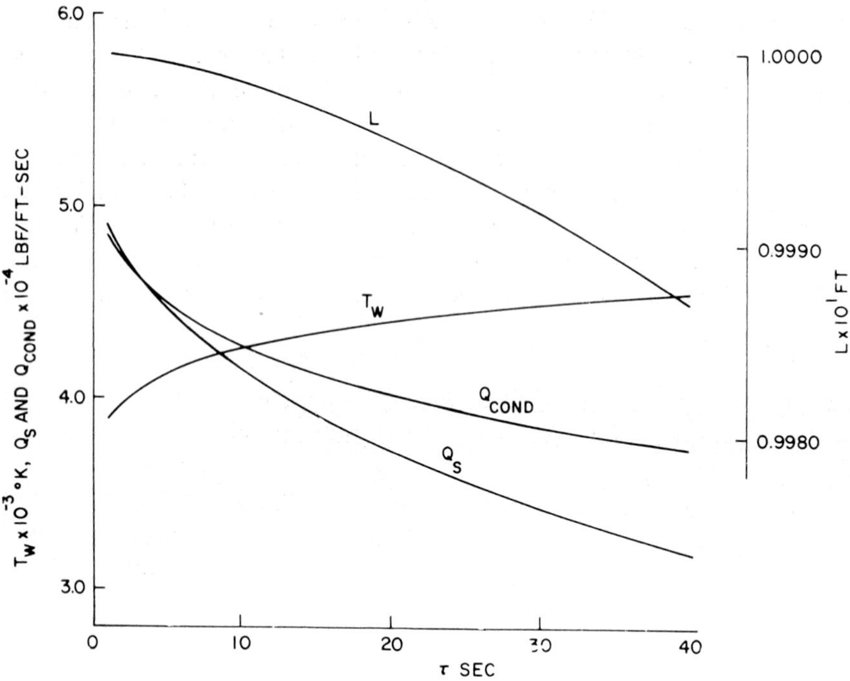

FIG. 2. Typical time variation of shield thickness, surface temperature and heating rate at cone station, $x = 3.0$ ft

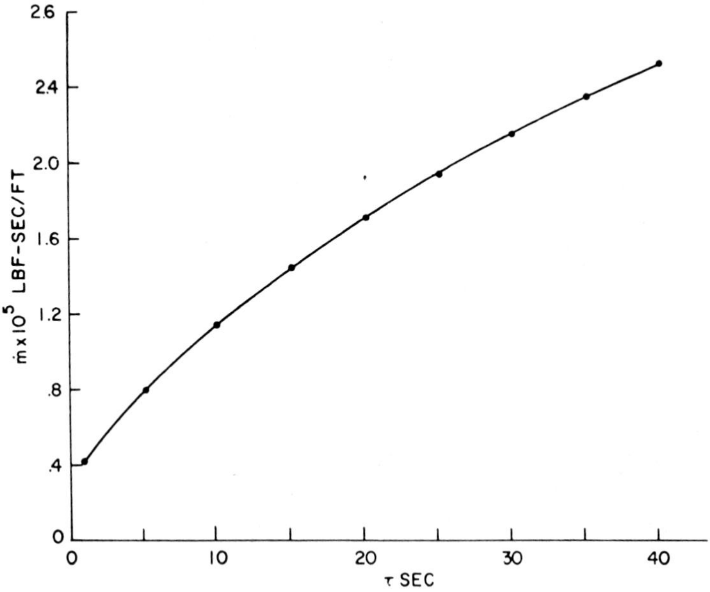

FIG. 3. Mass loss rate as a function of time for cone position, $x = 3.0$ ft

rate, and heating rate. The heating rate is given by two curves: one, the total heat being transferred through the solid (which is the same as the total heat being transferred from the gas to the solid—for continuity); and two, that part of the total heat transfer due to conduction in the gas phase. The difference between these two is the heat associated with the heterogeneous reactions and the blockage due to mass transfer. It is seen that, first, the heat transfer due to conduction is approximately 80% higher than the heat transfer actually seen by the shield. This points out that, for the present problem, heat blockage due to mass transfer is significant. Second, it is seen that the heat transfer due

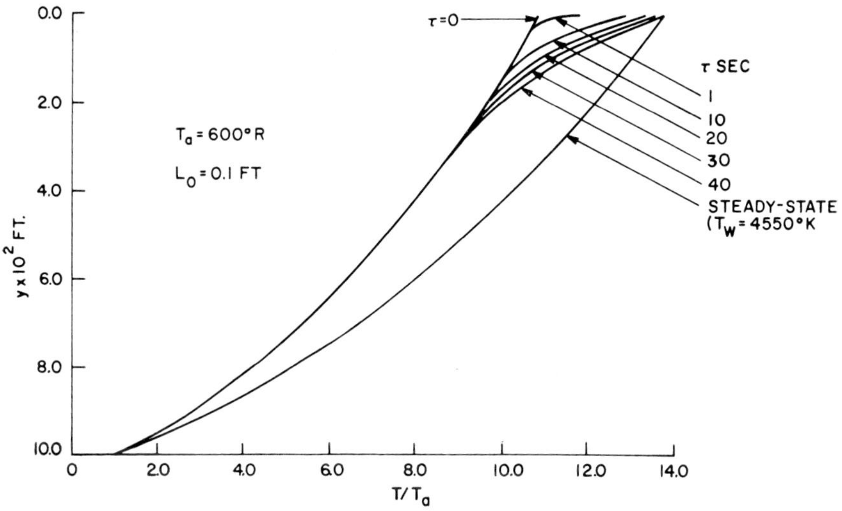

FIG. 4. Temperature variation through shield at a cone position, $x = 3.0$ ft

to gas phase conduction is reduced the further back we go along the cone. This is caused by the increased boundary layer thickness. This thickening means that the temperature change, from its surface value to its free stream value, is distributed over a larger distance. This, in turn, results in a reduced local gradient and, therefore, reduced heat conduction value.

It is also seen that the recession rate and, therefore, mass loss rate decrease as we proceed further back along the cone. The effect of this reduced mass loss rate on heat transfer rate is clearly observed. Though the heat due to conduction decreases, the total heat transfer remains relatively constant. That is, the heat being blocked by mass transfer is reduced to a sufficient extent to overcome the reduced heat conduction term.

In Fig. 2, the variation with time of shield thickness, surface temperature, and heat transfer rate are presented. Here again, it is seen that the heating rate due to conduction is greater than the overall heating rate (because of mass transfer blockage). Of significance, however, is the fact that the total heating rate is being reduced much faster than the heat conduction rate. One explanation is that the mass transfer blockage term is being reduced at a much lower rate than the conduction term. Thus, it would have a much greater effect on the total heat transfer rate than at earlier times. Or, another explanation could be that the mass transfer rate is still increasing and, therefore, it certainly would take on a more important part in the heat transfer process. As is evidenced by the increasing surface temperature (which is exponentially related to the mass loss rate) and the increasing rate of shield thickness change, it may be concluded that the total heat transfer rate is being effected by an increase in the mass blockage value. That is, it may be concluded that the mass loss rate is increasing with time. From Fig. 3, this is seen to be the case.

Figure 4 presents the results of the time variation of temperature through the shield. These results are typical for each body station. The shield temperature has been normalized with respect to its cold surface value ($T_a = 600$ °R = 333 °K). Because the shield's recession rate is low and its surface temperature is relatively constant with time, we would expect the temperature distribution to be approaching a steady state value. For comparison purposes, therefore, the steady state temperature distribution for a 0.1 ft structure with a hot-surface temperature of 4550 °K is also presented. This comparison shows that the shield's temperature does indeed appear to be approaching a steady value. It never will, however, because of the continuous reduction in shield thickness and resultant increase in surface temperature.

Not presented in figure form are the stress loads experienced by the shield's surface. The stress is composed of two components: a normal component and a tangential component. The normal component is given by the external pressure (p_{fs}) and is 4.5 lbf/in^2 along the entire shield surface (because the shield is conical). The tangential stress is given by the gas phase viscous shear stress ($\mu\, \partial u/\partial y$) and varies from 4×10^{-2} to 8×10^{-2} lbf/in^2 depending primarily on cone location (higher near the tip) and only slightly with time.

Based on the above unified ablation results (i.e., ablation rate and internal shield temperature distribution), the structure's mechanical response is obtained. Results, though presented for one particular body station ($x = 3.0$ ft), are typical for all, and are presented in figs. 5–7.

Due to the initial nonuniform temperature distribution, there exists an initial non-zero stress distribution. This distribution is shown in Fig. 5, and is based on the assumption that no elastic unloading has taken place previous to this time. It is seen that, even at the initial time, $\tau = 0$, a region of plastic flow exists adjacent to the hot surface. This, of course, is due primarily to the yield stress (which bounds the stress distribution both in tension and in compression) being at a relatively low level near this surface.

As heating of the structure continues and the temperature gradients become more severe in the vicinity of the hot surface, it is seen from Fig. 6 that the region of plastic flow existing initially through 11% of the structure's thickness grows uniformly until, at the end of the time period of interest, its depth has more than doubled to 24% of the structure's thickness.

The change in stress distribution resulting from these transient phenomena is illustrated in Fig. 7. The stress distribution at a typical intermediate time

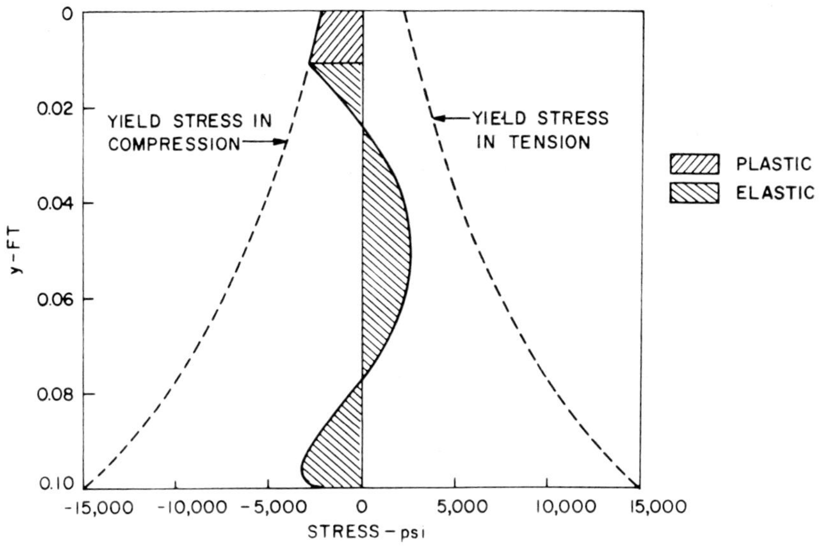

FIG. 5. Initial stress distribution in shield at $x = 3.0$ ft

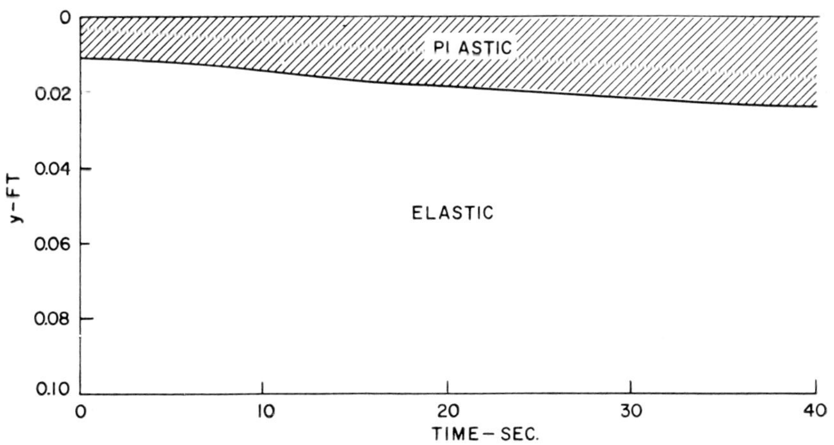

FIG. 6. Growth of plastic region

(τ = 30 sec) is presented. It is seen that the internal stresses at this time are virtually unaltered from those existing initially, through approximately 70% of the structure's thickness adjacent to the cold surface. This is the same region in which the temperature distribution is, for all intents and purposes, unaltered (Fig. 4). It is only in that region in which there is a significant variation of temperature with time and, therefore, a significant variation of yield stress with time, that the stress distribution is transient. It is this region which experiences a considerable degree of plastic flow, and it is primarily

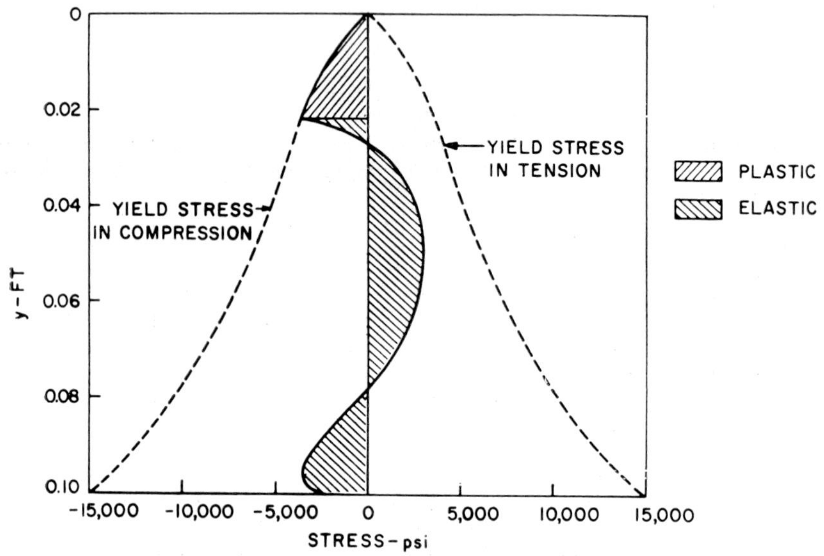

FIG. 7. Stress distribution at τ = 30 sec, x = 3.0 ft

the decreasing yield stress which results in a plastic region of increasing thickness (Fig. 6).

The two time-dependent quantities that determine the deformations (strains) throughout the structure, $F_1(\tau)$ and $F_2(\tau)$, are not presented here in figure form, since their variations with time for the problem and for the time period presently being considered are insignificant. It was found that both these quantities increase by approximately 1/2 of 1%. The internal deformations of the structure thus increase from their initial values by this same amount.

ACKNOWLEDGMENTS

The authors are pleased to acknowledge the assistance provided by several persons during the course of this study; in particular, Miss Sharon Smith for her most valuable contribution in coding, programming, debugging and, in general, putting the unified ablation theory into a numerically useful form; Mr. Octavio Winter for his programming of the thermostructural formulations and his suggestions with respect to obtaining the numerical output in concise form; and finally, Mr. Herbert Frey for his help in reducing some of the data and preparing the figures.

REFERENCES

1. SWANN, R. T., PITTMAN, C. M. and SMITH, J. C. One-Dimensional Numerical Analysis of the Transient Response of Thermal Protection Systems, NASA TN D-2976, September 1965.
2. GILBERT, L. M. and SCALA, S. M. Combustion and Sublimation of Cones, Spheres, and Wedges at Hypersonic Speeds, *AIAA Journal*, **3**, No. 11, November 1965.
3. SCALA, S. M. and GILBERT, L. M. Sublimation of Graphite at Hypersonic Speeds, *AIAA Journal*, **3**, No. 9, September 1965.
4. LEES, L. Convection Heat Transfer with Mass Addition and Chemical Reactions, *Third AGARD Combustion and Propulsion Panel Colloquium, Palermo, Sicily, March 1958*, published by Pergamon Press, New York.
5. BARON, J. R. The Binary-Mixture Boundary Layer Associated with Mass Transfer Cooling at High Speeds, MIT Naval Supersonic Laboratory, TR-160, May 1956.
6. CARSLAW, H. S. and JAEGER, J. C. *Conduction of Heat in Solids*, Oxford University Press, London, 1959.
7. PRICE, E. W., BRADLEY, H. H., JR., DEHORITY, G. L. and IBIRICU, M. M. Theory of Ignition of Solid Propellants, *AIAA Journal*, **4**, No. 7, July 1966.
8. SHARMA, O. P., ROTENBERG, M. and PERNER, S. S. Phase-Change Problems with Variable Surface Temperatures, *AIAA Journal*, **5**, No. 4, April 1967.
9. ALLEN, R. W. and SUESS, R. P. A Fortran Computer Program for the Solution of Multi-Dimensional, Transient Ablation Problems, Applied Physics Laboratory, The Johns Hopkins University, Report TG-648, February 1965, DDC-AD462 335.
10. OSTRACH, S. Heat Transfer with Receding Boundaries and Other Complications, Lecture series reproduced as NASA document N64-25108, August 1964.
11. SCHUYLER, F. L. and TORDA, T. P. An Aerothermochemical Analysis of Solid Propellant Combustion, *AIAA Journal*, **4**, No. 12, December 1966.
12. FREUDENTHAL, A. M. Effect of Rheological Behavior on Thermal Stresses, *J. Appl. Physics*, **25**, No. 9, 1110–1117, September 1954.
13. FREUDENTHAL, A. M. On Inelastic Thermal Stresses in Flight Structures, *J. of the Aeronautical Sciences*, **21**, 772–778 (1954).
14. WEINER, J. H. An Elastoplastic Thermal-Stress Analysis of a Free Plate, *J. Appl. Mech.*, **23**, 395–402 (1956).
15. LANDAU, H. G. and WEINER, J. H. Transient and Residual Stresses in Heat-Treated Plates, *J. Appl. Mech.*, **25**, 459–465, December 1958.
16. YÜKSEL, H. Elastic, Plastic Stresses in Free Plate with Periodically Varying Surface Temperature, *J. Appl. Mech.*, **25**, 603–606, December 1958.
17. LANDAU, H. G., WEINER, J. H. and ZWICKY, E. E. Thermal Stress in a Viscoelastic-Plastic Plate with Temperature-Dependent Yield Stress, *J. Appl. Mech.*, **27**, No. 2, 297–302, June 1960.

18. WEINER, J. H. and BOLEY, B. A. Elasto-Plastic Thermal Stresses in a Solidifying Body, *J. Mech. Phys. Solids*, **11**, 145–154 (1963).
19. FRIEDMAN, E. and BOLEY, B. A. Design Estimates of Stresses and Deformations in Melting Plates, Tech. Report No. 29, Contract No. Nonr-4259(07), Dept. of Civil Engr. and Engr. Mech., Columbia Univ., June 1965.
20. HILL, R. A Theory of the Yielding and Plastic Flow of Anisotropic Metals, *Proc. Royal Soc. of London*, **A193**, 281–297 (1948).
21. BOLEY, B. A. and WEINER, J. H. *Theory of Thermal Stresses*, Chaps. 14 and 16 J. Wiley and Sons, New York, 1960.

FABRICATION AND ENVIRONMENTAL INTERACTION EFFECTS OF FILAMENT-WOUND COMPOSITES

S. Brelant and I. Petker

Aerojet-General Corporation,
San Bernardino, California

I. INTRODUCTION

In order to understand the interaction of fabrication variables and untoward environments, it is necessary to delineate the effect of each separately upon structural performance. Further, an understanding of the failure mechanism is required in order to quantitatively evaluate the importance of process and environmental effects. For example, the strength of filament-wound internal pressure vessels is relatively insensitive to many process variables, whereas the interlaminar shear of filament-wound cylinders is a very critical function of the same processing variables. Again, the presence of moisture has been shown to markedly decrease the strength of some composites, whereas it is of little consequence in the structural performance of others. Accordingly, it appears that the geometry of the structure and the type of stress application determines the structural performance of the filament-wound structure, and the effect of processing defects and environmental interaction must be considered for each type of structure in a singular manner.

In this paper two types of filament-wound composites are considered. The first is the internal pressure vessel with filaments oriented along the principal axes of stress such that the composite is subjected to tensile loads only. Second, are small arc segments of N.O.L. rings loaded such that the principal stress is one of shear.

II. TEST SPECIMENS

Figure 1 illustrates the internal pressure vessel used as a test specimen in this study. It is 4 in. in diameter and slightly over 5 in. in length. Its criterion is the burst pressure when the bottle is pressurized hydrostatically and it affords excellent testimony that a filament-wound structure can be fabricated to perform in a highly reproducible manner. When fabricated with a standard

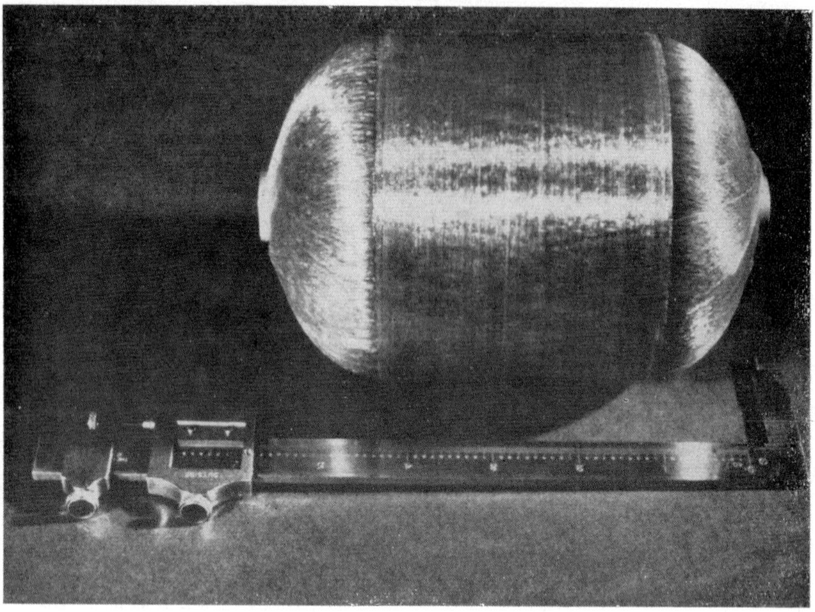

FIG. 1.

FIG. 2.

resin† and a constant optimized process, these bottles burst with a coefficient of variation of less than 2%.

In Fig. 2 an arc segment of an N.O.L. ring is shown in its test apparatus. This test is referred to as interlaminar or horizontal shear. Two configurations have been used in this test: one is a 1 in. long chord, $\frac{1}{4}$ in. wide and thick; the other is a 0.6 in. chord, $\frac{1}{4}$ in. wide and $\frac{1}{8}$ in. thick.

III. FAILURE MECHANISM

As the internal pressure vessel is loaded hydrostatically, growth occurs in two directions, approximately 3% in both length and diameter at failure. The pattern of stress is shown in Fig. 3. Although most epoxy materials used

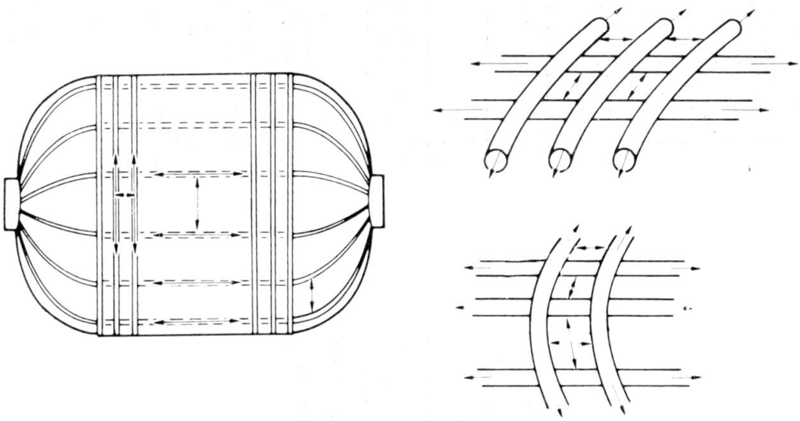

FIG. 3. Filament and resin stresses during pressurization

in bottle fabrication have an ultimate elongation between 3% and 6%, local strain incompatibilities exceed the strain capability of the resins and visible cracks (often called crazing) appear in the matrix. The crazing phenomenon occurs at levels of stress as low as 10% of the ultimate strength of the chamber.

More flexible resins may raise the threshold at which crazing initiates, but even with resins having ultimate elongation of 100%, crazing still occurs well below the ultimate failure stress of the bottle and little effect is observed in the actual burst pressure. Since bottle failure occurs well above the stress level at which crazing initiates, it is clear that crazing is not an immediate precursor of structural failure. The composite continues to carry load until

† Most of the composite data reported in this study is for 58-68R resin. This is a designation applied to an impregnating resin system by Aerojet-General Corporation which consists of Epon 828, Epon 1031, methyl nadic anhydride and benzylidmethylamine in the ratio of 50 : 50 : 84 : 0.5.

the ultimate strength of the filaments is exceeded wherein catastrophic failure occurs. The controlling factor, therefore, in the mechanism of chamber failure is filament strength and influences upon burst pressure must be viewed within the context of their influence upon the tensile strength of the filament. Two general categories of such influence can be identified, those which are active and directly reduce the strength of the filaments, and those which are passive

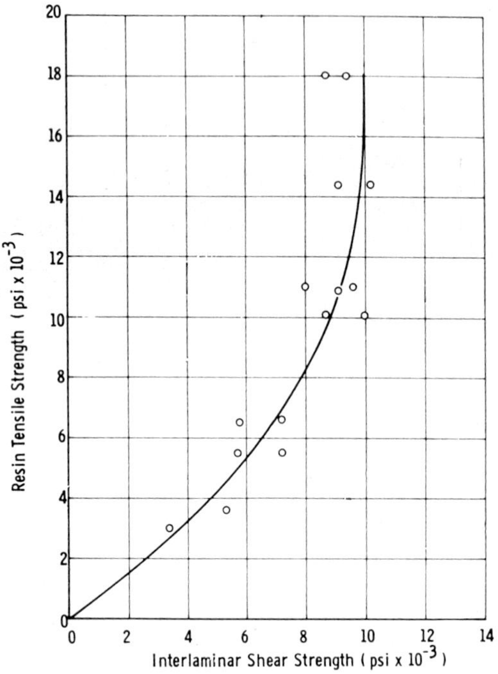

FIG. 4. Influence of resin strength on interlaminar shear strength (HTS finish)

and diminish the ability of the filament to realize its full inherent strength. In the category of passive influence the fabrication process can exert a profound effect[(1)] whereas the environment has little, if any, effect. In the category of active influence, the environment can be a key factor with crazing playing a direct and important role in the failure mechanism[(2)] and the fabrication process contributing little.[(3)]

For arc segments of N.O.L. rings, it has been shown that both the strength of the matrix and voids contribute directly to the failure mechanism in shear.[(4)] In Fig. 4 a correlation is shown between the cast resin tensile strength of a variety of resin systems and interlaminar shear strength. The nature of the curve clearly indicates that at least two distinct factors are controlling the failure level. At low and intermediate resin strength the correlation is essentially linear with resin strength indicating that the strength of the resin is controlling. At high resin strength, it appears that a limiting value of interlaminar shear strength has been reached, and further increase in resin strength has little or no effect upon interlaminar shear strength. The data in Fig. 4 was found to correlate with data from other laboratories, including those which has been taken at elevated temperature, as shown in Fig. 5.

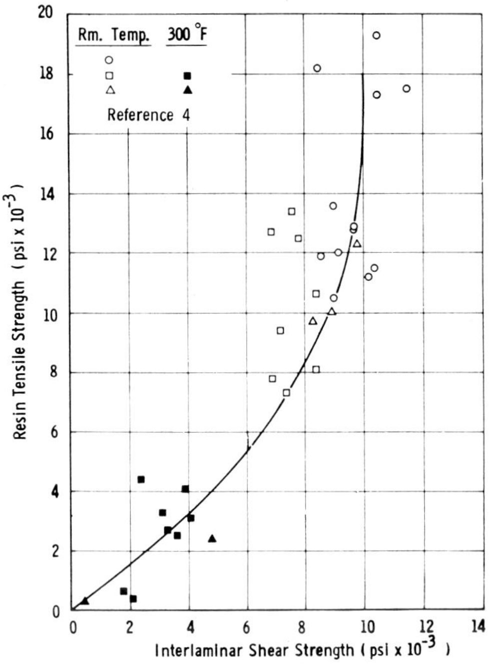

FIG. 5. Influence of temperature on resin tensile strength and interlaminar shear strength

Although it is possible to postulate a number of explanations for the apparent limit of interlaminar shear strength, evidence was presented in ref. 4 and corroborated by others[5, 6] that the void content of composites studied was the responsible, causative agent. Data were developed for a resin with a tensile strength of 11,000 psi (from Fig. 4) to show how the interlaminar shear strength of composites fabricated with the resin varied with the void content of the composites. The results are presented in Fig. 6. Due to inherent errors in void content measurements and the assumption of uniform distribution of voids, it is difficult to assign an exact value of interlaminar shear strength for a zero void content composite. However, it is possible to identify reasonable limits from the figure and assign a limiting range of 12 to 14,000 psi. These values were used with the previous data as shown in Fig. 7 and a linear correlation between resin strength and interlaminar shear strength was predicted for void free composites.[4]

This prediction has been verified by others[6] who successfully achieved minimal or void free composites by filament-winding in a vacuum chamber. Fortunately, two of the high strength resins used in this study were identical to those used to obtain the data of Fig. 4. When these new points are plotted

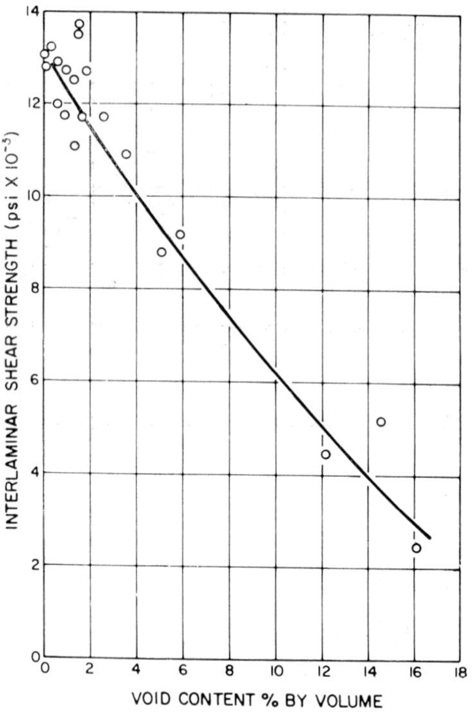

FIG. 6. Void content vs. interlaminar shear

with the previous data in Fig. 8, the linear relationship between resin strength and interlaminar shear is clearly indicated.

From the foregoing failure mechanism for shear it is extremely important to note that all influences upon shear strength must be considered from the standpoint of their influence upon the matrix and upon void content. This tensile strength and shear strength can be considered the poles of composite mechanical behavior, with one pole determined by filament performance and the other by matrix performance.

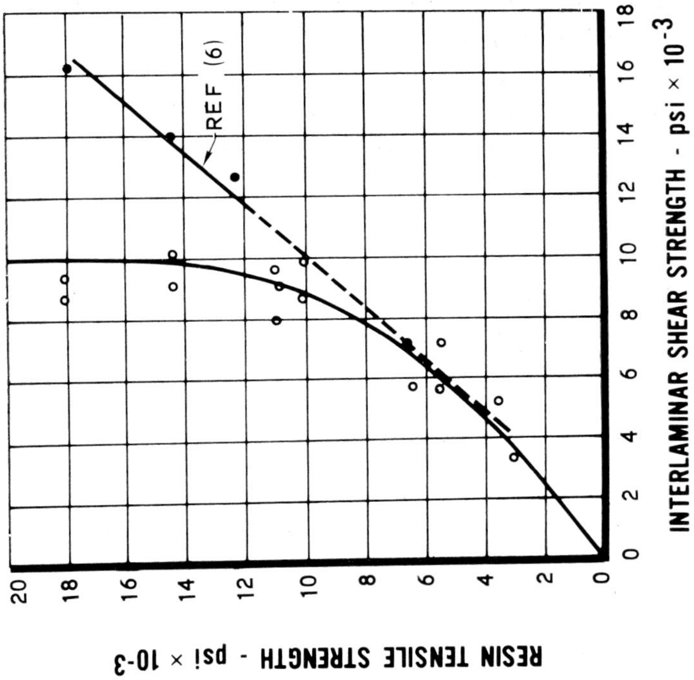

FIG. 8.

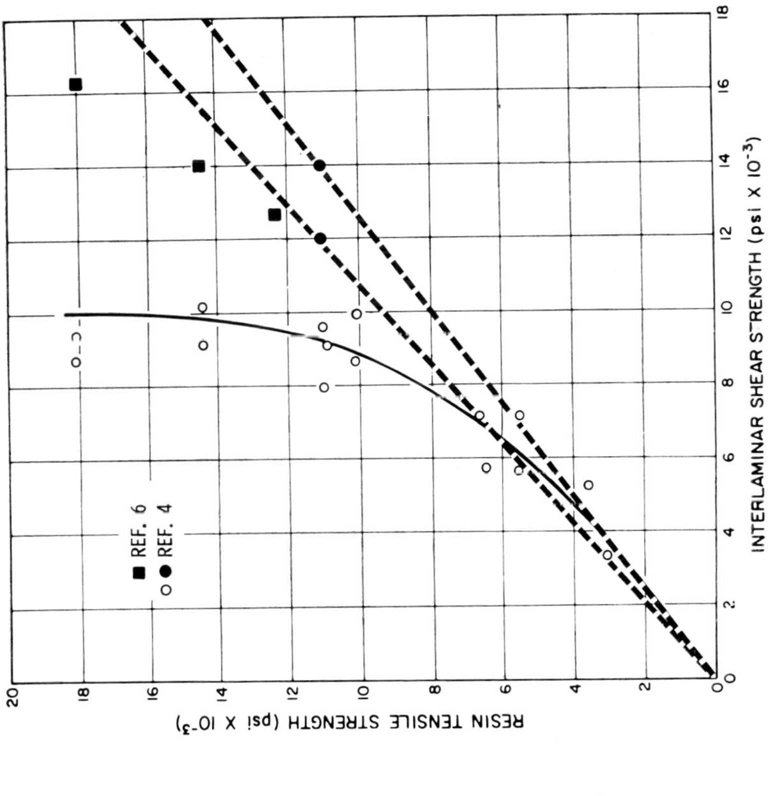

FIG. 7. Cast resin strength vs. interlaminar shear

IV. PROCESS CONSIDERATIONS

The process variables which affect the structural performance of both of the composites are: the ability to collimate and accurately space the filaments which in turn relates to the tension on the roving and the precision of the winding machine; the viscosity of the resin or the degree of "B" staging of a preimpregnated roving; and the cure cycle. All of these factors contribute to the degree of homogeneity which can be achieved in the final composite. The process should be such that poor alignment of the filaments does not occur and that both resin-rich areas and voids are minimized. These

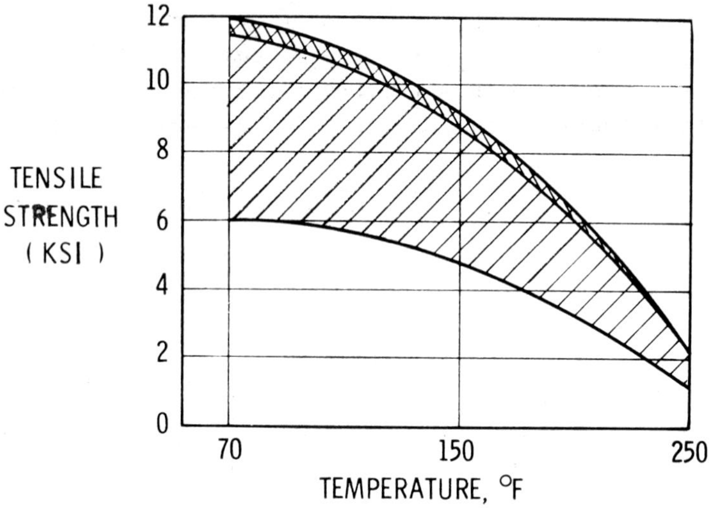

FIG. 9. Combined effects of voids and temperature on cast resin tensile strength

process variables may influence the interlaminar shear strength of the composite by a factor of as much as 50% or more as shown in Fig. 5. Influences of the same general nature can be observed in the behavior of unreinforced resin castings. In Fig. 9 tensile strength is shown as a function of temperature for a typical epoxy resin. In ambient temperature testing a very high variability was observed in the data which was attributed to voids and other defects. With very carefully prepared castings the variability was reduced to less than 5% as shown in the cross-hatched region of the curves. Also, it is shown that the effect of the defects was reduced as temperature was increased and the resin became more ductile.

The relation between voids and interlaminar shear strength is extremely important in composite technology, and the sensitivity of this parameter to processing cannot be overemphasized. Recently it was noted that a change of less than 25 °F in winding temperature could be responsible for void content differences of 3 to 5% by volume in otherwise identical composites. In this context there are also often very subtle interactions between small changes in the processability of the prepreg and the process which may influence composite performance much more seriously than often assumed.[1]

Damage to glass filaments incurred during the winding process and resulting in a reduction in filament strength has been shown to be either extremely small or nonexistent.[3] This is particularly true of modern prepregs where in the resin acts as a lubricant against interfilament abrasion and abrasion at other contact points. Again, the major contributions of recent finishes for glass fibers has been their ability to provide superior protection from abrasion and moisture. For these reasons it is difficult to identify active degradative factors on filament strengths in the nominal filament winding process. The fabrication process itself limits the realization of potential strength to the extent that it does or does not produce homogeneity in the final composites as discussed previously.

V. ENVIRONMENTAL CONSIDERATIONS

The environment which has received the greatest attention in respect to composite degradation is water, either as a vapor because of extended storage periods under varying humidity levels, or total immersion as would be required by submersible structures. Standard tests in the laboratory include immersion in boiling water for periods of 2 hr to 100 hr, or exposure to environmental conditions as specified in MIL-E-5272-272-C. Both of these tests supposedly accelerate the effects of a moisture environment on the composite.

It has been shown that moisture can seriously degrade the performance of an internal pressure vessel[2]. Under the conditions of MIL-E-5272-C as much as a 40% reduction in burst pressure has been identified, but only if the chamber has been crazed due to a previous pressurization cycle. Chambers without crazing and exposed to the same accelerated environment may degrade 5%, or at most 10%. In Fig. 10, data is shown for the burst pressure of crazed chambers as a function of time and humidity at 160 °F. It is clear that as the humidity level increases, degradation accelerates, and becomes more severe. In Table 1, strength data is shown for monofilaments immediately tested in a pristine condition, and after submersion in water for ten days. Bare filaments were severely degraded, whereas filaments coated with an appropriate finish do not degrade.

From this latter observation an important inference may be drawn regarding the failure site during crazing. It appears that debonding occurs and, moreover, that this debonding must occur at the glass-finish interface, thus removing the protective coating together with the resin. Thus reduction in

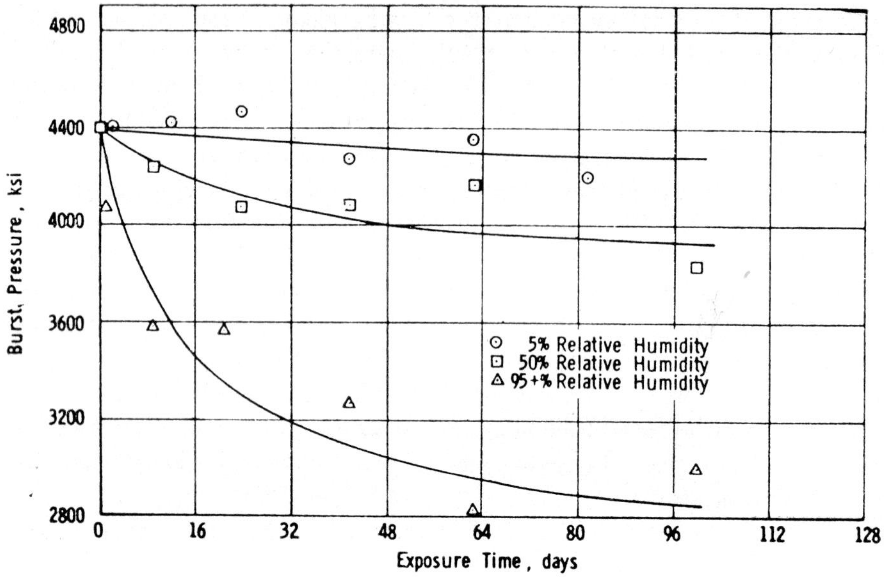

FIG. 10. Relationship of burst pressure to exposure time at 160°F

TABLE 1. EFFECT OF COATING AND MOISTURE EXPOSURE ON FILAMENT TENSILE STRENGTH

Environment	Tensile strength—psi after storage		
	1 hr	5 days	10 days
Controls			
Uncoated filaments stored in normal air	750	621	689
Uncoated filaments stored in desiccator 0% humidity	747	694	717
Uncoated filaments immersed in water	646	563	442
Coated filaments stored in water	707	687	695

filament strength, the necessary condition for degradation in chamber strength, can occur since the glass is no longer protected from direct attack by moisture.

In the case of interlaminar shear a very different set of circumstances prevails. In order for a reduction in interlaminar shear strength to be observed one necessary condition would be for the environment to effect a reduction in matrix strength. One possible route to such an affect would be the plasticization of a cured matrix by absorbed water and it should be possible to observe such a mechanism in the behavior of an isolated resin. In Table 2, data are shown for several epoxy systems which have been used in filament winding in which the compressive yield strength of the resin was measured

TABLE 2. EFFECT OF 5000 psi WATER IMMERSION ON CAST RESIN COMPRESSIVE YIELD STRENGTH

Immersion time (days)	Compressive yield strength (ksi)			
	Resin system A		Resin system B	
	Avg str.	Mean deviation	Avg str.	Mean deviation
0	19.10	0.07	22.42	0.17
1	19.17	0.26	21.65	0.07
7	18.22	0.06	20.36	0.55
30	20.26	0.08	21.35	0.17

Resin system A–Resorcinol diglycidyl ether /HHPA/BDMA
Resin system B–ERL 2256/ZZ0820

as a function of time during submersion in water at 5000 psi pressure. It is clear from this data that no degradation effect due to the water can be identified.

When N.O.L. ring composites were prepared from some of the same resins shown in Table 2 and exposed to water under the same conditions, similar results were observed and no degradation could be identified. This is shown in Tables 3 and 4 for two conditions; one in which arc segments were cut from the rings prior to submersion so that cut fiber ends were exposed to the water, and the other, presumably less severe, in which uncut rings were submerged. In essence, the data predicted by the performance of the unreinforced castings were corroborated by the composites.

TABLE 3. EFFECT OF 5000 psi WATER IMMERSION ON INTERLAMINAR SHEAR STRENGTH CUT FILAMENTS

Immersion time (days)	Interlaminar shear strength (ksi)								
	Resin system A			Resin system B			Resin system C		
	Avg str.	Mean deviation	% chg.	Avg str.	Mean deviation	% chg.	Avg str.	Mean deviation	% chg.
0	8.44	1.00	–	10.27	0.35	–	8.66	0.41	–
2	8.53	0.66	+1.0	9.83	0.56	−4.3	8.65	0.24	−0.1
5	8.53	0.34	+1.0	10.03	0.17	−2.4	9.24	0.56	+6.2
7	8.34	0.29	−1.2	10.10	0.20	−1.7	9.00	0.17	+3.8
14	8.52	0.26	+1.0	10.39	0.25	+1.2	8.77	0.68	+1.2
30	8.20	0.16	−2.8	9.90	0.25	−3.7	8.23	0.59	−5.0

Resin system A–Resorcinol diglycidyl ether /HHPA/BDMA
Resin system B–ERL 2256/ZZ0820
Resin system C–Epon 828/Epon 1031/MNA/BDMA

TABLE 4. EFFECT OF 5000 psi WATER IMMERSION ON INTERLAMINAR SHEAR STRENGTH (UNCUT FILAMENTS)

Immersion time (days)	Interlaminar shear strength (ksi)								
	Resin system A			Resin system B			Resin system C		
	Avg str.	Mean deviation	% chg.	Avg str.	Mean deviation	% chg.	Avg str.	Mean deviation	% chg.
0	8.42	0.33	–	10.38	0.41	–	8.55	0.21	–
2	8.31	0.14	−1.3	10.22	0.16	−1.5	8.73	0.34	+4.2
5	8.95	0.20	+5.9	9.86	0.43	−5.0	8.76	0.32	+2.4
7	8.51	0.05	+1.1	9.81	0.34	−5.5	8.01	0.32	−6.3
14	8.41	0.22	−0.2	10.14	0.29	−2.3	8.55	0.44	−117
30	8.47	0.33	+0.6	10.55	0.39	+1.6	8.73	0.24	+2.1

Resin system A – Resorcinol diglycidyl ether /HHPA/BDMA
Resin system B – ERL 2256/ZZ0820
Resin system C – Epon 828/Epon 1031/MNA/BDMA

An additional experiment was conducted to identify a possible interaction between voids and a severe water environment. In this work chords of N.O.L. rings were submerged in boiling water for periods of time up to about 100 hr. Specimens were chosen so that a wide variation in void content was represented. The data from these tests are given in Fig. 11. Each curve represents a

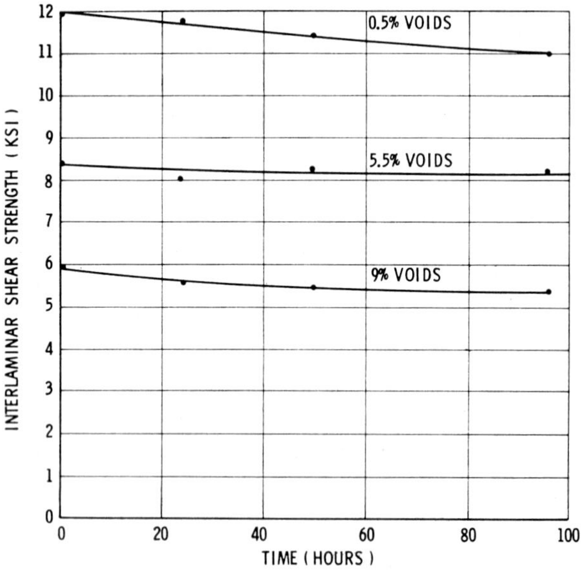

FIG. 11. Combined effects of water boil and voids on interlaminar strength

discrete level of void content and the initial displacement of the curves clearly shows the effect of voids on interlaminar shear strength. It is clear that the degradation caused by the boiling water was small in all cases and that the slopes of all the curves with time were very shallow and essentially the same.

This and the previous data appear to be in conflict with numerous reports that have identified severe degradation of interlaminar shear strength in a water environment. The fact that in uncrazed chambers and in rings prepared with efficient finishes small, but apparently real, water degradation is observed, tends to indicate that some diffusion of moisture to the interface does occur with untoward effects. The phenomenon is rate dependent and when severe degradation does occur the efficiency and contiguity of the coating affording moisture protection to the filaments must be suspect, as well as the specimen geometry and manner of loading such that the filament strength becomes an important factor in the failure mechanism.

VI. CONCLUSIONS

In years past it was common to attribute a low burst pressure to the fact that it had rained during fabrication or that the bottle had been stored at a relatively high humidity for some time before testing. This is contrary to the findings of the authors. For over four years involving production of hundreds of large rocket cases and thousands of small pressure vessels, we have never seen any seasonal variations or correlation with humidity levels in the burst pressure of internal pressure vessels. Bottles have been stored for up to one year immersed in salt water with retention of over 90% of their original strength based on controls tested before immersion. These data are very consistent with the very good performance of glass-reinforced plastic in water environments found during this study.

It is apparent, also, from the studies that no interaction between fabrication and environmental effects can be identified, at least for the materials, processes, and types of composites treated in this study. Individually, however, the fabrication process, the environment, and the materials can all influence the structural performance of the composite. Although it is now possible to describe the behavior of many of these factors with reasonable precision and confidence, much remains to be accomplished, particularly in regard to a more intimate description of behavior of the interface and of the role of adhesion in the failure mechanism and environmental stability of composites.

REFERENCES

1. PETKER, I. *et al.* Processability of Preimpregnated Materials, *Modern Plastics*, September and October 1965.

2. BRELANT, S. *et al.* Combined Effects of Prestress and HumidityCycling upon Filament-Wound Internal Pressure Vessels, *S.P.E. Journal*, September 1964.
3. WEINGART, O. Development of Improved Processes for Filament-Wound Reinforced Plastic Structures, *Proceedings 20th Annual Technical and Management Conference on Reinforced Plastics*, Society of Plastics Industry, 1965.
4. PETKER, I. The Influence of Resin Strength and Defects on the Interlaminar Shear Strength of Filament-Wound Composites, *Polymer Engineering and Science (Form S.P.E. Transactions)*, January 1965.
5. HAND, W. Quality Control of Filament-Wound Materials for Deep Submergence Vehicles, *Proceedings 20th Annual Technical and Management Conference on Reinforced Plastics*, Society of Plastics Industry, 1965.
6. PAUL, J. T. and THOMPSON, J. B. The Importance of Voids in the Filament-Wound Structure, *Ibid.*

DEGRADATION OF COMPOSITE MATERIALS: THE EFFECT OF WATER ON GLASS-REINFORCED PLASTICS†

N. FRIED

Materials Sciences Division,
U.S. Naval Applied Science Laboratory,
Brooklyn, New York

I. INTRODUCTION

Composite materials, in common with materials in general, are subject to degradation under a variety of chemical, physical, and biological influences. The significance of these degradative effects is accentuated in today's technology in which composites may be required to function under extreme conditions, from deep outer space to the abysmal depths of the ocean, and from the searing temperature of re-entry, through the heat and humidity of the tropics and the subfreezing temperatures of the Arctic, down to the cryogenic temperatures of liquid hydrogen.

The scope of the subject of degradation of composites is further broadened when the variety of materials is added to the variety of conditions. Thus the term "composites" could encompass such widely diverse materials as fiber-reinforced matrices, from wood (a low modulus matrix reinforced with high strength, high modulus cellulosic fibers), through fabric-reinforced materials such as rubber tires and plastic boat hull materials, to high performance glass or boron filament wound epoxy laminates; particulate reinforced materials, such as syntactic foams (resinous matrices containing glass microballoons), whisker reinforced metals, and concrete; and lamellar materials from double-layered packaging films, through safety glass, to clad steel armor plating.[1, 2]

Although in some instances it may be possible to obtain an indication of composite degradation from a knowledge of the response of the individual components to particular environments, it is generally not possible to arrive at precise predictions of behavior on this basis. Because of this uncertainty, the design engineer generally resorts to very specific experimental investigations, in which the material is exposed to laboratory ambients which simulate

† The opinions or assertions contained herein are those of the author and should not be construed as reflecting the views of the Naval Service at large.

expected service conditions as closely as possible, and degradative effects are observed over a particular time period. The technical literature abounds in descriptions of this type of investigation, which are so specific, and at the same time so diversified, as to defy codification. No attempt in this direction will be undertaken herein. Instead, a detailed study of a particular case, the degradation of the mechanical properties of glass-reinforced plastic composites under the influence of water, will be presented to illustrate the complexities of the processes of degradation and to provide the basis for an attempt to arrive at some generalizations regarding these processes as related to composite materials of this type, as a class.

II. DEGRADATION OF GLASS-REINFORCED PLASTIC COMPOSITES IN WATER

A. *Background*

It has long been recognized that glass-reinforced plastics (GRP) may be subject to degradation in mechanical properties upon exposure to water. This is discussed in general treatises dealing with GRP composites[3, 4] and is taken note of in specifications applicable to these materials.[5, 6] Degradation of GRP materials in water is also taken into consideration by marine engineers, who use "wet" rather than "dry" strengths in the design of GRP boat hulls and other marine structures.[7, 8]

The degree of degradation of different GRP composites under the influence of water varies considerably, depending on the component materials employed and the quality of the finished laminate. While earlier types of materials were frequently degraded significantly by water, considerable improvements have been accomplished over the years to the point that present-day GRP materials are relatively insensitive to degradation by water. Some degree of sensitivity still exists, however, and there remains room for further improvement. The extent to which degradation occurs, and the various pertinent factors relating thereto are discussed in the following paragraphs.

B. *Experimental Results*

1. Degradation in strength. The effects of extended seawater immersion on the strength of a number of earlier types of glass reinforced polyester laminates (1954), are shown in Fig. 1.[9] Although the several different materials were affected to different extents, the curves are similar in that there is a period of relatively rapid degradation, followed by a period of diminution of degradation rate, and finally a flattening-out of the curve, indicating essen-

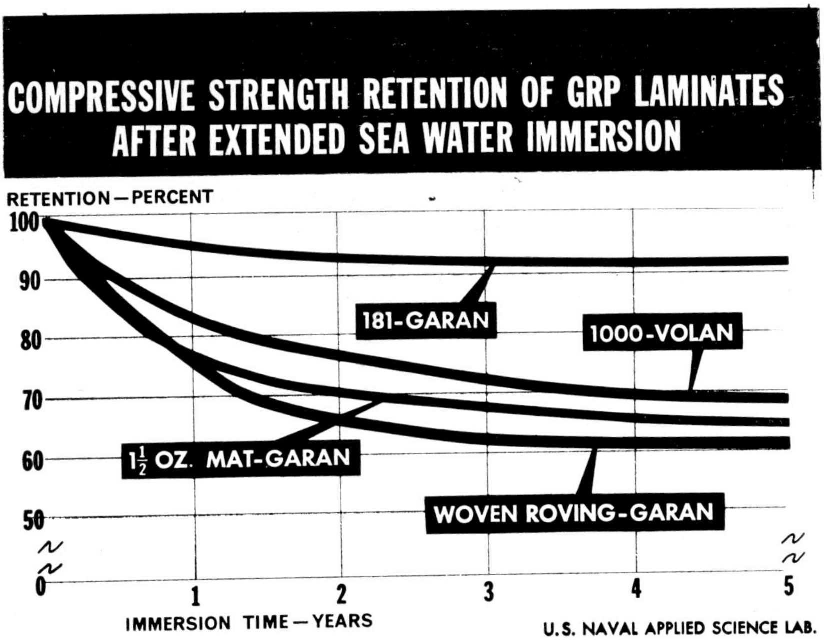

FIG. 1. Effect of extended immersion in the ocean on the compressive strength (wet) of glass-reinforced polyester laminates[9]

tially no further degradation with time of exposure. Although exceptions have been noted, this type of response is quite typical of GRP materials. Thus, any degradation which will occur takes place in a relatively short time and further exposure results in essentially no additional degradation.[10, 11]

2. DEGRADATION IN STIFFNESS. While the effect of water immersion on the strength of GRP is well defined, effects on stiffness are much less clearly established. As will be shown (see, for example, Figs. 4 and 5) there is evidence that for many materials elastic moduli are essentially unaffected by water immersion.[11] On the other hand there are materials which do show a degradation in stiffness, to a greater or lesser extent.[12] In general, however, it has been found that stiffness of GRP composites is much less sensitive to degradation by water than are strength characteristics.

3. STRESS-RUPTURE EFFECTS. As might be expected from experience with other sensitive materials, degradation of GRP composites is accentuated when the material is exposed to water while under stress. This is typified in Fig. 2, which presents stress-rupture data for an orthogonal filament wound glass reinforced epoxy composite.[11] Although the data shown are for interlaminar shear, similar behavior has been noted for specimens loaded in compression.[11]

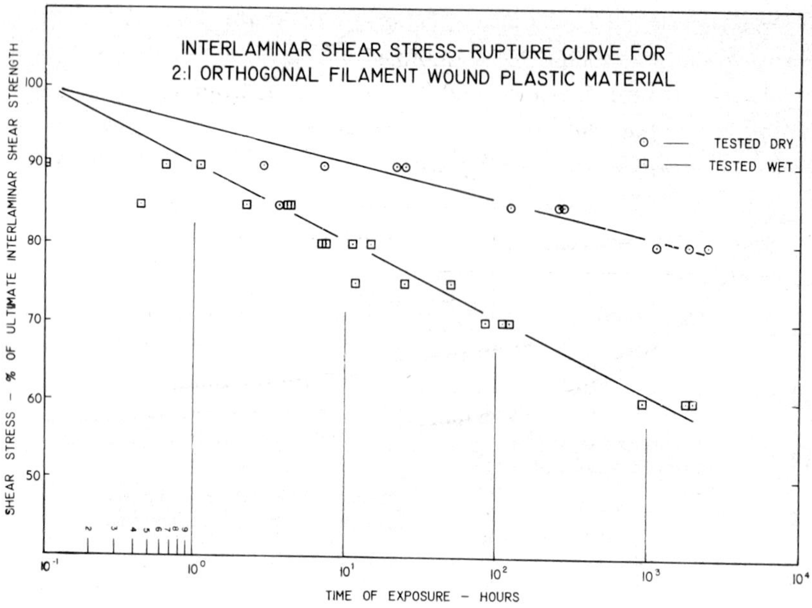

FIG. 2. Stress-rupture data for a filament wound epoxy laminate tested in air and under water[11]

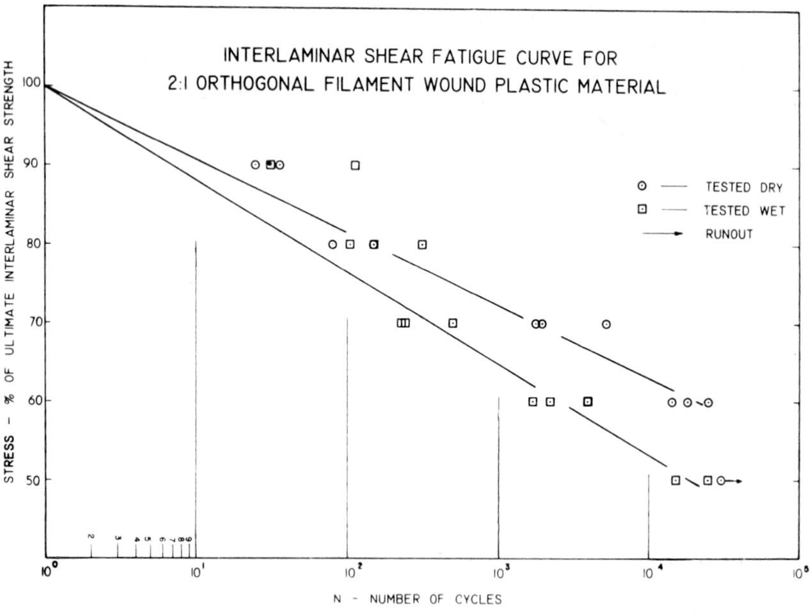

FIG. 3. Fatigue data for a filament wound epoxy laminate tested in air and under water[11]

4. Fatigue effects. Fatigue characteristics of GRP materials are also affected by water exposure, as exemplified by the S–N curves shown in Fig. 3.[11] These data were also developed for the filament wound material described above and, again, similar data were obtained for compression fatigue.[11]

5. Effects of environmental variables. The observation cited above applied to the effect of water exposure under normal conditions of pressure and temperature. An assessment of the effects of these two important environmental variables is given below.

(a) *Effect of pressure.* The effects of exposure to water at high hydrostatic pressure on the properties of an orthogonal filament wound epoxy laminate are shown in Figs. 4 and 5.[11] As is evident from the data, exposure at high pressure affected this material no more adversely than exposure at atmospheric

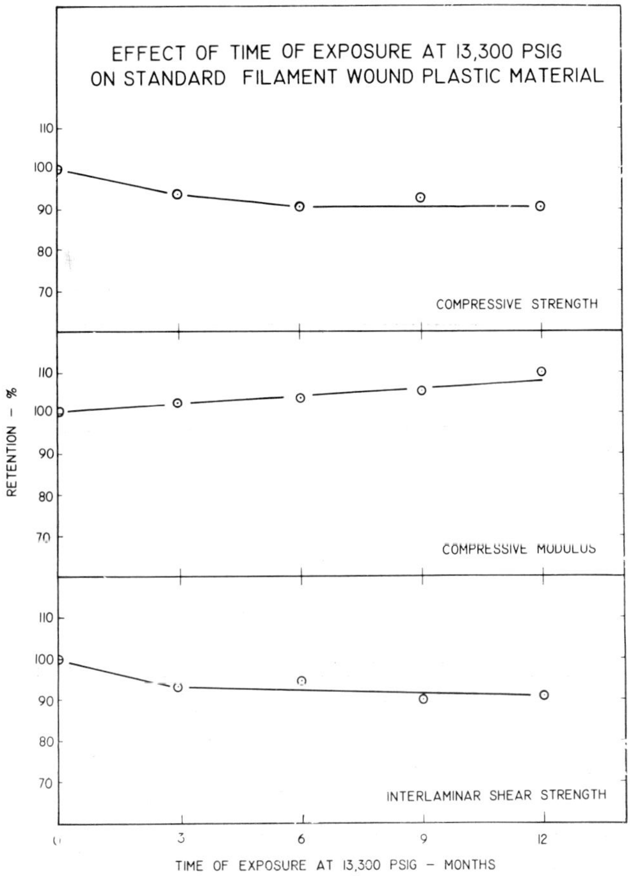

Fig. 4. Effect of extended exposure to water at 13,300 psi (30,000 ft) on the properties of a filament wound epoxy laminate[11]

pressure. It is also notable from the data presented that modulus of elasticity was unaffected by exposure, regardless of the pressure. A similar insensitivity to hydrostatic pressure has been noted for other high quality, low void materials but this behavior is by no means universal and marked pressure sensitivity has been observed in a number of GRP materials, several instances of which are discussed hereinafter.

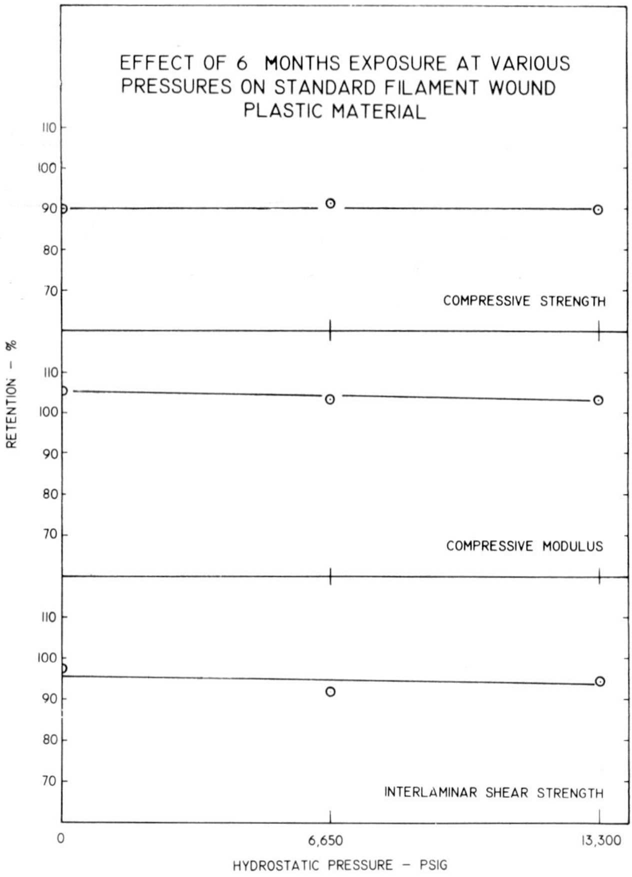

FIG. 5. Effect of six months immersion in water at various pressures on the properties of a filament wound epoxy laminate[11]

(b) *Effect of temperature*. The effect of exposure to water at elevated temperatures on the properties of GRP composites is illustrated in Fig. 6.[10] Increasing the temperature of the water tends to accelerate effects of degradation, and this provides the basis for short time tests to obtain an estimate of the effects of extended immersion at normal temperatures. It is generally accepted, for example, that a "two hour boil" is equivalent to about one month's exposure to water at room temperature.[10] This "two hour boil" technique is commonly employed in the laboratory as an accelerated test and, as such, has been included in various specifications.[5, 6] In this connection, it is important to note that exposure to boiling water can have the additional effect of further curing the resinous matrix, sometimes to the point that specimens are stronger and stiffer after the boil test. For this reason "two hour boil" data should be considered judiciously as a measure of the degradation of GRP due to extended water immersion.

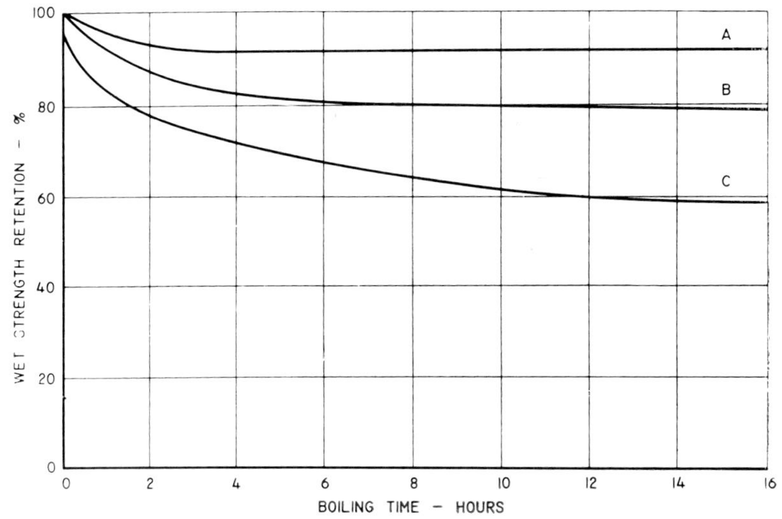

FIG. 6. Effect of exposure to boiling water on the strength of glass-reinforced plastic materials[10]

At the other end of the scale, it has been found that the mechanical properties of GRP are enhanced at lower temperatures.[13]

6. EFFECT OF CONSTRUCTIONAL VARIABLES. As would be expected, the degradation of GRP in water is markedly dependent upon constructional variables in the composite. A discussion of a number of the more important variables is presented below.

(a) *Effect of type of resin.* The type of resinous matrix in GRP can have a considerable influence on the sensitivity of the composite to water. An example of this is shown in Fig. 7.[11] These data are for an orthogonal filament wound laminate similar to the material for which data are shown in Figs. 4 and 5 except that in this case the matrix was an epoxy resin cured with a boron trifluoride (BF_3) complex hardener while the resin in the former material was cured with an anhydride hardener. As can be seen from the cited figures, not only is the BF_3 material more sensitive to water exposure but it is also sensitive to variations in hydrostatic pressure. Differences in water sensitivity have also been noted among epoxy, polyester, phenolic and other types of GRP composites.[14]

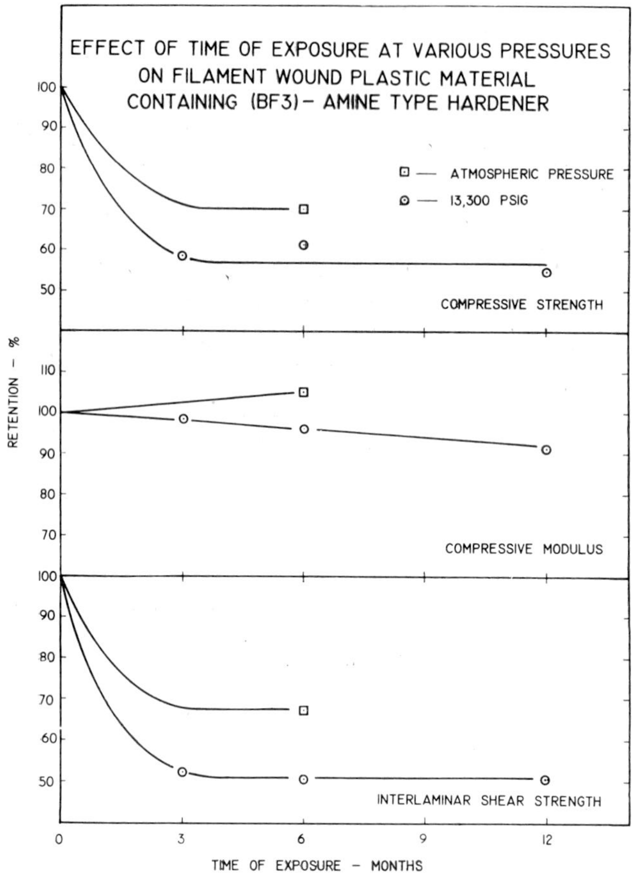

FIG. 7. Effect of extended exposure to water at atmospheric pressure and at 13,300 psi on the properties of a BF_3 hardened epoxy filament wound laminate[11]

(b) *Effect of type of reinforcement.* The type (and form) of the reinforcement can also influence the water sensitivity of a GRP composite. Differences due to variations in glass composition are typified in the data shown in Table 1,[15] which indicate the effects of forced conditions (100 hr exposure to boiling water) on the properties of polyester laminates reinforced with E (borosilicate) glass fibers and with A (alkali) type fibers, in mat form. The markedly increased water sensitivity incurred in using hollow glass fibers as reinforcement is shown in Fig. 8.[16] Of course, changing from glass fibers to

TABLE 1. EFFECT OF WATER ON POLYESTER LAMINATES REINFORCED WITH DIFFERENT TYPES OF GLASS FIBERS(15)

	E Glass	*A* Glass
Flexural strength (psi)		
Initial	42,000	43,000
100 hr boil	34,000	27,000
Tensile strength (psi)		
Initial	25,000	24,000
100 hr boil	23,000	18,000

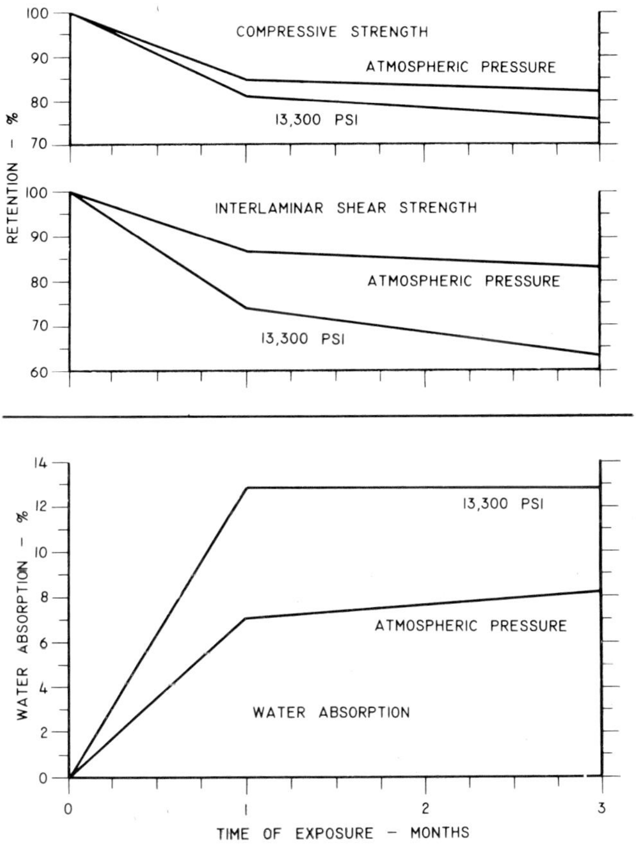

FIG. 8. Effect of extended exposure to water at atmospheric pressure and at 13,300 psi on the properties of a hollow glass fiber filament wound epoxy laminate(11)

other types of reinforcement can likewise have pronounced effects on water sensitivity of the composite. The use of paper reinforcement in a phenolic laminate, for example, results in markedly increased water sensitivity as against glass reinforced materials.[14]

(c) *Effect of coupling agent.* Variations in water sensitivity of GRP are also induced by the presence (or absence) of different coupling agents, which are applied to the glass fibers (or added to the resinous matrix) to improve bonding and to increase water resistance. Typical data illustrating the effects of a coupling agent are presented in Table 2.[17] While it is sufficient to present only the observed data at this point, coupling agents and reactions at the glass-resin interface are matters of considerable importance and will be discussed at some length hereinafter.

TABLE 2. EFFECTS OF CHEMICAL COUPLING AGENTS ON WATER RESISTANCE OF REINFORCED PLASTIC LAMINATES

Reinforcement and finish	Polyester laminates: Flexural strength, 1000 psi				Epoxy laminates: Flexural strength, 1000 psi			
	Original	Boiling water, 7 days	Room temp. water, 6 days	Room temp. water, 30 days	Original	Boiling water, 7 days	Room temp. water, 6 days	Room temp. water, 30 days
Style 181 glass cloth with silane finish	76.3	66.1	80.7	79.0	78.4	54.0	74.9	71.4
Style 112 glass cloth, heat cleaned with silane finish	71.1	54.4	64.5	62.4	79.6	41.0	78.3	74.0
(Same as above, but without silane finish)	49.6	18.2	20.5	19.0	79.7	19.8	72.8	65.9

(d) *Effect of resin content and voids.* The water sensitivity of GRP composites is also affected by resin content. In general, it has been found that increasing resin content tends to decrease water sensitivity.[10] Perhaps even more important than this factor (and probably related thereto) is void content which, if excessive, can seriously increase the water sensitivity of a composite. An indication of this is shown for an orthogonal filament wound material in Fig. 9.[11] These factors, too, will be discussed further in later sections.

7. RECOVERY. Up to this point, experimental evidence pertaining to degradation of GRP in water, and to the several factors which exert an influence on the degree of degradation, has been presented. In general, the foregoing has applied to properties of the composites in the wet state. It is of some interest

to consider the effects of drying of the composite following exposure to water, to determine whether the effects noted in "wet" material constitute permanent degradation. Data exemplifying this "recovery" effect are shown in Fig. 10.[14] As is evident in the figure, there is generally some recovery of properties upon extended drying following immersion. The degree of recovery

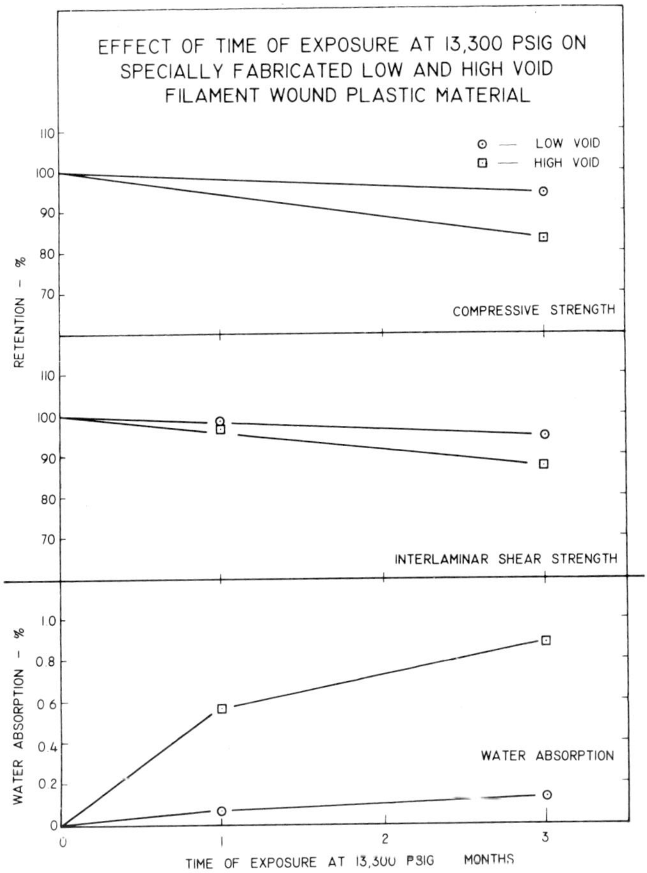

FIG. 9. Effect of exposure to water at 13,300 psi on the properties of low and high void content filament wound epoxy laminates[11]

may vary considerably but for many GRP laminates of the more common polyester and epoxy types, recovery is considerable and frequently approaches one hundred percent, indicating that little or no permanent degradation has occurred. In this connection, some recent data obtained in the author's laboratory (Table 3) are notable. These data were obtained in an investigation to determine the effects of extended exposure in the Tongue of the Ocean (5700 ft) on a number of different classes of materials. The polyester material was a representative Style 1000—Volan glass cloth laminate fabricated in house; the epoxy laminate was a parallel filament composite obtained from a commercial source. Upon removal of the samples from the ocean, they were stored for some time in a normal ambient and were essentially dry when received at the laboratory. The results of the evaluation of these two composite materials exemplify the extremes in behavior which may be encountered. The polyester material, upon drying after extended immersion, showed

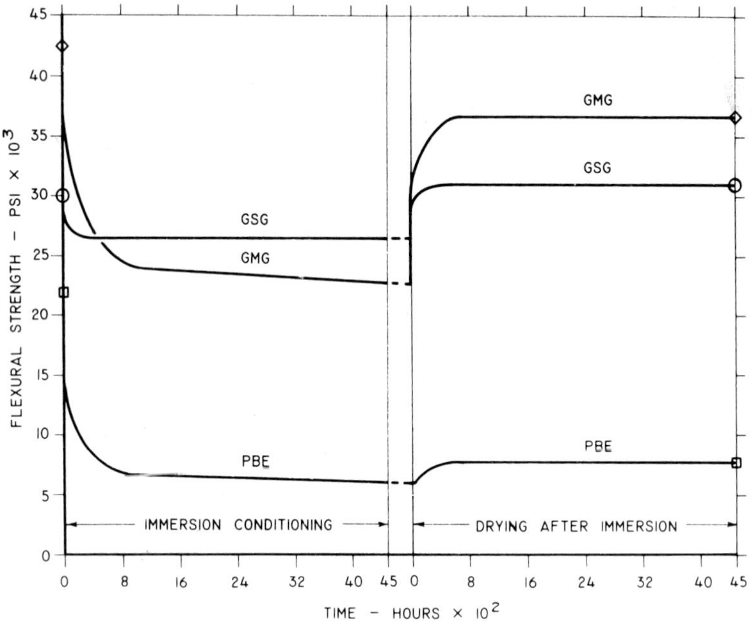

FIG. 10. Effect of extended exposure to water, and of subsequent drying, on the flexural strength of reinforced plastic laminates[14]

no deterioration, whereas the epoxy material had suffered severe, progressive, permanent degradation. The latter behavior is by no means typical of glass reinforced epoxy laminates, as is evident from the data presented in Figs. 4 and 5 which indicate that, even in the wet state, strength retention of this type of material exceeds 90% after extended exposure under severe conditions. Indeed, it has generally been found that epoxy laminates are less sensitive to water than polyester materials. An attempt was made to determine details of fabrication of the epoxy composite which had been immersed in the ocean but the supplier could offer no information other than to indicate that the glass used was of an unidentified composition and was not treated with a high wet strength finish or coupling agent.

TABLE 3. EFFECTS OF DEEP OCEAN EXPOSURE ON GLASS-REINFORCED PLASTIC LAMINATES

	Epoxy laminate			Polyester laminate		
	Initial	After exposure in the ocean at 5700 ft		Initial	After exposure in the ocean at 5700 ft	
		111 days	1045 days		111 days	1045 days
Compressive strength, psi	136,800	115,600	71,600	33,000	31,100	33,500
Compressive modulus, psi	5.97×10^6	5.84×10^6	6.13×10^6	2.72×10^6	2.61×10^6	2.90×10
Flexural strength, psi	–	–	–	38,800	39,500	38,100
Flexural modulus, psi	–	–	–	2.60×10^6	2.59×10^6	2.50×10^6
Interlaminar shear strength, psi	5900	4000	1760	1810	1680	1880

III. EFFECT OF WATER ON COMPONENT MATERIALS

Having presented a brief survey of the observed effects of water on GRP laminates, the several factors which influence the degree of degradation of these materials, and the extent to which effects of degradation are reversible, an attempt may now be made to explain the observed phenomena. Investigations directed to this end will therefore be considered, commencing with information relating to the effect of water on the individual components of GRP composites.

(a) *Effect of water on resins.* To a varying extent, resins which are commonly used in GRP composites tend to be water sensitive. In general, resins of this type (eg., polyesters and epoxies) absorb water and suffer some degree of degradation both in strength and stiffness under extended exposure, although at room temperature these effects are generally slight. The results of an accelerated test (9 hr boil plus extended exposure to water at room temperature) to determine the effects of water on an epoxy casting are shown in Table 4.[18] It is notable that, even after this severe conditioning, strength retention was 82% and stiffness retention 97%. There is some evidence that, at normal temperatures, the water absorption process may continue for long periods of time and that for some resin systems, equilibrium may not be attained for months or even years.[12] Interestingly enough, strength degradation effects appear to level off, just as in the case of the composite, despite continued water absorption.

TABLE 4. EFFECT OF WATER IMMERSION AND OF SUBSEQUENT DRYING ON EPOXY CASTINGS[18]

Condition	Modulus (psi$\times 10^5$)	Flexural strength (psi$\times 10^3$)
1. Dry	5.04	19.1
2. 9 hr boil +1570 hr in water at 72°F	4.87	15.6
3. 9 hr boil +560 hr in water at 72°F +900 hr dry in air at 140°F	5.43	19.2

Water may also have an effect on stress-rupture characteristics of resin castings, as shown in Table 5.[19] In this respect the resin again follows the behavior pattern of the composite. Finally, it has been noted that many types of resin castings may also show "recovery" effects in that samples which show degradation in properties upon water immersion recover from this deterioration upon drying (Table 4).[18]

TABLE 5. FLEXURAL STRESS-RUPTURE DATA FOR EPOXY CASTINGS[19]

Casting type	% of ult. load	Tested in air	Tested in salt water
		Time to fail (hr)	Time to fail (hr)
A	79.3	2690	454
	60.4	over 3600	over 3650
	51.8	over 5540	over 4600
B	79.9	5200	529
	56.6	over 5500	1025
C	49.3	over 6110	793
D	60.1	1783	over 6045
	49.6	2281	over 4078
E	59.8	607	less than 1

(b) *Effect of water on glass fibers.* Glass fibers may also be degraded by water immersion, particularly if exposed to the medium while under stress.[20, 21] Cameron[20] indicates that immersion in water acts to reduce the strength of glass fibers by reducing the effective cross section. Since the attack progresses inward from the surface, the smaller the fiber diameter, the more marked the effect. Since this type of degradation, attributed to a corrosive effect, would not be reversible, recovery of strength upon drying would not be expected. Otto[21] found that "E" glass fibers were unaffected by water immersion for a period of about one week but thereafter showed progressive degradation in strength. Coating of the fibers with a coupling agent, an epoxy resin, or both, did not protect the fibers against degradation. Although the general effects noted by Otto have been found by other investigators, there is also evidence to the contrary. Weisbart,[15] for example, found no degradation in strength of bundles of "A" (alkali) glass and "NON–A" (alkali-free) glass fibers after 80 days immersion in water. The preponderance of evidence, however, indicates that the glass fibers will show some degradation upon extended immersion in water.

(c) *The interface.* In addition to the two principal components of GRP composites which have been discussed above, there is a third, and most important, "phase" which must be considered in assessing the effect of water on these materials, the glass-resin interface. While a complete discussion of this subject is beyond the scope of this paper, a brief treatment of essentials will be presented to provide a basis for discussions which follow.

A primary function of the matrix in a GRP material is to transfer stress among the glass fibers, which are the principal load carrying elements in the composite. In order to accomplish this efficiently, and for optimum performance in general, it is important that the resin be properly adhered or "bonded" to the reinforcing elements. It has long been recognized also that conditions at the glass-resin interface exert a strong influence on wet-strength retention of the composite. As has already been shown, composites fabricated with untreated glass fibers are particularly sensitive to water immersion (Table 2). For this reason, work was undertaken to improve the glass-resin bond and at the same time to provide improved water resistance. The approach taken was to employ a "coupling agent," i.e., a molecule which, at least theoretically, would have both resin reactive and glass surface reactive functionality. An example of this might be a molecule of the type:

$$(EtO)_3Si—R—x$$

in which the ethoxide groups (or the hydroxyl hydrolysis products thereof) could react with silanol ($\equiv$Si—OH) groups on the glass surface to form Si—O—Si bonds. The x moiety could be a vinyl group (—CH=CH_2) for reaction with unsaturated polyester resins, or an amino group (—C—NH_2) if the resin employed were an epoxy type.

Table 2 shows that the use of such coupling agents, which are generally applied to the glass fibers, but which may also be added to the resin system, does indeed improve wet strength retention of composites. A considerable amount of data to substantiate this has been developed in recent years.[22, 23] However, there is no clear cut evidence that such coupling agents provide improved strength initially, i.e., in the dry state.

Although the theory of action of the coupling agent implies direct chemical reaction with the glass surface on the one hand, and the resin on the other, whether either of these reactions does in fact occur is still in question.

In point of fact different investigators at various times have presented evidence at the two extremes, tending to show varying degrees of interaction from covalent bonding to no chemical interaction at all. Evidence for the formation of an Si—O—Si bond between the glass surface and coupling agent has been presented in recent papers by White[24] and by Vogel[25] who also found evidence of reaction between the coupling agent and resin. Schrader[26] has also shown that an Si—O—P bond can be formed by interaction between glass and a phosphate ester coupling agent. Islinger[27] indicates the possibility of such reactions at the glass surface but found no evidence that bonding occurs between the coupling agent and the resin phase. At the other extreme is the opinion that bonding to the glass may not occur at all and that the coupling agent is effective in improving wet strength because of hydrophobicity. A pertinent example in this regard is the work of McGarry[28] which indicates that it is not necessary to posit chemical bonding between resin and glass to explain observed behavior, since this can be done on the basis of friction between the phases induced by shrinkage of the resin around the glass fibers.

Between these two extremes there are investigators who indicate the possibility of intermediate stages of interaction between resin and reinforcement, principally through hydrogen bonding or Van der Waal's forces.[29, 30] It is conceivable that all of these types of interaction may occur in combination, which would make for a very complex situation indeed. In any event, although there may be uncertainties regarding the precise nature of interactions at the glass-resin interface, there is no question that the effects of water at this site may have a profound influence on the process of degradation of GRP.

IV. MECHANISMS OF WATER PENETRATION AND ATTACK

Having reviewed the experimental evidence concerning the effects of water on GRP laminates and on the components thereof, three fundamental questions remain to be considered: how does the water penetrate the composite, and how does it react with the material thereafter, and what are the effects of this action on composite properties. These questions will be discussed in the paragraphs which follow.

A. *Mechanism of Water Penetration*

Penetration of GRP materials by water can occur by two general processes: diffusion through the resinous phase, and flow through cracks, pinholes, or other flaws in the composite.

1. DIFFUSION. The transport of a permeant through a polymeric material, free of defects which permit mass flow, normally occurs by a process of activated diffusion.[31] Mechanistically, the penetrant dissolves in the surface layers of the polymer and migrates into the bulk of the material under a concentration gradient. Rogers[31] describes the process as a movement of the penetrant molecules through transient "holes" which occur in the polymer structure at a rate dependent upon the nature of the polymer and penetrant, as well as the concentration gradient.

The permeation rate in a process of activated diffusion increases exponentially with temperature but is essentially independent of hydrostatic pressure, although some pressure sensitivity may occur if the polymer structure undergoes a change.[32] Lebovits[32] cites a number of instances to show this hydrostatic pressure insensitivity. McNeil[33] found no essential difference between rate of water permeation into epoxy castings at atmospheric pressure and at 10,000 psi. Indeed, in this case, the rate at atmospheric pressure seemed to be somewhat higher, leading to the supposition that the higher hydrostatic pressure tends to "close up" the structure and thus restrict diffusion, an effect also found by Eakins[34] in specimens subjected to mechanical compressive stress.

2. POISEUILLE FLOW. The penetration of water into a GRP composite through cracks or pinholes in the matrix is generally considered to occur by a capillary or Poiseuille type of flow.[35] It is important to note that this type of flow could also occur through defects at the glass-resin interface, resulting in the so-called "wicking effect". The Poiseuille equation for capillary flow is given by

$$\frac{V}{t} = \frac{\pi P r^4}{8L\eta}$$

where V/t is the rate of flow, P the pressure, r the radius of the capillary, L the length, and η the viscosity of the liquid. As is evident from this equation, the rate of water penetration by this mechanism would be a direct function of hydrostatic pressure.

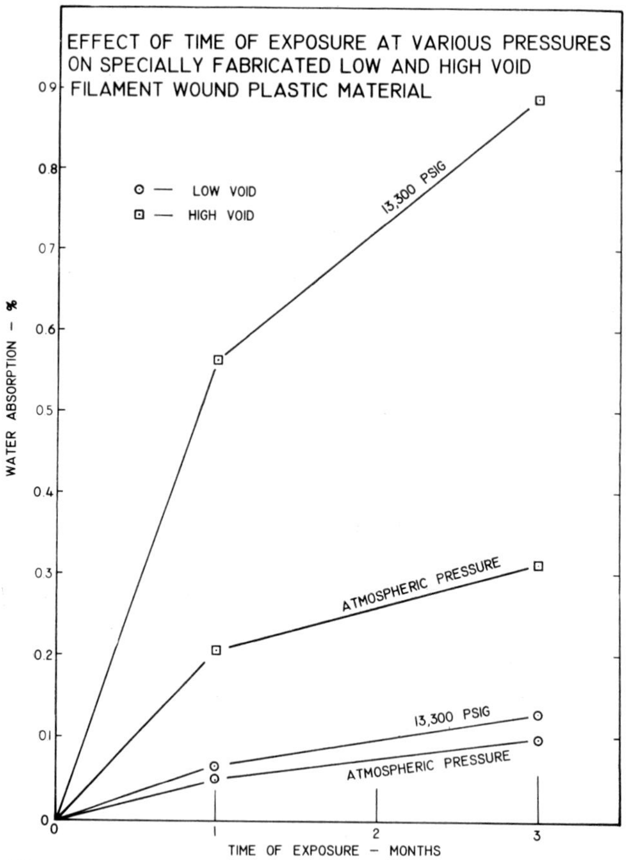

FIG. 11. Effect of void content on the water absorption of filament wound epoxy laminates[11]

It is generally believed that water penetrates GRP composites more readily at the resin-glass interface than by diffusion through the matrix. Laird, for example, found that the rate of water permeation through the resin-glass interface in an E glass-epoxy composite was 450 times more rapid than through the resin.[36] Glowacki[18] found considerable penetration in tests in which the cut edge of a glass reinforced epoxy laminate was exposed to water under high hydrostatic pressure.

The effects of high void content on water absorption of a composite are illustrated in Fig. 11. The data indicate not only increased water permeation due to voids but a pressure dependency as well, McGarry[37] has also noted increased permeation due to cracks in the matrix. In this work, GRP samples were subjected to a tensile stress well below the ultimate, but sufficient to cause resin cracking. Water absorption of specimens thus treated was markedly higher than for unstressed specimens.

B. *Mechanism and Effect of Water Attack*

The mechanism of degradation of GRP composites by water is complex, and probably multifarious, but a number of relevant mechanistic theories have been posited and these will be discussed below.

As a prelude to these considerations, it should be noted that any effect of water attack on a specific component would only be expected to affect the properties of the composite if this effect has an influence on the mode of failure under the particular condition of stress. Thus, if in a hypothetical case water tends to weaken the matrix but has no effect on the glass coupling agent bond, and if shear failure occurs at the latter site, water immersion may have little effect on interlaminar shear strength. On the other hand, weakening of the resin might have a pronounced effect if failure occurred within the resin phase.

1. THE RESIN PHASE. As has been indicated, water immersion can cause resin swelling, plasticization, and degradation in strength and stiffness. In instances in which no chemical changes occur due to water interaction (e.g. hydrolysis of chemical bonds) these effects are essentially reversible and resin properties are recovered upon drying.[18]

Degradation in resin properties would be expected to result in a parallel degradation in composite properties. For example, in the simple rule of mixtures equation:

$$E_c = A_f E_f + A_r E_r$$

where the subscripts c, f and r refer to the composite, filament, and resin respectively, A is fraction of cross sectional area, and E is elastic modulus; it is apparent that if E_r decreases E_c will likewise decrease. But the effect could be more marked than calculated from this equation since the decrease in resin stiffness could cause a loss in efficiency of stress transfer between filaments, resulting in unequal stress distributions therein. In compression, the effect could be further accentuated in that a lower modulus matrix could allow filament bending under load and thus cause a further decrease in modulus.[38] It is obvious, too, that flexural and shear moduli of the composite could likewise be adversely affected.

Degradation in matrix strength and modulus would also be expected to result in decreased composite strength in direct tension and compression, again, due to decreased stress transfer efficiency. In interlaminar shear, a weakening of the resin would cause a parallel decrease in composite strength if the locus of failure is the resin phase; or the locus of failure could be changed from, for example, the glass-coupling agent interface to the weakened resin phase. As indicated above, all of these effects on composite properties would be expected to be reversed upon drying if the resin "recovers" upon the removal of water.

2. THE REINFORCEMENT PHASE. The effect of degradation of the reinforcement on properties of the composite needs little discussion. It is obvious that any influence which decreases the strength of the primary load bearing elements, i.e., the glass fibers, will degrade the strength of the composite. Further, if there is a net decrease in the effective cross-sectional area of the totality of fibers, not only will strength decrease but, from the rule of mixtures equation, modulus will decrease as well. It would be expected that this type of degradation, if it occurred, would not be reversible upon drying of the composite.

3. THE INTERFACE. As has been indicated, it is the general consensus of investigators in the field that the most important factor in degradation of GRP composites in water is the action of the water at the glass-resin interface. In this respect a number of ideas have been advanced, as discussed below.

It has been proposed that a primary effect of water entering GRP composites is chemical interaction at the interface. Thus, if a coupling agent acts to form an Si—O—Si bond at the glass surface, water could cause bond scission by, for example, a base catalyzed redistribution due to alkalinity at cationic sites on the glass surface.[39] This could result in essentially complete debonding between the matrix and the glass fibers with the consequent adverse effects of decreased stress transfer efficiency discussed above. It is notable that this type of mechanism would require reformation of the Si—O bonds between the coupling agent and the glass, or at least formation of secondary bonds of some type, to explain recovery of composite properties on drying.

It has also been postulated that water entering the composite at the interface does not, in fact cause bond scission at this site but rather penetrates underneath the coupling agent and attacks the glass directly, with consequent weakening effects on the material.[25] Such a change would also be expected to be irreversible, so that properties could not be recovered on drying.

If it is assumed that interaction at the interface does not involve primary chemical bonding to the glass surface, but rather secondary bonding, such as H bonding or a multiplicity of Van der Waal's force interactions, it is conceivable that the interposition of a medium of high dielectric constant, such as water, could cause effective lessening of the interaction by a weakening of Coulomb forces. Further, the entering water molecules might offer a competitive interaction in which H bonding could occur between the water molecules and silanol groups on the glass on the one hand, and polar groups in the resin phase on the other. This hypothetical mechanism would also tend to weaken the resin-glass bond and result in degradation of composite properties. Interactions of this type would also require reversibility, e.g. reforming of glass-coupling agent H bonds, upon removal of water, to explain recovery effects.

As previously indicated, a theory has also been advanced which presupposes no bonding between the resin and reinforcement. In this case, it is postulated that water which penetrates at the interface simply acts as a lubricant, reducing stress transfer efficiency in this manner.[40] Obviously, recovery effects would occur upon removal of water. In this connection, it is notable that oil immersion has been found to affect the strength and stiffness of GRP composites to a much lesser extent than exposure to water.[40] While this may not be conclusive evidence, due to the generally lower absorption rate for oil, the results do cast some doubt on the simple lubricity theory.

V. SUMMARY

Glass reinforced plastic composites are subject to degradation in mechanical properties upon exposure to water. Degradation may occur in both strength and stiffness, the latter characteristic generally being less affected. Water exposure can also adversely affect stress-rupture and fatigue characteristics of GRP laminates.

The extent of degradation may vary with variations in environmental conditions. Increasing the temperature of the water generally accelerates effects of degradation. Increasing hydrostatic pressure may also accelerate degradation but many GRP materials suffer no more adverse effects under high pressure exposure than at atmospheric pressure.

The degree of degradation of GRP materials in water is a function of constructional variables and of the quality of the composite. The type of glass and resin employed in fabricating the laminate and particularly the efficiency and water resistance of the glass-resin bond will exert important influences in this respect. Defects such as voids, cracks, and pinholes will tend to make the material more susceptible to water attack.

Glass reinforced plastic materials tend to recover from the effects of water degradation upon drying. While there are exceptions, the degree of recovery is frequently substantial, some materials recovering essentially all of their strength and stiffness, when the water absorbed during immersion is removed.

Water can penetrate a GRP composite by two general routes: activated diffusion through the matrix, and flow through cracks, pinholes, voids, and other defects.

The components of GRP materials are, individually, subject to degradation by water. Resinous matrices tend to absorb water and to decrease in strength and stiffness, a process which is reversible to a degree since many resin systems tend to recover their properties upon drying. Glass fibers may show corrosion effects, with a net decrease in effective cross-section; this process, if it occurs, would not be expected to be reversible. In addition to the effects on static properties, water exposure may also adversely affect the stress-rupture characteristics of both resins and glass fibers. Degradation of the properties of the resin in a composite can result in reduced stress-transfer efficiency among the reinforcement fibers and consequent loss of strength and stiffness of the composite. Any decrease in the effective cross-section, or in the strength, of the glass fibers will also be reflected in decreased strength and stiffness of a composite.

Conditions at the glass-resin interface are of principal importance in determining the response of GRP materials to water exposure since stress transfer must occur across this boundary. In order to improve glass-resin bonding, and particularly to provide water resistance, coupling agents, which are designed to bond chemically to both the glass and resin, are employed at this site. There is some question whether primary chemical bonds are formed between coupling agents and either the glass or the resin phase. Alternate theories postulate secondary bonding or no bonding at all, but rather interaction through friction. Degradation at the interface has been variously ascribed to primary bond scission, weakening of secondary bonds, or lubricity effects. Regardless of the mechanism, degradation in the efficiency of the glass-resin bond may occur under the influence of water exposure. To the extent that this occurs, the composite will suffer a corresponding degradation in strength and stiffness.

VI. CONCLUDING REMARKS

The foregoing serves to illustrate the complexity of the processes of degraration of GRP composites in water. Although the principles related to these processes manifestly cannot be extended to apply to composites in general particularly under widely differing environmental conditions, they may be applicable in broad terms to degradation processes in generalized composites of the filament-reinforced matrix type under the influence of fluid media.

Thus, the importance of employing inherently inert components in the fabrication of a composite to resist a particular environment is obvious. The principles of employing a matrix with low permeability, and of eliminating defects such as cracks, voids, and pinholes would likewise apply to any composite material. But perhaps the most significant result of the considerable amount of investigational work which has been done on the degradation of GRP in water has been the pointing up of the extreme importance of conditions at the matrix-reinforcement interface.

A growing understanding of the principles of degradation of GRP in water has led to the development of resin systems and glass composition with improved water resistance, and particularly to the use of coupling agents at the glass–resin interface which has also resulted in major improvements. Further, improved fabrication processes have been developed with emphasis on elimination of voids and other defects. In sum, GRP materials of today, while still somewhat sensitive to water, frequently are affected insignificantly, and certainly are vastly more resistant to water degradation than earlier materials.

Knowledge gained in studies on GRP materials is also being applied in other areas in which some of the newer composites have shown evidence of water sensitivity. Thus, the principle of employing a coupling agent at the reinforcement–matrix interface has been used successfully to improve the water resistance of glass-reinforced thermoplastics[(41)] and is also being applied in syntactic foams (glass microballoons in a plastic matrix). Water sensitivity has also been observed in carbon filament reinforced plastics, and here again efforts are being made to improve bonding by use of coupling agents.[(42)] Water sensitivity in boron filament-epoxy composites has also been ascribed to sensitivity at the interface, specifically boron oxide on the fiber surface, and marked improvements have been attained by removal of this oxide layer.[(43)]

In closing, it must be emphasized that our knowledge of degradation processes remains fragmentary and indeed in many areas rudimentary. While broad guidelines are available which may enable a rational initial selection of materials for the fabrication of resistant composites, it is still frequently impossible to predict precisely the response of a particular composite to a specific environment. For this reason specific simulative laboratory studies continue to be made and may extend into the future, until further, more intensive research efforts delineate degradation processes more definitively.

REFERENCES

1. Holliday, L. (Ed.). *Composite Materials*, Elsevier Publishing Co., New York 1966.
2. Dietz, A. G. H. (Ed). *Engineering Laminates*, John Wiley & Sons, Inc., New York 1949.
3. MIL-HDBK-17. Plastics for Flight Vehicles—Part I, Reinforced Plastics (Nov. 1959) (under revision).
4. Bjorksten Res. Labs. *Polyesters and Their Applications*, Reinhold Pub. Co., New York 1956.
5. Spec MIL-P-17549. Plastic Laminates, Fibrous Glass Reinforced, Marine Structural.
6. Spec MIL-R-7575. Resin, Polyester, Low Pressure Laminating.
7. Tech. Res. Bull. No. 212. *Guide for the Selection of Fiberglass Reinforced Plastics for Marine Structures*, Soc. of Naval Architects and Marine Engrs., New York, Mar. 1965.
8. Engineers of Gibbs & Cox, Inc. *Marine Design Manual for Fiberglass Reinforced Plastics*, McGraw-Hill Book Co., Inc., New York 1960.
9. Goldfarb, P. M. U.S. Naval Applied Science Laboratory Lab. Project 4860-J-4, Final Report (21 Dec. 1959).
10. Rawe, A. W. *J. Plastics Inst. (Brit.) (Transactions)*, **30**, No. 85, p. 27 (1962).
11. Fried, N., Kaminetsky, J. and Silvergleit, M. *Proc. 21st Ann. Tech. Conf., Reinforced Plastics Div., SPI*, Sec. 14-A (1966).
12. Loewenstein, K. L. Glass systems, in *Composite Materials, loc. cit.*
13. Fontana, M. G. Bishop, S. M., and Spretnak, J. W. Air Force Tech. Rept. 5662, Part 5 (WADC) (Dec. 1953).
14. Winans, R. R., Fried, N., and Hand, W. *Electrical Manufacturing*, July 1955, p. 106.
15. Weisbart, H. *Proc. 17th Ann. Tech. Conf., Reinforced Plastics Div., SPI.*, Sec. 10-B (1962).
16. U.S. Naval Applied Science Laboratory. Unpublished data.
17. Vanderbilt, B. M. *Proc. 22nd Ann. Tech. Conf., SPE*, Montreal (March 1966).
18. Glowacki, A. S. *Proc. 21st Tech. Conf., Reinforced Plastics Div., SPI*, Sec. 14–1D (1966).
19. Burhaus, A. S. *et al. Ibid*, Sec. 14–C.
20. Cameron, N. M., T. & A.M. Report No. 274, Univ. of Illinois (Jan. 1965).
21. Otto, W. H. Final Summary Report, Contract NOnr 4522(00)(X), Narmco Division, the Whittaker Corp. (June 1966).
22. Plueddemann, E. P. *et al. Proc. 17th Ann. Tech. Conf., Reinforced Plastics Div., SPI*, Sec. 14–A (1962).
23. Clark, H. A., and Plueddemann, E. P., *Modern Plastics*, June 1963, p. 163.
24. White, T. E. *Proc. 20th Ann. Tech. Conf., Reinforced Plastics Div., SPI*, Sec. 3–B (1965).
25. Vogel, G. E. *et al. Proc. 22nd Ann. Tech. Conf., Reinforced Plastics Div., SPI*, Sec. 13-B (1967).
26. Schrader, M. E. and Lerner, I. *Proc. 20th Ann. Tech. Conf., Reinforced Plastics Div., SPI*, Sec. 3-C (1965).
27. Islinger, J. S., Gutfreund, K., Maguire, R. G. and Olson, O. H. WADC Tech. Report 59–600, Part I (Mar. 1964).
28. McGarry, F. J. *ASTM Bulletin*, No. 63 (Jan. 1959).
29. Weisbart, H. *Proc. 20th Ann. Tech. Conf., Reinforced Plastics Div., SPI*, Sec 3-D (1965).
30. Zisman, W. A. *Proc. 19th Ann. Tech. Conf., Reinforced Plastics Div., SPI*, Sec. 21-B (1964).
31. Rogers, C. E. In *Engineering Design for Plastics*, Ed. E. Baer, Reinhold Pub. Corp., New York 1964.
32. Lebovits, A. *Modern Plastics*, March 1966 (p. 139).
33. McNeil, D. W., Bennett, B. and Leininger, R. I. *Proc. 19th Ann. Tech. Conf., Reinforced Plastics Div., SPI*, Sec. 11-B (1964).

34. EAKINS, W. J. *Proc. 20th Ann. Tech. Conf., Reinforced Plastics Div., SPI*, Sec 15-C (1965).
35. REGESTER, R. F. *Proc. 22nd Ann. Tech. Conf., Reinforced Plastics Div., SPI*, Sec. 16-D (1967).
36. LAIRD, J. A. and NELSON, F. W. *Proc. 19th Ann. Tech. Conf., Reinforced Plastics Div., SPI*, Sec. 11-C (1964).
37. MCGARRY, F. J. Tech. Report AFML-RT-66-288 (MIT) (Sept. 1966).
38. FRIED, N. *Modern Plastics*, p. 163 (Sept. 1963).
39. CHAMBERLAIN, D. L., JR. *Proc. 22nd Ann. Tech. Conf. Reinforced Plastics Div., SPI*, Sec. 13-E (1967).
40. SHARP, W. H. and WEBER, M. K. *Corrosion (Nat'l Assn. of Corrosion Engrs.)*, **12,** 27 (1955).
41. STERMAN, S. and MARSDEN, J. G. *21st Ann. Tech. Conf., Reinforced Plastics Div., SPI*, Sec 3-A (Feb. 1966).
42. PROSEN, S. P. *et al. Ibid*, Sec. 8-D.
43. GUTFREUND, K., BROUTMAN, J. J. and JAFFE, E. H. In *Advanced Fibrous Reinforced Composites*, SAMPE Science of Advanced Materials and Process Eng., Vol. 10 (1966).

INTERNATIONAL FORUM

SUMMARY OF COMPOSITES RESEARCH IN JAPAN

TADASHI TAO[†]

ABSTRACT

FRP has been used in Japan for commercial purposes for many years. It was introduced in space and aeronautics applications, with the production of aircraft. However, only in the past few years have any research studies on this material been initiated by universities research institutes and industries in Japan.

The most current findings in this research were delivered before the symposium on "FRP in Space and Aeronautics" sponsored by the Institute of Space and Aeronautical Science at the University of Tokyo in January 1967.

This paper reviews briefly how FRP is applied to the Japanese aircraft industries, and then an outline of developmental and fundamental research accomplished on this material in this country.

I. FABRICATION OF AIRCRAFT PARTS

Aircraft parts are fabricated by means of hot press, vacuum forming, spray mat deicing, honeycomb sandwich construction, break away forming and hand lay-up (see Figs. 1 to 3).

FIG. 1. Hot press in aircraft work

[†] Chief Scientist, Nagoya Aircraft Factory, Mitsubishi Heavy Industries, Ltd., Japan.

FIG. 2. Spry mat deicing

FIG. 3. Oven in aircraft work

These parts have been used for building F-104J jet fighters, Vertol or Sikorsky helicopters and domestically designed YS-11 transports (Fig. 4) and MU-2 utility planes (Fig. 5). There are 148 parts, large and small, in the YS-11 and 174 parts in the MU-2. FRP was also used to construct a glider having an overall length of 7.5 m (25 ft).

FIG. 4. YS-11 transport

FIG. 5. MU-2 utility plane

II. DEVELOPMENTAL RESEARCH

A. *S–B Sounding Rocket by Science and Technology Agency, Japan (Fig. 6)*

This project was directed by Dr. Y. Takenaka. Overall length was 2.5 m (98 in.), outside diameter 165 mm (6.2 in.) gross weight 60 kg (135 lb). So far seven rounds have been launched for observing the weather in an altitude up to 60 km (37 miles) with success.

FRP parts	Construction and process
Nose cone	Phenolic resin, glass cloth, hot press
Chamber	Epoxy resin, combined circumferential FW with unidirectional cloth
Nozzle	Throat: phenolic resin and silica cloth. Skirt: phenolic resin and silica tape.

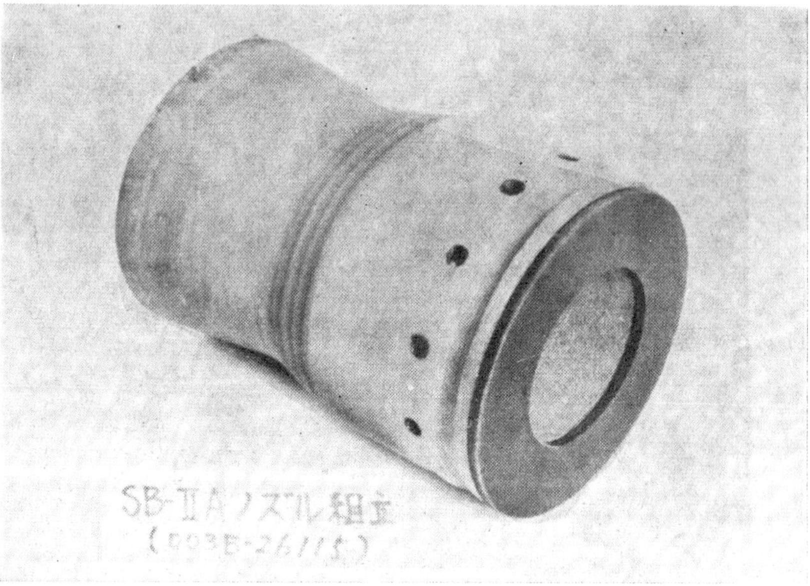

FIG. 6. S–B sounding rocket nozzle

B. *Kappa, Lambda and Mu Rocket by Institute of Space and Aeronautical Science, University of Tokyo, Japan (Figs. 7 and 8)*

This project was directed by Profs. D. Mori and R. Akiba.

FIG. 8. Mu-rocket nose cone

FIG. 7. Mu-rocket launching

FRP parts	Remarks
Nose cone including split type	Applied to all rockets, especially to small sounding rocket by epoxy mixed with phenolic powder
Nozzle	To Lambda 735 engine and Mu 1400 engine skirts by FRP, throat by graphite solid

C. *Combustion Part of Rocket Engine by Science and Technology Agency Japan (Fig. 9)*

This project was directed by Dr. Y. Takenaka. Combustion test is expected in the near future.

Parts	Remarks
Aft end closure insulator	Synthetic rubber molding
Nozzle	Inside: throat by graphite solid, skirt by phenolic Outside: epoxy circumferential FW and glass cloth
Chamber	By circumferential epoxy FW and unidirectional cloth length 1 m (39 in), outside dia. 500 mm (20 in) and length 3 m (14 in), outside dia. 350 mm (14 in)

FIG. 9. Phenolic prepreg tape winding

D. *480 ∅ mm In-plane Filament-wound Spherical Chamber by the Institute of Space and Aeronautical Science, University of Tokyo (Figs. 10–13)*

A metal has been used for the spherical motor chamber of the fourth stage of sounding rocket Lambda, developed by the Institute. However, FRP has been used in this application at a great saving in weight. This

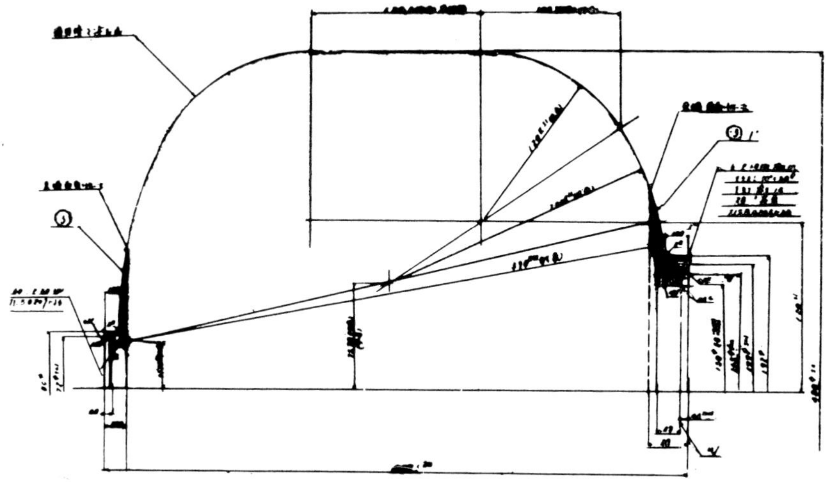

FIG. 10. 480 ∅ mm In-plane FW spherical chamber

FIG. 11. Spherical chamber In-plane FW

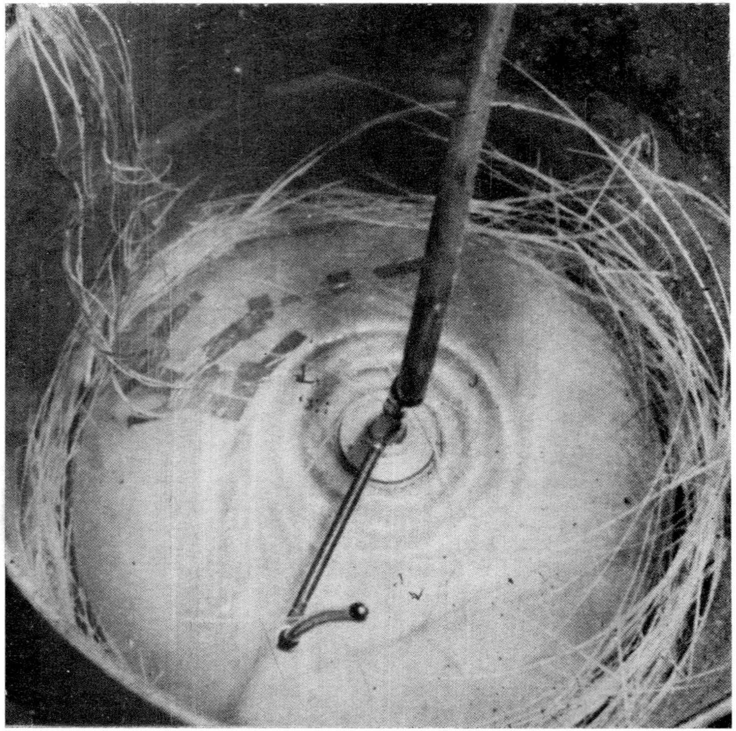

FIG. 12. Spherical chamber after hydraulic burst test

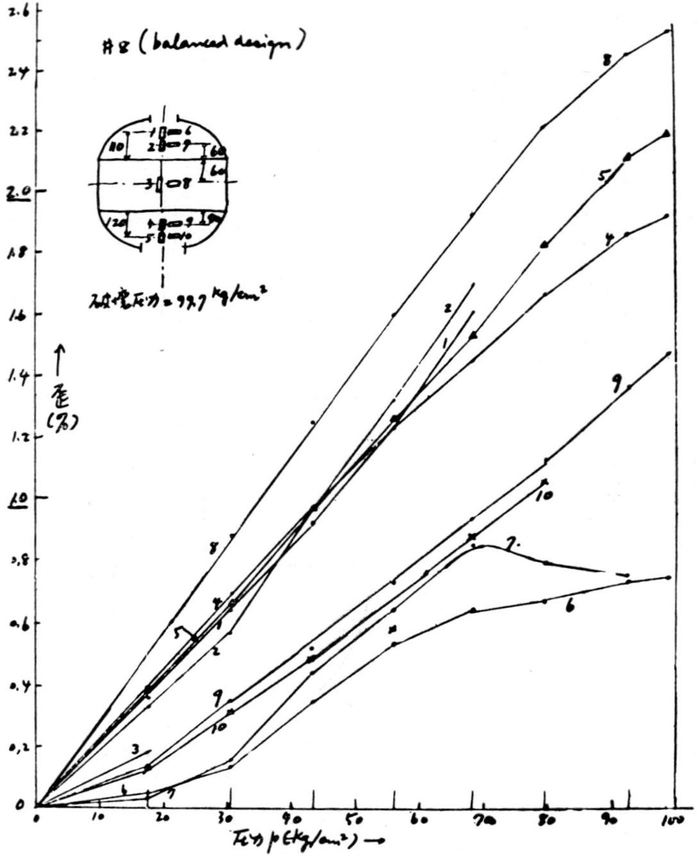

FIG. 13. Spherical chamber stress–strain curve

inplane filament-wound spherical chamber has a total length of 422 mm, and inner diameter of 480 mm, with weight of 6.5 kg including Al insert (nozzle side 1.37 kg and ignition side 0.55 kg). It has been manufactured and hydraulically burst-tested to be shown at 89–100 kg/cm^2. At the same time, fiber and matrix stress distributions are resolved by analysis and compared with experimental values. The analysis proves useful but frequently gives higher matrix stresses than observed in actual fracture of such vessels. Elastic moduli distributions are also calculated. The results show the promising application of FRP to the rocket component as a weight-saving material.

A static firing test will be conducted as well as actual employment in the sounding rocket series. As a conclusion, the in-plane filament-wound chamber has proved to be a powerful tool when the weight-saving is important. This project was directed by Professors D. Mori, K. Kawata and M. Uemura. This technique will be applied to larger cylindrical chambers.

E. *500 ∅ mm Filament-wound Cylindrical Chamber by the Institute of Space and Aeronautical Science, University of Tokyo (Figs. 14–16)*

This chamber, with 500 ∅ mm in diameter and 1700 mm in length, is produced as the third stage component of Lambda rocket, also for the purpose of weight-saving. The chamber is composed of a cylindrical part with glass-cloth and a filament hoop winding in addition to the molded end closures rivet-jointed. A considerable amount of weight-saving will be expected. This project was directed by Professors D. Mori, M. Uemura and K. Kawata.

Rapid progress is being made in the study of FW material and processes in many industrial concerns in cooperation with universities and national research institutes. Accomplishments in these fields of activities will lead to increased applications.

III. FUNDAMENTAL RESEARCH

A. *Newly Proposed ISAS Ring Test (Fig. 17)*

The ISAS ring test was proposed by Professors K. Kawata and S. Fukui in 1965. This test takes account of so-called ineffective width inherent to the actual filament-wound cylinder with arbitrary winding angle since the glass filament is cut away at ends. The experimental verification is made and the results assure its validity. This new test is quite unique and useful for actual hoop strength evaluation, since the famous N.O.L. ring test aims only at the unidirectional filament composite strength evaluation.

FIG. 14. 500 ∅ FW cylindrical chamber

FIG. 15. 500 ∅ FW cylindrical chamber after burst test

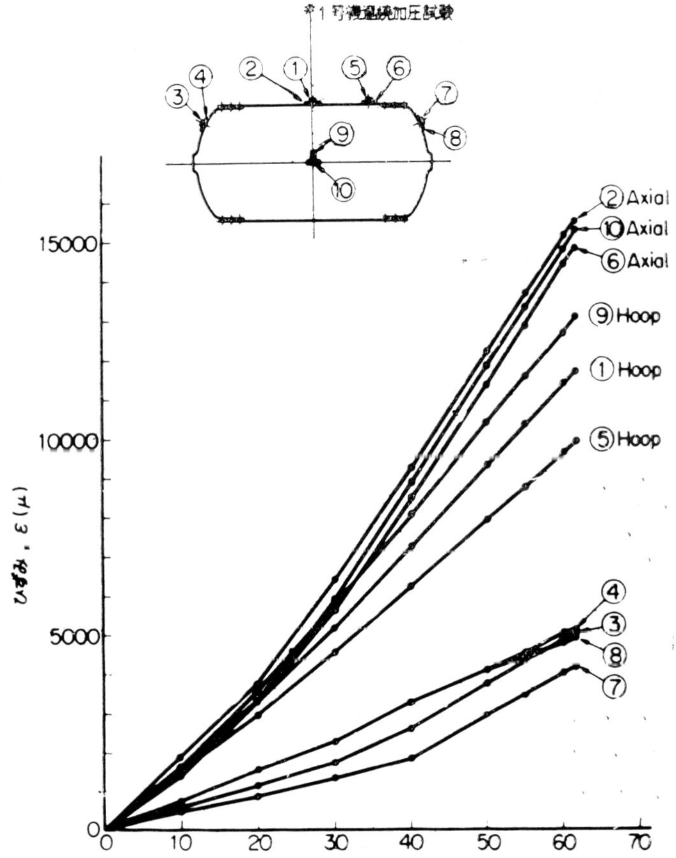

FIG. 16. 500 ∅ FW cylindrical chamber stress–strain curve

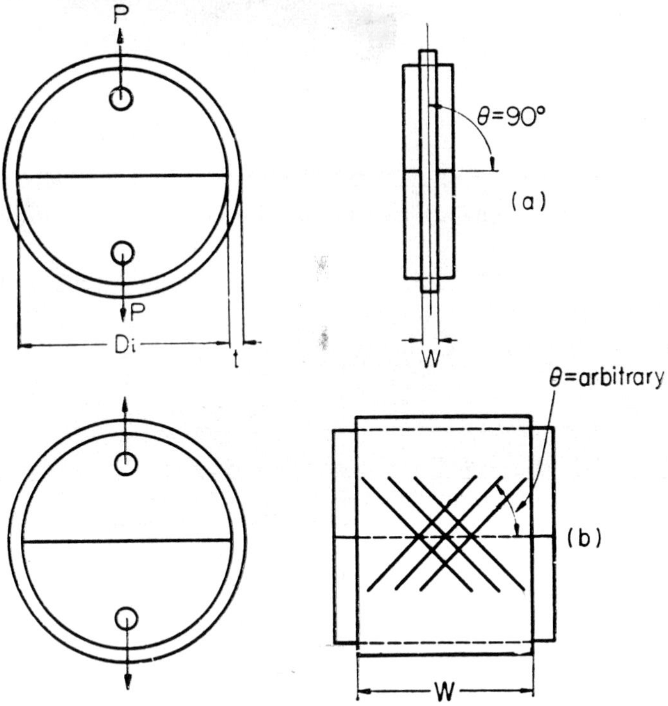

FIG. 17. NOL and ISAS ring tests

FIG. 18. Test piece of joint of FW cylinders

B. *Strength of Joint of Filament-wound Cylinders (Figs. 18 and 19)*

The joint of filament-wound cylinder is a critical item in the construction of the actual filament-wound structure. The filament-wound cylinder or spherical chamber should be tightly connected with the end closures in order to convey the external and internal forces and to keep the structural continuity. The mechanical joint problem is, therefore, very important. Professor M. Uemura, Institute of Space and Aeronautical Science, University of Tokyo, is working on this problem.

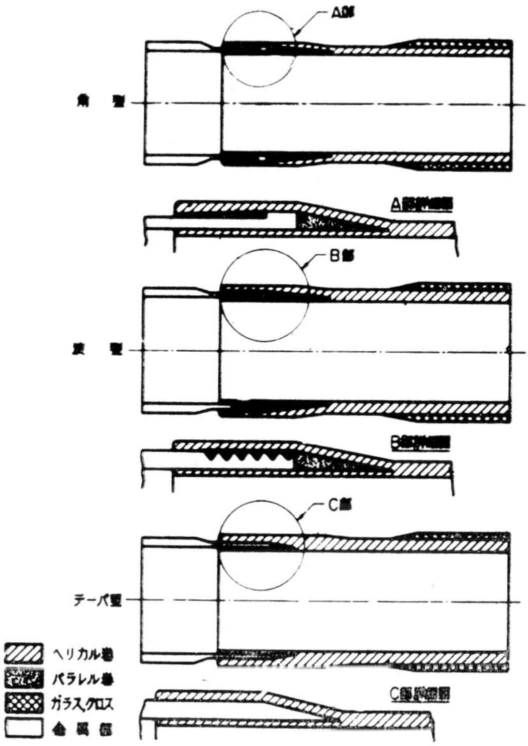

FIG. 19. Various joint types of FW cylinder

C. *Interlaminar Shear of Laminated Composite Plates*

Both the elasticity and strength of FRP laminate and Filament-wound Composite materials depend considerably upon the bonding condition between filaments and matrix, and between laminated layers. Despite the fact that much research has been conducted on these problems, so far no satisfactory theories have been established. Dr. T. Hayashi, Department of Aeronautics, University of Tokyo, has been studying this subject. He has found even in the most simple and fundamental case of rectangular specimens under tension test, the maximum interlaminar shear stress in the transverse direction may reach a considerable order of magnitude.

D. *Low-cycle Fatigue Testing of Filament-wound Materials (Figs. 20 and 21)*

The proof test by internal pressure is prerequisite to the actual rocket chamber at the completion. Therefore the effect of such preloading on the strength degradation must be studied. A series of low-cycle fatigue tests

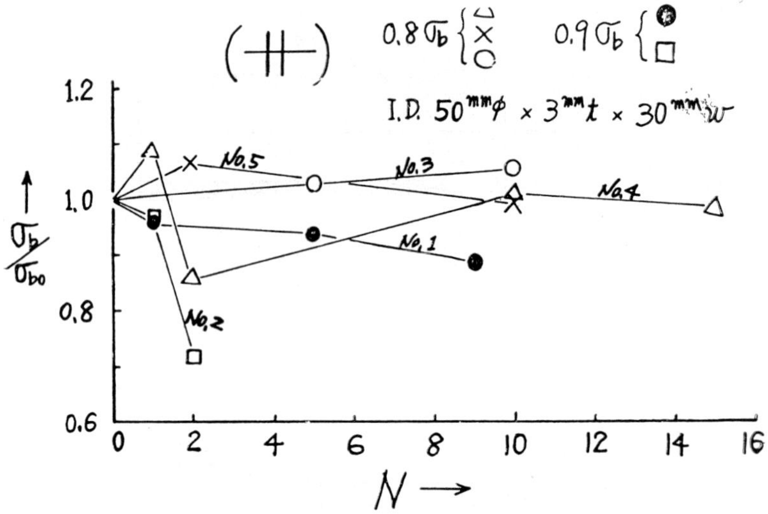

FIG. 20. Low cycle fatigue test of circumferential FW

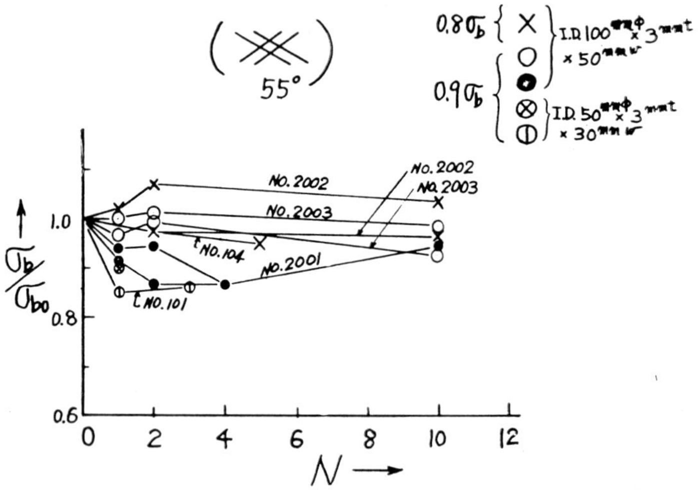

FIG. 21. Low cycle fatigue test of helical angle 55°FW

have been worked out by the solid mechanics group, headed by Professors S. Fukui and K. Kawata, and Associate Professor A. Kobayashi in charge, Institute of Space and Aeronautical Science, University of Tokyo. The preloads considered are 80% and 90% of virgin burst load. The portion of test results is shown on the slide. The low-cycle fatigue effects are conspicuous especially for 90% preloading.

E. *General Fatigue Test (Fig. 22)*

Mr. Shimamura, Chief Scientist, Government Mechanical Laboratory, Japan has been studying fatigue of composite material continuously for several years. His results are as follows:

1. Rigidity of specimen decreased as frequency was increased. When rigidity decreased to 90 or 80% of original rigidity, the reduction developed more rapidly until at last rigidity was nearly lost under the constant deformation type fatigue test, and resulted in complete fracture under constant load type fatigue test.
2. Rise in temperature of specimen was proportional to range of stress to the nth power ($1 < n \leqq 2$) and to the test rpm to the mth power ($0 < m \leqq 1$). As a result fatigue strength of ten million cycles was not affected by the testing speed, as long as it was around 3000 rpm. Cooling the specimen in air was not effective. Fatigue strength below one million cycles was affected by the testing speed and cooling the specimen in air was found effective.
3. Most of FRP did not reach fatigue limit at ten million cycles. Prof. D. Fujii, Municipal University of Osaka, Japan is also conducting research on the mechanism of fatigue of composite materials.

Fatigue strength under completely reversed plain bending σ_w''

Matrix	Reinforcement (content)	Process	σ_w'' Kg/mm²
Polyester	glass roving (–)	unidirectional FW	28.0
	" (–)	cross direction FW	14.0
	SC (0.52)	press	9.0
	PC (–)	"	7.0
	SC (0.57)	vacuum bag	7.0+
	M (0.55)	press	7.0
	M (0.22)	hand lay up	3.0
	SC+M+PC (–)	"	3.5
	PC+M+RC+M+PC (–)	"	2.7
epoxy	SC (0.57)	press	15.0
	SC+roving (–)	"	25.0
phenolic	SC (0.54)	press	12.0
	cotton cloth (–)	"	2.5
	paper (–)	"	2.5
polyacetal	glass chop (0.30)	injection	3.2

+ premature value under pulsating plain bending

SC: satin cloth
PC: plain cloth
RC: roving cloth
M: mat
+

FIG. 22. Fatigue strength of various construction FRP

F. *Rigidity and Strength of Filament-wound Cylinders (Fig. 23)*

A series of tests was developed to obtain mechanical properties such as Young's modulus and ultimate strength for the various winding angles of filament-wound cylinders. The relevant theoretical analysis will closely approximate the experimental verification for moduli. The work is done by Professor M. Uemura.

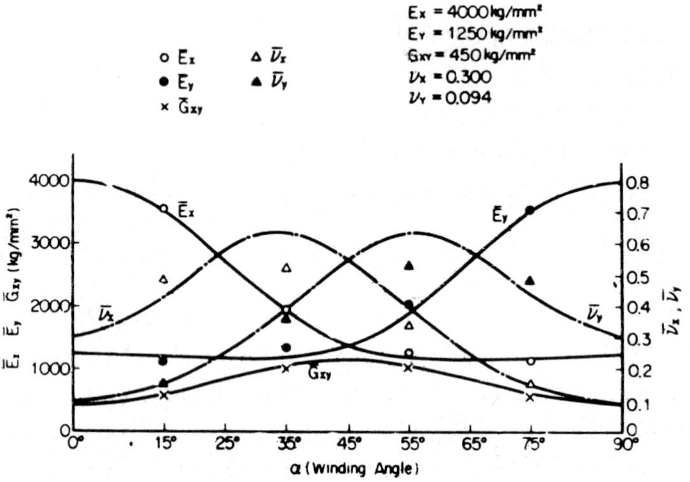

FIG. 23. Rigidity and strength of FW cylinders

G. *Fracture Toughness Evaluation of Filament-wound Materials*

The fracture toughness of orthotropic materials such as the filament-wound materials is being studied at the Institute of Space and Aeronautical Science, University of Tokyo, by K. Kawata and A. Kobayashi, Professor and Associate Professor, respectively. The results of preliminary investigations show that (1) the orthotropy is seen in the fracture toughness of filament-wound material, and (2) the fracture toughness of filament-wound material is superior to that of pure resin matrix. The glass filament-wound material is composed of glass filament and epoxy or TAC polyester. The absolute value in fracture toughness in the case of epoxy series is always higher than that of polyester. Further investigation is now urged.

H. *Photoelastic Method of Experimentation for Orthotropic Materials*

This project was directed by Prof. T. Hayashi, Department of Aeronautics, University of Tokyo, Japan. A new photoelastic method of experimentation for stress analysis of orthotropic materials made of FRP is presented.

Required orthotropy and good photoelastic properties can be obtained by proper choice of resin and fiberglass cloth. The test apparatus is the same as the usual one for isotropic test materials.

To test the availability of this method and to study the problem of stress concentration in orthotropic plates, experiments have been made on specimens of rectangular shape with a circular hole in the center. These were subjected to uniform tension applied at both ends, with the result that their practical usefulness appears promising and that the circumferential stress distribution of the hole obtained experimentally agrees fairly well with that obtained theoretically.

J. *Resin*

Polyimide resin has been tentatively manufactured as heat resisting matrix. The strength of the resin hardly changes up to a temperature of about 200°C.

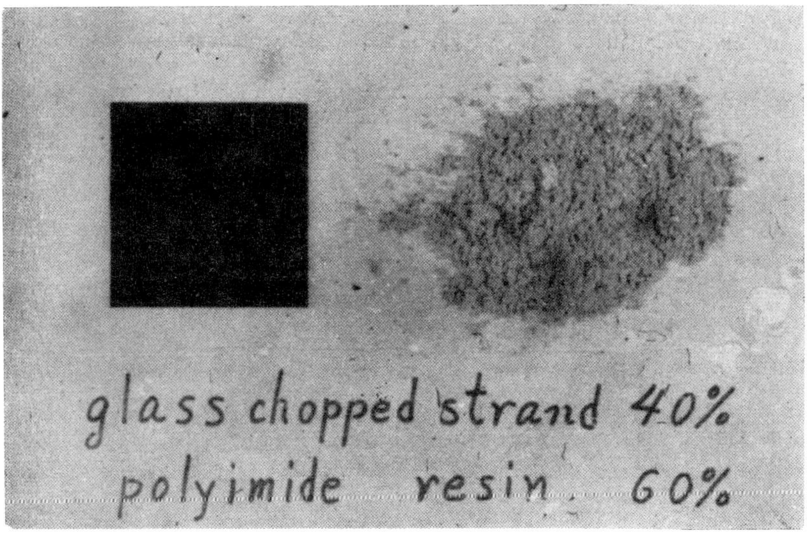

FIG. 24. Polyimide resin mixed glass chopped strand

Its forming properties are so poor that it is very difficult to perform compression molding. Through repeated test, bending strength has been improved to a level of 25 kg/mm^2 (35,000 psi) by mixing glass chopped strand to 40% (see Fig. 24).

K. *Carbon and Graphite Cloth*

Carbon and graphite cloth was experimentally manufactured by dehydrogenation of acrylonitrile fiber process and was subjected to various tests. This material has found application in rocket nozzles.

IV. CONCLUSIONS

Application of composite material to the manufacture of space vehicles is rapidly spreading owing to its advantages of high ablation heat resistance and high specific strength. It is desirable that better reinforcing material such as whiskers, boron fiber, etc., be used properly.

REFERENCES

1. KAWATA, K. and FUKUI, S.: ISAS Ring Test, Report No. 409, March 1967; University of Tokyo, *Proceedings of the Sixth International Symposium on Space Technology and Science, Tokyo*, (1965).
2. SHIMAMURA, A. Fatigue tests of FRP, JSME Semi-International Symposium, Tokyo (1967).
3. HAYASHI, T. Photoelastic Method of Experimentation for Orthotropic Materials, *Applied Physics*, **31**, No. 10 (1962), Japan Soc. Applied Physics.

SUMMARY OF SOME COMPOSITES APPLICATIONS AND RESEARCH IN SWEDEN

KARL-AXEL OLSSON

Royal Institute of Technology, Stockholm, Sweden

IT IS a great pleasure to be invited by the General Chairman, Dr., F. W. Wendt, to talk at this conference.

The most common composite material in Sweden is glass fiber reinforced plastic. Different types of this material have been produced in the past twenty years, much of it for motorboats and sailboats. The Swedish Aircraft Company (Saab) has used it in radomes and in other aircraft applications. A sports car, the Saab Sonett II, is now in production and is made of glass fiber reinforced plastic. Other uses include: tubes of different types, oil tanks, fans for the chemical industry, transport packings, swimming pools, etc.

Sandwich design has been used for example by AB Malmö Flygindustri for shelters, a smaller aircraft and a 32 ft sailboat and by AB Centrifugalplast, Sundsvall, for a hydraulic liftable cover of a truck (Fig. 1) and shelters.

Lack of time has made it impossible for me to collect information from all of the Swedish research centers and factories engaged in this field. Therefore, I will concentrate my presentat on on what has been done at the Division of Aircraft Structures at KTH in Stockholm under the direction of Professor T. Rand during the last three or four years. The main work has consisted of investigations on the structures of shelters for the Royal Swedish Air Force Board, Department of Telecommunication (see Figs. 2 to 6). These shelters were designed by the Air Force Board and are used mostly for electronic equipment. They can be carried by helicopter, suspended by wires and are built with sandwich panels of three dimensions, 1600×2100 mm, 1600×2350 mm and 2100×2350 mm. This module-building makes it possible to construct shelters with varying dimensions to suit any purpose. The shelter consists of an inner and outer steel frame to which sandwich panels are glued and riveted. The panels are built up on a core of polyvinyl chloride, PVC, with closed cells, Divinycell H-55-HT, made in Sweden by AB Bertildi on license from Klébér-Colombes, France. The core thickness is 50 mm and its density 55 kg/m^3 (3.4 lb/ft^3). The skins are made of glass fiber reinforced polyester with the thickness 2 mm (0.08 in.). The reinforcements consist of

FIG. 1. Hydraulic liftable cover over a truck

FIG. 2. Shelter, type 1

FIG. 3. Shelter, type 2

FIG. 4. Interior view of a shelter

300 g/m² (10 oz/yd²) mat, 650 g/m² (20 oz/yd²) rowing and 300 g/m² mat. The resin is self-extinguishing and non-flame-spreading as the core-material. The gelcoat on the outer face is IR-reflecting. The panels are made by a hand lay-up process in special steel forms under pressure and then taken out from the forms and heat cured in an oven. Until now about 5000 panels have been produced by AB Centrifugalplast in Sundsvall.

Around the door opening in each gable is the tightening border, mounted so that the shelters will be air tight when put together in a chain.

FIG. 5. Helicopter-carried shelter

FIG. 6. Helicopter delivering shelter to test site

We have tested this type of panel very carefully and we are still taking sample panels from production for testing. We believe this control is necessary as the product consists of so many raw products and the fabrication is a handicraft.

The tests consist mainly of:

Panel test in compression (see Figs. 7 and 8). The requirements specify that the load before failure on the panel shall be more than 10,000 kg on a panel with the dimensions 300×500 mm and a thickness of 52 mm which will give a strength of more than 800 kg/cm^2 (11,000 psi) in the skin.

Tension test of the skin in different directions. The tensile failure strength shall be more than 1000 kg/cm^2 (14,000 psi).

Tension test perpendicular against the skin with 50 mm cubes of the panel (Fig. 9). The tensile adhesion shall be higher than 7 kg/cm^2 (100 psi) and is now usually about 12–14 kg/cm^2.

This testing has resulted in several changes in production methods and choice of raw materials to obtain an optimal end product.

Other projects we have worked on are special tests with this earlier-mentioned PVC-core material, with densities from 35 to 250 kg/m^3 (2.2 to 15 lb/ft^3). For example these tests enabled us to determine the influence of curing time at different densities of the material, investigation of water absorption depending on time and pressure, volume stability at different tempera-

FIG. 7. Compression testing machine at Division of Aircraft Structure, KTH

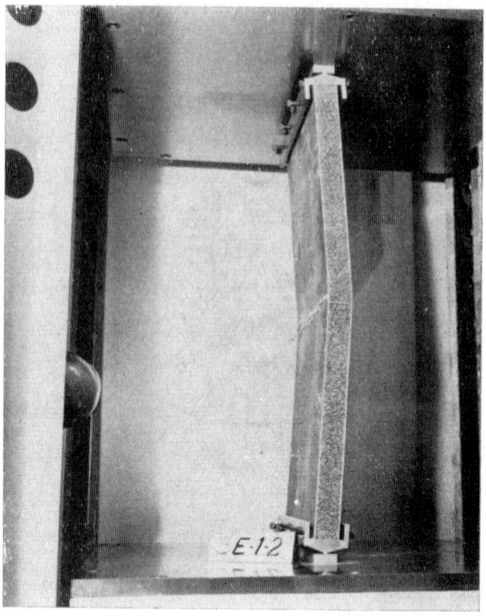

FIG. 8. Compression test with sandwich panel

FIG. 9. Tensile adhesion test

tures, maximum allowable hydraulic pressure on the material and, of course, investigations on the working curve, stress-strain curve, in tension and compression at room temperature. The working curve in compression is especially interesting for this material (see Fig. 10). We hope to get facilities to continue this work at higher and lower temperatures and in shear.

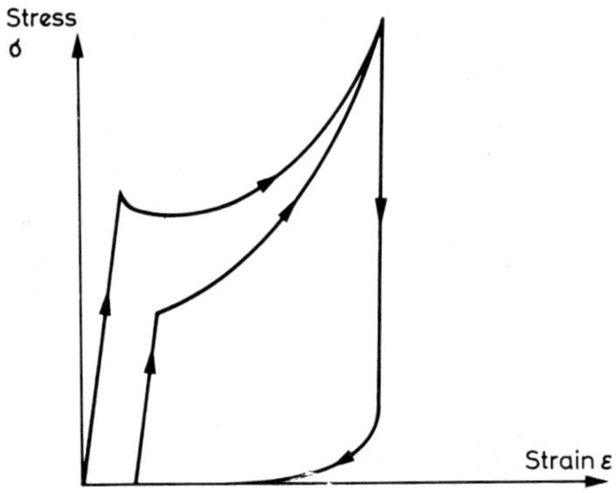

FIG. 10. Working curve compression for the PVC material

DINNER ADDRESS

SIDNEY R. GALLER

Smithsonian Institution, Assistant Secretary for Science

MR. CHAIRMAN, distinguished guests, ladies and gentlemen. When I was invited to address this group I must confess, very honestly, I did not pay much attention to the composition of the group, but I was intrigued with the fact that it was going to deal with composite engineering and composite materials. I assumed, therefore, that it was going to be a composite group. Lo and behold, it is the most homogeneous group of composite engineers I have ever encountered. However, I feel very comfortable, because that places me back in the role that I am used to, a minority of one, a biologist speaking to a group of engineers. I have had 15 years of experience trying to convince the engineers that there is something to biology. I must confess I have not fully succeeded.

Before I begin I would like to make one point, that is I am addressing these comments not only to engineers, but to scientists and technologists. The title of my talk this evening is "Ecological Engineering". Without attempting to talk down to this audience, I would like to say that I define ecology as the interrelationships of the physical, chemical, and biological factors of the environment, and I would like to add one other factor, the sociological. I maintain, and it is not novel with me, that it is not possible either in a military organization or in the civilian economy to separate man from the total complex of his physical environment—they are intertwined. Man is a part of the environment. He has an impact on the environment. However, by virtue of separate backgrounds and experiences, man tends to look at the environment and his interrelationship with it through the rose colored lenses of whatever academic background he comes from.

To illustrate this I would like to tell a story about a young physician fresh out of his internship, who for the first time in many years took a long and much needed vacation in a resort area. One early morning he decided he would take a stroll by a small lake. It was beautiful, the sun was rising, mist was over the water, and he was having a wonderful esthetic experience. All of a sudden he heard someone call for help. He looked out over the lake: obviously someone was drowning. So he did what any self-respecting young physician fresh out of medical school with a course of life saving would do:

he swam out, pulled this fellow half way up on shore and immediately started administering artificial respiration. This, by the way, was by the old fashioned Schäfer method, not mouth-to-mouth. He compressed the rib cage and a stream of water came out, he compressed the rib cage again and another stream of water came out, he did it once more and another stream of water came out. By this time the doctor was getting pretty perplexed. At that moment another chap came walking by; he stopped and examined the situation very carefully. Then he looked disapprovingly and shook his head. The young physician became quite upset; he looked up and said, "Sir, are you trying to tell me how to administer artificial respiration? I am a physician." The second chap said, "Oh no, sir Doc, I am not trying to tell you how to administer artificial respiration, but let me introduce myself. I am a hydraulics engineer, and I was about to suggest that if you don't pull this fellow all the way up on shore you're going to pump that lake dry." So you see there are different ways of looking at the environment.

I am really before you in a more serious vein to deliver not a sermon, not a lecture, but hopefully an invocation. This is really an expression of my own feeling towards the society, or that segment of the society that we call the engineering sciences or the engineering community. I, as a biologist, with perhaps a special interest in ecology, blended, mellowed, and kind of punched into shape by a long association with a hard nosed group in the Navy, am convinced that it is ultimately the responsibility of the engineer to serve as more than a responder but rather as the central component in a dynamic feedback mechanism that will raise certain critical questions to which the basic researcher must address himself. I do not know how many of you will agree with me, but my experiences with engineers is a kind of bivalent one. Individually, I find that they are extremely resourceful, extremely imaginative, and indeed socially responsible members of the total cultural, as well as the scientific community. But as a group, I feel that engineers almost by tradition have not been practicing leadership, they have been practicing what I call sheepsmanship. They have been perfectly content as a group to follow the dicta, to respond to the specifications and requirements that are established by other segments of the total scientific and technological community, whether it be a mission oriented agency, such as one of the Navy bureaus, or a private industrial outfit. In doing so, the engineering community is taking on the responsibility for manipulating man and his total environment. This may very well turn out to be not only deleterious, but irreversible. To be specific, I have seen many examples of engineers who have risen to the challenge of designing and constructing a major engineering project that is from an engineering point of view, very sound, in fact even an expression of a remarkable series of advances in technology. However, there is one thing wrong, not infrequently, the engineer being all too responsive, insufficiently critical, a poor communicator and lacking a feedback system designs and constructs an edifice which is self-defeating. This occurs not within the definition of economic feasibility over a finite period of time, in fact it may be a very profitable structure, but in the long run, it may turn out to be self-defeating in the sense that it reduces the effective environment in which man must live.

What I have said so far is a kind of preamble. Let me, if I may, carry on from some written lecture notes that I hope will be not only challenging and controversial, but I hope you will disagree with them, because I would like to debate this with you.

The birth of a technological society is for me synonymous with the harnessing of nonbiological energy to fulfill the roles of man's imagination. I suspect that seagoing man was the first discoverer of the fact that work could be done without sweat, as he used the wind for propulsion. However, little-by-little the gravitational energy of water, the chemical energy of wood and fossil fuels, and finally the atomic energy bound in radioactive materials were discovered and brought to the service of man. This was not a plotting course of discovery, tying together time and change in a regular sequence. The entire course of technological revolution has been run in 2000 years, even by reaching back to include the earliest germs of thought.

Democritus was the expounder of the hypothesis that the birth of axi-symmetrically constructed entities, as opposed to purely observed reality, was indeed the turning point that led through to the thinking processes of mathematics, physics and chemistry ultimately to the empirical understanding of man and nature that today distinguishes sharply between the technological and the nontechnological man. I wish to emphasize that within the society of man today, within the prisoner that we call man, who is both the prisoner within the biosphere as well as the warden of that biosphere, there has developed this dichotomy in thinking and in philosophy between the so-called technological man, primarily the engineering scientist and the practicing engineer, and the nontechnological man, the scholar, the intellectual, the humanitarian, the humanist. I want you to think about that for a moment, because indeed in the social community, there is this dichotomy between the nuts and bolts guy and the socially élite humanist.

The merging of science and technology that instituted the modern revolution in human competence is really an occidental invention little more than a century old. It was scarcely a century ago that man turned his talents to vital inventions, accepted surgery, vaccinations, public health measures, antibiotics, chemotherapy, etc. Of course, these are examples in an area that I feel somewhat conversant with; these are examples of inventions that introduced a new era in which man emerged as the first creature to directly influence the answers to the ageless questions of who dies, when and of what. The outcome of these discoveries are clearly revealed in the burgeoning human populations. It was the thought processes of science that consolidated the era of vital invention and started on the harried course to large scale environmental manipulation. The scientific image emerged.

In sharp conflict with the folk image of the living world a cult of engineering and technology has appeared. As has been pointed out on many occasions, even a relatively imperfect shift from the folk image of man and society to a scientific image involves man in at least two large irreversible and related changes. The first of these is the increase in self-consciousness, not of the individual himself, but of the society in which he has been placed. The second change is the development of the integrated systems point of view, where the world is seen as functioning in an orderly and predictable way, where imposed changes have predictable consequences. If I may digress for a moment it has been my experience that this almost epitomizes the thinking of what I call the hard scientist, the physical scientist, and the engineer; change impact will bring about a predictable response, but I submit that this does not have to be.

Man is no longer the fragile primate surrounded on every side by mysterious forces, wild beasts, and pestilence. He glories in his new capacity to go where he will, when he wishes, to conquer all natural obstacles, to guide his own star. We have man as the spoiler, the casual pursuer of short term goals, the arbiter of survival of the world's creatures, including man. At the same time, this man is the creator of majestic works, he is self-conscious, he is the only moral creature, we think, he is the conservator, and he is ready to answer for his errors and to extend the umbrella of his competence over many lesser forms of life with which he shares this limited environment that we call Earth. This to me is one of the great revolutions in attitude of all times. Since his beginning, man had lived in a world of natural realities, he sensed the cycles of the seasons and for himself, as well as for all other living organisms, there was birth, youth, maturity and death. Thus he was part of the human cycle inseparable from the interconnected observable cycles of nature. He was a part of his natural and physical environment.

Although it was the development of empirical science that set the stage for the era based upon molecular and nuclear energy, the great advances in the biological sciences, and in many other areas of human experience, have arisen from a concomitant and, frequently, dependent explosion in the observational sciences. Together they have altered human society beyond recognition. The world's resources are knit into a single entity and, in theory at least, local deficiencies have been cancelled out.

Progressive agriculture with its breeding programs and scientifically based culture of domesticated organisms has greatly increased the efficiency of converting solar energy into human food. Medicine, public health and, most recently, urban engineering have made possible the effective congregations of larger numbers of people than ever before into complexes that we call cities. There are no climates in which man cannot live with reasonable comfort if he so wishes. High speed travel, mass education and effective media of mass communication have expanded the horizons of technological societies and with them the experiences, imaginations, and the consciousness, as well as the consciences, of their citizens.

Inherent in the preoccupation of a technological society with the powerful concepts arising from the axisymmetrically derived sector of its science, there is a strong tendency to overlook its impact on the living world of demonstrable entities. The devastation of the living resources and the land that supports them is not new. This occurred in China, India, and the old inhabited lands of Asia Minor, all of them bearing witness to the destructiveness of the pretechnological culture.

The folk images of the world and its laws were inadequate when understanding of the larger issues were involved. It is paradoxical that the technological societies of today, affluent beyond previous experience, competent and equipped with the intellectual system to attack any problem, also stand on the brink of destroying the natural environments that nurture them. The very complexity of our technology has magnified and multiplied the roots of our impact on the environment, while the sophistication of many of our processes leads to consequences so subtle and so remote that they easily escape detection. The recent dominance of the molecular approach to biological knowledge has diverted funds, space and staff in universities into this area and has led to increasing neglect of the complex natural biological systems, especially those that cannot be easily studied within the laboratory but must instead be met outside of the laboratory in the living environment. So I submit that even though I am offering a hopefully constructive and friendly critique of the engineering society, I recognize that even at the basic biological level where the scientific community should be taken up with modeling living organisms in a natural environment, the fashions of the present have tended to swing the pendulum away from the holistic approach, away from the investigations of the interactions between the living organisms, including man, and his environment and into the molecular, biophysical, how many angels on the head of a pin kind of biology.

One of the concepts pressed upon us by the elusory success of molecular biology is the idea that life is, after all, nothing but a mixture of chemical reactions. Our failure to appreciate the profound connection between ignorance and death has led us increasingly to condone or even encourage the release into the environment of powerful chemicals distinguished by their capacity to kill. Now, mind you, I am not talking about chemical warfare, or biological warfare, or germ warfare, or radiological warfare: I am talking about the deliberate release of pesticides and other types of biocides into the environment supposedly to increase man's resources. In the long run these pollutants will serve as one of the elements in a synergistic system that very likely will bring the human community to the brink of an irreversible reaction within the next 50 years or so. Science, technology and the sociological elements of the community interjoin together in developing a model of the total environment that can be quantitated, replicated, and serve as a "master plan" to which any engineering group that is concerned with a major project with unknown consequences on the environment can refer to determine what the ultimate effect might be over a period of 1 year, 10 years, 50 years, and 100 years or more.

The most insidious influence of man in the environment arises from the disposal of wastes and from the purposeful distribution of biocide chemicals to destroy plants, insects, and related organisms regarded as inimical to certain human activities. These forms of activity have the common denominator of so altering the environment chemically or physically that it is no longer a suitable habitat for many native forms of life, and is often damaged as a habitat for man himself. Hercum C. Rains of the Rand Development Corporation states, "I agree that there is no good present technology for the treatment of certain pollutants, but I disagree that our environment is being increasingly polluted because of a technology gap, rather, and this is the critical point, it is a management gap. The major sources of pollution of man's biosphere are the consequences of what takes place in the management offices of industrial and municipal plants and in the offices of our political leaders, managers of the modern community." I don't want to dwell on pesticides or the individual technological, deleterious products that are going into man's total environment. The hour is growing late and I think that I would like to curtail the formal presentation. However, I do want to make a couple of points. What is happening to the atmosphere of the earth as a result of really uncontrolled, poorly managed technology which results as a consequence of management encouragement? There is evidence that in the past century our consumption of organic fuels has increased the carbon dioxide level in the earth's atmosphere by 14%. It is believed that steady increase in carbon dioxide will give rise to a gradual warming on earth. This might be considered beneficial particularly to the northern nations. However, there may be consequences in the gradual melting of the Antarctic and Greenland ice masses. Such melting could raise the level of the oceans as much as 400 feet above the present level with catastrophic inundation of coastal lowlands. Estimates suggest that carbon dioxide accumulation will be sufficient to begin this process by the year 2000. The probable speed of the melt is indefinable; however, by the time it starts the process will in all likelihood be irreversible. The time to interrupt the accumulation of CO_2 is before the critical level is reached, beginning now.

You must forgive me for speaking as a theologian or a preacher; as a matter of fact, last week I participated in a panel discussion that was sponsored by the National Council of the Churches of Christ in Chicago, and I confess that I was as critical of the theologians as I am of the friends of the engineers, but for a very different reason. The theologians are an integral part of management leadership, at least on social issues; and they have chosen to separate the social issues from the physical and environmental factors, and from the important role of technology. They have set up technology as the community's bugaboo.

You guys are the bad guys, and the reason you are is because through your technology you are the guys who are making atmospheric sewers of the earth. To use the vernacular, the monkey is not on the backs of the engineers, the monkey is on the backs of all of us, all of us who have the intelligence and the social responsibilities to recognize that the scientists and engineers, that science and engineering, at least, are neutral. It is what we do with the inventions and discoveries that decides whether man will advance as a social creature, or be destroyed as just one of the animals of the earth.

Within the last hundred years more than 180 species of living creatures have disappeared from the face of the earth. They have become extinct. This appeals to our esthetic sense; we decry the fact that the right whale and the blue whale are on the road to extinction. But I wish to point out that there is something much more basic than the esthetics. These creatures are essentially banks of genetic material, which, once gone, are irreplacable. One of our most essential resources is the genetic material found in the plants and animals of the earth. We use this material for hybridizing today in a way that we couldn't dream of 50 years ago, and what we do in this field 100 years hence will depend on these banks of germ plasms. As long as we have the banks, we can take out what we need. Once those banks are gone, they are gone forever.

Let me conclude by pinpointing what I consider to be, if not a solution, at least a direction towards a solution. I believe that engineers and physical scientists, in general, for various reasons, some of which I sympathize with, some of which I do not understand, have been all too smug in looking upon the life sciences as something that if you have nothing better to do you support. It is a kind of social varnish. An organization supports some of the basic life sciences just to be able to show the academic community, at least, that it is a distinguished organization, it is farsighted, it has a great breadth of purpose. I would like to submit that the day when we can make disciplinary, arbitrary distinctions between the usefulness of a technological speciality, and the usefulness of a fundamental science such as biology, is gone; it is long gone, we had better recognize it. Today the objectives are interdisciplinary, they are not simple objectives. Urban planning, environmental engineering, and more complex ecological engineering, necessitates an investment on the part of the engineers, the scientists, the physicists, and the biologists. We are all striving for one ultimate objective, and that is to develop stochastic models of ecology that we can use to predict the consequences of a major environmental manipulation. Today it is all too easy for man to manipulate his environment on a global scale from a technological point of view; but our intellectual appreciation and comprehension of the consequences are still on a paradoxical scale. I would like to see ultimately the establishment of at least one national center for comparative environmental manipulation. A center where any engineer, any industrial group or government group can go and say now here is what I would like to do. Put this into your system and give me some statistically valid probabilities on what is likely to happen if we build a bridge, or dam a river or blast a canal in region latitude x, longitude y, depth or altitude z. This is what we should be shooting for, this is what we must have. Without it we will continue to be as a Tower of Babel in technology, science and social issues, never to meet until ultimately some irreversible environmental manipulation will take place. Then perhaps we will meet, but who knows it may be too late.

Now I do not want to close on a somber note. I honestly believe that if the engineers and, I am using this in a very loose way, the technologists will recognize that they have two responsibilities: one, their professional responsibility to be responsive and convert basic discovery into application; two, and perhaps an even more subtle responsibility, to be leaders, to force the issues and to raise the questions that will in turn force some of our "long hairs", if you please, to address themselves to certain questions that indeed take on a mission oriented character. This can be done at a very basic level, where the scientist can start coming up with fundamental data, that will develop the stochastic ecological model that we need. Ultimately he will give us options and a basis for predicting engineering consequences; we will then have what I call "Ecological Engineering".

AUTHOR INDEX

SUBJECT INDEX